Genomics of Chloroplasts and Mitochondria

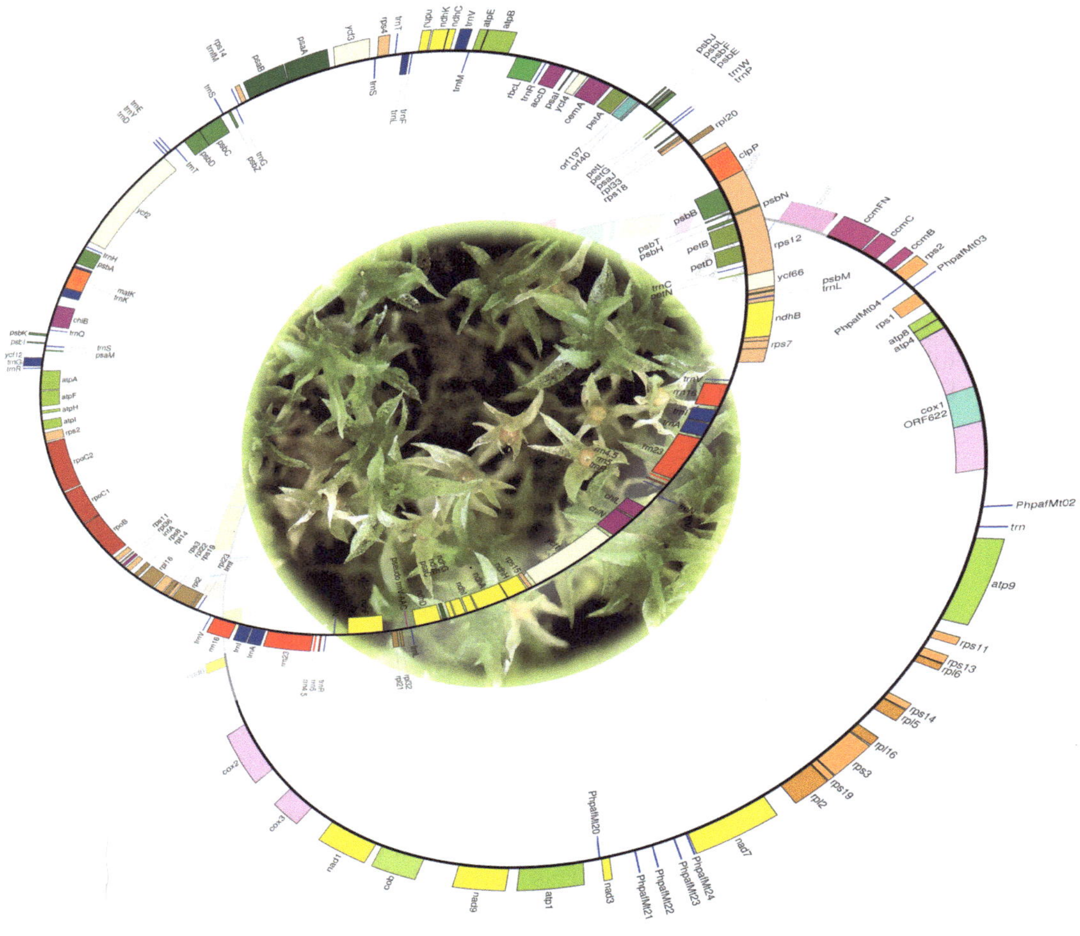

This illustration is a collage of a photograph of the model moss *Physcomitrella patens* and the graphic maps of its plastid (top/front) and mitochondrial (bottom/back) genomes. The *Physcomitrella* photograph (kindly provided by Anika Nicolaudius, Bonn and modified with Adobe Photoshop Elements including the water color style filter option) shows plantlets with developing light-brown spherical sporophytes. The organelle genome maps (modified with the Adobe tilting and shifting options) were constructed from the retrieved sequence files for the plastid genome (GenBank accession number AP005672) and the mitochondrial genome (GenBank accession number AB251495) using the freely available drawing tool for organellar genomes OGDRAW (http://ogdraw.mpimp-golm.mpg.de/). For details on nomenclature and color coding of gene classes, see: Lohse M, Drechsel O and Bock R (2007) OrganellarGenomeDRAW (OGDRAW) - a tool for the easy generation of high-quality custom graphical maps of plastid and mitochondrial genomes. Curr Genet 52: 267–274.

Advances in Photosynthesis and Respiration
Including Bioenergy and Related Processes

VOLUME 35

The book series ADVANCES IN PHOTOSYNTHESIS AND RESPIRATION *Including Bioenergy and Related Processes* provides a comprehensive and state-of-the-art account of research in photosynthesis, respiration and related processes. Virtually all life on our planet Earth ultimately depends on photosynthetic energy capture and conversion to energy-rich organic molecules. These are used for food, fuel, and fiber. Photosynthesis is the source of almost all bioenergy on Earth. The fuel and energy uses of photosynthesized products and processes have become an important area of study and competition between food and fuel has led to resurgence in photosynthesis research. This series of books spans topics from physics to agronomy and medicine; from femtosecond processes through season-long production to evolutionary changes over the course of the history of the Earth; from the photophysics of light absorption, excitation energy transfer in the antenna to the reaction centers, where the highly-efficient primary conversion of light energy to charge separation occurs, through the electrochemistry of intermediate electron transfer, to the physiology of whole organisms and ecosystems; and from X-ray crystallography of proteins to the morphology of organelles and intact organisms. In addition to photosynthesis in natural systems, genetic engineering of photosynthesis and artificial photosynthesis is included in this series. The goal of the series is to offer beginning researchers, advanced undergraduate students, graduate students, and even research specialists, a comprehensive, up-to-date picture of the remarkable advances across the full scope of research on photosynthesis and related energy processes. The purpose of this series is to improve understanding of photosynthesis and plant respiration at many levels both to improve basic understanding of these important processes and to enhance our ability to use photosynthesis for the improvement of the human condition.

For further volumes:
www.springer.com/series/5599

Genomics of Chloroplasts and Mitochondria

Edited by

Ralph Bock
Max Planck Institute of Molecular Plant Physiology
Potsdam-Golm
Germany

and

Volker Knoop
University of Bonn
Germany

Springer

Editors
Ralph Bock
Max Planck Institute of Molecular
Plant Physiology
Am Mühlenberg 1
D-14476 Potsdam-Golm
Germany
rbock@mpimp-golm.mpg.de

Volker Knoop
University of Bonn
IZMB - Institut für Zelluläre
und Molekulare Botanik
Kirschallee 1
D-53115 Bonn
Germany
volker.knoop@uni-bonn.de

ISBN 978-94-007-2919-3 (HB) ISBN 978-94-007-2920-9 (e-book)
DOI 10.1007/978-94-007-2920-9
Springer Dordrecht Heidelberg New York London

Library of Congress Control Number: 2012934278

Printed on acid-free paper

Springer is part of Springer Science+Business Media (www.springer.com)

From the Series Editors

Advances in Photosynthesis and Respiration Including Bioenergy and Related Processes

Volume 35: Genomics of Chloroplasts and Mitochondria

We are delighted to announce the publication of Volume 35 in this series. With this volume we are making some changes to keep the books a leading source of information on photosynthesis and related energy processes. The series title is now updated to ***Advances in Photosynthesis and Respiration*** *Including Bioenergy and Related Processes*. The front cover, which had a distinctive white background and color palette has been changed to a web-friendly green background. The series publisher, Springer, now makes the table of contents of all of the volumes freely available online. Links to each volume are given below. Readers may also see that this volume and the past few volumes have had significantly more color and the color figures are now better integrated into the chapters, instead of being collected in one section of the book. This improvement was possible because of changes in how the books are produced. Another change is that references to chapters in books will soon be tracked by bibliographic services. This will help authors provide evidence of the importance of their work. We hope that these updates will maintain the importance of these edited volumes in the dissemination of the science of photosynthesis and bioenergy.

This Book

This volume, *Genomics of Chloroplasts and Mitochondria* was conceived and edited by Ralph Bock (Director and Scientific Member at the Max Planck Institute for Molecular Plant Physiology, Postdam-Golm, Germany) and Volker Knoop (Head of the Department of Molecular Evolution at Bonn University, Germany). Professors Bock and Knoop are leading experts on the genomics of plant organelles. Mitochondria and chloroplasts are the energy organelles of plant cells and have their own genomes, reflecting their evolutionary origins as once free-living bacteria. While mitochondria are a hallmark of all eukaryotes, chloroplasts define plants and related photosynthetic organisms. This volume consists of 19 chapters of up-to-date information on the genomics of these fascinating organelles. Both organelles exchange genetic information with the host nucleus and at the same time retain some genes critical for the bioenergetics in these organelles. Further, both organelles have unique roles in photosynthesis and in plants more generally. Both organelles exhibit very interesting genomic behaviors. Perhaps most puzzling and worth learning about is RNA editing, in which plant organelles literally change the RNA sequence after transcription such that the protein sequence is different from that coded by the DNA. Changes in the mitochondrial genome over the course of evolution of plants and algae is also a fascinating theme woven through the book. We hope the readers will find this volume enlightening and fascinating. We are grateful to the editors for their timely submission of the book and to all the 52 authors who contributed to this look at the genomics of the energy organelles that underlie life on Earth.

Authors

The current book contains 19 chapters written by 52 authors from 8 countries (Australia (5); Belgium (2); Canada (6); France (4); Germany (13); Italy (2); UK (1); and USA (19)). We thank all the authors for their valuable contribution to this book; their names (arranged alphabetically) are:

John M. **A**rchibald (Canada; Chap. 1); Michael A. Ayliffe (Australia; Chap. 9); Sidonie **B**ellot (Germany; Chap. 10); Debashish Bhattacharya (USA; Chap. 2); Alexandra-Viola Bohne (Germany; Chap. 16); Nathalie Bonnefoy (France; Chap. 19); Hans-Peter Braun (Germany; Chap. 15); Gertraud Burger (Canada; Chap. 6); Teodoro **C**ardi (Italy; Chap. 14); Anil **D**ay (UK; Chap. 18); Holger **E**ubel (Australia; Chap. 15); Sabrina **F**inster (Germany; Chap. 13); Susan **G**abay-**L**aughnan (USA; Chap. 12); Philippe Giegé (France; Chap. 14); Michael W. Gray (Canada; Chap. 1); Stephan Greiner (Germany; Chap. 11); Jeferson Gross (USA; Chap. 2); Patrice **H**amel (USA; Chap. 19); Robert K. **J**ansen (USA; Chap. 5); Sabine **K**ahlau (Germany; Chap. 14); Kenneth G. Karol (USA; Chap. 4); Frank Kempken (Germany; Chap. 19); Volker Knoop (Germany; Chap. 8); B. Franz **L**ang (Canada; Chap. 3); Veronique Larosa (Belgium; Chap. 19); Julia Legen (Germany; Chap. 13); Libo Li (USA; Chap. 7); Yang Liu (USA; Chap. 7); Andrew H. Lloyd (Australia; Chap. 9); Pal **M**aliga (USA; Chap. 17); Aurora M. **N**edelcu (Canada; Chaps. 3 and 6); Kathleen J. Newton (USA; Chap. 12); Jörg Nickelsen (Germany; Chap. 16); Karen N. **P**elletreau (USA; Chap. 2); Yin-Long **Q**iu (USA; Chap. 7); Yujiao Qu (Germany; Chap. 13); Claire **R**emacle (Belgium; Chap. 19); Susanne S. Renner (Germany; Chap. 10); Adrian Reyes-Prieto (USA; Chap. 2); Mathieu Rousseau-Gueutin (France; Chap. 9); Tracey A. Ruhlman (USA; Chap. 5); Mary E. Rumpho (USA; Chap. 2); Thalia **S**alinas (France; Chap. 19); Christian Schmitz-Linneweber (Germany; Chap. 13); Nunzia Scotti (Italy; Chap. 14); Anna E. Sheppard (Australia; Chap. 9); Nitya Subramanian (USA; Chap. 19); Jeremy N. **T**immis (Australia; Chap. 9); Bin **W**ang (USA; Chap. 7); Paul G. Wolf (USA; Chap. 4); Jiayu **X**ue (USA; Chap. 7).

Our Books: 35 Volumes

We list below information on all the 35 volumes that have been published thus far (see http://www.springer.com/series/5599 for the series web site). We are pleased to note that Springer, our publisher, is now producing complete *Tables of Contents* of these books. Electronic access to individual chapters depends on subscription (ask your librarian) but Springer provides free downloadable front matter as well as indexes. As of July, 2011, Tables of Contents are available for all volumes. The available web sites of the books in the Series are listed below.

- **Volume 34 (2012) Photosynthesis – Plastid Biology, Energy Conversion and Carbon Assimilation**, edited by Julian Eaton-Rye, Baishnab C. Tripathy, and Thomas D. Sharkey, from New Zealand, India, and USA. Thirty-three chapters, 854 pp., Hardcover, ISBN: 978-94-007-1578-3 (HB) ISBN 978-94-007-1579-0 (e-book) [http://www.springerlink.com/content/978-94-007-1578-3/]
- **Volume 33 (2012): Functional Genomics and Evolution of Photosynthetic Systems**, edited by Robert L. Burnap and Willem F.J. Vermaas, from USA. Fifteen chapters, 428 pp., ISBN: 978-94-007-1532-5 [http://www.springerlink.com/content/978-94-007-1532-5/]
- **Volume 32 (2011): C4 Photosynthesis and Related CO_2 Concentrating Mechanisms,** edited by Agepati S. Raghavendra and Rowan Sage, from India and Canada. Nineteen chapters, 425 pp., Hardcover, ISBN: 978-90-481-9406-3 [http://www.springerlink.com/content/978-90-481-9406-3/]
- **Volume 31 (2010): The Chloroplast: Basics and Applications**, edited by Constantin Rebeiz (USA), Christoph Benning (USA), Hans J. Bohnert (USA), Henry Daniell (USA), J. Kenneth Hoober (USA), Hartmut K. Lichtenthaler (Germany), Archie R. Portis (USA), and Baishnab C. Tripathy (India). Twenty-five chapters, 451 pp., Hardcover, ISBN: 978-90-481-8530-6 [http://www.springerlink.com/content/978-90-481-8530-6/]

- **Volume 30 (2009): Lipids in Photosynthesis: Essential and Regulatory Functions**, edited by Hajime Wada and Norio Murata, both from Japan. Twenty chapters, 506 pp., Hardcover, ISBN: 978-90-481-2862-4; e-book, ISBN: 978-90-481-2863-1 [http://www.springerlink.com/content/978-90-481-2862-4/]
- **Volume 29 (2009): Photosynthesis in Silico: Understanding Complexity from Molecules**, edited by Agu Laisk, Ladislav Nedbal, and Govindjee, from Estonia, The Czech Republic, and USA. Twenty chapters, 525 pp., Hardcover, ISBN: 978-1-4020-9236-7 [http://www.springerlink.com/content/978-1-4020-9236-7/]
- **Volume 28 (2009): The Purple Phototrophic Bacteria**, edited by C. Neil Hunter, Fevzi Daldal, Marion C. Thurnauer and J. Thomas Beatty, from UK, USA and Canada. Forty-eight chapters, 1053 pp., Hardcover, ISBN: 978-1-4020-8814-8 [http://www.springerlink.com/content/978-1-4020-8814-8/]
- **Volume 27 (2008): Sulfur Metabolism in Phototrophic Organisms**, edited by Christiane Dahl, Rüdiger Hell, David Knaff and Thomas Leustek, from Germany and USA. Twenty-four chapters, 551 pp., Hardcover, ISBN: 978-4020-6862-1 [http://www.springerlink.com/content/978-1-4020-6862-1/]
- **Volume 26 (2008): Biophysical Techniques Photosynthesis**, Volume II, edited by Thijs Aartsma and Jörg Matysik, both from The Netherlands. Twenty-four chapters, 548 pp., Hardcover, ISBN: 978-1-4020-8249-8 [http://www.springerlink.com/content/978-1-4020-8249-8/]
- **Volume 25 (2006): Chlorophylls and Bacteriochlorophylls: Biochemistry, Biophysics, Functions and Applications**, edited by Bernhard Grimm, Robert J. Porra, Wolfhart Rüdiger, and Hugo Scheer, from Germany and Australia. Thirty-seven chapters, 603 pp., Hardcover, ISBN: 978-1-40204515-8 [http://www.springerlink.com/content/978-1-4020-4515-8/]
- **Volume 24 (2006): Photosystem I: The Light-Driven Plastocyanin:Ferredoxin Oxidoreductase**, edited by John H. Golbeck, from USA. Forty chapters, 716 pp., Hardcover, ISBN: 978-1-40204255-3 [http://www.springerlink.com/content/978-1-4020-4255-3/]
- **Volume 23 (2006): The Structure and Function of Plastids**, edited by Robert R. Wise and J. Kenneth Hoober, from USA. Twenty-seven chapters, 575 pp., Softcover, ISBN: 978-1-4020-6570–6; Hardcover, ISBN: 978-1-4020-4060-3 [http://www.springerlink.com/content/978-1-4020-4060-3/]
- **Volume 22 (2005): Photosystem II: The light-Driven Water:Plastoquinone Oxidoreductase**, edited by Thomas J. Wydrzynski and Kimiyuki Satoh, from Australia and Japan. Thirty-four chapters, 786 pp., Hardcover, ISBN: 978-1-4020-4249-2 [http://nwww.springerlink.com/content/978-1-4020-4249-2/]
- **Volume 21 (2005): Photoprotection, Photoinhibition, Gene Regulation, and Environment**, edited by Barbara Demmig-Adams, William W. Adams III and Autar K. Mattoo, from USA. Twenty-one chapters, 380 pp., Hardcover, ISBN: 978-14020-3564-7 [http://www.springerlink.com/content/978-1-4020-3564-7/]
- **Volume 20 (2006): Discoveries in Photosynthesis**, edited by Govindjee, J. Thomas Beatty, Howard Gest and John F. Allen, from USA, Canada and UK. One hundred and eleven chapters, 1,304 pp., Hardcover, ISBN: 978-1-4020-3323-0 [http://www.springerlink.com/content/978-1-4020-3323-0/]
- **Volume 19 (2004): Chlorophyll *a* Fluorescence: A Signature of Photosynthesis**, edited by George C. Papageorgiou and Govindjee, from Greece and USA. Thirty-one chapters, 820 pp., Hardcover, ISBN: 978-1-4020-3217-2 [http://www.springerlink.com/content/978-1-4020-3217-2/]
- **Volume 18 (2005): Plant Respiration: From Cell to Ecosystem**, edited by Hans Lambers and Miquel Ribas-Carbo, from Australia and Spain. Thirteen chapters, 250 pp., Hardcover, ISBN: 978-14020-3588-3 [http://www.springerlink.com/content/978-1-4020-3588-3/]
- **Volume 17 (2004): Plant Mitochondria: From Genome to Function**, edited by David Day, A. Harvey Millar and James Whelan, from Australia. Fourteen chapters, 325 pp., Hardcover, ISBN: 978-1-4020-2399-6 [http://www.springerlink.com/content/978-1-7923-2399-6/]
- **Volume 16 (2004): Respiration in Archaea and Bacteria: Diversity of Prokaryotic Respiratory Systems**, edited by Davide Zannoni, from Italy. Thirteen chapters, 310 pp., Hardcover, ISBN: 978-14020-2002-5 [http://www.springerlink.com/content/978-1-4020-2002-5]

- **Volume 15 (2004): Respiration in Archaea and Bacteria: Diversity of Prokaryotic Electron Transport Carriers**, edited by Davide Zannoni, from Italy. Thirteen chapters, 350 pp., Hardcover, ISBN: 978-1-4020-2001-8 [http://www.springerlink.com/content/978-0-7923-2001-8/]
- **Volume 14 (2004): Photosynthesis in Algae**, edited by Anthony W. Larkum, Susan Douglas and John A. Raven, from Australia, Canada and UK. Nineteen chapters, 500 pp., Hardcover, ISBN:978-0-7923-6333-0[http://www.springerlink.com/content/978-0-7923-6333-0/]
- **Volume 13 (2003): Light-Harvesting Antennas in Photosynthesis**, edited by Beverley R. Green and William W. Parson, from Canada and USA. Seventeen chapters, 544 pp., Hardcover, ISBN: 978-07923-6335-4 [http://www.springerlink.com/content/978-0-7923-6335-4/]
- **Volume 12 (2003): Photosynthetic Nitrogen Assimilation and Associated Carbon and Respiratory Metabolism**, edited by Christine H. Foyer and Graham Noctor, from UK and France. Sixteen chapters, 304 pp., Hardcover, ISBN: 978-07923-6336-1 [http://www.springerlink.com/content/978-0-7923-6336-1/]
- **Volume 11 (2001): Regulation of Photosynthesis**, edited by Eva-Mari Aro and Bertil Andersson, from Finland and Sweden. Thirty-two chapters, 640 pp., Hardcover, ISBN: 978-0-7923-6332-3 [http://www.springerlink.com/content/978-0-7923-6332-3/]
- **Volume 10 (2001): Photosynthesis: Photobiochemistry and Photobiophysics**, edited by Bacon Ke, from USA. Thirty-six chapters, 792 pp., Softcover, ISBN: 978-0-7923-6791-8; Hardcover: ISBN: 978-0-7923-6334-7 [http://www.springerlink.com/content/978-0-7923-6334-7/]
- **Volume 9 (2000): Photosynthesis: Physiology and Metabolism**, edited by Richard C. Leegood, Thomas D. Sharkey and Susanne von Caemmerer, from UK, USA and Australia. Twenty-four chapters, 644 pp., Hardcover, ISBN: 978-07923-6143-5 [http://www.springerlink.com/content/978-0-7923-6143-5/]
- **Volume 8 (1999): The Photochemistry of Carotenoids**, edited by Harry A. Frank, Andrew J. Young, George Britton and Richard J. Cogdell, from USA and UK. Twenty chapters, 420 pp., Hardcover, ISBN: 978-0-7923-5942-5 [http://www.springerlink.com/content/978-0-7923-5942-5/]
- **Volume 7 (1998): The Molecular Biology of Chloroplasts and Mitochondria in *Chlamydomonas***, edited by Jean David Rochaix, Michel Goldschmidt-Clermont and Sabeeha Merchant, from Switzerland and USA. Thirty-six chapters, 760pp., Hardcover, ISBN: 978-0-7923-5174-0 [http://www.springerlink.com/content/978-0-7923-5174-0/]
- **Volume 6 (1998): Lipids in Photosynthesis: Structure, Function and Genetics**, edited by Paul-André Siegenthaler and Norio Murata, from Switzerland and Japan. Fifteen chapters, 332 pp., Hardcover, ISBN: 978-0-7923-5173-3 [http://www.springerlink.com/content/978-0-7923-5173-3/]
- **Volume 5 (1997): Photosynthesis and the Environment**, edited by Neil R. Baker, from UK. Twenty chapters, 508 pp., Hardcover, ISBN: 978-07923-4316-5 [http://www.springerlink.com/content/978-0-7923-4316-5/]
- **Volume 4 (1996): Oxygenic Photosynthesis: The Light Reactions**, edited by Donald R. Ort and Charles F. Yocum, from USA. Thirty-four chapters, 696 pp., Softcover: ISBN: 978-0-7923- 3684–6; Hardcover, ISBN: 978-0-7923-3683-9 [http://www.springerlink.com/content/978-0-7923-3683-9/]
- **Volume 3 (1996): Biophysical Techniques in Photosynthesis**, edited by Jan Amesz and Arnold J. Hoff, from The Netherlands. Twenty-four chapters, 426 pp., Hardcover, ISBN: 978-0-7923-3642-6 [http://www.springerlink.com/content/978-0-7923-3642-6/]
- **Volume 2 (1995): Anoxygenic Photosynthetic Bacteria**, edited by Robert E. Blankenship, Michael T. Madigan and Carl E. Bauer, from USA. Sixty-two chapters, 1331 pp., Hardcover, ISBN:978-0-7923-3682-8[http://www.springerlink.com/content/978-0-7923-3681-5/]
- **Volume 1 (1994): The Molecular Biology of Cyanobacteria**, edited by Donald R. Bryant, from USA. Twenty-eight chapters, 916 pp., Hardcover, ISBN: 978-0-7923-3222-0 [http://www.springerlink.com/content/978-0-7923-3222-0/]

Further information on these books and ordering instructions can be found at http://www.springer.com/series/5599. Contents of volumes 1–29 can also be found at <http://www.life.uiuc.edu/govindjee/photosynSeries/ttocs.html>.

Special 25% discounts are available to members of the International Society of Photosynthesis Research, ISPR http://www.photosynthesisresearch.org/. See http://www.springer.com/ispr.

Future Advances in Photosynthesis and Respiration and Other Related Books

The readers of the current series are encouraged to watch for the publication of the forthcoming books (not necessarily arranged in the order of future appearance):

- Chloroplast Biogenesis: During Leaf Development and Senescence (Editors: Basanti Biswal, Karin Krupinska and Udaya Chand Biswal)
- The Structural Basis of Biological Energy Generation (Editor: Martin Hohmann-Marriott)
- Photosynthesis in Bryophytes and Early Land Plants (Editors: David T. Hanson and Steven K. Rice)
- Canopy Photosynthesis: From Basics to Applications (Editors: Kouki Hikosaka, Ülo Niinemets and Niels P.R. Anten)
- Saga of Non-Photochemical Quenching (NPQ) and Thermal Energy Dissipation In Plants, Algae and Cyanobacteria (Editors: Barbara Demmig-Adams,Györző Garab, William W. Adams III, and Govindjee)

In addition to the above contracted books, the following topics are under consideration:

- Algae, Cyanobacteria: Biofuel and Bioenergy
- Artificial Photosynthesis
- ATP Synthase and Proton Translocation
- Bacterial Respiration II
- Biohydrogen Production
- Carotenoids II
- Cyanobacteria II
- The Cytochromes
- Ecophysiology
- Evolution of Photosynthesis
- FACE Experiments
- Global Aspects of Photosynthesis
- Green Bacteria and Heliobacteria
- Interactions between Photosynthesis and other Metabolic Processes
- Limits of Photosynthesis: Where do we go from here
- Photosynthesis, Biomass and Bioenergy
- Photosynthesis under Abiotic and Biotic Stress
- Plant Respiration II

If you have any interest in editing/co-editing any of the above listed books, or being an author, please send an E-mail to Tom Sharkey (tsharkey@msu.edu) and/or to Govindjee at gov@illinois.edu. Suggestions for additional topics are also welcome.

In view of the interdisciplinary character of research in photosynthesis and respiration, it is our earnest hope that this series of books will be used in educating students and researchers not only in Plant Sciences, Molecular and Cell Biology, Integrative Biology, Biotechnology, Agricultural Sciences, Microbiology, Biochemistry, Chemical Biology, Biological Physics, and Biophysics, but also in Bioengineering, Chemistry, and Physics.

We take this opportunity to thank and congratulate Ralph Bock and Volker Knoop for their outstanding editorial work; they have done a fantastic job not only in editing, but also in organizing this book for all of us, and for their highly professional dealing with the reviewing process. We thank all the 52 authors of this book (see the list above): without their authoritative chapters, there would be no such volume. We give special thanks to Saravanan Purushothaman, SPi Global, India for directing the typesetting of this book; his expertise has been crucial in bringing this book to completion. We owe Jacco Flipsen, Ineke Ravesloot and André Tournois (of Springer) thanks for their friendly working relation with us that led to the production of this book.

Thomas D. Sharkey
Department of Biochemistry and
Molecular Biology,
Michigan State University,
East Lansing, MI, 48823, USA
e-mail: tsharkey@msu.edu

Govindjee
Department of Plant Biology
Department of Biochemistry and Center
of Biophysics & Computational Biology,
University of Illinois at Urbana-Champaign,
Urbana, IL 61801, USA
(e-mail: gov@life.illinois.edu)

Series Editors

Govindjee, who uses one name only, was born on October 24, 1932, in Allahabad, India. Since 1999, he has been Professor Emeritus of Biochemistry, Biophysics and Plant Biology at the University of Illinois at Urbana-Champaign (UIUC), Urbana, IL, USA. He obtained his B.Sc. (Chemistry and Biology) and M.Sc. (Botany; Plant Physiology) in 1952 and 1954, from the University of Allahabad. He studied 'Photosynthesis' at the UIUC, under two pioneers of photosynthesis Robert Emerson, and Eugene Rabinowitch, obtaining his Ph.D. in 1960, in Biophysics. He is best known for his research on the excitation energy transfer, light emission, the primary photochemistry and the electron transfer in "Photosystem II" (PS II, water-plastoquinone oxido-reductase). His research, with many collaborators, has included the discovery of a short-wavelength form of chlorophyll (Chl) *a* functioning in the Chl *b*-containing system, now called PS II; of the two-light effect in Chl *a* fluorescence; and, with his wife Rajni Govindjee, of the two-light effect (Emerson enhancement) in NADP reduction in chloroplasts. His major achievements, together with several other researchers, include an understanding of the basic relationships between Chl *a* fluorescence and photosynthetic reactions; an unique role of bicarbonate/carbonate on the electron acceptor side of PS II, particularly in the protonation events involving the Q_B binding region; the theory of thermoluminescence in plants; the first picosecond measurements on the primary photochemistry of PS II; and the use of Fluorescence Lifetime Imaging Microscopy (FLIM) of Chl *a* fluorescence in understanding photoprotection, by plants, against excess light. His current focus is on the 'History of Photosynthesis Research', in 'Photosynthesis Education', and in the 'Possible Existence of Extraterrestrial Life'. He has served on the faculty of the UIUC for ~40 years. Govindjee's honors include: Fellow of the American Association of Advancement of Science (AAAS); Distinguished Lecturer of the School of Life Sciences, UIUC; Fellow and Lifetime member of the National Academy of Sciences (India); President of the American Society for Photobiology (1980–1981); Fulbright Scholar and Fulbright Senior Lecturer; Honorary President of the 2004 International Photosynthesis Congress (Montréal, Canada); the first recipient of the Lifetime Achievement Award of the Rebeiz

Foundation for Basic Biology, 2006; Recipient of the Communication Award of the International Society of Photosynthesis Research, 2007; and the Liberal Arts & Sciences Lifetime Achievement Award of the UIUC, 2008. Further, Govindjee was honored (1) in 2007, through two special volumes of Photosynthesis Research, celebrating his 75th birthday and for his 50-year dedicated research in 'Photosynthesis' (Guest Editor: Julian Eaton-Rye); (2) in 2008, through a special International Symposium on 'Photosynthesis in a Global Perspective', held in November, 2008, at the University of Indore, India; and (3) Volume 34 of this Series "*Photosynthesis – Plastid Biology, Energy Conversion and Carbon Assimilation*", edited by Julian Eaton-Rye, Baishnab C. Tripathy, and Thomas D. Sharkey, was dedicated to him, celebrating his 80th year. Govindjee is coauthor of 'Photosynthesis' (John Wiley, 1969); and editor of many books, published by several publishers including Academic Press and Kluwer Academic Publishers (now Springer). Since 2007, each year a Govindjee and Rajni Govindjee Award is given to graduate students, by the Department of Plant Biology, at the UIUC, to recognize Excellence in Biological Sciences. For further information on Govindjee, see his web site at http://www.life.illinois.edu/govindjee.

Thomas D. (Tom) Sharkey obtained his Bachelor's degree in Biology in 1974 from Lyman Briggs College, a residential science college at Michigan State University, East Lansing, Michigan. After 2 years as a research technician, Tom entered a Ph.D. program in the Department of Energy Plant Research Laboratory at Michigan State University under the mentorship of Klaus Raschke and finished in 1979. Post-doctoral research was carried out with Graham Farquhar at the Australian National University, in Canberra, where he coauthored a landmark review on photosynthesis and stomatal conductance. For 5 years he worked at the Desert Research Institute, Reno, Nevada. After Reno, Tom spent 20 years as Professor of Botany at the University of Wisconsin in Madison. In 2008, Tom became Professor and Chair of the Department of Biochemistry and Molecular Biology at Michigan State University. Tom's research interests center on the exchange of gases between plants and the atmosphere. The biochemistry and biophysics underlying carbon dioxide uptake and isoprene emission from plants form the two major research topics in his laboratory. Among his contributions are measurement of the carbon dioxide concentration inside leaves, an exhaustive study of short-term feedback effects in carbon metabolism, and a significant contribution to elucidation of the pathway by which leaf starch breaks down at night. In the isoprene research field, Tom is recognized as the leading advocate for thermotolerance of photosynthesis as the explanation for why plants emit isoprene. In addition, his laboratory has cloned many of the genes that underlie isoprene synthesis and published many papers on the biochemical regulation of isoprene synthesis. Tom has co-edited three books, the first on trace gas emissions from plants in 1991 (with Elizabeth Holland and Hal Mooney) and then volume 9 of this series (with Richard Leegood and Susanne von Caemmerer) on the physiology of carbon metabolism of photosynthesis in 2000 and volume 34 (with Julian Eaton-Rye and Baishnab C. Tripathy) entitled *Photosynthesis: Plastid Biology, Energy Conversion and Carbon Assimilation*. Tom is listed in Who's Who and is a "Highly Cited Researcher" according to the Thomson Reuters Institute for Scientific Information.

Contents

Preface

The genomics era for plant cell organelles started in 1979, when Hans Kössel at the University of Freiburg, Germany, sequenced the first piece of chloroplast DNA: the 3′-terminal part of the 16S ribosomal RNA gene from maize (*Zea mays*). This was a remarkable feat, because, in addition to obtaining the very first DNA sequence from a plant, the study also provided strong molecular evidence for the endosymbiotic origin of plastids (which, at that time, was still hotly debated). By showing that the chloroplast 16S ribosomal RNA sequence displayed strong similarity to the (previously sequenced) 16S ribosomal RNA gene from the bacterium *Escherichia coli*, Hans and his post-doc Zsuzsanna Schwarz-Sommer provided compelling evidence for the plastid DNA stemming from a prokaryotic ancestor. In the following years, many more plastid genes were sequenced from many different plant species. At the same time, with the advent of new molecular tools, such as restriction enzymes and hybridization techniques, insights into the overall structure and organization of the chloroplast genome were obtained. Numerous restriction maps and physical maps were constructed and integrated with the growing amount of sequence information. In 1986, the mapping, cloning and sequencing efforts culminated in the determination of two complete chloroplast genome sequences: from a bryophyte, the liverwort *Marchantia polymorpha* (sequenced by Kanji Ohyama's group), and from a seed plant, the cultivated tobacco *Nicotiana tabacum* (sequenced by Masahiro Sugiura's group).

Due to its more complex genome structure and smaller copy number per cell, research on plant mitochondrial genomes followed suit with some delay. In 1992, it was again Kanji Ohyama's group who determined the first complete genome sequence of a plant mitochondrial genome, again from their favorite model plant, the liverwort *Marchantia polymorpha*. Five years later, the first genome sequence from a seed plant, the brassicaceous weed *Arabidopsis thaliana*, followed (sequenced by Axel Brennicke's laboratory).

The past decade has seen an explosion of completed organellar genome projects and the list of fully sequenced plastid and mitochondrial genomes (http://www.ncbi.nlm.nih.gov/genomes/GenomesHome.cgi?taxid=2759&hopt=html) is now growing almost every day. Importantly, this development has been accompanied by a steadily improving coverage of diverse taxonomic groups and, in this way, organellar genomics continues to contribute greatly to resolving unclear evolutionary links and phylogenetic relationships. Over the last few years, next-generation sequencing techniques have revolutionized the genomics field and, foreseeably, their ever increasing power will keep accelerating also genome research on plant organelles. At the same time, technologies for the genome-wide analysis of gene expression at the RNA and protein levels become more and more powerful and, while currently mainly applied in a handful of model plants, will likely become routine tools for monitoring the dynamic changes of organellar transcriptomes and proteomes in response to environmental stimuli and developmental cues in many more species. With that, plant organelles can potentially be at the forefront of future efforts to model gene expression networks and metabolite networks, which represents one of the cornerstones of the nowadays much-talked-about field of systems biology.

This volume of *Advances in Photosynthesis and Respiration* attempts to summarize the state of the art in genomics research on plant mitochondria and plastids. Its first two chapters are dedicated to the evolution of organelles by primary, secondary and higher-order endosymbioses. Both chapters illustrate Natures amazing fondness of experimenting with endosymbiotic associations and pinpoint important principles involved in the successful establishment of enduring endosymbiotic relationships. The following six chapters (Chaps. 3, 4, 5, 6, 7, 8) give an account of genomics research in the various taxonomic groups of algae and embryophyte plants. They describe the characteristic features of plastid and mitochondrial genomes in each taxonomic group, highlight recurring evolutionary patterns that underlie changes in genome structure and gene content and emphasize the enormous contributions of organellar genomics to constructing phylogenetic trees and resolving uncertain evolutionary relationships. Chapters 9 and 10 are dedicated to the gene transfer processes involving organellar DNA. While the intracellular transfer of plastid and mitochondrial DNA to the nucleus represents the initiating event in endosymbiotic gene transfer and, in its presumably non-functional form, generates so-called promiscuous DNA (Chap. 9), the transfer of organellar DNA between cells can result in marvelous examples of horizontal gene transfer between plants, but also between plants and other organisms (Chap. 10). Chapters 11 and 12 deal with mutants in plant organellar genomes, their isolation and experimental induction. In both plastids and mitochondria, the study of mutants has contributed greatly to our understanding of organelle physiology and the mechanisms of organellar gene expression. In addition, mutations in organellar genomes are responsible for important agronomic traits, such as cytoplasmic male sterility and resistances to herbicides. Chapters 13, 14 and 15 illuminate general aspects of the expression of plant organellar genomes. Chapter 13 summarizes our current knowledge about RNA editing, a still largely enigmatic RNA processing step that post-transcriptionally alters the identity of individual nucleotide positions in organellar transcripts and may represent a mechanism to correct DNA mutations at the RNA level (Chap. 13). The two following chapters address the expression of plastid and mitochondrial genomes at the RNA (Chap. 14) and protein (Chap. 15) levels. The authors describe methods for expression profiling (transcriptomics) and proteomics and the exciting insights that these studies have provided into the dynamic changes in organellar gene expression in response to environmental cues, developmental stimuli and genetic perturbations of the organellar genetic system. The final four chapters (Chaps. 16, 17, 18, 19) are dedicated to the genetic transformation of organellar genomes. The current status in transformation methods for plastid genomes of algae (Chap. 16) and higher plants (Chap. 17) as well as methods for mitochondrial transformation and transfection of isolated organelles (Chap. 19) are reviewed and the various applications that organelle transformation technologies have in basic research and biotechnology are discussed. A special chapter (Chap. 18) is devoted to reverse genetics in plastids, which has not only been instrumental in elucidating gene functions and structure-function relationships in chloroplast proteins, but also has provided fundamental insights in principles of organelle transformation, recombination and genome sorting and, moreover, led to the development of novel tools for plastid transformation that are widely applicable.

This volume of *Advances in Photosynthesis and Respiration* is written primarily for researchers working in the fields of organelle genetics, gene expression and biotechnology as well as plant systematics and evolutionary biology. In addition to providing a comprehensive overview of their topic, the authors of the individual chapters have tried to discuss concepts, appraise current hypotheses and emphasize general principles. Inevitably, there is some overlap between the contributions, which, however, has been largely limited to the extent needed to ensure that the individual chapters can be read in isolation.

Authors and editors hope that this book will also serve as a stepping stone for students becoming interested in organelle biology, genomics and evolution and for new researchers entering these fields.

Last but not least, we wish to express our sincere thanks to the authors of each chapter. We have been very fortunate to win preeminent experts for all chapters – their thoroughness and commitment made this book possible. We are also grateful to the Series Editor Tom Sharkey and the staff at Springer Publishers for their valuable help in editing and formatting this volume of *Advances in Photosynthesis and Respiration*.

Ralph Bock
Max Planck Institute of Molecular Plant Physiology
Potsdam-Golm, Germany
rbock@mpimp-golm.mpg.de

Volker Knoop
University of Bonn
Germany
volker.knoop@uni-bonn.de

The Editors

Ralph Bock

Ralph Bock was born on October 8, 1967 in Wolfen, Germany. He is a Director at the Max Planck Institute of Molecular Plant Physiology (MPI-MP) in Potsdam-Golm, Germany, and a Professor of Plant Biology at the University of Potsdam. Ralph received an M.Sc. (Diplom) degree in Genetics in 1993 from the University of Halle, Germany, under the supervision of Professor Rudolf Hagemann. He did his Ph.D. work at the Waksman Institute, Rutgers, The State University of New Jersey (with Professor Pal Maliga), and at the University of Freiburg, Germany (with Professor Hans Kössel), and received his Ph.D. in 1996 from the University of Freiburg. He was an Assistant Professor at the Institute of Biology III, University of Freiburg from 1996 to 2001, and a Full Professor at the University of Münster, Germany, from 2001 to 2004, before joining the May Planck Society as a Director at MPI-MP in 2004. Ralph is distinguished for his manifold contributions to chloroplast genetics, biochemistry and biotechnology. His early work was focused on chloroplast RNA processing and, especially, on the study of RNA editing, a curious transcript maturation step by which individual nucleotides in the messenger RNA are post-transcriptionally altered. Later, he broadened his research interests to include the study of the mechanisms and regulation of other transcriptional and post-transcriptional processes in plastid gene expression, with the ultimate goal to obtain a systems-level understanding of organellar function in the context of the genetic and biochemical networks operating in the plant cell. He has also made seminal contributions to the development of tools for the genetic engineering of higher plant chloroplast genomes and their application in functional genomics, photosynthesis research and biotechnology. Using experimental evolution approaches, Ralph's research team successfully reconstructed endosymbiotic gene transfer processes in laboratory experiments and, recently, also discovered a path for horizontal gene transfer between plants. Among many other community activities, Ralph currently serves on the editorial boards of *Current Genetics*, *Transgenic Research* and *Eukaryotic Cell*. His pioneering research contributions were recognized, inter alia, by his election as a Member of the National Academy of Science (Leopoldina).

Volker Knoop

Volker Knoop was born on July 13, 1963 in Dortmund, Germany. He is Professor of Botany at the University of Bonn, Germany and head of the Molecular Evolution group at the IZMB (Institute for Cellular and Molecular Botany). Volker received his Diplom (M.Sc. equivalent) in Biochemistry from the Free University Berlin (West-Berlin at that time) in 1989 under the supervision of Professor Lothar Willmitzer at the Institut für Genbiologische Forschung GmbH (IGF) in Berlin-Dahlem in the independent research group of Dr. Ulla Bonas (now Professor at the University of Halle). He did his Ph.D. work on trans-splicing and RNA editing in plant mitochondria at the IGF under the supervision of Prof. Axel Brennicke and received his Ph.D. (Dr. rer. nat.) in 1992. He continued his work on molecular evolution of plant mitochondrial DNA as an independent group leader in the department of Axel Brennicke, first at the IGF in Berlin and, from 1996 to 2002, at the University of Ulm. In 2002, Volker was appointed as a professor at the Institute of Botany at the University of Bonn and, in 2003, he became the founding director of the newly established IZMB. Volker is best known for his work on the "deep green" molecular evolution of mitochondrial DNA in early land plants, with a strong focus on peculiarities in RNA processing, such as the origin of RNA editing and the evolutionary history of trans-splicing introns. More recently, his group also investigated the particularly unique mitochondrial DNAs of lycophytes. Volker received the Merckle research prize in 2000. Aside from his evolutionary interests, his group investigates a gene family of membrane transport proteins for magnesium in the model plants *Arabidopsis thaliana* and *Physcomitrella patens*. Volker was in charge of masterminding the Plant Sciences master course program at the University of Bonn – the first study program exclusively dedicated to plant sciences in Germany, which is now starting into its fourth year. Volker currently serves on the editorial board of *Journal of Systematics and Evolution*. Together with Kai Müller (now Professor at the University of Münster), he has authored the textbook "Gene und Stammbäume" (Genes and Phylogenetic Trees) on phylogenetic analyses, the second edition of which was published in 2009. Volker is married and has four children.

Contributors

John M. Archibald
Department of Biochemistry and Molecular Biology, Sir Charles Tupper Medical Building, Dalhousie University, Halifax, NS B3H 1X5, Canada

Michael A. Ayliffe
Division of Plant Industry, CSIRO, Canberra, ACT 2601, Australia

Sidonie Bellot
Systematic Botany and Mycology, University of Munich, D-80638 Munich, Germany

Debashish Bhattacharya
Department of Ecology, Evolution and Natural Resources and Institute of Marine and Coastal Sciences, Rutgers University, New Brunswick, NJ 08901, USA

Alexandra-Viola Bohne
Molekulare Pflanzenwissenschaften, Biozentrum LMU München, D-82152 Planegg-Martinsried, Germany

Nathalie Bonnefoy
Centre National de la Recherche Scientifique, Centre de Génétique Moléculaire FRE3144, FRC3115, Gif-sur-Yvette Cedex 91198, France

Hans-Peter Braun
Department V: Plant Proteomics, Institute for Plant Genetics, Leibniz Universität Hannover, D-30419 Hannover, Germany

Gertraud Burger
Department of Biochemistry, Robert-Cedergren Centre for Bioinformatics and Genomics, Université de Montréal, Montreal, QC H3T 1J4, Canada

Teodoro Cardi
CRA-ORT, Agricultural Research Council, Research Centre for Vegetable Crops, Pontecagnano, SA 84098, Italy

Anil Day
Faculty of Life Sciences, The University of Manchester, Manchester M13 9PT, UK

Holger Eubel
Department V: Plant Proteomics, Institute for Plant Genetics, Leibniz Universität Hannover, D-30419 Hannover, Germany

Sabrina Finster
Institut für Biologie, Humboldt-Universität Berlin, D-10115 Berlin, Germany

Susan Gabay-Laughnan
Department of Plant Biology, University of Illinois, Urbana, IL 61801, USA

Philippe Giegé
IBMP-CNRS, Strasbourg 67084, France

Michael W. Gray
Department of Biochemistry and Molecular Biology, Sir Charles Tupper Medical Building, Dalhousie University, Halifax, NS B3H 1X5, Canada

Stephan Greiner
Max-Planck-Institut für Molekulare Pflanzenphysiologie, Wissenschaftspark Golm, D-14476 Potsdam-Golm, Germany

Jeferson Gross
Department of Ecology, Evolution and Natural Resources and Institute of Marine and Coastal Sciences, Rutgers University, New Brunswick, NJ 08901, USA

Patrice Hamel
Department of Molecular Genetics and Department of Molecular and Cellular Biochemistry, The Ohio State University, 500 Aronoff Laboratory, Columbus, OH 43210, USA

Robert K. Jansen
Section of Integrative Biology and Institute of Cellular and Molecular Biology, University of Texas at Austin, Austin, TX 78712, USA

Sabine Kahlau
QIAGEN GmbH, D-40724 Hilden, Germany

Kenneth G. Karol
The Lewis B. and Dorothy Cullman Program for Molecular Systematics Studies, The New York Botanical Garden, Bronx, NY 10458, USA

Frank Kempken
Abteilung Botanische Genetik und Molekularbiologie, Botanisches Institut, Christian-Albrechts-Universität zu Kiel, D-24098 Kiel, Germany

Volker Knoop
Abteilung Molekulare Evolution, Institut für Zelluläre und Molekulare Botanik, Universität Bonn, D-53115 Bonn, Germany

B. Franz Lang
Department of Biochemistry, Robert-Cedergren Centre for Bioinformatics and Genomics, Université de Montréal, Montreal, QC H3T 1J4, Canada

Veronique Larosa
Laboratory of Genetics of Microorganisms, Institute of Botany B22, University of Liège, Liège B-4000, Belgium

Julia Legen
Institut für Biologie, Humboldt-Universität Berlin, D-10115 Berlin, Germany

Libo Li
Department of Ecology & Evolutionary Biology, University of Michigan, Ann Arbor, MI 48109-1048, USA

Yang Liu
Department of Ecology & Evolutionary Biology, University of Michigan, Ann Arbor, MI 48109-1048, USA

Andrew H. Lloyd
School of Molecular and Biomedical Science, The University of Adelaide, Adelaide, SA 5005, Australia

Pal Maliga
Waksman Institute, Rutgers University, Piscataway, NJ, 08854, USA
Department of Plant Biology, Rutgers University, New Brunswick, NJ 08901, USA

Aurora M. Nedelcu
Department of Biology, University of New Brunswick, Fredericton, NB E3B 5A3, Canada

Kathleen J. Newton
Division of Biological Sciences, University of Missouri, Columbia, MO 65211, USA

Jörg Nickelsen
Molekulare Pflanzenwissenschaften, Biozentrum LMU München, D-82152 Planegg-Martinsried, Germany

Karen N. Pelletreau
Department of Molecular and Biomedical Sciences, University of Maine, Orono, ME 04469, USA

Yujiao Qu
Institut für Biologie, Humboldt-Universität Berlin, D-10115 Berlin, Germany

Yin-Long Qiu
Department of Ecology & Evolutionary Biology, University of Michigan, Ann Arbor, MI 48109-1048, USA

Claire Remacle
Laboratory of Genetics of Microorganisms, Institute of Botany B22, University of Liège, Liège B-4000, Belgium

Susanne S. Renner
Systematic Botany and Mycology, University of Munich, D-80638 Munich, Germany

Adrian Reyes-Prieto
Biology Department, University of New Brunswick, Fredericton, NB E3B 5A3, Canada

Mathieu Rousseau-Gueutin
School of Molecular and Biomedical Science, The University of Adelaide, Adelaide, SA 5005, Australia

Tracey A. Ruhlman
Section of Integrative Biology and Institute of Cellular and Molecular Biology, University of Texas at Austin, Austin, TX 78712, USA

Mary E. Rumpho
Department of Molecular and Biomedical Sciences, University of Maine, Orono, ME 04469, USA

Thalia Salinas
Laboratory of Genetics of Microorganisms, Institute of Botany B22, University of Liège, Liège B-4000, Belgium

Christian Schmitz-Linneweber
Institut für Biologie, Humboldt-Universität Berlin, D-10115 Berlin, Germany

Nunzia Scotti
CNR – IGV, Institute of Plant Genetics, Res. Div. Portici, Portici, NA 0055, Italy

Anna E. Sheppard
School of Molecular and Biomedical Science, The University of Adelaide, Adelaide, SA 5005, Australia

Nitya Subramanian
Department of Molecular Genetics and Department of Molecular and Cellular Biochemistry, The Ohio State University, 500 Aronoff Laboratory, Columbus, OH 43210, USA

Jeremy N. Timmis
School of Molecular and Biomedical Science, The University of Adelaide, Adelaide, SA 5005, Australia

Bin Wang
Department of Ecology & Evolutionary Biology, University of Michigan, Ann Arbor, MI 48109-1048, USA

Paul G. Wolf
Department of Biology, Utah State University, Logan, UT 84322, USA

Jiayu Xue
School of Life Sciences, Nanjing University, Nanjing 210093, China

Author Index

Chapter 1

Origins of Mitochondria and Plastids

Michael W. Gray* and John M. Archibald
Canadian Institute for Advanced Research, Program in Integrated Microbial Biodiversity, Department of Biochemistry and Molecular Biology, Sir Charles Tupper Medical Building, Dalhousie University, Halifax, NS B3H 4R2, Canada

Summary

The evolution of mitochondria and plastids (chloroplasts) by endosymbiosis is a central tenet of modern eukaryotic cell biology. Evidence in support of a prokaryotic ancestry for these textbook organelles is now stronger than ever, but despite decades of genomics-enabled research, fundamental questions about the earliest events leading to their establishment remain unanswered. Foremost among them is the precise nature of the cells involved in these endosymbiotic mergers. Mitochondria and plastids are traditionally considered to be the products of independent, but fundamentally similar, endosymbioses involving eukaryotic hosts and bacterial endosymbionts. Such a model still holds true for plastids, but increasingly it appears that mitochondria could have evolved in a fashion quite different from this 'classical' scenario. In this introductory chapter we provide an overview of the primary endosymbiotic origins of mitochondria and plastids, focusing on advances coming from the latest comparative genomic and proteomic investigations. In the case of plastids, the recently evolved photosynthetic 'organelles' of the testate amoeba *Paulinella* provide a possible window on the ancient origin of canonical plastids and are thus also discussed in detail.

*Author for correspondence, e-mail: M.W.Gray@dal.ca

R. Bock and V. Knoop (eds.), Genomics of Chloroplasts and Mitochondria, Advances in Photosynthesis and Respiration 35, pp. 1–30,
DOI 10.1007/978-94-007-2920-9_1,

I. Introduction

The classical endosymbiont hypothesis for the origin of mitochondria and plastids (chloroplasts) posits that on two separate occasions, free-living prokaryotes – from within the bacterial groups α-Proteobacteria and Cyanobacteria, respectively – entered into increasingly intimate associations with eukaryotic host cells, ultimately evolving into the well-studied sub-cellular organelles they are today. Over time and with much data accumulated, this hypothesis has risen to the status of theory; that these quintessential membrane-bound organelles are derived from prokaryotes is now considered textbook 'fact'. In this era of high-throughput genomics and proteomics, how much do the data still support this classical view?

This chapter provides an overview of the 'primary' endosymbiotic origins of mitochondria and plastids and in so doing sets the stage for the chapters that follow. As we shall see, many aspects of the classical endosymbiont hypothesis still hold true for both organelles, but others have changed significantly in response to a wealth of comparative data from diverse eukaryotic lineages. This is especially true in the case of mitochondria, where recent years have seen numerous alternative evolutionary scenarios proposed. Mitochondria are still believed to have evolved before plastids, but there is no evidence supporting the existence of an amitochondriate phase in eukaryotic evolution. Indeed, it is possible that the origin of the mitochondrion was contemporaneous with the origin of the eukaryotic cell itself.

Abbreviations: CSS – Cellular signature structure; ER – Endoplasmic reticulum; ESP – Eukaryotic signature protein; EST – Expressed sequence tag; GTA – Gene transfer agent; HGT – Horizontal gene transfer; ISC – Iron–sulfur cluster; LECA – Last eukaryotic common ancestor; MRO – Mitochondrion-related organelle; MS – Mass spectrometry; PFO – Pyruvate: ferredoxin oxidoreductase; PGK – Phosphoglycerate kinase; TCA – Tricarboxylic acid; Tic – Translocater of the inner chloroplast membrane; Toc – Translocater of the outer chloroplast membrane

II. Mitochondria

A single origin of mitochondria from within the domain Bacteria (eubacteria) is now widely accepted and overwhelmingly supported by a variety of evidence, in particular molecular evidence that has accumulated over the past four decades. At the heart of this conclusion is the demonstration that the mitochondrial genome is clearly the remnant of a eubacterial genome, arising specifically from within the α-class of the phylum Proteobacteria (*Alphaproteobacteria*, also known as α-Proteobacteria). Data underlying this conclusion have been extensively reviewed elsewhere (Gray and Doolittle 1982; Gray 1989, 1992, 1993, 1999; Gray and Spencer 1996; Lang et al. 1999a; Gray et al. 2001), and are consistent with the long-standing endosymbiont hypothesis for the evolution of mitochondria. Indeed, the most gene-rich and ancestral (least derived) mitochondrial genome known, that of the protozoon *Reclinomonas americana*, resembles to a striking degree a miniaturized eubacterial genome, retaining distinct traces of its α-proteobacterial genomic heritage (Lang et al. 1997).

Although the α-proteobacterial ancestry of the mitochondrial genome seems firmly established, our understanding of the evolutionary route from eubacterium to mitochondrion is still murky. It is somewhat disconcerting to realize that the appearance of new molecular data, particularly genomic, phylogenomic and proteomic, has muddied the waters with respect to mitochondrial origin and evolution, rather than clarifying the issue. The bottom line is that we are rather less certain today than we were two decades ago that the 'classical' endosymbiont hypothesis, which posits an amitochondriate eukaryotic host cell taking up a bacterial symbiont (Margulis 1970), provides a correct – or at least fully accurate – description of the origin of mitochondria. One reason for this reservation is that although the mitochondrial genome is indisputably α-proteobacterial in evolutionary origin, most of the constituents of the mitochondrial proteome – the collection of proteins that constitute the functional organelle – are not (Gray et al. 2001). Thus, accepting that an α-proteobacterial endosymbiont was the ancestor of mitochondria, the evolutionary re-modeling that occurred subsequently has obviously been so extensive that the contemporary mitochondrion would best be described as a genetic and functional mosaic (Szklarczyk and Huynen 2010).

Another complication is that no extant eukaryotic lineages have been discovered that can convincingly be shown to have diverged before the putative endosymbiotic acquisition of mitochondria (i.e., there are no known eukaryotic lineages that are primitively amitochondriate, in the sense of never having harbored mitochondria during their evolution). Hence, we cannot currently point to any existing eukaryotic candidates that could serve as examples of the sort of host cell that the classical endosymbiont hypothesis requires. As we shall see, these complications considerably constrain our understanding of the origin and evolution of mitochondria.

A. Genetic, Genomic and Phylogenomic Data Bearing on Mitochondrial Origins

Following the discovery that mitochondria contain a genome and carry out DNA replication, transcription and translation, various biochemical and molecular biological data based on these findings were marshaled in support of a eubacterial, endosymbiotic origin of mitochondria (a xenogenous origin), as opposed to an autogenous origin (an origin from within the eukaryotic cell itself; Gray and Doolittle 1982). For example, the fact that mitochondrial protein synthesis is sensitive to chloramphenicol but not to cycloheximide was an early indication that the mitochondrial ribosome is functionally eubacterial in character, and not an evolutionary derivative of the cytoplasmic ribosome. Molecular data stemming from comparative studies of mitochondrial and bacterial genomes have been particularly informative when it comes to assessing the origin of mitochondria. Arguments based on these data, and supporting a single origin of the mitochondrial genome from within the eubacterial class *Alphaproteobacteria*, are of three sorts (Gray et al. 1999; Lang et al. 1999a). First, mitochondrial genomes in different eukaryotes encode relatively few genes (<100), but in all studied mitochondrial genomes, these genes are essentially sub-sets of the ones found in the *R. americana* mtDNA. Accepting that mitochondrial genomes in different eukaryotic taxa are radically (but differently) reduced versions of a much larger eubacterial genome carrying a substantially greater number of genes, it is highly improbable that independently acquired α-proteobacterial genomes would have undergone convergent reduction to the same small set of residual genes. Second, although mitochondrial genome organization and gene order vary markedly among eukaryotes, a number of minimally derived mitochondrial genomes (principally found among single-celled eukaryotic microbes, or protists) retain vestiges of eubacterial operons. These operon-like clusters are invariably missing some of the genes that are found in the corresponding eubacterial operons, and these specific deletions are shared among mitochondrial genomes. For example, in the mitochondrial version of the eubacterial S10 operon, comprising a cluster of 11 ribosomal protein genes, the same six genes (*rpl3-rpl4-rpl23, rpl22, rpl29-rps17*) are missing in all characterized mitochondrial genomes that encode clustered ribosomal protein genes. The inference is that these mitochondrion-specific deletions must have been present in the mitochondrial genome of a common ancestor of the taxa in question. Finally, in phylogenetic reconstructions based on alignments of mtDNA-encoded genes, both rRNA and protein-coding, mitochondria appear as a monophyletic clade branching within *Alphaproteobacteria*. This is the case even when alignments include nucleus-encoded homologs that are encoded in the mtDNA of some eukaryotes but in the nuclear DNA of others (e.g., Burger et al. 1996), additionally providing support for the concept of mitochondrion-to-nucleus gene transfer that has shaped the evolution of the mitochondrial and nuclear genomes.

Initial phylogenetic reconstructions based on rRNA sequence data identified the *Alphaproteobacteria* as the probable evolutionary source of mitochondria (Yang et al. 1985), a satisfying conclusion in view of the fact that a member of this class (*Paracoccus denitrificans*) had earlier been proposed as the bacterium whose electron transport chain bears an especially strong resemblance to the mitochondrial one in its biochemical properties (John and Whatley 1975). Subsequently, in phylogenetic trees based on protein as well as rRNA sequences, mitochondria were found to branch together with Rickettsiales (Gupta 1995; Lang et al. 1999a), one of six or more orders within *Alphaproteobacteria* (Williams et al. 2007). Again, this result was intellectually pleasing because members of Rickettsiales (genera such as *Rickettsia*, *Anaplasma*, *Ehrlichia* and *Wolbachia*) are obligate, intracellular parasites of eukaryotic cells, superficially resembling mitochondria in their dependence on a host: and, like mitochondria, they harbor markedly reduced genomes compared to typical eubacteria (Sällström and Andersson 2005). However, the genomes of the mitochondrion and members of the Rickettsiales are clearly the products of independent evolutionary reduction (Andersson et al. 1998; Gray 1998), which argues that mitochondria cannot have been derived directly from a Rickettsiales taxon; rather, these two groups share a more distant common ancestor.

Although a specific evolutionary connection between mitochondria and Rickettsiales has been repeatedly demonstrated (Viale and Arakaki 1994; Gupta 1995; Sicheritz-Pontén et al. 1998; Lang et al. 1999b), it is still not certain whether the two are sister groups, or whether mitochondria actually branch within Rickettsiales, which comprises two distinct families, *Rickettsiaceae* and *Anaplasmataceae* (Williams et al. 2007). A number of studies have concluded that mitochondria are more closely related to the former family (containing various Rickettsia species) than to the latter (comprising the genera *Anaplasma*, *Ehrlichia* and *Wolbachia*) (Karlin and Brocchieri 2000; Emelyanov 2001a, b, 2003a, b).

The specific affiliation of mitochondria and Rickettsiales in phylogenetic trees has been questioned on the grounds that this rooting, although robust, may represent a phylogenetic artifact attributable to the high rate of sequence divergence and high A+T content of the genomes of Rickettsiales taxa and mitochondria (in other words, a long-branch-attraction artifact). Indeed, Esser et al. (2004) were not able to ascertain with certainty the placement of mitochondria within their phylogenetic trees, pointing out that *Rhodosprillum rubrum*, a member of a different α-proteobacterial order (Rhodospirillales), "came as close to mitochondria as any α-proteobacterium investigated". Accordingly, there has been considerable interest in expanding the availability of α-proteobacterial genome sequences, in particular ones from free-living members of Rickettsiales. These phylogenetic analyses have used different combinations of data sets and methods of phylogenetic inference to first show that it is possible to generate a robust phylogeny of *Alphaproteobacteria*, despite concerns about possible complications such as base composition, codon bias, variable rate of sequence divergence, and horizontal gene transfer (HGT) compromising the underlying phylogenetic signal (Williams et al. 2007). These analyses, based on expanded taxon sampling, uniformly place mitochondria within *Alphaproteobacteria*, although they still vary somewhat in the specific branching position of mitochondria within this subdivision. Wu et al. (2004), for example, reported strong support for a grouping of *Wolbachia* (family *Anaplasmataceae*) + Rickettsia (family *Rickettsiaceae*) within Rickettsiales, to the exclusion of mitochondria. Fitzpatrick et al. (2006), however, concluded that the Rickettsiales as a whole constitutes the sister group to mitochondria. On balance, there is strong convergence on an evolutionary affiliation of mitochondria with the order Rickettsiales within the class *Alphaproteobacteria*, but still no compelling consistency as to whether mitochondria branch within the Rickettsiales or as a sister group to this order.

A potentially discriminating addition to this debate has been the recent discovery of a large group of predominantly marine members of *Alphaproteobacteria*, the so-called SAR11 clade, which constitute new representatives of the Rickettsiales (Morris et al. 2002; Williams et al. 2007). Unlike other members of this order, the SAR11 clade comprises free-living species that nevertheless share a number of genomic features with their parasitic cousins, such as streamlined genome and limited metabolic capacity. The genome of one SAR11 taxon, *Pelagibacter ubique*, is the smallest currently known for a free-living microorganism, contains the lowest number of predicted open reading frames, and exhibits the shortest intergenic spacers yet observed for any cell, with no evidence of pseudogenes, introns, mobile or extrachromosomal elements or inteins (Giovannoni et al. 2005). The phylogenetic placement of the SAR11 clade with respect to mitochondria within *Alphaproteobacteria* is still uncertain, but we may anticipate that the addition of more genome sequence data from this group will augment phylogenetic reconstructions aimed at answering this question.

While debate about the origin of mitochondria has centered on wholesale acquisition of an α-proteobacterial genome, in which case the various genes encoded by its mitochondrial descendent are assumed to have had the same evolutionary origin, a new wrinkle has been introduced by the discovery that gene transfer agents (GTAs) are pervasive in the genomes of *Alphaproteobacteria* taxa, including within the various Rickettsiales genera (although not, apparently, *Pelagibacter*; McDaniel et al. 2010). GTAs are virus-like elements that seem to function solely in the high-frequency transfer of DNA (and therefore genes) between cells, with no apparent adverse effects on the recipient. Recently, Richards and Archibald (2011) raised the possibility that the α-proteobacterial ancestor of mitochondria might have had a genome that had already undergone GTA-mediated gene exchange with other proteobacteria as well as non-proteobacterial species, so that the available phylogenetic signal is to some extent scrambled. Such a situation could conceivably compromise our ability to identify with certainty the precise α-proteobacterial lineage from which mitochondria originated "by generating incongruent tree topologies with 'mitochondrial' genes branching in different places within the α-proteobacterial phylogeny and prokaryotes as a whole". Some analyses (e.g., Esser et al. 2007) have reported discordant phylogenetic affinities for some mtDNA-encoded genes, although apparent discordance is much less pronounced in other studies (e.g., Fitzpatrick et al. 2006). It remains to be seen how much of this discrepancy can be attributed to the vagaries of single-gene tree reconstruction, where the number of phylogenetically informative characters is likely to be limited. In any event, potential gene transfer into the α-proteobacterial proto-mitochondrial ancestor (Esser et al. 2007), perhaps GTA-mediated at least in part (Richards and Archibald 2011), is yet another complication that will have to be taken into account in the continuing quest to delineate more precisely the evolutionary origin of the mitochondrial genome and the genes it contains.

B. Nature of the Host

Although we have a good albeit still imprecise idea of the phylogenetic provenance of the organism that ultimately gave rise to mitochondria, our picture of the host cell is far less clear, particularly regarding whether it was a full-fledged but amitochondriate eukaryotic cell, as classical endosymbiotic theory suggests. Our uncertainty about the host reflects two fundamental but still unanswered questions about the process of eukaryotic cell evolution (eukaryogenesis): (1) What is the evolutionary relationship of eukaryotes to the two prokaryotic groups, archaebacteria and eubacteria? (2) Were the formation of the eukaryotic cell per se and the formation of mitochondria congruent or sequential processes? In other words, was the prior emergence of an amitochondriate eukaryotic cell a sine qua non for the subsequent formation of mitochondria, or did the emergence of the defining structural and biochemical complexity of the eukaryotic cell depend on the prior establishment of the mitochondrion?

Debate about the phylogenetic relationships among eukaryotes, eubacteria and archaebacteria has centered on two opposing views of the so-called 'tree of life'. Supporters of the **three-domains tree** accept three separate and phylogenetically distinct (monophyletic) primary divisions, or domains, of life: Eucarya (eukaryotes), Archaea (archaebacteria) and Bacteria (eubacteria; Pace et al. 1986; Woese et al. 1990). On the basis of phylogenetic analyses of ancient paralogous genes – products of a gene duplication event that is presumed to have occurred before the separation of the three groups – Eucarya and Archaea are often considered to be sister groups, to the exclusion of Bacteria (e.g., Gogarten et al. 1989; Iwabe et al. 1989). On the other hand, proponents of the **eocyte tree** argue that eukaryotes branch *within* archaebacteria, with a specific group called Chrenarchaeota (eocytes), to the exclusion of the other major group of archaebacteria, Euryarchaeota (Lake et al. 1984; Rivera and Lake 1992). In the eocyte tree, therefore, archaebacteria are paraphyletic, not monophyletic.

Irrespective of which of the above alternatives is correct (Archibald 2008; Cox et al. 2008), what has become increasingly evident is that the nuclear genome of eukaryotes is a genetic mosaic: some of its genes are clearly more similar to archaeal homologs, others to eubacterial homologs, and still others appear to be eukaryote-specific inventions, having recognizable homologs in neither of the other two groups. Koonin (2010) has pointed out a number of examples where components of key functional systems and molecular machines of eukaryotes show evidence of varied phylogenetic ancestry. For example, chromatin/nucleosome proteins and protein constituents of the RNA interference machinery and the endomembrane/endoplasmic reticulum all appear to be complex mixes of archaeal and bacterial origin. Archaeal homologs tend to be 'informational', i.e., involved in genetic information transfer and processing (principally components of the replication, transcription and translation machineries), whereas eubacterial homologs are largely 'operational', i.e., involved in biosynthesis and metabolism (Rivera et al. 1998; Jain et al. 1999).

This dichotomy has prompted a variety of prokaryote–prokaryote fusion models for the origin of the nuclear genome, whereby an archaeal-type genome is combined with a eubacterial-type genome (e.g., Rivera and Lake 2004). Differential gene loss in the resulting hybrid then results in retention of mostly archaeal informational genes and eubacterial operational genes, although how the initial genome fusion and subsequent gene re-assortment occur (and why) is largely unspecified in these models. Alternatively, in models that invoke symbiosis-type interactions, an archaea-like cell is often seen as serving as host to a eubacteria-like symbiont. The resulting combination subsequently evolves the various hallmarks of the eukaryotic cell, such as membrane-bounded nucleus, endocytosis, complex endomembrane system and cytoskeleton. Importantly, however, the archaea-like and eubacteria-like gene complements of eukaryotic nuclear genomes show a diversity of origins within these two domains, to the extent that it has not been possible, as it has in the case of mitochondria, to pinpoint specific archaeal and eubacterial taxa as 'founding' lineages of a putative chimeric proto-eukaryotic cell. In rationalizing these observations, Koonin (2010) has suggested "that the archaeal ancestor of eukaryotes combined a variety of features found separately in diverse archaea".

Complicating efforts to untangle the evolutionary history of eukaryotes is the phenomenon of horizontal gene transfer (HGT), which is a major contributor to prokaryotic genome evolution (Doolittle 1999; Ochman et al. 2000), but also operates in eukaryotic genome evolution (e.g., Archibald et al. 2003; Andersson 2005; Keeling and Palmer 2008). If one accepts that the nuclear genome arose through a fusion of archaeal-type and eubacterial-type cells, the genomes of these progenitor cells may have themselves already been mosaic to some extent, added to which subsequent HGT from diverse archaeal and eubacterial sources could have further scrambled the nuclear genome's underlying phylogenetic signal. Thus, we should perhaps not be surprised that the three-domains tree is robustly supported in some analyses while the eocyte tree is strongly supported in others: in the case of a highly mosaic genome, the concept of a unique single origin simply does not apply.

Given the limitations attending reconstruction of deep phylogenies based on single-gene and multiple-gene alignments (even using concatenates of >100 genes), an alternative 'supertree' approach shows considerable promise. Supertree methods take individual phylogenetic trees as their input, synthesizing a single consensus tree from the collection of separate trees (Steel et al. 2000; Bininda-Emonds 2004; Wilkinson et al. 2005). Employing a supertree approach, Pisani et al. (2007) generated phylogenetic trees for 5,741 single-copy genes contained in 165 sequenced genomes, including >10 eukaryotic nuclear ones. After rigorously pruning the underlying conservative alignments to reduce sequence-based methodological artifacts (see Esser and Martin 2007), the results suggested that the nuclear genome of eukaryotes is dominated by genes of cyanobacterial and α-proteobacterial origin, attributable to the bacterial symbionts that gave rise to plastids (see below) and mitochondria, respectively, as well as a third component of archaeal origin. These intriguing results suggest that supertree methods may be able to recover signals due to symbiosis/genome melding events even in the face of considerable HGT 'noise'. Whether this approach is able to further tease apart the key events underlying the formation of the nuclear genome, to the degree we have been able to do so with mitochondrial and plastid genomes, remains to be seen.

A radically different view of eukaryogenesis has been championed by Kurland et al. (2006), who argue that the modern eukaryotic cell is the evolutionary descendent of "a unique primordial lineage", and that prokaryotes are derived from a common ancestor with eukaryotes through a process of genomic streamlining (reductive genome evolution) that has operated minimally in the eukaryotic lineage, affecting only select groups. These authors dispute the evidence that has been marshaled in support of the idea that the nuclear genome comprises a hybrid of archaea-derived and eubacteria-derived genes; they further posit that proteins unique to eukaryotes (so-called 'eukaryotic signature proteins', or ESPs) are retained primitive traits, rather than derived, eukaryote-specific inventions.

While it is increasingly clear that the Last Eukaryotic Common Ancestor (LECA) was already a highly complex cell containing most or all of the cellular signature structures (CSSs) that distinguish eukaryotic cells from archaeal and bacterial cells (Koonin 2010), compelling data are lacking to support the thesis that these features *trace* back to the last common ancestor of all cells. This particular eukaryogenesis scenario remains controversial (Kurland et al. 2007; Martin et al. 2007; Koonin 2010). It does not, for example, account for the large number of proteins of apparent archaeal origin in some of the key functional systems and molecular machines of eukaryotes (particularly DNA replication and repair, transcription and translation; Koonin 2010). Instead, Kurland et al. (2006) view archaea-like and eubacteria-like genes in the nuclear genome largely as retained primitive traits present in the last common ancestor of the three domains. The implication of this assumption is that ancestral genes shared specifically between Archaea and Eucarya were selectively lost from Bacteria, whereas those specifically shared between Bacteria and Eucarya (save those attributable to mitochondrial and plastid symbioses) were selectively lost from Archaea. At the moment, the accumulated genomic and phylogenomic data are better accommodated by eukaryogenesis scenarios in which the nuclear genome is initially formed through contributions from (an) archaeal and (a) eubacterial genome(s), however that mixing occurred (via cell/genome fusion or more indirectly; see below), with ESPs evolving as eukaryote-specific inventions within the resulting hybrid cell.

C. How Did It Happen?

A surprisingly large number of endosymbiotic models have been proposed over the years to account for the origin of mitochondria (see Martin et al. 2001 for an excellent and comprehensive overview). In essence, these models can be seen as variations on two fundamentally different themes that have been referred to, respectively, as the **archezoan scenario** and the **symbiogenesis scenario** (Koonin 2010). The archezoan scenario holds that "the host of the proto-mitochondrial endosymbiont was a hypothetical primitive amitochondrial eukaryote, termed archezoan". In contrast, the symbiogenesis scenario proposes that "a single endosymbiotic event involving the uptake of an α-proteobacterium by an archaeal cell led to the generation of the mitochondria", followed subsequently "by the evolution of the nucleus and compartmentalization of the eukaryotic cell." The archezoan scenario most closely approximates the classical endosymbiotic hypothesis of mitochondrial origin (Margulis 1970; Doolittle 1980). The hydrogen hypothesis of Martin and Müller (1998) exemplifies the symbiogenesis scenario. A fundamental difference between these two scenarios is whether the α-proteobacterial endosymbiosis that provided the proto-mitochondrion occurred at the same time as (and was integral to) the formation of the eukaryotic cell, or occurred subsequent to the formation of a primitive, amitochondriate cell that was already essentially eukaryotic.

1. Archezoan Scenario

The archezoan scenario received a major boost with the discovery of eukaryotes that not only lacked identifiable mitochondria but that also appeared to be the earliest branching taxa in phylogenetic trees based on rRNA sequences. The purportedly amitochondriate lineages included microsporidia, diplomonads and parabasalids, protists living as parasites of other eukaryotes, in anaerobic environments. These 'amitochondriate' parasites (collectively termed Archezoa) were initially assumed to represent contemporary examples of the sort of primitive eukaryote that might have served as host to an α-proteobacterial symbiont, which would subsequently become the mitochondrion.

In recent years, the archezoan scenario has been substantially weakened by two key findings. First, a number of studies have convincingly demonstrated that the apparent early branching of archezoan taxa is a long-branch artifact, due to an unusually rapid rate of sequence divergence of the genes selected for phylogenetic analysis (e.g., Inagaki et al. 2004). These long-branch sequences cluster at the base of eukaryotic phylogenetic trees, closest to the outgroup (prokaryotic) sequences used to root the trees. Microsporidia, for example, are now known to be evolutionarily degenerate fungi rather than 'early-branching' eukaryotes (Hirt et al. 1999; Keeling et al. 2000). In fact, the root of the eukaryotic tree has been notoriously difficult to discern and is not yet established, with six or so eukaryotic supergroups appearing to diverge from one another virtually simultaneously, on an evolutionary timescale. As a result, the eukaryotic tree more closely resembles a bush, and no one lineage can be clearly identified as earliest diverging (Keeling et al. 2005; Koonin 2010).

A second notable nail in the coffin of the archezoan scenario was the discovery in one archezoan species after another of organelles that were eventually recognized as highly reduced mitochondria. These 'mitochondrion-related organelles' (MROs) are of two basic types, distinguished by whether or not they retain any capacity for energy (ATP) generation. The first of these MROs to be discovered was the hydrogenosome (Lindmark and Müller 1973). This double-membrane-bound organelle, found in certain anaerobic protists such as the parabasalid *Trichomonas vaginalis*, lacks a number of the defining features of a conventional mitochondrion: it has no genome, no complete tricarboxylic acid (TCA) cycle, no cytochromes and lacks a complete electron transport chain. Although the *T. vaginalis* hydrogenosome does not have a capacity to produce ATP through coupled electron transport-oxidative phosphorylation, it is still able to generate ATP from pyruvate via a substrate-level pathway, through the combined activities of a set of enzymes characteristic of this organelle, including an iron–iron hydrogenase. Molecular hydrogen (H_2) is one of the end products of this pathway, accounting for the organelle's name. Because of its unique anaerobic metabolism, it was originally supposed that the hydrogenosome might be the evolutionary product of a separate endosymbiosis – in this case with an anaerobic-type eubacterium such as a *Clostridium* (Whatley et al. 1979) – than the one that gave rise to the mitochondrion. However, subsequent studies have demonstrated that the *T. vaginalis* hydrogenosome contains a number of proteins typical of mitochondria, such as chaperonins (Bui et al. 1996), the NADH dehydrogenase module of electron transport Complex I (Hrdy et al. 2004), and components of the mitochondrial machinery for synthesis of iron–sulfur (Fe-S) clusters, the ISC biosynthesis pathway (Sutak et al. 2004). These results strongly support the view that the *T. vaginalis* hydrogenosome is an evolutionary derivative of a conventional mitochondrial ancestor.

A second group of double membrane-bound MROs, in this case lacking any capacity to generate ATP, has been found in a number of anaerobic, parasitic protists, including the amoebozoans *Entamoeba histolytica* (Clark and Roger 1995; Mai et al. 1999; Tovar et al. 1999) and *Mastigamoeba balamuthi* (Gill et al. 2007), the microsporidians *Trachipleistophora hominis* (Williams et al. 2002) and *Encephalitozoon cuniculi* (Goldberg et al. 2008; Tsaousis et al. 2008), and the diplomonad *Giardia lamblia* (Tovar et al. 2003). Collectively, the term 'mitosome' is most often applied to these particular MROs (Embley et al. 2003; Embley 2006; Hjort et al. 2010). Again, molecular data identifying typical mitochondrial proteins in these organelles has solidified the view that mitosomes, also, are derived mitochondria, but even more highly reduced than hydrogenosomes (see Hjort et al. 2010 for a detailed listing and discussion of relevant data). The limited metabolic capacity of mitosomes has focused attention on what functionality has been retained in these organelles, suggesting that Fe-S cluster formation, rather than oxidative phosphorylation, may be the essential *raison d' être* of the mitochondrion and its evolutionary derivatives.

More recently, the distinction between mitochondria, 'classical' hydrogenosomes and mitosomes has become blurred by the discovery of what appear to be transitional evolutionary forms that retain a reduced genome, lacking a number of typical mtDNA-encoded genes. Like genome-deficient hydrogenosomes, these novel genome-containing MROs are able to generate H_2 via a hydrogenase-mediated reaction; however, they also carry out a more complex biochemistry than genome-deficient hydrogenosomes. Two such genome-containing MROs that have been studied in some detail are those in the anaerobic ciliate *Nyctotherus ovalis* (Boxma et al. 2005) and the anaerobic stramenopile *Blastocystis* sp. (Pérez-Brocal and Clark 2008; Stechmann et al. 2008; Wawrzyniak et al. 2008), a relative of brown algae, diatoms and oomycetes (see Chap. 2). In both cases, the MRO genome encodes components of an organellar translation system (rRNAs, tRNAs, ribosomal proteins) as well as components of electron transport complexes I and II, suggesting the presence of a partial electron transport chain.

The punctuate phylogenetic distribution of MROs of various types, and their interspersion with aerobic taxa within particular lineages, strongly indicate that MROs have arisen independently a number of times from a conventional mitochondrial ancestor (Embley et al. 2003; Embley 2006; Hjort et al. 2010), and that many of the seemingly shared characteristics of MROs (e.g., between those of *Nyctotherus* and *Blastocystis*) are due to convergent evolution rather than vertical inheritance. The continued study of variously evolved MROs will not only be key to elucidating both the pathways and mechanisms involved in the evolutionary conversion of conventional to relict mitochondrion, but will also give us a better appreciation of the evolutionary flexibility of mitochondria: a theme considered below in the discussion of mitochondrial proteome evolution.

Significantly, with regard to models of mitochondrial origin, the discovery of MROs has greatly weakened the concept of primitively amitochondriate protists. Although a number of 'amitochondrial' eukaryotes (i.e., lacking conventional mitochondria) obviously exist, we can point to no convincing examples of primitively 'amitochondriate' eukaryotes (i.e., ones whose evolutionary ancestors never had mitochondria). Accordingly, we are forced to conclude that if such organisms ever existed, their descendent lineages must all have become extinct.

2. Symbiogenesis Scenario

As support for the archezoan scenario has waned, the alternative view – that the host cell for the mitochondrial endosymbiosis was a prokaryote (specifically an archaeon), not a eukaryote – has correspondingly gained prominence (Koonin 2010). Perhaps the best-known symbiogenesis scenario is the hydrogen hypothesis of Martin and Müller (1998), which suggests that eukaryotes have arisen "through symbiotic association of an anaerobic, strictly hydrogen-dependent, strictly autotrophic archaebacterium (the host) with a eubacterium (the symbiont) that was able to respire, but generated molecular hydrogen as a waste product of anaerobic heterotrophic metabolism. The host's dependence upon molecular hydrogen produced by the symbiont is put forward as the selective principle that forged the common ancestor of eukaryotic cells."

Assuming that the symbiont was an α-proteobacterium that was capable of both anaerobic and aerobic energy metabolism, the hydrogen hypothesis can account for the origins of eukaryotic energy metabolism, assuming that respiration machinery genes (Krebs cycle and oxidative phosphorylation) and genes for anaerobic energy metabolism (PFO, hydrogenase) were both retained in the hybrid cell but differentially expressed under the relevant environmental conditions, and that genes for aerobic respiration were differentially lost in those eukaryotic lineages, in which the mitochondrion was converted to an anaerobic MRO. Thus, the hydrogen hypothesis "posits that the origins of the heterotrophic organelle (the symbiont) and the origins of the eukaryotic lineage are identical". A corollary of the hydrogen hypothesis and other symbiogenesis scenarios is that the complexity of the eukaryotic cell and its defining features developed *after* the mitochondrial symbiosis, rather than *before*.

As noted by Koonin (2010), several arguments can be advanced against a symbiogenesis scenario for the origin of mitochondria. For example, endocytosis (a hallmark eukaryotic character) has long been considered to be an essential capacity for uptake of a bacterial endosymbiont. Cases of bacterial endosymbioses (e.g., γ-proteobacteria inside β-proteobacteria) have, however, been documented (von Dohlen et al. 2001; Thao et al. 2002). Also, as noted earlier, it has not been possible to trace the archaeal and eubacterial contributions to the nuclear genome to single extant prokaryotic lineages: although an α-proteobacterial signal does predominate (Pisani et al. 2007), in any given eukaryotic taxon collectively more eubacterial-type genes appear to derive from a diversity of non-α-proteobacterial lineages (or to branch within Bacteria as a whole, but not robustly with any specific group). Nevertheless, it is possible that ancestral lineages contributing to a eubacterial-archaeal symbiogenesis might have had more complex genomes than their contemporary relatives: genomes already affected to a certain extent by HGT. The hydrogen hypothesis does make a number of testable predictions, for example, that genes of anaerobic energy metabolism (such as PFO and hydrogenase) should form monophyletic clades in phylogenetic reconstructions, branching together with α-Proteobacteria. However, a rigorous study of the phylogenetic distributions and histories of proteins involved in anaerobic pyruvate metabolism in eukaryotes has not provided support for this prediction (Hug et al. 2010).

Very recently, a new hypothesis, based on a consideration of the energetics of genome complexity, has added fuel to the eukaryogenesis fire. Lane and Martin (2010) argue that the increase in the number of proteins that eukaryotes encode and express, compared to prokaryotes, required an increase in cellular energy that only the mitochondrion could have provided. Accordingly, this hypothesis views mitochondria as the sine qua non to eukaryotic genomic and cellular complexity. The authors conclude, rather definitively, that "the host for mitochondria was a prokaryote".

On balance, a symbiogenesis scenario (eubacterial endosymbiont in an archaeal host) better accommodates the accumulated data that address the origin of the mitochondrion than does an archezoan scenario (eubacterial endosymbiont in an amitochondriate but essentially eukaryotic host). However, the latter scenario cannot be ruled out absolutely at this stage. Each scenario raises complications and objections that are difficult to rationalize without resorting to ad hoc explanations – e.g., that true archezoan eukaryotes may exist but simply have not yet been discovered, or that all such lineages have become extinct – and each is complicated by the fact that there is no obvious way to discern how similar the genomes of the proposed prokaryotic ancestors of the eukaryotic cell were compared to their extant descendants.

D. Evolution of the Mitochondrial Proteome

Given that even the most gene-rich mitochondrial genomes retain only a small fraction of the genes that are assumed to have been contained in the genome of its α-proteobacterial ancestor (Gray 1999), gene loss has evidently played a major role in the evolution of the mitochondrial genome. Many of these 'lost' genes have been transferred to the nuclear genome from where they are now expressed, with import of only a minority of the resulting proteins back into the organelle; in fact, most of these transferred 'proto-mitochondrial' genes now function in other subcellular compartments (Gabaldón and Huynen 2003). Because functional mitochondria are composed of hundreds or even thousands of nucleus-encoded proteins, many of which belong to the category of ESPs, re-tailoring of the mitochondrial proteome through addition of new proteins and functions has been extensive in the course of evolution. Hence, just as comparative mitochondrial genomics, based on complete sequencing of mtDNAs, has proven to be a powerful approach for discerning the nature of the ancestral mitochondrial genome and revealing patterns and mechanisms of mitochondrial genome evolution (Gray et al. 1998; Gray 1999), so is comparative mitochondrial proteomics, based on mass spectrometric analysis of whole mitochondria or sub-mitochondrial fractions and complexes (Dreger 2003; Yan et al. 2009), proving to be an equally powerful method for elucidating the evolution of the mitochondrial proteome.

Initially, the composition of the mitochondrial proteome and the phylogenetic origins of mitochondrial proteins were assessed from complete genome sequence data via bioinformatics analyses of proteins possessing N-terminal mitochondrial targeting peptides. A number of algorithms have been developed to identify such targeting sequences (e.g., Claros and Vincens 1996; Emanuelsson et al. 2000), although not all imported mitochondrial proteins possess mitochondrial import signals identifiable in this way, and the algorithms are variably accurate and may have limited sensitivity in cases where protein sequences are highly divergent (Richly et al. 2003). Early estimates of the number of proteins in the yeast mitochondrial proteome ranged from ~400 to ~800, or between ~7% and ~13% of the total yeast proteome of ~6,100 proteins (Karlberg et al. 2000; Marcotte et al. 2000; Kumar et al. 2002). More broadly applied predictions suggest that functional mitochondria could harbor as few as several hundred proteins in *Plasmodium falciparum*, the malaria parasite, to >3,000 in vertebrate animals (Richly et al. 2003).

Such studies provided a first, and surprising, overview of the evolutionary origins of proteins constituting the yeast mitochondrion (Karlberg et al. 2000; Marcotte et al. 2000; Kumar et al. 2002): surprising because a much smaller proportion (only ~10–15%) of the mitochondrial proteome than might have been anticipated proved to originate clearly from the α-proteobacterial lineage. A larger, generically 'prokaryotic' proportion (~40–50%) contained proteins whose origins appear to be outside α-Proteobacteria but without necessarily a robust affiliation to any particular bacterial or archaeal lineage. Members of another, 'eukaryotic' fraction (~20–30%) have no obvious homologs in either Archaea or Bacteria and so are, by definition, ESPs. A final, 'unique' subset (~20%) comprises seemingly species-specific proteins having no identifiable homologs in other eukaryotes or in prokaryotes. These results indicate that the yeast mitochondrial proteome has multiple evolutionary origins, and a complex evolutionary history (Kurland and Andersson 2000; Gray et al. 2001), a conclusion now firmly established for the mitochondria of other eukaryotes (Gabaldón and Huynen 2004; Szklarczyk and Huynen 2010). A small contribution of bacteriophage-like proteins (notably the mitochondrial RNA polymerase in most eukaryotes) has also been added to the evolutionary mix (Shutt and Gray 2006).

Direct proteomics surveys relying on mass spectrometry (Aebersold and Mann 2003; Yan et al. 2009) have confirmed and extended the initial, bioinformatics-based findings that pointed to a mosaic evolutionary origin of the mitochondrial proteome. This approach, while not biased toward proteins containing N-terminal mitochondrial targeting sequences, has its own limitations, most particularly a bias in favor of the most abundant, soluble targets. Nevertheless, mass spectrometry (MS) has afforded a powerful means of uncovering novel mitochondrial proteins that cannot be identified on the basis of sequence similarity with known mitochondrial proteins. For example, in an MS study of mitochondria from the ciliate protozoon, *Tetrahymena thermophila*, ~30% of identified proteins were found to have no demonstrable sequence homologs outside of the ciliate lineage, while a further ~10% are unique to *T. thermophila* (Smith et al. 2007). At least 13 of the novel, ciliate-specific proteins have subsequently been found as components of the purified mitochondrial F_1F_O-ATP synthase (Complex V) of this protist (Nina et al. 2010), illustrating an emerging theme in mitochondrial research: taxon-specific re-tooling of mitochondrial complexes such as electron transport chain assemblies and ribosomes, only the core components of which derive from the α-proteobacterial ancestor of mitochondria. This re-tailoring occurs by addition of novel proteins of generally unknown function, sometimes accompanied by loss of otherwise conserved components. One such example is the ATP synthase of *Chlamydomonas reinhardtii*, a chlorophycean green alga, in which nine novel 'Asa' subunits of unknown evolutionary origin replace eight subunits that are otherwise conserved in the ATP synthase of other non-chlorophycean green algae, as well as in plants, animals and fungi (Lapaille et al. 2010).

This re-tailoring theme can also be seen in other well-studied mitochondrial respiratory complexes, such as Complex I (CI; NADH:ubiquinone oxidoreductase), the multi-subunit proton pump that carries out the first step in the canonical respiratory chain – the oxidation of NADH and subsequent reduction of ubiquinone. Bacterial CI comprises 14 subunits, all of which are present in the corresponding mammalian complex, with seven of the subunits encoded in the mammalian mitochondrial genome. A further 18 subunits that are present in mammalian CI are ubiquitous throughout eukaryotes but are not found in bacteria, and so are assumed to be eukaryote-specific additions already present in the last eukaryotic common ancestor. Thirteen other subunits of mammalian CI appear to have a narrow phylogenetic distribution, having so far been found only in metazoan animals (Brandt 2006).

Attempts have been made to reconstruct the proteins contributed to the eukaryotic cell by the proto-mitochondrial endosymbiont, through comparisons of proteins encoded in sequenced α-proteobacteria with those specified by sequenced eukaryotic genomes. This approach has identified at least 840 orthologous groups that are considered to bear a clear α-proteobacterial signature – i.e., a close and specific evolutionary relationship to α-proteobacterial homologs, without any evidence of recent HGT (Gabaldón and Huynen 2003, 2007; Szklarczyk and Huynen 2010). Comparisons among α-proteobacterial genomes suggest that the free-living bacterial ancestor of mitochondria contained ~3,000–5,000 genes (Boussau et al. 2004), with an upper bound of ~1,700 ancestral clusters of orthologous genes in the proto-mitochondrial genome (Szklarczyk and Huynen 2010). These estimates imply that upwards of 1,000–3,000 genes were lost in the transition from bacterial symbiont to proto-organelle. Significantly, of the >800 human genes that display an α-proteobacterial signature, only ~200 comprise part of the human mitochondrial proteome, clearly implying that the proto-mitochondrial contribution to eukaryotic cell evolution and function extends well beyond the mitochondrion itself.

Pathways that are considered to have been complete in the proto-mitochondrion include the full electron transport chain and β-oxidation of fatty acids (providing NADH and $FADH_2$ to the former), indicating that the mitochondrial endosymbiont had an aerobic metabolism. Also prominently represented are pathways for the synthesis of lipids, biotin, heme and iron-sulfur clusters, as well an abundance of cation transporters. In all, the reconstructed metabolism suggests that the proto-mitochondrion was capable of at least facultative aerobic respiration (Szklarczyk and Huynen 2010). More than half of what remains of this proto-mitochondrial metabolism in modern mitochondria comprises functions involved in energy metabolism and translation, including post-translational modifications: a veritable "hijacking of mitochondrial protein synthesis and metabolism" (Gabaldón and Huynen 2007).

In attempts to elucidate in more detail the ancestral state of selected mitochondrial components and pathways, several groups have initiated comparative analyses of emerging eukaryotic genome data. As noted earlier, mitochondrial CI has an additional 18 subunits that are not present in its bacterial counterpart, and that are considered to have been incorporated at the earliest stages of mitochondrial CI evolution (Gabaldón et al. 2005; Brandt 2006). In plants (Heazlewood et al. 2003; Perales et al. 2004) and green algae (Cardol et al. 2004), mitochondrial CI has also been found to contain multiple proteins with high similarity to γ-type carbonic anhydrases (γCAs), with comparative studies initially suggesting that these proteins represented specific additions in the plant lineage (Parisi et al. 2004). However, a more recent study focusing on protists has revealed a much broader distribution of mitochondrial γCAs, either demonstrated or presumed to be associated with mitochondrial CI (Gawryluk and Gray 2010), than previously suspected. It appears likely that γCAs were ancestral components of mitochondrial CI, and that they were subsequently lost from CI specifically in the evolutionary line leading to animals and fungi (opisthokonts), rather than added to CI specifically in the line leading to plants and algae. These results emphasize the importance of comprehensive taxon coverage in drawing conclusions about mitochondrial proteome evolution.

Other studies have demonstrated that the ancestral mitochondrial ribosome in the last eukaryotic common ancestor was already much larger than its bacterial ancestor, containing some 19 additional eukaryote-specific proteins (Smits et al. 2007; Desmond et al. 2011). The fact that these novel mitochondrial ribosomal proteins are found throughout the eukaryotic domain, in all of the currently recognized eukaryotic supergroups, is yet another strong argument in favor of a monophyletic origin of contemporary mitochondria: a conclusion in this case based on eukaryote-specific rather than prokaryote-specific features.

The mitochondrial ribosome presents a particularly dramatic example of mitochondrial re-tailoring, with both the RNA and protein components varying markedly in size and number among eukaryotes. For example, the 55S human mitochondrial ribosome contains rRNA species that are about half the size of their bacterial 23S and 16S counterparts; however, it has 29 different small subunit and 48 different large subunit proteins, compared to values of 21 and 34, respectively, in *E. coli* (O'Brien 2003). Clearly, the human mitochondrial ribosome has lost substantial RNA and gained substantial protein in the course of its evolution from a bacterial progenitor, reversing the usual protein:RNA ratio (33:67) to become protein-rich (69:31) (O'Brien 2002).

An even more extreme situation is seen in the kinetoplastid protozoa, such as *Trypanosoma brucei* (Ziková et al. 2008) and *Leishmania tarentolae* (Sharma et al. 2009). Here, rRNA shrinkage is even more pronounced than in the human mitochondrial ribosome whereas protein content has been further expanded, with the *Trypanosoma* mitochondrial ribosome containing 56 small subunit and 77 large subunit proteins. Notably, the novel mitoribosomal proteins identified in these analyses do not have detectable homologs outside of the kinetoplastid protozoa, and display only a low degree of sequence conservation within this lineage. These observations reinforce the importance of direct mass spectrometric analyses of isolated mitochondrial complexes in order to accurately determine their composition, given that so many of these components appear to be new, lineage-specific inventions. Overall, the mitochondrial proteome has proven to be surprisingly malleable, a situation that is mirrored by the picture emerging from investigations of the plastid proteome in photosynthetic eukaryotes.

III. Plastids

The notion that plastids are of endosymbiotic origin is more than a century old. The Russian botanist Konstantin Mereschkowsky (1855–1921) is generally credited as having been the first to elaborate on the significance of similarities between '*Cyanophyceae*' (cyanobacteria) and the 'chromatophores' (chloroplasts or plastids) of plants and unicellular algae such as diatoms (Mereschkowsky 1905; Martin and Kowallik 1999). Mereschkowsky developed the concept of symbiogenesis – the evolution of new life forms from the amalgamation of two separate organisms – which was championed and rendered 'mainstream' by Margulis (1970) as the endosymbiont hypothesis for the evolution of mitochondria and plastids. In the sections that follow we discuss the wealth of data brought to bear on the origin and early evolution of 'primary' plastids, i.e., those that have been inherited in a vertical fashion since their inception. The following chapter by Bhattacharya and colleagues deals with so-called 'secondary' and 'tertiary' endosymbioses, whereby plastids have spread horizontally by mergers between two eukaryotes. Molecular evidence in support of a classical endosymbiotic origin for primary plastids is (and has always been) stronger than that for mitochondria. Nevertheless, as is the case for mitochondrial evolution, genomic and proteomic investigations continue to expose layer upon layer of unexpected complexity.

A. Cyanobacterial Endosymbiont, Complex Eukaryotic Host

With several decades worth of ultrastructural, biochemical and molecular phylogenetic data in hand, it can now be concluded that (1) plastids evolved after mitochondria, (2) the endosymbiont was an ancestor of modern-day cyanobacteria capable of oxygenic photosynthesis, and (3) the host was a 'complex', fully formed eukaryote with the ability to phagocytose prey (Gray and Spencer 1996; Reyes-Prieto et al. 2007; Gould et al. 2008). The precise ecological and physiological conditions present at the time of the evolution of plastids are unknown, but eukaryotic heterotrophs would presumably have benefited greatly from the ingestion of organisms capable of generating energy from sunlight. Like today's cyanobacteria, primary plastids are characterized by the presence of two membranes (Gould et al. 2008). This suggests that the endosymbiont somehow 'escaped' from its phagocytic vacuole, perhaps allowing it to persist for progressively longer periods of time without being digested. Regardless, the cyanobacterial progenitor of the plastid gradually became one with its eukaryotic host: non-essential genes were lost, scores of essential genes were transferred to the nuclear genome, a protein import machinery evolved, and mechanisms for metabolite transport were 'invented', allowing the proto-alga to reap the benefits of cyanobacterial carbon fixation (Martin and Herrmann 1998; McFadden 1999, 2001; Soll and Schleiff 2004; Weber et al. 2006; Howe et al. 2008). Both endosymbiont- and host-derived components appear to have contributed to the integration of the two cells.

How derived are plastids relative to cyanobacteria? Hundreds of plastid genomes have now been sequenced, and even the most gene-rich among them contain only ~250 genes (Stoebe and Kowallik 1999; Martin et al. 2002; Hagopian et al. 2004). This coding capacity stands in stark contrast to that of cyanobacteria, which have at least ~1,700 genes (Rocap et al. 2003). The plastid genomes of all photosynthetic organisms retain a very similar core set of genes encoding proteins primarily involved in photosynthesis, transcription and translation (Turmel et al. 1999; Martin et al. 2002; Howe et al. 2003; Kim and Archibald 2009). In a situation analogous to the retention of mitochondrion-related organelles in anaerobic protists, virtually all known secondarily non-photosynthetic eukaryotes, including parasitic plants (Krause 2008), the green algal parasites *Helicosporidium* (de Koning and Keeling 2006) and *Prototheca* (Borza et al. 2005), and the malaria parasite *Plasmodium* (Waller and McFadden 2005; see Chap. 2) retain a plastid. This is because the plastid is the site of essential biochemical processes entirely unrelated to photosynthesis, including the synthesis of heme precursors, fatty acids and certain amino acids (Borza, et al. 2005; Mazumdar et al. 2006).

B. Single or Multiple Origins?

While mitochondria (and their derivatives) are part-and-parcel with the eukaryotic condition, plastid-bearing organisms exhibit a 'patchy' phylogenetic distribution. Primary plastids bearing two membranes are restricted to three lineages, the glaucophyte (or glaucocystophyte) algae, red algae and green algae (Bhattacharya et al. 2003; Keeling 2010). Glaucophytes are poorly studied fresh-water unicells that are of particular interest due to the fact that their plastid envelopes possess a layer of peptidoglycan, as do the cell walls of cyanobacteria (Graham and Wilcox 2000). Despite retention of this ancestral feature, glaucophyte plastids have a genome that is as reduced as those of green and red algae and that shares many features in common with them (Löffelhardt et al. 1997). Red algae are a diverse lineage comprising both unicellular and multicellular forms, some of which are capable of living in highly acidic environments and at temperatures greater than 50°C (Ciniglia et al. 2004; Reeb and Bhattacharya 2010; Yoon et al. 2010). Green algae are a speciose assemblage of terrestrial and aquatic (both freshwater and marine) phototrophs that are divided into two distinct lines, the chlorophytes (e.g., the model laboratory alga *Chlamydomonas*) and the streptophytes. It is from within this latter group that multicellular land plants evolved (Karol et al. 2001; Lewis and McCourt 2004; Finet et al. 2010). Together, glaucophytes, red algae and green algae plus land plants belong to the eukaryotic 'supergroup' Archaeplastida or Plantae (Adl et al. 2005; Keeling et al. 2005). Whether Archaeplastida represents a monophyletic assemblage is a topic of ongoing debate.

Molecular phylogenetic analyses of plastid rRNA and protein genes from red, green and glaucophyte algae almost always show a clear connection to cyanobacteria (e.g., Douglas and Gray 1991; Delwiche et al. 1995; Turner et al. 1999), but no particular extant cyanobacterial lineage has yet emerged as an unambiguous, specific relative of plastids. The topologies of individual protein and rRNA gene trees have proven frustratingly sensitive to phylogenetic artifacts and taxon representation of both algae and cyanobacteria (Lockhart et al. 1992a, b; Sato 2006; Larkum et al. 2007). Nevertheless, the current trend towards multi-gene and whole-genome-scale analyses has improved matters somewhat.

For example, a recent analysis of combined 16S rRNA and ribulose 1,5-bisphosphate carboxylase/oxygenase (*rbcL*) genes by Falcón et al. (2010) resolved a monophyletic primary plastid clade and suggested a specific association between plastids and nitrogen-fixing unicellular cyanobacteria belonging to the Chroococcales. This result is consistent with a phylogenomics-based analysis carried out by Deusch et al. (2008). These authors compared the complete nuclear genomes of *Arabidopsis thaliana*, rice, *Chlamydomonas reinhardtii*, and the red alga *Cyanidioschyzon merolae* to nine cyanobacterial genomes. They found that, in terms of gene presence/absence and overall sequence similarity, the cyanobacterial-derived gene sets contained in the algal genomes were most similar to those of the nitrogen-fixing, heterocyst-forming cyanobacteria *Nostoc* sp. and *Anabaena variabilis* (Deusch et al. 2008). While certainly not definitive, these studies are nevertheless significant in positing a specific role for nitrogen fixation in the early stages of plastid evolution.

As noted above, the plastid genomes of all three of the primary plastid-harboring lineages are highly reduced compared to those of known cyanobacteria. Their gene contents overlap to a substantial degree and when structural similarities are taken into consideration, such as the near-universal presence of rDNA-containing inverted repeats, conserved gene clusters (e.g., the *atpA* operon) and an unusual $tRNA^{Leu}$ intron, it seems improbable that green, red and glaucophyte plastid genomes could have evolved to such similar 'endpoints' from different (but closely related) cyanobacterial endosymbionts (Kowallik 1997; Martin and Herrmann 1998; Stoebe and Kowallik 1999; Besendahl et al. 2000; Palmer 2003). Nevertheless, convergent evolution of plastid genome structure and content, as could occur if there were serious gene-specific constraints on the process of endosymbiont-to-nucleus gene transfer, cannot be dismissed outright (Palmer 2003; Stiller et al. 2003). Furthermore, as emphasized recently by Larkum et al. (2007), the recovery of monophyletic plastid sequences in phylogenetic trees does not necessarily mean that the organelle was acquired in a common ancestor shared exclusively by the organisms that harbor them (Howe et al. 2003). Much recent attention has thus been given to answering the question of whether phylogenies of mitochondrial, plastid and nuclear genes of red, green and glaucophyte algae agree with one another. The answer appears to be a qualified 'maybe.'

It is now common practice to try and maximize the extraction of ancient phylogenetic signal from molecular data by analyzing dozens to hundreds of loci together in the context of a single supermatrix (Delsuc et al. 2005). Applied to the question of primary plastid monophyly versus polyphyly, the first such phylogenomic analyses of concatenated mitochondrial-, plastid- and nucleus-encoded proteins yielded results consistent with the hypothesis that red and green algae are each other's closest relatives (e.g., Burger et al. 1999; Moreira et al. 2000; Rodríguez-Ezpeleta et al. 2005; Burki and Pawlowski 2006). When available, glaucophyte sequences were also found to branch specifically with those of red and green algae, although the relative branching order of the three groups was not resolved (Moreira et al. 2000; Rodríguez-Ezpeleta et al. 2005).

Unfortunately, with more data and increased analytical sophistication, the interrelationships between primary plastid-containing algae have become less and less clear. Individual protein trees sometimes do not agree with one another, even when they correspond to different genes from the same genome (e.g., Stiller and Hall 1997; Longet et al. 2003; Kim and Graham 2008), and analyses of particular subsets of the data in isolation (e.g., slowly evolving proteins) sometimes yield trees that do not show red, green and glaucophyte algae as specific sister lineages (Nozaki et al. 2007, 2009). Such results have spawned alternate hypotheses, such as the idea that a truly ancient primary endosymbiotic event occurred in a common ancestor shared between members of the Archaeplastida and other eukaryotic groups that currently lack plastids, including members of the supergroup Excavata (Nozaki 2005; Nozaki et al. 2007). Parfrey et al. (2010) recently carried out a comprehensive 'taxon-rich', multi-gene analysis designed to resolve higher-order relationships amongst eukaryotes. These authors concluded that "…there is no support in any analysis for 'Archaeplastida' ('Plantae')". In 2003, Palmer provided the following synopsis of the state of knowledge on early plastid evolution, which, as it has turned out, still fits today: "There is universal consensus that all well-recognized types of primary plastid-containing organisms fall into three groups, each clearly monophyletic: the green algae (including, of course, land plants), red algae and glaucophytes…. There is also broad consensus, based on many lines of evidence, that all three of these lineages 'probably' trace back to the same cyanobacterial endosymbiosis; that is, primary plastids arose once and only once. I say 'probably', because some authors regard the issue as settled and others see a need for more evidence" (Palmer 2003). One of the few points on which there is unanimous agreement is the notion that more data are sorely needed from glaucophytes and red algae: only expressed sequence tag (EST) data are currently available for glaucophyte algae, and red algae are at present represented by only a single (and apparently highly reduced) nuclear genome sequence (Matsuzaki et al. 2004). Fortunately, such data will soon be forthcoming and will undoubtedly give rise to another wave of phylogenomic analyses.

C. Primary Endosymbiosis and Genome–Proteome Mosaicism

One of the most profound recent advances in the field of plastid evolution has been recognition of the huge extent to which the cyanobacterial progenitor of the plastid appears to have contributed to the biochemistry and cell biology of the earliest photosynthetic eukaryotes. Endosymbiotic gene transfer has long been recognized as the mechanism by which endosymbionts surrender genetic material to their hosts (Martin et al. 1993; Martin and Herrmann 1998; Timmis et al. 2004; Kleine et al. 2009). Together with the evolution of an import apparatus for targeting the products of transferred genes, endosymbiotic gene transfer is an essential step in the transition from endosymbiont to organelle (Cavalier-Smith and Lee 1985; Theissen and Martin 2006; Cavalier-Smith 2007). In the days before whole genome-scale analyses, Weeden's 'product specificity corollary' (Weeden 1981) posited that the products of transferred genes remain faithful to their subcellular compartment of origin: proteins functioning in the plastid that are not currently encoded in its genome are the product of cyanobacterial-derived nuclear genes that *were* present in the plastid progenitor (Weeden 1981). This has turned out to be true in many cases but it is by no means the rule (Martin and Cerff 1986; Martin and Schnarrenberger 1997; Martin 2010). Conversely, few would have predicted that the cyanobacterial 'footprint' on the nuclear genome of algae and plants would extend so far beyond the plastid and photosynthesis.

Pioneering work in the 1980s and 1990s by Martin, Cerff and colleagues provided the first glimpses of the remarkable degree of mosaicism now known to exist in plant metabolic pathways. For example, both the plastid-targeted and cytosol-localized isoforms of the Calvin cycle/glycolytic enzyme phosphoglycerate kinase (PGK) are of cyanobacterial ancestry (Brinkmann and Martin 1996). Such gene duplication-enabled functional reassignments are known as endosymbiotic gene replacements (Martin and Schnarrenberger 1997) and can also happen 'in reverse': in the case of plant fructose-1,6-bisphosphatase, the plastid-localized and cytosolic enzymes are derived from duplicated genes of *cytosolic* (i.e., eukaryotic host) origin (Martin et al. 1996). Endosymbiotic and reverse endosymbiotic gene replacements are now well recognized as generators of metabolic complexity and innovation, and are also useful markers for testing evolutionary hypotheses (e.g., Fast et al. 2001; Nowitzki et al. 2004; Patron et al. 2004; Rogers and Keeling 2004). The phenomenon has been aptly summarized as follows: "there is no evolutionary 'homing device' that automatically directs the product of a transferred gene back to the organelle of its provenance, the products of genes that are acquired by endosymbionts are free to explore any and all targeting possibilities within the cell; they can and do replace pre-existing host genes, or even whole pathways, and sometimes pre-existing host genes can be duplicated to provide organelle-targeted copies of host enzymes that can replace organelle-encoded functions" (Martin 2010).

The potential full scope of genome and proteome mosaicism in photosynthetic eukaryotes was revealed in 2002 with an analysis of the flowering plant, *Arabidopsis thaliana*. Martin et al. (2002) compared the complete set of ~25,000 genes in the *A. thaliana* genome to the gene sets of yeast, archaea, bacteria, and cyanobacteria. Approximately 1,700 of the 9,368 *A. thaliana* genes whose ancestry could be inferred were deemed to be of cyanobacterial origin. Extrapolated to the whole genome, this amounts to ~4,500 genes, or 18% of the complete gene complement. Unexpectedly, fewer than half of the genes of putative cyanobacterial ancestry were predicted to encode plastid-targeted proteins. Those that did not could be assigned to a wide range of predicted functional categories having nothing to do with the plastid, including cell division and intracellular transport (Martin et al. 2002). Conversely, Suzuki and Miyagishima (2010) recently estimated that ~40% of the plastid-targeted proteins thought to have been present in the common ancestor of red algae and plants are not of cyanobacterial ancestry but are derived from the eukaryotic host and various bacterial groups. As amply demonstrated on a case-by-case basis for metabolic enzymes in plants (above), whole genome-scale analyses suggest that there is no strict correlation between the evolutionary origin of a given protein and the cellular compartment or biological process in which it presently functions (Martin 2010).

As striking as these numbers are, there are reasons to tread cautiously. Reyes-Prieto et al. (2006) carried out an analysis of cyanobacterial genes in an EST-based dataset assembled for the glaucophyte alga *Cyanophora paradoxa*, concluding that ~10% (~1,500) of the estimated 12,000–15,000 genes in the genome are cyanobacterial in origin. In contrast to the predictions for *A. thaliana* (Martin et al. 2002), these authors found that >90% of these proteins were predicted to be plastid-targeted, i.e., <10% of the cyanobacterial proteins in *C. paradoxa* appear to have plastid-independent functions (Reyes-Prieto et al. 2006). The reasons for these differences are not clear but could be both biological and methodological in nature (Archibald 2006b; Reyes-Prieto et al. 2006). In addition, there is growing evidence for the existence of non-canonical protein import pathways in algae. Primary plastids utilize an evolutionarily conserved import apparatus comprised of the Toc and Tic super-complexes (translocators of the outer and inner chloroplast membranes, respectively; Soll and Schleiff 2004; Gutensohn et al. 2006). Nucleus-encoded pre-proteins destined for the plastid possess a characteristic N-terminal transit peptide extension (McFadden 1999; Gould et al. 2008), and it is these extensions that are the target of in silico screens (Emanuelsson et al. 2000, 2007). Modern biochemical analyses have, however, revealed that we currently have a quite limited understanding of the biochemical determinants of plastid targeting. For example, only ~60% of a set of 604 *A. thaliana* plastid proteins identified by proteomics contained plastid targeting signals that could actually be identified using bioinformatics (Kleffmann et al. 2004). Examples of ER-to-Golgi-to-plastid targeting have also been uncovered (Radhamony and Theg 2006). The take-home message is that our inferences about the extent to which endosymbiotic gene transfers and replacements have shaped the biology of photosynthetic eukaryotes are ultimately only as good as our ability to accurately determine where in the cell proteins actually function.

Several additional points are worthy of mention from the perspective of non-cyanobacterial contributions to the establishment of primary plastids. Evolutionary analyses have revealed that most – but apparently not all – of the protein components of the Tic and Toc import machinery are demonstrably cyanobacterial in nature (Gould et al. 2008). One such exception is Tic110, a protein found in both red and green algae, which lacks a cyanobacterial counterpart and has been suggested to represent a host-derived contribution to the plastid protein import apparatus (McFadden and van Dooren 2004; Kalanon and McFadden 2008). In addition, phylogenetic analyses of plastid metabolite transporters reveal that they are of host, not endosymbiont, origin (Weber et al. 2006; Tyra et al. 2007). The primary endosymbiosis that gave rise to the plastid was clearly "…a period of considerable evolutionary experimentation, facilitated on one hand by functional redundancy at the level of enzymes and metabolic pathways, and on the other by combining the genetic potential of two very different cell types" (Archibald 2005).

Did other cells play a role as well? Some authors believe so. Huang and Gogarten (2007) uncovered 21 instances of apparent gene transfer from members of the bacterial genus *Chlamydia* into the algal nuclear genome and proposed that such genes are the remnants of a chlamydial endosymbiont that was somehow involved in cementing the relationship between the cyanobacterial progenitor of the plastid and its eukaryotic host. Interestingly, one of the chlamydial genes in the genome of the red alga *Cyanidioschyzon merolae* encodes an ATP/ADP translocase, which could have allowed the cyanobacterial endosymbiont to acquire energy from its host. There is little to go on in terms of confirming or refuting this hypothesis: on balance the data are also consistent with an ancestral relationship between cyanobacteria and Chlamydiae (Brinkman et al. 2002) or the presence of chlamydial genes in the cyanobacterial genome prior to the evolution of plastids. Regardless, the results of Huang and Gogarten are interesting in that most of the chlamydial genes in algal nuclear genomes encode proteins with predicted plastid targeting sequences, suggesting that they now contribute to the function of the organelle regardless of their origin (Huang and Gogarten 2007).

D. 'Recent' Cyanobacterial Endosymbioses: A Window on Plastid Evolution?

Evolutionary biologists work on the assumption that understanding processes taking place in modern-day organisms can shed light on events that have occurred in the past. Understanding the ancient origin of plastids is no exception. This section is devoted to discussion of two examples of recently established associations between microbial eukaryotes and cyanobacterial 'endosymbionts'. We say 'endosymbionts' because it is often far from clear whether the term 'endosymbiont' or 'organelle' is most appropriate. When does an endosymbiont become an organelle? As noted in previous sections, gene transfer from endosymbiont to host is a major part of this process, and the evolution of a mechanism for importing protein products of such transferred genes is often considered to be the tipping point (e.g., Cavalier-Smith and Lee 1985; Theissen and Martin 2006). There is seemingly no end to the number of recent host-endosymbiont relationships with the potential to improve our understanding of symbiogenesis and the origin of organelles (Nowack and Melkonian 2010).

Arguably the most striking example is that of the rhizarian testate amoeba *Paulinella chromatophora*, first discovered by the German biologist Robert Lauterborn in 1894. Lauterborn (1869–1952) noticed that *P. chromatophora* possesses one or two blue/green-pigmented bodies – chromatophores – in its cytoplasm and was clearly struck by their resemblance to cyanobacteria (Lauterborn 1895; Melkonian and Mollenhauer 2005). More than 100 years later, this organism has become the focus of intense genomic investigations to understand the precise nature of the chromatophore-host relationship. Preliminary molecular data indicated that the chromatophore was clearly *not* specifically related to canonical plastids, but rather to a specific sub-lineage of cyanobacteria, the *Synechococcus/Prochloroccus* group (Marin et al. 2005; Yoon et al. 2006). This connection has been firmly established by complete sequencing of the chromatophore genome (Nowack et al. 2008). At ~1 Mbp in size and with only 867 protein genes, it is the smallest cyanobacterial genome yet sequenced. The genes it retains – and has lost – provide a fascinating window into the biology of the chromatophore and the extent to which it has integrated with its host (Keeling and Archibald 2008).

First and foremost, the chromatophore is clearly all about phototrophy: its genome contains a near-complete set of genes for photosynthesis. It also lacks many genes that would be predicted to be dispensable for an obligate endosymbiont, in particular, those encoding membrane transporters and proteins involved in certain amino acid and cofactor biosynthetic pathways (Nowack et al. 2008; Nowack and Melkonian 2010). With only a quarter of the coding capacity inferred to have been present in its free-living cyanobacterial progenitor, the chromatophore is obviously no longer an autonomous entity. But is it an organelle? Chromatophore division is known to happen in concert with its host (Hoogenraad 1927; Kies 1974; Johnson et al. 1988), an observation that fueled speculation that genes encoding division proteins had been transferred to the *P. chromatophora* nuclear genome (Archibald 2006a; Yoon et al. 2006). Indeed, while most of the 'usual suspects' for cell division are encoded by the chromatophore (e.g., FtsZ and MinD), *sulA*, a gene encoding an FtsZ polymerization inhibitor, is absent (Nowack et al. 2008). This gene might now reside in the nucleus.

The first definitive evidence for endosymbiotic gene transfer in *P. chromatophora* came not from cell division protein genes but for a core photosystem gene. Nakayama and Ishida (2009) showed that a cyanobacterial-derived, spliceosomal intron-containing *psaE* gene encoding subunit IV of the PSI reaction centre is located in the host nuclear genome. Exactly how a nucleus-encoded PsaE protein would make its way to the chromatophore was not immediately obvious. A canonical N-terminal plastid targeting signal was not detected (Nakayama and Ishida 2009) but a follow-up investigation revealed the presence of a signal peptide of the sort that directs co-translational insertion of proteins into the eukaryotic secretory pathway (Mackiewicz and Bodyl 2010). Bodyl and colleagues have now presented compelling data and arguments for the existence of a bona fide protein import apparatus in *P. chromatophora*, one that could involve divergent chromatophore-encoded Tic-Toc components and the host cell signal peptide secretion system (Bodyl et al. 2010; Mackiewicz and Bodyl 2010). Most recently, Nowack et al. (2011) used next-generation transcriptome sequencing to expand the number of endosymbiotic gene transfer candidates to 32, most of which encode small photosynthetic proteins. Combined with information gleaned from another chromatophore genome sequence from a second species (Reyes-Prieto et al. 2010), these authors speculate on the existence of a minimum of several dozen to perhaps as many as 100 chromatophore-derived genes in the *Paulinella* nuclear genome (Nowack et al. 2011). Whether the term 'plastid' should be used to describe the chromatophores of *Paulinella* species is perhaps a matter of taste, but 'organelle' would now seem to be entirely appropriate.

A second interesting example of recent endosymbiosis involving a eukaryote and a cyanobacterium is in the diatom *Rhopalodia*. This case is very different from the situation in *Paulinella*, and indeed from what is believed to have occurred in the primary endosymbiotic origin of canonical plastids, because the host was already photosynthetic. Diatoms are environmentally significant marine algae that acquired their plastids by secondary endosymbiosis, i.e., the engulfment of a primary plastid-bearing alga (in this case a red alga) by a eukaryotic heterotroph (see Chap. 2). Therefore, the selection pressures that would have driven the establishment of a permanent connection between the photosynthetic host and photosynthetic endosymbiont are not as clear-cut. In the case of *R. gibba*, the so-called 'spheroid bodies' reside within cytoplasmic vacuoles (Geitler 1977) and have been shown to be most closely related to members of the cyanobacterial genus *Cyanothece* (Prechtl et al. 2004), which are well known for carrying out nitrogen fixation. As was the case with *Paulinella*, genome sequencing has provided important insight into the *raison d'etre* of the *Rhopalodia* host-spheroid body association. Large genomic fragments from the *R. gibba* spheroid body genome reveal the presence of a complete set of N_2-fixation enzymes and, interestingly, recent pseudogenization of numerous photosynthetic genes (Kneip et al. 2008). There is still much to learn about this fascinating system, but for now it is possible that *Rhopalodia* is well on it way to becoming a nitrogen-fixing eukaryotic organelle.

IV. Conclusion

Three decades ago we viewed the evolutionary origin of mitochondria and plastids as two sides of the same coin. At that time, the existing evidence supported the view that the two organelles had a 'classical' endosymbiotic origin from different eubacterial groups (α-Proteobacteria in the case of mitochondria, Cyanobacteria in the case of plastids) within an initially organelle-less but essentially eukaryotic host cell, with the mitochondrion emerging first and the plastid some time later. Endosymbiosis was followed by pronounced genome reduction as the transition from free-living bacterium to organelle progressed, with the chloroplast genome retaining a greater number of genes and a more pronounced resemblance to a bacterial ancestor than the mitochondrial genome. Endosymbiotic gene transfer from organelle to nucleus contributed in a major way to the evolution of the resulting composite cell, with many initially proto-organellar proteins now performing their functions elsewhere in the cell, and/or acquiring new functions. At the same time, newly minted proteins, novel inventions within the eukaryotic lineage, were acquired by the organelles and assumed essential roles in their biogenesis and function. Particularly prominent among these acquired proteins are membrane components that allow the regulated flow of both small metabolites and macromolecules (proteins but also RNA) across the double-membrane-bound organelles.

In the ensuing years, the evolutionary scenario for the plastid has changed little from that summarized above, but our understanding of the origin and subsequent evolution of the mitochondrion has undergone a substantial shift. With the recognition that mitochondria or mitochondrion-related organelles (MROs) are present in all eukaryotes that have been studied, the archezoan scenario has been severely challenged. It is still possible that the mitochondrion originated in a eukaryotic host cell populating an amitochondriate lineage ('archezoan') that has since become extinct; however, recent evidence and argument are turning the tide in favour of a symbiogenesis scenario, in which the host organism for the α-proteobacteria-like endosymbiont was a prokaryotic cell (archaeon?) rather than a eukaryote. Such a scenario raises the possibility that the origin of the mitochondrion was not only concurrent with the origin of the eukaryotic cell, but was in fact the sine qua non of eukaryogenesis.

Comparative analysis of mitochondrial and plastid proteomes has shown that these organelles are genetically highly mosaic, with organellar proteins having evolutionary origins well beyond the specific eubacterial lineages from which the organelles originated. Such studies are increasingly emphasizing how evolutionarily malleable organelles are, with a limited set of universally conserved core proteins and functions and a much larger assemblage of proteins that are phylogenetically diverse. Determining the functions of these lineage-specific proteins constitutes a formidable challenge.

As always, more data from phylogenetically strategic groups will be required to address many of the questions still outstanding about organelle evolution. For example, comprehensive genomic data from red algae and glaucophytes will be critical to resolving once and for all the question of primary plastid monophyly versus polyphyly. Additional data will undoubtedly yield many more examples of biochemical 'tinkering' in the course of mitochondrial and plastid evolution. We confidently expect that 'Origins of Mitochondria and Plastids' will continue to be a subject of debate for the foreseeable future, and we will not be at all surprised if our understanding of this evolutionary process takes a few more unexpected twists and turns as relevant new information continues to challenge currently accepted ideas.

Acknowledgements

The authors thank the Canadian Institute for Advanced Research, Programs in Evolutionary Biology and Integrated Microbial Biodiversity, for long-term research support and collaborative opportunities. MWG gratefully acknowledges past support from the Canadian Institutes of Health Research (CIHR) and the Canada Research Chairs Program, and current support from the Tula Foundation. JMA gratefully acknowledges CIHR and the Natural Sciences and Engineering Research Council of Canada for research funding and the CIHR New Investigator Program for salary support.

References

Adl SM, Simpson AGB, Farmer MA, Andersen RA, Anderson OR, Barta JR, Bowser SS, Brugerolle G, Fensome RA, Fredericq S, James TY, Karpov S, Kugrens P, Krug J, Lane CE, Lewis LA, Lodge J, Lynn DH, Mann DG, McCourt RM, Mendoza L, Moestrup Ø, Mozley-Standridge SE, Nerad TA, Shearer CA, Smirnov AV, Spiegel FW, Taylor MFJR (2005) The new higher level classification of eukaryotes with emphasis on the taxonomy of protists. J Eukaryot Microbiol 52:399–451

Aebersold R, Mann M (2003) Mass spectrometry-based proteomics. Nature 422:198–207

Andersson JO (2005) Lateral gene transfer in eukaryotes. Cell Mol Life Sci 62:1182–1197

Andersson SGE, Zomorodipour A, Andersson JO, Sicheritz-Pontén T, Alsmark UCM, Podowski RM, Näslund AK, Eriksson A-S, Winkler HH, Kurland CG (1998) The genome sequence of *Rickettsia prowazekii* and the origin of mitochondria. Nature 396:133–140

Archibald JM (2005) Jumping genes and shrinking genomes – probing the evolution of eukaryotic photosynthesis using genomics. IUBMB Life 57:539–547

Archibald JM (2006a) Endosymbiosis: double-take on plastid origins. Curr Biol 16:R690–R692

Archibald JM (2006b) Algal genomics: exploring the imprint of endosymbiosis. Curr Biol 16:R1033–R1035

Archibald JM (2008) The eocyte hypothesis and the origin of eukaryotic cells. Proc Natl Acad Sci USA 105:20049–20050

Archibald JM, Rogers MB, Toop M, Ishida K, Keeling PJ (2003) Lateral gene transfer and the evolution of plastid-targeted proteins in the secondary plastid-containing alga *Bigelowiella natans*. Proc Natl Acad Sci USA 100:7678–7683

Besendahl A, Qiu Y-L, Lee J, Palmer JD, Bhattacharya D (2000) The cyanobacterial origin and vertical transmission of the plastid tRNA^{Leu} group-I intron. Curr Genet 37:12–23

Bhattacharya D, Yoon HS, Hackett JD (2003) Photosynthetic eukaryotes unite: endosymbiosis connects the dots. Bioessays 26:50–60

Bininda-Emonds ORP (2004) The evolution of supertrees. Trends Ecol Evol 19:315–322

Bodyl A, Mackiewicz P, Stiller JW (2010) Comparative genomic studies suggest that the cyanobacterial endosymbionts of the amoeba *Paulinella chromatophora* possess an import apparatus for nuclear-encoded proteins. Plant Biol (Stuttg) 12:639–649

Borza T, Popescu CE, Lee RW (2005) Multiple metabolic roles for the nonphotosynthetic plastid of the green alga *Prototheca wickerhamii*. Eukaryot Cell 4:253–261

Boussau B, Karlberg EO, Frank AC, Legault B-A, Andersson SGE (2004) Computational inference of scenarios for α-proteobacterial genome evolution. Proc Natl Acad Sci USA 101:9722–9727

Boxma B, de Graaf RM, van der Staay GWM, van Alen TA, Ricard G, Gabaldón T, van Hoek AHAM, Moon-van der Staay SY, Koopman WJH, van Hellemond JJ, Tielens AGM, Friedrich T, Veenhuis M, Huynen MA, Hackstein JHP (2005) An anaerobic mitochondrion that produces hydrogen. Nature 434:74–79

Brandt U (2006) Energy converting NADH:quinone oxidoreductase (complex I). Annu Rev Biochem 75:69–92

Brinkman FSL, Blanchard JL, Cherkasov A, Av-Gay Y, Brunham RC, Fernandez RC, Finlay BB, Otto SP, Ouellette BFF, Keeling PJ, Rose AM, Hancock

REW, Jones SJM, Greberg H (2002) Evidence that plant-like genes in *Chlamydia* species reflect an ancestral relationship between Chlamydiaceae, cyanobacteria, and the chloroplast. Genome Res 12:1159–1167
Brinkmann H, Martin W (1996) Higher-plant chloroplast and cytosolic 3-phosphoglycerate kinases: a case of endosymbiotic gene replacement. Plant Mol Biol 30:65–75
Bui ET, Bradley PJ, Johnson PJ (1996) A common evolutionary origin for mitochondria and hydrogenosomes. Proc Natl Acad Sci USA 93:9651–9656
Burger G, Lang BF, Reith M, Gray MW (1996) Genes encoding the same three subunits of respiratory complex II are present in the mitochondrial DNA of two phylogenetically distant eukaryotes. Proc Natl Acad Sci USA 93:2328–2332
Burger G, Saint-Louis D, Gray MW, Lang BF (1999) Complete sequence of the mitochondrial DNA of the red alga *Porphyra purpurea*. Cyanobacterial introns and shared ancestry of red and green algae. Plant Cell 11:1675–1694
Burki F, Pawlowski J (2006) Monophyly of Rhizaria and multigene phylogeny of unicellular bikonts. Mol Biol Evol 23:1922–1930
Cardol P, Vanrobaeys F, Devreese B, Van Beeumen J, Matagne RF, Remacle C (2004) Higher plant-like subunit composition of mitochondrial complex I from *Chlamydomonas reinhardtii*: 31 conserved components among eukaryotes. Biochim Biophys Acta 1658:212–224
Cavalier-Smith T (2007) The chimaeric origin of mitochondria: photosynthetic cell enslavement, gene-transfer pressure, and compartmentation efficiency. In: Martin WF, Müller M (eds) Origin of mitochondria and hydrogenosomes. Springer, Berlin, pp 161–199
Cavalier-Smith T, Lee JJ (1985) Protozoa as hosts for endosymbioses and the conversion of symbionts into organelles. J Protozool 32:376–379
Ciniglia C, Yoon HS, Pollio A, Pinto G, Bhattacharya D (2004) Hidden biodiversity of the extremophilic Cyanidiales red algae. Mol Ecol 13:1827–1838
Clark CG, Roger AJ (1995) Direct evidence for secondary loss of mitochondria in *Entamoeba histolytica*. Proc Natl Acad Sci USA 92:6518–6521
Claros MG, Vincens P (1996) Computational method to predict mitochondrially imported proteins and their targeting sequences. Eur J Biochem 241:779–786
Cox CJ, Foster PG, Hirt RP, Harris SR, Embley TM (2008) The archaebacterial origin of eukaryotes. Proc Natl Acad Sci USA 105:20356–20361
de Koning AP, Keeling PJ (2006) The complete plastid genome sequence of the parasitic green alga *Helicosporidium* sp. is highly reduced and structured. BMC Biol 4:12
Delsuc F, Brinkmann H, Philippe H (2005) Phylogenomics and the reconstruction of the tree of life. Nat Rev Genet 6:361–375
Delwiche CF, Kuhsel M, Palmer JD (1995) Phylogenetic analysis of *tuf*A sequences indicates a cyanobacterial origin of all plastids. Mol Phylogenet Evol 4:110–128
Desmond E, Brochier-Armanet C, Forterre P, Gribaldo S (2011) On the last common ancestor and early evolution of eukaryotes: reconstructing the history of mitochondrial ribosomes. Res Microbiol 162:53–70
Deusch O, Landan G, Roettger M, Gruenheit N, Kowallik KV, Allen JF, Martin W, Dagan T (2008) Genes of cyanobacterial origin in plant nuclear genomes point to a heterocyst-forming plastid ancestor. Mol Biol Evol 25:748–761
Doolittle WF (1980) Revolutionary concepts in evolutionary biology. Trends Biochem Sci 5:146–149
Doolittle WF (1999) Phylogenetic classification and the universal tree. Science 284:2124–2129
Douglas SE, Gray MW (1991) Plastid origins. Nature 352:290
Dreger M (2003) Proteome analysis at the level of subcellular structures. Eur J Biochem 270:589–599
Emanuelsson O, Nielsen H, Brunak S, von Heijne G (2000) Predicting subcellular localization of proteins based on their N-terminal amino acid sequence. J Mol Biol 300:1005–1016
Emanuelsson O, Brunak S, von Heijne G, Nielsen H (2007) Locating proteins in the cell using TargetP, SignalP and related tools. Nat Protoc 2:953–971
Embley TM (2006) Multiple secondary origins of the anaerobic lifestyle in eukaryotes. Philos Trans R Soc Lond B Biol Sci 361:1055–1067
Embley TM, van der Giezen M, Horner DS, Dyal PL, Bell S, Foster PG (2003) Hydrogenosomes, mitochondria and early eukaryotic evolution. IUBMB Life 55:387–395
Emelyanov VV (2001a) Rickettsiaceae, rickettsia-like endosymbionts, and the origin of mitochondria. Biosci Rep 21:1–17
Emelyanov VV (2001b) Evolutionary relationship of Rickettsiae and mitochondria. FEBS Lett 501:11–18
Emelyanov VV (2003a) Mitochondrial connection to the origin of the eukaryotic cell. Eur J Biochem 270:1599–1618
Emelyanov VV (2003b) Common evolutionary origin of mitochondrial and rickettsial respiratory chains. Arch Biochem Biophys 420:130–141
Esser C, Martin W (2007) Supertrees and symbiosis in eukaryote genome evolution. Trends Microbiol 15:435–437

Esser C, Ahmadinejad N, Wiegand C, Rotte C, Sebastiani F, Gelius-Dietrich G, Henze K, Kretschmann E, Richly E, Leister D, Bryant D, Steel MA, Lockhart PJ, Penny D, Martin W (2004) A genome phylogeny for mitochondria among α-proteobacteria and a predominantly eubacterial ancestry of yeast nuclear genes. Mol Biol Evol 21:1643–1660

Esser C, Martin W, Dagan T (2007) The origin of mitochondria in light of a fluid prokaryotic chromosome model. Biol Lett 3:180–184

Falcón LI, Magallón S, Castillo A (2010) Dating the cyanobacterial ancestor of the chloroplast. ISME J 4:777–783

Fast NM, Kissinger JC, Roos DS, Keeling PJ (2001) Nuclear-encoded, plastid-targeted genes suggest a single common origin for apicomplexan and dinoflagellate plastids. Mol Biol Evol 18:418–426

Finet C, Timme RE, Delwiche CF, Marlétaz F (2010) Multigene phylogeny of the green lineage reveals the origin and diversification of land plants. Curr Biol 20:2217–2222

Fitzpatrick DA, Creevey CJ, McInerney JO (2006) Genome phylogenies indicate a meaningful α-proteobacterial phylogeny and support a grouping of the mitochondria with the Rickettsiales. Mol Biol Evol 23:74–85

Gabaldón T, Huynen MA (2003) Reconstruction of the proto-mitochondrial metabolism. Science 302:609

Gabaldón T, Huynen MA (2004) Shaping the mitochondrial proteome. FEBS Lett 1659:212–220

Gabaldón T, Huynen MA (2007) From endosymbiont to host-controlled organelle: the hijacking of mitochondrial protein synthesis and metabolism. PLoS Comput Biol 3:2209–2218

Gabaldón T, Rainey D, Huynen MA (2005) Tracing the evolution of a large protein complex in the eukaryotes, NADH: ubiquinone oxidoreductase (complex I). J Mol Biol 348:857–870

Gawryluk RMR, Gray MW (2010) Evidence for an early evolutionary emergence of γ-type carbonic anhydrases as components of mitochondrial respiratory complex I. BMC Evol Biol 10:176

Geitler L (1977) On the life history of the *Epithemiaceae Epithemia*, *Rhopalodia* and *Denticula* (*Diatomophyceae*) and their presumably symbiotic spheroid bodies. Plant Syst Evol 128:259–275

Gill EE, Diaz-Triviño S, Barberà MJ, Silberman JD, Stechmann A, Gaston D, Tamas I, Roger AJ (2007) Novel mitochondrion-related organelles in the anaerobic amoeba *Mastigamoeba balamuthi*. Mol Microbiol 66:1306–1320

Giovannoni SJ, Tripp HJ, Givan S, Podar M, Vergin KL, Baptista D, Bibbs L, Eads J, Richardson TH, Noordewier M, Rappé MS, Short JM, Carrington JC, Mathur EJ (2005) Genome streamlining in a cosmopolitan oceanic bacterium. Science 309: 1242–1245

Gogarten JP, Kibak H, Dittrich P, Taiz L, Bowman EJ, Bowman BJ, Manolson MF, Poole RJ, Date T, Oshima T, Konishi J, Denda K, Yoshida M (1989) Evolution of the vacuolar H^+-ATPase: implications for the origin of eukaryotes. Proc Natl Acad Sci USA 86:6661–6665

Goldberg AV, Molik S, Tsaousis AD, Neumann K, Kuhnke G, Delbac F, Vivares CP, Hirt RP, Lill R, Embley TM (2008) Localization and functionality of microsporidian iron-sulphur cluster assembly proteins. Nature 452:624–629

Gould SB, Waller RF, McFadden GI (2008) Plastid evolution. Annu Rev Plant Biol 59:491–517

Graham LE, Wilcox LW (2000) Algae. Prentice-Hall, Upper Saddle River

Gray MW (1989) The evolutionary origins of organelles. Trends Genet 5:294–299

Gray MW (1992) The endosymbiont hypothesis revisited. Int Rev Cytol 141:233–357

Gray MW (1993) Origin and evolution of organelle genomes. Curr Opin Genet Dev 3:884–890

Gray MW (1998) Rickettsia, typhus and the mitochondrial connection. Nature 396:109–110

Gray MW (1999) Evolution of organellar genomes. Curr Opin Genet Dev 9:678–687

Gray MW, Doolittle WF (1982) Has the endosymbiont hypothesis been proven? Microbiol Rev 46:1–42

Gray MW, Spencer DF (1996) Organellar evolution. In: Roberts DM, Sharp P, Alderson G, Collins MA (eds) Evolution of microbial life. Cambridge University Press, Cambridge, pp 109–126

Gray MW, Lang BF, Cedergren R, Golding GB, Lemieux C, Sankoff D, Turmel M, Brossard N, Delage E, Littlejohn TG, Plante I, Rioux P, Saint-Louis D, Zhu Y, Burger G (1998) Genome structure and gene content in protist mitochondrial DNAs. Nucleic Acids Res 26:865–878

Gray MW, Burger G, Lang BF (1999) Mitochondrial evolution. Science 283:1476–1481

Gray MW, Burger G, Lang BF (2001) The origin and early evolution of mitochondria. Genome Biol 2:Reviews 1018.1011–1018.1015

Gupta RS (1995) Evolution of the chaperonin families (Hsp60, Hsp10 and Tcp-1) of proteins and the origin of eukaryotic cells. Mol Microbiol 15:1–11

Gutensohn M, Fan E, Frielingsdorf S, Hanner P, Hou B, Hust B, Klösgen RB (2006) Toc, Tic, Tat et al.: structure and function of protein transport machineries in chloroplasts. J Plant Physiol 163:333–347

Hagopian JC, Reis M, Kitajima JP, Bhattacharya D, de Oliveira MC (2004) Comparative analysis of the complete plastid genome sequence of the red alga

Gracilaria tenuistipitata var. *liui* provides insights into the evolution of rhodoplasts and their relationship to other plastids. J Mol Evol 59:464–477
Heazlewood JL, Howell KA, Millar AH (2003) Mitochondrial complex I from Arabidopsis and rice: orthologs of mammalian and fungal components coupled with plant-specific subunits. Biochim Biophys Acta 1604:159–169
Hirt RP, Logsdon JM Jr, Healy B, Dorey MW, Doolittle WF, Embley TM (1999) Microsporidia are related to fungi: evidence from the largest subunit of RNA polymerase II and other proteins. Proc Natl Acad Sci USA 96:580–585
Hjort K, Goldberg AV, Tsaousis AD, Hirt RP, Embley TM (2010) Diversity and reductive evolution of mitochondria among microbial eukaryotes. Philos Trans R Soc Lond B Bio Sci 365:713–727
Hoogenraad HR (1927) Zur Kenntnis der Fortpflanzung von *Paulinella chromatophora* Lauterb. Zool Anz 72:140–150
Howe CJ, Barbrook AC, Koumandou VL, Nisbet ER, Symington HA, Wightman TF (2003) Evolution of the chloroplast genome. Philos Trans R Soc Lond B Biol Sci 358:99–107
Howe CJ, Barbrook AC, Nisbet RE, Lockhart PJ, Larkum AW (2008) The origin of plastids. Philos Trans R Soc Lond B Biol Sci 363:2675–2685
Hrdy I, Hirt RP, Dolezal P, Bardonová L, Foster PG, Tachezy J, Embley TM (2004) *Trichomonas* hydrogenosomes contain the NADH dehydrogenase module of mitochondrial complex I. Nature 432:618–622
Huang J, Gogarten JP (2007) Did an ancient chlamydial endosymbiosis facilitate the establishment of primary plastids? Genome Biol 8:R99
Hug LA, Stechmann A, Roger AJ (2010) Phylogenetic distributions and histories of proteins involved in anaerobic pyruvate metabolism in eukaryotes. Mol Biol Evol 27:311–324
Inagaki Y, Susko E, Fast NM, Roger AJ (2004) Covarion shifts cause a long-branch attraction artifact that unites microsporidia and archaebacteria in EF-1α phylogenies. Mol Biol Evol 21:1340–1349
Iwabe N, Kuma K, Hasegawa M, Osawa S, Miyata T (1989) Evolutionary relationship of archaebacteria, eubacteria, and eukaryotes inferred from phylogenetic trees of duplicated genes. Proc Natl Acad Sci USA 86:9355–9359
Jain R, Rivera MC, Lake JA (1999) Horizontal gene transfer among genomes: the complexity hypothesis. Proc Natl Acad Sci USA 96:3801–3806
John P, Whatley FR (1975) *Paracoccus denitrificans* and the evolutionary origin of the mitochondrion. Nature 254:495–498
Johnson PW, Hargraves PE, Sieburth JM (1988) Ultrastructure and ecology of *Calycomonas ovalis* Wulff, 1919, (Chrysophyceae) and its redescription as a testate rhizopod, *Paulinella ovalis* n. comb. (Filosea: Euglyphina). J Protozool 35:618–626
Kalanon M, McFadden GI (2008) The chloroplast protein translocation complexes of *Chlamydomonas reinhardtii*: a bioinformatic comparison of Toc and Tic components in plants, green algae and red algae. Genetics 179:95–112
Karlberg O, Canbäck B, Kurland CG, Andersson SGE (2000) The dual origin of the yeast mitochondrial proteome. Yeast 17:170–187
Karlin S, Brocchieri L (2000) Heat shock protein 60 sequence comparisons: duplications, lateral transfer, and mitochondrial evolution. Proc Natl Acad Sci USA 97:11348–11353
Karol KG, McCourt RM, Cimino MT, Delwiche CF (2001) The closest living relatives of land plants. Science 294:2351–2353
Keeling PJ (2010) The endosymbiotic origin, diversification and fate of plastids. Philos Trans R Soc Lond B Biol Sci 365:729–748
Keeling PJ, Archibald JM (2008) Organelle evolution: what's in a name? Curr Biol 18:R345–R347
Keeling PJ, Palmer JD (2008) Horizontal gene transfer in eukaryotic evolution. Nat Rev Genet 9:605–618
Keeling PJ, Luker MA, Palmer JD (2000) Evidence from beta-tubulin phylogeny that microsporidia evolved from within the fungi. Mol Biol Evol 17:23–31
Keeling PJ, Burger G, Durnford DG, Lang BF, Lee RW, Pearlman RE, Roger AJ, Gray MW (2005) The tree of eukaryotes. Trends Ecol Evol 20:670–676
Kies L (1974) Electron microscopical investigations on *Paulinella chromatophora* Lauterborn, a thecamoeba containing blue-green endosymbionts (Cyanelles) (author's transl). Protoplasma 80:69–89
Kim E, Archibald JM (2009) Diversity and evolution of plastids and their genomes. In: Aronsson H, Sandelius AS (eds) The chloroplast – interactions with the environment. Springer, Berlin, pp 1–39
Kim E, Graham LE (2008) EEF2 analysis challenges the monophyly of Archaeplastida and Chromalveolata. PLoS One 3:e2621
Kleffmann T, Russenberger D, von Zychlinski A, Christopher W, Sjölander K, Gruissem W, Baginsky S (2004) The *Arabidopsis thaliana* chloroplast proteome reveals pathway abundance and novel protein functions. Curr Biol 14:354–362
Kleine T, Maier UG, Leister D (2009) DNA transfer from organelles to the nucleus: he idiosyncratic genetics of endosymbiosis. Annu Rev Plant Biol 60:115–138

Kneip C, Voss C, Lockhart PJ, Maier UG (2008) The cyanobacterial endosymbiont of the unicellular algae *Rhopalodia gibba* shows reductive genome evolution. BMC Evol Biol 8:30

Koonin EV (2010) The origin and early evolution of eukaryotes in the light of phylogenomics. Genome Biol 11:209

Kowallik KV (1997) Origin and evolution of chloroplasts: current status and future perspectives. In: Schenk HE, Herrmann RG, Jeon KW, Müller NE, Schwemmler W (eds) Eukaryotism and symbiosis: intertaxonic combination versus symbiotic adaptation. Springer, Berlin, pp 3–23

Krause K (2008) From chloroplasts to "cryptic" plastids: evolution of plastid genomes in parasitic plants. Curr Genet 54:111–121

Kumar A, Agarwal S, Heyman JA, Matson S, Heidtman M, Piccirillo S, Umansky L, Drawid A, Jansen R, Liu Y, Cheung K-H, Miller P, Gerstein M, Roeder GS, Snyder M (2002) Subcellular localization of the yeast proteome. Genes Dev 16:707–719

Kurland CG, Andersson SGE (2000) Origin and evolution of the mitochondrial proteome. Microbiol Mol Biol Rev 64:786–820

Kurland CG, Collins LJ, Penny D (2006) Genomics and the irreducible nature of eukaryotic cells. Science 312:1011–1014

Kurland CG, Collins LJ, Penny D (2007) The evolution of eukaryotes: response. Science 316:543

Lake JA, Henderson E, Oakes M, Clark MW (1984) Eocytes: a new ribosome structure indicates a kingdom with a close relationship to eukaryotes. Proc Natl Acad Sci USA 81:3786–3790

Lane N, Martin W (2010) The energetics of genome complexity. Nature 467:929–934

Lang BF, Burger G, O'Kelly CJ, Cedergren R, Golding GB, Lemieux C, Sankoff D, Turmel M, Gray MW (1997) An ancestral mitochondrial DNA resembling a eubacterial genome in miniature. Nature 387:493–497

Lang BF, Gray MW, Burger G (1999a) Mitochondrial genome evolution and the origin of eukaryotes. Annu Rev Genet 33:351–397

Lang BF, Seif E, Gray MW, O'Kelly CJ, Burger G (1999b) A comparative genomics approach to the evolution of eukaryotes and their mitochondria. J Eukaryot Microbiol 46:320–326

Lapaille M, Escobar-Ramírez A, Degand H, Baurain D, Rodríguez-Salinas E, Coosemans N, Boutry M, Gonzalez-Halphen D, Remacle C, Cardol P (2010) Atypical subunit composition of the chlorophycean mitochondrial F_1F_O-ATP synthase and role of Asa7 protein in stability and oligomycin resistance of the enzyme. Mol Biol Evol 27:1630–1644

Larkum AW, Lockhart PJ, Howe CJ (2007) Shopping for plastids. Trends Plant Sci 12:189–195

Lauterborn R (1895) Protozoenstudien II. *Paulinella chromatophora* nov. gen., nov. spec., ein beschalter Rhizopode des Süßwassers mit blaugrünen chromatophorenartigen Einschlüssen. Z Wiss Zool 59:537–544

Lewis LA, McCourt RM (2004) Green algae and the origin of land plants. Am J Bot 91:1535–1556

Lindmark DG, Müller M (1973) Hydrogenosome, a cytoplasmic organelle of the anaerobic flagellate *Tritrichomonas foetus*, and its role in pyruvate metabolism. J Biol Chem 248:7724–7728

Lockhart PJ, Howe CJ, Bryant DA, Beanland TJ, Larkum AW (1992a) Substitutional bias confounds inference of cyanelle origins from sequence data. J Mol Evol 34:153–162

Lockhart PJ, Penny D, Hendy MD, Howe CJ, Beanland TJ, Larkum AW (1992b) Controversy on chloroplast origins. FEBS Lett 301:127–131

Löffelhardt W, Bohnert HJ, Bryant DA (1997) The complete sequence of the *Cyanophora paradoxa* cyanelle genome. In: Bhattacharya D (ed) Origins of algae and their plastids. Springer, Wien, pp 142–162

Longet D, Archibald JM, Keeling PJ, Pawlowski J (2003) Foraminifera and Cercozoa share a common origin according to RNA polymerase II phylogenies. Int J Syst Evol Microbiol 53:1735–1739

Mackiewicz P, Bodyl A (2010) A hypothesis for import of the nuclear-encoded PsaE protein of *Paulinella chromatophora* (Cercozoa, Rhizaria) into its cyanobacterial endosymbionts/plastids via the endomembrane system. J Phycol 46:847–859

Mai Z, Ghosh S, Frisardi M, Rosenthal B, Rogers R, Samuelson J (1999) Hsp60 is targeted to a cryptic mitochondrion-derived organelle ("crypton") in the microaerophilic protozoan parasite *Entamoeba histolytica*. Mol Cell Biol 19:2198–2205

Marcotte EM, Xenarios I, van Der Bliek AM, Eisenberg D (2000) Localizing proteins in the cell from their phylogenetic profiles. Proc Natl Acad Sci USA 97:12115–12120

Margulis L (1970) Origin of eukaryotic cells. Yale University Press, New Haven

Marin B, Nowack ECM, Melkonian M (2005) A plastid in the making: evidence for a second primary endosymbiosis. Protist 156:425–432

Martin W (2010) Evolutionary origins of metabolic compartmentalization in eukaryotes. Philos Trans R Soc Lond B Biol Sci 365:847–855

Martin W, Cerff R (1986) Prokaryotic features of a nucleus-encoded enzyme. cDNA sequences for chloroplast and cytosolic glyceraldehyde-3-phosphate

dehydrogenases from mustard (*Sinapis alba*). Eur J Biochem 159:323–331
Martin W, Herrmann RG (1998) Gene transfer from organelles to the nucleus: how much, what happens, and why? Plant Physiol 118:9–17
Martin W, Kowallik KV (1999) Annotated English translation of Mereschkowsky's 1905 paper 'Über Natur und Ursprung der Chromatophoren im Pflanzenreiche'. Eur J Phycol 34:287–295
Martin W, Müller M (1998) The hydrogen hypothesis for the first eukaryote. Nature 392:37–41
Martin W, Schnarrenberger C (1997) The evolution of the Calvin cycle from prokaryotic to eukaryotic chromosomes: a case study of functional redundancy in ancient pathways through endosymbiosis. Curr Genet 32:1–18
Martin W, Brinkmann H, Savonna C, Cerff R (1993) Evidence for a chimeric nature of nuclear genomes: eubacterial origin of eukaryotic glyceraldehyde-3-phosphate dehydrogenase genes. Proc Natl Acad Sci USA 90:8692–8696
Martin W, Mustafa AZ, Henze K, Schnarrenberger C (1996) Higher-plant chloroplast and cytosolic fructose-1,6-bisphosphatase isoenzymes: origins via duplication rather than prokaryote-eukaryote divergence. Plant Mol Biol 32:485–491
Martin W, Hoffmeister M, Rotte C, Henze K (2001) An overview of endosymbiotic models for the origins of eukaryotes, their ATP-producing organelles (mitochondria and hydrogenosomes), and their heterotrophic lifestyle. Biol Chem 382:1521–1539
Martin W, Rujan T, Richly E, Hansen A, Cornelsen S, Lins T, Leister D, Stoebe B, Hasegawa M, Penny D (2002) Evolutionary analysis of *Arabidopsis*, cyanobacterial, and chloroplast genomes reveals plastid phylogeny and thousands of cyanobacterial genes in the nucleus. Proc Natl Acad Sci USA 99:12246–12251
Martin W, Dagan T, Koonin EV, Dipippo JL, Gogarten JP, Lake JA (2007) The evolution of eukaryotes. Science 316:542–543
Matsuzaki M, Misumi O, Shin-i T, Maruyama S, Takahara M, Miyagishima S, Mori T, Nishida K, Yagisawa F, Nishida K, Yoshida Y, Nishimura Y, Nakao S, Kobayashi T, Momoyama Y, Higashiyama T, Minoda A, Sano M, Nomoto H, Oishi K, Hayashi H, Ohta F, Nishizaka S, Haga S, Miura S, Morishita T, Kabeya Y, Terasawa K, Suzuki Y, Ishii Y, Asakawa S, Takano H, Ohta N, Kuroiwa H, Tanaka K, Shimizu N, Sugano S, Sato N, Nozaki H, Ogasawara N, Kohara Y, Kuroiwa T (2004) Genome sequence of the ultrasmall unicellular red alga *Cyanidioschyzon merolae* 10D. Nature 428:653–657
Mazumdar J, Wilson EH, Masek K, Hunter CA, Striepen B (2006) Apicoplast fatty acid synthesis is essential for organelle biogenesis and parasite survival in *Toxoplasma gondii*. Proc Natl Acad Sci USA 103:13192–13197
McDaniel LD, Young E, Delaney J, Ruhnau F, Ritchie KB, Paul JH (2010) High frequency of horizontal gene transfer in the oceans. Science 330:50
McFadden GI (1999) Plastids and protein targeting. J Eukaryot Microbiol 46:339–346
McFadden GI (2001) Chloroplast origin and integration. Plant Physiol 125:50–53
McFadden GI, van Dooren GG (2004) Evolution: red algal genome affirms a common origin of all plastids. Curr Biol 14:R514–R516
Melkonian M, Mollenhauer D (2005) Robert Lauterborn (1869–1952) and his *Paulinella chromatophora*. Protist 156:253–262
Mereschkowsky C (1905) Über Natur und Ursprung der Chromatophoren im Pflanzenreiche. Biol Centralbl 25:593–604
Moreira D, Le Guyader H, Phillippe H (2000) The origin of red algae and the evolution of chloroplasts. Nature 405:69–72
Morris RM, Rappé MS, Connon SA, Vergin KL, Slebold WA, Carlson CA, Giovannoni SJ (2002) SAR11 clade dominates ocean surface bacterioplankton communities. Nature 420:806–810
Nakayama T, Ishida K-I (2009) Another acquisition of a primary photosynthetic organelle is underway in *Paulinella chromatophora*. Curr Biol 19:R284–R285
Nina PB, Dudkina NV, Kane LA, van Eyk JE, Boekema EJ, Mather MW, Vaidya AB (2010) Highly divergent mitochondrial ATP synthase complexes in *Tetrahymena thermophila*. PLoS Biol 8:1–15
Nowack ECM, Melkonian M (2010) Endosymbiotic associations within protists. Philos Trans R Soc Lond B Biol Sci 365:699–712
Nowack ECM, Melkonian M, Glöckner G (2008) Chromatophore genome sequence of *Paulinella* sheds light on acquisition of photosynthesis by eukaryotes. Curr Biol 18:410–418
Nowack ECM, Vogel H, Groth M, Grossman AR, Melkonian M, Glöckner G (2011) Endosymbiotic gene transfer and transcriptional regulation of transferred genes in *Paulinella chromatophora*. Mol Biol Evol 28:407–422
Nowitzki U, Gelius-Dietrich G, Schwieger M, Henze K, Martin W (2004) Chloroplast phosphoglycerate kinase from *Euglena gracilis*. Endosymbiotic gene replacement going against the tide. Eur J Biochem 271:4123–4131
Nozaki H (2005) A new scenario of plastid evolution: plastid primary endosymbiosis before the divergence of the "Plantae," emended. J Plant Res 118: 247–255

Nozaki H, Iseki M, Hasegawa M, Misawa K, Nakada T, Sasaki N, Watanabe M (2007) Phylogeny of primary photosynthetic eukaryotes as deduced from slowly evolving nuclear genes. Mol Biol Evol 24:1592–1595

Nozaki H, Maruyama S, Matsuzaki M, Nakada T, Kato S, Misawa K (2009) Phylogenetic positions of Glaucophyta, green plants (Archaeplastida) and Haptophyta (Chromalveolata) as deduced from slowly evolving nuclear genes. Mol Phylogenet Evol 53:872–880

O'Brien TW (2002) Evolution of a protein-rich mitochondrial ribosome: implications for human genetic disease. Gene 286:73–79

O'Brien TW (2003) Properties of human mitochondrial ribosomes. IUBMB Life 55:505–513

Ochman H, Lawrence JG, Groisman EA (2000) Lateral gene transfer and the nature of bacterial innovation. Nature 405:299–304

Pace NR, Olsen GJ, Woese CR (1986) Ribosomal RNA phylogeny and the primary lines of evolutionary descent. Cell 45:325–326

Palmer JD (2003) The symbiotic birth and spread of plastids: how many times and whodunit? J Phycol 39:4–11

Parfrey LW, Grant J, Tekle YI, Lasek-Nesselquist E, Morrison HG, Sogin ML, Patterson DJ, Katz LA (2010) Broadly sampled multigene analyses yield a well-resolved eukaryotic tree of life. Syst Biol 59:518–533

Patron NJ, Rogers MB, Keeling PJ (2004) Gene replacement of fructose-1,6-bisphosphate aldolase supports the hypothesis of a single photosynthetic ancestor of chromalveolates. Eukaryot Cell 3:1169–1175

Parisi G, Perales M, Fornasari MS, Colaneri A, González-Schain N, Gómez-Casati D, Zimmermann S, Brennicke A, Araya A, Ferry JG, Echave J, Zabaleta E (2004) Gamma carbonic anhydrases in plant mitochondria. Plant Mol Biol 55:193–207

Perales M, Parisi G, Fornasari MS, Colaneri A, Villarreal F, González-Schain N, Echave J, Gómez-Casati D, Braun HP, Araya A, Zabaleta E (2004) Gamma carbonic anhydrase like complex interact with plant mitochondrial complex I. Plant Mol Biol 56:947–957

Pérez-Brocal V, Clark AG (2008) Analysis of two genomes from the mitochondrion-like organelle of the intestinal parasite *Blastocystis*: complete sequences, gene content and genome organization. Mol Biol Evol 25:2475–2482

Pisani D, Cotton JA, McInerney JO (2007) Supertrees disentangle the chimerical origin of eukaryotic genomes. Mol Biol Evol 24:1752–1760

Prechtl J, Kneip C, Lockhart P, Wenderoth K, Maier U-G (2004) Intracellular spheroid bodies of *Rhopalodia gibba* have nitrogen-fixing apparatus of cyanobacterial origin. Mol Biol Evol 21:1477–1481

Radhamony RN, Theg SM (2006) Evidence for an ER to Golgi to chloroplast protein transport pathway. Trends Cell Biol 16:385–387

Reeb V, Bhattacharya D (2010) The thermo-acidophilic Cyanidiophyceae (Cyanidiales). In: Seckback J (ed) Red algae in the genomic age. Springer, New York, pp 409–426

Reyes-Prieto A, Hackett JD, Soares MB, Bonaldo MF, Bhattacharya D (2006) Cyanobacterial contribution to algal nuclear genomes is primarily limited to plastid functions. Curr Biol 16:2320–2325

Reyes-Prieto A, Weber AP, Bhattacharya D (2007) The origin and establishment of the plastid in algae and plants. Annu Rev Genet 41:147–168

Reyes-Prieto A, Yoon HS, Moustafa A, Yang EC, Andersen RA, Boo SM, Nakayama T, Ishida K, Bhattacharya D (2010) Differential gene retention in plastids of common recent origin. Mol Biol Evol 27:1530–1537

Richards TA, Archibald JM (2011) Cell evolution: gene transfer agents and the evolution of mitochondria. Curr Biol 21(3):R112–R114

Richly E, Chinnery PF, Leister D (2003) Evolutionary diversification of mitochondrial proteomes: implications for human disease. Trends Genet 19:356–362

Rivera MC, Lake JA (1992) Evidence that eukaryotes and eocyte prokaryotes are immediate relatives. Science 257:74–76

Rivera MC, Lake JA (2004) The ring of life provides evidence for a genome fusion origin of eukaryotes. Nature 431:152–155

Rivera MC, Jain R, Moore JE, Lake JA (1998) Genomic evidence for two functionally distinct gene classes. Proc Natl Acad Sci USA 95:6239–6244

Rocap G, Larimer FW, Lamerdin J, Malfatti S, Chain P, Ahlgren NA, Arellano A, Coleman M, Hauser L, Hess WR, Johnson ZI, Land M, Lindell D, Post AF, Regala W, Shah M, Shaw SL, Steglich C, Sullivan MB, Ting CS, Tolonen A, Webb EA, Zinser ER, Chisholm SW (2003) Genome divergence in two *Prochlorococcus* ecotypes reflects oceanic niche differentiation. Nature 424:1042–1047

Rodríguez-Ezpeleta N, Brinkmann H, Burey SC, Roure B, Burger G, Löffelhardt W, Bohnert HJ, Philippe H, Lang BF (2005) Monophyly of primary photosynthetic eukaryotes: green plants, red algae, and glaucophytes. Curr Biol 15:1325–1330

Rogers M, Keeling PJ (2004) Lateral transfer and recompartmentalization of Calvin cycle enzymes of plants and algae. J Mol Evol 58:367–375

Sällström B, Andersson SGE (2005) Genome reduction in the α-Proteobacteria. Curr Opin Microbiol 8:579–585

Sato N (2006) Origin and evolution of plastids: genomic view on the unification and diversity of plastids. In: Wise RR, Hoober JK (eds) The structure and function of plastids. Springer, Dordrecht, pp 75–102

Sharma MR, Booth TM, Simpson L, Maslov DA, Agrawal RK (2009) Structure of a mitochondrial ribosome with minimal RNA. Proc Natl Acad Sci USA 106:9637–9642

Shutt TE, Gray MW (2006) Bacteriophage origins of mitochondrial replication and transcription proteins. Trends Genet 22:90–95

Sicheritz-Pontén T, Kurland CG, Andersson SG (1998) A phylogenetic analysis of the cytochrome *b* and cytochrome *c* oxidase I genes supports an origin of mitochondria from within the Rickettsiaceae. Biochim Biophys Acta 1365:545–551

Smith DG, Gawryluk RMR, Spencer DF, Pearlman RE, Siu KWM, Gray MW (2007) Exploring the mitochondrial proteome of the ciliate protozoon *Tetrahymena thermophila*: direct analysis by tandem mass spectrometry. J Mol Biol 374:837–863

Smits P, Smeitink JAM, van den Heuvel LP, Huynen MA, Ettema TJG (2007) Reconstructing the evolution of the mitochondrial ribosomal proteome. Nucleic Acids Res 35:4686–4703

Soll J, Schleiff E (2004) Protein import into chloroplasts. Nat Rev Mol Cell Biol 5:198–208

Stechmann A, Hamblin K, Pérez-Brocal V, Gaston D, Richmond GS, van der Giezen M, Clark CG, Roger AJ (2008) Organelles in *Blastocystis* that blur the distinction between mitochondria and hydrogenosomes. Curr Biol 18:580–585

Steel M, Dress AW, Böcker S (2000) Simple but fundamental limitations on supertree and consensus tree methods. Syst Biol 49:363–368

Stiller JW, Hall BD (1997) The origin of red algae: implications for plastid evolution. Proc Natl Acad Sci USA 94:4520–4525

Stiller JW, Reel DC, Johnson JC (2003) A single origin of plastids revisited: convergent evolution in organellar genome content. J Phycol 39:95–105

Stoebe B, Kowallik KV (1999) Gene-cluster analysis in chloroplast genomics. Trends Genet 15:344–347

Sutak R, Dolezal P, Fiumera HL, Hrdy I, Dancis A, Delgadillo-Correa M, Johnson PJ, Müller M, Tachezy J (2004) Mitochondrial-type assembly of FeS centers in the hydrogenosomes of the amitochondriate eukaryote *Trichomonas vaginalis*. Proc Natl Acad Sci USA 101:10368–10373

Suzuki K, Miyagishima SY (2010) Eukaryotic and eubacterial contributions to the establishment of plastid proteome estimated by large-scale phylogenetic analyses. Mol Biol Evol 27:581–590

Szklarczyk R, Huynen M (2010) Mosaic origin of the mitochondrial proteome. Proteomics 10:4012–4024

Thao ML, Gullan PJ, Baumann P (2002) Secondary (γ-*Proteobacteria*) endosymbionts infect the primary (β-*Proteobacteria*) endosymbionts of mealybugs multiple times and coevolve with their hosts. Appl Environ Microbiol 68:3190–3197

Theissen U, Martin W (2006) The difference between organelles and endosymbionts. Curr Biol 16:R1016–R1017, author reply R1017–8

Timmis JN, Ayliffe MA, Huang CY, Martin W (2004) Endosymbiotic gene transfer: organelle genomes forge eukaryotic chromosomes. Nat Rev Genet 5:123–135

Tovar J, Fischer A, Clark CG (1999) The mitosome, a novel organelle related to mitochondria in the amitochondrial parasite *Entamoeba histolytica*. Mol Microbiol 32:1013–1021

Tovar J, Léon-Avila G, Sánchez LB, Sutak R, Tachezy J, van der Giezen M, Hernández M, Müller M, Lucocq JM (2003) Mitochondrial remnant organelles of *Giardia* function in iron-sulphur protein maturation. Nature 426:172–176

Tsaousis AD, Kunji ERS, Goldberg AV, Lucocq JM, Hirt RP, Embley TM (2008) A novel route for ATP acquisition by the remnant mitochondria of *Encephalitozoon cuniculi*. Nature 453:553–557

Turmel M, Otis C, Lemieux C (1999) The complete chloroplast DNA sequence of the green alga *Nephroselmis olivacea*: insights into the architecture of ancestral chloroplast genomes. Proc Natl Acad Sci USA 96:10248–10253

Turner S, Pryer KM, Miao VP, Palmer JD (1999) Investigating deep phylogenetic relationships among cyanobacteria and plastids by small subunit rRNA sequence analysis. J Eukaryot Microbiol 46:327–338

Tyra HM, Linka M, Weber AP, Bhattacharya D (2007) Host origin of plastid solute transporters in the first photosynthetic eukaryotes. Genome Biol 8:R212

Viale AM, Arakaki AK (1994) The chaperone connection to the origins of the eukaryotic organelles. FEBS Lett 341:146–151

von Dohlen CD, Kohler S, Alsop ST, McManus WR (2001) Mealybug β-proteobacterial endosymbionts contain γ-proteobacterial symbionts. Nature 412:433–436

Waller RF, McFadden GI (2005) The apicoplast: a review of the derived plastid of apicomplexan parasites. Curr Issues Mol Biol 7:57–79

Wawrzyniak I, Roussel M, Diogon M, Couloux A, Texier C, Tan KS, Vivarès CP, Delbac F, Wincker P, El Alaoui H (2008) Complete circular DNA in the mitochondria-like organelles of *Blastocystis hominis*. Int J Parasitol 38:1377–1382

Weber AP, Linka M, Bhattacharya D (2006) Single, ancient origin of a plastid metabolite translocator family in Plantae from an endomembrane-derived ancestor. Eukaryot Cell 5:609–612

Weeden NF (1981) Genetic and biochemical implications of the endosymbiotic origin of the chloroplast. J Mol Evol 17:133–139

Whatley JM, John P, Whatley FR (1979) From extracellular to intracellular: the establishment of mitochondria and chloroplasts. Proc R Soc Lond B Biol Sci 204:165–187

Wilkinson M, Cotton JA, Creevey C, Eulenstein O, Harris SR, Lapointe FJ, Levasseur C, McInerney JO, Pisani D, Thorley JL (2005) The shape of supertrees to come: tree shape related properties of fourteen supertree methods. Syst Biol 54:419–431

Williams BAP, Hirt RP, Lucocq JM, Embley TM (2002) A mitochondrial remnant in the microsporidian *Trachipleistophora hominis*. Nature 418:865–869

Williams KP, Sobral BW, Dickerman AW (2007) A robust species tree for the *Alphaproteobacteria*. J Bacteriol 189:4578–4586

Woese CR, Kandler O, Wheelis ML (1990) Towards a natural system of organisms: proposal for the domains Archaea, Bacteria, and Eucarya. Proc Natl Acad Sci USA 87:4576–4579

Wu M, Sun LV, Vamathevan J, Riegler M, Deboy R, Brownlie JC, McGraw EA, Martin W, Esser C, Ahmadinejad N, Wiegand C, Madupu R, Beanan MJ, Brinkac LM, Daugherty SC, Durkin AS, Kolonay JF, Nelson WC, Mohamoud Y, Lee P, Berry K, Young MB, Utterback T, Weidman J, Nierman WC, Paulsen IT, Nelson KE, Tettelin H, O'Neill SL, Eisen JA (2004) Phylogenomics of the reproductive parasite *Wolbachia pipientis* *w*Mel: a streamlined genome overrun by mobile genetic elements. PLoS Biol 2:0327–0341

Yan W, Aebersold R, Raines EW (2009) Evolution of organelle-associated protein profiling. J Proteomics 72:4–11

Yang D, Oyaizu Y, Oyaizu H, Olsen GJ, Woese CR (1985) Mitochondrial origins. Proc Natl Acad Sci USA 82:4443–4447

Yoon HS, Reyes-Prieto A, Melkonian M, Bhattacharya D (2006) Minimal plastid genome evolution in the *Paulinella* endosymbiont. Curr Biol 16:R670–R672

Yoon HS, Zuccarello G, Bhattacharya D (2010) Evolutionary history and taxonomy of red algae. In: Seckback J, Chapman DJ (eds) Red algae in the genomic age. Springer, New York, pp 25–42

Ziková A, Panigrahi AK, Dalley RA, Acestor N, Anupama A, Ogata Y, Myler PJ, Stuart K (2008) *Trypanosoma brucei* mitochondrial ribosomes. Affinity purification and component identification by mass spectrometry. Mol Cell Proteomics 7:1286–1296

Chapter 2

Secondary and Tertiary Endosymbiosis and Kleptoplasty

Jeferson Gross* and Debashish Bhattacharya**
Department of Ecology, Evolution and Natural Resources and Institute of Marine and Coastal Sciences, Rutgers University, New Brunswick, NJ 08901, USA

Karen N. Pelletreau* and Mary E. Rumpho
Department of Molecular and Biomedical Sciences, University of Maine, Orono, ME 04469, USA

Adrian Reyes-Prieto
Biology Department, University of New Brunswick, 4400, Fredericton, NB E3B 5A3, Canada

*Equal contribution made by these authors.
**Author for correspondence, e-mail: bhattacharya@aesop.rutgers.edu

R. Bock and V. Knoop (eds.), Genomics of Chloroplasts and Mitochondria, Advances in Photosynthesis and Respiration 35, pp. 31–58,
DOI 10.1007/978-94-007-2920-9_2, © Springer Science+Business Media B.V. 2012

Summary

Algae refers to a diverse group of photosynthetic eukaryotes that have a polyphyletic origin in the tree of life. Although genomics has provided powerful tools for understanding the evolution of algal photosynthesis many issues remain unresolved. These include explaining the intermingling of plastid-lacking taxa such as ciliates and oomycetes among plastid-containing groups of chromalveolates. Does this pattern reflect a single ancient endosymbiosis in the chromalveolate ancestor followed by independent plastid losses or multiple secondary endosymbioses? Here we review current knowledge about chromalveolate evolution and phylogeny with a focus on secondary and tertiary endosymbiosis and survey recent genome-wide analyses to assess the potentially broad and lasting impacts of plastid transfer on eukaryote evolution. We assess the evidence for "footprints" of photosynthetic pasts that remain even when the plastid is lost. These data comprise remnant algal genes in the nucleus of plastid-lacking taxa that have putatively originated via intracellular gene transfer from the former endosymbiont. We also provide a survey of recent work done in the field of protein import (i.e., via translocons) into chromalveolate and other plastids derived from secondary endoysmbiosis. We contrast the similarities and differences between primary and secondary plastid protein import machineries and speculate on the key innovations that led to their establishment. And finally, we take a careful look at the remarkable case of sea slug (*Elysia chlorotica*) kleptoplasty and photosynthesis and review recent work aimed at explaining this phenomenon in different metazoa. In particular, we critically assess support for the hypothesis that sea slug photosynthesis is explained by massive horizontal gene transfer (HGT) from the genome of the captured alga.

I. Plastid Origin

A. Plastids Acquired via Eukaryote–Eukaryote Endosymbiosis

Algae is a widely used informal name that refers to a diverse group of photosynthetic eukaryotes such as euglenids and diatoms that have a polyphyletic origin (Reyes-Prieto et al. 2007). The Plantae and the Chromalveolata comprise the two largest eukaryote supergroups of presumed photosynthetic ancestry. Current data clearly show that the primary plastid traces its origin to primary (i.e., eukaryote–prokaryote) endosymbiosis, in which a unicellular protist (the "host") engulfed and retained a photosynthetic cyanobacterium (the endosymbiont) (Chap. 1). The resulting photosynthetic eukaryote is the putative common ancestor of the Plantae (Moreira et al. 2000; Rodriguez-Ezpeleta et al. 2005; Hackett

Abbreviations: BTS – Bipartite topogenic signal; DM – Dry mass; ERAD – Endoplasmic reticulum associated degradation; EGT – Endosymbiotic gene transfer; E/HGT – Endosymbiotic and/or horizontal gene transfer; ER – Endoplasmic reticulum; HGT – Horizontal gene transfer; IEM – Inner envelope membrane; MAA – Microsporine-like amino acids; MAST – Marine stramenopile; OEM – Outer envelope membrane; PPC – Periplastid compartment; PPM – Periplastid membrane; PRK – Phosphoribulokinase; RDRP – RNA-dependent RNA polymerase; RT – Reverse transcriptase; SP – Signal pepide; SAR – Stramenopiles Alveolata, Rhizaria; TOC/TIC – Translocon on the outer/inner envelope of chloroplasts; ToL – Tree of life; TP – Transit peptide

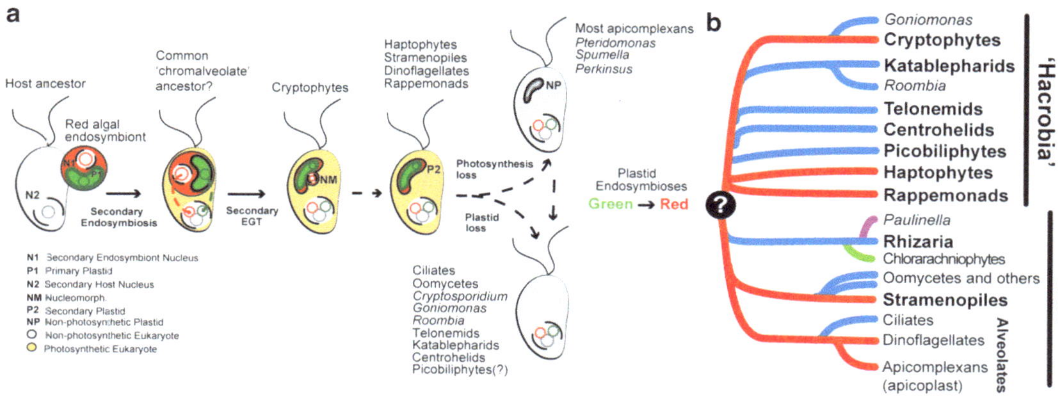

Fig. 2.1. Secondary endosymbiosis in eukaryote evolution. (**a**) Schematic representation of major events that presumably occurred during plastid evolution in chromalveolate lineages. Red algal secondary plastids are present in most photosynthetic chromalveolates (*yellow* cells). Secondary endosymbiotic gene transfer (EGT) into the host (host ancestor) genome (N2) from both nucleus (N1; *red dotted line*) and plastid (*green dotted line*) of the captured red alga are shown. There are several cases (*black dotted lines*) of photosynthesis and plastid loss in different chormalveolate lineages. Representative genera (or groups) are indicated close to each known non-photosynthetic variant (*grey* cells). (**b**) Mock phylogeny showing what is currently known about the origin of plastids and the interrelationships of the 'Hacrobia' and 'SAR' clades. The common photosynthetic origin of the chromalveolate red-algal plastid (e.g., the chromalveolate hypothesis; secondary endosymbiosis) is a contentious scenario, and recent genome analyses suggest an ancestral green algal endosymbiosis (*green line* at base) that predates the red algal capture. Photosynthetic lineages (*red lines*) are intermingled with non-photosynthetic (*blue lines*). Some of the non-photosynthetic groups, such as the ciliates and oomycetes, presumably evolved after independent losses of the red algal plastid. The Rhizaria regained two different types of plastids (also presumably after loss of the ancestral red algal plastid), the green algal plastid in chlorarachniophytes and the cyanobacterium-derived plastid in *Paulinella* species (Yoon et al. 2006; Reyes-Prieto et al. 2010) are indicated in *green* and *purple*, respectively.

et al. 2007) that is comprised of red, green (including plants), and glaucophyte algae (Cavalier-Smith 1981; Chan et al. 2011). Upon establishment, the primary plastid was apparently maintained in all extant Plantae. There are many colorless (non-photosynthetic) Plantae known, for example in parasitic plants and pathogenic algae (e.g., *Epifagus virginiana*, *Cuscuta* spp., *Aneura mirabilis*, *Polytomella* sp., *Helicosporidium* spp; Wolfe et al. 1992; Bungard 2004; McNeal et al. 2007; Wickett et al. 2008; Tartar et al. 2002), however, each of these taxa retains a vestigial plastid to perform other organelle functions, such as carbohydrate storage and heme biosynthesis (Atteia et al. 2005). In contrast, the chromalveolate ancestor gained its plastid through secondary (i.e., eukaryote–eukaryote) endosymbiosis (see Fig. 2.1a), whereby under the original hypothesis (Cavalier-Smith 1992), a red alga was engulfed and reduced to a secondary plastid. Phylogenomic data have shown, however, that a number of genes shared by chromalveolates are of green algal origin (e.g., Li et al. 2006; Nosenko and Bhattacharya 2007). Of particular significance is the finding of five enzymes of green algal origin involved in carotenoid biosynthesis in "chromist" (stramenopile, cryptophyte, and haptophyte) algae (Frommolt et al. 2008). Three of these genes branch deeply within prasinophyte green algae in phylogenetic analyses. This suggests they may have originated via endosymbiotic gene transfers (EGTs) or extensive horizontal gene transfer (HGT) from an ancient green algal endosymbiont (i.e., prasinophytes form a basal split among green algae; e.g., Steinkoetter et al. 1994; Fawley et al. 2000) that predates the widespread red algal plastid in chromalveolates (Fig. 2.1b).

This hypothesis gained further support when Moustafa et al. (2009) found evidence of hundreds of genes of green algal origin in the diatoms (stramenopiles, Bacillariophyta) *Thalassiosira pseudonana* and *Phaeodactylum tricornutum*. Similarly, phylogenomic analysis of the brown alga (stramenopiles, Phaeophyta) *Ectocarpus siliculosus* turned up ca. 2,600 genes of putative green algal origin, that contrast with only 611 genes of red algal provenance in this species (Cock et al. 2010). Whether these "green genes" trace their origin to a single cryptic endosymbiosis, to repeated HGTs, or more likely a combination of the two, these data highlight the complex nature of chromalveolate genome evolution that is only now being fully appreciated (see Fig. 2.1b and Baurain et al. 2010). The working hypothesis favored by Moustafa et al. (2009) is that the presence of a red algal-derived plastid in many chromalveolates conceals a past endosymbiosis with the "green" nuclear encoded genes acting as footprints of ancient E/HGT (see also Elias and Archibald 2009; Dagan and Martin 2009).

Two other eukaryote supergroups, the Rhizaria and Excavata also contain photosynthetic members (Chlorarachniophyta and Euglenozoa, respectively) but these algae are derived branches of what are believed to be anciently plastid-lacking lineages (hereafter, plastid [−]; in contrast to plastid [+]). Recent molecular studies have unveiled phylogenetic ties (see Fig. 2.1) between Rhizaria, some groups of chromalveolates, and nonphotosynthetic lineages such as telonemids and katablepharids (Shalchian-Tabrizi et al. 2006b, Okamoto and Inouye 2005; Hackett et al. 2007; Burki et al. 2007, 2009; Reeb et al. 2009). If these hypotheses are substantiated with genome data, then it will be important to trace the evolution of plastids via secondary endosymbioses across trees with intermingled plastid [−] and plastid [+] lineages whose origins extend back hundreds of millions of years to the time of eukaryote origin (see below and Moustafa et al. 2009; Cavalier-Smith 2010).

Understanding the convoluted history of secondary plastid evolution is aided significantly by a well-resolved nuclear host phylogeny. The chromalveolates however pose a great challenge in this respect. Plastid data usually support the monophyly of photosynthetic chromalveolates (e.g., Yoon et al. 2002; Khan et al. 2007), however nuclear gene trees disagree with the chromalveolate hypothesis in two key respects. First, recent multi-gene analyses support the inclusion of the Rhizaria within chromalveolates (Hackett et al. 2007; Burki et al. 2007; 2009; the 'SAR' clade in Fig. 2.1b), and second, chromists as originally proposed by Cavalier-Smith (1992) are polyphyletic with cryptophytes often found sister to haptophytes in a major lineage that also may include telonemids (Shalchian-Tabrizi et al. 2006b), katablepharids, centrohelids (Okamoto and Inouye 2005; Okamoto et al. 2009), picobiliphytes (Not et al. 2007), and the recently described photosynthetic rappemonads (Kim et al. 2011). This assemblage is sometimes referred to as the 'Hacrobia' (Okamoto et al. 2009) and is putatively sister to the SAR clade (Hackett et al. 2007; Burki et al. 2007; Patron et al. 2007; Burki et al. 2008; Okamoto et al. 2009). Clearly much more work has to be done to resolve the origins of chromalveolate-affiliated taxa.

B. How Is the Nuclear Genome Affected by Plastid Origin and Loss?

A key aspect of organelle evolution is the transfer of endosymbiont genes to the host nucleus, followed by import of the gene products into the organelle (Herrmann 1997; Martin et al. 1998; Martin and Herrmann 1998). Over time, EGT enriches the host genome with hundreds of transferred genes (Martin and Herrmann 1998; Moustafa et al. 2008a, b, 2009). The magnitude of EGT in Plantae genomes was estimated for *Arabidopsis thaliana* (Martin et al. 2002; Sato et al. 2005), and the unicellular algae *Chlamydomonas reinhardtii* (Moustafa and Bhattacharya 2008), *Cyanophora paradoxa* (Reyes-Prieto et al. 2006) and *Cyanidioschyzon merolae* (Sato et al. 2005). The results of these studies suggest that unicellular algae contain ca. 600–900 genes of cyanobacterial origin in their nucleus, and the vast majority encode proteins with plastid functions (Reyes-Prieto et al. 2006).

1. Secondary Endosymbiotic Gene Transfer

In contrast to primary plastid evolution, the acquisition of secondary plastids involves a more complex scenario for EGT. In these cases, genes are transferred both from the nucleus of the eukaryotic endosymbiont (Robertson and Tartar 2006; Li et al. 2006) into the host genome as well as directly from the plastid (Sanchez-Puerta et al. 2005; Oudot-Le Secq et al. 2007; see Fig. 2.1a). Nucleus–nucleus EGT facilitates the transfer of eukaryotic genes required for plastid function and maintenance (e.g., Archibald et al. 2003; Li et al. 2006), as well as other genes to provide redundancy and/or perform novel non-plastid functions (e.g., Stibitz et al. 2000), or replace the existing host gene copies (e.g., Hackett et al. 2007). Eukaryote–eukaryote endosymbiosis provides therefore the potential for extensive nuclear gene transfers that can significantly alter the host gene pool. Given the clear evidence for green algal genes in photosynthetic chromalveolates (e.g., Frommolt et al. 2008; Moustafa et al. 2009; Cock et al. 2010), it is likely that secondary EGT and/or HGT in these lineages encompasses both red and green algal (as well as potentially other) sources. This leads to an important point to consider with regard to anciently phagotrophic lineages such as chromalveolates: the evolutionary history of the plastid and nuclear genome may be uncoupled in these taxa. The red-algal plastid most likely represents the most recent organelle capture and associated EGT to the nucleus, whereas the nucleus contains not only this information but potentially evidence of all endosymbioses/EGTs and HGTs that have occurred in the history of the lineage (e.g., dinoflagellates; see Li et al. 2006; Patron et al. 2006; Nosenko et al. 2006). Nuclear genome data (in spite of its convoluted history) offers therefore a more accurate view of host evolution than does the plastid. Discriminating between competing evolutionary scenarios will however require extensive taxon sampling (that is not yet available) to understand better the impact of E/HGT on chromalveolate nuclear genome evolution (e.g., ortholog gene replacement, gene losses, impacts of homologous recombination, heterogeneous evolutionary rates; Harper and Keeling 2004; Richards et al. 2006), to ameliorate their misleading effects (e.g., phylogenetic artifacts) on multiprotein phylogenies.

2. Alveolate Plastids

Genome analysis of different apicomplexans (e.g., the parasitic protists *Plasmodium falciparum*, *Theileria parva*, and *Toxoplasma gondii*; Huang et al. 2004a) has turned up dozens of genes of algal (endosymbiotic) origin, with some of them encoding apicoplast-targeted proteins (Gardner et al. 2002). Even in the apicoplast-lacking apicomplexan *Cryptosporidium parvum*, several dozen genes of putative endosymbiotic origin have been identified (Huang et al. 2004b). The recent discovery of the photosynthetic relatives of apicomplexans, the coral-endosymbiont *Chromera velia* (Moore et al. 2008) and its relative *Chromerida* sp. CCMP3155 supports an algal ancestry for this group and likely as well for the common ancestor of dinoflagellates and apicomplexans (Moore et al. 2008; Janouskovec et al. 2010). Numerous examples of mixotrophic (Stoecker 1990) or non-photosynthetic (plastid-lacking or with a relic plastid) dinoflagellates (or closely related taxa) have been described in the past (e.g., *Noctiluca*, *Pfiesteria, Gymnodinium, Protoperidinium* and some *Dinophysis* species; Gaines and Elbrächter 1987; Jeong 1999), but until recently the genomic footprint of a past plastid (i.e., endosymbiont) has been described only in the heterotroph *Crypthecodinium cohnii* (Sanchez-Puerta et al. 2007), the early branching (plastid-lacking) dinoflagellate *Oxyrrhis marina* (Slamovits and Keeling 2008), and the bivalve parasite *Perkinsus marinus* (Matsuzaki et al. 2008).

The evolution of dinoflagellate plastids is marked by multiple independent examples of tertiary endosymbiosis involving the capture of algae harboring secondary plastids (e.g., photosynthetic stramenopiles, haptophytes, cryptophytes). The phylogeny of these taxa suggests that each of these events has involved independent replacements of the ancestral red algal plastid (Saldarriaga et al. 2001; Shalchian-Tabrizi et al. 2006a). Some gymnodiniacean dinoflagellates, such as *Karenia brevis* and *Karlodinium micrum,* harbor multi-membrane-bound plastids containing the typical haptophyte photopigment 19′ hexanoyl oxy-fucoxanthin. These fucoxanthin-containing plastids of tertiary origin presumably replaced the original secondary organelle of the *Karenia–Karlodinium* ancestor after an endosymbiosis involving a haptophyte alga. Phylogenies of nuclear-encoded plastid targeted proteins in *K. brevis* (Nosenko et al. 2006) and *K. micrum* (Patron et al. 2006) indicate that the proteome of these tertiary plastids accumulates a complex collection of proteins of bacterial, haptophyte, red and even green algal origin (Patron et al. 2006; Nosenko et al. 2006) as a consequence of multiple E/HGT events. The "recycling" of a fraction of the ancestral secondary-plastid proteome for the tertiary plastid suggests that both algal-derived organelles likely co-existed in the *Karenia–Karlodinium* ancestor during consolidation of the tertiary endosymbiosis (Patron et al. 2006).

Dinoflagellates have also recruited diatom endosymbionts on multiple occasions (Dodge 1971; Inagaki et al. 2000; Imanian and Keeling 2007). Relevant cases are *Durinskia baltica* and *Kryptoperidinium foliaceum* that harbor permanent tertiary plastids derived from a common pennate diatom ancestor (Inagaki et al. 2000; Imanian and Keeling 2007). The tertiary plastids of *D. baltica* and *K. foliaceum* maintain the endosymbiont nuclear membrane, endoplasmic reticulum surrounding the plastids, mitochondria, and ribosomes (Eschbach et al. 1990). Moreover, other dinoflagellate species have acquired independently plastids from centric diatoms (Imanian et al. 2010), and a putative case of successive replacement of diatom-derived plastids was reported recently (Takano et al. 2008). Some species of the genus *Dinophysis* harbor plastids (possibly kleptoplastids) of cryptophyte origin (Schnepf and Elbrachter 1988; Takishita et al. 2002). In this case, the two-membrane-bound plastids of tertiary origin contain phycobilins instead of peridinin as the main accessory photosynthetic pigment. Typical features of cryptophyte photosynthetic organelles, such as four bounding plastid membranes and the nucleomorph, are no longer present in *Dinophysis* plastids. It is plausible that the common ancestor of *Dinophysis* species lost the peridinin-containing plastid and some of these lineages are able to retain temporarily plastids from cryptophyte prey (Minnhagen and Janson 2006; see also *Elysia* section below). These results provide strong evidence that the ancestral red algal plastid has been lost repeatedly during dinoflagellate evolution.

Ciliates constitute the third branch of alveolates and form a non-photosynthetic sister to apicomplexans and dinoflagellates. In contrast to other alveolates, ciliates have no apparent physical remnant of a plastid, begging the question of a potential photosynthetic past for this group and for the alveolates as a whole. Until now, no *Chromera*-like taxon has turned up as an early diverging "ciliate" but a recent analysis done by our lab identified multiple examples of algal genes in this group (Reyes-Prieto et al. 2008; see also Archibald 2008).

3. Were Ciliates Once Algae?

Phylogenomic analysis can generate thousands of single-gene trees that are placed into categories reflecting phylogenetic origin (e.g., vertically inherited in eukaryotes or candidates for E/HGT). Single gene trees that address billion year-old splits are however particularly prone to stochastic behavior due to a paucity of signal in the limited data (e.g., Martin et al. 2002). Therefore, single

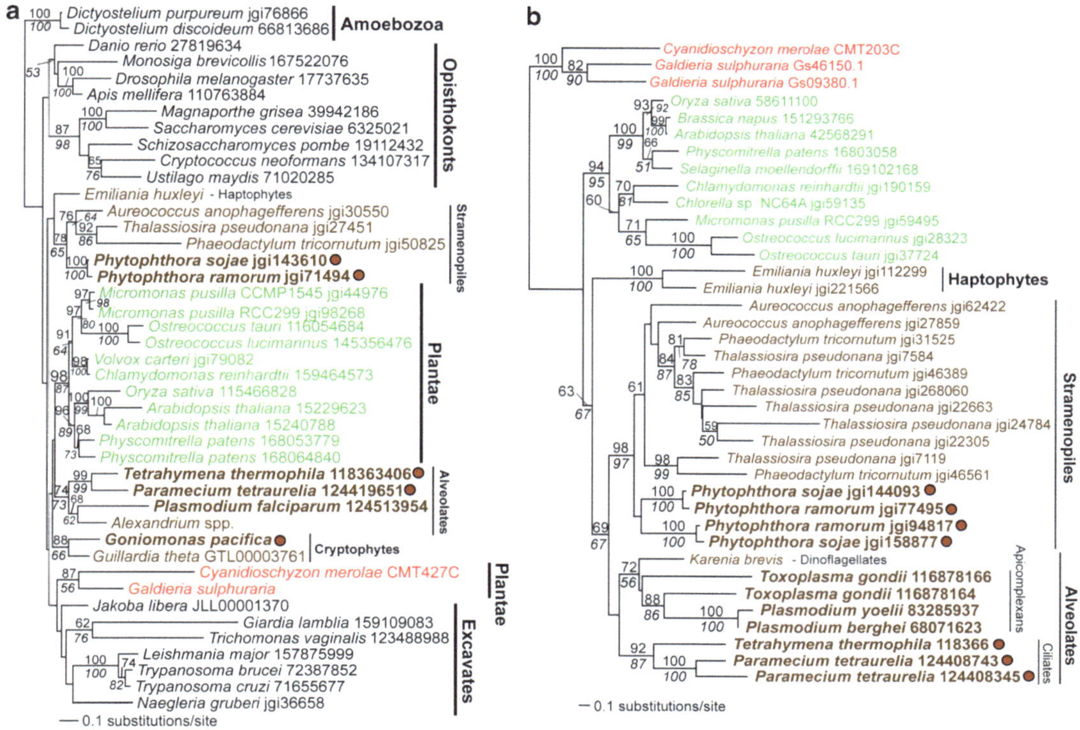

Fig. 2.2. RAxML phylogenetic trees of (**a**) RuvB-like 2 DNA helicase-like and (**b**) aspartyl protease family proteins. These trees represent typical output from phylogenomic analysis with the RuvB-like tree showing a putatively vertically inherited gene and the aspartyl protease tree showing a case of EGT from a red or green algal source into the putative common ancestor of chromalveolates. Bootstrap values (when ≥50%) from a RAxML analysis are indicated *above* the branches and PhyML bootstrap values are shown in *Italic* text below the branches. The branch lengths are proportional to the number of substitutions per site (see scales in the figure). Green algae and plants are shown in *green* text, red algae in *red* text, chromalveolates in *brown* text and all others in *black*. Non-photosynthetic chromalveolates are in *boldface* text and taxa that have lost the plastid are indicated with the filled *brown* circles. The tree shown in (**a**) was rooted on the branch leading to the Amoebozoa, whereas the tree shown in (**b**) was rooted arbitrarily on the branch leading to red algae.

trees need to be interpreted with great caution. As an example of the type of result produced by phylogenomics, Fig. 2.2 shows two trees generated by past analyses done in our lab (see Moustafa and Bhattacharya 2008; Moustafa et al. 2008b). The first (Fig. 2.2a) was inferred using RuvB-like 2 (DNA helicase-like) protein, a putative vertically inherited gene that provides a reasonable single-protein estimate of the eukaryote tree of life. The second example (Fig. 2.2b) is a tree inferred from an aspartyl protease family protein that is clearly of algal origin (i.e., the tree is arbitrarily rooted on the branch leading to red algae) and has been maintained in a broad diversity of plastid [+] and [−] chromalveolates. Note however that both single-protein trees fail to provide bootstrap support for deeper splits (e.g., for supergroups), but do substantiate the monophyly of most phyla (e.g., stramenopiles, cryptophytes, green algae).

Using this type of approach, we reported 16 genes of putative algal origin in the genome of the ciliate *Tetrahymena thermophila* that are shared with another distantly related ciliate, *Paramecium tetraurelia* (Reyes-Prieto et al. 2008). It should be noted that for some ciliate proteins, single homologous sequences are also returned from other taxa such as Amoebozoa, excavates or opisthokonts (the clade comprising fungi, animals and their immediate protist ancestors). We suggest these genes originated via independent HGTs in non-chromalveolate taxa. The complex topology of most of the phylogenies presented in Reyes-Prieto et al. (2008) and potential for recurrent HGTs from eukaryotic sources, however, renders it difficult to unambiguously distinguish between rare, ancient EGT and recurrent, independent HGTs as explanation for gene origin in ciliates. The wide phylogenetic distribution of some genes in both plastid [+] and plastid [−] chromalveolates does not prove, but is consistent with, ancient origin via EGT, under the dual endosymbiosis hypothesis described above (Nosenko et al. 2006; Frommolt et al. 2008; Moustafa et al. 2009). Most of the ciliate algal-derived genes are shared with at least one other chromalveolate group, and several are present in at least two other lineages. The presence of algal-derived genes in non-photosynthetic and photosynthetic chromalveolates strongly suggests therefore a common and ancient origin of these sequences (e.g., Fig. 2.2b). Although some of the plant (e.g., *Arabidopsis thaliana*) homologs are potentially plastid-targeted, most of the putative algal proteins in ciliates are not derived from plastid-targeted sequences in photosynthetic eukaryotes (Reyes-Prieto et al. 2008). Therefore genome analysis of ciliates provides tantalizing evidence of the footprints of endosymbiosis that have persisted over hundreds of millions of years in spite of presumptive plastid loss. It should be noted that the set of algal genes identified in ciliates (even when likely to grow in number with more sophisticated genome analysis) is by definition an underestimate of the true value given that over time, sequence divergence blurs the evolutionary history of some genes, making it impossible to determine their origin using standard molecular phylogenetic methods (e.g., Martin et al. 2002; Dagan and Martin 2006; Reyes-Prieto et al. 2008).

4. Stramenopile Plastids

The extraordinarily diverse stramenopiles comprise many ecologically relevant photosynthetic groups such as diatoms (Bacilariophyta), phaeophytes, and chrysophytes, but includes as well members with vestigial, non-photosynthetic plastids (e.g., the chrysophytes *Spumella* spp. and *Antophysa vegetans* and the dictyochophytes (axodines) *Pteridomonas danica* and *Ciliophrys infusionum*; Sekiguchi et al. 2002), and plastid [−] lineages or lineages with plastid-derived vestigial structures, such as oomycetes, bicosoecids, labyrinthulids and opalinids. Some studies suggest that outright plastid loss has occurred only twice, early in stramenopile evolution; i.e., in the ancestors of oomycetes and a putative monophyletic group formed by opalinids, labyrinthulids and bicosoecids (Cavalier-Smith and Chao 2006). The placement of the many heterotrophic environmental MAST (Marine Stramenopile) picoeukaryotes within the tree (e.g., the MAST-1 clade in Not et al. 2007) may however inflate the number of putative plastid losses in this group if they are intermingled with photosynthetic groups. Nevertheless, consistent with the idea of an algal past for plastid [−] stramenopiles, analysis of the complete nuclear genome sequence from the oomycetes *Phytophthora ramorum* and *P. sojae* revealed at least 30 (with up to several hundred) genes of putative cyanobacterial or algal (i.e., endosymbiotic) origin, including 12 *Phytophthora* genes with plant/algal homologs that encode plastid-targeted proteins (Tyler et al. 2006; see Stiller et al. 2009 for an alternative explanation for algal/cyanobacterial genes in oomycetes).

5. 'Hacrobia': Cryptophyte and Haptophyte Plastids

Most members of this clade contain a red algal-derived secondary plastid that was likely acquired by their putative common ancestor (Rice and Palmer 2006). However examples of non-photosynthetic haptophytes (Andersen

2004) and cryptophytes (Clay et al. 1999) are known. Cryptophytes are a notable case to highlight the history of secondary plastid evolution in chromalveolates given the persistence in most of these taxa of a reduced nucleus (nucleomorph) that can be traced back to the red algal endosymbiont (Douglas et al. 2001). Two fully sequenced cryptophyte nucleomorph genomes show the conservation of hundreds of protein-coding genes (ca. 470, mostly housekeeping), including some essential players in photosynthesis (Douglas et al. 2001; Lane et al. 2007; Kim et al. 2008). The cryptophytes comprise a number of non-photosynthetic lineages, such as members of the genus *Cryptomonas* (Hoef-Emden 2005) with a relic plastid and a nucleomorph and members of the genus *Goniomonas* that contain neither a plastid nor a nucleomorph (McFadden et al. 1994). Phylogenetic analyses show that *Goniomonas* diverges earliest from the branch leading to photosynthetic cryptophytes, suggesting to some that a red algal endosymbiosis may have occurred independently in cryptophytes after the *Goniomonas* split (McFadden et al. 1994; Deane et al. 2002), whereas an alternative interpretation is that the cryptophyte ancestor was photosynthetic and *Goniomonas* lost outright the organelle (Cavalier-Smith et al. 1996). This last scenario is the most parsimonious considering the likely common ancestry of cryptophytes and haptophytes (Burki et al. 2007, 2009; Hackett et al. 2007; Rice and Palmer 2006; Okamoto et al. 2009), strongly supported by a unique plastid gene (*rpl36*) replacement shared by these two lineages (Rice and Palmer 2006). Finally, the phylogenetic affiliation between cryptophytes and katablepharids, and the intermingling of haptophytes with other lineages, such as telonemids, picobiliphytes and centrohelid heliozoa (Burki et al. 2009; Okamoto et al. 2009), within the proposed Hacrobia, suggest the putative ancestor of this assembly was photosynthetic. Additional genome data are needed to confirm the existence of footprints of ancient endosymbioses in the Hacrobia that putatively constitutes a major eukaryotic lineage of photosynthetic ancestry.

C. Future Directions

A key question is raised as genome data accumulate for members of the SAR, Hacrobia and other under-studied microbial eukaryotes: what role did cryptic secondary endosymbiosis of red and/or green algae versus recurrent HGT from these sources play in the evolutionary history of these taxa? Here we stress that the question remains largely unanswered but some key insights can already be made. First, E/HGT is substantial in the genomes of taxa originally included in the Chromalveolata and presumably in recently erected groups (see Okamoto et al. 2009; Cavalier-Smith 2010). The magnitude of E/HGT remains a challenging issue for projects that aim to reconstruct organism and plastid history using multi-gene data. The inability to convincingly settle the issue of chromalveolate phylogenetic history, the relationship among major photosynthetic lineages (e.g., Plantae, SAR, Hacrobia) and the phylogenetic positions of novel taxa (e.g., the heterotophic biflagellate *Palpitomonas;* Yabuki et al. 2010), even when using large data sets is worrisome (e.g., Nozaki et al. 2007; Patron et al. 2007; Burki et al. 2008; Kim and Graham 2008; Yoon et al. 2008; Archibald 2009). What has however not yet happened is the combination of broad taxon sampling and a large data set (dozens of proteins) derived from analysis of complete genome data (e.g., Rodriguez-Ezpeleta et al. 2005; Hackett et al. 2007; Burki et al. 2008). It is this missing piece of the tree of life puzzle that needs to be filled to accurately infer the number of plastid endosymbioses that have occurred during eukaryote evolution.

II. The Evolution of Plastid Protein Topogenesis in Chromalveolates

Under the chromalveolate hypothesis, Cavalier-Smith emphasized that any event of organellogenesis leading to the transformation of a red algal endosymbiont into a plastid must include the evolution of an organized molecular system (e.g., protein translocons) to catalyze the transport of nuclear-encoded proteins into the endosymbiont subcompartments (Cavalier-Smith 1999). An additional "difficulty" was assumed to be that each gene transferred from the red algal chromosome to the host nucleus needed to acquire specific sorting signals to "inform" the final destination of the encoded product into the organelle. Cavalier-Smith argued that the de novo emergence of such protein translocons and topogenic signals is a complex, if not highly improbable, evolutionary leap. One should therefore favor a parsimonious scenario in which only a single endosymbiotic event and unique evolution of the organellar protein targeting explains the chromalveolate plastid origin (Cavalier-Smith 1999). Cell biologists are now constructing a picture of how protein topogenesis operates into secondary plastids. The new data open up the opportunity to revisit the original premises of the chromalveolate hypothesis and to approach the fundamental question of how a complex plastid can derive from a secondary endosymbiont.

A. Protein Targeting to Secondary Plastids

Organellogenesis involving a secondary red algal endosymbiont resulted in a four membrane-bound plastid in most members of the chromalveolate clade (one exemption is the plastid of peridinin-containing dinoflagellates which is surrounded by three membranes; Bolte et al. 2009; Keeling 2009). This defines new compartments not found in primary plastids; i.e., the periplastid compartment (PPC), corresponding to the remnant of the red algal cytosol, and the periplastid membrane (PPM), derived from the endosymbiont plasma membrane. The outermost membrane of cryptophyte, stramenopile and haptophyte plastids is contiguous with that of the endoplasmic reticulum (ER), indicating these complex organelles are embedded within the endomembrane system. This fact is compatible with the idea that the original red algal endosymbiont was acquired via phagocytosis (Patron and Waller 2007; Bolte et al. 2009). Accordingly, proteins directed to the complex chromalveolate plastids are first routed to the ER via the Sec61 translocon. Typically, proteins imported into secondary plastids have bipartite topogenic signals (BTS) that comprise an N-terminal signal peptide (SP) to cotranslationally direct the imported protein into the ER lumen, followed by a canonical transit peptide (TP) for plastid targeting (Waller et al. 2000; Apt et al. 2002; Patron et al. 2005; Gould et al. 2006; Patron and Waller 2007; Bolte et al. 2009). The SP is cleaved off upon substrate entrance into the ER, thereby exposing the TP-like leader sequence. In cryptophytes, stramenopiles, and haptophytes this TP further directs the pre-proteins across the PPM (Apt et al. 2002; Gould et al. 2006; Gruber et al. 2007; Bolte et al. 2009). In apicomplexans and peridinin-containing dinoflagellates, the imported proteins are presumably routed via vesicular transport from the ER to the outermost membranes of their respective complex plastids (Patron et al. 2005; Agrawal and Striepen 2010).

Recent data suggest that in chromalveolates, a molecular system originally derived from the endosymbiont ERAD (Endoplasmic Reticulum Associated Degradation) is responsible for protein translocation across the PPM. In eukaryotes, ERAD components are involved in an energy-depended retro-translocation of misfolded proteins from the ER lumen into the cytosol, where they are tagged with poly-ubiquitins to be routed to the degradosome (Xie and Ng 2010). The homologs of the ERAD subunits Der1-1, Der1-2, Hrd1, and Udf1 are still encoded in the nucleomorph genome of the cryptophyte *Guillardia theta* (Sommer et al. 2007). The nucleomorph Der gene (ORF201) can complement a yeast strain with a defective homolog allele, indicating a conserved function despite of the fact that the secondary red algal plastid is apparently devoid of a remnant ER (Sommer et al. 2007). In diverse chromalveolates, two sets of ERAD components are encoded in the nucleus. One group of homologs corresponds to the canonical ER retro-translocon of host origin. The other set of encoded ERAD homologs is phylogenetically distinct and contains standard BTSs capable of targeting reporter markers to the complex plastids in experiments of subcellular localization (Sommer et al. 2007; Hempel et al. 2009; Spork et al. 2009; Felsner et al. 2010b). A detailed understanding of the cell biology of the ERAD system in the ER is still lacking (Xie and Ng 2010). One model posits that oligomers of polytopic Der subunits may form the protein-conducting pore of the retro-translocon at the ER. Such an idea that Der components compose a protein channel may be applicable to translocation into the secondary plastid because the *Phaeodactylum tricornutum* Der1-1 and Der1-2 homologs form homo- and hetero-oligomers that are associated with the PPM (Hempel et al. 2009). In addition, Der complexes interact with TPs of imported intermediates directed to the PPC (but not with TPs of stromal-targeted proteins; Hempel et al. 2009). Genetic evidence that chromalveolate symbiont-derived Der components function in protein import to the complex plastids comes from an engineered conditional *Der*1 mutant of *Toxoplasma gondii* in which the decrease in protein translocation into the apicoplast is directly proportional to the ablation of the conditionally expressed Der1 protein (Agrawal et al. 2009). Empirical data in *P. tricornutum* also support the idea that the ubiquitin ligase (ptE3P) and the de-ubiquitinase (ptDUP) homologs have conserved enzymatic functions and are located in the PPM and PPC of the secondary plastid, respectively (Hempel et al. 2010). Although the hypothesis of ubiquitination-dependent protein translocation into the complex plastids still needs to be verified, current evidence favors the idea that the ERAD system was co-opted from the red algal endosymbiont to mediate protein import across the PPM.

The ERAD system likely represented an "immediate" evolutionary solution for protein import into the new organelle. Secondary plastids seem to have retained components of the TOC and TIC machineries (translocon at the outer/inner envelope of chloroplasts), which are responsible for protein import across the outer and inner envelope membranes (OEM and IEM, respectively) of the primary red algal plastid (Gross and Bhattacharya 2009a). Toc75 has been identified by bioinformatic analysis in the diatoms *P. tricornutum*, *Thalassiosira pseudonana*, the haptophyte *Emiliania huxleyi* and the apicomplexan parasites *Plasmodium falciparum* and *T. gondii* (Bullmann et al. 2010). The *P. tricornutum* Toc75 forms a channel with electrophysiological properties similar to cyanobacterial and land plant homologs and is targeted to the second innermost plastid membrane by a pathway that may involve its transient accumulation in the intermembrane space (Bullmann et al. 2010). This route of protein sorting is analogous to that observed for Pea Toc75 (Baldwin and Inoue 2006), indicating remarkable conservation of protein topogenesis in plastids across widely separated taxa and after remodeling of the organelle by secondary endosymbiosis. The TIC translocon also seems to be conserved during the evolution of secondary plastids. Tic110 and Tic22 are encoded in the nucleomorph of cryptophyte algae and together with Tic20 are also found in the in the nuclear genome of diatoms and *E. huxleyi* (McFadden and van Dooren 2004; Gross and Bhattacharya, personal observations). Tic20 and Tic22 are also encoded in the nucleus of apicomplexan parasites and are targeted to the plastid-derived compartment (van Dooren et al. 2008; Kalanon et al. 2009; Agrawal and Striepen 2010). A conditional null mutant of *T. gondii* Tic20 showed that ablation of Tic20 expression impacts protein import into the apicoplast and is lethal to the parasite. However protein import into the organelle is only extinguished after 2 days of the complete absence of immunological detection of Tic20 (van Dooren et al. 2008). This situation contrasts with an immediate pronounced drop in protein import rates into the apicoplast once the *T. gondii* Der homolog is depleted (Agrawal et al. 2009). The lag time between Tic20 knockout and apicoplast protein import decline may suggest that Tic20 does not represent a central protein-conducting pore (van Dooren et al. 2008), but rather is an accessory component, or a factor for biogenesis of the IEM translocon. Despite many remaining uncertainties, increasing evidence for the involvement of TOC and TIC components in protein import across the second and first innermost membranes of chromalveolate plastids, respectively, provides a novel perspective on the evolution of secondary endosymbiosis.

B. A Bottleneck to Evolve a Secondary Plastid?

What can recent experimental data tell us about the initial premises of the chromalveolate hypothesis? Phylogenetic trees of ERAD components Cdc48 and Uba1, and Toc75 tend to place plastid-targeted chromalveolate proteins in the same branch forming a sister group to red algal homologs (Agrawal et al. 2009; Bullmann et al. 2010; Felsner et al. 2010b). These important data lend strong support to key ideas of the chromalveolate hypothesis; i.e., the red algal secondary plastid had a single origin and was made possible by the unique emergence of an organelle protein sorting system (Cavalier-Smith 1999). Recent observations indicate that the chromalveolate plastids may conserve the TOC and TIC pathways (Agrawal and Striepen 2010; Bullmann et al. 2010). This raises an intriguing perspective in which the bottleneck to evolve protein targeting to the red algal endosymbiont captured within the endomembrane system was the rerouting of proteins from the ER lumen across the endosymbiont plasma membrane. The initial steps of organellogenesis probably included transfer to the nucleus of genes encoding TP-contained products that eventually were mistargeted to the host ER. The redirection of ERAD components (e.g., the Der subunits) from the red algal ER to its plasma membrane may have been an adaptation to retro-translocate ER dispersed TP-containing proteins into the endosymbiont cytosol. Once entering that compartment, proteins equipped with a TP would by default have been routed to the red algal primarily plastid via TOC and TIC translocons. This scenario is supported by the observation that ERAD components once relocated to the PPM would retain the same topology as in the ER membrane.

C. Co-option of Pre-existing Topogenic Signals

Another important assumption of the chromalveolate hypothesis is the "difficulty" to establish a system of topogenic signals for proteins directed to the new organelle (Cavalier-Smith 1999). Chromalveolate plastid-directed proteins are first targeted into the ER via a standard N-terminal SP contained in the BTS, and then further directed across the PPM by a canonical TP for targeting into the primary plastids (Bolte et al. 2009). The TP-like sequences of the chromalveolates tend to preserve features found in TPs of Plantae (Tonkin et al. 2006; Patron and Waller 2007; Felsner et al. 2010a); i.e., the bias for the hydroxylated amino acids serine and threonine and under-representation of acidic residues, conferring a net positive charge to the TP. More important is the tendency for conservation of a phenylalanine near the first amino acid of the TP, that is a hallmark of the Rhodophyta (Patron and Waller 2007). This observation points to a "less-difficult" scenario, whereby topogenic signals were not created de novo during the evolution of secondary plastids, but instead recycled from pre-existing systems, probably via exon shuffling of SPs and TPs to newly established protein-coding genes with a function in the organelle (Kilian and Kroth 2004). It is noteworthy that precursors of proteins with a final destination in the PPC are also equipped with TP-like sequences. Empirical and bioinformatic evidence in cryptophytes and stramenopiles indicates that absence of the critical N-terminal phenylalanine at the TP seems to be the topogenic determinant to retain import substrates in the PPC once the import intermediate crosses the PPM (Gould et al. 2006; Patron and Waller 2007; Felsner et al. 2010a). Such a feature suggests that a standard sorting system in the new organelle initially evolved to target precursors to the primary plastid, and a mechanism to halt proteins in the PPC was superposed on to this feature. It is likely that the translocon at the PPM was initially under selective pressure to evolve affinity for TP-containing substrates directed to the primary plastid, indicating that the functions of the red algal primary plastid (e.g., photosynthesis) were the target of selection. That TPs emerged as a canonical signal to cross the red algal former plasma membrane is in accordance with our previous hypothesis that topogenic signals tend to emerge from physical properties already present in the import substrate (see Gross and Bhattacharya 2009b).

D. Evolution of Secondary Plastids, an Insiders' Perspective?

In light of the ideas discussed above, can we draw comparisons between the evolution of protein targeting to the primary plastid in Plantae and to the secondary plastid in chromalveolates? We previously postulated that the evolution of the primary plastid and the mitochondrion was constrained by topological factors (Gross and Bhattacharya 2009b). Initially, host-encoded proteins that were synthesized in the cytosol could not easily cross the two membranes of the Gram-negative endosymbiont progenitors of mitochondria and plastids to directly gain access to their interior. Therefore, organellogenesis of plastid and mitochondria was hypothetically initiated by limited targeting of host-proteins constrained to the OM of the captive endosymbionts. Such a view, referred to as an "outsiders' perspective", suggests that organelle evolution then progressed by gradually establishing an inward organized topological system to finally direct proteins into the organelle lumen (Gross and Bhattacharya 2009b). However, organellogenesis of secondary plastids may be conceptually different from that of primary plastids and mitochondria. The fact that TPs seem to be the standard topological signal to move imported proteins across the PPM indicates that organelle protein sorting was initially selected to import substrates directly to the primary plastid located inside the endosymbiont (i.e., an "insiders' perspective"). The Toc and Tic translocons and pathways to further route proteins to the thylakoid membranes of Plantae plastids seem to be conserved in chromalveolates (Broughton et al. 2006; Gould et al. 2007; van Dooren et al. 2008; Bullmann et al. 2010). If the red algal endosymbiont was trapped within the host endomembrane system, it is then conceivable that the only physical obstacle for host proteins to reach the red algal innermost compartments was to cross the endosymbiont plasma membrane. The recruitment of the endosymbiont ERAD translocon to the PPM likely represented a one-step solution to overcome this topological constraint. Interestingly, ERAD homolgs are still encoded in the nucleomorph genome of *G. theta* indicating that these molecular components were directly co-opted from the endosymbiont genome (Sommer et al. 2007). This case represents a deviation from the pattern that most molecular components supporting the evolution of primary plastids and mitochondria arguably evolved in the host genome (Gross and Bhattacharya 2009b, 2011). Conceivably, direct recruitment of the ERAD components from the endosymbiont chromosome only involved minor modifications. Nonetheless, the overall tendency for shrinkage and disappearance of the former red algal nuclear genome and occurrence of EGT requires that the secondary plastid evolution should be interpreted as a result of events predominantly selected in the host nuclear genome (Gross and Bhattacharya 2009b. 2011).

E. Convergent Evolution of Secondary Plastids

Taxa belonging to Chlorarachniophyta and Euglenophyta also have a secondary plastid surrounded by four and three membranes, respectively (Bolte et al. 2009). These organelles are derived from green algal plastids via two independent secondary endosymbiotic events. Curiously, many overlapping features are observed between protein topogenesis of these green algal secondary plastids and chromalveolates. For example, proteins targeted to the chlorarachniophyte and euglenophyte plastids also have BTSs analogous to that of chromalveolates (Durnford and Gray 2006; Bolte et al. 2009; Hirakawa et al. 2010). The N-terminus of these BTSs is a SP that specifies routing through the ER. It is followed by a canonical TP that shares features with green plastid TPs, such as an overall positive charge and enrichment of hydroxylated amino acids. In accordance, components of the TOC and TIC translocons may also be conserved in green algal-derived secondary plastids. The Toc75 and Tic20 components are encoded in the nucleomorph genome of the chlorarachniophyte *Bigelowiella natans* (Gilson et al. 2006). In addition, as in chromalveolates, the routing of proteins destined to the PPC of chlorarachniophytes seems to rely on topogenic signals that retain TP-containing proteins in that compartment (Hirakawa et al. 2010). Finally, protein targeting into the complex plastids of euglenophytes most likely proceeds via vesicular transport and is determined by a BTS that contains an additional hydrophobic stop-transfer signal immediately downstream to the TP (Durnford and Gray 2006). This may serve to anchor import substrates into the vesicular membrane. Despite having a different origin, an analogous stop-transfer signal following the TP is also observed in proteins targeted to the plastid of peridinin-containing dinoflagellates (Patron et al. 2005). In light of all these examples implying convergent evolution between complex plastids of different origins in chromalveolates, chlorarachniophytes and euglenophytes it is tempting to speculate that there is a defined trajectory for the evolution of a secondary plastid. This may be the result of strong selection for increasing host control over the photosynthetic organelle. The establishment of a translocation system in the endosymbiont plasma membrane (e.g., the ERAD machinery) and conservation of pre-existing endosymbiont protein-sorting components (e.g., the TOC and TIC translocons) and topologic signals (e.g., SPs and TPs) may be recurrent evolutionary solutions to this problem. Convergent evolution of protein targeting principles in unrelated secondary plastids surrounded by multiple membranes also corroborates the notion that topological constrains are critical barriers that impose a trajectory to evolutionary processes, such as organellogenesis, as recognized by the outsiders' hypothesis (Gross and Bhattacharya 2009b). Finally, the idea that minimal topological innovations are required for establishment of a secondary plastid may reinforce the feasibility of a past existence of a cryptic green algal secondary organelle/endosymbiont that presumably left a phylogenetic footprint of EGT in the genome of the chromalveolate ancestor (Moustafa et al. 2009).

III. Kleptoplasty of a Secondary Endosymbiont in a Metazoan System

A. Introduction

Sacoglossan molluscs are marine invertebrates generally referred to as sea slugs. Many sacoglossans have evolved close evolutionary relationships with their algal food source (though not all are herbivorous) and, for some, the feeding mechanism and apparatus are highly specialized for suctorial feeding on their algal prey (Jensen 1997). These molluscs break down the algal components, except for the plastids which are retained intact in the animal digestive tissue for anywhere from 24 h to 10 months (Händeler et al. 2009; Yamomoto et al. 2009; Rumpho et al. 2011). This unique relationship between animal host and algal plastids is referred to as a symbiosis (more correctly, an endosymbiosis), because the host retains the organelle intracellularly and is endowed with a novel metabolic trait – photosynthesis. Photosynthate from the plastid has been traced to the host and the animals can be maintained in the laboratory for months with light and CO_2 alone; no additional energy or food sources are required (reviewed by Rumpho et al. 2011).

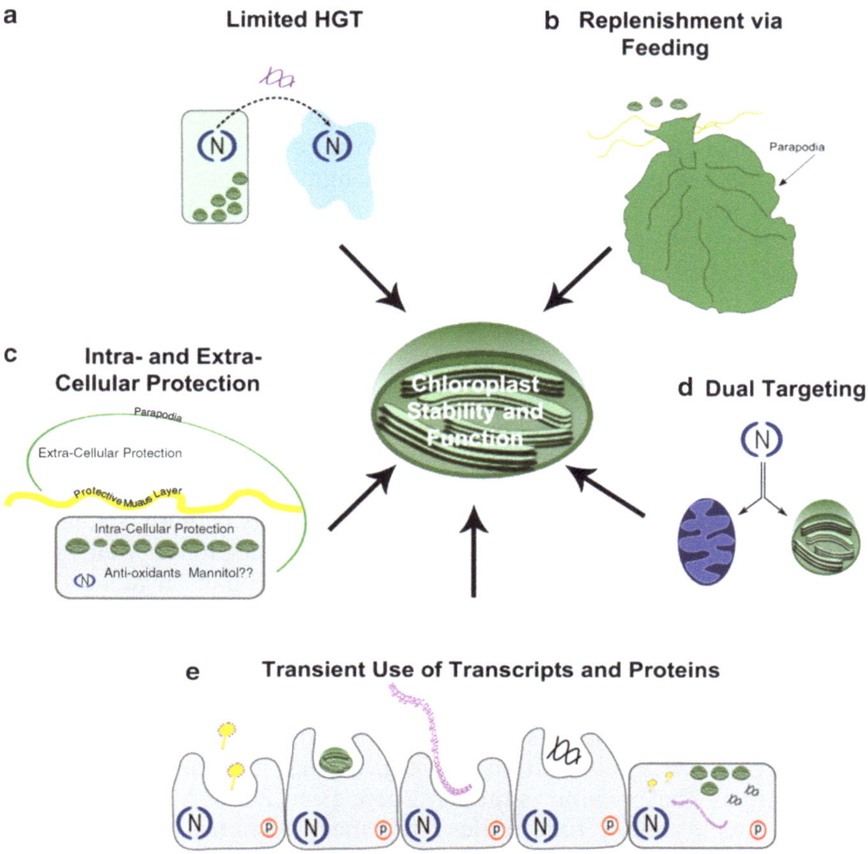

Fig. 2.3. Putative mechanisms that may work synergistically to support long-term plastid function in the sea slug *Elysia chlorotica*. (**a**) A limited amount of horizontal gene transfer (HGT) from the algal nucleus to the animal nucleus. (**b**) In the environment, the animal can replenish algal plastids, proteins and transcripts through feeding. (**c**) Multiple mechanisms are in place in the animal to provide protection of the plastids from photo-oxidation and free radicals, including the parapodial extensions covering the body when folded, protective muco-polysaccharides, and intracellular protective compounds. (**d**) Proteins encoded in the animal nuclei and targeted to the mitochondria may be co-opted and used in plastid function. (**e**) Cellular uptake and transient use of materials encountered in the environment or during feeding, such as nucleic acids and proteins (algal, bacterial or viral), may occur in addition to the integration of plastids into the digestive cells of *E. chlorotica*.

The vast majority of these specialized sacoglossans feed on chlorophyte algae possessing plastids of primary endosymbiotic origin (see Fig. 2.3; Jensen 1997; Händeler et al. 2009; Händeler et al. 2010; Wägele et al. 2010); thus, these host animals are models of kleptoplasty of a secondary endosymbiont. In contrast, *E. chlorotica* is unique in that its algal prey is the stramenopile alga, *Vaucheria litorea* (Xanthophytes, yellow–green algae). The plastids in *Vaucheria* are products of a secondary endosymbiosis involving the uptake of a red alga (Fig. 2.3). Stramenopile plastids, such as in *Vaucheria* sp., are typically surrounded by multiple bounding membranes (three or four) including the two original primary plastid membranes from the uptake of the cyanobacterial ancestor, a third membrane derived from the cellular membrane of the symbionts (red algal cell in the case of *Vaucheria*), and the fourth membrane which presumably originates from the plasma membrane of the phagocytic host and is continuous with the host's outer nuclear membrane. After *E. chlorotica* feeds on *Vaucheria*, the plastids within the animal cells definitively retain the two original plastid membranes. The presence of a third membrane surrounding the plastids is variable (but see Pierce et al. 2009); the fourth membrane appears to be lost through the mechanics of feeding and digestion. In cases where a third membrane has been visualized via electron microscopy, the source of the membrane (algal plastid membrane or host vacuolar/phagocytic membrane) remains to be determined.

Because the plastids are secondarily derived in the algal prey for *E. chlorotica*, this sea slug kleptoplasty is most analogous to tertiary endosymbiosis. The most common models of tertiary symbiosis are those observed in dinoflagellates (Yoon et al. 2005; Hackett et al. 2004; Palmer 2003), but the *E. chlorotica-Vaucheria* system has received much attention regarding the temporary establishment, sustainment and ultimately the evolution of photosynthesis in a multicellular heterotrophic host. Thus, both the source and stability of the plastids in this model render *E. chlorotica* unique from its sacoglossan relatives that exhibit similar abilities, but on much shorter time scales with plastids of primary endosymbiotic origin.

B. The Stability Dilemma

Although this unique example of functional photosynthesis in a metazoan has been researched for decades (reviewed by Trench 1975; Rumpho et al. 2006; Rumpho et al. 2011; Pelletreau et al. 2011), questions surrounding the mechanisms involved in obtaining and maintaining the foreign organelle, remain unanswered. Research into the intracellular sequestration of the plastids, recognition of the plastids by the digestive cells, and the host immune response (or lack thereof), are just beginning to receive attention. Perhaps the most intriguing aspect of the *E. chlorotica-Vaucheria* symbiosis is the lack of algal nuclei in the animal tissue. Repeated work using a variety of techniques (from microscopy to molecular tools) has failed to provide evidence for algal nuclei or algal housekeeping genes in the animal tissue. However, like in all other photosynthetic organisms, the *V. litorea* plastid genome has retained only a small fraction of the genes needed for the synthesis, maintenance and turnover of photosynthesis proteins, along with sigma factors and other transcriptional and translational modifiers. The *V. litorea* plastid genome shows no exceptional coding capacity that would facilitate the symbiosis (Rumpho et al. 2008). In the absence of an algal nucleus encoding for these presumably requisite proteins, how then does the plastid remain stable and functional in the animal?

The inherent energy-intensive processes involved in photosynthesis typically result in rapid turnover of the protein pool to ensure efficient energy capture. Yet, in *E. chlorotica*, plastids continue to function, synthesize photosynthate and transfer reduced carbon to the host without a known source for the replenishment of nuclear-encoded plastid proteins (reviewed by Rumpho et al. 2006; Rumpho et al. 2011). With the apparent widespread distribution of horizontal gene transfer (HGT; e.g., Olendzenski and Gogarten 2009; Bock 2010; Boto 2010; Moran and Jarvik 2010), this mechanism has been implicated in helping to sustain plastid stability and function in *E. chlorotica*. The sea slug shares a physically close relationship with filaments of *Vaucheria* during development. In fact, this association is essential for the sea slug to undergo metamorphosis from the veliger larva stage to juvenile sea slug, and for the juveniles to mature into adult sea slugs. This requisite physical association further supports the plausibility of transient HGT from the algal nucleus to the sea slug. PCR-based results support the presence of certain genes in aposymbiotic host tissue, i.e., the egg and larval stages that are not exposed to algal prey and do not sequester plastids (Pierce et al. 1996, 2003, 2007; Rumpho et al. 2001, 2008, 2009; Schwartz et al. 2010). However, partial sequencing of the sea slug genome and transcriptome of actively photosynthesizing adult *E. chlorotica*, has not revealed any photosynthesis-related genes, or genes specifically originating from *Vaucheria* (Pelletreau et al. 2011 [but see Pierce et al. 2011]). Similar results were also observed for two other sacoglossan species, *E. timida*, which harbors chlorophyte plastids for several weeks at a time, and *Plakobranchus ocellatus,* which harbors mixed plastids for several months (Evertsen et al. 2007; Händeler et al. 2009; Wägele et al. 2010). No photosynthetic genes were identified following 454 pyrosequencing of the transcriptomes from both organisms (Wägele et al. 2010). All of these results suggest that extensive HGT has not occurred between host and symbiont, despite the close physical relationship with the symbiont's nuclear genome during feeding, and/or HGT has not occurred uniformly among populations of sea slugs. Thus, at present, additional mechanisms must be proposed and explored to explain the sustained viability and stability of the plastids observed in the host tissue.

C. Alternate Mechanisms to Explain Plastid Stability

It is apparent that the mechanistic complexity underlying plastid stability in *E. chlorotica* is much greater than once presumed. Neither the algal plastid genome (Rumpho et al. 2008), nor our current understanding of the nuclear genome and transcriptome of the sea slug, provide a compelling explanation for plastid function. Although there are numerous potential explanations; here, four mechanisms will be discussed which may work synergistically with *limited* HGT to contribute to plastid function and stability: (1) Plastid replenishment; (2) Plastid durability and protection; (3) Transient transcript expression and protein function; and (4) Dual targeting of animal proteins (Fig. 2.1). The relative contribution of each of these mechanisms remains unknown and all avenues warrant further investigation in order to fully comprehend the processes involved in animal photosynthesis.

1. Limited HGT

Although emerging sequencing studies do not, at present, support the presence of expressed transferred genes, numerous studies employing a variety of methods have provided evidence for the presence of genes for nuclear-encoded plastid-targeted photosynthesis proteins in *E. chlorotica* (reviewed by Rumpho et al. 2006, 2011; Schwartz et al. 2010). The majority of these studies have employed single gene investigation of genomic and complementary DNA from *E. chlorotica* during both the aposymbiotic and symbiotic phases of the animal's life history (veliger larvae and eggs vs. 'green' adults). To date, the available evidence supports the transfer of six algal nuclear genes related to energy capture and photosynthetic electron transport. These genes encode the following proteins: light harvesting complex proteins (Pierce et al. 2007), the manganese stabilizing protein of photosystem II (Rumpho et al. 2008), the Calvin-Benson cycle enzyme phosphoribulokinase (PRK; Rumpho et al. 2009; Schwartz et al. 2010; Soule 2010), and three proteins involved in chlorophyll synthesis (Pierce et al. 2009; Schwartz et al. 2010 [see also Pierce et al. 2011]). In addition, evidence for the synthesis of chlorophyll in *E. chlorotica* has been reported using ^{14}C radiolabeling, suggesting the presence of additional nuclear encoded algal genes functioning in the animal (Pierce et al. 2009).

2. Plastid Replenishment

In its natural environment, *E. chlorotica* encounters *Vaucheria* specimens sporadically throughout the year. Observations of individuals fed algae in the lab suggest that, when available, plastids from *Vaucheria* are continually incorporated into the digestive cells of the animals. It is therefore reasonable to assume that proteins, transcripts and other materials that can contribute to plastid stability would likewise be available to the animal when feeding in nature or when provided algal prey in the laboratory. In the natural environment, food availability would presumably be the limiting factor for plastid sustainability for any herbivorous sacoglossan. In the marsh habitat for *E. chlorotica*, growth of *Vaucheria* is limited during the winter seasons and during this time the animals would need to support plastid functions without the introduction of new algal materials through feeding. Of interest, the animals are not observed in the field through the winter, and some speculation exists of "hibernation" in sediment or deep water. If this is the case, photosynthesis would be minimal during this time. This behavioral explanation works to explain how in nature the animals could sustain plastid function; however, in the laboratory, animals are maintained in well lit and aerated aquaria for 9–10 months after collection from the field without any additional *Vaucheria* to feed on ("starved" conditions). In this scenario, it is apparent that replenishment of materials via feeding is not required for long-term plastid function. As a result, one must still seek explanations to explain how plastid proteins are maintained for an extended period of time in the absence of the algal nuclei.

3. Plastid Durability and Protection

Many of the sacoglossan molluscs that are able to exploit plastid function feed on coenocytic (lacking cross walls) algae. Trench et al. (1973) first suggested that the inherent nature of the plastids and the morphology of coenocytic algae may play a role in the evolution of the sacoglossan-plastid symbiosis. *Vaucheria* plastids exhibit a unique "robustness" when isolated from the algal filament or from *E. chlorotica*. In comparison to spinach plastids, *Vaucheria litorea* plastids remain structurally intact and able to fix CO_2 for 3 days after isolation from the alga when simply suspended in a buffered iso-osmotic medium. Conversely, spinach plastids showed a rapid deterioration of shape and function within 24 h of isolation (Green et al. 2005). The *V. litorea* plastids are also resistant to varying osmotic concentrations and able to translate proteins throughout the 3 day isolation period. When isolated, plastids of land plants and other algal species typically exhibit a precipitous drop in photosynthetic activity and protein translation, ceasing within hours of isolation (Kirk and Tilney-Bassett 1967; Morgenthaler and Morgenthaler 1976; Mayfield et al. 1995). Therefore, the physical properties of *V. litorea* plastids may facilitate the successful establishment of the symbiosis. It is important to remember that the vast majority of sacoglossan species feed on chlorophyte algae derived from the green algal lineage, which do not have additional envelope membranes, and there is very little known about the physical characteristics of plastids from these other algal species or if the additional membranes "protect" secondary plastids.

Photoprotection of the plastids within the animal, in addition to the robust nature of the plastids, would synergistically aid in long-term photosynthesis. Photoprotection can result from physical shading, sunscreens, antioxidants, and enzymes that counteract the effects of free radicals. All of these mechanisms may be involved in this symbiosis. *Elysia* species are members of the family Placobranchiodea, a distinguishing feature of which is the presence of parapodia. These wing-like extensions on the animal open and close in response to light, movement, and other environmental cues. The parapodia are thought to have contributed to the evolution of the symbiosis and the longevity of plastid function within these animals (Trench 1975; Rahat and Monselise 1979; Händeler et al. 2009; Wägele et al. 2010). Additionally, sacoglossans produce copious amounts of mucus and polysaccharides that, in other marine invertebrates, contain many UV absorbing compounds and sunscreens such as microsporine-like amino acids (MAAs; reviewed by Karentz 2001). The presence of MAAs in *E. chlorotica* has not been investigated. These physical characteristics of the host would presumably generate a protective environment for the plastids that could ameliorate potentially damaging effects of light absorption.

Of equal importance are mechanisms involved in intracellular photoprotection of the organelle after it has been sequestered into the digestive cells of the animal. *Vaucheria litorea* contains relatively high concentrations (~2.7% dry mass) of mannitol within its tissues. Mannitol is important to marine organisms in variable saline environments, where it functions as a compatible solute (Munda 1964; Reed et al. 1985; Iwamoto and Shiraiwa 2005). In algae, the synthesis of mannitol from glycolytic intermediates only requires two additional enzymes, mannitol-1-P dehydrogenase and mannitol-1-P specific phosphatase (Iwamoto and Shiraiwa 2005; Rousvoal et al. 2011). More recently, mannitol has shown importance as an antioxidant. Plastids from tobacco plants were genetically modified to carry mannitol-1-P dehydrogenase, which synthesizes mannitol-1-P from fructose-6-P, and the transgenic progeny exhibited a greater ability to scavenge free radicals (Shen et al. 1997a). Furthermore, the Calvin-Benson cycle enzyme PRK was protected from inactivation normally caused by free radicals (Shen et al. 1997b). Preliminary investigation of mannitol concentrations in *E. chlorotica* showed the presence of mannitol in the animal 2 months after removal from its algal prey, *V. litorea* (the presumed source of mannitol; Rumpho ME and W Loescher, unpublished data). Thus, it is possible that upon feeding on *V. litorea* and during the uptake of plastids, the host animal also sequesters mannitol (or synthesizes it itself) and the presence of mannitol may play a role in stabilizing and protecting the intracellular photosynthetic machinery.

Preliminary investigation into the transcriptome of actively photosynthesizing *E. chlorotica* is revealing an abundance of genes which play a role in anti-oxidant function, including catalase, peroxisomal biogenesis factor 16, caspase 7, cytosolic phospholipase A2 beta, Cu/Zn superoxide dismutase, ferritin, manganese superoxide dismutase, peroxiredoxin 6, Ser/Thr-protein phosphatase 2A catalytic subunit beta isoform, selenium-dependent glutathione peroxidase and thioredoxin peroxidase (Pelletreau et al. 2011). Further investigation into expression levels and timing of expression of these various anti-oxidants will clarify their role in plastid protection and stability.

4. Transient Transcript Expression and Protein Function

As discussed earlier, transcriptome data do not support HGT as a sole or major explanation for plastid function and stability; yet, several studies have provided evidence for the presence of algal nuclear genes in varied phases of the animal life history. This discrepancy may be reconciled via transient processes (rather than permanent integration into the host genome). Several mechanisms may allow host cells to take up and use "foreign" proteins, nucleic acids or other molecules. One involves microvesicles or smaller exosomes as vehicles for transferring proteins and other molecules between cells in a variety of organisms from human and mouse to fungi (see reviews by Valadi et al. 2007; Casadevall et al. 2009; Mansfield and Keene 2009; Feng et al. 2010). Valadi et al. (2007) first demonstrated that exosomes from human and mouse mast cell lines also contain functional mRNAs and regulatory microRNAs, and these RNAs can be transferred in vitro to other cells, translated, and new proteins are observed in the recipient cells (but also see Smalheiser 2007). More recently, reports of similar biologically active vesicles akin to exosomes have been characterized in several fungi facilitating transport across the cell wall and stimulating macrophage activity in animal hosts (Casadevall et al. 2009; Regente et al. 2009; Oliveira et al. 2010).

A second mechanism facilitating transfer and expression of foreign DNA or RNA involves RNA and reverse transcriptase (RT)-mediated inheritance of novel traits. This is well documented in zygotes and spermatozoa of mice (Sciamanna et al. 2003; Rassoulzadegan et al. 2006; Cuzin, et al. 2008; Spadafora 2008; Sciamanna et al. 2009; Garcia-Olmo et al. 2010), cow (Canovas, et al. 2010; Feitosa et al. 2010), pig (Garcia-Vazquez et al. 2010) and fish (Collares et al. 2010), and is now a common mechanism employed in generating transgenic animals (Sciamanna et al. 2009). In these systems, RNA injected directly into zygotes, or sperm containing foreign DNA and RNA and incubated with cells, or cells bathed in free nucleic acids all took up foreign nucleic acids with subsequent expression of the encoded traits in the embryos. In these cases, expression of the DNA or RNA was mosaic in nature; i.e., differential expression was observed between individuals and/or among cells of one individual, and transient over time (Sciamanna et al. 2003, 2009; Rassoulzadegan et al. 2006). Genes transferred *via* RT/RNA mediation in *E. chlorotica* would presumably be expressed in a mosaic fashion, which may explain the often confounding results obtained using PCR to amplify gene products, whereby the reproducibility of the PCR reaction among animals or samples can at times be less than 25% (Pierce et al. 2007; Schwartz et al. 2010).

Development and establishment of irreversible plastid endosymbiosis (or kleptoplasty) in *E. chlorotica* takes place during the development of the animal gut tissue and other advanced features. For this transition to occur, the animal must feed on *V. litorea* resulting in its digestive tract being bathed in algal-derived nucleic acids and protein, establishing a prime environment for RT-mediated inheritance. Mechanistically, direct injection of RNA requires an RNA-dependent RNA polymerase (RDRP) for amplification, whereas cells bathed in free nucleic acids rely on RT and retrovirus activity (Alleman et al. 2006; Sciamanna et al. 2009). It is interesting to note that viral particles have been observed to increase in density and measurable RT activity increases in older senescing specimens of *E. chlorotica* (Pierce et al. 1999; Mondy and Pierce 2003). Additionally, the partial transcriptome library obtained to date is replete with top hits to transposases, reverse transcriptases, RDRP and retroviral Gag-Pol (polyprotein-reverse transcriptase) sequences; perhaps even more intriguing are several foreign viral signatures for RDRP (Pelletreau et al. 2011). The possibility exists for RT activity in *E. chlorotica* that is analogous to that observed in spermatozoa, enabling transient expression of non-animal RNA-derived cDNA in affected cells.

5. Dual Targeting of Cytosolic Host Proteins

Transient gene expression and DNA encountered upon feeding provide plausible mechanisms for the provision of requisite proteins in the absence of algal nuclei and massive HGT. However, this model fails to account for animals which are maintained without exposure to food for months in the laboratory, yet retain photosynthetic ability. Replenishment of these essential components could be provided through co-option of native cytosolic and mitochondrial proteins to provide analogous functions in the plastid. Several processes are shared by mitochondria and plastids including DNA replication and repair, gene expression, protein processing and proteolysis, and generation of ATP (Mackenzie 2005), and several metabolic enzymes are "shared" between pathways in the cytosol and plastid. For example, in *V. litorea* all of the enzymes of the Calvin-Benson photosynthetic carbon reduction cycle are nuclear encoded except for ribulose-1,5-bisphosphate carboxylase/oxygenase, Rubisco (Rumpho et al. 2008). However, all but two of these enzymes (sedoheptulose-1,7-bisphosphatase and PRK) have cytosolic counterparts in *E. chlorotica*. There is increasing evidence supporting dual targeting of some proteins to mitochondria and plastids (reviewed by Carrie et al. 2009) and this possibility should be considered as a contributor to long-term plastid functioning in *E. chlorotica*.

D. Future Directions

The mechanisms supporting this unique model of long-lasting kleptoplasty of a secondary endosymbiont in a metazoan remain elusive and the data are not reconciled. The theory of massive HGT enabling plastid function is questionable, and investigation into other explanations and mechanisms of gene expression are required, if we are to fully understand the evolution of such a novel and highly complex metabolic trait in an animal. It is likely that a combination of factors, such as those outlined here, enable the plastids to remain functional in the animal cells for long periods of time. Greater understanding of how these varied mechanisms work and their relative contributions to plastid stability and function will enhance the overall understanding of the evolution of photosynthesis and the novel acquisition of such an important metabolic function among Metazoa.

Acknowledgments

This research was partially supported by grants from the National Science Foundation awarded to DB (EF 08-27023, DEB 09-36884, and MCB 09-46528) and to MER (IOS-0726178). MER was also supported by the Maine Agricultural and Forest Experiment Station (ME08361-08MRF, NC 1168). ARP acknowledges support from the Canadian Institute for Advanced Research's Integrated Microbial Biodiversity Program.

References

Agrawal S, Striepen B (2010) More membranes, more proteins: complex protein import mechanisms into secondary plastids. Protist 161:672–687

Agrawal S, van Dooren GG, Beatty WL, Striepen B (2009) Genetic evidence that an endosymbiont-derived endoplasmic reticulum-associated protein degradation (ERAD) system functions in import of apicoplast proteins. J Biol Chem 284:33683–33691

Alleman M, Sidorenko L, McGinnis K, Seshadri S, Dorweiler JE, White J, Sikkink K, Chandler VL (2006) An RNA-dependent RNA polymerase is required for paramutation in maize. Nature 442:295–298

Andersen RA (2004) Biology and systematics of heterokont and haptophyte algae. Am J Bot 91:1508–1522

Apt KE, Zaslavkaia L, Lippmeier JC, Lang M, Kilian O, Wetherbee R, Grossman AR, Kroth PG (2002) In vivo characterization of diatom multipartite plastid targeting signals. J Cell Sci 115:4061–4069

Archibald JM (2008) Plastid evolution: remnant algal genes in ciliates. Curr Biol 18:R663–R665

Archibald JM (2009) The puzzle of plastid evolution. Curr Biol 19:R81–R88

Archibald JM, Rogers MB, Toop M, Ishida K, Keeling PJ (2003) Lateral gene transfer and the evolution of plastid-targeted proteins in the secondary plastid-containing alga *Bigelowiella natans*. Proc Natl Acad Sci USA 100:7678–7683

Atteia A, van Lis R, Beale SI (2005) Enzymes of the heme biosynthetic pathway in the nonphotosynthetic alga *Polytomella* sp. Eukaryot Cell 4:2087–2097

Baldwin AJ, Inoue K (2006) The most C-terminal tri-glycine segment within the polyglycine stretch of the pea Toc75 transit peptide plays a critical role for targeting the protein to the chloroplast outer envelope membrane. FEBS J 273:1547–1555

Baurain D, Brinkmann H, Petersen J, Rodríguez-Ezpeleta N, Stechmann A, Demoulin V, Roger AJ, Burger G, Lang BF, Philippe H (2010) Phylogenomic evidence for separate acquisition of plastids in cryptophytes, haptophytes, and stramenopiles. Mol Biol Evol 27:1698–1709

Bock R (2010) The give-and-take of DNA: horizontal gene transfer in plants. Trends Plant Sci 15:11–22

Bolte K, Bullmann L, Hempel F, Bozarth A, Zauner S, Maier UG (2009) Protein targeting into secondary plastids. J Eukaryot Microbiol 56:9–15

Boto L (2010) Horizontal gene transfer in evolution: facts and challenges. Proc R Soc B Biol Sci 277:819–827

Broughton MJ, Howe CJ, Hiller RG (2006) Distinctive organization of genes for light-harvesting proteins in the cryptophyte alga *Rhodomonas*. Gene 369:72–79

Bullmann L, Haarmann R, Mirus O, Bredemeier R, Hempel F, Maier UG, Schleiff E (2010) Filling the gap, evolutionarily conserved Omp85 in plastids of chromalveolates. J Biol Chem 285:6848–6856

Bungard RA (2004) Photosynthetic evolution in parasitic plants: insight from the chloroplast genome. Bioessays 26:235–247

Burki F, Shalchian-Tabrizi K, Minge M, Skjaeveland A, Nikolaev SI, Jakobsen KS, Pawlowski J (2007) Phylogenomics reshuffles the eukaryotic supergroups. PLoS One 2:e790

Burki F, Shalchian-Tabrizi K, Pawlowski J (2008) Phylogenomics reveals a new 'megagroup' including most photosynthetic eukaryotes. Biol Lett 4:366–369

Burki F, Inagaki Y, Bråte J, Archibald JM, Keeling PJ, Cavalier-Smith T, Sakaguchi M, Hashimoto T, Horak A, Kumar S, Klaveness D, Jakobsen KS, Pawlowski J, Shalchian-Tabrizi K (2009) Large-scale phylogenomic analyses reveal that two enigmatic protist lineages, telonemia and centroheliozoa, are related to photosynthetic chromalveolates. Genome Biol Evol 1:231–238

Canovas S, Gutierrez-Adan A, Gadea J (2010) Effect of exogenous DNA on bovine sperm functionality using the sperm mediated gene transfer (SMGT) technique. Mol Reprod Dev 77:687–698

Carrie C, Giraud E, Whelan J (2009) Protein transport in organelles: dual targeting of proteins to mitochondria and chloroplasts. FEBS J 276:1187–1195

Casadevall A, Nosanchuk JD, Williamson P, Rodrigues ML (2009) Vesicular transport across the fungal cell wall. Trends Microbiol 17:158–162

Cavalier-Smith T (1981) Eukaryote kingdoms: seven or nine? Biosystems 14:461–481

Cavalier-Smith T (1992) The number of symbiotic origins of organelles. Biosystems 28:91–106

Cavalier-Smith T (1999) Principles of protein and lipid targeting in secondary symbiogenesis: euglenoid, dinoflagellate, and sporozoan plastid origins and the eukaryote family tree. J Eukaryot Microbiol 46:347–366

Cavalier-Smith T (2010) Kingdoms Protozoa and Chromista and the eozoan root of the eukaryotic tree. Biol Lett 6:342–345

Cavalier-Smith T, Chao EE (2006) Phylogeny and megasystematics of phagotrophic heterokonts (kingdom Chromista). J Mol Evol 62:388–420

Cavalier-Smith T, Couch JA, Thorsteinsen KE, Gilson P, Deane JA, Hill DRA, McFadden GI (1996) Cryptomonad nuclear and nucleomorph 18S rRNA phylogeny. Eur J Phycol 31:315–328

Chan CX, Yang EC, Banerjee T, Yoon HS, Martone PT, Estevez JM, Bhattacharya D (2011) Red and green algal monophyly and extensive gene sharing found in a rich repertoire of red algal genes. Curr Biol 21:328–333

Clay BL, Kugrens P, Lee RE (1999) A revised classification of Cryptophyta. Bot J Linn Soc 131:131–151

Cock JM, Sterck L, Rouzé P, Scornet D, Allen AE, Amoutzias G, Anthouard V, Artiguenave F, Aury JM, Badger JH, Beszteri B, Billiau K, Bonnet E, Bothwell JH, Bowler C, Boyen C, Brownlee C, Carrano CJ, Charrier B, Cho GY, Coelho SM, Collén J, Corre E, Da Silva C, Delage L, Delaroque N, Dittami SM, Doulbeau S, Elias M, Farnham G, Gachon CM, Gschloessl B, Heesch S, Jabbari K, Jubin C, Kawai H, Kimura K, Kloareg B, Küpper FC, Lang D, Le Bail A, Leblanc C, Lerouge P, Lohr M, Lopez PJ, Martens C, Maumus F, Michel G, Miranda-Saavedra D, Morales J, Moreau H, Motomura T, Nagasato C, Napoli CA, Nelson DR, Nyvall-Collén P, Peters AF, Pommier C, Potin P, Poulain J, Quesneville H, Read B, Rensing SA, Ritter A, Rousvoal S, Samanta M, Samson G, Schroeder DC, Ségurens B, Strittmatter M, Tonon T, Tregear JW, Valentin K, von Dassow P, Yamagishi T, Van de Peer Y, Wincker P (2010) The *Ectocarpus* genome and the independent evolution of multicellularity in brown algae. Nature 465:617–621

Collares T, Campos VF, Seixas FK, Cavalcanti PV, Dellagostin OA, Moreira HLM, Deschamps JC (2010) Transgene transmission in South American catfish (*Rhamdia quelen*) larvae by sperm-mediated gene transfer. J Biosci 35:39–47

Cuzin F, Grandjean V, Rassoulzadegan M (2008) Inherited variation at the epigenetic level: paramutation from the plant to the mouse. Curr Opin Genet Dev 18:193–196

Dagan T, Martin W (2006) The tree of one percent. Genome Biol 7:118

Dagan T, Martin W (2009) Microbiology. Seeing green and red in diatom genomes. Science 324:1651–1652

Deane JA, Strachan IM, Saunders GW, Hill DRA, McFadden GI (2002) Cryptomonad evolution: nuclear 18S rDNA phylogeny versus cell morphology and pigmentation. J Phycol 38:1236–1244

Dodge JD (1971) A dinoflagellate with both a mesokaryotic and a eukayotic nucleus. I. Fine structure of the nuclei. Protoplasma 73:145–157

Douglas S, Zauner S, Fraunholz M, Beaton M, Penny S, Deng LT, Wu X, Reith M, Cavalier-Smith T, Maier UG (2001) The highly reduced genome of an enslaved algal nucleus. Nature 410:1091–1096

Durnford DG, Gray MW (2006) Analysis of *Euglena gracilis* plastid-targeted proteins reveals different classes of transit sequences. Eukaryot Cell 5:2079–2091

Elias M, Archibald JM (2009) Sizing up the genomic footprint of endosymbiosis. Bioessays 31:1273–1279

Eschbach S, Speth V, Hansmann P, Sitte P (1990) Freeze-fracture study of the single membrane between host cell and endocytobiont in the dinoflagellates *Glenodinium foliaceum* and *Peridinium balticum*. J Phycol 26:324–328

Evertsen J, Burghardt I, Johnsen G, Wägele H (2007) Retention of functional chloroplasts in some sacoglossans from the Indo-Pacific and Mediterranean. Mar Biol 151:2159–2166

Fawley MW, Yun Y, Qin M (2000) Phylogenetic analyses of 18s rDNA sequences reveal a new coccoid lineage of the prasinophyceae (Chlorophyta). J Phycol 36:387–393

Feitosa WB, Mendes CM, Milazzotto MP, Rocha AM, Martins LF, Simoes R, Paula-Lopes FF, Visintin JA, Assumpcao M (2010) Exogenous DNA uptake by bovine spermatozoa does not induce DNA fragmentation. Theriogenology 74:563–568

Felsner G, Sommer MS, Maier UG (2010a) The physical and functional borders of transit peptide-like sequences in secondary endosymbionts. BMC Plant Biol 10:223

Felsner G, Sommer MS, Gruenheit N, Hempel F, Moog D, Zauner S, Martin W, Maier UG (2010b) ERAD components in organisms with complex red plastids suggest recruitment of a preexisting protein transport pathway for the periplastid membrane. Genome Biol Evol 3:140–150

Feng D, Zhao WL, Ye YY, Bai XC, Liu RQ, Chang LF, Zhou Q, Sui SF (2010) Cellular internalization of exosomes occurs through phagocytosis. Traffic 11:675–687

Frommolt R, Werner S, Paulsen H, Goss R, Wilhelm C, Zauner S, Maier UG, Grossman AR, Bhattacharya D, Lohr M (2008) Ancient recruitment by chromists of green algal genes encoding enzymes for carotenoid biosynthesis. Mol Biol Evol 25:2653–2667

Gaines G, Elbrächter M (1987) Heterotrophic nutrition. In: Taylor FJR (ed) The biology of dinoflagellates. Blackwell Scientific, Oxford, pp 224–268

Garcia-Olmo DC, Dominguez C, Garcia-Arranz M, Anker P, Stroun M, Garcia-Verdugo JM, Garcia-

Olmo D (2010) Cell-free nucleic acids circulating in the plasma of colorectal cancer patients induce the oncogenic transformation of susceptible cultured cells. Cancer Res 70:560–567

Garcia-Vazquez FA, Ruiz S, Matas C, Izquierdo-Rico MJ, Grullon LA, De Ondiz A, Vieira L, Aviles-Lopez K, Gutierrez-Adan A, Gadea J (2010) Production of transgenic piglets using ICSI-sperm-mediated gene transfer in combination with recombinase RecA. Reproduction 140:259–272

Gardner MJ, Hall N, Fung E, White O, Berriman M, Hyman RW, Carlton JM, Pain A, Nelson KE, Bowman S, Paulsen IT, James K, Eisen JA, Rutherford K, Salzberg SL, Craig A, Kyes S, Chan MS, Nene V, Shallom SJ, Suh B, Peterson J, Angiuoli S, Pertea M, Allen J, Selengut J, Haft D, Mather MW, Vaidya AB, Martin DM, Fairlamb AH, Fraunholz MJ, Roos DS, Ralph SA, McFadden GI, Cummings LM, Subramanian GM, Mungall C, Venter JC, Carucci DJ, Hoffman SL, Newbold C, Davis RW, Fraser CM, Barrell B (2002) Genome sequence of the human malaria parasite *Plasmodium falciparum*. Nature 419:498–511

Gilson PR, Su V, Slamovits CH, Reith ME, Keeling PJ, McFadden GI (2006) Complete nucleotide sequence of the chlorarachniophyte nucleomorph: nature's smallest nucleus. Proc Natl Acad Sci USA 103:9566–9571

Gould SB, Sommer MS, Hadfi K, Zauner S, Kroth PG, Maier UG (2006) Protein targeting into the complex plastid of cryptophytes. J Mol Evol 62:674–681

Gould SB, Fan E, Hempel F, Maier UG, Klösgen RB (2007) Translocation of a phycoerythrin alpha subunit across five biological membranes. J Biol Chem 282:30295–30302

Green BJ, Fox TC, Rumpho ME (2005) Stability of isolated algal chloroplasts that participate in a unique mollusc/kleptoplast association. Symbiosis 40:31–40

Gross J, Bhattacharya D (2009a) Revaluating the evolution of the Toc and Tic protein translocons. Trends Plant Sci 14:13–20

Gross J, Bhattacharya D (2009b) Mitochondrial and plastid evolution in eukaryotes: an outsiders' perspective. Nat Rev Genet 10:495–505

Gross J, Bhattacharya D (2011) Endosymbiont or host: who drove mitochondrial and plastid evolution? Biol Direct 6:12

Gruber A, Vugrinec S, Hempel F, Gould SB, Maier UG, Kroth PG (2007) Protein targeting into complex diatom plastids: functional characterization of a specific targeting motif. Plant Mol Biol 64:519–530

Hackett JD, Anderson DM, Erdner DL, Bhattacharya D (2004) Dinoflagellates: a remarkable evolutionary experiment. Am J Bot 91:1523–1534

Hackett JD, Yoon HS, Li S, Reyes-Prieto A, Rummele SE, Bhattacharya D (2007) Phylogenomic analysis supports the monophyly of cryptophytes and haptophytes and the association of rhizaria with chromalveolates. Mol Biol Evol 24:1702–1713

Händeler K, Grzymbowski YP, Krug PJ, Wägele H (2009) Functional chloroplasts in metazoan cells – a unique evolutionary strategy in animal life. Front Zool 6:28

Händeler K, Wägele H, Wahrmund U, Rüdinger M, Knoop V (2010) Slugs' last meals: molecular identification of sequestered chloroplasts from different algal origins in Sacoglossa (Opisthobranchia, Gastropoda). Mol Ecol Res 10:968–978

Harper JT, Keeling PJ (2004) Lateral gene transfer and the complex distribution of insertions in eukaryotic enolase. Gene 340:227–235

Hempel F, Bullmann L, Lau J, Zauner S, Maier UG (2009) ERAD-derived preprotein transport across the second outermost plastid membrane of diatoms. Mol Biol Evol 26:1781–1790

Hempel F, Felsner G, Maier UG (2010) New mechanistic insights into pre-protein transport across the second outermost plastid membrane of diatoms. Mol Microbiol 76:793–801

Herrmann RG (1997) Eukaryotism, towards a new interpretation. In: Schenk HEA, Herrmann RG, Jeon KW, Mueller NE, Schwemmler W (eds) Eukaryotism and symbiosis. Springer, Hedelberg, pp 73–118

Hirakawa Y, Gile GH, Ota S, Keeling PJ, Ishida K (2010) Characterization of periplastidal compartment-targeting signals in chlorarachniophytes. Mol Biol Evol 27:1538–1545

Hoef-Emden K (2005) Multiple independent losses of photosynthesis and differing evolutionary rates in the genus *Cryptomonas* (Cryptophyceae): combined phylogenetic analyses of DNA sequences of the nuclear and the nucleomorph ribosomal operons. J Mol Evol 60:183–195

Huang J, Mullapudi N, Sicheritz-Ponten T, Kissinger JC (2004a) A first glimpse into the pattern and scale of gene transfer in the Apicomplexa. Int J Parasitol 34:265–274

Huang J, Mullapudi N, Lancto CA, Scott M, Abrahamsen MS, Kissinger JC (2004b) Phylogenomic evidence supports past endosymbiosis, intracellular and horizontal gene transfer in *Cryptosporidium parvum*. Genome Biol 5:R88

Imanian B, Keeling PJ (2007) The dinoflagellates *Durinskia baltica* and *Kryptoperidinium foliaceum* retain functionally overlapping mitochondria from two evolutionarily distinct lineages. BMC Evol Biol 7:172

Imanian B, Pombert JF, Keeling PJ (2010) The complete plastid genomes of the two 'dinotoms'

Durinskia baltica and *Kryptoperidinium foliaceum*. PLoS One 5:e10711

Inagaki Y, Dacks JB, Doolittle WF, Watanabe KI, Ohama T (2000) Evolutionary relationship between dinoflagellates bearing obligate diatom endosymbionts: insight into tertiary endosymbiosis. Int J Syst Evol Microbiol 50:2075–2081

Iwamoto K, Shiraiwa Y (2005) Salt-regulated mannitol metabolism in algae. Mar Biotechnol 7:407–415

Janouskovec J, Horák A, Oborník M, Lukes J, Keeling PJ (2010) A common red algal origin of the apicomplexan, dinoflagellate, and heterokont plastids. Proc Natl Acad Sci USA 107:10949–10954

Jensen KR (1997) Evolution of the sacoglossa (Mollusca, Opisthobranchia) and the ecological associations with their food plants. Evol Ecol 11:301–335

Jeong HJ (1999) The ecological roles of heterotrophic dinoflagellates in marine planktonic community. J Eukaryot Microbiol 46:390–396

Kalanon M, Tonkin CJ, McFadden GI (2009) Characterization of two putative protein translocation components in the apicoplast of *Plasmodium falciparum*. Eukaryot Cell 8:1146–1154

Karentz D (2001) Chemical defenses of marine organisms against solar radiation exposure: UV-absorbing mycrosporine-like amino acids and scytonemin. In: McClintock JB, Baker BJ (eds) Marine chemical ecology. CRC Press, Boca Raton, pp 481–520

Keeling PJ (2009) Chromalveolates and the evolution of plastids by secondary endosymbiosis. J Eukaryot Microbiol 56:1–8

Khan H, Parks N, Kozera C, Curtis BA, Parsons BJ, Bowman S, Archibald JM (2007) Plastid genome sequence of the cryptophyte alga *Rhodomonas salina* CCMP1319: lateral transfer of putative DNA replication machinery and a test of chromist plastid phylogeny. Mol Biol Evol 24:1832–1842

Kilian O, Kroth PG (2004) Presequence acquisition during secondary endocytobiosis and the possible role of introns. J Mol Evol 58:712–721

Kim E, Graham LE (2008) EEF2 analysis challenges the monophyly of Archaeplastida and Chromalveolata. PLoS One 3:e2621

Kim E, Lane CE, Curtis BA, Kozera C, Bowman S, Archibald JM (2008) Complete sequence and analysis of the mitochondrial genome of *Hemiselmis andersenii* CCMP644 (Cryptophyceae). BMC Genomics 9:215

Kim E, Harrison JW, Sudek S, Jones MD, Wilcox HM, Richards TA, Worden AZ, Archibald JM (2011) Newly identified and diverse plastid-bearing branch on the eukaryotic tree of life. Proc Natl Acad Sci USA 108:1496–1500

Kirk JT, Tilney-Bassett RA (eds) (1967) The plastids; their chemistry, structure, growth and inheritance. WH Freeman, San Francisco

Lane CE, van den Heuvel K, Kozera C, Curtis BA, Parsons BJ, Bowman S, Archibald JM (2007) Nucleomorph genome of *Hemiselmis andersenii* reveals complete intron loss and compaction as a driver of protein structure and function. Proc Natl Acad Sci USA 104:19908–19913

Li S, Nosenko T, Hackett JD, Bhattacharya D (2006) Phylogenomic analysis identifies red algal genes of endosymbiotic origin in the chromalveolates. Mol Biol Evol 23:663–674

Mackenzie SA (2005) Plant organellar protein targeting: a traffic plan still under construction. Trends Cell Biol 10:548–554

Mansfield KD, Keene JD (2009) The ribonome: a dominant force in co-ordinating gene expression. Biol Cell 101:169–181

Martin W, Herrmann RG (1998) Gene transfer from organelles to the nucleus: how much, what happens, and why? Plant Physiol 118:9–17

Martin W, Stoebe B, Goremykin V, Hapsmann S, Hasegawa M, Kowallik KV (1998) Gene transfer to the nucleus and the evolution of chloroplasts. Nature 393:162–165

Martin W, Rujan T, Richly E, Hansen A, Cornelsen S, Lins T, Leister D, Stoebe B, Hasegawa M, Penny D (2002) Evolutionary analysis of *Arabidopsis*, cyanobacterial, and chloroplast genomes reveals plastid phylogeny and thousands of cyanobacterial genes in the nucleus. Proc Natl Acad Sci USA 99:12246–12251

Matsuzaki M, Kuroiwa H, Kuroiwa T, Kita K, Nozaki H (2008) A cryptic algal group unveiled: a plastid biosynthesis pathway in the oyster parasite *Perkinsus marinus*. Mol Biol Evol 25:1167–1179

Mayfield SP, Yohn CB, Cohen A, Danon A (1995) Regulation of chloroplast gene expression. Ann Rev Plant Physiol Plant Mol Biol 46:147–166

McFadden GI, van Dooren GG (2004) Evolution: red algal genome affirms a common origin of all plastids. Curr Biol 14:R514–R516

McFadden GI, Gilson PR, Hill DRA (1994) *Goniomonas*: rRNA sequences indicate that this phagotrophic flagellate is a close relative of the host component of cryptomonads. Eur J Phycol 29:29–32

McNeal JR, Arumugunathan K, Kuehl JV, Boore JL, Depamphilis CW (2007) Systematics and plastid genome evolution of the cryptically photosynthetic parasitic plant genus *Cuscuta* (Convolvulaceae). BMC Biol 5:55

Minnhagen S, Janson S (2006) Genetic analyses of *Dinophysis* spp. support kleptoplastidy. FEMS Microbiol Ecol 57:47–54

Mondy WL, Pierce SK (2003) Apoptotic-like morphology is associated with annual synchronized death in kleptoplastic sea slugs (*Elysia chlorotica*). J Invertebr Biol 122:126–137

Moore RB, Oborník M, Janouskovec J, Chrudimský T, Vancová M, Green DH, Wright SW, Davies NW, Bolch CJ, Heimann K, Slapeta J, Hoegh-Guldberg O, Logsdon JM, Carter DA (2008) A photosynthetic alveolate closely related to apicomplexan parasites. Nature 451:959–963

Moran NA, Jarvik T (2010) Lateral transfer of genes from fungi underlies carotenoid production in aphids. Science 328:624–627

Moreira D, Le Guyader H, Philippe H (2000) The origin of red algae and the evolution of chloroplasts. Nature 405:69–72

Morgenthaler JJ, Morgenthaler L (1976) Synthesis of soluble, thylakoid and envelope membrane proteins by spinach chloroplasts purified from gradients. Arch Biochem Biophys 172:51–58

Moustafa A, Bhattacharya D (2008) PhyloSort: a user–friendly phylogenetic sorting tool and its application to estimating the cyanobacterial contribution to the nuclear genome of Chlamydomonas. BMC Evol Biol 8:6

Moustafa A, Reyes-Prieto A, Bhattacharya D (2008a) Chlamydiae has contributed at least 55 genes to Plantae with predominantly plastid functions. PLoS One 3:e2205

Moustafa A, Chan CX, Danforth M, Zear D, Ahmed H, Jadhav N, Bhattacharya D (2008b) A phylogenomic approach for studying plastid endosymbiosis. Genome Inform 21:165–176

Moustafa A, Beszteri B, Maier UG, Bowler C, Valentin K, Bhattacharya D (2009) Genomic footprints of a cryptic plastid endosymbiosis in diatoms. Science 324:1724–1726

Munda I (1964) The quantity and chemical composition of *Ascophyllum nodosum* (L) Le Jol along the coast between the rivers Ölfusá and Thjorsá (Southern Iceland). Bot Mar 7:76–89

Nosenko T, Bhattacharya D (2007) Horizontal gene transfer in chromalveolates. BMC Evol Biol 7:173

Nosenko T, Lidie KL, Van Dolah FM, Lindquist E, Cheng JF, Bhattacharya D (2006) Chimeric plastid proteome in the Florida "red tide" dinoflagellate *Karenia brevis*. Mol Biol Evol 23:2026–2038

Not F, Valentin K, Romari K, Lovejoy C, Massana R, Tobe K, Vaulot D, Medlin LK (2007) Picobiliphytes: a marine picoplanktonic algal group with unknown affinities to other eukaryotes. Science 315:253–255

Nozaki H, Iseki M, Hasegawa M, Misawa K, Nakada T, Sasaki N, Watanabe M (2007) Phylogeny of primary photosynthetic eukaryotes as deduced from slowly evolving nuclear genes. Mol Biol Evol 24:1592–1595

Okamoto N, Inouye I (2005) The katablepharids are a distant sister group of the Cryptophyta: a proposal for Katablepharidophyta divisio nova/Kathablepharida phylum novum based on SSU rDNA and beta-tubulin phylogeny. Protist 156:163–179

Okamoto N, Chantangsi C, Horák A, Leander BS, Keeling PJ (2009) Molecular phylogeny and description of the novel katablepharid *Roombia truncata* gen. et sp. nov., and establishment of the Hacrobia taxon nov. PLoS ONE 4:7080

Olendzenski L, Gogarten JP (2009) Evolution of genes and organisms the tree/web of life in light of horizontal gene transfer. In: Witzany G (ed) Natural genetic engineering and natural genome editing. Blackwell, Oxford, pp 137–145

Oliveira DL, Freire-de-Lima CG, Nosanchuk JD, Casadevall A, Rodrigues ML, Nimrichter L (2010) Extracellular vesicles from *Cryptococcus neoformans* modulate macrophage functions. Infect Immun 78:1601–1609

Oudot-Le Secq MP, Grimwood J, Shapiro H, Armbrust EV, Bowler C, Green BR (2007) Chloroplast genomes of the diatoms *Phaeodactylum tricornutum* and *Thalassiosira pseudonana*: comparison with other plastid genomes of the red lineage. Mol Genet Genomics 277:427–439

Palmer JD (2003) The symbiotic birth and spread of plastids: how many times and whodunit? J Phycol 39:4–11

Patron NJ, Waller RF (2007) Transit peptide diversity and divergence: a global analysis of plastid targeting signals. Bioessays 29:1048–1058

Patron NJ, Waller RF, Archibald JM, Keeling PJ (2005) Complex protein targeting to dinoflagellate plastids. J Mol Biol 348:1015–1024

Patron NJ, Waller RF, Keeling PJ (2006) A tertiary plastid uses genes from two endosymbionts. J Mol Biol 357:1373–1382

Patron NJ, Inagaki Y, Keeling PJ (2007) Multiple gene phylogenies support the monophyly of cryptomonad and haptophyte host lineages. Curr Biol 17:887–891

Pelletreau KN, Bhattacharya D, Price DB, Worful JM, Moustafa A, Rumpho ME (2011) Update on sea slug kleptoplasty and plastid maintenance in a metazoan. Plant Physiol 155:1561–1565

Pierce SK, Biron R, Rumpho ME (1996) Endosymbiotic chloroplasts in molluscan cells contain proteins synthesized after plastid capture. J Exp Biol 199:2323–2330

Pierce SK, Maugel TK, Rumpho ME, Hanten JJ, Mondy WL (1999) Annual viral expression in a sea

slug population: life cycle control and symbiotic chloroplast maintenance. Biol Bull 197:1–6
Pierce SK, Massey SE, Hanten JJ, Curtis NE (2003) Horizontal transfer of functional nuclear genes between multicellular organisms. Biol Bull 204:237–240
Pierce SK, Curtis NE, Hanten JJ, Boerner SL, Schwartz JA (2007) Transfer, integration and expression of functional nuclear genes between multicellular species. Symbiosis 43:57–64
Pierce SK, Curtis NE, Schwartz JA (2009) Chlorophyll a synthesis by an animal using transferred algal nuclear genes. Symbiosis 49:121–131
Pierce SK, Fang X, Schwartz JA, Jiang X, Zhao W, Curtis NE, Kocot K, Yang B, Wang J (2011) Transcriptomic evidence for the expression of horizontally transferred algal nuclear genes in the photosynthetic sea slug, *Elysia chlorotica*. Mol Biol Evol, epub ahead of print
Rahat M, Monselise EB (1979) Photobiology of the chloroplast hosting mollusc *Elysia timida* (Opisthobranchia). J Exp Biol 79:225–233
Rassoulzadegan M, Grandjean V, Gounon P, Vincent S, Gillot I, Cuzin F (2006) RNA-mediated non-mendelian inheritance of an epigenetic change in the mouse. Nature 441:469–474
Reeb VC, Peglar MT, Yoon HS, Bai JR, Wu M, Shiu P, Grafenberg JL, Reyes-Prieto A, Rümmele SE, Gross J, Bhattacharya D (2009) Interrelationships of chromalveolates within a broadly sampled tree of photosynthetic protists. Mol Phylogenet Evol 53:202–211
Reed RH, Davison IR, Chudek JA, Foster R (1985) The osmotic role of mannitol in the Phaeophyta: an appraisal. Phycologia 24:35–47
Regente M, Corti-Monzon G, Maldonado AM, Pinedo M, Jorrin J, de la Canal L (2009) Vesicular fractions of sunflower apoplastic fluids are associated with potential exosome marker proteins. FEBS Lett 583:3363–3366
Reyes-Prieto A, Hackett JD, Soares MB, Bonaldo MF, Bhattacharya D (2006) Cyanobacterial contribution to algal nuclear genomes is primarily limited to plastid functions. Curr Biol 16:2320–2325
Reyes-Prieto A, Weber AP, Bhattacharya D (2007) The origin and establishment of the plastid in algae and plants. Ann Rev Genet 41:147–168
Reyes-Prieto A, Moustafa A, Bhattacharya D (2008) Multiple genes of apparent algal origin suggest ciliates may once have been photosynthetic. Curr Biol 18:956–962
Reyes-Prieto A, Yoon HS, Moustafa A, Yang EC, Andersen RA, Boo SM, Nakayama T, Ishida K, Bhattacharya D (2010) Differential gene retention in plastids of common recent origin. Mol Biol Evol 27:1530–1537
Rice DW, Palmer JD (2006) An exceptional horizontal gene transfer in plastids: gene replacement by a distant bacterial paralog and evidence that haptophyte and cryptophyte plastids are sisters. BMC Biol 4:31
Richards TA, Dacks JB, Campbell SA, Blanchard JL, Foster PG, McLeod R, Roberts CW (2006) Evolutionary origins of the eukaryotic shikimate pathway: gene fusions, horizontal gene transfer, and endosymbiotic replacements. Eukaryot Cell 5:1517–1531
Robertson DL, Tartar A (2006) Evolution of glutamine synthetase in heterokonts: evidence for endosymbiotic gene transfer and the early evolution of photosynthesis. Mol Biol Evol 23:1048–1055
Rodriguez-Ezpeleta N, Brinkmann H, Burey SC, Roure B, Burger G, Löffelhardt W, Bohnert HJ, Philippe H, Lang BF (2005) Monophyly of primary photosynthetic eukaryotes: green plants, red algae, and glaucophytes. Curr Biol 15:1325–1330
Rousvoal S, Groisillier A, Dittami SM, Michel G, Boyon C, Tonon T (2011) Mannitol-1-phosphate dehydrogenase activity in *Ectocarpus siliculosus*, a key role for mannitol synthesis in brown algae. Planta 233:261–273
Rumpho ME, Summer EJ, Green GJ, Fox TC, Manhart JR (2001) Mollusc/algal chloroplast symbiosis: how can isolated chloroplasts continue to function for months in the cytosol of a sea slug in the absence of an algal nucleus? Zoology 104:303–312
Rumpho ME, Dastoor FP, Manhart JR, Lee J (2006) The kleptoplasty. In: Wise RR, Hoober JK (eds) Advances in photosynthesis and respiration – the structure and function of plastids. Springer, Dordrecht, pp 451–473
Rumpho ME, Worful JM, Lee J, Kannan K, Tyler MS, Bhattacharya D, Moustafa A, Manhart JR (2008) Horizontal gene transfer of the algal nuclear gene *psb*O to the photosynthetic sea slug *Elysia chlorotica*. Proc Natl Acad Sci USA 105:17867–17871
Rumpho ME, Pochareddy S, Worful JM, Summer EJ, Bhattacharya D, Pelletreau KN, Tyler MS, Lee J, Manhart JR, Soule KM (2009) Molecular characterization of the Calvin cycle enzyme phosphoribulokinase in the stramenopile alga *Vaucheria litorea* and the plastid hosting mollusc *Elysia chlorotica*. Mol Plant 2:1384–1396
Rumpho ME, Pelletreau KN, Moustafa A, Bhattacharya D (2011) The making of a photosynthetic animal. J Exp Biol 214:301–311
Saldarriaga JF, Taylor FJR, Keeling PJ, Cavalier-Smith T (2001) Dinoflagellate nuclear SSU rRNA phylogeny suggests multiple plastid losses and replacements. J Mol Evol 53:204–213
Sanchez-Puerta MV, Bachvaroff TR, Delwiche CF (2005) The complete plastid genome sequence of

the haptophyte *Emiliania huxleyi*: a comparison to other plastid genomes. DNA Res 12:151–156

Sanchez-Puerta MV, Lippmeier JC, Apt KE, Delwiche CF (2007) Plastid genes in a non-photosynthetic dinoflagellate. Protis 158:105–117

Sato N, Ishikawa M, Fujiwara M, Sonoike K (2005) Mass identification of chloroplast proteins of endosymbiont origin by phylogenetic profiling based on organism-optimized homologous protein groups. Genome Inform 16:56–68

Schnepf E, Elbrachter M (1988) Cryptophycean-like double membrane-bound chloroplast in the dinoflagellate, *Dinophysis* ehrenb – evolutionary, phylogenetic and toxicological implications. Bot Acta 101:196–203

Schwartz JA, Curtis NE, Pierce SK (2010) Using algal transcriptome sequences to identify transferred genes in the sea slug, *Elysia chlorotica*. Evol Biol 37:29–37

Sciamanna I, Barberi L, Martire A, Pittoggi C, Beraldi R, Giordano R, Magnano AR, Hogdson C, Spadafora C (2003) Sperm endogenous reverse transcriptase as mediator of new genetic information. Biochem Biophys Res Commun 312:1039–1046

Sciamanna I, Vitullo P, Curatolo A, Spadafora C (2009) Retrotransposons, reverse transcriptase and the genesis of new genetic information. Gene 448:180–186

Sekiguchi H, Moriya M, Nakayama T, Inouye I (2002) Vestigial chloroplasts in heterotrophic stramenopiles *Pteridomonas danica* and *Ciliophrys infusionum* (Dictyochophyceae). Protist 153:157–167

Shalchian-Tabrizi K, Minge MA, Cavalier-Smith T, Nedreklepp JM, Klaveness D, Jakobsen KS (2006a) Combined heat shock protein 90 and ribosomal RNA sequence phylogeny supports multiple replacements of dinoflagellate plastids. J Eukaryot Microbiol 53:217–224

Shalchian-Tabrizi K, Eikrem W, Klaveness D, Vaulot D, Minge MA, Le Gall F, Romari K, Throndsen J, Botnen A, Massana R, Thomsen HA, Jakobsen KS (2006b) Telonemia, a new protist phylum with affinity to chromist lineages. Proc Biol Sci 273:1833–1842

Shen B, Jensen RG, Bohnert HJ (1997a) Increased resistance to oxidative stress in transgenic plants by targeting mannitol biosynthesis to chloroplasts. Plant Physiol 113:1177–1183

Shen B, Jensen RG, Bohnert HJ (1997b) Mannitol protects against oxidation by hydroxyl radicals. Plant Physiol 115:527–532

Slamovits CH, Keeling PJ (2008) Plastid-derived genes in the nonphotosynthetic alveolate *Oxyrrhis marina*. Mol Biol Evol 25:1297–1306

Smalheiser NR (2007) Exosomal transfer of proteins and RNAs at synapses in the nervous system. Biol Dir 2:35

Sommer MS, Gould SB, Lehmann P, Gruber A, Przyborski JM, Maier UG (2007) Der1-mediated preprotein import into the periplastid compartment of chromalveolates? Mol Biol Evol 24:918–928

Soule KM (2010) Light-related photosynthetic gene expression and enzyme activity in the heterokont alga *Vaucheria litorea* and its symbiotic partner the sacoglossan mollusc *Elysia chlorotica*. MS Thesis, University of Maine, Orono

Spadafora C (2008) Sperm-mediated 'reverse' gene transfer: a role of reverse transcriptase in the generation of new genetic information. Hum Reprod 23:735–740

Spork S, Hiss JA, Mandel K, Sommer M, Kooij TW, Chu T, Schneider G, Maier UG, Przyborski JM (2009) An unusual ERAD-like complex is targeted to the apicoplast of *Plasmodium falciparum*. Eukaryot Cell 8:1134–1145

Steinkoetter J, Bhattacharya D, Semmelroth I, Bibeau C, Melkonian M (1994) Prasinophytes form independent lineages within the Chlorophyta: evidence from ribosomal RNA sequence comparisons. J Phycol 30:340–345

Stibitz TB, Keeling PJ, Bhattacharya D (2000) Symbiotic origin of a novel actin gene in the cryptophyte Pyrenomonas helgolandii. Mol Biol Evol 17:1731–1738

Stiller JW, Huang J, Ding Q, Tian J, Goodwillie C (2009) Are algal genes in nonphotosynthetic protists evidence of historical plastid endosymbioses? BMC Genomics 10:484

Stoecker DE (1990) Mixotrophy among dinoflagellates. J Eukaryot Microbiol 46:397–401

Takano Y, Hansen G, Fujita D, Horiguchi T (2008) Serial replacement of diatom endosymbionts in two freshwater dinoflagenates, *Peridiniopsis* spp. (Peridiniales, Dinophyceae). Phycologia 47:41–53

Takishita K, Koike K, Maruyama T, Ogata T (2002) Molecular evidence for plastid robbery (Kleptoplastidy) in *Dinophysis*, a dinoflagellate causing diarrhetic shellfish poisoning. Protist 153:293–302

Tartar A, Boucias DG, Adams BJ, Becnel JJ (2002) Phylogenetic analysis identifies the invertebrate pathogen *Helicosporidium* sp. as a green alga (Chlorophyta). Int J Syst Evol Microbiol 52:273–279

Tonkin CJ, Roos DS, McFadden GI (2006) N-terminal positively charged amino acids, but not their exact position, are important for apicoplast transit peptide fidelity in *Toxoplasma gondii*. Mol Biochem Parasitol 150:192–200

Trench RK (1975) Of 'leaves that crawl': functional chloroplasts in animal cells. In: Jennings DH (ed) Symposia of the society for experimental biology. Cambridge University Press, London, pp 229–265

Trench RK, Boyle EJ, Smith DC (1973) The association between chloroplasts of *Codium fragile* and the mollusc *Elysia viridis*. I. Characteristics of isolated *Codium* chloroplasts. Proc R Soc Lond B Biol Sci 184:51–61

Tyler BM, Tripathy S, Zhang X, Dehal P, Jiang RH, Aerts A, Arredondo FD, Baxter L, Bensasson D, Beynon JL, Chapman J, Damasceno CM, Dorrance AE, Dou D, Dickerman AW, Dubchak IL, Garbelotto M, Gijzen M, Gordon SG, Govers F, Grunwald NJ, Huang W, Ivors KL, Jones RW, Kamoun S, Krampis K, Lamour KH, Lee MK, McDonald WH, Medina M, Meijer HJ, Nordberg EK, Maclean DJ, Ospina-Giraldo MD, Morris PF, Phuntumart V, Putnam NH, Rash S, Rose JK, Sakihama Y, Salamov AA, Savidor A, Scheuring CF, Smith BM, Sobral BW, Terry A, Torto-Alalibo TA, Win J, Xu Z, Zhang H, Grigoriev IV, Rokhsar DS, Boore JL (2006) *Phytophthora* genome sequences uncover evolutionary origins and mechanisms of pathogenesis. Science 313: 1261–1266

Valadi H, Ekstrom K, Bossios A, Sjostrand M, Lee JJ, Lotvall JO (2007) Exosome-mediated transfer of mRNAs and microRNAs is a novel mechanism of genetic exchange between cells. Nat Cell Biol 9:654–672

van Dooren GG, Tomova C, Agrawal S, Humbel BM, Striepen B (2008) *Toxoplasma gondii* Tic20 is essential for apicoplast protein import. Proc Natl Acad Sci USA 105:13574–13579

Wägele H, Deusch O, Händeler K, Martin R, Schmitt V, Christa G, Pinzger B, Gould SB, Dagan T, Klussmann-Kolb A, Martin W (2010) Transcriptomic evidence that longevity of acquired plastids in the photosynthetic slugs *Elysia timida* and *Plakobrachus ocellatus* does not entail lateral transfer of algal nuclear genes. Mol Biol Evol 28:699–706

Waller RF, Reed MB, Cowman AF, McFadden GI (2000) Protein trafficking to the plastid of *Plasmodium falciparum* is via the secretory pathway. EMBO J 19:1794–1802

Wickett NJ, Zhang Y, Hansen SK, Roper JM, Kuehl JV, Plock SA, Wolf PG, DePamphilis CW, Boore JL, Goffinet B (2008) Functional gene losses occur with minimal size reduction in the plastid genome of the parasitic liverwort *Aneura mirabilis*. Mol Biol Evol 25:393–401

Wolfe KH, Morden CW, Palmer JD (1992) Function and evolution of a minimal plastid genome from a nonphotosynthetic parasitic plant. Proc Natl Acad Sci USA 89:10648–10652

Xie W, Ng DT (2010) ERAD substrate recognition in budding yeast. Semin Cell Dev Biol 21:533–539

Yabuki A, Inagaki Y, Ishida K (2010) *Palpitomonas bilix* gen. et sp. nov.: a novel deep-branching heterotroph possibly related to Archaeplastida or Hacrobia. Protist 161:523–538

Yamomoto YY, Yusa Y, Yamamoto S, Hirano Y, Yoshiaki H, Motomura T, Tanemura T, Obokata J (2009) Identification of photosynthetic sacoglossans from Japan. Endocytobiosis Cell Res 19: 112–119

Yoon HS, Hackett JD, Bhattacharya D (2002) A single origin of the peridinin- and fucoxanthin-containing plastids in dinoflagellates through tertiary endosymbiosis. Proc Natl Acad Sci USA 99:11724–11729

Yoon HS, Hackett JD, Van Dolah FM, Nosenko T, Lidie KL, Bhattacharya D (2005) Tertiary endosymbiosis driven genome evolution in dinoflagellate algae. Mol Biol Evol 22:1299–1308

Yoon HS, Reyes-Prieto A, Melkonian M, Bhattacharya D (2006) Minimal plastid genome evolution in the *Paulinella* endosymbiont. Curr Biol 16: R670–R672

Yoon HS, Grant J, Tekle YI, Wu M, Chaon BC, Cole JC, Logsdon JM Jr, Patterson DJ, Bhattacharya D, Katz LA (2008) Broadly sampled multigene trees of eukaryotes. BMC Evol Biol 8:14

Chapter 3

Plastid Genomes of Algae

B. Franz Lang*
Department of Biochemistry, Robert-Cedergren Centre for Bioinformatics and Genomics, Université de Montréal, 2900 Edouard-Montpetit, Montreal, QC H3T 1J4, Canada

Aurora M. Nedelcu
Department of Biology, University of New Brunswick, Fredericton, NB E3B 5A3, Canada

*Author for correspondence, e-mail: Franz.Lang@Umontreal.ca

R. Bock and V. Knoop (eds.), Genomics of Chloroplasts and Mitochondria, Advances in Photosynthesis and Respiration 35, pp. 59–87,
DOI 10.1007/978-94-007-2920-9_3,

Summary

Algae are characterized by the presence of plastids (chloroplasts), which are organelles of cyanobacterial origin. Plastids have their own genome, machineries for replication, transcription and translation, and are the site of photosynthesis (except in secondarily non-photosynthetic species) and a variety of other biological functions. Algae are subdivided into those whose plastids can be traced back to a common cyanobacterial endosymbiont (algae with primary plastids), and others in which plastids are second-hand acquisitions that were introduced by eukaryote-eukaryote endosymbioses.

Only a fraction of plastid components is encoded in plastid DNA; the majority of genes coding for plastid proteins are in the nucleus, many of which originated through transfers (in some cases still ongoing) from the organelle to the nuclear genome. Despite the broad phylogenetic affiliation of algae, most plastid genomes are fairly homogenous, coding for about 100–250 genes, except in non-photosynthetic algae that rapidly lose genes involved in photosynthesis. The most gene-rich and cyanobacteria-like plastid genomes are in red algae, followed by glaucophyte and green algae. Genomes in secondary or higher-order plastids usually have a reduced gene count, compared to their primary photosynthetic donors. In this chapter, we provide an overview on the evolutionary history, organization and coding properties of algal plastid genomes, for which complete (or almost complete) sequences are publicly available.

I. Introduction

The term ‘algae’ as used here includes all plastid-containing eukaryotes, except land plants and ‘blue-green algae’ (a popular misnomer for Cyanobacteria). Algae cover a large variety of about 20 taxonomic groups (among the best-known ones are green, red, brown and golden algae, diatoms, glaucophytes, raphidophytes, cryptophytes, haptophytes, chlorarachniophytes, dinoflagellates and euglenids). Some of these groups include both unicellular and multi-cellular species (e.g., the large-size brown algal kelp, various red and green algal taxa). In rare instances, algae are secondarily non-photosynthetic, carrying a plastid genome with reduced coding capacity; these include the colorless green algae *Prototheca* and *Helicosporidium* (Knauf and Hachtel 2002; Pombert and Keeling 2010); the euglenid *Euglena (Astasia) longa* (Knauf and Hachtel 2002); and *Plasmodium* and its apicomplexan relatives (McFadden and Waller 1997; Wilson and Williamson 1997).

Plastid genomes are best described and compared within an evolutionary framework (a phylogenetic tree based on plastid

Abbreviations: aa – Amino acid; CASH – Cryptophyta Alveolata, Stramenopila plus Haptophyta their plastids are of red algal origin and pt genomes are closely related (which is incompatible with respective nuclear genome phylogenies). CW – ‘Clockwise’ arrangement of flagellar basal bodies in Chlamydomonadales; DO – ‘Directly opposed’ arrangement of flagellar basal bodies in Sphaeropleales; IR – Inverted genomic repeat region occurs in a large number of ptDNAs; LBA – Long Branch Attraction phylogenetic artifact that leads to the incorrect grouping of fast-evolving species or attraction to distant outgroups, due to evolutionary model violations and under-estimation of repeated sequence change; mtDNA – Mitochondrial DNA protists – eukaryotes other than fungi animals and plants; pt – Plastid (chloroplast); ptDNA – Plastid DNA; SC – Single-copy regions separating large inverted repeats in ptDNAs; tmRNA – Transfer mRNA occurs in bacterial some plastid and jakobid mitochondrial genomes typically contains a tRNA-like and a protein-coding domain involved in releasing ribosomes that are stalled by degraded mRNAs without in-frame stop codons

protein sequences is shown in Fig. 3.1), which is however more easily said than done. This is because phylogenetic placement of reduced or fast-evolving plastid sequences is challenging due to lack of phylogenetic signal. Another difficulty arises from the different evolutionary routes followed by plastids: vertical descent from a Cyanobacterium, and lateral acquisition from other eukaryotes. The latter entails the transfer of both the complete plastid DNA (ptDNA) plus an often undetermined number of nuclear genes from the symbiont to the host nucleus, leading to potential phylogenetic misinterpretations. For instance, the plastid tree in Fig. 3.1 groups dinoflagellates such as *Kryptoperidinium* and *Durinskia* with diatoms, apicomplexans with stramenopiles, cercozoans with green algae, and so on. One may indeed wonder which of the shown phylogenetic relationships represent vertical evolutionary descent at all. The only notable exception are primary photosynthetic eukaryotes (green, red, glaucophyte algae and land plants – collectively known as 'Plantae'; (Cavalier-Smith 1981); see also Chap. 1), whose plastids derive directly from a cyanobacterial origin, and which are therefore expected to form a monophyletic group with nuclear, plastid and mitochondrial genes in phylogenetic analyses (Baurain et al. 2010).

A. Origin and Evolution of Primary Photosynthetic Algae and Their Plastids

The origin of primary plastids represents a relatively late step in eukaryotic evolution, well after the endosymbiosis with the α-Proteobacterium that evolved into the mitochondrion. Most plastid genomes retain many more features of their (cyano) bacterial ancestor than do mitochondrial genomes, such as large conserved bacterial operons and bacteria-like RNA polymerases (but see the notable exception in jakobid mitochondria; Lang et al. 1997). Although plastids came in relatively late, the exact nature of the eukaryotic group which acquired plastids remains vague, as primary (ancestrally) non-photosynthetic members belonging to Plantae are unknown. In fact, even phylogenomic evidence for the monophyly (i.e. divergence from a single, common origin) of Plantae varies with taxon and gene sampling, with significant statistical support in some cases (e.g., Rodriguez-Ezpeleta et al. 2005 and references therein) but not in others (e.g., Burki et al. 2009; Baurain et al. 2010; Parfrey et al. 2010; Chan et al. 2011). Likewise, the branching order of primary photosynthetic lineages has been elusive, depending much on the choice of genes and species included in phylogenies (Rodriguez-Ezpeleta et al. 2005; Reyes-Prieto and Bhattacharya 2007; Deschamps and Moreira 2009). Taken together, much remains to be done in terms of resolving the origin and the evolutionary divergence of Plantae. Apparently, the resolution of the deepest branches of the eukaryotic tree remains unsatisfying, as deep eukaryotic (protist) diversity continues to be poorly sampled at the genome level. Yet, for sake of simplicity, we will assume in the following that Plantae is a valid taxonomic grouping, and therefore discuss plastids in two major subdivisions, (1) those derived from a primary endosymbiotic event and (2) those that have been acquired by higher-order (secondary, tertiary …) endosymbioses among eukaryotes.

B. Algae with Second-Hand Plastids: Eukaryote-Eukaryote Endosymbioses

In contrast to Plantae – which are characterized by plastids with two surrounding membranes – there are three or four membranes in algae that have undergone eukaryote-eukaryote endosymbiosis, the focus of most reviews on plastid DNAs (ptDNAs; e.g., Douglas and Gray 1991; Wolfe et al. 1991; Douglas 1998; McFadden 1999; Moreira and Philippe 2001; Archibald and Keeling 2002; Stoebe and Maier 2002; Bhattacharya et al. 2004; Reyes-Prieto et al. 2007; Gould et al. 2008; Archibald 2009; Keeling 2009; Keeling 2010). These plastids are in most instances

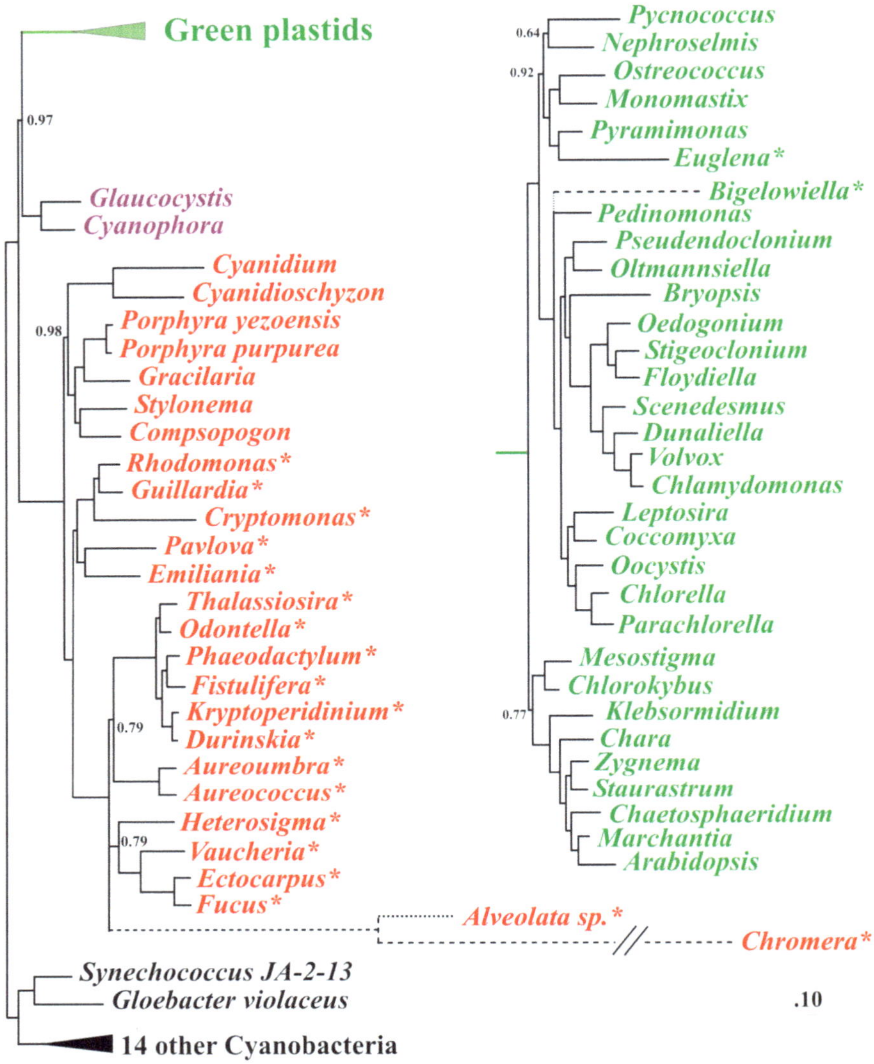

Fig. 3.1. Phylogeny based on ptDNA-encoded proteins. The phylogeny was inferred with a set of 76 derived ptDNA-encoded protein sequences that are most common across algae. Non-photosynthetic species such as *Plasmodium* or *Helicosporidium* were excluded, as their ptDNAs encode only small subsets of genes that evolve at elevated rates (i.e., problematic for phylogenetic inference). Sequences were aligned with software developed in-house. Briefly, derived protein sequences are pre-aligned with Muscle (Edgar 2004), and alignments are iteratively refined with HMMalign (S. Eddy; http://hmmer.janelia.org) using E-values obtained with respective HMM models as an optimization criterion. Sequence positions that are not aligned with a posterior probability value of 1.0 are discarded, and alignments are concatenated. The resulting dataset including 48 ptDNAs (plus a cyanobacterial outgroup) has 17,409 amino acid positions. It was analyzed using PhyloBayes that implements Bayesian inference and the CAT model, known to be least sensitive to LBA artifacts (Lartillot and Philippe 2004; Rodriguez-Ezpeleta et al. 2007b; Lartillot and Philippe 2008, and references therein). It is best suited for the placement of rapidly-evolving species. All branches are supported by posterior probability values of 1.0 except where indicated. Species are color-coded according to their plastid origins as *green*, *red* or *magenta* (glaucocystophytes). Plastids that were transferred by secondary or higher-order eukaryote-eukaryote endosymbiosis are marked with an *asterisk*. The branch length of *Chromera* (*dotted broken line*) is about three times as long as indicated. Its association with *Alveolata* could be due to LBA; if analyzed alone, *Chromera* tends to group weakly with stramenopiles.

retraced to either a red or green algal origin (Fig. 3.1), but whether the endosymbiotic event is secondary, tertiary or higher-order often remains speculative. In particular, the source of the highly reduced 'apicoplast' plastids in alveolates (e.g., *Plasmodium*, *Eimeria*, etc.) remains uncertain, believed to be of either red (Williamson et al. 1994; Fast et al. 2001; Foth and McFadden 2003) or green algal origin (Kohler et al. 1997; Funes et al. 2002, 2004). Even gene transfer from mitochondrial DNA (mtDNA) to apicoplast DNA has been proposed (Obornik et al. 2002), and there is currently no convincing avenue for overcoming the massive phylogenetic artifacts (long-branch-attraction artifacts or LBA) that are the likely cause of the unsettled dispute. Phylogenetic analyses including these species are so questionable because of both the small number of remaining plastid genes and their extreme evolutionary rates.

Another confounding factor in these analyses is the number of symbiotic events that took place across eukaryotes. Plastids in cryptophytes, alveolates, stramenopiles plus haptophytes (collectively, CASH) likely arose from a single secondary endosymbiosis with a red alga because of a unique, shared feature of their plastids, the presence of chlorophyll c, and because phylogenies based on plastid sequences (plus a few nuclear genes involved in plastid function) clearly regroup CASH with red algae. However, the consensus in interpretation stops here. Based on the idea that eukaryote-eukaryote endosymbiosis is a very rare event, proponents of the 'chromalveolate hypothesis' (Cavalier-Smith 2002; Keeling 2009, 2010) postulate a single, ancient secondary endosymbiosis with a red alga. This supposition is contested by others predicting much more frequent (higher-order), serial plastid transfers (Sanchez-Puerta et al. 2007; Baurain et al. 2010; Gray 2010 and references therein). We share the interpretation of frequent transfers because of cumulating evidence in this direction. In dinoflagellates, for instance, there is compelling evidence for a number of subsequent plastid replacements (e.g., Minge et al. 2010). Another (contentious) example is in two presumed photosynthetic relatives of Apicomplexa, *Chromera velia* and *Alveolata sp.* (CCMP3115). Phylogenies with several concatenated nuclear genes confirm that they represent taxonomically deep divergences to Apicomplexa, and a 34 plastid gene phylogeny associates their plastids close to (but outside of) stramenopiles (heterokonts; Janouskovec et al. 2010). According to the authors' interpretation, this represents support for the chromalveolate hypothesis. Yet, the plastid phylogeny that we performed for the purpose of this review, with an extended number of species (79) and proteins (76) comes to a different conclusion, placing *Chromera* and *Alveolata* plastids together within stramenopiles (Fig. 3.1), indicative of a higher-order endosymbiosis. This example indicates that phylogenetic analyses with data from fast-evolving genomes have to be interpreted with extreme prudence, in particular when these diverge deeply in a tree, an indicator for a potential phylogenetic reconstruction artifact (Philippe et al. 2005). In turn, when broader taxon sampling and/or the use of a superior (more realistic) evolutionary model, such as CAT (Lartillot and Philippe 2004; Lartillot et al. 2007), leads to an alternative tree topology favoring the regrouping of rapidly with slowly evolving species, even with limited statistical support (as in Fig. 3.1), it is more likely the correct one. Clearly, further investigation of the given example is needed, which falls outside the mission of this review.

Given the confusion in distinguishing secondary and higher-order eukaryote-eukaryote endosymbionts, we will only refer to the following five well established taxa: (1) golden, brown, diatom and raphidophyte algae (Stramenopila; Patterson 1989), (2) Alveolata plus Stramenopila and Rhizaria (SAR group; Burki et al. 2007; Hackett et al. 2007; Rodriguez-Ezpeleta et al. 2007a), (3) haptophytes, (4) cryptophytes, and (5) euglenids (belonging to the 'JEH group' uniting jakobids, Euglenozoa plus Heterolobosea; Rodriguez-Ezpeleta et al. 2007a). It is noteworthy that there is only one major eukaryotic supergroup without photosynthetic members (and without evident genetic remnants of eukaryote-eukaryote endosymbioses), the Unikonta. This group comprises Opisthokonta (animals, fungi and their protist relatives), Amoebozoa, and arguably, Apusozoa.

In the following, we will review plastid genome organization in the various groups of algae. The highly reduced alveolate ptDNAs (apicoplasts) will not be discussed in detail as they have been well described elsewhere (Wilson and Williamson 1997; McFadden 2011).

II. Plastid Genome Organization, Genes and Functions

We will start with a short introduction on the structure of plastid genomes and the type of genes they encode, across all eukaryotes. For sequence records we refer to the plastid genome section at GenBank, and two curated databases, GOBASE (O'Brien et al. 2009) and ChloroplastDB (Cui et al. 2006). Note that (1) the catalogue of complete ptDNAs in GenBank's genome section is currently incomplete (e.g., most records of the reduced apicomplexan and several green algal ptDNAs are missing and have to be retrieved from the nucleotide section), and gene and intron information is only validated as to consistency; (2) information in GOBASE is no longer being updated as of 2010, and (3) ChloroplastDB's last update (at the time of writing this review) was in 2007, lacks taxonomical grouping of species, and data on certain structural RNAs (RNase P, tmRNA and signal recognition particle RNAs).

A. Plastid Genome Structure

Generally, plastids contain a single type of chromosome in multiple copies. Restriction analysis and sequencing revealed that most ptDNAs are circular-mapping (not to be confused with truly circular DNA molecules), likely representing linear head-to-tail concatemers, plus subgenome-size fragments that tend to occur in genomes carrying repeat regions (Bendich 2004, 2007; Oldenburg and Bendich 2004). A similar genome structure is observed in mitochondria (Oldenburg and Bendich 2001; Ling and Shibata 2004). The mechanism of replication remains essentially an open question. Small organelle genomes might be replicated by a rolling circle mechanism, but the presence of a substantial fraction of subgenome-size fragments suggests a more complex mechanism, likely including recombination, or template switching in other instances. Experimental evidence for this or any other type of organization and replication is available for only a small fraction of known plastid genomes. A notable curiosity exists in dinoflagellates, where several genes are separately encoded on DNA minicircles (Zhang et al. 1999, 2002; Howe et al. 2008). However, whether these circles represent the principal genome organization or highly abundant subgenomic molecules (e.g., replicative rolling circle DNAs) remains to be demonstrated. For more information on dinoflagellate plastids, we refer to recent publications (e.g., Zhang et al. 1999, 2002; Stoebe and Maier 2002; Hackett et al. 2004; Laatsch et al. 2004; Howe et al. 2008; Keeling 2010).

A widespread feature of pt genomes is a large inverted repeat (IR) region that contains genes for rRNAs and a variable number of tRNAs and proteins (e.g., Gardner et al. 1993; Douglas and Penny 1999; Sanchez Puerta et al. 2005; Belanger et al. 2006; Cattolico et al. 2008; Tanaka et al. 2011). The biological role of the IR region is likely increased gene dosage for ribosomal components (ribosomes are among the most abundant sub-cellular structures). The IR region may be present or not in related species (e.g., *Pedinomonas minor*, *Parachlorella kessleri* and *Oocystis solitaria* have this trait, whereas *Chlorella vulgaris* does not; Turmel et al. 2009b). Similarly, the two ptDNAs of photosynthetic cryptomonads have large inverted repeat regions containing rDNA genes in contrast to the direct repeats in *Porphyra* species, and no repeat in *Cyanidium* (Glockner et al. 2000), *Cyanidioschyzon* (Ohta et al. 2003) and *Gracilaria* (Hagopian et al. 2004). The pt genome of the secondarily non-photosynthetic cryptomonad *C. paramecium* has single-copy rRNAs as most red algae, which is best explained as a secondary loss of the repeat. This comparison shows that, although repeat features are given high attention in many publications on complete plastid genomes, they are not well conserved across eukaryotes and are of undefined value for understanding the evolution of genome structure and function.

B. Plastid-Encoded Functions, Genes and Introns

Plastids perform numerous biological functions that rely to a large extent on nuclear genes, and that are translated in the cytoplasm and transported into plastids. For detailed information on protein import see (McFadden 1999; Wastl et al. 2000; Wastl and Maier 2000; van Dooren et al. 2001; Foth et al. 2003; Nassoury et al. 2003; Patron et al. 2005; Durnford and Gray 2006; Chaal and Green 2007; Patron and Waller 2007; Kessler and Schnell 2009; Ma et al. 2009; Felsner et al. 2010; Hempel et al. 2010; Kovacs-Bogdan et al. 2010; Li and Chiu 2010; Strittmatter et al. 2010).

Biological processes that involve at least some ptDNA-encoded genes are translation and photosynthesis. Only species that lost their photosynthetic capacity gradually eliminate the corresponding genes (Gockel and Hachtel 2000; de Koning and Keeling 2006). Additional biological processes that rely on pt-encoded genes involve transcription, protein transport and plastid division. Further, in a more restricted number of cases, ptDNAs code for components for tRNA processing (RNase P RNA), quality control of protein translation (tmRNAs; Gueneau de Novoa and Williams 2004), the signal recognition particle RNA (Rosenblad and Samuelsson 2004; Schunemann 2004), plus several other functions that are limited to the most gene-rich ptDNAs, in particular from red algae (Glockner et al. 2000; Ohta et al. 2003; Hagopian et al. 2004). Currently recognized pt genes and their functions are compiled in Table 3.1. This list is expected to extend as the functions of *ycf* genes and additional ORFs are being identified. All of the above processes are directly derived from the cyanobacterial ancestor of plastids (only a few genes/functions were acquired by lateral transfer). The pattern of genes and functions represented by ptDNA-encoded genes often does not correspond with phylogenetic affinities (i.e., gene presence/absence is an unreliable phylogenetic marker), as gene migration to the nucleus or complete gene loss has occurred numerous times independently across various eukaryotic lineages.

The most reduced ptDNAs of photosynthetically active species are those in dinoflagellates that are organized in minicircles (Howe et al. 2008), encoding a bit more than a dozen identified genes, followed by the apicomplexans *Alveolata sp.* (CCMP3115; 124 genes) and *Chromera velia* (112 genes; Janouskovec et al. 2010). At the other side of the spectrum, red algae have the most gene-rich, densely packed pt genomes (with up to ~254 genes).

Intron counts for ptDNAs are most variable: none in almost all red algae and in plastids of red algal origin (e.g., Douglas and Penny 1999; Glockner et al. 2000; Hagopian et al. 2004; Sanchez Puerta et al. 2005; Oudot-Le Secq et al. 2007), but 26 in the green alga *Floydiella terrestris*, and contrary to expectations, more than 100 in the red alga *Compsopogon caeruleus* (B.F.L. unpublished). Introns in pt genomes belong to group I and II, and are sometimes difficult to classify, because distinct secondary structure features are highly derived; i.e. such introns are only detected because coding regions are discontinuous. The most derived introns (group III, some organized in 'twintrons') are present in *Euglena gracilis* and *E. longa* ptDNAs (Copertino and Hallick 1993; Hallick et al. 1993), and are likely derived from group II introns (Copertino and Hallick 1993). The >100 introns in *Compsopogon sp.* (staghorn alga) ptDNA are also group

Table 3.1. Plastid-encoded genes and their products[a]

Gene	Product
accA	Acetyl-CoA carboxylase, carboxyl transferase subunit alpha
accB	Acetyl-CoA carboxylase, biotin carboxyl carrier protein
accD	Acetyl-CoA carboxylase, carboxyl transferase subunit beta
acpP	Acyl carrier protein
apcA	Allophycocyanin alpha subunit
apcB	Allophycocyanin beta subunit
apcD	Allophycocyanin gamma subunit
apcE	Phycobilisome linker polypeptide
apcF	Allophycocyanin beta 18 subunit
argB	Acetylglutamate kinase
atpA	ATP synthase CF1 subunit alpha
atpB	ATP synthase CF1 subunit beta
atpC	ATP synthase CF1 subunit gamma
atpD	ATP synthase CF1 subunit delta
atpE	ATP synthase CF1 subunit epsilon
atpF	ATP synthase CF0 subunit I
atpG	ATP synthase CF0 subunit II
atpH	ATP synthase CF0 subunit III
atpI	ATP synthase CF0 subunit IV
bas1	2-Cys peroxiredoxin
carA	Carbamoyl phosphate synthase small subunit
cbbX	Rubisco expression protein
ccsA	Heme attachment to plastid cytochrome c
cemA	Envelope membrane protein
chlB	Protochlorophyllide reductase subunit B
chlI	Mg chelatase subunit e
chlL	Protochlorophyllide reductase iron protein subunit
chlN	Protochlorophyllide reductase subunit N
clpC	Clp protease ATP binding subunit
clpP	Clp protease proteolytic subunit
cobA	Uroporphyrin-III C-methyltransferase
cpcA	Phycocyanin alpha subunit
cpcB	Phycocyanin beta subunit
cpcG	Phycobilisome rod-core linker polypeptide
cpeA	Phycoerythrin subunit a
cpeB	Phycoerythrin subunit b
crtE	Geranylgeranyl pyrophosphate synthase
cysA	Sulfate ABC transporter ATP-binding subunit
cysT	Sulfate ABC transporter permease subunit CysT
cysW	Sulfate ABC transporter permease subunit CysW
desA	Fatty-acid desaturase
dfr	Drug sensory protein A
dnaB	Replication helicase subunit
dnaK	Hsp70-type chaperone
dsbD	Thiol:disulfide interchange protein
fabH	Beta-ketoacyl-acyl carrier protein synthase III
ffs	Signal recognition particle RNA
ftrB	Ferredoxin thioreductase subunit b
ftsH	Cell division protein FTSH
ftsI	Peptidoglycan synthetase
ftsW	Cell division protein FTSW
glmS	glutamine-fructose-6-phosphate amidotransferase
glnB	Nitrogen regulatory protein PII
gltB	Glutamate synthase
groEL	60-kDa chaperonin
groES	GroES chaperonin
hemA	Glutamyl-tRNA reductase
hisH	Glutamine amidotransferase
hlp	DNA-binding protein Hu homolog
ilvB	Acetohydroxyacid synthetase large subunit
ilvH	Acetohydroxyacid synthetase small subunit
infA	Translation initiation factor 1
infB	Translation initiation factor 2
infC	Translation initiation factor 3
leuC	3-isopropylmalate dehydratase large subunit
leuD	3-isopropylmalate dehydratase small subunit
lipB	Lipoate-protein ligase B
lpxA	Acyl-UDP-N-acetylglucosamine O-acyltransferase
lpxC	UDP-3-0-acyl N-acetylglucosamine deacetylase
matK	Intron maturase
matR	Intron maturase
menA	1,4-dihydroxy-2-naphthoate octaprenyltransferase
menB	Naphthoate synthase
menC	O-succinylbenzoate synthase
menD	2-succinyl-6-hydroxy-2,4-cyclohexadiene-1-carboxylate synthase
menE	O-succinylbenzoic acid-CoA ligase
menF	Menaquinone-specific isochorismate synthase
minD	Organelle division inhibitor factor
minE	Septum site-determining protein
moeB	Molybdopterin biosynthesis protein
nadA	Quinolinate synthetase
nblA	Phycobilisome degradation protein
ndhA	NADH-plastoquinone oxidoreductase subunit 1
ndhB	NADH-plastoquinone oxidoreductase subunit 2
ndhC	NADH-plastoquinone oxidoreductase subunit 3

(continued)

Table 3.1. (continued)

Gene	Product
ndhD	NADH-plastoquinone oxidoreductase subunit 4
ndhE	NADH-plastoquinone oxidoreductase subunit 4L
ndhF	NADH-plastoquinone oxidoreductase subunit 5
ndhG	NADH-plastoquinone oxidoreductase subunit 6
ndhH	NADH-plastoquinone oxidoreductase subunit 7
ndhI	NADH-plastoquinone oxidoreductase subunit I
ndhJ	NADH-plastoquinone oxidoreductase subunit J
ndhK	NADH-plastoquinone oxidoreductase subunit K
ntcA	Global nitrogen transcriptional regulator
odpA	Pyruvate dehydrogenase E1 component, alpha subunit
odpB	Pyruvate dehydrogenase E1 component, beta subunit
ompR	Probable transcriptional regulator ompR
pbsA	Heme oxygenase
petA	Apocytochrome f
petB	Cytochrome b6
petD	Cytochrome b6/f complex subunit 4
petF	Ferredoxin
petG	Cytochrome b6/f complex subunit 5
petJ	Cytochrome c553
petL	Cytochrome b6-f complex subunit 6
petM	Cytochrome b6-f complex subunit 7
petN	Cytochrome b6-f complex subunit 8
pgmA	Phosphoglycerate mutase
preA	Prenyl transferase
psaA	Photosystem I P700 apoprotein A1
psaB	Photosystem I P700 apoprotein A2
psaC	Photosystem I subunit VII (iron-sulfur center)
psaD	Photosystem I reaction center subunit II (ferredoxin-binding)
psaE	Photosystem I reaction center subunit IV
psaF	Photosystem I reaction center subunit III (plastocyanin-binding)
psaI	Photosystem I reaction center subunit VIII
psaJ	Photosystem I reaction center subunit IX
psaK	Photosystem I reaction center subunit X
psaL	Photosystem I reaction center subunit XI
psaM	Photosystem I reaction center subunit M
psbA	Photosystem II reaction center protein D1
psbB	Photosystem II CP47 chlorophyll apoprotein
psbC	Photosystem II CP43 chlorophyll apoprotein
psbD	Photosystem II reaction center protein D2
psbE	Photosystem II cytochrome b559 alpha subunit
psbF	Photosystem II cytochrome b559 beta subunit
psbH	Photosystem II 10 kDa phosphoprotein
psbI	Photosystem II protein I
psbJ	Photosystem II protein J
psbK	Photosystem II protein K
psbL	Photosystem II protein L
psbM	Photosystem II protein M
psbN	Photosystem II protein N
psbT	Photosystem II protein T
psbV	Photosystem II cytochrome c550
psbW	Photosystem II protein W
psbX	Photosystem II protein X
psbY	Photosystem II protein Y
psbZ	Photosystem II protein Z
rbcL	Rubisco large subunit
rbcR	Transcription regulator of Rubisco operon
rbcS	Rubisco small subunit
rdrp	Viral RNA-dependent RNA polymerase
rne	Ribonuclease E
rnl	Large subunit ribosomal RNA
rnpB	Ribonuclease P RNA
rns	Small subunit ribosomal RNA
rpl1	Ribosomal protein L1
rpl11	Ribosomal protein L11
rpl12	Ribosomal protein L12
rpl13	Ribosomal protein L13
rpl14	Ribosomal protein L14
rpl16	Ribosomal protein L16
rpl18	Ribosomal protein L18
rpl19	Ribosomal protein L19
rpl2	Ribosomal protein L2
rpl20	Ribosomal protein L20
rpl21	Ribosomal protein L21
rpl22	Ribosomal protein L22
rpl23	Ribosomal protein L23
rpl24	Ribosomal protein L24
rpl27	Ribosomal protein L27
rpl29	Ribosomal protein L29
rpl3	Ribosomal protein L3
rpl31	Ribosomal protein L31
rpl32	Ribosomal protein L32
rpl33	Ribosomal protein L33
rpl34	Ribosomal protein L34
rpl35	Ribosomal protein L35
rpl36	Ribosomal protein L36
rpl4	Ribosomal protein L4
rpl5	Ribosomal protein L5
rpl6	Ribosomal protein L6

(continued)

Table 3.1. (continued)

Gene	Product
rpl7	Ribosomal protein L7
rpl9	Ribosomal protein L9
rpoA	RNA polymerase α-subunit
rpoB	RNA polymerase β-subunit
rpoC	RNA polymerase β'-subunit
rpoC1	RNA polymerase β'-subunit
rpoC2	RNA polymerase β''-subunit
rps1	Ribosomal protein S1
rps10	Ribosomal protein S10
rps11	Ribosomal protein S11
rps12	Ribosomal protein S12
rps13	Ribosomal protein S13
rps14	Ribosomal protein S14
rps15	Ribosomal protein S15
rps16	Ribosomal protein S16
rps17	Ribosomal protein S17
rps18	Ribosomal protein S18
rps19	Ribosomal protein S19
rps2	Ribosomal protein S2
rps20	Ribosomal protein S20
rps3	Ribosomal protein S3
rps4	Ribosomal protein S4
rps5	Ribosomal protein S5
rps6	Ribosomal protein S6
rps7	Ribosomal protein S7
rps8	Ribosomal protein S8
rps9	Ribosomal protein S9
rrn4.5	4.5S rRNA
rrn5	5S rRNA
secA	Preprotein-translocase subunit a
secY	SecY-type transporter protein
sprA	Small plastid RNA
ssrA	tmRNA
sufB	SufB protein
syfB	Phenylalanine-tRNA ligase beta subunit
syh	Histidine-tRNA ligase
tatC	Sec-independent protein translocase component TatC
thdF	Thiophen and furan oxidation protein
thiG	thiG protein
tilS	tRNA Ile-lysidine synthetase
trnA...Y	Transfer RNA Alanine...Tyrosine
trpA	Tryptophan synthase alpha subunit
trpG	Anthranilate synthase component II
trxA	Thioredoxin
tscA	*psaA* trans-splicing trans-acting factor
tsf	Translation elongation factor Ts
tufA	Translation elongation factor Tu
upp	Uracil phosphoribosyltransferase
ycf16	ABC transporter subunit
ycf3	Photosystem I assembly protein Ycf3
ycf4	Photosystem I assembly protein Ycf4
ycf59	Leucine zipper-containing protein
ycf65	Plastid-specific 30S ribosomal protein 3
ycf79	Photosystem II 13 kDa protein
orf#	Hypothetical protein of # aa
ycf#	Conserved hypothetical protein

[a]Taken from GOBASE at http://gobase.bcm.umontreal.ca/

II-related, some typical but others barely recognizable (B.F.L., unpublished). Finally, in some instances of group II intron-mediated trans-splicing, exons are located in distant genomic regions, transcribed separately and ligated to give rise to functional mRNAs (e.g., Goldschmidt-Clermont et al. 1991; Rochaix 1996; Rivier et al. 2001; Turmel et al. 2002; Belanger et al. 2006; Brouard et al. 2008, 2010; Jacobs et al. 2010). In conclusion, identification of introns can be difficult and some may be missed, even when applying most sophisticated search algorithms.

Currently, only few tools are available for automated intron recognition plus classification (Eddy 2008; Beck and Lang 2009; Gardner et al. 2009), and ptDNA annotation in general (Wyman et al. 2004; Jansen et al. 2005; Beck and Lang 2010). Tools developed by us (MFannot, RNAweasel; Beck and Lang 2009, 2010), although not fine-tuned for ptDNAs, appear to be most effective and miss only a few genes, small exons, and complex gene structures due to trans-splicing. Identification of structured RNAs is an area that needs improvements, including precise delineation of rRNA gene extremities and intron/exon boundaries. RNAse P RNA can be identified with RNAweasel or MFannot, but search models for tmRNAs and signal recognition particle RNAs remain to be added, together with an update of structural models that allow prediction of the whole range of plastid introns. Given the rapidly increasing number of genome sequences produced by new sequencing technologies, we will have to develop increasingly effective, semi-automated ways of genome annotation and GenBank submission to keep pace with data production.

III. Plastids Derived from Primary Endosymbiosis with Cyanobacteria

Plantae is a potentially monophyletic assemblage of photosynthetic (and some secondarily non-photosynthetic) lineages with primary plastids, i.e. derived directly from an endosymbiotic cyanobacterium. This large and diverse group is divided into the glaucophytes, rhodophytes (red algae) and Viridiplantae (green algae and land plants). To date, plastid genomes are available for only two glaucophytes and seven red algae (two of which unpublished; B.F.L.), but a large and rapidly growing number of green algae. The reason for this bias may be related to the difficulty of growing sufficient quantities of cell material for red and glaucophyte algae, a difficulty that no longer exists with the new sequencing technologies that require only small quantities of total DNA.

A. Rhodophyta

Rhodophyta is a morphologically diverse group with several thousand described species, both unicellular and multicellular ones. Red algal cells are characterized by the lack of centrioles and a flagellar apparatus, and the presence of phycoerythrin-containing plastids with unstacked thylakoids. Resolution of phylogenetic relationships among red algal lineages is currently limited by taxon and gene sampling (e.g., Le Gall and Saunders 2007; Verbruggen et al. 2010) and references therein), and may also be due to unequal rates of sequence evolution among red algae.

Complete plastid genomes are available from only five species, These include the multicellular taxa *Porphyra purpurea* and *Porphyra yeozensis* (Bangiales; Reith and Munholland 1995), the unicellular Cyanidales *Cyanidioschyzon merolae* (Ohta et al. 2003) and *Cyanidium caldarium* (Glockner et al. 2000), and the florideophycean *Gracilaria tenuistipitata* (Hagopian et al. 2004). Two additional ptDNAs are currently being sequenced in our laboratory (*Stylonema alsidii*, UTEX LB1424 and *Compsopogon caeruleus*, UTEX LB1553).

The first sequenced red algal ptDNA (*P. purpurea*; Reith and Munholland 1995) turned out to be more cyanobacterial-like than any other alga, based on features such as gene count, a large tRNA set, genes encoding transcriptional regulators and bacteria-like operons. This conclusion also applies to other red algal pt genomes. Whereas land plant and green algal ptDNAs encode 88–138 genes (Lemieux et al. 2007; Turmel et al. 2007), this number is close to double in red algae (230–254). Many of these genes are unique to red algae or rare in other ptDNAs, and include RNase P RNA (present in all red algal ptDNAs including *Cyandium*; otherwise only present in a few green plastids including *Nephroselmis*, *Pycnococcus*, *Monomastix*, *Ostreococcus* and in cyanelles of the two glaucophytes; (Shevelev et al. 1995; Turmel et al. 2009a); our own analysis), tmRNA (http://www.indiana.edu/~tmrna/; Andersen et al. 2006) and signal recognition particle RNA (Andersen et al. 2006). In contrast, genes for components of the NADPH dehydrogenase complex (in ptDNAs of some prasinophyte and most land plant lineages) are absent.

B. Glaucophyta

Glaucophytes (glaucocystophytes) are freshwater algae that are particularly important for understanding the origin and evolution of photosynthesis in eukaryotes. Plastids of these organisms are unique in having retained two cyanobacterial features: a true, bacterial-type peptidoglycan cell wall (Pfanzagl et al. 1996), and carboxysomes – polyhedral micro-compartments involved in CO_2 fixation (Kaplan and Reinhold 1999). The presence of these unique features strongly suggests that glaucocystophyte plastids originated directly from a symbiosis with a Cyanobacterium, and there has been a perception that this algal group might therefore have emerged early in the evolution of photosynthetic eukaryotes. However, more recent phylogenetic analyses with broad species sampling and a large number of genes do not support this idea, placing the origin of the glaucophyte plastid close to the divergence point of green and red plastids (see for instance Fig. 3.1).

The only complete plastid genome sequence from glaucophytes is that of *Cyanophora paradoxa* (Löffelhardt and Bohnert 1994). Recently, we have sequenced most of the *Glaucocystis nostochinearum* ptDNA (Lang et al. unpublished). Despite their evolutionary distance (see the deep divergence in Fig. 3.1), the two genomes are similar in terms of genome organization and gene content (a potential inverted repeat region remains to be confirmed for *Glaucocystis* ptDNA). The number of genes in glaucophyte ptDNAs (a total of 191 in *Cyanophora*, including protein, tRNA and rRNA genes; Cui et al. 2006) is relatively low compared to that of red algae (between 230 and 254). This might seem unexpected when considering that glaucophyte plastids still have a bacterial cell wall and other 'primitive' cyanobacterial features. In addition, in phylogenetic analyses with plastid data, glaucophyte branches are amongst the shortest ones, whereas the red algal plastids are among the more rapidly evolving ones. Evidently, gene counts do not correlate with evolutionary rates in this example.

C. Viridiplantae

Viridiplantae (green plants) is a morphologically and ecologically diverse group including the Streptophyta (land plants and their closest green algal relatives, the charophytes) and Chlorophyta (i.e., the rest of the green algae; Lewis and McCourt 2004; Sluiman 1985). Based on flagellar apparatus ultrastructure and features related to cytokinesis, Chlorophyta is further divided into four classes: Prasinophyceae (a paraphyletic group of unicellular species thought to be descendants of the ancestral flagellates from which the main green algal lineages evolved), Trebouxiophyceae, Chlorophyceae and Ulvophyceae (Lewis and McCourt 2004; Mattox and Stewart 1984). Although molecular data support the early divergence of prasinophytes (e.g., Guillou et al. 2004), the branching order of Trebouxiophyceae, Ulvophyceae and Chlorophyceae within Chlorophyta remains uncertain (see Pombert et al. 2004, 2006 for discussion and references), which is also consistent with our analysis (Fig. 3.1).

To date, 28 green algal plastid genomes (22 from Chlorophyta and 6 from Streptophyta) have been fully sequenced (Table 3.2), and they revealed an unexpected diversity both within and between algal groups. Overall, green algal ptDNAs differ in many respects from the well characterized plastid genomes of land plants (see Chaps. 4, 5). The latter typically share the same quadripartite structure (characterized by the presence of two copies of a large inverted repeat sequence separating a small single-copy and a large single-copy region) and have the same gene partitioning pattern between the two single copies. Their genes are densely packed and most of them are organized in conserved clusters. In contrast, green algal ptDNAs are "hotbeds" for chloroplast genome evolution (Belanger et al. 2006), exhibiting great diversity in genome and gene organization, including loss or inversion of the inverted repeat, gene rearrangements, intergenic expansions, invasion by repeat elements and introns, gene loss, gene expansion and gene fragmentation.

1. Prasinophytes

Prasinophytes are primarily marine unicellular algae that show great variation in terms of cell size and shape, flagella number, membrane covering (i.e., with our without scales) and biochemical features (Graham and Wilcox 2000). Seven prasinophyte clades are currently recognized; however, the exact

Table 3.2. Features of green algal plastid genomes for which complete genome sequences are available

	Genome size (Kbp)	AT content%	Gene number[a]	Intergenic space (%)	Intron numbers (group I/ group II)	IR size (Kbp)
Prasinophytes						
Nephroselmis olivacea	200.8	57.9	128	31.3	0/0	46.1
Pycnococcus provasolli	80.2	60.5	98	11.6	0/1	–
Ostreococcus tauri	71.6	60.1	88	15.1	0/1	6.8
Pyramimonas parkae	101.6	65.3	110	19.6	0/1	13.0
Monomastix sp.	114.5	61.0	94	43.9	5/1	–
Pedinomonas minor	98.3	65.2	105	25.6	0/0	10.3
Trebouxiophyceae						
Chlorella vulgaris	150.6	68.4	112	45.4	3/0	–
Oocystis solitaria	96.3	71.0	111	22.2	1/1	nd
Leptosira terrestris	195.1	72.7	106	51.5	4/0	–
Parachlorella kessleri	123.9	70.0	112	35.2	1/0	10.9
Coccomyxa sp.	175.7	50	115	61	1/0	nd
Helicosporidium sp.	37.5	73.1	54	5.1	1	–
Chlorophyceae						
Chlamydomonas reinhardtii	203.8	65.5	94	49.2	5/2	22.2
Volvox carteri	~525	~57	>91	>77	>3/>5	>16
Dunaliella salina	269	67.9	97	52	35/8	14.4
Scenedesmus obliquus	161.4	73.1	96	32.8	7/2	12.0
Stigeoclonium helveticum	223.9	71.1	97	46.7	16/5	–
Floydiella terrestris	521.1	65.5	97	77.8	19/7	–
Oedogonium cardiacum	196.5	70.5	99	22.6	17/4	35.5
Ulvophyceae						
Oltmannsiellopsis viridis	151.9	59.5	104	40.8	5/0	18.5
Pseudendoclonium akinetum	195.8	62.3	105	37.7	27/0	6.0
Bryopsis hypnoides	153.4	66.9	111[b]	59.8	5/6[c]	–
Charophyceae						
Mesostigma viride	118.3	69.9	137	26.8	0	6.0
Chlorokybus atmophyticus	152.2	63.8	138	41.2	1/0	7.6
Staurastrum punctulatum	157.1	67.5	121	42	1/7	–
Zygnema circumcarinatum	165.3	68.9	125	42	1/12	–
Chara vulgaris	184.9	73.8	127	38.8	2/16	10.9
Chaetosphaeridium globosum	131.1	70.4	125	23	1/17	12.4

[a]Only known genes and conserved ORFs are included
[b]Our preliminary analyses suggest a higher number of genes
[c]Our preliminary analyses suggest a different intron composition

relationships between these lineages and their affiliation with other green algal groups remain unresolved (Marin and Melkonian 2010).

The six currently available plastid genome sequences belong to: (1) *Nephroselmis olivacea* (Pseudoscourfieldiales) – a flagellate unicellular alga; (2) *Pycnococcus provasolli* (Pseudoscourfieldiales, Pycnococcaceae) – a coccoid picoplanktonic alga; (3) *Ostreococcus tauri* (Mamiellales) – the smallest known eukaryotic organism; (4) *Pyramimonas* (Pyramimonadales) – a scaly quadriflagellate alga; (5) *Monomastix* – a scaly flagellate of unknown affiliation; and (6) *Pedinomonas minor* (Pedinomonadales) – a small naked uniflagellate with no clear affiliation to the other prasinophyte clades probably related to, or ancestral to, Trebouxiophyceae (Turmel et al. 2009b; see also Fig. 3.1). Overall, prasinophyte ptDNAs show extreme diversity in size (an almost 3-fold variation), gene repertoire and genome organization. On the other hand, these genomes are similar in base composition and harbor no or just a few introns (Table 3.2).

Interestingly, both ancestral and derived types of genome organization (relative to the presumed plastid genome in the most recent common ancestor of green plants; Turmel et al. 1999) have been reported among the plastid genomes described in this group. Ancestral types are characterized by large gene complements, ancestral gene clusters and a quadripartite genome structure (i.e., two identical copies of a large inverted repeat (IR), separated by single-copy (SC) regions), whereas derived types have reduced and rearranged genomes. With 128 conserved genes, the 200.8 Kbp plastid genome of *Nephroselmis* has the largest gene complement yet reported for a chlorophyte alga and has retained many ancestral gene clusters (Turmel et al. 1999). Its quadripartite architecture resembles that of streptophyte counterparts in displaying (1) unequal SC regions – a large and a small one – that contain highly conserved sets of genes and (2) IR-encoded rRNA operons transcribed towards the small SC region. The ptDNA of *Nephroselmis* codes for several genes with limited phylogenetic distribution; for instance, *ftsI* (involved in peptidoglycan synthesis) has not been reported in other ptDNAs, and *ndh* genes (coding for subunits of the NADH:ubiquinone oxidoreductase) are absent from chlorophyte ptDNAs, but are present in other prasinophytes and land plants. At the other extreme is the plastid genome of *Ostreococcus*, with 88 genes highly scrambled over 71.6 Kbp, representing the smallest genome with the most reduced gene complement among photosynthetic green plants (Robbens et al. 2007). Both the small size and overall low proportion of intergenic spacers (representing 15% of the genome and varying from 1 to 476 nt length) as well as the presence of three cases of overlapping genes make this genome one of the most compact green plant ptDNAs (Table 3.2). Moreover, in contrast to *Nephroselmis*, its SC regions – although different in size – have the same number of genes, and the rRNA operons are transcribed away from the SC regions.

Reductions in plastid genome size and gene complement as well as the loss of the inverted repeat took place independently in several other prasinophyte lineages, leading to a variety of distinct genome configurations. For instance, the *Pycnococcus* plastid genome resembles the *Ostreococcus* counterpart in being small and highly compact (with two cases of overlapping genes and only ~11% intergenic regions). However, it lacks the IR, and its gene complement is more similar to that of chlorophycean plastid genomes (Turmel et al. 2009a). On the other hand, the plastid genome of *Monomastix* has a larger size but a slightly lower number of genes (Table 3.2; Turmel et al. 2009a). The ptDNA of *Pyramimonas* displays intermediate genome size, compactness and gene repertoire (including six *ndh* genes present only in *Nephroselmis* and land plants, and two other genes – *rpl22* and *ycf65* – not reported in other chlorophytes; Turmel et al. 2009a). Lastly, the ptDNA of *Pedinomonas*, although very small, compact, and with a low gene count (Table 3.2), has retained the highest degree of ancestral gene linkages among all chlorophyte algae (i.e., linkages that predate the divergence of chlorophytes and streptophytes; Turmel et al. 2009b).

2. *Trebouxiophyceae*

Trebouxiophyceae (sensu Friedl 1995) are a group of morphologically heterogeneous algae (unicellular non-flagellated or filamentous) that inhabit mostly soil and freshwaters. Most phycobionts of lichens, ciliates and animals are also included in this class (Booton et al. 1998; Graham and Wilcox 2000; Lewis and McCourt 2004). To date, five plastid genomes from four photosynthetic species (*Chlorella vulgaris* and

Oocystis solitaria – Chlorellales; *Parachlorella kessleri* and *Leptosira terrestris* – Ctenocladales) and one non-photosynthetic relative (*Helicosporidium sp.* – Chlorellales) have been published (Wakasugi et al. 1997; de Cambiaire et al. 2007; Turmel et al. 2009a). In addition, nearly complete ptDNAs are available from *Coccomyxa* sp C-169 (Coccomyxaceae; GenBank accession number HQ693844), *Chlorella ellipsoidea* and the colorless *Prototheca wickerhamii* (Knauf and Hachtel 2002; Yamada 1991). All trebouxiophyte ptDNAs sequenced so far are rather AT-rich, with *Helicosporidium* and *Leptosira* being among the most AT-rich green algal genomes (Table 3.2).

Although the plastid genomes from the four fully characterized photosynthetic species have similar gene contents, they vary significantly in size (a twofold variation). Most of this variation is accounted for by size differences in intergenic regions (Table 3.2). Gene order also varies considerably. For instance, the *Chlorella* plastid genome has retained many of the gene clusters present in streptophytes and prasinophytes. On the other hand, *Leptosira* shares little similarity in gene order with other plastid genomes and exhibits derived traits reminiscent of evolutionary patterns described for the ulvophyte and chlorophycean lineages (Turmel et al. 2009b).

The IR is missing in both *Chlorella* and *Leptosira* pt genomes, which is a feature also shared with the non-photosynthetic *Helicosporidium* (Table 3.2). Nevertheless, it is believed that the last common ancestor of trebouxiophytes possessed a plastid genome with a quadripartite structure (very similar to that of *Nephroselmis* and streptophytes) and that the IR was lost independently on at least two occasions. These suggestions are based on the finding of IRs in other trebouxiophyte plastid genomes (including that of *Chlorella ellipsoidea*, which has a large IR with a split rRNA operon; Yamada and Shimaji 1987) and on the presence of an IR remnant in *Chlorella vulgaris* (de Cambiaire et al. 2007).

The ptDNAs of the non-photosynthetic trebouxiophytes *Helicosporidium* and *Prototheca* are both highly reduced in size (partially sequenced; ~37.5 and 45 Kbp, respectively). Based on its structure and compactness, the *Helicosporidium* genome is more similar to that described in the non-photosynthetic plastids of apicomplexan parasites. As expected, it lacks all genes for photosynthesis (de Koning and Keeling 2006), but its size reduction is due to both gene loss and reduced non-coding regions, overlapping genes, and the loss of the IR. Notable is the loss of the rRNA operon structure – an event that is thought to have taken place independently in several other lineages (including the trebouxiophyte *C. ellipsoidea,* several ulvophytes and charopytes as well as other non-photosynthetic algae; de Koning and Keeling 2006).

3. Chlorophyceae

The Chlorophyceae (sensu Mattox and Stewart 1984) comprise mostly freshwaters species, but several marine species are also known. Species in this group show diverse morphologies – from unicellular (flagellated or coccoid) to complex multicellular (colonial or filamentous) forms – and distinct configurations of their flagellar apparatus. Based on the arrangement of the flagellar basal bodies in their motile cells, two sister clades are generally described in this group. They are commonly referred to as CW ("clockwise"; Chlamydomonadales) and DO ("directly opposed"; Sphaeropleales) groups (Booton et al. 1998). Three additional lineages (Oedogoniales, Chaetopeltidales and Chaetophorales) are basal to these clades, but their divergence order is not well understood (Brouard et al. 2010; Buchheim et al. 2001; Shoup and Lewis 2003; Turmel et al. 2008). To date, seven plastid genomes from representatives of the five main chlorophycean lineages have been completely sequenced: (1) Chlamydomonadales – *Chlamydomonas reinhardtii* (Maul et al. 2002), *Volvox carteri* (Smith and Lee 2009, 2010), and *Dunaliella salina* (Smith et al.

2010); (2) Sphaeropleales – *Scenedesmus obliquus* (de Cambiaire et al. 2006); (3) Chaetophorales – *Stigeoclonium helveticum* (Belanger et al. 2006); (4) Oedogoniales – *Oedogonium cardiacum* (Brouard et al. 2008); and (5) Chaetopeltidales – *Floydiella terrestris* (Brouard et al. 2010).

Overall, plastid genomes in this group show tremendous variation in terms of genome size, intergenic spacers and intron numbers (Table 3.2). At the same time, the number of genes encoded in these genomes has been kept remarkably constant, within the range of derived prasinophyte pt genomes (*Pycnococcus* and *Monomastix*; Table 3.2). In terms of general genome organization, both types – with or without inverted repeats – are found among chlorophycean ptDNAs.

In cases where ptDNAs maintained the quadripartite structure, the organization of the IR and SC regions as well as the gene distribution within these regions differ among lineages. For instance, in *Chlamydomonas*, the two SC regions have similar sizes and differ radically in both gene content and gene organization from their counterparts in ancestral prasinophyte plastid genomes (Maul et al. 2002). Interestingly, although the *Scenedesmus* ptDNA shares with its *Chlamydomonas* counterpart a similar quadripartite structure, the sets of genes in the SC regions are very different between the two species, which indicates that genes were shuffled since the divergence of the DO and CW lineages (de Cambiaire et al. 2006). On the other hand, the *Oedogonium* plastid genome revealed an atypical structure with an IR significantly larger than in most of its green algal counterparts (with the notable exception of *Nephroselmis*) and two SC regions of vastly unequal size. Furthermore, the partitioning of genes among the two SC regions is distinctly different from that in *Chlamydomonas* and *Scenedesmus* (de Cambiaire et al. 2006).

Consistent with the situation among trebouxiophytes, the IR-lacking ptDNAs of *Stigeoclonium* and *Floydiella* also have loosely packed genes and intergenic regions rich in short repeats (Brouard et al. 2010). The most re-arranged chlorophycean plastid genome appears to be that of *Stigeoclonium*, which completely lacks the ancestral gene partitioning pattern displayed by *Nephroselmis* and streptophytes, and overall, exhibits the fewest ancestral features among all plastid genomes completely sequenced to date (Belanger et al. 2006).

Chlorophycean ptDNAs differ substantially in the amount of short repeated sequences. At one extreme, there are *Oedogonium* and *Scenedesmus*, in which such sequences occupy only 1.3% and 3% of genomes, respectively. At the other extreme, there are the ptDNAs of *Chlamydomonas, Stigeoclonium, Volvox,* and *Floydiella,* which are extremely rich in repeated sequences. For instance, short palindromic repeats (potentially acquired *via* mitochondria-to-plastid transfers involving mobile introns) constitute ~64% of the *Volvox* plastid genome. Repeats larger than 30 bp account for half of the *Floydiella* pt genome (almost three times more than in *Chlamydomonas* and *Stigeoclonium*; Brouard et al. 2010; Smith and Lee 2009).

Several atypical features have also been described in this group, including: (1) strong bias in the distribution of genes between the two DNA strands (in *Stigeoclonium* and *Scenedesmus*), (2) breakup of protein-coding genes by putatively trans-spliced group II introns (*rbcL, psaC, petD, psaA*) (in *Stigeoclonium* and *Floydiella*); (3) fragmentation of protein-coding genes into distinct open reading frames (contiguous or distant from each other) that are not associated with any introns (*rpoC1, rps2, rpoB*); (4) the substantial expansion (over fivefold increase) of many protein-coding genes (e.g., *cemA, clpP, ftsH, rpoB, rpoC1, rpoC2, rps3, rps4, and ycf1*) due to the presence of insertions whose post-transcriptional fate (i.e., excised or not) or biological significance are mostly unknown; (5) intergenic intron-like sequences of unknown origin and function in *Dunaliella*; and (6) genes (*int* and *dpoB*, coding for a tyrosine recombinase and a DNA-dependent DNA polymerase, respectively) potentially acquired via horizontal gene transfer from a mitochondrial genome donor in *Oedogonium* (Belanger et al. 2006; Brouard et al. 2008, 2010; Smith et al. 2010).

Overall, the plastid genome in this group of algae has experienced major changes, and it displays the lowest degree of ancestral traits relative to other chlorophytes. Some of the most eccentric ptDNAs among all Viridiplantae are also found in this group: over 520 Kbp and over 77% intergenic spacers in *Floydiella* and *Volvox*; 73% AT-content in *Scenedesmus*; and 43 introns in *Dunaliella* (Table 3.2).

4. Ulvophyceae

Ulvophyceae are unicellular (including macroscopic forms composed of a single, large multinucleate cell) and multicellular species that are common in rocky intertidal coasts of temperate regions, but secondarily freshwater species are also known. The flagellar basal bodies in their motile cells are arranged in a counterclockwise (CCW) orientation (Floyd and Okelly 1984). To date, complete plastid genome sequences are available from three unicellular ulvophyte species: *Oltmannsiellopsis viridis* (Oltmannsiellopsidales; Pombert et al. 2006), *Pseudendoclonium akinetum* (Ulotrichales; Pombert et al. 2005) and *Bryopsis hypnoides* (Bryopsidales; Lu et al. 2010). The first two species belong to lineages believed to occupy a basal position within the group, whereas the phylogenetic position of the latter is uncertain (Lu et al. 2010 and Fig. 3.1). Partial sequence information is also available from *Codium fragile* (Ulvales; Manhart et al. 1989) and *Caulerpa sertularoides* (Bryopsidales; Lehman and Manhart 1997).

Although different in size, the *Oltmannsiellopsis* and *Pseudendoclonium* plastid genomes share a similar number of genes and coding density (Table 3.2). The difference in genome size is mostly accounted for by a difference in intron numbers (Table 3.2). The 27 introns in *Pseudendoclonium* make up for 14.8% of the genome and are thought to have arisen from the intragenomic proliferation of a few founding introns in this lineage (Pombert et al. 2005). Both genomes share a quadripartite structure that deviates from the ancestral type. Nevertheless, the IR sequences in the two genomes differ in size (with that of *Oltmannsiellopsis* being ~12 Kbp larger) and gene content (the *Pseudendoclonium* IR encodes only the rRNA operon, while the *Oltmannsiellopsis* IR contains five additional genes). Also, *Pseudendoclonium* shows evidence of inter-organellar lateral transfer (involving some dispersed repeats and one intron) between its plastid and mitochondrial genomes (Pombert et al. 2005).

The plastid genome of *Bryopsis* differs significantly from those of *Oltmannsiellopsis* and *Pseudendoclonium* in several important ways. These include the absence of IRs (also lacking in the two other ulvophytes for which partial information is available; *Caulerpa* and *Codium*) and the presence of multimeric forms of ptDNA (including monomer, dimer, trimer, tetramer, and even higher-order multimers), which is a trait that has only been reported in land plants (Lu et al. 2010). Furthermore, this genome is unique in possessing 10 tRNA genes that have not been found in other completely sequenced chlorophyte ptDNAs. Note that while five of them are known in embryophytes the other five have only been reported in some bacterial genomes. Also, its rRNA locus consists of five (rrn23, rrn16, rrn7, rrn5, and rrn3) instead of the usual four coding regions; a similar situation is only found in *C. reinhardtii* ptDNA (Maul et al. 2002). The number of genes reported for this ptDNA is similar to that of the other two ulvophytes (Table 3.2). However, our preliminary analyses indicate a larger gene complement for this genome; likewise, the number of introns in this genome might prove to be different than listed in Table 3.2. Overall, although ulvophyte ptDNAs feature an atypical quadripartite structure, they maintained a relatively large gene complement and the degree of remodeling is intermediate relative to those seen in their trebouxiophyte and chlorophycean counterparts.

5. Charophyceae

Charophytes comprise thousands of mainly freshwater algal species exhibiting great variability in morphology and reproduction. They are subdivided into six monophyletic lineages: (1) Mesostigmatales represented by the scaly biflagellate *Mesostigma viride* (previously regarded as a member of the Prasinophyceae), (2) Chlorokybales also represented by a single species (the sarcinoid *Chlorokybus atmophyticus*), (3) Klebsormidiales, (4) Zygnematales, (5) Coleochaetales and (6) Charales. Phylogenetic analyses indicate Mesostigmatales and Chlorokybales as the earliest-diverging charophycean lineages (forming a distinct clade; Turmel et al. 2007). The branching order among the other groups remains debatable. Charales are the closest relatives of plants in some studies, while other analyses favor that Charales diverged prior to Coleochaetales and Zygnematales (see Turmel et al. 2006 for discussion and references; see also Fig. 3.1).

Complete plastid genome sequences are available from six species belonging to five of the six main charophycean lineages: *Mesostigma viride* (Mesostigmatales), *Chlorokybus atmophyticus* (Chlorokybales), *Staurastrum punctulatum* and *Zygnema circumcarinatum* (Zygnematales), *Chaetosphaeridium globosum* (Coleochaetales) and *Chara vulgaris* (Charales; Lemieux et al. 2007; Turmel et al. 2002, 2005, 2006). In addition, the almost complete ptDNA of *Klebsormidium flaccidum* has been sequenced (Fig. 3.1; BFL unpublished). Overall, charophycean ptDNAs vary in size, gene content, intron content, gene order and include the most gene-rich green plastid genomes (Table 3.2).

Consistent with their basal position among charophytes, the plastid genomes of *Mesostigma* and *Chlorokybus* are gene-rich and feature a typical quadripartite structure (Turmel et al. 2007). The two genomes are similar in gene content and gene order, with the notable presence in each of the two genomes of genes that have not been identified in other green algal and land plant pt genomes. Genes are loosely packed in *Chlorokybus* (the average size of intergenic spacers in *Chlorokybus* is twice that of *Mesostigma)*, which also reflects in the larger genome size (Table 3.2; Turmel et al. 2007). Nevertheless, relative to the gene order in *Nephroselmis* and Streptophyta ptDNAs, the *Chlorokybus* plastid genome is more rearranged than its *Mesostigma* counterpart. Both genomes are intron-poor, with none in *Mesostigma* and a single intron in *Chlorokybus* (Table 3.2).

Relative to *Mesostigma* and *Chlorokybus*, the plastid genomes of the two zygnematalean lineages, *Staurastrum* and *Zygnema*, have a slightly reduced gene repertoire (Table 3.2) and lack the rRNA-encoding IR typical of other charophytes and streptophytes. Notably, the lack of IR is also shared with *Spirogyra maxima* – another zygnematalean species for which partial genome information is available (Manhart et al. 1990). Furthermore, both these genomes are loosely packed with genes (due to the expansion of their intergenic spacers), and feature a larger number of introns (which have also expanded in size). However, the two genomes differ extensively from one another in gene order. Also, many intergenic regions in the *Staurastrum* ptDNA harbour tandem repeats while such sequences are virtually absent in the *Zygnema* counterpart (Turmel et al. 2005).

On the other hand, the pt genomes of *Chaetosphaeridium globosum* (Coleochaetales) and *Chara vulgaris* (Charales) exhibit the typical quadripartite structure found in streptophytes, and resemble their land plant counterparts more closely than do other charophycean relatives. Although the two genomes have similar coding capacities (Table 3.2), *Chara* features four genes (*rpl12, trnL(gag), rpl19,* and *ycf20*) that are entirely missing from other charophycean and land plant ptDNAs. Furthermore, despite similarities in genome organization, gene content and intron composition, the two genomes differ in size, gene density and AT content, with the *Chara* genome representing the largest and most AT-rich streptophyte ptDNA (Table 3.2). Notably, *Chara*'s increased genome size and AT-content is mainly accounted for by increased AT-rich intergenic spacers and introns, which represent 38.8% and 13.4% of the total genome, respectively (Turmel et al. 2006). Overall, among streptophyte green algae, the ptDNAs of the charophytes *Mesostigma* and *Chlorokybus* exhibit the most ancestral features (including the largest gene complement among Viridiplantae; 137–138 genes), while the genomes of *Chara* and *Chaetosphaeridium* resemble most their land plant counterparts.

IV. Plastids Acquired via Eukaryote-Eukaryote Endosymbiosis

According to the chromalveolate hypothesis, chlorophyll c-containing plastids originated from a single photosynthetic ancestor, which acquired its plastids only once by secondary endosymbiosis with a red alga (Cavalier-Smith 2002; Keeling 2009, 2010). However, phylogenetic studies suggest a much higher incidence of plastid transfer among eukaryotes, favoring complex evolutionary scenarios involving multiple eukaryote-eukaryote endosymbioses (Sanchez-Puerta et al. 2007; Archibald 2009). The arguably most rigorous analysis in this sense is by Baurain and co-workers (Baurain et al. 2010), who find that monophyly of Cryptophytes, Alveolates, Stramenopiles, and Haptophytes (CASH) is seen neither with mitochondrial nor nuclear sequence data. This means that the very strongly supported phylogenetic relationships in trees constructed with plastid proteins (plastid-encoded as in Fig. 3.1; as well as nucleus-encoded genes of cyanobacterial origin) do not represent the evolution of CASH species but more likely multiple plastid transfers. In some instances, higher-order eukaryote-eukaryote endosymbioses are in fact evident, for instance, the grouping of plastids from the dinoflagellates (*Durinskia* and *Kryptoperidinium*; Imanian et al. 2010) with diatoms (Fig. 3.1), and the (weak) association of *Alveolata* sp. (Apicomplexa) plastids with stramenopiles.

A shared characteristic of 'second hand' plastid genomes is their reduced coding capacity relative to that of the plastid donor, which is in most instances a red and only in rare cases a green alga (i.e., in the rhizarian *Bigelowiella* and relatives, and the euglenozoan *Euglena*). Plastids of red origin are in general remarkably similar in gene content, despite their turbulent evolutionary past. In the following we will focus on the few main differences, and refer the reader otherwise to the corresponding original publications. It should be noted that gene counts and identifications differ slightly across different papers and database compilations (Cui et al. 2006; O'Brien et al. 2009). Although minor (up to about ten), these differences need to be resolved in the future, by establishing gene identification based on the same criteria. Eventually, all ptDNAs should be reannotated by using the same tools, a task that was unfortunately out of reach for this review.

A. Stramenopila

Stramenopiles is the largest group among CASH protists whose monophyly is well supported (e.g., (Baurain et al. 2010). A sizable portion of stramenopile taxa are non-photosynthetic and without plastid relicts, such as oomycetes (*Phytophthora*) and bicosoecids (*Cafeteria*). Whether or not the stramenopile ancestor had plastids, and of which origin, has been the subject of heated debates. The controversy is in part due to over-interpretation of BLAST analyses and lack of resolution in single-gene phylogenies (Stiller et al. 2009) and references therein). The few clear examples pointing to a plastid origin of genes in plastid-less stramenopiles may in fact be explained by transfer of individual genes, rather than endosymbiotic events.

PtDNA sequences are available from bacillariophytes (diatoms), phaeophytes (brown algae), raphidophytes, pelagophytes, xanthophytes, but curiously not from chrysophytes (golden algae).

1. Diatoms

Bacillariophyta are most diverse (>250 genera), unicellular, silica-walled algae that live either attached to surfaces or are planktonic. Complete ptDNAs have been sequenced from four phylogenetically relatively distant species: *Phaeodactylum tricornutum*, *Thalassiosira pseudonana* (Oudot-Le Secq et al. 2007), *Odontella sinensis* (Kowallik et al. 1995) and *Fistulifera* sp. (Tanaka et al. 2011).

These ptDNAs are relatively uniform, coding for a similar set of 160–170 genes. A putative serine recombinase gene (*serC2*) is potentially of plasmid origin. It also occurs in the diatom plastids residing in certain dinoflagellates (Imanian et al. 2010).

2. Phaeophytes

Brown algae are a large group of multicellular organisms (~250 genera) that occur mostly in marine habitats and grow attached to surfaces. Complete ptDNAs are published from two representatives of distinct orders, *Ectocarpus siliculosus* and *Fucus vesiculosus* (Le Corguille et al. 2009). Their gene counts are similar to those of diatoms, with only minor differences.

3. Raphidophytes

Raphidophytes is a small group (four genera) of flagellated unicellular organisms that occur in both marine and fresh water habitats, and that lack a rigid cell wall. A complete ptDNA sequence is available for two strains of *Heterosigma akashiwo* (Cattolico et al. 2008). The number of ptDNA-encoded genes (197) is relatively high compared to other algae with plastids from secondary or higher-order endosymbioses, and a putative serine recombinase gene is present as in diatoms. Another unusual ORF codes for a potential G-protein-coupled receptor. Again, the functionality and biological role of these extra genes remain to be demonstrated. Several protein-coding genes and their mRNAs contain large, in-frame inserts, when compared to orthologs in other plastids. These inserts likely represent derived forms of protein introns (inteins; Liu 2000; Gogarten and Hilario 2006) that may have lost their capacity for splicing. In fact, one typical *bona fide* intein has been identified in the *dnaB* gene of *H. akashiwo* ptDNA (Cattolico et al. 2008).

4. Pelagophytes

This group of algae known for causing algal blooms was originally included in the Chrysophyceae, but based on biochemical, physiological and phylogenetic criteria it now forms its own class Pelagophyceae. Complete ptDNAs are available from *Aureococcus anophagefferens* and *Aureoumbra lagunensis* (Ong et al. 2010). The large inverted repeat, otherwise common in other second-hand red plastids is missing, and the two genomes code for only 137 and 141 genes, respectively. About 20 genes that are usually present in stramenopile ptDNAs are absent from both pelagophytes. According to our phylogenetic analysis with plastid data, pelagophytes branch deeply within stramenopiles, but their placement relative to the raphidophytes and xanthophytes is unresolved (Fig. 3.1).

5. Xanthophytes

The *Vaucheria litorea* plastid genome has been characterized during the course of a most unusual investigation of the green sea slug *Elysia chlorotica*. This animal acquires plastids ("kleptoplasts", see Chap. 2) by ingesting *Vaucheria litorea* as food, and sequestrating the organelles into the digestive epithelium, where photosynthesis occurs for several months (Rumpho et al. 2008). As it turns out, the plastid genome sequence is typical for stramenopiles (167 genes), and contains the common inverted repeat. According to the authors, some nuclear gene products that have to be imported and are required for plastid function are likely encoded in the animal's nuclear genome (the algal nucleus is digested during the organelle sequestration process). So far, horizontal gene transfer from the algal genome to the mollusk genome has been demonstrated only for a few nuclear genes. Evidently, nuclear genome sequences of the sea slug and of *Vaucheria* are required to substantiate this unusual case of horizontal gene transfer (see Chap. 2).

B. Alveolata

Alveolates comprise ciliates, apicomplexans and dinoflagellates, but only the two latter ones contain photosynthetic plastids.

1. Dinoflagellata

In most dinoflagellates, the ptDNA consists of multiple minicircles that code for a total of about a dozen genes. Here we will only discuss the pt genomes of *Kryptoperidinium foliaceum* and *Peridinium quinquecorne* that possess a conventional genome organization, since their ptDNAs derive from a higher-order endosymbiosis with diatoms (Imanian et al. 2010; see also Fig. 3.1). These dinoflagellate ptDNAs possess IR regions similar to those in diatoms, and *K. foliaceum* has as a putative serine recombinase gene that is characteristic for diatom and raphidophyte ptDNAs. According to the authors' interpretation (Imanian et al. 2010), the larger size of the *K. foliaceum* ptDNAs may be due to the insertion of numerous plasmid-derived genes that are dispensable for plastid function.

2. Apicomplexa

As already mentioned in the introduction, ptDNAs have been sequenced from two photosynthetic relatives of Apicomplexa, *Chromera velia* and *Alveolata sp.* (CCMP3115; Janouskovec et al. 2010). The *Chromera* plastid DNA is very rapidly evolving, and therefore difficult to place in phylogenetic analyses. Its genome is larger than that of *Alveolata* sp., and translates UGA stop codons as tryptophan as is otherwise common for (in most cases also rapidly evolving) mtDNAs.

The gene count of both ptDNAs is modest (124 and 112 genes, respectively) compared to other second-hand red algal ptDNAs. A gene for a horizontally transferred phosphonopyruvate decarboxylase is inserted into the rRNA operon of *Alveolata*. According to our phylogenetic analysis (Fig. 3.1), plastids of the two species could have a common origin by vertical descent, yet the positioning of the *Chromera* ptDNA alone is unresolved, somewhere close to stramenopiles. According to our phylogenetic results with *Alveolata*, its plastids may stem from a tertiary endosymbiosis with a photosynthetic stramenopile rather than from a unique secondary acquisition, as proposed by the chromalveolate hypothesis. In fact, the authors of the original genome paper state that '*comparing gene content among alveolate plastids reveals the nearly mutually-exclusive gene sets of apicomplexans and dinoflagellates*', which can be interpreted as further evidence against their common origin.

C. Cercozoa (Rhizaria)

Chlorarachniophytes are a small group of photosynthetic marine flagellates with two recognized genera *Chlorarachnion* and *Bigelowiella*. Similar to cryptophytes (for details on cryptomonads see below) they carry a second reduced nucleus (nucleomorph), but of green algal origin (not precisely identified according to our analyses presented in Fig. 3.1 and those published by others; Rogers et al. 2007). A complete ptDNA sequence is available for *Bigelowiella natans*. The genome has a small size (69.2 Kbp), a highly compact gene organization, and a nearly full complement of photosynthesis-related genes that is similar to those in some of the less gene-rich green algae such as *Chlamydomonas* (Rogers et al. 2007). Most of the reduction in gene content comes from the loss of *ycf* and tRNA genes.

D. Cryptomonada

Cryptomonads are unicellular flagellates that are mostly photosynthetic, containing chlorophyll c and phycobilins as photosynthetic pigments. They carry direct physical evidence for eukaryote-eukaryote endosymbiosis in form of a second, remnant eukaryotic nucleus, the ‘nucleomorph’ (for a recent review see Moore and Archibald 2009) of evidently red algal origin. Non-photosynthetic cryptomonad species include *Cryptomonas paramecium* that contains plastids with a secondarily reduced plastid genome (Donaher et al. 2009), and heterotrophic *Goniomonas* species that have no plastids. Whether *Goniomonas* is indeed primarily without plastids (e.g., Keeling et al. 1999) and may thus represent the ancestral group that engulfed an alga with red plastids, remains to be demonstrated with nuclear genome sequence data.

The three completely sequenced cryptomonad pt DNAs are from *Guillardia theta* (Douglas and Penny 1999), *Rhodomonas (Pyrenomonas) salina* (Khan et al. 2007) and the non-photosynthetic *C. parasiticum* (Donaher et al. 2009). The gene count of cryptomonad ptDNAs is >180, more than in green algae but about a quarter less than in red algae. The non-photosynthetic *C. parasiticum* has about 70 genes less in its plastid genome\, including only a few remaining members of the *pet*, *psa* and *psb* photosynthetic gene families (Donaher et al. 2009). An interesting acquisition in *R. salina* ptDNA is a gene for the tau/gamma subunit of DNA polymerase III (*dnaX*) that was likely acquired by lateral gene transfer from a firmicute bacterium (Khan et al. 2007). Whether or not this gene is transcribed, translated, and functional in plastids, remains to be shown.

E. Haptophyta

Haptophytes (prymnesiophytes) are unicellular photosynthetic flagellates (some are colonial), and unlike in cryptophytes, heterotrophic taxa are unknown in this clade. Currently, pt genomes of only two species are available, those of *Emiliana huxleyi* (Sanchez Puerta et al. 2005) and *Pavlova lutheri* (Burger et al. unpublished). Their genomes have about the same size and gene content (105 Kbp and 155 genes in *E. huxleyi*), and carry few notable features. Phylogenetic analyses based on pt data sometimes (but not always) unite haptophytes and cryptophytes (Bachvaroff et al. 2005; Keeling 2009; Le Corguille et al. 2009; Fig. 3.1).

F. Euglenids

Euglenids are unicellular flagellates, some of which contain plastids (chlorophyll a and b, β-carotene and xanthophylls), which were acquired via secondary endosymbiosis with a green alga. Euglenid ptDNA sequences are available from two species, *Euglena gracilis* (Copertino and Hallick 1993; Hallick et al. 1993) and the non-photosynthetic *Euglena (Astasia) longa* (Knauf and Hachtel 2002). In both instances, plastid genes are loaded with a large number of unusual introns (see above). At only 73 Kbp, the *A. longa* ptDNA has about half the size of its photosynthetic relatives, with all photosynthesis-related protein genes missing except for *rbcL*. According to published phylogenetic analyses based on pt sequences (Turmel et al. 2009a), *Euglena* plastids derive from a relative of the green alga *Pyramimonas*, which is clearly corroborated by our phylogenetic analysis (Fig. 3.1).

V. Conclusions

The availability of information on plastid genomes has increased over the last few years at an almost disquieting pace, in particular in green algae (as well as in land plants that are not covered in this chapter). Unfortunately, from the standpoint of evolutionary biology, the traditional bias in attention to green algae and plants remains. In particular, we have sequence data from just a handful of red algal pt genomes, a skimpy two from glaucophytes, and similarly low coverage for the numerous groups of algae with second-hand plastids. In fact, we are surprised that sequencing of almost identical flowering plant ptDNAs appears to be more important than sequencing those for which we know so little.

During the course of writing this review, we have come across several issues that touch on data production and analysis. For most pt genome projects underway, sequencing is performed with new technologies, some of which are fraught with systematic error (e.g., pyrosequencing technology suffers from frameshifts in homopolymer stretches among other, less well understood sequencing artifacts). This shortcoming may lead to mistaking genes for pseudogenes with great confidence (based on high coverage of systematic error). In a few cases, we have seen omission of gene annotation that may be due to such frameshifts. Further, as new genome data are pouring in at an unprecedented rate, detailed genome annotation by the end user (typically manual intervention) becomes increasingly challenging. The best solution to both issues, detecting erroneous gene features and potential sequencing error, and keeping up with high standards of genome annotation, is the development of automated genome annotation pipelines. We are aware of only one published tool for organelle genome annotation (DOGMA; (Wyman et al. 2004), and the currently unpublished but freely available tools developed by ourselves (MFannot, RNAweasel; Lang et al. 2007; Beck and Lang 2009, 2010). These are still far from perfect, justifying a continued time investment that should ideally be integrated with ongoing large scale sequencing projects. In this context we noticed that plastid gene identification is relatively straightforward, based on a wide consensus on gene names and functions (which cannot be said for mitochondrial genes). Yet, it seems that renaming *ycf* genes with now known functions would be timely, so would be a systematic identification and renaming of conserved ORFs as *ycf*, as long as they are present in distant species. Identification of weakly conserved genes is best achieved by HMM searches (http://hmmer.janelia.org; Eddy 1996, 1998) that are as fast and by far more sensitive and reliable than BLAST.

Acknowledgments

We are grateful to Dr. G. Burger (Université de Montréal, Montreal, Canada) for insightful discussion and comments on the manuscript. This work was supported by the Canadian Research Chair Program and the Natural Sciences and Engineering Research Council (NSERC; 194560–2011) of Canada.

References

Andersen ES, Rosenblad MA, Larsen N, Westergaard JC, Burks J, Wower IK, Wower J, Gorodkin J, Samuelsson T, Zwieb C (2006) The tmRDB and SRPDB resources. Nucleic Acids Res 34:D163–D168

Archibald JM (2009) The puzzle of plastid evolution. Curr Biol 19:R81–R88

Archibald JM, Keeling PJ (2002) Recycled plastids: a 'green movement' in eukaryotic evolution. Trends Genet 18:577–584

Bachvaroff TR, Sanchez Puerta MV, Delwiche CF (2005) Chlorophyll c-containing plastid relationships based on analyses of a multigene data set with all four chromalveolate lineages. Mol Biol Evol 22:1772–1782

Baurain D, Brinkmann H, Petersen J, Rodriguez-Ezpeleta N, Stechmann A, Demoulin V, Roger AJ, Burger G, Lang BF, Philippe H (2010) Phylogenomic evidence for separate acquisition of plastids in cryptophytes, haptophytes, and stramenopiles. Mol Biol Evol 27:1698–1709

Beck N, Lang BF (2009) RNAweasel, a webserver for identification of mitochondrial, structured RNAs. http://megasun.bch.umontreal.ca/RNAweasel

Beck N, Lang BF (2010) MFannot, organelle genome annotation websever. http://megasun.bch.umontreal.ca/papers/MFannot

Belanger AS, Brouard JS, Charlebois P, Otis C, Lemieux C, Turmel M (2006) Distinctive architecture of the chloroplast genome in the chlorophycean green alga *Stigeoclonium helveticum*. Mol Genet Genomics 276:464–477

Bendich AJ (2004) Circular chloroplast chromosomes: the grand illusion. Plant Cell 16:1661–1666

Bendich AJ (2007) The size and form of chromosomes are constant in the nucleus, but highly variable in bacteria, mitochondria and chloroplasts. Bioessays 29:474–483

Bhattacharya D, Yoon HS, Hackett JD (2004) Photosynthetic eukaryotes unite: endosymbiosis connects the dots. Bioessays 26:50–60

Booton AS, Floyd GL, Fuerst PA (1998) Polyphyly of tetrasporalean green algae inferred from nuclear small subunit rDNA. J Phycol 34:306–311

Brouard JS, Otis C, Lemieux C, Turmel M (2008) Chloroplast DNA sequence of the green alga *Oedogonium cardiacum* (Chlorophyceae): unique genome architecture, derived characters shared with the Chaetophorales and novel genes acquired through horizontal transfer. BMC Genomics 9:290

Brouard JS, Otis C, Lemieux C, Turmel M (2010) The exceptionally large chloroplast genome of the green alga *Floydiella terrestris* illuminates the evolutionary history of the Chlorophyceae. Genome Biol Evol 2:240–256

Buchheim MA, Michalopulos EA, Buchheim JA (2001) Phylogeny of the Chlorophyceae with special references to the Sphaeropleales. J Phycol 37:819–835

Burki F, Shalchian-Tabrizi K, Minge M, Skjaeveland A, Nikolaev SI, Jakobsen KS, Pawlowski J (2007) Phylogenomics reshuffles the eukaryotic supergroups. PLoS One 2:e790

Burki F, Inagaki Y, Brate J, Archibald JM, Keeling PJ, Cavalier-Smith T, Sakaguchi M, Hashimoto T, Horak A, Kumar S, Klaveness D, Jakobsen KS, Pawlowski J, Shalchian-Tabrizi K (2009) Large-scale phylogenomic analyses reveal that two enigmatic protist lineages, telonemia and centroheliozoa, are related to photosynthetic chromalveolates. Genome Biol Evol 1:231–238

Cattolico RA, Jacobs MA, Zhou Y, Chang J, Duplessis M, Lybrand T, McKay J, Ong HC, Sims E, Rocap G (2008) Chloroplast genome sequencing analysis of *Heterosigma akashiwo* CCMP452 (West Atlantic) and NIES293 (West Pacific) strains. BMC Genomics 9:211

Cavalier-Smith T (1981) Eukaryote kingdoms: seven or nine? Biosystems 14:461–481

Cavalier-Smith T (2002) The phagotrophic origin of eukaryotes and phylogenetic classification of Protozoa. Int J Syst Evol Microbiol 52:297–354

Chaal BK, Green BR (2007) Protein targeting in "secondary" or "complex" chloroplasts. Methods Mol Biol 390:207–217

Chan CX, Yang EC, Banerjee T, Yoon HS, Martone PT, Estevez JM, Bhattacharya D (2011) Red and green algal monophyly and extensive gene sharing found in a rich repertoire of red algal genes. Curr Biol 21:328–333

Copertino DW, Hallick RB (1993) Group II and group III introns of twintrons: potential relationships with nuclear pre-mRNA introns. Trends Biochem Sci 18:467–471

Cui L, Veeraraghavan N, Richter A, Wall K, Jansen RK, Leebens-Mack J, Makalowska I, dePamphilis CW (2006) ChloroplastDB: the chloroplast genome database. Nucleic Acids Res 34:D692–D696

de Cambiaire JC, Otis C, Lemieux C, Turmel M (2006) The complete chloroplast genome sequence of the chlorophycean green alga *Scenedesmus obliquus* reveals a compact gene organization and a biased distribution of genes on the two DNA strands. BMC Evol Biol 6:37

de Cambiaire JC, Otis C, Turmel M, Lemieux C (2007) The chloroplast genome sequence of the green alga *Leptosira terrestris*: multiple losses of the inverted repeat and extensive genome rearrangements within the Trebouxiophyceae. BMC Genomics 8:213

de Koning A, Keeling P (2006) The complete plastid genome sequence of the parasitic green alga *Helicosporidium* sp. is highly reduced and structured. BMC Biol 4:12

Deschamps P, Moreira D (2009) Signal conflicts in the phylogeny of the primary photosynthetic eukaryotes. Mol Biol Evol 26:2745–2753

Donaher N, Tanifuji G, Onodera NT, Malfatti SA, Chain PS, Hara Y, Archibald JM (2009) The complete plastid genome sequence of the secondarily nonphotosynthetic alga *Cryptomonas paramecium*: reduction, compaction, and accelerated evolutionary rate. Genome Biol Evol 1:439–448

Douglas SE (1998) Plastid evolution: origins, diversity, trends. Curr Opin Genet Dev 8:655–661

Douglas SE, Gray MW (1991) Plastid origins. Nature 352:290

Douglas SE, Penny SL (1999) The plastid genome of the cryptophyte alga, *Guillardia theta*: complete sequence and conserved synteny groups confirm its common ancestry with red algae. J Mol Evol 48:236–244

Durnford DG, Gray MW (2006) Analysis of Euglena gracilis plastid-targeted proteins reveals different classes of transit sequences. Eukaryot Cell 5:2079–2091

Eddy SR (1996) Hidden Markov models. Curr Opin Struct Biol 6:361–365

Eddy SR (1998) Profile hidden Markov models. Bioinformatics 14:755–763

Eddy S (2008) Infernal website. http://infernal.janelia.org

Edgar RC (2004) MUSCLE: a multiple sequence alignment method with reduced time and space complexity. BMC Bioinformatics 5:113

Fast NM, Kissinger JC, Roos DS, Keeling PJ (2001) Nuclear-encoded, plastid-targeted genes suggest a single common origin for apicomplexan and dinoflagellate plastids. Mol Biol Evol 18:418–426

Felsner G, Sommer MS, Maier UG (2010) The physical and functional borders of transit peptide-like sequences in secondary endosymbionts. BMC Plant Biol 10:223

Floyd GL, Okelly CJ (1984) Motile cell ultrastructure and the circumscription of the orders Ulotrichales and Ulvales (Ulvophyceae, Chlorophyta). Am J Bot 71:111–120

Foth BJ, McFadden GI (2003) The apicoplast: a plastid in *Plasmodium falciparum* and other Apicomplexan parasites. Int Rev Cytol 224:57–110

Foth BJ, Ralph SA, Tonkin CJ, Struck NS, Fraunholz M, Roos DS, Cowman AF, McFadden GI (2003) Dissecting apicoplast targeting in the malaria parasite *Plasmodium falciparum*. Science 299:705–708

Friedl T (1995) Inferring taxonomic positions and testing genus level assignments in coccoid green lichen algae – a phylogenetic analysis of 18S ribosomal RNA sequences from *Dictyochloropsis reticulata* and from members of the genus *Myrmecia* (Chlorophyta, Trebouxiohyceae Cl-Nov). J Phycol 31:632–639

Funes S, Davidson E, Reyes-Prieto A, Magallón S, Herion P, King MP, González-Halphen D (2002) A green algal apicoplast ancestor. Science 298:2155

Funes S, Reyes-Prieto A, Pérez-Martínez X, González-Halphen D (2004) On the evolutionary origins of apicoplasts: revisiting the rhodophyte vs. chlorophyte controversy. Microbes Infect 6:305–311

Gardner MJ, Feagin JE, Moore DJ, Rangachari K, Williamson DH, Wilson RJ (1993) Sequence and organization of large subunit rRNA genes from the extrachromosomal 35 kb circular DNA of the malaria parasite *Plasmodium falciparum*. Nucleic Acids Res 21:1067–1071

Gardner PP, Daub J, Tate JG, Nawrocki EP, Kolbe DL, Lindgreen S, Wilkinson AC, Finn RD, Griffiths-Jones S, Eddy SR, Bateman A (2009) Rfam: updates to the RNA families database. Nucleic Acids Res 37:D136–D140

Glockner G, Rosenthal A, Valentin K (2000) The structure and gene repertoire of an ancient red algal plastid genome. J Mol Evol 51:382–390

Gockel G, Hachtel W (2000) Complete gene map of the plastid genome of the nonphotosynthetic euglenoid flagellate *Astasia longa*. Protist 151:347–351

Gogarten JP, Hilario E (2006) Inteins, introns, and homing endonucleases: recent revelations about the life cycle of parasitic genetic elements. BMC Evol Biol 6:94

Goldschmidt-Clermont M, Choquet Y, Girard-Bascou J, Michel F, Schirmer-Rahire M, Rochaix JD (1991) A small chloroplast RNA may be required for trans-splicing in *Chlamydomonas reinhardtii*. Cell 65:135–143

Gould SB, Waller RF, McFadden GI (2008) Plastid evolution. Annu Rev Plant Biol 59:491–517

Graham LE, Wilcox LW (2000) Algae. Prentice-Hall, Upper Saddle River

Gray MW (2010) Rethinking plastid evolution. EMBO Rep 11:562–563

Gueneau de Novoa P, Williams KP (2004) The tmRNA website: reductive evolution of tmRNA in plastids and other endosymbionts. Nucleic Acids Res 32:D104–D108

Guillou L et al (2004) Diversity of picoplanktonic prasinophytes assessed by direct nuclear SSU rDNA sequencing of environmental samples and novel isolates retrieved from oceanic and coastal marine ecosystems. Protist 155:193–214

Hackett JD, Yoon HS, Soares MB, Bonaldo MF, Casavant TL, Scheetz TE, Nosenko T, Bhattacharya D (2004) Migration of the plastid genome to the nucleus in a peridinin dinoflagellate. Curr Biol 14:213–218

Hackett JD, Yoon HS, Li S, Reyes-Prieto A, Rummele SE, Bhattacharya D (2007) Phylogenomic analysis supports the monophyly of cryptophytes and haptophytes and the association of rhizaria with chromalveolates. Mol Biol Evol 24:1702–1713

Hagopian JC, Reis M, Kitajima JP, Bhattacharya D, de Oliveira MC (2004) Comparative analysis of the complete plastid genome sequence of the red alga *Gracilaria tenuistipitata* var. liui provides insights into the evolution of rhodoplasts and their relationship to other plastids. J Mol Evol 59:464–477

Hallick RB, Hong L, Drager RG, Favreau MR, Monfort A, Orsat B, Spielmann A, Stutz E (1993) Complete sequence of *Euglena gracilis* chloroplast DNA. Nucleic Acids Res 21:3537–3544

Hempel F, Felsner G, Maier UG (2010) New mechanistic insights into pre-protein transport across the second outermost plastid membrane of diatoms. Mol Microbiol 76:793–801

Howe CJ, Nisbet RE, Barbrook AC (2008) The remarkable chloroplast genome of dinoflagellates. J Exp Bot 59:1035–1045

Imanian B, Pombert JF, Keeling PJ (2010) The complete plastid genomes of the two 'dinotoms' *Durinskia baltica* and *Kryptoperidinium foliaceum*. PLoS One 5:e10711

Jacobs J, Glanz S, Bunse-Grassmann A, Kruse O, Kuck U (2010) RNA trans-splicing: identification of components of a putative chloroplast spliceosome. Eur J Cell Biol 89:932–939

Janouskovec J, Horak A, Obornik M, Lukes J, Keeling PJ (2010) A common red algal origin of the apicomplexan, dinoflagellate, and heterokont plastids. Proc Natl Acad Sci USA 107:10949–10954

Jansen RK, Raubeson LA, Boore JL, de Pamphilis CW, Chumley TW, Haberle RC, Wyman SK, Alverson AJ, Peery R, Herman SJ, Fourcade HM, Kuehl JV, McNeal JR, Leebens-Mack J, Cui L (2005) Methods for obtaining and analyzing whole chloroplast genome sequences. Methods Enzymol 395:348–384

Kaplan A, Reinhold L (1999) CO2 concentrating mechanisms in photosynthetic microorganisms. Annu Rev Plant Physiol Plant Mol Biol 50:539–570

Keeling PJ (2009) Chromalveolates and the evolution of plastids by secondary endosymbiosis. J Eukaryot Microbiol 56:1–8

Keeling PJ (2010) The endosymbiotic origin, diversification and fate of plastids. Philos Trans R Soc Lond B Biol Sci 365:729–748

Keeling PJ, Deane JA, Hink-Schauer C, Douglas SE, Maier UG, McFadden GI (1999) The secondary endosymbiont of the cryptomonad *Guillardia theta* contains alpha-, beta-, and gamma-tubulin genes. Mol Biol Evol 16:1308–1313

Kessler F, Schnell D (2009) Chloroplast biogenesis: diversity and regulation of the protein import apparatus. Curr Opin Cell Biol 21:494–500

Khan H, Parks N, Kozera C, Curtis BA, Parsons BJ, Bowman S, Archibald JM (2007) Plastid genome sequence of the cryptophyte alga *Rhodomonas salina* CCMP1319: lateral transfer of putative DNA replication machinery and a test of chromist plastid phylogeny. Mol Biol Evol 24:1832–1842

Knauf U, Hachtel W (2002) The genes encoding subunits of ATP synthase are conserved in the reduced plastid genome of the heterotrophic alga *Prototheca wickerhamii*. Mol Genet Genomics 267:492–497

Kohler S, Delwiche CF, Denny PW, Tilney LG, Webster P, Wilson RJ, Palmer JD, Roos DS (1997) A plastid of probable green algal origin in Apicomplexan parasites. Science 275:1485–1489

Kovacs-Bogdan E, Soll J, Bolter B (2010) Protein import into chloroplasts: the Tic complex and its regulation. Biochim Biophys Acta 1803:740–747

Kowallik KV, Stoeb B, Schaffran I, Kroth-Pancic P, Freier U (1995) The chloroplast genome of a chlorophyll a+c-containing alga, *Odontella sinenesis*. Plant Mol Biol Rep 13:336–342

Laatsch T, Zauner S, Stoebe-Maier B, Kowallik KV, Maier UG (2004) Plastid-derived single gene minicircles of the dinoflagellate *Ceratium horridum* are localized in the nucleus. Mol Biol Evol 21:1318–1322

Lang BF, Burger G, O'Kelly CJ, Cedergren R, Golding GB, Lemieux C, Sankoff D, Turmel M, Gray MW (1997) An ancestral mitochondrial DNA resembling a eubacterial genome in miniature. Nature 387:493–497

Lang BF, Laforest MJ, Burger G (2007) Mitochondrial introns: a critical view. Trends Genet 23:119–125

Lartillot N, Philippe H (2004) A Bayesian mixture model for across-site heterogeneities in the amino-acid replacement process. Mol Biol Evol 21:1095–1109

Lartillot N, Philippe H (2008) Improvement of molecular phylogenetic inference and the phylogeny of Bilateria. Philos Trans R Soc Lond B Biol Sci 363:1463–1472

Lartillot N, Brinkmann H, Philippe H (2007) Suppression of long-branch attraction artefacts in the animal phylogeny using a site-heterogeneous model. BMC Evol Biol 7(Suppl 1):S4

Le Corguille G, Pearson G, Valente M, Viegas C, Gschloessl B, Corre E, Bailly X, Peters AF, Jubin C, Vacherie B, Cock JM, Leblanc C (2009) Plastid genomes of two brown algae, *Ectocarpus siliculosus* and *Fucus vesiculosus*: further insights on the evolution of red-algal derived plastids. BMC Evol Biol 9:253

Le Gall L, Saunders GW (2007) A nuclear phylogeny of the Florideophyceae (Rhodophyta) inferred from combined EF2, small subunit and large subunit ribosomal DNA: establishing the new red algal subclass Corallinophycidae. Mol Phylogenet Evol 43:1118–1130

Lehman RL, Manhart JR (1997) A preliminary comparison of restriction fragment patterns in the genus Caulerpa (Chlorophyta) and the unique structure of the chloroplast genome of *Caulerpa sertularioides*. J Phycol 33:1055–1062

Lemieux C, Otis C, Turmel M (2007) A clade uniting the green algae *Mesostigma viride* and *Chlorokybus atmophyticus* represents the deepest branch of the Streptophyta in chloroplast genome-based phylogenies. BMC Biol 5:2

Lewis LA, McCourt RM (2004) Green algae and the origin of land plants. Am J Bot 91:1535–1556

Li HM, Chiu CC (2010) Protein transport into chloroplasts. Annu Rev Plant Biol 61:157–180

Ling F, Shibata T (2004) Mhr1p-dependent concatemeric mitochondrial DNA formation for generating yeast mitochondrial homoplasmic cells. Mol Biol Cell 15:310–322

Liu XQ (2000) Protein-splicing intein: genetic mobility, origin, and evolution. Annu Rev Genet 34:61–76

Löffelhardt W, Bohnert HJ (1994) Structure and function of the cyanelle genome. Int Rev Cytol 151:29–65

Lu F, Xu W, Tian C, Wang G, Niu J, Pan G, Hu S (2010) The *Bryopsis hypnoides* plastid genome: multimeric forms and complete nucleotide sequence. PLoS One 6:e14663

Ma Y, Jakowitsch J, Deusch O, Henze K, Martin W, Löffelhardt W (2009) Transketolase from *Cyanophora paradoxa*: in vitro import into cyanelles and pea chloroplasts and a complex history of a gene often, but not always, transferred in the context of secondary endosymbiosis. J Eukaryot Microbiol 56:568–576

Manhart JR, Hoshaw RW, Palmer JD (1990) Unique chloroplast genome in *Spirogyra maxima* (Chlorophyta) revealed by physical and gene mapping. J Phycol 26:490–494

Manhart JR, Kelly K, Dudock BS, Palmer JD (1989) Unusual characteristics of *Codium fragile* chloroplast DNA revealed by physical and gene mapping. Mol Gen Genet 216:417–421

Marin B, Melkonina M (2010) Molecular phylogeny and classification of the Mamiellophyceae class. nov (Chlorophyta) based on sequence comparisons of the nuclear- and plastid-encoded rRNA operons. Protist 161:304–336

Mattox KR, Stewart KD (1984) Classification of the green algae: a concept based on comparative ecology. In: Irvine DEG, John DM (eds) The systematics of the green algae. Academic Press, London, pp 29–72

Maul JE et al. (2002) The *Chlamydomonas reinhardtti* plastid chromosome: islands of genes in a sea of repeats. Plant Cell 14:2659–2679

McFadden GI (1999) Plastids and protein targeting. J Eukaryot Microbiol 46:339–346

McFadden GI (2010) The apicoplast. Protoplasma. 248:641–650

McFadden GI, Waller RF (1997) Plastids in parasites of humans. Bioessays 19:1033–1040

Minge MA, Shalchian-Tabrizi K, Torresen OK, Takishita K, Probert I, Inagaki Y, Klaveness D, Jakobsen KS (2010) A phylogenetic mosaic plastid proteome and unusual plastid-targeting signals in the green-colored dinoflagellate *Lepidodinium chlorophorum*. BMC Evol Biol 10:191

Moore CE, Archibald JM (2009) Nucleomorph genomes. Annu Rev Genet 43:251–264

Moreira D, Philippe H (2001) Sure facts and open questions about the origin and evolution of photosynthetic plastids. Res Microbiol 152:771–780

Nassoury N, Cappadocia M, Morse D (2003) Plastid ultrastructure defines the protein import pathway in dinoflagellates. J Cell Sci 116:2867–2874

Obornik M, Van de Peer Y, Hypsa V, Frickey T, Slapeta JR, Meyer A, Lukes J (2002) Phylogenetic analyses suggest lateral gene transfer from the mitochondrion to the apicoplast. Gene 285:109–118

O'Brien EA, Zhang Y, Wang E, Marie V, Badejoko W, Lang BF, Burger G (2009) GOBASE: an organelle genome database. Nucleic Acids Res 37: D946–D950

Ohta N, Matsuzaki M, Misumi O, Miyagishima SY, Nozaki H, Tanaka K, Shin IT, Kohara Y, Kuroiwa T (2003) Complete sequence and analysis of the plastid genome of the unicellular red alga *Cyanidioschyzon merolae*. DNA Res 10:67–77

Oldenburg DJ, Bendich AJ (2001) Mitochondrial DNA from the liverwort *Marchantia polymorpha*: circularly permuted linear molecules, head-to-tail concatemers, and a 5' protein. J Mol Biol 310:549–562

Oldenburg DJ, Bendich AJ (2004) Most chloroplast DNA of maize seedlings in linear molecules with defined ends and branched forms. J Mol Biol 335:953–970

Ong HC, Wilhelm SW, Gobler CJ, Bullerjahn G, Jacobs MA, McKay J, Sims EH, Gillett WG, Zhou Y, Haugen E, Rocap G, Cattolico RA (2010) Analysis of the complete chloroplast genome sequences of two members of the Pelagophyceae: *Aureococcus anophagefferens* and *Aureoumbra lagunensis*. J Phycol 46:602–615

Oudot-Le Secq MP, Grimwood J, Shapiro H, Armbrust EV, Bowler C, Green BR (2007) Chloroplast genomes of the diatoms *Phaeodactylum tricornutum* and *Thalassiosira pseudonana*: comparison with other plastid genomes of the red lineage. Mol Genet Genomics 277:427–439

Parfrey LW, Grant J, Tekle YI, Lasek-Nesselquist E, Morrison HG, Sogin ML, Patterson DJ, Katz LA (2010) Broadly sampled multigene analyses yield a well-resolved eukaryotic tree of life. Syst Biol 59:518–533

Patron NJ, Waller RF (2007) Transit peptide diversity and divergence: a global analysis of plastid targeting signals. Bioessays 29:1048–1058

Patron NJ, Waller RF, Archibald JM, Keeling PJ (2005) Complex protein targeting to dinoflagellate plastids. J Mol Biol 348:1015–1024

Patterson DJ (1989) Stramenopiles: chromophyte from a protistan perspective. In: Green JC, Leadbeater ESC, Diver WL (eds) The chromophyte algae: problems and perspectives. Clarendon, Oxford, pp 357–379

Pfanzagl B, Zenker A, Pittenauer E, Allmaier G, Martinez-Torrecuadrada J, Schmid ER, De Pedro MA, Löffelhardt W (1996) Primary structure of cyanelle peptidoglycan of *Cyanophora paradoxa*: a prokaryotic cell wall as part of an organelle envelope. J Bacteriol 178:332–339

Philippe H, Delsuc F, Brinkmann H, Lartillot N (2005) Phylogenomics. Annu Rev Ecol Evol Syst 36: 541–562

Pombert JF, Keeling PJ (2010) The mitochondrial genome of the entomoparasitic green alga *Helicosporidium*. PLoS One 5:e8954

Pombert JF, Lemieux C, Turmel M (2006) The complete chloroplast DNA sequence of the green alga *Oltmannsiellopsis viridis* reveals a distinctive quadripartite architecture in the chloroplast genome of early diverging ulvophytes. BMC Biol 4:3

Pombert JF, Otis C, Lemieux C, Turmel M (2004) The complete mitochondrial DNA sequence of the green alga *Pseudendoclonium akinetum* (Ulvophyceae) highlights distinctive evolutionary trends in the chlorophyta and suggests a sister-group relationship between the Ulvophyceae and Chlorophyceae. Mol Biol Evol 21:922–935

Pombert JF, Otis C, Lemieux C, Turmel M (2005) The chloroplast genome sequence of the green alga *Pseudendoclonium akinetum* (Ulvophyceae) reveals unusual structural features and new insights into the branching order of chlorophyte lineages. Mol Biol Evol 22:1903–1918

Reith ME, Munholland J (1995) Complete nucleotide sequence of the *Porphyra pupurea* chloroplast. Plant Mol Biol Rep 13:333–335

Reyes-Prieto A, Bhattacharya D (2007) Phylogeny of nuclear-encoded plastid-targeted proteins supports an early divergence of glaucophytes within Plantae. Mol Biol Evol 24:2358–2361

Reyes-Prieto A, Weber AP, Bhattacharya D (2007) The origin and establishment of the plastid in algae and plants. Annu Rev Genet 41:147–168

Rivier C, Goldschmidt-Clermont M, Rochaix JD (2001) Identification of an RNA-protein complex involved in chloroplast group II intron trans-splicing in *Chlamydomonas reinhardtii*. EMBO J 20:1765–1773

Robbens S, Derelle E, Ferraz C, Wuyts J, Moreau H, Van de Peer Y (2007) The complete chloroplast and mitochondrial DNA sequence of *Ostreococcus tauri*: organelle genomes of the smallest eukaryote are examples of compaction. Mol Biol Evol 24:956–968

Rochaix JD (1996) Post-transcriptional regulation of chloroplast gene expression in *Chlamydomonas reinhardtii*. Plant Mol Biol 32:327–341

Rodriguez-Ezpeleta N, Brinkmann H, Burey SC, Roure B, Burger G, Löffelhardt W, Bohnert HJ, Philippe H, Lang BF (2005) Monophyly of primary photosynthetic eukaryotes: green plants, red algae, and glaucophytes. Curr Biol 15:1325–1330

Rodriguez-Ezpeleta N, Brinkmann H, Burger G, Roger AJ, Gray MW, Philippe H, Lang BF (2007a) Toward resolving the eukaryotic tree: the phylogenetic positions of jakobids and cercozoans. Curr Biol 17:1420–1425

Rodriguez-Ezpeleta N, Brinkmann H, Roure B, Lartillot N, Lang BF, Philippe H (2007b) Detecting and overcoming systematic errors in genome-scale phylogenies. Syst Biol 56:389–399

Rogers MB, Gilson PR, Su V, McFadden GI, Keeling PJ (2007) The complete chloroplast genome of the chlorarachniophyte *Bigelowiella natans*: evidence for independent origins of chlorarachniophyte and euglenid secondary endosymbionts. Mol Biol Evol 24:54–62

Rosenblad MA, Samuelsson T (2004) Identification of chloroplast signal recognition particle RNA genes. Plant Cell Physiol 45:1633–1639

Rumpho ME, Worful JM, Lee J, Kannan K, Tyler MS, Bhattacharya D, Moustafa A, Manhart JR (2008) Horizontal gene transfer of the algal nuclear gene psbO to the photosynthetic sea slug *Elysia chlorotica*. Proc Natl Acad Sci USA 105:17867–17871

Sanchez Puerta MV, Bachvaroff TR, Delwiche CF (2005) The complete plastid genome sequence of the haptophyte *Emiliania huxleyi*: a comparison to other plastid genomes. DNA Res 12:151–156

Sanchez-Puerta MV, Bachvaroff TR, Delwiche CF (2007) Sorting wheat from chaff in multi-gene analyses of chlorophyll c-containing plastids. Mol Phylogenet Evol 44:885–897

Schunemann D (2004) Structure and function of the chloroplast signal recognition particle. Curr Genet 44:295–304

Shevelev EL, Bryant DA, Löffelhardt W, Bohnert HJ (1995) Ribonuclease-P RNA gene of the plastid chromosome from *Cyanophora paradoxa*. DNA Res 2:231–234

Shoup S, Lewis LA (2003) Polyphyletic origin of parallel basal bodies in swimming cells of chlorophycean green algae (Chlorophyta). J Phycol 39:789–796

Sluiman HJ (1985) A cladistic evaluation of the lower and higher green plants (Viridiplantae). Plant Syst Evol 149:217–232

Smith DR, Lee RW (2009) The mitochondrial and plastid genomes of *Volvox carteri*: bloated molecules rich in repetitive DNA. BMC Genomics 10:132

Smith DR, Lee RW (2010) Low nucleotide diversity for the expanded organelle and nuclear genomes of *Volvox carteri* supports the mutational-hazard hypothesis. Mol Biol Evol 27:2244–2256

Smith DR, Lee RW, Cushman JC, Magnuson JK, Tran D, Polle JE (2010) The *Dunaliella salina* organelle genomes: large sequences, inflated with intronic and intergenic DNA. BMC Plant Biol 10:83

Stiller JW, Huang J, Ding Q, Tian J, Goodwillie C (2009) Are algal genes in nonphotosynthetic protists evidence of historical plastid endosymbioses? BMC Genomics 10:484

Stoebe B, Maier UG (2002) One, two, three: nature's tool box for building plastids. Protoplasma 219:123–130

Strittmatter P, Soll J, Bolter B (2010) The chloroplast protein import machinery: a review. Methods Mol Biol 619:307–321

Tanaka T, Fukuda Y, Yoshino T, Maeda Y, Muto M, Matsumoto M, Mayama S, Matsunaga T (2011) High-throughput pyrosequencing of the chloroplast genome of a highly neutral-lipid-producing marine pennate diatom, Fistulifera sp. strain JPCC DA0580. Photosynth Res 109:223–229

Turmel M, Otis C, Lemieux C (1999) The complete chloroplast DNA sequence of the green alga *Nephroselmis olivacea*: insights into the architecture of ancestral chloroplast genomes. Proc Natl Acad Sci USA 96:10248–10253

Turmel M, Otis C, Lemieux C (2002) The complete mitochondrial DNA sequence of *Mesostigma viride* identifies this green alga as the earliest green plant divergence and predicts a highly compact mitochondrial genome in the ancestor of all green plants. Mol Biol Evol 19:24–38

Turmel M, Otis C, Lemieux C (2005) The complete chloroplast DNA sequences of the charophycean green algae *Staurastrum* and *Zygnema* reveal that the chloroplast genome underwent extensive changes during the evolution of the Zygnematales. BMC Biol 3:22

Turmel M, Otis C, Lemieux C (2006) The chloroplast genome sequence of *Chara vulgaris* sheds new light into the closest green algal relatives of land plants. Mol Biol Evol 23:1324–1338

Turmel M, Otis C, Lemieux C (2007) An unexpectedly large and loosely packed mitochondrial genome in the charophycean green alga *Chlorokybus atmophyticus*. BMC Genomics 8:137

Turmel M, Brouard JS, Gagnon C, Otis C, Lemieux C (2008) Deep division in the Chlorophyceae (Chlorophyta) revealed by chloroplast phylogenomic analyses. J Phycol 44:739–750

Turmel M, Gagnon MC, O'Kelly CJ, Otis C, Lemieux C (2009a) The chloroplast genomes of the green algae *Pyramimonas*, *Monomastix*, and *Pycnococcus* shed new light on the evolutionary history of prasinophytes and the origin of the secondary chloroplasts of euglenids. Mol Biol Evol 26:631–648

Turmel M, Otis C, Lemieux C (2009b) The chloroplast genomes of the green algae *Pedinomonas minor*, *Parachlorella kessleri*, and *Oocystis solitaria* reveal a shared ancestry between the Pedinomonadales and Chlorellales. Mol Biol Evol 26:2317–2331

van Dooren GG, Schwartzbach SD, Osafune T, McFadden GI (2001) Translocation of proteins across the multiple membranes of complex plastids. Biochim Biophys Acta 1541:34–53

Verbruggen H, Maggs CA, Saunders GW, Le Gall L, Yoon HS, De Clerck O (2010) Data mining approach identifies research priorities and data requirements for resolving the red algal tree of life. BMC Evol Biol 10:16

Wakasugi T, Nagai T, Kapoor M, Sugita M, Ito M, Ito S, Tsudzuki J, Nakashima K, Tsudzuki T, Suzuki Y, Hamada A, Ohta T, Inamura A, Yoshinaga K, Sugiura M (1997) Complete nucleotide sequence of the chloroplast genome from the green alga *Chlorella vulgaris*: the existence of genes possibly involved in chloroplast division. Proc Natl Acad Sci USA 94:5967–5972

Wastl J, Maier UG (2000) Transport of proteins into cryptomonads complex plastids. J Biol Chem 275:23194–23198

Wastl J, Duin EC, Iuzzolino L, Dorner W, Link T, Hoffmann S, Sticht H, Dau H, Lingelbach K, Maier UG (2000) Eukaryotically encoded and chloroplast-located rubredoxin is associated with photosystem II. J Biol Chem 275:30058–30063

Williamson DH, Gardner MJ, Preiser P, Moore DJ, Rangachari K, Wilson RJ (1994) The evolutionary origin of the 35 kb circular DNA of *Plasmodium falciparum*: new evidence supports a possible rhodophyte ancestry. Mol Gen Genet 243:249–252

Wilson RJ, Williamson DH (1997) Extrachromosomal DNA in the Apicomplexa. Microbiol Mol Biol Rev 61:1–16

Wolfe KH, Morden CW, Palmer JD (1991) Ins and outs of plastid genome evolution. Curr Opin Genet Dev 1:523–529

Wyman SK, Jansen RK, Boore JL (2004) Automatic annotation of organellar genomes with DOGMA. Bioinformatics 20:3252–3255

Yamada T (1991) Repetitive sequence-mediated rearrangements in *Chlorella ellipsoidea* chloroplast DNA – completion of nucleotide-sequence of the large inverted repeat. Curr Genet 19:139–147

Yamada T, Shimaji M (1987) Splitting of the ribosomal-RNA operon on chloroplast DNA from *Chlorella ellipsoidea*. Mol Gen Genet 208:377–383

Zhang Z, Green BR, Cavalier-Smith T (1999) Single gene circles in dinoflagellate chloroplast genomes. Nature 400:155–159

Zhang Z, Cavalier-Smith T, Green BR (2002) Evolution of dinoflagellate unigenic minicircles and the partially concerted divergence of their putative replicon origins. Mol Biol Evol 19:489–500

Chapter 4

Plastomes of Bryophytes, Lycophytes and Ferns

Paul G. Wolf*
Department of Biology, Utah State University, Logan, UT 84322, USA

Kenneth G. Karol
The Lewis B. and Dorothy Cullman Program for Molecular Systematics Studies, The New York Botanical Garden, Bronx, NY 10458, USA

Summary

We review current progress in our understanding of chloroplast genomes (plastomes) of liverworts, mosses, hornworts, lycophytes and monilophytes. We briefly cover some of the methods used to obtain complete nucleotide sequences of plastomes and we summarize the published sequences from the plant groups above. We explore some of the evolutionary changes that have occurred in terms of gene content, introns and position of the inverted repeat boundaries. We also discuss RNA editing, which is especially high in plastome genes of some non-seed land plants. We finish with a phylogenetic analysis of available plastome genes and we suggest some possible directions for future research.

I. Introduction

Land plants have a chloroplast (plastid) genome (plastome) with a basic canonical organization that is similar to that of their algal ancestors (see Chap. 3). This represents one of the most evolutionary conserved genomic structures in nature. However, from this basic organization, several structural changes have occurred on

*Author for correspondence, e-mail: wolf@biology.usu.edu

R. Bock and V. Knoop (eds.), Genomics of Chloroplasts and Mitochondria, Advances in Photosynthesis and Respiration 35, pp. 89–102,
DOI 10.1007/978-94-007-2920-9_4,

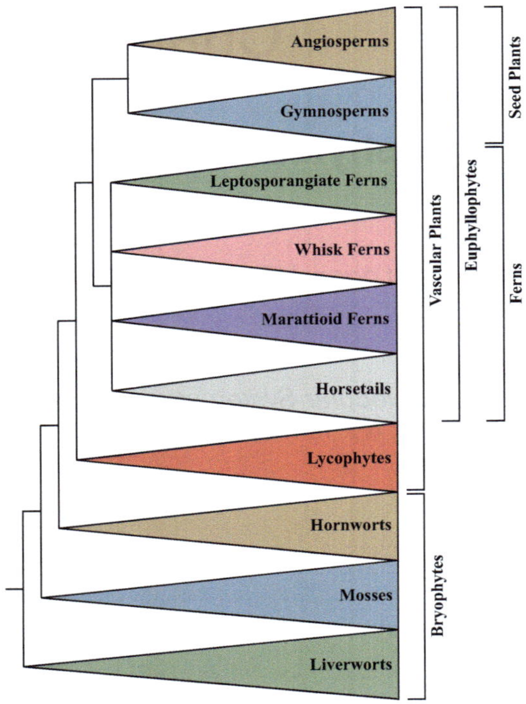

Fig. 4.1. Our current understanding of relationships among major land plant lineages. The extant bryophytes represent a grade of three lineages with liverworts shown sister to all other extant land plants and hornworts shown sister to extant vascular plants. Vascular plants include the lycophytes, monilophytes and seed plants. Four major monilophyte lineages are shown as an unresolved polytomy sister to seed plants.

various evolutionary branches. Here we review aspects of plastomes of extant land plants, except for seed plants (see next chapter). The main lineages include the nonvascular bryophyte lineages (hornworts, liverworts and mosses), the lycophytes and monilophytes. The latter, which include leptosporangiate ferns and horsetails, are also referred to elsewhere as 'ferns' (e.g., Pryer et al. 2004; Schneider et al. 2009). Seed plants appear to be the sister to monilophytes (Pryer et al. 2001). Our current understanding of relationships among these lineages is depicted in Fig. 4.1. We begin with an overview of the taxa and structural aspects of plastomes. We then summarize the major events of gene and intron loss in plastomes of non-seed land plants. Next we discuss the phenomenon of RNA editing, a process that occurs at much higher rates in non-seed land plants than in seed plants.

Abbreviations: IR – Inverted repeat; kb – Kilobases; LSC – Large single copy; mya – Million years ago; PCR – Polymerase chain reaction; PPR – Pentatricopeptide repeat; SSC – Small single copy

II. Techniques and Overall Plastome Organization

Until about the mid-1990s, restriction site mapping was the main approach to inferring plastome organization. The technique involves digesting DNA with restriction endonucleases, separating the DNA fragments on an agarose gel and then transferring them to a membrane. The fragments on this membrane are then probed with labeled plastid DNA from a well-characterized species, or fragments of the same species cut with a different restriction enzyme. After careful analysis, a coarse-scale map of the plastome can be constructed. The first such physical map was that of the *Zea mays* plastome (Bedbrook and Kolodner 1979). Mapping studies also indicated that, within plant cells, the plastome exists in two orientations (Palmer 1983), a pattern that is maintained by a form of homologous recombination (so-called flip-flop recombination; Stein et al. 1986). Subsequently, plastomes of many species were mapped (reviewed by Palmer 1985), verifying that in most (but not all) lineages, plastomes map to a circle with a large single copy region (LSC) and a small single copy region (SSC) separated by two copies of an inverted repeat (IR), which include the ribosomal RNA genes (Palmer 1985). Fine-scale mapping requires nucleotide sequencing, which is easier and cheaper with today's techniques. The first two plastomes to be completely sequenced were those of the flowering plant tobacco (*Nicotiana tabacum*; Shinozaki et al. 1986), and the liverwort *Marchantia polymorpha* (Ohyama et al. 1986). These data confirmed the earlier inferences on overall plastome organization that had been deduced from mapping studies.

Most green plant plastomes map to a circle of about 150 kb. However, the largest reported plastome, that of the green alga *Floydiella terrestris*, is more than 500 kb (Brouard et al. 2010). Most plant cells contain many copies of the plastome; even plants with a single plastid (e.g., the unicellular green alga *Chlamydomonas reinhardtii*) can contain many copies of the plastome. At the other extreme, wheat cells have more than 50 plastids per cell and more than 300 plastome copies per plastid (Boffey and Leech 1982). Thus, although the plastome is a small genome compared to its nuclear counterpart, plastid DNA makes up a significant proportion of total cellular DNA, as much as 20% in some species (Boffey and Leech 1982).

Plastid DNA is not assembled into chromosomes and it does not reside in the plastid as a population of free circular molecules. Rather, several plastomes are organized, with proteins and RNA, into structures known as nucleoids (Sato et al. 2003). Most nucleoids are attached to the envelope membrane, but mature chloroplasts can also have nucleoids associated with the thylakoid membrane (Sato et al. 2003). It is likely that nucleoid structure plays an important role in plastome replication, transcription and post-transcriptional modification. However, the general relationships between plastome packaging and these processes remain poorly understood (Bock 2007).

Although plastomes are typically depicted as circles, most plastid DNA is not in this form in a living plant cell (Bendich 2004; Bock 2007). Researchers have found linear plastomes, concatenated pieces representing multiple plastomes (sometimes circular) (Bendich 2004), and even branched forms (Oldenburg and Bendich 2004a). This variety of possible conformations is likely a function of both phylogenetic divergence and stage of plastome replication. The plastome replication process itself is also poorly understood (Bock 2007), and several mechanisms have been proposed. Early models involved bidirectional replication similar to that in bacteria, resulting in displacement (D) loops (Kolodner and Tewari 1975b). Rolling circle amplification (RCA) could also be used to achieve additional replication (Kolodner and Tewari 1975a). A double D-loop mechanism has also been proposed (Kunnimalaiyaan and Nielsen 1997). However, these models have been challenged, based on the degree of linear DNA observed (Bendich 2004), and a recombination-dependent mechanism was instead proposed (Oldenburg and Bendich 2004b). The challenge of studying replication is making observations during the actual process. Alternatively, researchers can examine the signature of replication, which can be deduced from variation in base composition. Studies of mitochondrial genomes found that regions accumulate adenine-to-guanine transitions due to deamination during the single-stranded phase of replication. This is because $A \rightarrow G$ transitions accumulate evenly over time whereas the accumulation of $C \rightarrow T$ substitutions is complex and asymptotic (Krishnan et al. 2004). Thus, gradients in A/G composition, especially for non-coding DNA, is a function of total amount of time spent in the single-stranded phase, and therefore can reveal origins and directions of replication. This approach was used recently to show that A/G composition gradients are most consistent with the earlier models (bidirectional and RCA) across a wide range of published green plant plastomes (Krishnan and Rao 2009). Direct testing of these models is now needed. Meanwhile, evidence continues to accumulate for a role of recombination-dependent replication in *Arabidopsis*, especially as a repair process for maintaining plastome integrity (Rowan et al. 2010). Clearly, the evidence suggests that more than one replication process appears to be operating, and the result is a complex population of molecules representing the plastome. Regardless, most land plant plastomes map to a circle and have a fairly conserved set of protein and RNA encoding genes. The map of the plastome of the whisk fern *Psilotum nudum* is depicted in linear fashion in Fig. 4.2 as a guide to this overall structure.

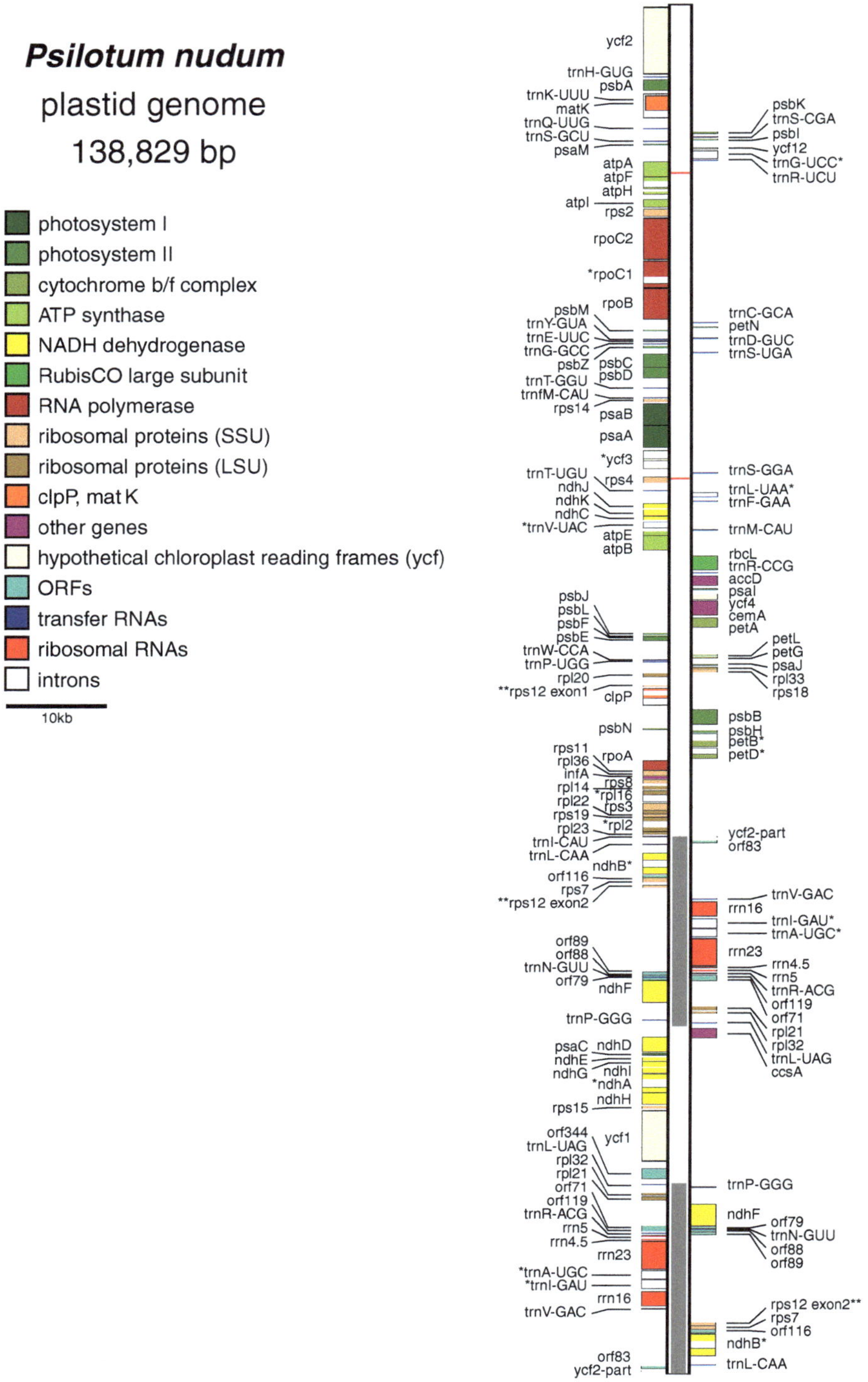

Fig. 4.2. Psilotum nudum plastid genome structure. Genes (*colored boxes*) on the *right side* of the map are transcribed in the *top down* direction, whereas those on the *left side* are transcribed *bottom up*. The tRNA genes are indicated by the three-letter amino acid code followed by the anticodon. Intron-containing genes are show with an *asterisk* (*); the trans-spliced gene *rps12* is shown with *two asterisks* (**). The two horizontal *red lines* along the genome indicate the insertion/deletion events unique to all monilophytes and the two *grey* boxes along the genome indicate the inverted repeats. Note a fragment of *ycf2* is found in the inverted repeat.

A. Bryophytes

The bryophytes represent a grade of three extant lineages (Fig. 4.1, Mishler and Churchill 1984; Nickrent et al. 2000; Renzaglia et al. 2007; Shaw and Renzaglia 2004). Several phylogenetic analyses lead to the hypothesis that liverworts are sister to all other extant land plants (Qiu et al. 1998) and hornworts are sister to extant vascular plants (e.g., Groth-Malonek et al. 2005; Qiu et al. 2006, 2007; Qiu 2008). Recent findings of cryptospores from the early Middle Ordovician (c. 473–471 mya Rubinstein et al. 2010) may represent liverworts or at least their ancestors. A broad-scale phylogenetic analysis of liverworts reveals several key lineages. The earliest branching lineage, Haplomitriopsida, is the sister to all remaining extant liverworts. There is then a major split between the complex thalloid liverworts (Marchantiopsida) and a heterogenous clade (Jungermanniopsida) which includes two clades (Metzgeriidae and Pellidae) of simple thalloid taxa (which is therefore a paraphyletic group) and a monophyletic "leafy" clade (Jungermanniidae) which excludes a few taxa previously considered as leafy (Forrest et al. 2006). The complex-thalloid liverwort *Marchantia polymorpha* was the first plant for which the chloroplast genome was sequenced (Ohyama et al. 1986). Later, the complete mitochondrial genome of *M. polymorpha* was also sequenced (Oda et al. 1992), providing yet another important genomic resource for nonvascular plants. A second liverwort plastome was recently sequenced (Wickett et al. 2008b), that of the only known parasitic bryophyte, the simple-thalloid liverwort *Aneura mirabilis*. Nonphotosynthetic plants often lose plastid genes that are associated with photosynthetic functions (Wickett et al. 2008b; Wolfe et al. 1992). Indeed, *A. mirabilis* has lost some of the same genes as has the parasitic angiosperm *Epifagus virginiana* (Wolfe et al. 1992). However, the loss of only a subset of these genes in *A. mirabilis* suggests that this liverwort is in an earlier stage of acquiring a parasitic life history stage (Wickett et al. 2008a, b).

Mosses are a diverse clade of more than 12,000 species, representing about eight main extant lineages (Cox et al. 2004; Goffinet and Buck 2004; Newton et al. 2000; Wahrmund et al. 2009, 2010). Two complete moss plastomes have been sequenced: that of the model species for molecular genetic studies, *Physcomitrella patens* (Sugiura et al. 2003) and the desiccation-tolerant species *Syntrichia (=Tortula) ruralis* (Oliver et al. 2010). These plastomes differed by a large (71 kb) inversion in the large single copy (LSC) region, with *S. ruralis* possessing the apparently ancestral organization. Further analysis revealed that the inversion is unique to the Funariidae (Goffinet et al. 2007). This inversion is the largest plastome reorganization reported to date for land plant plastomes, and appears to represent a single evolutionary event (Goffinet et al. 2007).

Hornworts represent the third main clade of nonvascular land plants, with about 400 extant species (Bateman et al. 1998). Hornworts are probably sister group to the vascular plants (Groth-Malonek et al. 2005; Qiu et al. 2007). Phylogeny within the hornworts has been examined by Duff and coworkers (Duff et al. 2004; 2007). Currently, there is only a single published complete plastome sequence of a hornwort, *Anthoceros formosae* (Kugita et al. 2003b). This plastome has a very high level of RNA editing (Kugita et al. 2003a), as do several mitochondrial and plastid genes in most hornworts studied (Duff and Moore 2005; Duff 2006). More details on RNA editing are provided later in this chapter.

B. Lycophytes

The lycophytes include a large assemblage of both extant and extinct lineages. Extant groups include the heterosporous Isoetopsida with about 150 species of *Isoetes* (quillworts) and about 700 species of *Selaginella* (spikemosses). The remaining extant lineage is the homosporous Lycopodiopsida (clubmosses) of which about 300 species are known, including

Lycopodium, *Huperzia* and related genera. Extinct lineages include many fossil species, especially from the late Silurian (about 420 mya) through the Carboniferous (about 300 mya, Kenrick and Crane 1997). Ancient representative of this group of plants formed many of the fossil coal beds. Photosynthesis in these plants harnessed the sun's energy, which is now used as one major source of fossil fuels. These extinct lycophytes were large plants; some reached 30 m, whereas today's species are less than 1 m. As a group the lycophytes appear to be a sister group to Euphyllophytes (monilophytes plus seed plants, see below). This early split is supported both by analysis of morphology in fossil taxa (Kenrick and Crane 1997) and extant taxa (Kranz and Huss 1996). However, an additional convincing piece of evidence comes from analysis of plastome organization. Monilophytes and seed plants possess a 30 kb inversion in the LSC relative to lycophytes and bryophytes (Raubeson and Jansen 1992). Further details of the organization of lycophyte plastomes came from restriction site mapping of an *Isoetes* plastome (Duff and Schilling 2000), which confirmed the overall similarity of the lycophyte and bryophyte plastomes. The first complete plastome sequence of a lycophyte was that of *Huperzia lucidula* (Wolf et al. 2005). Since then, additional plastomes have been sequenced from the heterosporous genera, *Selaginella moellendorffii, S. uncinata* and *Isoetes flaccida* (Karol et al. 2010; Tsuji et al. 2007). Although lycophytes share structural similarities with bryophytes, the former do have some unique features. For example, *ycf2* normally resides in the LSC in most plastomes, but has been translocated to the SSC in *I. flaccida*, with the 5′ end now incorporated into the IR. In addition, the *chlL/chlN* gene cluster has been inverted in *I. flaccida* so that it is now adjacent to *ycf2* rather than *ycf1* as in *H. lucidula*. The *ycf2* translocation and the *chlL/chlN* inversion occur in neither of the *Selaginella* plastomes. Both *Selaginella* plastomes differ considerably in gene order from other plastomes (Karol et al. 2010). An approximately 14-kb region has been translocated from the LSC to the IR/SSC in both *Selaginella* plastomes. The genes included in this translocation differ slightly between *S. uncinata* and *S. moellendorffii*. In addition, *rps4* is in the IR in *Selaginella* and marks one endpoint of the translocated segment. The other endpoint resides in the SSC and is marked by *psbD* in *S. moellendorffii*. In *S. uncinata*, the same endpoint includes three additional genes *(trnE-UUC, trnY-GUA and trnD-GUC*), which remain in the LSC adjacent to *ycf2* in *S. moellendorffii*. *Selaginella uncinata* also has a ~20-kb LSC inversion (*psbI to rpoB-trnC-GAC*), a duplication of the *psbK/trnQ-UUG* region, and translocation of *petN* from the LSC to the SSC. These features appear to be unique to *S. uncinata* (Karol et al. 2010). Because complete plastome sequences are available from only four species of lycophytes, it is not yet possible to infer the phylogenetic extent of all plastome changes. Additional taxon sampling will be needed to understand more fully how recent and extensive these changes are.

C. Monilophytes (Ferns)

Monilophytes represent another group of vascular plants with an extensive fossil history. Here we consider four main extant lineages: (1) leptosporangiate ferns (about 11,000 species), (2) a clade that includes whisk ferns (*Psilotum* and *Tmesipteris*) and the Ophioglossales, (3) Marattioid ferns and (4) Horsetails (*Equisetum*). Data from plastid and nuclear gene sequences (Pryer et al. 2001) and morphology (Kenrick and Crane 1997; Schneider et al. 2009) find support for monophyly of a clade that includes these four lineages. Together the clade is called moniliformopses (Kenrick and Crane 1997), monilophytes (Pryer et al. 2004), or ferns *sensu lato* (Schneider et al. 2009). Further resolution of relationships among these four groups has not yet been achieved. Although monophyly of monilophytes has support from analyses of extant taxa, analyses that include fossil taxa has questioned this idea (Rothwell and Nixon 2006).

The first monilophyte to have a plastome sequenced was *Psilotum nudum* (GenBank accession #AP004638 from 2002, see Fig. 4.2). Several phylogenetic studies support inclusion of the ophioglossoid ferns with the whisk ferns (Pryer et al. 2001, 2004; Qiu et al. 2007), but so far no complete plastome from the ophioglossoid ferns has been published. Complete plastome sequences are available from one horsetail (Karol et al. 2010), one marattioid fern (Roper et al. 2007) and four leptosporangiate ferns (Der 2010; Gao et al. 2009; Wolf et al. 2003, 2011).

An inversion in the LSC involving *trnG-GCC* to *trnT-GGU* is found in all fern plastomes and no other land plant plastomes (Karol et al. 2010), thus providing further evidence for monilophyte monophyly. Within the leptosporangiate ferns, a series of additional inversions has occurred, two of which (18 kb and 21 kb respectively) result in a reverse gene order within the IR (Wolf et al. 2010). An additional pair of inversions occurred more recently in the LSC of a large clade of ferns (the "polypods", Wolf et al. 2010).

III. The Inverted Repeat Boundaries

Plastome IRs from most plants typically house a similar gene content, which includes primarily rRNA and tRNA genes (Jansen et al. 2007; Palmer and Stein 1986; Turmel et al. 2007). This is seen also in some leptosporangiate ferns where, except for a few early-diverging clades, the IR itself is inverted (Wolf et al. 2003). Most of the variation in IR gene content occurs at the ends of the IR. This "ebb and flow" of the IR boundaries into and out of the LSC and SSC regions has been attributed to effects of recombination and gene conversion (Goulding et al. 1996). Effects of these positional changes have been seen in related species at the nucleotide level in several species of *Nicotiana* (Goulding et al. 1996). Furthermore, when comparing distantly related lineages of land plants, several plastomes exhibit unique IR boundaries that differ from the basic theme (Karol et al. 2010). But this is not always the case: other distantly related taxa have very similar IR boundaries. For example, *Marchantia polymorpha*, two mosses and *Equisetum arvense* were identical in gene content at both ends of the IR. This suggests that whereas the ends of the IR clearly ebb and flow in some lineages, in other lineages they appear to be rather stable, at least at the scale of gene order (Karol et al. 2010).

IV. Changes in Gene and Intron Content

Most plastomes sequenced to date contain a very similar repertoire of genes. The most significant exceptions are plastomes from parasitic plants in which many photosynthetic genes are lost or pseudogenized (Wickett et al. 2008b; Wolfe et al. 1992). Overlaid on the basic pattern are found a few genes that are absent in some sequenced plastomes. Some of these genes seem to have been lost multiple times based on their phylogenetic distribution (See Fig. 4.3). These include *infA* and *ycf1*. Other genes appear to be distinctly present or absent in particular clades. Here we briefly list these latter patterns based on what we know is a very limited sample of plastomes (especially for non-seed land plant clades). We ignore many that are specific to only one plastome, except where that plastome is the sole representative (such as the single published hornwort plastome).

The genes *ccsA* and *rpoA* are absent from the plastomes of two mosses (*Syntrichia ruralis* and *Physcomitrella patens*, Oliver et al. 2010; Sugiura et al. 2003), *petN* is lacking in *S. ruralis*, and *cemA* is absent from both *Selaginella* plastomes. Mosses and liverworts lack *rps16*, but the gene is present in hornworts and some vascular plants. The genes *matK* and *rps15* are pseudogenes in the hornwort (Kugita et al. 2003b). The gene cluster *chlB*, *chlL*, and *chlN* is absent from *Psilotum nudum* and angiosperms. The gene *psaM* is lacking from the three polypod ferns (*Adiantum capillus-veneris, Cheilanthes lindheimeri* and *Pteridium aquilinum*), as

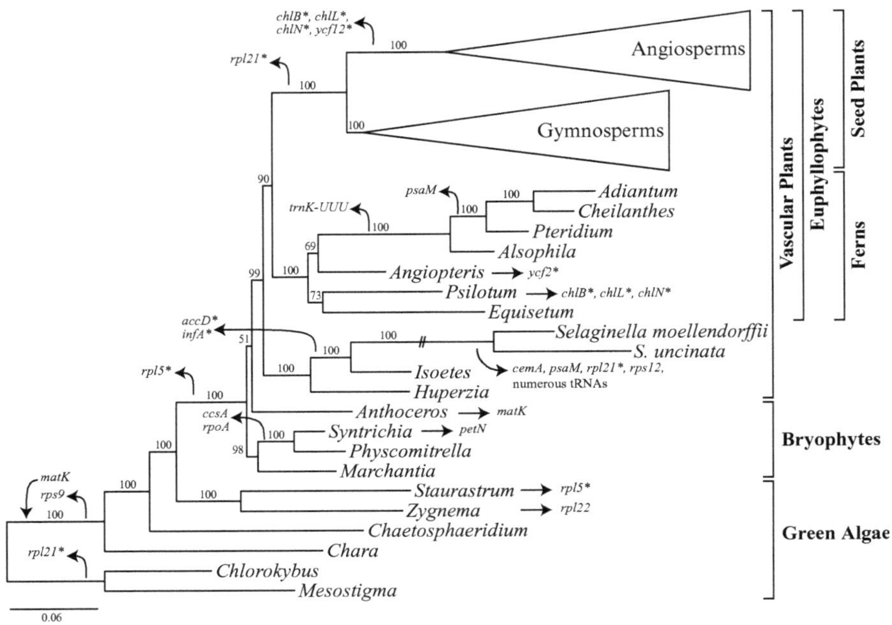

Fig. 4.3. Phylogenetic results using nucleotide data. Phylogenetic analyses were performed using 49 plastome gene sequences from 45 completely sequenced plastomes, including 39 land plants and six charophycean algae. The nucleotide alignment from Karol et al. (2010) was used as a starting point (49 genes from 43 taxa). To this we incorporated into the alignment sequence data from two new leptosporangiate fern plastomes: *Cheilanthes lindheimeri* (Wolf et al. 2011) and *Pteridium aquilinum* (Der 2010). Maximum likelihood analyses were performed on the Cyberinfrastructure for Phylogenetic Research (CIPRES) Portal (v.3.1, Miller et al. 2009) using RAxML-HPC (v. 7.2.7, Stamatakis 2006, 2008) with 200 bootstrap replicates. Third codon positions were excluded to avoid problems associated with relatively rapidly evolving sites. The best tree (−ln = 195205.737395) is shown with bootstrap proportions drawn above branches. The relationships among major fern lineages are weakly supported, though monophyly of the ferns is strongly supported. The branch leading to *Selaginella* was drawn to one-half scale to accommodate this figure. Note that the sister relationship of the liverwort and mosses is strongly supported and is in contrast to the cladogram shown in Fig. 4.1. This relationship was also recovered by Karol et al. (2010) when divergent taxa (*Selaginella* spp.) were excluded from phylogenetic analyses. Furthermore, Renzaglia and Garbary (2001) concluded that characters related to sperm cell development were compelling evidence for the monophyly of liverworts plus mosses, a clade they referred to as Setaphytes. Names of lost plastid genes are shown with *arrows*. An *asterisk* (*) following a gene name indicates that this gene has been lost in at least two lineages independently. The rare gain of a plastid gene (*matK*) is also indicated in the *green algae*.

well as from the two *Selaginella* plastomes and the majority of seed plant plastomes. Seed plant plastomes lack *rpl21,* as do the two *Selaginella* plastomes. The parasitic liverwort *A. mirabilis* has lost several genes (including several *ndh* genes) and many others exist as possibly recent pseudogenes (Wickett et al. 2008a).

A group II intron, along with its encoded maturase gene (*matK*) invaded the *trnK-UUU* gene in charophycean algae after the divergence of chlorophytes and charophytes. All chloropycean algae and some early diverging charophycean algae (Mesostigmatophyceae and Chlorokybophyceae) do not contain this intron. More derived charophycean algae (Charophyceae, Coleocheatophyceae and Zygnematophyceae) have the intron. There is one lineage (Klebsormidiophyceae, which is sister to Charophyceae, Coleocheatophyceae, Zygnematophyceae and land plants) where we do not yet fully know the condition of *trnK*. A large clade of leptosporangiate ferns has subsequently lost *trnK-UUU* and its intron (Wolf et al. 2010, 2011), yet *matK* remains. The introns of *clpP* are variable across land plants, with some plastomes having two, and others having one intron in this gene, but there appears to be no distinct phylogenetic pattern (Karol et al. 2010).

Thus, although plastome gene content tends to be well-conserved among land plant lineages, several clade-specific gene losses are apparent.

V. RNA Editing

The central dogma of molecular genetics requires conservation of information from genomic DNA through messenger RNA to the final amino acid sequence of a protein. However, detailed studies of the various products of transcription and translation have found exceptions to this conservation. Considerable post-translational modification occurs to proteins. In addition to the various aspects of RNA processing that occur, an independent post-transcriptional stage is RNA editing. This process alters the nucleotides in the primary transcript so that the messenger RNA differs from the genomic encoding sequence (See Chap. 13). RNA editing is found throughout eukaryotes, and is especially common in organellar genomes (reviewed by Tillich et al. 2006). In plastome genes from seed plants, the process occurs at fewer than 40 sites and about ten times that number have been reported in ferns and hornworts. In most cases, cytosines are edited to uracils, but in hornworts and ferns, additional uracil-to-cytosine edits have been reported (Kugita et al. 2003a; Wolf et al. 2004).

RNA editing requires both cis- and trans-acting factors. Cis-acting factors include the actual site to be edited. Other cis-acting factors include upstream and downstream recognition sequences (Kobayashi et al. 2008). However, the latter appear to have no obvious pattern across sites. This might be because the trans-acting factors (nuclear-encoded proteins) are likely to be of several types (Hammani et al. 2009). To date, over 20 different nuclear factors have been associated with RNA editing in *Arabidopsis* (see Stern et al. 2010), most of which are pentatricopeptide repeat (PPR) proteins (Kotera et al. 2005; Okuda and Shikanai 2008). These proteins are characterized by tandem repeats of a degenerate 35 amino acid motif, and several PPR gene subfamilies are found across eukaryotic lineages.

The functions of RNA editing are not obvious. Several authors have argued that RNA editing repairs errors in genomic sequences (Jobson and Qiu 2008; Stern et al. 2010). However, this seems far less efficient than a simple nucleotide substitution at the DNA level of the genome, which would require no further action. An additional role has been implicated in gene regulation, whereby RNA editing varies with developmental stage and could be used to restore correct translation when the gene product is needed (Hirose et al. 1999). This has been observed in a few cases in animals, but seems to play a minor role in plants (Stern et al. 2010). It seems more likely that the enzymes that edit RNA have evolved for other cellular functions and their editing ability then releases selective constraints for the edited sites in genes. In fact, some of these other functions of editing enzymes are known. In primates, the APOBEC family of RNA editing enzymes includes cytosine deaminases that act to restrict infection from retroviruses (Bransteitter et al. 2009). Further research is needed on the RNA editing factors of *Arabidopsis* and other plants if we are to understand further the function and cellular significance of RNA editing.

RNA editing can cause problems for comparative analyses of nucleotide sequences. Most phylogenetic analyses are based on alignment of orthologous genomic sequences. However, if RNA editing occurs, these DNA sequences represent the unedited versions. Should one use the genomic sequences or the edited versions? The latter can only be inferred accurately by using mature RNA transcripts to generate cDNA. Until this is done, one does not know which sites have been edited. For analyses of seed plants, this dilemma is trivial because RNA editing rates are so low. But in ferns, lycophytes and some bryophytes, the effect on the outcomes of analyses can be significant. In hornworts, RNA editing rates are so high that the same site can be C to U edited in some taxa and U to C edited in other taxa (Duff and Moore 2005). When phylogenetic analyses of hornworts use cDNA sequences, the results are different from those from genomic sequences (Duff and Moore 2005; Duff 2006). Removal of edited sites does not help, because that reduces the amount of potentially useful phylogenetic signal. The solution can only be attained once we know the evolutionary stability of RNA editing itself. If relatively stable, then the fact that a site is edited provides an evolutionary marker. If sites come and go rapidly, then RNA editing sites are homoplastic and the results of phylogenetic analysis of cDNA sequences will be misleading. The answer will depend on the relative levels of homoplasy in genomic sequences versus RNA editing sites, and this is likely to vary across clades of land plants.

VI. Phylogenetic Analyses

Over the last few decades single gene phylogenetic analyses have served as powerful tools for reconstructing the evolutionary history of every major lineage of life on Earth (Donoghue and Cracraft 2004). Reduced costs and improvements in sequencing technologies have allowed several genes to be sequenced across a broad range of taxa for phylogenetic reconstruction (Holton and Pisani 2010; Nickrent et al. 2000; Qiu et al. 2007; Shalchian-Tabrizi et al. 2008). Indeed, with new second-generation sequencing technologies, complete plastome sequences are now being generated at an ever increasing rate (Cronn et al. 2008; Wolf et al. 2011). We reanalyzed the plastome alignment of Karol et al. (2010) and included two new leptosporangiate fern taxa (*Cheilanthes lindheimeri* and *Pteridium aquilinum*). This analysis included 49 plastome genes from 45 green plant taxa and the results are shown in Fig. 4.3. The overall topology is consistent with results presented in Karol et al. (2010), with the two new fern taxa found in a monophyletic leptosporangiate clade. Relationships among the major monilophyte lineages remained weakly supported. Most of the currently available land plant plastome sequences are from seed plants, with very few available from the presumed sister clade, monilophytes. With additional data from other fern representatives, including ophioglossoid ferns, it will become possible to gain further insight into early land plant evolution as well as the patterns and processes that shape the evolution of plastomes.

VII. Future Directions

Currently, the distribution of complete plastome sequence data is biased toward angiosperms. In general, clades more distantly related to angiosperms are less well sampled. There are especially critical clades in the algae for which no representative plastome sequence is available (e.g., Klebsormidiophyceae, *Coleochaete*). Although obtaining the actual DNA sequence is relatively easy, limiting steps in plastome sequencing mostly involve isolating plastome DNA. Although this can be done through various centrifugation and other procedures (Jansen et al. 2005), there are some alternative approaches. If the plastome component of total DNA is high then a total genomic shotgun sequence can provide sufficient data from which the plastome sequence can be assembled (Wolf et al. 2011). A more cost-effective approach involves multiplex sequencing-by-synthesis on the Illumina platform (Cronn et al. 2008). In this protocol, more than a hundred plastomes can be sequenced simultaneously. However, custom probes or PCR-primers will be needed for each major clade, the range of these depending on sequence divergence levels. One problem with the shotgun genome approach is that it may not be possible to distinguish genuine reads of plastome DNA from those that are plastid DNA that has been transferred to the nucleus (Bock and Timmis 2008). To some extent, this is a problem for all approaches to plastome studies, but the problem is exacerbated by short reads and the use of total genomic DNA extractions. Regardless, the prospects seem good for filling many of the critical clade gaps in the next few years. This should ease the trend away from recent exemplar studies (with a few, though critical taxa) toward more taxon-dense studies with broad phylogenetic breadth. Although such a trend may not always uncover much new in terms of phylogenetic hypotheses, it is sure to show us more details of the evolution of plastomes themselves.

Acknowledgements

This work was supported by NSF DEB1020660 and DEB1036466 to KGK and NSF DEB-0228432 to PGW.

References

Bateman RM, Crane PR, DiMichele WA, Kenrick PR, Rowe NP, Speck T, Stein WE (1998) Early evolution of land plants: phylogeny, physiology, and ecology of the primary terrestrial radiation. Annu Rev Ecol Syst 29:263–292

Bedbrook JR, Kolodner R (1979) Structure of chloroplast DNA. Annu Rev Plant Physiol Plant Mol Biol 30:593–620

Bendich AJ (2004) Circular chloroplast chromosomes: the grand illusion. Plant Cell 16:1661–1666

Bock R (2007) Structure, function, and inheritance of plastid genomes. Cell Mol Biol Plastids 19:29–63

Bock R, Timmis JN (2008) Reconstructing evolution: gene transfer from plastids to the nucleus. Bioessays 30:556–566

Boffey SA, Leech RM (1982) Chloroplast DNA levels and the control of chloroplast division in light-grown wheat leaves. Plant Physiol 69:1387–1391

Bransteitter R, Prochnow C, Chen XJS (2009) The current structural and functional understanding of APOBEC deaminases. Cell Mol Life Sci 66:3137–3147

Brouard JS, Otis C, Lemieux C, Turmel M (2010) The exceptionally large chloroplast genome of the green alga *Floydiella terrestris* illuminates the evolutionary history of the Chlorophyceae. Genome Biol Evol 2:240–256

Cox CJ, Goffinet B, Shaw AJ, Boles SB (2004) Phylogenetic relationships among the mosses based on heterogeneous Bayesian analysis of multiple genes from multiple genomic compartments. Syst Bot 29:234–250

Cronn R, Liston A, Parks M, Gernandt DS, Shen R, Mockler T (2008) Multiplex sequencing of plant chloroplast genomes using Solexa sequencing-by-synthesis technology. Nucleic Acids Res 36:e122

Der JP (2010) Genomic perspectives on evolution in bracken fern. Utah State University, Logan

Donoghue MJ, Cracraft J (2004) Charting the tree of life. In: Donoghue MJ, Cracraft J (eds) Assembling the tree of life. Oxford University Press, New York, pp 1–4

Duff RJ (2006) Divergent RNA editing frequencies in hornwort mitochondrial *nad5* sequences. Gene 366:285–291

Duff RJ, Moore FBG (2005) Pervasive RNA editing among hornwort *rbcL* transcripts except *Leiosporceros*. J Mol Evol 61:571–578

Duff RJ, Schilling EE (2000) The chloroplast genome structure of the vascular plant *Isoetes* is similar to that of the liverwort *Marchantia*. Am Fern J 90:51–59

Duff RJ, Cargill DC, Villarreal JC, Renzaglia KS (2004) Phylogenetic relationships of the hornworts based on *rbcL* sequence data: novel relationships and new insights. Monogr Syst Bot Ann Missouri Bot Gard 98:41–58

Duff RJ, Villarreal JC, Caagill DC, Renzaglia KS (2007) Progress and challenges toward developing a phylogeny and classification of the hornworts. Bryologist 110:214–243

Forrest LL, Davis EC, Long DG, Crandall-Stotler BJ, Clark A, Hollingsworth ML (2006) Unraveling the evolutionary history of the liverworts (Marchantiophyta): multiple taxa, genomes and analyses. Bryologist 109:303–334

Gao L, Yi X, Yang YX, Su YJ, Wang T (2009) Complete chloroplast genome sequence of a tree fern *Alsophila spinulosa*: insights into evolutionary changes in fern chloroplast genomes. BMC Evol Biol 9:130

Goffinet B, Buck WR (2004) Systematics of the Bryophyta (Mosses): from molecules to a revised classification. Monog Syst Bot Missouri Bot Gard 98:203–223

Goffinet B, Wickett NJ, Werner O, Ros RM, Shaw AJ, Cox CJ (2007) Distribution and phylogenetic significance of the 71-kb inversion in the plastid genome in Funariidae (Bryophyta). Ann Bot 99:747–753

Goulding SE, Olmstead RG, Morden CW, Wolfe KH (1996) Ebb and flow of the chloroplast inverted repeat. Mol Gen Genet 252:195–206

Groth-Malonek M, Pruchner D, Grewe F, Knoop V (2005) Ancestors of trans-splicing mitochondrial introns support serial sister group relationships of hornworts and mosses with vascular plants. Mol Biol Evol 22:117–125

Hammani K, Okuda K, Tanz SK, Chateigner-Boutin AL, Shikanai T, Small I (2009) A study of new *Arabidopsis* chloroplast RNA editing mutants reveals general features of editing factors and their target sites. Plant Cell 21:3686–3699

Hirose A, Kusumegi T, Tsudzuki T, Sugiura M (1999) RNA editing sites in tobacco chloroplast transcripts: editing as a possible regulator of chloroplast RNA polymerase activity. Mol Gen Genet 262:462–467

Holton TA, Pisani D (2010) Deep genomic-scale analyses of the Metazoa reject Coelomata: evidence from single- and multigene families analyzed under a supertree and supermatrix paradigm. Genome Biol Evol 2:310–324

Jansen RK, Raubeson LA, Boore JL, dePamphilis CW, Chumley TW, Haberle RC, Wyman SK, Alverson AJ, Peery R, Herman SJ (2005) Methods for obtaining and analyzing whole chloroplast genome sequences. Methods Enzymol 395:348–384

Jansen RK, Cai Z, Raubeson LA, Daniell H, dePamphilis CW, Leebens-Mack J, Müller KF, Guisinger-Bellian M, Haberle RC, Hansen AK, Chumley TW, Lee S-B, Peery R, McNeal JR, Kuehl JV, Boore JL (2007) Analysis of 81 genes from 64 plastid genomes resolves relationships in angiosperms and identifies genome-scale evolutionary patterns. Proc Natl Acad Sci USA 104:19369–19374

Jobson RW, Qiu YL (2008) Did RNA editing in plant organellar genomes originate under natural selection or through genetic drift? Biol Direct 3:43

Karol KG, Arumuganathan K, Boore JL, Duffy AM, Everett KDE, Hall JD, Hansen SK, Kuehl JV, Mandoli DF, Mishler BD, Olmstead RG, Renzaglia KS, Wolf PG (2010) Complete plastome sequences of *Equisetum arvense* and *Isoetes flaccida*: implications for phylogeny and plastid genome evolution of early land plant lineages. BMC Evol Biol 10:321

Kenrick P, Crane PR (1997) The origin and early diversification of land plants: a cladistic study. Smithsonian Press, Washington, DC

Kobayashi Y, Matsuo M, Sakamoto K, Wakasugi T, Yamada K, Obokata J (2008) Two RNA editing sites with cis-acting elements of moderate sequence identity are recognized by an identical site-recognition protein in tobacco chloroplasts. Nucleic Acids Res 36:311–318

Kolodner R, Tewari KK (1975a) Chloroplast DNA from higher plants replicates by both the Cairns and the rolling circle mechanism. Nature 256:708–711

Kolodner R, Tewari KK (1975b) Presence of displacement loops in covalently closed circular chloroplast deoxyribonucleic-acid from higher-plants. J Biol Chem 250:8840–8847

Kotera E, Tasaka M, Shikanai T (2005) A pentatricopeptide repeat protein is essential for RNA editing in chloroplasts. Nature 433:326–330

Kranz HD, Huss VAR (1996) Molecular evolution of pteridophytes and their relationship to seed plants: evidence from complete 18S rRNA gene sequences. Plant Syst Evol 202:1–11

Krishnan NM, Rao BJ (2009) A comparative approach to elucidate chloroplast genome replication. BMC Genomics 10:237

Krishnan NM, Seligmann H, Raina SZ, Pollock DD (2004) Detecting gradients of asymmetry in site-specific substitutions in mitochondrial genomes. DNA Cell Biol 23:707–714

Kugita M, Yamamoto Y, Fujikawa T, Matsumoto T, Yoshinaga K (2003a) RNA editing in hornwort chloroplasts makes more than half the genes functional. Nucleic Acids Res 31:2417–2423

Kugita M, Kaneko A, Yamamoto Y, Takeya Y, Matsumoto T, Yoshinaga K (2003b) The complete nucleotide sequence of the hornwort (*Anthoceros formosae*) chloroplast genome: insight into the earliest land plants. Nucleic Acids Res 31:716–721

Kunnimalaiyaan M, Nielsen BL (1997) Chloroplast DNA replication: mechanism, enzymes and replication origins. J Plant Biochem Biotech 6:1–7

Miller MA, Holder MT, Vos R, Midford PE, Liebowitz T, Chan L, Hoover P, Warnow T (2009) The CIPRES Portals. http://www.phylo.org/sub_sections/portal

Mishler BD, Churchill SP (1984) A cladistic approach to the phylogeny of the bryophytes. Brittonia 36:406–424

Newton AE, Cox CJ, Duckett JG, Wheeler JA, Goffinet B, Hedderson TAJ, Mishler BD (2000) Evolution of the major moss lineages: phylogenetic analyses based on multiple gene sequences and morphology. Bryologist 103:187–211

Nickrent DL, Parkinson CL, Palmer JD, Duff RJ (2000) Multigene phylogeny of land plants with special reference to bryophytes and the earliest land plants. Mol Biol Evol 17:1885–1895

Oda K, Yamato K, Ohta E, Nakamura Y, Takemura M, Nozato N, Akashi K, Kanegae T, Ogura Y, Kohchi T, Ohyama K (1992) Gene organization deduced from the complete sequence of liverwort *Marchantia polymorpha* mitochondrial DNA: a primitive form of plant mitochondrial genome. J Mol Biol 223:1–7

Ohyama K, Fukuzawa H, Kohchi T, Shirai H, Sano T, Sano S, Umesono K, Shiki Y, Takeuchi M, Chang Z, Aota S, Inokuchi H, Ozeki H (1986) Chloroplast gene organization deduced from complete sequence of liverwort *Marchantia polymorpha* chloroplast DNA. Nature 322:572–574

Okuda K, Shikanai T (2008) PPR proteins function as a trans-factor in chloroplast RNA editing. In: Allen JF, Gantt E, Golbeck J (eds) Photosynthesis energy from the sun. Springer, New York, pp 1211–1214

Oldenburg DJ, Bendich AJ (2004a) Most chloroplast DNA of maize seedlings in linear molecules with defined ends and branched forms. J Mol Biol 335:953–970

Oldenburg DJ, Bendich AJ (2004b) Changes in the structure of DNA molecules and the amount of DNA per plastid during chloroplast development in maize. J Mol Biol 344:1311–1330

Oliver MJ, Murdock AG, Mishler BD, Kuehl JV, Boore JL, Mandoli DF, Everett KDE, Wolf PG, Duffy AM, Karol KG (2010) Chloroplast genome sequence of the moss *Tortula ruralis*: gene content, polymorphism, and structural arrangement relative to other green plant chloroplast genomes. BMC Genomics 11:143

Palmer JD (1983) Chloroplast DNA exists in 2 orientations. Nature 301:92–93

Palmer JD (1985) Comparative organization of chloroplast genomes. Annu Rev Genet 19:325–354

Palmer JD, Stein DB (1986) Conservation of chloroplast genome structure among vascular plants. Curr Genet 10:823–833

Pryer KM, Schneider H, Smith AR, Cranfill R, Wolf PG, Hunt JS, Sipes SD (2001) Horsetails and ferns are a monophyletic group and the closest living relatives to seed plants. Nature 409:618–622

Pryer KM, Schuettpelz E, Wolf PG, Schneider H, Smith AR, Cranfill R (2004) Phylogeny and evolution of ferns (monilophytes) with a focus on the early leptosporangiate divergences. Amer J Bot 91:1582–1598

Qiu YL (2008) Phylogeny and evolution of charophytic algae and land plants. J Syst Evol 46:287–306

Qiu YL, Cho YR, Cox JC, Palmer JD (1998) The gain of three mitochondrial introns identifies liverworts as the earliest land plants. Nature 394:671–674

Qiu YL, Li LB, Wang B, Chen ZD, Knoop V, Groth-Malonek M, Dombrovska O, Lee J, Kent L, Rest J, Estabrook GF, Hendry TA, Taylor DW, Testa CM, Ambros M, Crandall-Stotler B, Duff RJ, Stech M, Frey W, Quandt D, Davis CC (2006) The deepest divergences in land plants inferred from phylogenomic evidence. Proc Natl Acad Sci USA 103:15511–15516

Qiu YL, Li LB, Wang B, Chen ZD, Dombrovska O, Lee J, Kent L, Li RQ, Jobson RW, Hendry TA, Taylor DW, Testa CM, Ambros M (2007) A nonflowering land plant phylogeny inferred from nucleotide sequences of seven chloroplast, mitochondrial, and nuclear genes. Int J Plant Sci 168:691–708

Raubeson LA, Jansen RK (1992) Chloroplast DNA evidence on the ancient evolutionary split in vascular land plants. Science 255:1697–1699

Renzaglia KS, Garbary DJ (2001) Motile gametes of land plants: diversity, development, and evolution. Crit Rev Plant Sci 20:107–213

Renzaglia KS, Schuette S, Duff RJ, Ligrone R, Shaw AJ, Mishler BD, Duckett JG (2007) Bryophyte phylogeny: advancing the molecular and morphological frontiers. Bryologist 110:179–213

Roper JM, Hansen SK, Wolf PG, Karol KG, Mandoli DF, Everett KDE, Kuehl J, Boore JL (2007) The complete plastid genome sequence of *Angiopteris evecta* (G. Forst.) Hoffm. (Marattiaceae). Am Fern J 97:95–106

Rothwell GW, Nixon KC (2006) How does the inclusion of fossil data change our conclusions about the phylogenetic history of euphyllophytes? Int J Plant Sci 167:737–749

Rowan BA, Oldenburg DJ, Bendich AJ (2010) *RecA* maintains the integrity of chloroplast DNA molecules in *Arabidopsis*. J Exp Bot 61:2575–2588

Rubinstein CV, Gerrienne P, de la Puente GS, Astini RA, Steemans P (2010) Early middle Ordovician evidence for land plants in Argentina (eastern Gondwana). New Phytol 188:365–369

Sato N, Terasawa K, Miyajima K, Kabeya Y (2003) Organization, developmental dynamics, and evolution of plastid nucleoids. Int Rev Cytol Surv Cell Biol 232:217–262

Schneider H, Smith AR, Pryer KM (2009) Is morphology really at odds with molecules in estimating fern phylogeny? Syst Bot 34:455–475

Shalchian-Tabrizi K, Minge MA, Espelund M, Orr R, Ruden T, Jakobsen KS, Cavalier-Smith T (2008) Multigene phylogeny of Choanozoa and the origin of animals. PLoS One 3:e2098

Shaw J, Renzaglia K (2004) Phylogeny and diversification of bryophytes. Amer J Bot 91:1557–1581

Shinozaki K, Ohme M, Tanaka M, Wakasugi T, Hayashida N, Matsubayashi T, Zaita N, Chunwongse J, Obokata J, Yamaguchi-Shinozaki K, Ohto C, Torazawa K, Meng BY, Sugita M, Deno H, Kamogashira T, Yamada K, Kusuda J, Takaiwa F, Kato A, Tohdoh N, Shimada H, Sugiura M (1986) The complete nucleotide sequence of tobacco chloroplast genome: its gene organization and expression. EMBO J 5:2043–2049

Stamatakis A (2006) RAxML-VI-HPC: maximum likelihood-based phylogenetic analyses with thousands of taxa and mixed models. Bioinformatics 22:2688–2690

Stamatakis A, Hoover P, Rougemont J (2008) A rapid bootstrap algorithm for the RAxML web-servers. Syst Biol 57:758–771

Stein DB, Palmer JD, Thompson WF (1986) Structural evolution and flip-flop recombination of chloroplast DNA in the fern genus *Osmunda*. Curr Genet 10:835–841

Stern DB, Goldschmidt-Clermont M, Hanson MR (2010) Chloroplast RNA metabolism. Annu Rev Plant Biol 61:125–155

Sugiura C, Kobayashi Y, Aoki S, Sugita C, Sugita M (2003) Complete chloroplast DNA sequence of the moss *Physcomitrella patens*: evidence for the loss and relocation of *rpoA* from chloroplast to the nucleus. Nucleic Acids Res 31:5324–5331

Tillich M, Lehwark P, Morton BR, Maier UG (2006) The evolution of chloroplast RNA editing. Mol Biol Evol 23:1912–1921

Tsuji S, Ueda K, Nishiyama T, Hasebe M, Yoshikawa S, Konagaya A, Nishiuchi T, Yamaguchi K (2007) The chloroplast genome from a lycophyte (microphyllophyte), *Selaginella uncinata*, has a unique inversion, transpositions and many gene losses. J Plant Res 120:281–290

Turmel M, Pombert J-F, Charlebois P, Otis C, Lemieux C (2007) The green algal ancestry of land plants as revealed by the chloroplast genome. Int J Plt Sci 168:679–689

Wahrmund U, Rein T, Müller KF, Groth-Malonek M, Knoop V (2009) Fifty mosses on five trees: comparing phylogenetic information in three types of non-coding mitochondrial DNA and two chloroplast loci. Plant Syst Evol 282:241–255

Wahrmund U, Quandt D, Knoop V (2010) The phylogeny of mosses – addressing open issues with a new mitochondrial locus: group I intron cobi420. Mol Phylogen Evol 54:417–426

Wickett NJ, Fan Y, Lewis PO, Goffinet B (2008a) Distribution and evolution of pseudogenes, gene losses, and a gene rearrangement in the plastid genome of the nonphotosynthetic liverwort, *Aneura mirabilis* (metzgeriales, jungermanniopsida). J Mol Evol 67:111–122

Wickett NJ, Zhang Y, Hansen SK, Roper JM, Kuehl JV, Plock SA, Wolf PG, dePamphilis CW, Boore JL, Goffinet B (2008b) Functional gene losses occur with minimal size reduction in the plastid genome of the parasitic liverwort *Aneura mirabilis*. Mol Biol Evol 25:393–401

Wolf PG, Rowe CA, Sinclair RB, Hasebe M (2003) Complete nucleotide sequence of the chloroplast genome from a leptosporangiate fern, *Adiantum capillus-veneris* L. DNA Res 10:59–65

Wolf PG, Rowe CA, Hasebe M (2004) High levels of RNA editing in a vascular plant chloroplast genome: analysis of transcripts from the fern *Adiantum capillus-veneris*. Gene 339:89–97

Wolf PG, Karol KG, Mandoli DF, Kuehl J, Arumuganathan K, Ellis MW, Mishler BD, Kelch DG, Olmstead RG, Boore JL (2005) The first complete chloroplast genome sequence of a lycophyte, *Huperzia lucidula* (Lycopodiaceae). Gene 350:117–128

Wolf PG, Roper JM, Duffy AM (2010) The evolution of chloroplast genome structure in ferns. Genome 53:731–738

Wolf PG, Der JP, Duffy AM, Jacobson JB, Grusz AL, Pryer KM (2011) The evolution of chloroplast genes and genomes in ferns. Plant Mol Biol 76: 251–261

Wolfe KH, Morden CW, Palmer JD (1992) Functions and evolution of a minimal plastid genome from a nonphotosynthetic parasitic plant. Proc Natl Acad Sci USA 89:10648–10652

Chapter 5

Plastid Genomes of Seed Plants

Robert K. Jansen* and Tracey A. Ruhlman
Section of Integrative Biology, University of Texas at Austin, 1 University Station C0930, Austin, TX 78712, USA
and Department of Biological Sciences, Genomics and Biotechnology Section, King Abdulaziz University, Jeddah 21589, Saudi Arabia

Summary

The field of comparative plastid genomics has burgeoned during the past decade, largely due to the availability of rapid, less expensive genome sequencing technologies. Currently there are 200 plastid genomes (plastomes) publicly available with 65% of these from seed plants. Comparative analyses have demonstrated that there is an overall stability in plastome architecture, gene and intron content, and gene order across seed plants. However, a number of unrelated lineages of both gymnosperms and angiosperms do not follow this pattern and have experienced considerable genomic upheaval. Within angiosperms these lineages with highly rearranged plastomes exhibit three other phenomena: highly accelerated rates of nucleotide substitutions, an increase in the number of dispersed repeats, many of which are associated with rearrangement endpoints, and biparental plastid inheritance. The correlation

*Author for correspondence, e-mail: jansen@mail.utexas.edu

R. Bock and V. Knoop (eds.), Genomics of Chloroplasts and Mitochondria, Advances in Photosynthesis and Respiration 35, pp. 103–126,
DOI 10.1007/978-94-007-2920-9_5,

between genomic upheaval and these other phenomena suggest that aberrant DNA repair mechanisms may be involved in destabilizing these plastid genomes. Experimental studies support this idea because knocking out DNA repair genes destabilizes plastomes. Further studies of nuclear-plastid interactions, especially in seed plant lineages with highly rearranged plastomes, are needed to clarify the causes of the plastome instability. The large number of plastome sequences has also provided valuable data for resolving phylogenetic relationships among seed plants. This is especially true for angiosperms where these data have been instrumental in clarifying relationships among the early diverging clades, an endeavor that had stymied plant biologists for over a decade. The most recent plastome phylogenies clearly identify *Amborella* as the earliest diverging lineage of flowering plants and provide strong support for the position of magnoliids as sister to a large clade that includes eudicots and monocots. This robust phylogenetic estimate provides an evolutionary framework for examining patterns and rates of change in plastid genomes across angiosperms.

I. Introduction

The field of comparative plastid genomics began in 1986 with the publication of the first two land plant plastid genome sequences (plastomes) for *Nicotiana tabacum* (Shinozaki et al. 1986) and *Marchantia polymorpha* (Ohyama et al. 1986). Following these landmark papers there was a slow and steady increase in the number of completed plastome sequences for about 15 years (Fig. 5.1). The development of less expensive, high-throughput DNA sequencing methods resulted in a rapid rise in the number of publicly available plastome sequences during the past decade. Currently (as of February 16, 2011) there are 205 plastome sequences available on Genbank representing many major lineages of photosynthetic organisms. The vast majority (175) represent green plants (Viridiplantae) with most of these from the two major lineages of seed plants, angiosperms (118) and gymnosperms (16). Within seed plants many of the major lineages of gymnosperms (Wu et al. 2007, 2009; McCoy et al. 2008; Lin et al. 2010; Zhong et al. 2010) and angiosperms (Jansen et al. 2007, 2011; Moore et al. 2007) are now represented, although there is still limited sampling for some clades, especially among gymnosperms. The increased availability of plastome sequences has provided a wealth of new comparative data for understanding patterns of genome organization, rates of sequence evolution, mechanisms of evolutionary change, and phylogenetic relationships among seed plants.

During the past decade there have been several reviews of plastid genome organization and evolution and the phylogenetic implications of the newly acquired plastome data (Odintsova and Yurina 2003; Raubeson and Jansen 2005; Bock 2007; Ravi et al. 2008; Khan et al. 2010; Gao et al. 2010; Wolf et al. 2011) but many of these were published when there were limited plastome sequences available and none of them focused exclusively on seed plants. In this chapter, we summarize the current knowledge of the organization and evolution of seed plant plastid genomes with a focus on their genome organization, inheritance, rate of nucleotide substitution and genomic rearrangements, and the utility of plastome data for resolving phylogenetic relationships.

Abbreviations: bp – Basepair; indel – Insertion/deletion; IR – Inverted repeat; IRLC – Inverted repeat lacking clade; kb – Kilobase; LSC – Large single copy region; PEP – Plastid encoded polymerase; SSC – Small single copy region

II. Plastid Genome Organization

A. Overall Organization

In general plastome organization is highly conserved among seed plants with most having a quadripartite structure with two

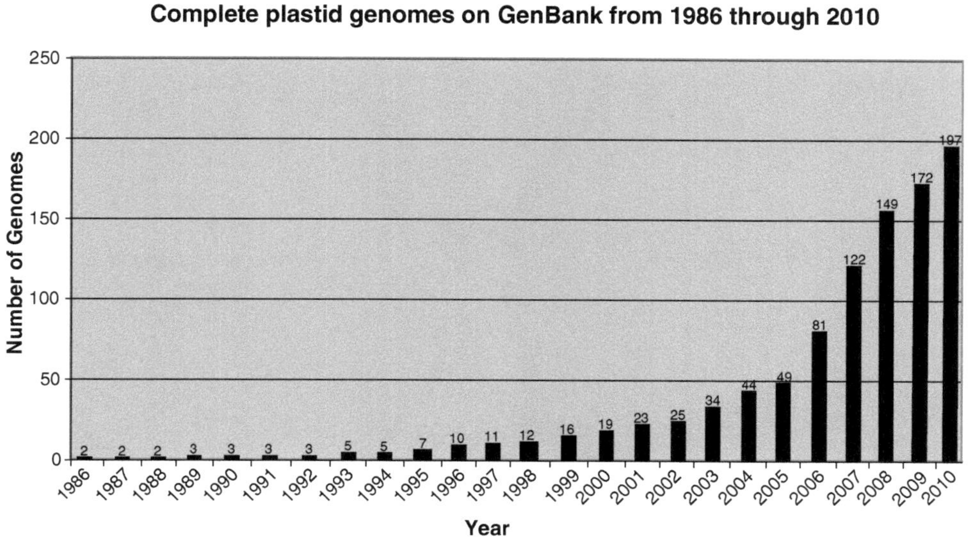

Fig. 5.1. Histogram showing number of plastid genomes available on GenBank from 1986 to December 1, 2010. (http://www.ncbi.nlm.nih.gov/genomes/GenomesGroup.cgi?taxid=2759&opt=plastid)

copies of a large inverted repeat (IR) separated by small (SSC) and large (LSC) single copy regions (Fig. 5.2a, b). The two copies of the IR facilitate flip–flop recombination resulting in the presence of isoforms that differ in the orientation of the single copy regions (Palmer 1983). The prevailing view has been that plastid genomes are circular, and this was supported by early electron microscopic comparisons that revealed circular genomes in either monomeric or multimeric form (Kolodner and Tewari 1979). More recently considerable evidence has accumulated that suggests a much more complex structure, with circular, linear, branched, and multimeric configurations that vary during plastid development (Lilly et al. 2001; Bendich 2004; Oldenburg and Bendich 2004; Shaver et al. 2006).

The majority of plastid genes are contained in operons and transcribed as polycistronic units, a feature that reflects the endosymbiotic origin of plastids from a cyanobacterial ancestor (see Fig. 5.2b for operon organization). Among seed plant plastomes there are very few instances of disruption of operons. Exceptions occur in the three angiosperm families Campanulaceae (Cosner et al. 1997; Haberle 2006; Haberle et al. 2008), Geraniaceae (Chumley et al. 2006; Guisinger et al. 2011), and Fabaceae (Milligan et al. 1989; Cai et al. 2008; Palmer et al. 1988; Perry et al. 2002). In the Geraniaceae, the *rps2-atpA* operon is disrupted in the most recent common ancestor of *Erodium texanum* and *Geranium palmatum* but two other plastomes from this family, *Pelargonium hortorum* and *Monsonia speciosa*, have this operon intact. The highly conserved S10 operon is also disrupted in Geraniaceae. This operon is split into two groups of genes, *rpl23-rps3* and *rpl16-rpoA,* in *E. texanum* and *M. speciosa*, whereas it is split into four pieces (*rpl23*, *rpl2*, *rps19-rpl22*, and *rps3-rpoA*) in *G. palmatum*. *Pelargoniuim hortorum* has an intact S10 operon except that *rpoA* is so divergent that its functionality is in question. In Campanulaceae, the plastomes have two disrupted operons, *rps2–atpI–atpH-atpF–atpA* and *clpP-5′rps12-rpl20* (Haberle 2006; Haberle et al. 2008). Two operons, *rpoB-rpoC1-rpoC2* and *clpP-5′rps12-rpl20*, are disrupted in *Trifolium* but they are intact in the plastomes of the seven other sequenced legumes, including the closely related genus *Medicago*. The disruption of

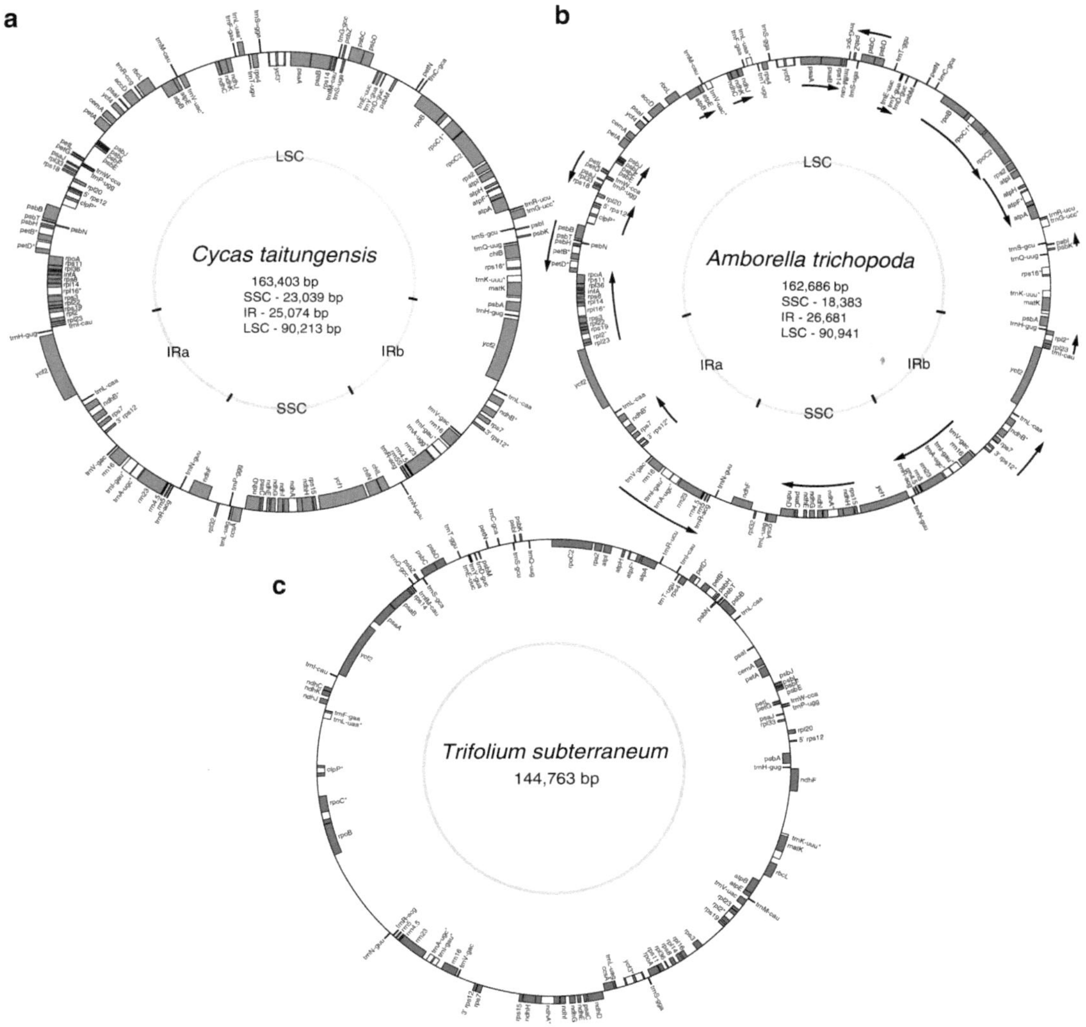

Fig. 5.2. Physical maps of the three seed plant plastid genomes: (**a**) *Cycas taitungensis* (NC_009618); (**b**) *Amborella trichopoda* (NC_005086); and (**c**) *Trifolium subterraneum* (NC_011828). Maps were constructed using GenomeVX (Conant and Wolfe 2008; http://wolfe.gen.tcd.ie/GenomeVx/). Genes annotated inside the circle are transcribed clockwise while those outside are transcribed counterclockwise. *Arrows* in (**b**) indicate polycistrionic transcription units. Introns are annotated as open boxes and genes containing introns are marked with *asterisks*.

the *rpoB* operon is notable because it includes three of four genes for the plastid-encoded RNA polymerase (PEP), the multi-subunit enzyme that transcribes many plastid genes. In all of these cases, relocated segments of operons must have acquired new promoters to drive gene transcription but experimental studies have not been performed to determine how these segments are transcribed in their new location.

B. Genome Size, Gene/Intron Content, and GC Content

Genome size varies considerably among photosynthetic seed plant plastomes (Table 5.1),

Table 5.1. Characteristics of plastid genomes of representative photosynthetic seed plants and the parasitic, non-photosynthetic plant *Epifagus virginica*. Data in this table came from Wu et al. 2009; Lin et al. 2010, and Guisinger et al. 2011 or from analyses using sequences on GenBank (see a–g below)

Genome characteristic	*Cycas taitungensis*[a]	*Ephedra equisetina*[b]	*Pinus thunbergii*[c]	*Cryptomeria japonica*[d]	*Amborella trichopoda*[e]	*Pelargonium hortorum*[f]	*Epifagus virginica*[g]
Size (bp)	163,403	109,518	119,707	131,810	162,686	217,942	70,028
LSC length (bp)	90,216	59,906	65,696	93,958	90,941	59,710	19,799
SSC length (bp)	23,039	8,104	53,021	37,624	18,383	6,750	4,759
IR length (bp)	25,074	20,754	495	114	26,681	75,741	22,735
Number of different genes/total number of genes	109/124	101/117	114/115	116/118	114/132	109/160	42/53
Number of different protein-coding genes (duplicated in IR)	74 (4)	66 (6)	66 (0)	82 (0)	80 (7)	76 (39)	21 (3)
Number of different tRNA genes (duplicated in IR)	31 (7)	31 (6)	32	32 (1)	30 (7)	29 (8)	17 (4)
Number of different rRNA genes (duplicated in IR)	4 (4)	4 (4)	4 (0)	4 (0)	4 (4)	4 (4)	4 (4)
Number of genes duplicated in IR	15	16	1	2	18	51	11
Number of different genes with introns	18	12	12	17	18	16	4
Percent of genome coding for genes	57.2	72.3	56.5	60.8	56.6	51.5	43
Gene density – total number of genes/genome length including IR (genes/kb)	0.759	1.068	1.01	0.91	0.81	0.734	0.757
GC content (%)	39.4	36.6	38.5	35.4	38.3	39.6	36.0

[a]NC_009618 (Wu et al. 2007)
[b]NC_011954 (Wu et al. 2009)
[c]NC_001631 (Wakasugi et al. 1994)
[d]NC_010548 (Hirao et al. 2008)
[e]NC_005086 (Goremykin et al. 2003)
[f]NC_008454 (Chumley et al. 2006)
[g]NC_001568 (Wolfe et al. 1992)

ranging from 107,122 bp (*Cathaya argophylla* [NC_014589]) to 217,942 bp (*Pelargonium hortorum* [NC_008454]) with an average length of 144,824 bp. The most remarkable example of plastome size variation within a single family occurs in the Geraniaceae with the smallest genome, *Erodium carvifolium*, at 116,935 bp and the largest, *Pelargonium hortorum* at 217,942 bp (Guisinger et al. 2011; Blazier et al. 2011). Plastomes from a few non-photosynthetic, parasitic plants have been sequenced and these genomes are greatly reduced in size ranging from 59,190 bp in *Rizanthella gardneri* (Delannoy et al. 2011) to 86,744 bp in *Cuscuta gronovii* (Funk et al. 2007). Several factors contribute to this wide variation in genome size. First, expansion/contraction and loss of the IR is one of the most evident causes; it has been recognized for some time that small changes in the extent of the IR are very common due to shifting of the IR/SC boundaries (Goulding et al. 1996). There is considerable variation in IR size across seed plants ranging from absent in the

IR loss clade of legumes (Lavin et al. 1990) and *Erodium texanum* (Guisinger et al. 2011) to 75,741 bp in *Pelargonium hortorum* (Chumley et al. 2006). The second factor contributing to genome size variation is gene loss and additional gene duplications outside of the IR. In gnetophytes (McCoy et al. 2008; Wu et al. 2009) the loss of up to 18 genes has resulted in a more compact genome with gene densities lower than other gymnosperms and most angiosperms (Table 5.1). In Geraniaceae, there has been considerable partial or complete duplication of genes, which could be partly responsible for the larger plastomes of some members of this angiosperm family (Guisinger et al. 2011). The third factor involves downsizing of introns and intergenic spacer regions as was observed for the gnetophyte plastomes (McCoy et al. 2008; Wu et al. 2009).

Seed plant plastomes usually contain 101–118 different genes (Table 5.1) with the majority of these (66–82) coding for proteins involved in photosynthesis and gene expression and several others with miscellaneous functions, 29–32 transfer RNAs, and four ribosomal RNA genes (Fig. 5.2, Table 5.1). The range in the total number of genes is higher (115–160), largely due to duplication of genes in the IR (Table 5.1). The highest number (160 in *Pelargonium hortorum*) is due to the duplication of 39 genes in the 76 kilobase (kb) IR (Chumley et al. 2006). Among photosynthetic seed plants the gnetophytes have the most reduced gene content with up to 18 gene losses, including the absence of all 11 NADH dehydrogenase genes (Wu et al. 2009). The most highly reduced plastomes in terms of gene content are from parasitic plants that have varying degrees of capacity for photosynthesis. The completely non-photosynthetic plants *Epifagus virginica* and *Rhizanthella gardneri* have only 42 and 33 intact genes, respectively (Wolfe et al. 1992; Delannoy et al. 2011), and most of these represent ribosomal genes (tRNAs, rRNAs, and ribosomal proteins). The genus *Cuscuta* has wide variation in adaptations to a parasitic life history. Some species are fully photosynthetic, others are intermediate with limited photosynthetic capacity and others are completely non-green and parasitic. Recent examinations of plastomes from four *Cuscuta* species with varying photosynthetic capacity have demonstrated a progressive loss of plastid genes with increasing parasitism (Funk et al. 2007; McNeal et al. 2007b).

Intron content of seed plant plastomes is highly conserved; most have 18 genes with introns, six in tRNAs and 12 in protein coding genes (Fig. 5.2a, b). Fifteen of the 18 intron-containing genes have a single intron and three genes, *ycf3*, *clpP*, and *rps12*, have two introns, resulting in a total of 21 introns in most seed plants. Twenty of the introns are group II, whereas *trnL*-uaa is the only group I intron in seed plant plastomes. Splicing of exons for 17 of these genes involves cis-splicing. The single exception is *rps12*, which has cis-splicing for exons 2 and 3 and trans-splicing for exon 1 (Hildebrand et al. 1988). Variation in the number of genes with introns occurs among photosynthetic seed plants, ranging from 12 in gnetophytes to 18 in most species (Table 5.1). Similar to the situation in gene content, parasitic plants have a reduced number of introns with only four and six different genes with introns in the holoparasites *Epifagus virginica* and *Rhizanthella gardneri* and 12 in hemiparasitic *Cuscuta* species (Wolfe et al. 1992; Funk et al. 2007; McNeal et al. 2007b; Delannoy et al. 2011).

GC content among seed plant plastomes ranges between 34% and 40% (Goremykin et al. 2003; Kim and Lee 2004; Cai et al. 2006; Raubeson et al. 2007; Guisinger et al. 2011; Table 5.1). There is an uneven distribution of GC content over the plastid genome and this pattern is due primarily to three factors (Cai et al. 2006). First, coding regions have a significantly higher GC content than non-coding regions. Second, the distribution of GC content by regions of the genome varies with the highest in the IR and the lowest in the SSC. The higher GC content in the IR is due primarily to the presence of the four rRNA genes that have highest GC content of any coding regions. The lowest GC content

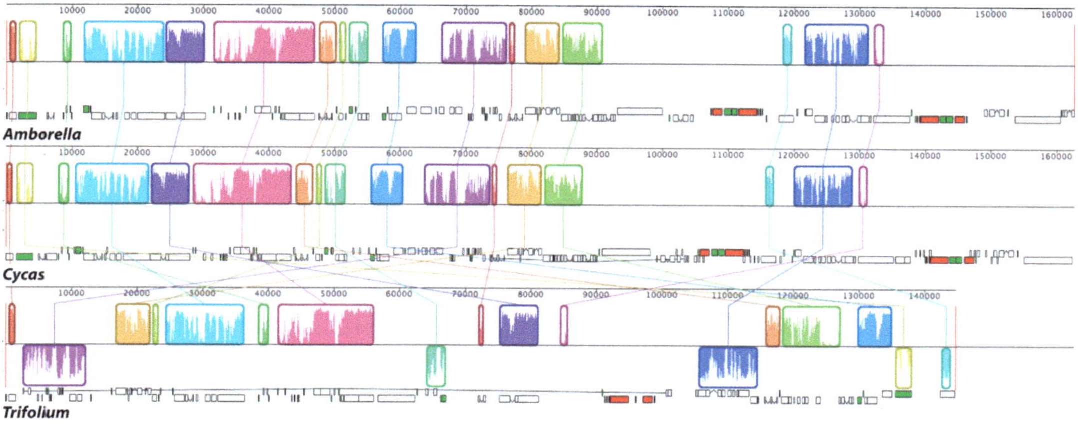

Fig. 5.3. Gene order comparison of two highly conserved and one rearranged plastid genomes. Whole plastid genome sequences were downloaded from Genbank for *Cycas taitungensis* (NC_009618), *Amborella trichopoda* (NC-005086) and *Trifolium subterraneum* (NC_011828). Alignments were performed in Geneious Pro (Drummond et al. 2010) with the mauveAligner algorithm (Darling et al. 2010), which aligns synthetic blocks of genes and predicts inversions relative to a reference genome.

in the SSC is caused by the presence of 8 of the 11 NADH dehydrogenase genes, which have the lowest GC content of any functional group. Third, GC content varies by functional groups. Among protein-coding genes, GC content is highest for photosynthetic genes, lowest for NADH genes, with genetic system genes having intermediate values. GC content also varies by codon position in protein-coding genes (Cai et al. 2006; Raubeson et al. 2007; Guisinger et al. 2011). For each of the three classes of genes (photosynthetic, genetic system, and NADH) the third position in the codon has a significant AT bias. This pattern has been attributed to codon usage bias (Shimada and Sugiura 1991; Kim and Lee 2004; Chaw et al. 2004; Liu and Xue 2005). Several studies have examined codon usage of plastid genes to determine if these biases can be attributed to nucleotide compositional bias, selection for translational efficiency, or a balance among mutational biases, natural selection, and genetic drift (Morton 1993, 1994, 1998; Wall and Herbeck 2003). The recent investigation of codon usage and GC content in Geraniaceae and related rosids concluded that codon usage in plastid genes is generally driven by selection and not GC content (Guisinger et al. 2011).

C. Gene Order

Gene order among seed plant plastomes is generally highly conserved. This is evident by comparison of the gene order of *Cycas* with the basal angiosperm *Amborella* (Fig. 5.3, top two genomes). These two genomes are co-linear, suggesting that the ancestral gene order for seed plants was similar to *Cycas*. Although there is no published plastome sequence for *Ginkgo*, gene mapping studies (Palmer and Stein 1986) indicate that this genome is also co-linear with *Cycas* and basal angiosperms. Despite the high level of conservation in gene order across seed plants, a number of groups, including gnetophytes, conifers, and several lineages of angiosperms have experienced considerable change (see Fig. 5.3 and Sect. IV below).

Three different mechanisms have been suggested to cause gene order changes in seed plant plastomes. First, inversion facilitated by recombination is considered the most common mechanism of plastome rearrangement (Palmer 1991; Raubeson and Jansen 2005). Intramolecular recombination of plastid DNA has been documented in *Oryza*; in this case repeats <15 bp recombine and generate deletions in both coding and non-coding regions (Kanno et al. 1993; Kawata et al. 1997). Intermolecular recombination between tRNA sequences in *Oryza* was also shown to result in gene order change (Hiratsuka et al. 1989). Recombination between repeats has generated genome rearrangements in transplastomic plants, providing experimental evidence for this mechanism in generating inversions (Rogalski et al. 2006; Gray et al. 2009). Several studies in angiosperms have documented the presence of a large number of repetitive sequences in highly rearranged plastomes, with the highest concentration of repeats occurring at rearrangement endpoints (Haberle et al. 2008; Chumley et al. 2006; Guisinger et al. 2011). The most extensive comparisons are in the Geraniaceae where the size and number of repeats is correlated with the degree of genomic rearrangement (Guisinger et al. 2011). Also, shared families of repeats flank rearrangement endpoints across the four genomes examined in this family. Second, transposition was suggested as a mechanism of plastome rearrangement in *Trachelium* (Cosner et al. 1997) and *Trifolium* (Milligan et al. 1989). However, plastid genome sequences for these species have not confirmed transposition in either of these plastomes (Haberle et al. 2008; Cai et al. 2008). The only case of a plastome transposable element is the degenerate "Wendy" element of the alga *Chlamydomonas* (Fan et al. 1995). Third, expansion and contraction of the IR has been suggestion as the cause of gene order changes in the green alga *Chlamydomonas* (Boudreau and Turmel 1995) and the angiosperm families Fabaceae (Perry et al. 2002) and Geraniaceae (Chumley et al. 2006; Guisinger et al. 2011). The Geraniaceae plastomes are the best example of this because of their incredible variation in the size of the IR, ranging from absent to 76 kb.

III. Plastid Inheritance

Considerable progress has been made in recent years to improve our understanding of modes of plastid inheritance in seed plants (reviewed in Hagemann 2004; Bock 2007; Hu et al. 2008; Nagata 2010; Kuroiwa 2010). Gymnosperms have not been investigated as extensively as angiosperms and they have been erroneously considered to have almost exclusively paternal inheritance. It turns out that all three modes (biparental, maternal and paternal) have been documented, with cycads, *Ginkgo*, and gnetophytes having maternal inheritance and conifers having substantial variation in mode of inheritance. Most studies have reported paternal plastid inheritance in conifers (Mogensen 1996) but biparental inheritance is known in *Cryptomeria* (Ohba et al. 1971) and progeny from crosses in *Larix* detected a mixture of maternal and paternal plastids (Szmidt et al. 1987). Overall, the prevailing mode of plastid inheritance in conifers is paternal but some species may have maternal or biparental inheritance. Examination of a broader phylogenetic diversity of gymnosperms is needed to fully understand the extent of variation in their mode of plastid inheritance.

Plastids have historically been thought to have largely maternal inheritance among angiosperms (Corriveau and Coleman 1988; Birky 1995; Mogensen 1996; Zhang et al. 2003; Hagemann 2004). The proportion of angiosperms with maternal inheritance is currently estimated at about 80% with the remaining species having biparental inheritance. The only known case of exclusively paternal inheritance is *Actinidia speciosa* (Testolin and Cipriani 1997). The phylogenetic distribution of mode of plastid inheritance in angiosperms indicates that maternal inheritance is ancestral and that there have been repeated conversions to biparental inheritance scattered among more derived lineages (Fig. 5.4; Hu et al. 2008). Furthermore, the phylogenetic distribution also indicates that changes in mode of inheritance are unidirectional because there are no cases of the derivation of maternal inheritance from a biparental ancestor.

Several different mechanisms are known to prevent paternal plastids from being transmitted during fertilization (Hagemann 2004; Bock 2007). Most angiosperms with maternal inheritance lack plastids in the generative cell, which results in their exclusion in the sperm cells. The presence of plastids in the generative cell does not necessarily result in their transmission to the embryo. Mechanisms to prevent paternal plastid transmission occur in multiple post-fertilization stages from exclusion just prior to fertilization to differential replication of maternal and paternal plastids in the embryo. Several studies have documented a surprising amount of variation in inheritance patterns, including situations where the mode of inheritance varied in progeny from crosses depending on whether the cross is inter- or intraspecific (Cruzan et al. 1993; Soliman et al. 1987; Yang et al. 2000; Lee et al. 1988; Hansen et al. 2007a).

There has been some discussion on why biparental inheritance of plastids has evolved multiple times from maternal inheritance. Zhang and Sodmergen (2010) suggested that biparental inheritance evolved as a mechanism to overcome defective maternal plastids in angiosperms with nuclear plastid incompatibility. This is a tantalizing hypothesis and is consistent with two lines of evidence. First, a number of angiosperms with nuclear plastid incompatibility systems have biparental inheritance, including *Oenothera* (Chiu and Sears 1993), *Passiflora* (Mrácek 2005), *Pelargonium* (Metzlaff et al. 1982), *Trifolium* (Pandey et al. 1987), and *Zantedeschia* (Snijder et al. 2007). Second, crossing studies demonstrate that more distant crosses (i.e., interspecific), which are more likely to cause genomic incompatibilities, result in progeny with paternal plastids, whereas progeny from crosses within species have maternal plastids.

IV. Genomic Rearrangements

A. IR Loss or Expansion/Contraction

The IR is present in the vast majority of seed plant plastomes, and some have argued that this structure promotes stability for the rest of the genome, largely via intramolecular recombination between the two IR copies, which limits recombination between the single copy regions (Palmer et al. 1987; Palmer 1991). This idea was supported earlier by the fact that plastomes known to lack the IR experienced more genomic rearrangements, especially some legumes. However, this correlation has not held up as more plastomes have been sequenced. In fact, some of the most highly rearranged seed plant plastomes have retained their IR, including the angiosperm families Campanulaceae (Cosner et al. 2004; Haberle et al. 2008), Lobeliaceae (Knox and Palmer 1999), Oleaceae (Lee et al. 2007), Geraniaceae (Chumley et al. 2006; Guisinger et al. 2011), and the gnetophytes among gymnosperms (McCoy et al. 2008; Wu et al. 2009).

The IR has been reported lost at least five times independently in seed plants. Within angiosperms IR loss has occurred at least two times within rosids (Fig. 5.4). The first reported loss was in a large, monophyletic group of papilionoid Fabaceae referred to as the inverted repeat lacking clade (IRLC; Wojciechowski et al. 2004). There have been two independent losses reported in the Geraniaceae in *Erodium texanum* and *Monsonia vanderietieae* (Downie and Palmer 1992) but only one of these has been verified (Guisinger et al. 2011; Blazier et al. 2011). Complete plastome sequences confirmed the IR loss in *Erodium texanum* (Guisinger et al. 2011) and *E. carvifolium* (Blazier et al. 2011), and draft genome sequences of 12 other *Erodium* species indicate that the IR has been lost throughout the genus (C. Blazier and R. Jansen, unpublished). The situation in *Monsonia* is not fully resolved. The complete genome sequence of *Monsonia speciosa* has an IR, although it is greatly reduced to 7 kb (Guisinger et al. 2011). A draft genome sequence of *M. vanderietieae* suggests that there may be a small IR of at least 3 kb, although assembly of this genome is complicated by the large number of rearrangements and repeats (M. Guisinger and R. Jansen, unpublished). Thus, it is likely that there has only been a single IR loss in Geraniaceae. IR losses have been suggested for two genera of Orobanchaceae, *Conophilis* and *Striga*

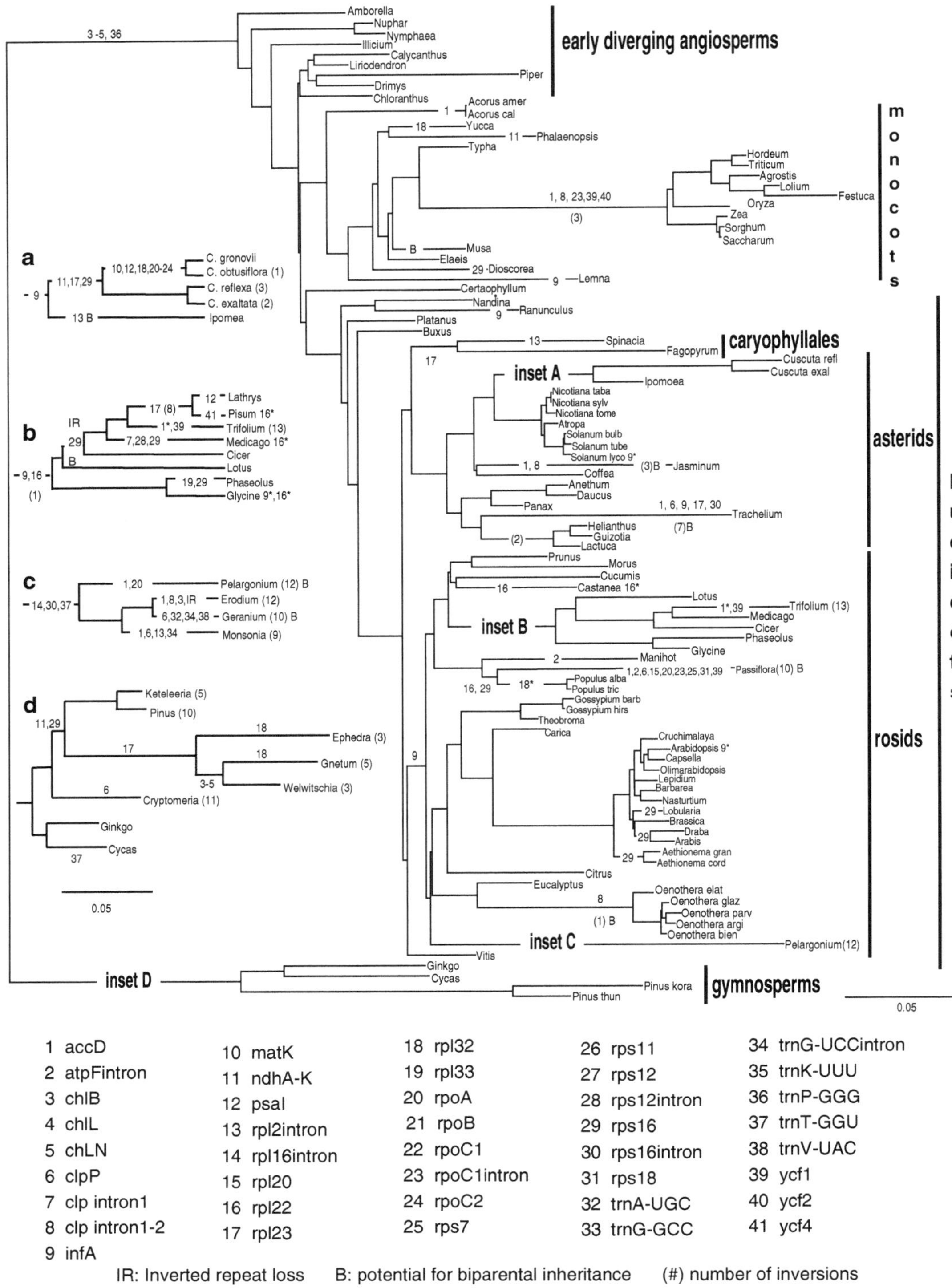
3 -5, 36
Amborella
Nuphar
Nymphaea
Illicium
Calycanthus
Liriodendron
Piper
Drimys
Chloranthus
early diverging angiosperms
1
Acorus amer
Acorus cal
18
Yucca
11
Phalaenopsis
Typha
Hordeum
Triticum
Agrostis
Lolium
Festuca
1, 8, 23,39,40
(3)
Oryza
Zea
Sorghum
Saccharum
B
Musa
Elaeis
29 Dioscorea
9
Lemna
monocots
Certaophyllum
Nandina
9
Ranunculus
Platanus
Buxus
13
Spinacia
Fagopyrum
caryophyllales
17
inset A
Cuscuta refl
Cuscuta exal
Ipomoea
Nicotiana taba
Nicotiana sylv
Nicotiana tome
Atropa
Solanum bulb
Solanum tube
Solanum lyco 9*
1, 8
(3)B
Jasminum
Coffea
Anethum
Daucus
Panax
1, 6, 9, 17, 30
Trachelium
(7)B
Helianthus
Guizotia
Lactuca
(2)
asterids
Prunus
Morus
Cucumis
16
Castanea 16*
inset B
Lotus
1*,39
Trifolium (13)
Medicago
Cicer
Phaseolus
Glycine
2
Manihot
1,2,6,15,20,23,25,31,39
Passiflora(10) B
16, 29
18*
Populus alba
Populus tric
Gossypium barb
Gossypium hirs
Theobroma
Carica
Cruchimalaya
Arabidopsis 9*
Capsella
Olimarabidopsis
Lepidium
Barbarea
Nasturtium
29
Lobularia
Brassica
Draba
Arabis
29
Aethionema gran
Aethionema cord
Citrus
Eucalyptus
8
(1) B
Oenothera elat
Oenothera glaz
Oenothera parv
Oenothera argi
Oenothera bien
inset C
Pelargonium(12)
9
Vitis
rosids
Eudicots
inset D
Ginkgo
Cycas
Pinus kora
Pinus thun
gymnosperms
0.05
a
9
11,17,29
10,12,18,20-24
C. gronovii
C. obtusiflora (1)
C. reflexa (3)
C. exaltata (2)
13 B
Ipomea
b
9,16
(1)
IR
29
B
17 (8)
12
Lathrys
41
Pisum 16*
1*,39
Trifolium (13)
7,28,29
Medicago 16*
Cicer
Lotus
19,29
Phaseolus
Glycine 9*,16*
c
14,30,37
1,20
Pelargonium (12) B
1,8,3,IR
Erodium (12)
6,32,34,38
Geranium (10) B
1,6,13,34
Monsonia (9)
d
11,29
Keteleeria (5)
Pinus (10)
17
18
Ephedra (3)
18
Gnetum (5)
3-5
Welwitschia (3)
6
Cryptomeria (11)
Ginkgo
Cycas
37
0.05
1 accD
2 atpFintron
3 chlB
4 chlL
5 chLN
6 clpP
7 clp intron1
8 clp intron1-2
9 infA
10 matK
11 ndhA-K
12 psaI
13 rpl2intron
14 rpl16intron
15 rpl20
16 rpl22
17 rpl23
18 rpl32
19 rpl33
20 rpoA
21 rpoB
22 rpoC1
23 rpoC1intron
24 rpoC2
25 rps7
26 rps11
27 rps12
28 rps12intron
29 rps16
30 rps16intron
31 rps18
32 trnA-UGC
33 trnG-GCC
34 trnG-UCCintron
35 trnK-UUU
36 trnP-GGG
37 trnT-GGU
38 trnV-UAC
39 ycf1
40 ycf2
41 ycf4
IR: Inverted repeat loss
B: potential for biparental inheritance
(#) number of inversions

(Downie and Palmer 1992; Palmer 1991). Draft plastid genome sequences for species from both of these genera confirm the IR loss in *Conophilis* but the situation in *Striga* remains uncertain because of assembly issues (C. dePamphilis, personal communication, 2011). If it turns out that both of these genera lack an IR these would be independent events because phylogenetic analyses of Orobanchaceae indicate that *Striga* and *Conophilis* are not sister genera (Bennett and Mathews 2006). Thus, there would be four independent IR losses in angiosperms, one in Fabaceae, one in Geraniaceae, and two in Orobanchaceae.

The fifth putative IR loss was reported in gymnosperms. Early work on *Pinus thunbergii* suggested that the IR was lost in this genus and that the loss was shared by all conifers (Raubeson and Jansen 1992a). However, it turns out that the IR in *Pinus* has been greatly reduced and consists of a 495 bp repeat that includes *trnI*-cau and a portion of *psbA* (Tsudzuki et al. 1992). Plastome sequences of other Pinaceae (Lin et al. 2010) have identified short IRs in three other genera, *Cathaya* – 429 bp, *Cedrus* – 236 bp, and *Keteleeria* – 267 bp. A similar situation occurs in another family of conifers, Cupressaceae (Hirao et al. 2008). The plastome sequence of *Cryptomeria japonica* has a residual IR that is only 114 bp and includes the *trnI*-cau gene (Lin et al. 2010).

B. Gene and Intron Loss

The ancestral genome organization represented by *Cycas* among gymnosperms and *Amborella* among angiosperms includes the full complement of genes and introns but there have been scattered gene and intron losses across seed plants based on the phylogenetic distribution of these events (Fig. 5.4). Within gymnosperms (Fig. 5.4d) most of these losses occurred in the gnetophytes and Pinaceae, including the loss of all 11 plastid-encoded subunits of NADH dehydrogenase. In the case of gnetophytes these losses are part of an overall downsizing of plastid genomes (McCoy et al. 2008; Wu et al. 2009). In angiosperms there is a high level of conservation of gene and intron content among the basal lineages with repeated bursts of losses in mostly unrelated lineages of monocot and eudicots (Fig. 5.4). In most cases, the causes of these losses or the fate of the genes has not been determined. In the case of intron loss, one mechanism that has been proposed involves reverse transcription of an edited RNA intermediate, followed by homologous recombination between an intron-less cDNA and the original intron-containing copy. This mechanism was suggested for the *atpF* intron loss in the angiosperm order Malphigiales (Daniell et al. 2008).

Although there have been many gene losses documented among seed plants (Fig. 5.4; Raubeson and Jansen 2005; Jansen et al. 2007; Magee et al. 2010) very few of these events have been investigated rigorously. It is widely known that plastid DNA transfer to the nucleus occurs at a high rate (Timmis et al. 2004; Matsuo et al. 2005; Noutsos et al. 2005) but only a few functional gene transfers to the nucleus have been

Fig. 5.4. Angiosperm and gymnosperm phylogenetic trees based on complete plastid genome sequences. The large maximum likelihood phylogram was constructed from 97 taxa based on 81 plastid gene sequences (adapted from Jansen et al. 2011; see this publication for details of the phylogenetic analyses). Scale bars for large tree and inset *d* indicate 0.05 substitutions per site. Inset *a* is a phylogram of gymnosperms adapted from Zhong et al. (2010). Insets *a*, *b*, and *c* are adapted from McNeal et al. (2007b), Magee et al. (2010), and Guisinger et al. (2011), respectively. Gene and intron losses and IR losses plotted on branches are based on Jansen et al. 2007, Magee et al. 2010, and on published papers on each of the sequenced genomes. The number of estimated inversions (in parentheses) is based on Jansen et al. (2007) and on GRIMM (http://grimm.ucsd.edu/GRIMM/; Bourque and Pevzner 2002) comparisons with *Cycas* (for gymnosperms) and *Amborella* (for angiosperms) for those taxa not included in Jansen et al. Asterisks indicate reported cases of nuclear transfer. *B* indicates those taxa that have biparental inheritance or the potential for biparental inheritance based on Corriveau and Coleman (1988), Birky (1995), Mogensen (1996), Zhang et al. (2003), Hagemann (2004), and Hu et al. (2008).

characterized in seed plants. The reason for the paucity of documented examples is twofold: (1) once a gene is transferred it must acquire the required sequences to properly regulate nuclear transcription and a transit peptide to target the product back to the plastid; and (2) there have been very few experimental studies to search for nuclear copies. Successful gene transfers to the nucleus in seed plants have been documented for only four genes (Fig. 5.4): *infA* in rosids (Millen et al. 2001), independent transfers of *rpl22* in Fabaceae (Gantt et al. 1991) and Fagaceae (Jansen et al. 2011), *rpl32* in some Salicaceae (Cusack and Wolfe 2007; Ueda et al. 2007), and *accD* in *Trifolium* (Magee et al. 2010). The loss of *rps16* from the plastomes of *Medicago* and *Populus* was determined to be a gene substitution because a nuclear-encoded, mitochondrial-targeted copy is now also targeted to the plastid (Ueda et al. 2008). The acetyl-CoA carboxylase (ACC) subunit D gene (*accD*) has been lost at least seven times among angiosperm plastid genomes (Fig. 5.4) and in one case (*Trifolium*) a copy was found in the nucleus (Magee et al. 2010). The fate of *accD* in grasses is different. In this case, the prokaryotic multisubunit enzyme has been replaced by plastid-targeted eukaryotic ACC (Konishi et al. 1996; Gornicki et al. 1997). A similar situation occurs in *Spinacia oleracea* where the prokaryotic plastid *rpl23* has been replaced by a eukaryotic cytosolic copy of this ribosomal protein (Bubunenko et al. 1994). Therefore, in the few examined cases of gene loss in seed plants, three different pathways have been detected: gene transfer to the nucleus (*infA*, *rpl22*, *rpl32* and *accD*), substitution by a nuclear-encoded, mitochondrial targeted gene product (RPS16), and substitution by a nuclear-encoded protein for a plastid gene product (ACC, RPL23).

C. Gene Order Changes

As mentioned earlier the majority of seed plant plastomes lack any changes in gene order as is evident by the comparison of the *Cycas* and *Amborella* genomes (Fig. 5.3). Thus, gene order has been highly conserved over long periods of time during the evolutionary history of seed plants. However, both gymnosperms and angiosperms have experienced multiple bursts of gene order change (Fig. 5.4); in some cases this has resulted from one or few inversions while in others there is evidence for more severe genomic upheaval. This pattern is most evident among angiosperms, partly because the amount of plastome sequence data available is much greater. The phylogenetic distribution of inferred inversions (Fig. 5.4) indicates a long period of genomic stability starting from the early diverging angiosperms, monocots, and eudicots, followed by isolated instances of gene order changes in more derived lineages, especially among eudicots. Three of the most striking examples of extensive gene order changes among photosynthetic angiosperm lineages occur in the Campanulaceae, Fabaceae, and Geraniaceae and are summarized briefly below.

Campanulaceae (sensu APG III 2009, including Lobeliaceae) have experienced a high degree of gene order change. Although only one plastome sequence for *Trachelium caeruleum* has been published (Haberle et al. 2008), draft genomes have been completed for several other genera (Haberle 2006) and restriction site and gene maps have been published for many others (Knox and Palmer 1999; Cosner et al. 2004). The most extensive comparisons included gene maps for 18 genera of Campanulaceae (Cosner 1993; Cosner et al. 2004) and these authors estimated that the gene order changes were due to a minimum of 42 inversions, 18 large insertions (>5 kb) of unknown origin, five IR expansions and contractions, and several putative transpositions. The complete genome sequence for *Trachelium* (Haberle et al. 2008), the least rearranged taxon examined by Cosner et al. (2004), confirmed that at least seven inversions are present in this genome, but it did not identify any evidence for transposition as a mechanism for gene order changes.

Fabaceae are known to exhibit a number of unusual phenomena in their plastomes (Fig. 5.4b), including the loss of the IR in a large clade of papilionoids (Wojciechowski et al. 2004), transfer of genes to the nucleus (Gantt et al. 1991; Magee et al. 2010), intron losses (Doyle et al. 1995; Jansen et al. 2008), and inversions (Doyle et al. 1996; Bruneau et al. 1990). *Trifolium* has experienced the most extensive genomic reconfigurations within the family, including the loss of the IR, 14–18 inversions, duplication of parts or all of nine genes, and insertions of 20 kb of novel DNA (Milligan et al. 1989; Cai et al. 2008; Figs. 5.3, 5.4).

Geraniaceae have been examined more extensively in terms of complete plastome sequences, which are now available for five species from four of the five genera in the family (Chumley et al. 2006; Guisinger et al. 2011; Blazier et al. 2011). Like Campanulaceae, two mechanisms are responsible for gene order changes in this group, inversions and expansion/contraction of the IR. For *Pelargonium hortorum* Chumley et al. (2006) developed an evolutionary scenario that required a minimum of 12 inversions and eight IR boundary changes. The situation in Geraniaceae is so complex among the five sequenced plastomes that it is as yet not possible to reconstruct an evolutionary model to explain the gene order differences among these plastomes. This will require sequencing many more genomes within each genus so that intermediate stages in gene order can be reconstructed more reliably.

V. Patterns and Rates of Nucleotide Substitutions

A. Sequence Evolution in Coding Regions Versus Intergenic Regions and Introns

The relative frequency of base substitutions and insertions/deletions (indels) in plastid genomes has been examined among both closely related and distantly related species. Most studies (e.g., Golenberg et al. 1993; Ingvarsson et al. 2003) compared selected regions of the genome and concluded that indels occur at an equal or slightly higher rate than nucleotide substitutions, especially in comparisons among closely related species. The availability of complete plastome sequences opened up the opportunity to perform genome-wide comparisons, and several such studies have been completed in different angiosperm families, including Asteraceae (Timme et al. 2007), Poaceae (Masood et al. 2004; Saski et al. 2007; Yamane et al. 2006), Ranunculaceae (Kim et al. 2009), and Solanaceae (Kahlau et al. 2006; Chung et al. 2006; Daniell et al. 2006). All of these comparisons confirmed that the relative frequencies of indels are similar or slightly higher than nucleotide substitutions. In Poaceae, genome-wide comparisons demonstrated that most indels occurred in intergenic spacers (56–64%), with coding regions (10–19%) and introns (25–26%) having less than half the number of indels. In terms of genomic region most indels are concentrated in the LSC (84%), followed by the SSC (12%) and IR (4%). The vast majority of the indels represent single or few bp changes that are likely caused by slip-strand mispairing during DNA replication.

A number of genome-wide comparisons have also examined sequence divergence across plastomes by partitioning the genome into different regions; coding, intron, and intergenic spacer (Timme et al. 2007; Saski et al. 2007; Kim et al. 2009; Daniell et al. 2006). Rates of change varied considerably within these regions in different angiosperm families, however, coding regions were most highly conserved, followed by introns, and intergenic spacers. For example, in the Asteraceae the average p- distance, the proportion of substitution changes between two sequences, was 0.057 for intergenic spacers, 0.030 for introns, 0.022 for protein coding genes, and 0.008 for RNA genes (Timme et al. 2007). Such comparisons have been valuable for deciding which genes or regions to utilize for phylogenetic studies within angiosperms (Shaw et al. 2007; Timme et al. 2007). Comparisons of sequence divergence across intergenic spacers also have important implications for plastid biotechnology; it has been demonstrated that using homologous flanking and regulatory sequences for plastid transformation significantly increases transgene integration and expression of foreign proteins, respectively (Ruhlman et al. 2010).

B. Rates of Sequence Evolution in Protein Coding Genes

Early evolutionary rate comparisons of plastid coding regions of photosynthetic seed plants were based on a limited number of genes and/or genomes. Several general observations were made from these analyses: synonymous substitution rates are low in plastid DNA relative to nuclear DNA (Wolfe et al. 1987; Gaut 1998); rates can vary among lineages (Wolfe et al. 1987), among codon positions (Gaut et al. 1993), and among genes in different functional groups (Palmer 1991; Gaut et al. 1993); substitution rates in the three regions of the plastome vary, with genes in the IR having a lower rate of synonymous substitutions relative to those in the SSC and LSC regions (Clegg et al. 1984; Wolfe et al. 1987; Gaut 1998; Perry and Wolfe 2002); base composition often plays an important role in plastid DNA sequence evolution (Olmstead et al. 1998; Decker-Walters et al. 2004) resulting in mutations that are spatially biased across the genome. Several earlier comparisons of non-photosynthetic seed plants also focused on a limited number of genes and taxa (dePamphilis et al. 1997; Wolfe et al. 1992; Young and dePamphilis 2000, 2005). Not surprisingly, these studies demonstrated that non-photosynthetic plants have elevated rates of nucleotide substitution, largely due to relaxed selection. For some genes both synonymous (*dS*) and nonsynonymous (*dN*) rates increased, suggesting that other forces, including generation time, speciation rate, and population size, may be affecting rates, especially at synonymous sites. More recently there have been a number of genome-wide rate comparisons performed and these have provided a much more comprehensive view of rates of plastome sequence evolution. The most extensive comparisons have been performed in selected gymnosperms (Wu et al. 2007, 2009), the angiosperm families Poaceae (Chang et al. 2006; Zhong et al. 2009; Guisinger et al. 2010) and Geraniaceae (Guisinger et al. 2008), and the non-photosynthetic genus *Cuscuta* (McNeal et al. 2007a). Some notable observations from two angiosperm families, Poaceae and Geraniaceae, are described briefly below.

Genome-wide comparisons (Chang et al. 2006; Zhong et al. 2009; Guisinger et al. 2010) confirmed that rates of change were accelerated on the branch leading to Poaceae, while internal Poaceae branches have experienced a significant rate deceleration. Furthermore, genes involved in gene expression and photosynthesis metabolism have higher values of *dN*, and several genes appear to be under positive selection, i.e. the *dN*/*dS* ratio is greater than one. The precise timing of this rate acceleration is not clear since only two of the 16 families of Poales (Poaceae and Typhaceae) have complete plastome sequences available. Rate heterogeneity in Poaceae could be due to one or more factors, including relaxed or positive selection, mutational bias, altered DNA repair, or differences in levels of gene expression.

The situation in the Geraniaceae is novel for two reasons. First, this is the only seed plant lineage where extreme rate acceleration has been documented in both the mitochondrial (Parkinson et al. 2005; Mower et al. 2007) and plastid genomes (Guisinger et al. 2008). Second, plastid genomes in this family are among the most highly rearranged of any seed plant lineage. Analyses of 72 protein coding genes for nine Geraniaceae and 38 other angiosperms detected both locus and lineage specific rate heterogeneity (Guisinger et al. 2008). Values of *dN* were highly accelerated in the branch leading to the Geraniaceae as well as within several lineages within the family for ribosomal protein and RNA polymerase genes. In addition, *dN*/*dS* ratios were significantly higher for these two functional classes of genes and for ATPase genes. It was hypothesized that these unusual phenomena were caused by a combination of aberrant DNA repair and altered levels of gene expression.

C. Correlation Between Rates of Nucleotide Substitutions and Genomic Rearrangements

A significant positive correlation between rates of nucleotide substitutions and genomic rearrangements (indels, gene/intron losses, and inversions) was previously identified across angiosperms (Jansen et al. 2007). This pattern is evident in Fig. 5.4, which plots the distribution of genomic changes on a phylogram that was constructed using sequences of 81 genes from 97 seed plant plastomes. It is evident that early diverging lineages of angiosperms, eudicots, and monocots had very stable plastomes even though there was rapid diversification in morphology, anatomy, and reproductive biology among these lineages. This plastomic stasis was followed by repeated bursts of change in both rates of nucleotide substitution, gene order and gene content in disparate and more derived eudicot and monocot lineages. More extensive studies of two unrelated angiosperm families, Geraniaceae and Poaceae (Guisinger et al. 2008, 2010, 2011), have identified similar positive correlations between rates of nucleotide substitutions and genomic rearrangements.

A correlation between rates of nucleotide substitution and genomic rearrangements has been previously identified in bacterial (Belda et al. 2005) and animal mitochondrial genomes (Shao et al. 2003; Xu et al. 2006). The mitochondrial studies suggested several possible mechanisms to explain this correlation, but argued that accuracy of DNA replication is the most likely cause. In the case of plastid genomes, it was suggested that accelerated rates of genome rearrangements and nucleotide substitutions were possibly caused by aberrant DNA repair mechanisms (Jansen et al. 2007; Guisinger et al. 2008, 2010, 2011). Four classes of nuclear-encoded genes have been implicated in DNA repair in plastids of angiosperms: chloroplast mutator (CHM/MSH1), RecA-like homologs, OSBs (organellar single-stranded DNA-binding proteins), and the Whirlies (reviewed in Maréchal and Brisson 2010). These genes produce proteins that suppress recombination between repeated DNA sequences, and thus provide stability to the genome by preventing illegitimate recombination. Mutations in genes encoding either Whirly (Maréchal et al. 2009) or RecA (Rowan et al. 2010) have been shown to generate plastome rearrangements. We suggest that plastomes with accelerated rates of nucleotide substitutions and genomic rearrangements may result from mutations in nuclear-encoded DNA repair and/or replication genes. The prevalence of large numbers of dispersed repeats in highly rearranged plastomes (Chumley et al. 2006; Haberle et al. 2008; Cai et al. 2008; Guisinger et al. 2011) is consistent with this idea.

There is also a correlation between lineages with accelerated rates of change and biparental inheritance. Although this correlation has not been tested rigorously, it is evident from the distribution of biparental inheritance on the plastome phylogram for angiosperms (Fig. 5.4). The evolutionary significance of this correlation is not obvious. One possibility is that biparental inheritance provides a mechanism for bringing together multiple plastid types, and these could undergo intermolecular recombination to produce plastomes with novel organization. There is limited evidence available demonstrating plastid recombination (e.g., Medgyesy et al. 1985). Clearly more detailed investigations are needed to confirm this correlation and to examine its possible role in enhancing plastome diversity.

VI. Phylogenetic Utility of Plastome Data for Resolving Relationships Among Seed Plants

Most molecular phylogenetic investigations of seed plant relationships have relied on features of the plastid genome (reviewed in Raubeson and Jansen 2005). Early studies from 1985 to 1995 used restriction site and gene mapping comparisons to examine phylogenetic relationships at a wide range of taxonomic levels. Restriction site/fragment analyses were mostly utilized at the generic level or below, whereas gene map comparisons were valuable for defining major clades. Several early examinations of gene order identified one or a few inversions that were extremely valuable in defining major lineages of seed plants. This included a 22 kb inversion that identified the subfamily Barnadesioideae as the earliest diverging lineage in the largest angiosperm family Asteraceae (Jansen and Palmer 1987), a 30 kb inversion that placed lycophytes as the earliest clade of land plants (Raubeson and Jansen 1992b), and 50 kb inversion that supports the monophyly of a major clade of papilionoid legumes (Doyle et al. 1996). In addition to inversions, several other structural changes of plastomes were utilized to define major groups of seed plants. The loss of the IR was used to support the monophyly of conifers (Raubeson and Jansen 1992a) and the legume IRLC (Lavin et al. 1990; Wojciechowski et al. 2004). A number of gene and intron losses were also identified and in some cases these were powerful phylogenetic markers (e.g., *rpl22* gene loss in all legumes; Doyle et al. 1995), whereas in other instances such changes were shown to occur multiple times (*rpoC1* intron loss multiple times across angiosperms; Downie et al. 1996), limiting their utility as phylogenetic characters. In a few groups extensive structural changes in their plastomes have been utilized for phylogenetic analyses. The best example is the angiosperm family Campanulaceae. Gene maps for 18 genera identified a total of 84 structural changes, including inversions, putative transpositions, insertions, and gene and intron losses. Despite the extreme genomic complexity phylogenetic trees generated from these data exhibited very little homoplasy and were congruent with trees generated from DNA sequence data for the same taxa (Cosner et al. 2004).

During the past 10 years the field of plastid molecular phylogenetics has changed dramatically due the availability of rapid, less expensive methods for amassing large quantities of DNA sequence data. Thus, rather than relying on generating sequences for only a handful of markers or using the limited data from restriction site and gene mapping, it is now possible to produce large amounts of genomic data for phylogenetic studies. This has resulted in the production of very large data sets both in terms of number of genes and taxa for examining phylogenetic relationships (Jansen et al. 2007, 2011; Moore et al. 2007, 2010; Lin et al. 2010; Zhong et al. 2010). These studies focused entirely on plastome sequencing by utilizing isolated plastid DNA and either standard Sanger sequencing or 454 pyrosequencing. More recently, plastid genome sequencing for phylogeny reconstruction has shifted to sequencing platforms, such as Ilumina, that utilizes shorter reads of up to 75–100 bp (Cronn et al. 2008; Parks et al. 2009). This approach, combined with multiplexing samples, has greatly reduced the cost for generating draft plastid genome sequences. Another recent development in plastome sequencing is the use of total genomic DNA as template for next generation sequencing (Nock et al. 2010; Atherton et al. 2010). The outcome of these new developments is that we have made huge improvements in our understanding of phylogenetic relationship among seed plants and some of these are described below.

One of the most controversial, remaining issues in seed plant phylogeny concerns the position of the three morphologically unique genera of gnetophytes, *Ephedra*, *Gnetum*, and *Welwitschia* (reviewed in Burleigh and Mathews 2004; Mathews 2009). Morphological studies suggested that gnetophytes were sister to angiosperms (Anthophyte hypothesis; Doyle and Donoghue 1986) but most molecular phylogenetic studies do not support this relationship. The situation became more contentious because molecular phylogenetic studies supported three different hypotheses of relationships depending on which genes and taxa were included in the analyses: gnetophytes sister to conifers (Gnetifer hypothesis); gnetophytes sister to Pinaceae (Gnepine hypothesis); gnetophytes sister to Cupressaceae (Gnecup hypothesis). Several recent papers have utilized plastome sequences to try to resolve this issue but in all cases the number of genomes available was limited and the issue remains unresolved (Wu et al. 2007; McCoy et al. 2008). The major problem with such limited sampling is that it can cause artifacts in phylogenetic tree construction, often arising from long branch attraction. Previous phylogenetic studies using complete plastome sequences in angiosperms has shown that increased taxon sampling can alleviate issues associated with long branch attraction (Leebens-Mack et al. 2005).

The most comprehensive phylogenetic analysis of complete plastome sequences to resolve the position of gnetophytes included only eight gymnosperms and five outgroups (see inset in Fig. 5.4; Zhong et al. 2010). This analysis involved 56 protein coding genes shared among these 13 taxa. Initial results from analyses of these data supported the Gnecup hypothesis but the relationship was the result of long branch attraction between the single Cupressaceae genome and gnetophytes. Removal of the fastest evolving proteins from the dataset, many of which had parallel amino acid substitutions between gnetophytes and Cupressaceae, resulted in trees that supported the Gnepine hypothesis. This hypothesis is also supported by the fact that both gnetophytes and Pinaceae have lost all 11 NADH dehydrogenase genes (Braukmann et al. 2009) and *rps16* (Wu et al. 2007, 2009). Clearly additional gymnosperm plastome sequences are needed, especially from conifers, before the position of gnetophytes can be resolved.

Plastome sequences have been utilized extensively to examine phylogenetic relationships among the major clades of angiosperms. The earliest studies utilized 61 protein coding genes from a limited number of plastome sequences and suffered from some of the same issues mentioned above, especially long branch attraction (Goremykin et al. 2003, 2004, 2005). As more plastome sequences were completed it became evident that these data provide a valuable resource for resolving phylogenetic relationships among angiosperms (Leebens-Mack et al. 2005; Cai et al. 2006; Ruhlman et al. 2006; Hansen et al. 2007b). The most comprehensive studies examined up to 97 plastid genomes using 81–83 genes. These studies have provided strong support for resolving relationships among all major clades of angiosperms (Fig. 5.4; Jansen et al. 2007, 2011; Moore et al. 2007, 2010), including the placement of *Amborella* as the earliest diverging lineage, the position of magnoliids as sister to Chloranthaceae and this group sister to a large clade that includes both eudicots and monocots, placement of *Ceratophyllum* sister to eudicots, sister relationship between monocots and eudicots, and resolution of relationships among many of the orders within both monocots and eudicots. A number of ongoing plastome sequencing projects in these major clades will provide much new data for resolving the angiosperm tree of life.

VII. Conclusions and Future Directions

Rapid improvement in DNA sequencing technology at a much lower cost has generated a glut of plastome data for plant biologists. For plant evolutionary biologists, plastome sequences have provided reams of data for resolving phylogenetic relationships among the major clades of seed plants and for examining rates and patterns of sequence evolution. These two endeavors are closely intertwined since an understanding of how sequences evolve is essential for using them correctly for making phylogenetic inferences. Some lineages, especially among gymnosperms, are still underrepresented but projects are underway that will fill these gaps and ultimately generate a tree of life for seed plants. The knowledge we have gained about the diversity of plastome organization among seed plants is providing the framework for examining the mechanisms of change in these genomes. We have confirmed that across most seed plant lineages there is an incredible stability of genome organization in terms of overall architecture, size, gene/intron content, and gene order. However, several unrelated groups have experienced genomic upheaval and these taxa are positioned to illuminate the mechanisms of change in plastomes. Future investigations should focus on these extraordinary lineages by sequencing more representative in these groups, and by examining their nuclear-plastid interactions. Such studies will reveal critical insights into how these genomes have co-evolved to control the many biochemical processes that are coordinated between nuclear and plastid genomes in seed plants. Comparisons of these natural mutant lineages, combined with experimental studies of plastid-nuclear interactions using plastid genetic engineering, will lead to new insights into compartmental crosstalk, which is critical for plant cells to function properly.

Acknowledgments

Support was provided by the National Science Foundation (DEB-0717372 to RKJ and IOS-1027259 to RKJ and TAR) and the Fred G. Gloeckner Foundation (to TAR and RKJ). The authors thank J. Chris Blazier and Mao-Lun Weng for comments on an earlier version of the manuscript.

References

Angiosperm Phylogeny Group [APG] (2009) An update of the Angiosperm Phylogeny Group classification for the orders and families of flowering plants: APG III. Bot J Linn Soc 161:105–121

Atherton RA, McComish BJ, Shepherd LD, Berry LA, Albert NW, Lockhart PJ (2010) Whole genome sequencing of enriched chloroplast DNA using the Illumina GAII platform. BMC Plant Methods 6:22

Belda E, Moya A, Silva FJ (2005) Genome rearrangement distances and gene order phylogeny in gamma-Proteobacteria. Mol Biol Evol 22:1456–1467

Bendich AJ (2004) Circular chloroplast chromosomes: the grand illusion. Plant Cell 16:1661–1666

Bennett JR, Mathews S (2006) Phylogeny of the parasitic plant family Orobanchaceae inferred from phytochrome A. Am J Bot 93:1039–1051

Birky CWJR (1995) Uniparental inheritance of mitochondrial and chloroplast genes: mechanisms and evolution. Proc Natl Acad Sci USA 92: 11331–11338

Blazier JC, Guisinger MM, Jansen RK (2011) Recent loss of plastid-encoded *ndh* genes within *Erodium* (Geraniaceae). Plant Mol Biol 76:263–272

Bock R (2007) Structure, function, and inheritance of plastid genomes. In: Bock R (ed) Cell and molecular biology of plastids. Springer, Berlin/Heidelberg, pp 1610–2096

Boudreau E, Turmel M (1995) Gene rearrangements in *Chlamydomonas* chloroplast DNAs are accounted for by inversions and by expansion/contraction of the inverted repeat. Plant Mol Biol 27:351–364

Bourque G, Pevzner P (2002) Genome-scale evolution: reconstructing gene orders in the ancestral species. Genome Res 12:26–36

Braukmann TW, Kuzmina M, Stefanovic S (2009) Loss of all plastid ndh genes in Gnetales and conifers: extent and evolutionary significance for the seed plant phylogeny. Curr Genet 55:323–337

Bruneau A, Doyle JJ, Palmer JD (1990) A chloroplast DNA inversion as a subtribal character in the Phaseoleae (Leguminosae). Syst Bot 15:378–386

Bubunenko MG, Schmidt J, Subramanian AR (1994) Protein substitution in chloroplast ribosome evolution. A eukaryotic cytosolic protein has replaced its organelle homologue (L23) in spinach. J Mol Biol 240:28–41

Burleigh JG, Mathews S (2004) Phylogenetic signal in nucleotide data from seed plants: implications for resolving the seed plant tree of life. Am J Bot 91:1599–1613

Cai Z, Penaflor C, Kuehl JV, Leebens-Mack J, Carlson J, dePamphilis CW, Jansen RK (2006) Complete plastid genome sequences of *Drimys*, *Liriodendron*, and *Piper*: implications for the phylogeny of magnoliids. BMC Evol Biol 6:77

Cai Z, Guisinger M, Kim HG, Ruck E, Blazier JC, McMurtry V, Kuehl JV, Boore J, Jansen RK (2008) Extensive reorganization of the plastid genome of *Trifolium subterraneum* (Fabaceae) is associated with numerous repeated sequences and novel DNA insertions. J Mol Evol 67:696–704

Chang C-C, Lin H-C, Lin I-P, Chow T-Y, Chen H-H, Chen W-H, Cheng C-H, Lin C-Y, Liu S-M, Chang C-C, Chaw S-M (2006) The chloroplast genome of *Phalaenopsis aphrodite* (Orchidaceae): comparative analysis of evolutionary rate with that of grasses and its phylogenetic implications. Mol Biol Evol 23:279–291

Chaw S-M, Chang C-C, Chen H-L, Li W-H (2004) Dating the monocot dicot divergence and the origin of core eudicots using whole chloroplast genomes. J Mol Evol 58:424–441

Chiu WL, Sears BB (1993) Plastome-genome interactions affect plastid transmission in *Oenothera*. Genetics 133:989–997

Chumley TW, Palmer JD, Mower JP, Fourcade HM, Calie PJ, Boore JL, Jansen RK (2006) The complete chloroplast genome sequence of *Pelargonium* x *hortorum*: organization and evolution of the largest and most highly rearranged chloroplast genome of land plants. Mol Biol Evol 23:2175–2190

Chung H-J, Jung JD, Park H-W, Kim J-H, Cha HW, Min SR, Jeong W-J, Liu JR (2006) The complete chloroplast genome sequences of *Solanum tuberosum* and comparative analysis with *Solanaceae* species identified the presence of a 241-bp deletion in cultivated potato chloroplast DNA sequence. Plant Cell Rep 25:1369–1379

Clegg MT, Brown ADH, Whitfield PR (1984) Chloroplast DNA diversity in wild and cultivated barley: implications for genetic conservation. Genet Res 43:339–343

Conant GC, Wolfe KH (2008) GenomeVx: simple web-based creation of editable circular chromosome maps. Bioinformatics 24:861–862

Corriveau JS, Coleman AW (1988) Rapid screening method to detect potential biparental inheritance of plastid DNA and results for over 200 angiosperm species. Am J Bot 75:1443–1458

Cosner ME (1993) Phylogenetic and molecular evolutionary studies of chloroplast DNA variation in the Campanulaceae. PhD thesis, The Ohio State University, Columbus

Cosner ME, Jansen RK, Palmer JD, Downie SR (1997) The highly rearranged chloroplast genome of *Trachelium caeruleum* (Campanulaceae): multiple inversions, inverted repeat expansion and contraction, transposition, insertions/deletions, and several repeat families. Curr Genet 31:419–429

Cosner ME, Raubeson LA, Jansen RK (2004) Chloroplast DNA rearrangements in Campanulaceae: phylogenetic utility of highly rearranged genomes. BMC Evol Biol 4:27

Cronn R, Liston A, Parks M, Gernandt D, Shen R, Mockler T (2008) Multiplex sequencing of plant chloroplast genomes using Solexa sequencing-by synthesis technology. Nucleic Acids Res 36:e122

Cruzan MB, Arnold ML, Carney SE, Wollenberg KR (1993) CpDNA inheritance in interspecific crosses and evolutionary inference in Louisiana irises. Am J Bot 80:344–350

Cusack BP, Wolfe KH (2007) When gene marriages don't work: divorce by subfunctionalization. Trends Genet 23:270–272

Daniell H, Lee S-B, Grevich J, Saski C, Guda C, Tompkins J, Jansen RK (2006) Complete chloroplast genome sequences of *Solanum tuberosum*, *Solanum lycopersicum* and comparative analyses with other Solanaceae genomes. Theor Appl Genet 112:1503–1518

Daniell H, Wurdack KJ, Kanagaraj A, Lee S-B, Saski C, Jansen RK (2008) The complete nucleotide sequence of the cassava (*Manihot esculenta*) chloroplast genome and multiple losses of the *atpF* intron in Malpighiales. Theor Appl Genet 116:723–737

Darling AE, Mau B, Perna NT (2010) Progressive mauve: multiple genome alignment with gene gain, loss, and rearrangement. PLoS One 5:e11147

Decker-Walters DS, Chung SM, Staub JE (2004) Plastid sequence evolution: a new pattern of nucleotide substitutions in the Cucurbitaceae. J Mol Evol 58:606–614

Delannoy E, Fujii S, des Frances CC, Brundrett M, Small I (2011) Rampant gene loss in the underground orchid *Rhizanthella gardneri* highlights evolutionary constraints on plastid genomes. Mol Biol Evol 28:2077–2086

dePamphilis CW, Young ND, Wolfe AD (1997) Evolution of plastid gene *rps2* in a lineage of hemiparasitic and holoparasitic plants: many losses of photosynthesis and complex patterns of rate variation. Proc Natl Acad Sci USA 94:7367–7372

Downie SR, Palmer JD (1992) Use of chloroplast DNA rearrangements in reconstructing plant phylogeny. In: Soltis PS, Soltis DE, Doyle JJ (eds) Molecular systematics of plants. Chapman and Hall, New York, pp 14–35

Downie SR, Llanas E, Katz-Downie DS (1996) Multiple independent losses of the *rpoC1* intron in angiosperm chloroplast DNAs. Syst Bot 21:135–151

Doyle JA, Donoghue MJ (1986) Seed plant phylogeny and the origin of angiosperms: an experimental cladistic approach. Bot Rev 52:321–431

Doyle JJ, Doyle JL, Palmer JD (1995) Multiple independent losses of two genes and one intron from legume chloroplast genomes. Syst Bot 20:272–294

Doyle JJ, Doyle JL, Palmer JD (1996) The distribution and phylogenetic significance of a 50-kb chloroplast DNA inversion in the flowering plant family Leguminosae. Mol Phylogenet Evol 5:429–438

Drummond AJ, Ashton B, Buxton S, Cheung M, Cooper A, Heled J, Kearse M, Moir R, Stones-Havas S, Sturrock S, Thierer T, Wilson A (2010) Geneious v5.1. http://www.geneious.com

Fan WH, Woelfle MA, Mosig G (1995) Two copies of a DNA element, 'Wendy', in the chloroplast chromosome of *Chlamydomonas reinhardtii* between rearranged gene clusters. Plant Mol Biol 29:63–80

Funk HT, Berg S, Krupinska K, Maier UG, Krause K (2007) Complete DNA sequences of the plastid genomes of two parasitic flowering plant species, *Cuscuta reflexa* and *Cuscuta gronovii*. BMC Plant Biol 7:45

Gantt JS, Baldauf SL, Calie PJ, Weeden NF, Palmer JD (1991) Transfer of *rpl22* to the nucleus greatly preceded its loss from the chloroplast and involved the gain of an intron. EMBO J 10:3073–3078

Gao L, Su Y-J, Wang T (2010) Plastid genome sequencing, comparative genomics, and phylogenomics: current status and prospects. J Syst Evol 48:77–93

Gaut BS (1998) Molecular clocks and nucleotide substitution rates in higher plants. In: Hecht MK (ed) Evolutionary biology. Plenum Press, New York, pp 93–120

Gaut BS, Muse SV, Clegg MT (1993) Relative rates of nucleotide substitution in the chloroplast genome. Mol Phylogenet Evol 2:89–96

Golenberg EM, Clegg MT, Durbin ML, Doebley J, Ma DP (1993) Evolution of a noncoding region of the chloroplast genome. Mol Phylogenet Evol 2:52–64

Goremykin VV, Hirsch-Ernst KI, Wolfl S, Hellwig FH (2003) Analysis of the *Amborella trichopoda* chloroplast genome sequence suggests that *Amborella* is not a basal angiosperm. Mol Biol Evol 20:1499–1505

Goremykin VV, Hirsch-Ernst KI, Wolfl S, Hellwig FH (2004) The chloroplast genome of *Nymphaea alba*: whole-genome analyses and the problem of identifying the most basal angiosperm. Mol Biol Evol 21:1445–1454

Goremykin VV, Holland B, Hirsch-Ernst KI, Hellwig FH (2005) Analysis of *Acorus calamus* chloroplast genome and its phylogenetic implications. Mol Biol Evol 22:1813–1822

Gornicki P, Faris J, King I, Podkowinski J, Gill B, Haselkorn R (1997) Plastid localized acetyl-Co-A carboxylase of bread wheat is encoded by a single gene on each of the three ancestral chromosome sets. Proc Natl Acad Sci USA 94:14179–14184

Goulding SE, Olmstead RG, Morden CW, Wolfe KH (1996) Ebb and flow of the chloroplast inverted repeat. Mol Gen Genet 252:195–206

Gray BN, Ahner BA, Hanson MR (2009) Extensive homologous recombination between introduced and native regulatory plastid DNA elements in transplastomic plants. Transgenic Res 18:559–572

Guisinger MM, Kuehl JV, Boore JL, Jansen RK (2008) Genome-wide analyses of Geraniaceae plastid DNA reveal unprecedented patterns of increased nucleotide substitutions. Proc Natl Acad Sci USA 105:18424–18429

Guisinger MM, Chumley TW, Kuehl JV, Boore JL, Jansen RK (2010) Implications of the plastid genome sequence of *Typha* (Typhaceae, Poales) for understanding genome evolution in Poaceae. J Mol Evol 70:149–166

Guisinger MM, Kuehl JV, Boore JL, Jansen RK (2011) Extreme reconfiguration of plastid genomes in the angiosperm family Geraniaceae: rearrangements, repeats, and codon usage. Mol Biol Evol 28:583–600

Haberle RC (2006) Phylogeny and comparative chloroplast genomics of the Campanulaceae. PhD thesis, University of Texas, Austin

Haberle RC, Fourcade HM, Boore JL, Jansen RK (2008) Extensive rearrangements in the chloroplast genome of *Trachelium caeruleum* are associated with repeats and tRNA genes. J Mol Evol 66:350–361

Hagemann R (2004) The sexual inheritance of plant organelles. In: Daniell H, Chase C (eds) Molecular biology and biotechnology of plant organelles. Springer, Heidelberg, pp 93–113

Hansen AK, Escobar LK, Gilbert LE, Jansen RK (2007a) Paternal, maternal, and biparental inheritance of the chloroplast genome in *Passiflora* (Passifloraceae): implications for phylogenetic studies. Am J Bot 94:42–46

Hansen DR, Dastidar SG, Cai Z, Penaflor C, Kuehl V, Boore JL, Jansen RK (2007b) Phylogenetic and evolutionary implications of complete chloroplast genome sequences of four early diverging angiosperms: *Buxus* (Buxaceae), *Chloranthus* (Chloranthaceae), *Dioscorea* (Dioscoreaceae), and *Illicium* (Schisandraceae). Mol Phylogenet Evol 45:547–563

Hildebrand M, Hallick RB, Passavant CW, Bourque DP (1988) Trans-splicing in chloroplasts: the *rps12* loci of *Nicotiana tabacum*. Proc Natl Acad Sci USA 85:372–376

Hirao T, Watanabe A, Kurita M, Kondo T, Takata K (2008) Complete nucleotide sequence of the *Cryptomeria japonica* D. Don. chloroplast genome and comparative chloroplast genomics: diversified genome structure of coniferous species. BMC Plant Biol 8:70

Hiratsuka J, Shimada H, Whittier R, Ishibashi T, Sakamoto M, Mori M, Kondo C, Honji Y, Sun CR, Meng BY, Li YQ, Kanno A, Nishizawa Y, Hirai A, Shinozaki K, Sugiura M (1989) The complete sequence of the rice *Oryza sativa* chloroplast genome – intermolecular recombination between distinct transfer-RNA genes accounts for a major plastid DNA inversion during the evolution of the cereals. Mol Gen Genet 217:185–194

Hu YC, Zhang Q, Rao GY, Sodmergen (2008) Occurrence of plastids in the sperm cells of Caprifoliaceae: biparental plastid inheritance in angiosperms is unilaterally derived from maternal inheritance. Plant Cell Physiol 49:958–968

Ingvarsson PK, Ribstein S, Taylor DR (2003) Molecular evolution of insertions and deletions in the chloroplast genome of *silene*. Mol Biol Evol 20:1737–1740

Jansen RK, Palmer JD (1987) A chloroplast DNA inversion marks an ancient evolutionary split in the sunflower family (Asteraceae). Proc Natl Acad Sci USA 84:5818–5822

Jansen RK, Cai Z, Raubeson LA, Daniell H, de Pamphilis CW, Leebens-Mack J, Müller KF, Guisinger-Bellian M, Haberle RC, Hansen AK, Chumley TW, Lee SB, Peery R, McNeal JR, Kuehl JV, Boore JL (2007) Analysis of 81 genes from 64 plastid genomes resolves relationships in angiosperms and identifies genome-scale evolutionary patterns. Proc Natl Acad Sci USA 104: 19369–19374

Jansen RK, Wojciechowski MF, Sanniyasi E, Lee S-B, Daniell H (2008) Complete plastid genome sequence of the chickpea (*Cicer arietinum*) and the phylogenetic distribution of *rps12* and *clpP* intron losses among legumes (Fabaceae). Mol Phylogenet Evol 48:1204–1207

Jansen RK, Saski C, Lee S-B, Hansen AK, Daniell H (2011) Complete plastid genome sequences of three rosids (*Castanea*, *Prunus*, *Theobroma*): evidence for at least two independent transfers of *rpl22* to the nucleus. Mol Biol Evol 28:835–847

Kahlau S, Aspinall S, Gray JC, Bock R (2006) Sequence of the tomato chloroplast DNA and evolutionary comparison of Solanaceous plastid genomes. J Mol Evol 63:194–207

Kanno A, Watanabe K, Nakamura I, Hirai A (1993) Variations in chloroplast DNA from rice (*Oryza sativa*): differences between deletions mediated by short direct-repeat sequences within a single species. Theor Appl Genet 86:579–584

Kawata M, Harada T, Shimamoto Y, Oono K, Takaiwa F (1997) Short inverted repeats function as hotspots of intermolecular recombination giving rise to oligomers of deleted plastid DNAs (ptDNAs). Curr Genet 31:179–184

Khan A, Khan IA, Asif H, Azim MK (2010) Current trends in chloroplast genome research. Afr J Biotechnol 9:3494–3500

Kim K-J, Lee H-L (2004) Complete chloroplast genome sequence from Korean Ginseng (*Panax schiseng* Nees) and comparative analysis of sequence evolution among 17 vascular plants. DNA Res 11:247–261

Kim Y-K, Park C-W, Kim K-J (2009) Complete chloroplast DNA sequence from a Korean endemic genus, *Megaleranthus saniculifolia*, and its evolutionary implications. Mol Cells 27:365–381

Knox EB, Palmer JD (1999) The chloroplast genome arrangement of *Lobelia thuliana* (Lobeliaceae): expansion of the inverted repeat in an ancestor of the Campanulales. Plant Syst Evol 214:49–64

Kolodner R, Tewari KK (1979) Inverted repeats in chloroplast DNA of higher plants. Proc Natl Acad Sci USA 76:41–45

Konishi T, Shinohara K, Yamada K, Sasaki Y (1996) Acetyl-CoA carboxylase in higher plants: most plants other than gramineae have both prokaryotic and eukaryotic forms of this enzyme. Plant Cell Physiol 37:117–122

Kuroiwa T (2010) Review of cytological studies on cellular and molecular mechanisms of uniparental (maternal or paternal) inheritance of plastid and mitochondrial genomes induced by active digestion of organelle nuclei (nucleoids). J Plant Res 123:207–230

Lavin M, Doyle JJ, Palmer JD (1990) Evolutionary significance of the loss of the chloroplast-DNA inverted repeat in the Leguminosae subfamily Papilionoideae. Evolution 44:390–402

Lee DJ, Blake TK, Smith SE (1988) Biparental inheritance of chloroplast DNA and the existence of heteroplasmic cells in alfalfa. Theor Appl Genet 76:545–549

Lee H-L, Jansen RK, Chumley TW, Kim K-J (2007) Gene relocations within chloroplast genomes of *Jasminum* and *Menodora* (Oleaceae) are due to multiple, overlapping inversions. Mol Biol Evol 24:1161–1180

Leebens-Mack J, Raubeson LA, Cui L, Kuehl JV, Fourcade MH, Chumley TW, Boore JL, Jansen RK, dePamphilis CW (2005) Identifying the basal angiosperm node in chloroplast genome phylogenies: sampling one's way out of the Felsenstein zone. Mol Biol Evol 22:1948–1963

Lilly JW, Havey MJ, Jackson SA, Jiang J (2001) Cytogenomic analyses reveal the structural plasticity of the chloroplast genome in higher plants. Plant Cell 13:245–254

Lin C-P, Huang J-P, Wu C-S, Hsu C-Y, Chaw S-M (2010) Comparative chloroplast genomics reveals the evolution of Pinaceae genera and subfamilies. Genome Biol Evol 2:504–517

Liu Q, Xue Q (2005) Comparative studies on codon usage pattern of chloroplasts and their host nuclear genes in four plant species. J Genet 84:55–62

Magee AM, Aspinall S, Rice DW, Cusack BP, Semon M, Perry AS, Stefanovic S, Milbourne D, Barth S, Palmer JD, Gray JC, Kavahagh TA, Wolfe KH (2010) Localized hypermutation and associated gene losses in legume chloroplast genomes. Genome Res 20:1700–1710

Maréchal A, Brisson N (2010) Recombination and the maintenance of plant organelle genome stability. New Phytol 186:299–317

Maréchal A, Parent J, Véronneau-Lafortune F, Joyeux A, Lang BF, Brisson N (2009) Whirly proteins

maintain plastid genome stability in *Arabidopsis*. Proc Natl Acad Sci USA 106:14693–14698
Masood MS, Nishikawa T, Fukuoka S, Njenga PK, Tsudzuki T, Kadowaki K (2004) The complete nucleotide sequence of wild rice (*Oryza nivara*) chloroplast genome: first genome wide comparative sequence analysis of wild and cultivated rice. Gene 340:133–139
Mathews S (2009) Phylogenetic relationships among seed plants: persistent questions and the limits of molecular data. Am J Bot 96:228–236
Matsuo M, Ito Y, Yamauchi R, Obokata J (2005) The rice nuclear genome continuously integrates, shuffles, and eliminates the chloroplast genome to cause chloroplast–nuclear DNA flux. Plant Cell 17:665–675
McCoy SR, Kuehl JV, Boore JL, Raubeson LA (2008) The complete plastid genome sequence of *Welwitschia mirabilis*: an unusually compact plastome with accelerated divergence rates. BMC Evol Biol 8:130
McNeal JR, Arumugunathan K, Kuehl JV, Boore JL, de Pamphilis CW (2007a) Systematics and plastid genome evolution of the cryptically photosynthetic parasitic plant genus *Cuscuta* (Convolvulaceae). BMC Biol 5:55
McNeal JR, Kuehl JV, Boore JL, dePamphilis CW (2007b) Complete plastid genome sequences suggest strong selection for retention of photosynthetic genes in the parasitic plant genus *Cuscuta*. BMC Plant Biol 7:57
Medgyesy P, Fejes E, Maliga P (1985) Interspecific chloroplast recombination in a *Nicotiana* somatic hybrid. Proc Natl Acad Sci USA 82:6960–6964
Metzlaff M, Pohlheim F, Börner T, Hagemann R (1982) Hybrid variegation in the genus *Pelargonium*. Curr Genet 5:245–249
Millen RS, Olmstead RG, Adams KL, Palmer JD, Lao NT, Heggie L, Kavanagh TA, Hibberd JM, Gray JC, Morden CW, Calie PJ, Jermlin LS, Wolfe KH (2001) Many parallel losses of *infA* from chloroplast DNA during angiosperm evolution with multiple independent transfers to the nucleus. Plant Cell 13:645–658
Milligan BG, Hampton JN, Palmer JD (1989) Dispersed repeats and structural reorganization in subclover chloroplast DNA. Mol Biol Evol 6:355–368
Mogensen HL (1996) The hows and whys of cytoplasmic inheritance in seed plants. Am J Bot 83:383–404
Moore MJ, Bell CD, Soltis PS, Soltis DE (2007) Using plastid genome-scale data to resolve enigmatic relationships among basal angiosperms. Proc Natl Acad Sci USA 104:19363–19368
Moore MJ, Soltis PS, Bell CD, Burleigh JG, Soltis DE (2010) Phylogenetic analysis of 83 plastid genes further resolves the early diversification of eudicots. Proc Natl Acad Sci USA 107:4623–4628
Morton BR (1993) Chloroplast DNA codon use: evidence for selection at the *psbA* locus based on tRNA availability. J Mol Evol 37:273–280
Morton BR (1994) Codon use and the rate of divergence of land plant chloroplast genes. Mol Biol Evol 11:231–238
Morton BR (1998) Selection on the codon bias of chloroplast and cyanelle genes in different plant and algal lineages. J Mol Evol 46:449–459
Mower JP, Touzet P, Gummow JS, Delph LF, Palmer JD (2007) Extensive variation in synonymous substitution rates in mitochondrial genes of seed plants. BMC Evol Biol 7:135
Mrácek J (2005) Investigation of interspecific genome-plastome incompatibility in *Oenothera* and *Passiflora*. PhD dissertation, University of Munich, Germany
Nagata N (2010) Mechanisms for independent cytoplasmic inheritance of mitochondria and plastids in angiosperms. J Plant Res 123:193–199
Nock C, Waters D, Edwards M, Bowen W, Rice N, Cordeiro G, Henry R (2010) Chloroplast genome sequences from total DNA for exploring plant relationships. Plant Biotechnol J 9:328–333
Noutsos C, Richly E, Leister D (2005) Generation and evolutionary fate of insertions of organelle DNA in the nuclear genomes of flowering plants. Genome Res 15:616–628
Odintsova MS, Yurina NP (2003) Plastid genomes of higher plants and algae: structure and functions. Mol Biol 37:649–662
Ohba K, Iwakawa M, Ohada Y, Murai M (1971) Paternal transmission of a plastid anomaly in some reciprocal crosses of Suzi, Cryptomeria japonica D. Don. Silvae Genet 210:101–107
Ohyama K, Fukuzawa H, Kohchi T, Shirai H, Sano T, Sano S, Umesono K, Shiki Y, Takeuchi M, Chang Z, Aota S, Inokuchi H, Ozeki H (1986) Chloroplast gene organization deduced from complete sequence of Liverwort *Marchantia-Polymorpha* chloroplast DNA. Nature 322:572–574
Oldenburg DJ, Bendich AJ (2004) Changes in the structure of DNA molecules and the amount of DNA per plastid during chloroplast development in maize. J Mol Biol 344:1311–1330
Olmstead RG, Reeves PA, Yen AC (1998) Patterns of sequence evolution and implications for parsimony analysis of chloroplast DNA. In: Soltis DE, Soltis PS, Doyle JJ (eds) Molecular systematics of plants II: DNA sequencing. Kluwer Academic, Norwell, pp 164–187
Palmer JD (1983) Chloroplast DNA exists in two orientations. Nature 301:92–93

Palmer JD (1991) Plastid chromosomes: structure and evolution. In: Hermann RG (ed) The molecular biology of plastids. Cell culture and somatic cell genetics of plants. Springer, Vienna, pp 5–53
Palmer JD, Stein DB (1986) Conservation of chloroplast genome structure among vascular plants. Curr Genet 10:823–833
Palmer JD, Aldrich J, Thompson WF (1987) Chloroplast DNA evolution among legumes: loss of a large inverted repeat occurred prior to other sequence rearrangements. Curr Genet 11:275–286
Palmer JD, Osorio B, Thompson WF (1988) Evolutionary significance of inversions in legume chloroplast DNAs. Curr Genet 14:65–74
Pandey K, Grant J, Williams E (1987) Interspecific hybridisation between *Trifolium repens* and *T. uniflorum*. Aust J Bot 35:171–182
Parkinson CL, Mower JP, Qiu Y-Q, Shirk AJ, Song K, Young ND, de Pamphilis CW, Palmer JD (2005) Multiple major increases and decreases in mitochondrial substitution rates in the plant family Geraniaceae. BMC Evol Biol 5:73
Parks M, Cronn R, Liston A (2009) Increasing phylogenetic resolution at low taxonomic levels using massively parallel sequencing of chloroplast genomes. BMC Biol 7:84
Perry AS, Wolfe KH (2002) Nucleotide substitution rates in legume chloroplast DNA depend on the presence of the inverted repeat. J Mol Evol 55:501–508
Perry AS, Brennan S, Murphy DJ, Kavanagh TA, Wolfe KH (2002) Evolutionary re-organization of a large operon in Adzuki Bean chloroplast DNA caused by inverted repeat movement. DNA Res 9:157–162
Raubeson LA, Jansen RK (1992a) A rare chloroplast-DNA structural mutation is shared by all conifers. Biochem Syst Ecol 20:17–24
Raubeson LA, Jansen RK (1992b) Chloroplast DNA evidence on the ancient evolutionary split in vascular land plants. Science 255:1697–1699
Raubeson LA, Jansen RK (2005) Chloroplast genomes of plants. In: Henry RJ (ed) Plant diversity and evolution: genotypic and phenotypic variation in higher plants. CABI, Cambridge, MA, pp 45–68
Raubeson LA, Peery R, Chumley TW, Dziubek C, Fourcade HM, Boore JL, Jansen RK (2007) Comparative chloroplast genomics: analyses including new sequences from the angiosperms *Nuphar advena* and *Ranunculus macranthus*. BMC Genomics 8:174
Ravi V, Khurana JP, Tyagi AK, Khurana P (2008) An update on chloroplast genomes. Plant Syst Evol 271:101–122
Rogalski M, Ruf S, Bock R (2006) Tobacco plastid ribosomal protein S18 is essential for cell survival. Nucleic Acids Res 34:4537–4545
Rowan BA, Oldenburg DJ, Bendich AJ (2010) RecA maintains the integrity of chloroplast DNA molecules in *Arabidopsis*. J Exp Bot 61:2575–2588
Ruhlman T, Lee S-B, Jansen RK, Hostetler JB, Tallon LJ, Town CD, Daniell H (2006) Complete chloroplast genome sequence of *Daucus carota*: implications for biotechnology and phylogeny of angiosperms. BMC Genomics 7:222
Ruhlman T, Verma D, Samson N, Daniell H (2010) The role of heterologous chloroplast sequence elements in transgene integration and expression. Plant Physiol 152:2088–2104
Saski C, Lee S-B, Fjellheim S, Guda C, Jansen RK, Luo H, Tomkins J, Rognli OA, Daniell H, Clarke JL (2007) Complete chloroplast genome sequences of *Hordeum vulgare*, *Sorghum bicolor* and *Agrostis stolonifera* and comparative analyses with other grass genomes. Theor Appl Genet 115:571–590
Shao R, Dowton M, Murrell A, Barker SC (2003) Rates of gene rearrangement and nucleotide substitution are correlated in the mitochondrial genomes of insects. Mol Biol Evol 20:1612–1619
Shaver JM, Oldenburg DJ, Bendich AJ (2006) The structure of chloroplast DNA molecules and the effects of light on the amount of chloroplast DNA during development in *Medicago truncatula*. Plant Physiol 146:1064–1074
Shaw J, Lickey EB, Schilling EE, Small RL (2007) Comparison of whole chloroplast genome sequences to choose noncoding regions for phylogenetic studies in angiosperms: the tortoise and the hare III. Am J Bot 94:275–288
Shimada H, Sugiura M (1991) Fine-structural features of the chloroplast genome – comparison of the sequenced chloroplast genomes. Nucleic Acids Res 19:983–995
Shinozaki K, Ohme M, Tanaka M, Wakasugi T, Hayashida N, Matsubayashi T, Zaita N, Chunwongse J, Obokata J, Yamaguchishinozaki K, Ohto C, Torazawa K, Meng BY, Sugita M, Deno H, Kamogashira T, Yamada K, Kusuda J, Takaiwa F, Kato A, Tohdoh N, Shimada H, Sugiura M (1986) The complete nucleotide-sequence of the tobacco chloroplast genome – its gene organization and expression. EMBO J 5:2043–2049
Snijder RC, Brown FS, van Tuyk JM (2007) The role of plastome-genome incompatibility and biparental plastid inheritance in interspecific hybrids in the genus *Zantedeschia* (Araceae). Floricul Ornam Biotechnol 1:150–157
Soliman K, Fedak G, Allard RW (1987) Inheritance of organelle DNA in barley and *Hordeum* X *Secale* intergeneric hybrids. Genome 29:867–872

Szmidt AE, Alden T, Hallgren JE (1987) Paternal inheritance of chloroplast DNA in *Larix*. Plant Mol Biol 9:59–64

Testolin R, Cipriani G (1997) Paternal inheritance of chloroplast DNA and maternal inheritance of mitochondrial DNA in the genus *Actinidia*. Theor Appl Genet 94:897–903

Timme RE, Kuehl JV, Boore JL, Jansen RK (2007) A comparison of the first two sequenced chloroplast genomes in Asteraceae: Lettuce and Sunflower. Am J Bot 94:302–312

Timmis JN, Ayliffe MA, Huang CY, Martin W (2004) Endosymbiotic gene transfer: organelle genomes forge eukaryotic chromosomes. Nat Rev Genet 5:123–135

Tsudzuki J, Nakashima K, Tsudzuki T, Hiratsuka J, Shibata M, Wakasugi T, Sugiura M (1992) Chloroplast DNA of black pine retains a residual inverted repeat lacking rRNA genes: nucleotide sequences of *trnQ*, *trnK*, *psbA*, *trnI* and *trnH* and the absence of *rps16*. Mol Gen Genet 232:206–214

Ueda M, Fujimoto M, Arimura S-I, Murata J, Tsutsumi N, Kadowaki K-I (2007) Loss of the *rpl32* gene from the chloroplast genome and subsequent acquisition of a preexisting transit peptide within the nuclear gene in *Populus*. Gene 402:51–56

Ueda M, Fujimoto M, Takanashi H, Arimura S-I, Tsutsumi N, Kadowaki K-I (2008) Substitution of the gene for chloroplast *rps16* was assisted by generation of dual targeting signal. Mol Biol Evol 25:1566–1575

Wakasugi T, Tsudzuki J, Ito S, Nakashima K, Tsudzuki T, Sugiura M (1994) Loss of all ndh genes as determined by sequencing the entire chloroplast genome of the black pine *Pinus thunbergii*. Proc Natl Acad Sci USA 91:9794–9798

Wall DP, Herbeck JT (2003) Evolutionary patterns of codon usage in the chloroplast gene *rbcL*. J Mol Evol 56:673–688

Wojciechowski MF, Lavin M, Sanderson MJ (2004) A phylogeny of legumes (Leguminosae) based on analysis of the plastid *matK* gene resolves many well supported subclades within the family. Am J Bot 91:1846–1862

Wolf PG, Der JP, Duffy AM, Davidson JB, Grusz AL, Pryer KM (2011) The evolution of chloroplast genes and genomes in ferns. Plant Mol Biol 76:251–261

Wolfe KH, Li WH, Sharp PM (1987) Rates of nucleotide substitution vary greatly among plant mitochondrial, chloroplast, and nuclear DNAs. Proc Natl Acad Sci USA 84:9054–9058

Wolfe KH, Morden CW, Palmer JD (1992) Function and evolution of a minimal plastid genome from a nonphotosynthetic parasitic plant. Proc Natl Acad Sci USA 89:10648–10652

Wu CS, Wang YN, Liu SM, Chaw SM (2007) Chloroplast genome (cpDNA) of *Cycas taitungensis* and 56 cp protein-coding genes of *Gnetum parvifolium*: insights into cpDNA evolution and phylogeny of extant seed plants. Mol Biol Evol 24:1366–1379

Wu CS, Lai YT, Lin CP, Wang YN, Chaw SM (2009) Evolution of reduced and compact chloroplast genomes (cpDNAs) in gnetophytes: selection toward a lower-cost strategy. Mol Phylogenet Evol 52:115–124

Xu W, Jameson D, Tang B, Higgs PG (2006) The relationship between the rate of molecular evolution and the rate of genome rearrangement in animal mitochondrial genomes. J Mol Evol 63:375–392

Yamane K, Yano K, Kawahara T (2006) Pattern and rate of indel evolution from whole chloroplast intergenic regions in sugarcane, maize and rice. DNA Res 13:197–204

Yang TW, Yang YA, Xiong Z (2000) Paternal inheritance of chloroplast DNA in interspecific hybrids in the genus *Larrea* (Zygophyllaceae). Am J Bot 87:1452–1458

Young ND, dePamphilis CW (2000) Purifying selection detected in the plastid gene *matK* and flanking ribozyme regions within a group II intron of nonphotosynthetic plants. Mol Biol Evol 17:1933–1941

Young ND, dePamphilis CW (2005) Rate variation in parasitic plants: correlated and uncorrelated patterns among plastid genes of different function. BMC Evol Biol 5:16

Zhang Q, Sodmergen (2010) Why does biparental plastid inheritance revive in angiosperms? J Plant Res 123:201–206

Zhang Q, Liu Y, Sodmergen (2003) Examination of the cytoplasmic DNA in male reproductive cells to determine the potential for cytoplasmic inheritance in 295 angiosperm species. Plant Cell Physiol 44:941–951

Zhong BJ, Yonezawa T, Zhong Y, Hasegawa M (2009) Episodic evolution and adaptation of chloroplast genomes in ancestral grasses. PLoS One 4:e5297

Zhong B, Yonezawa T, Zhong Y, Hasegawa M (2010) The position of Gnetales among seed plants: overcoming pitfalls of chloroplast phylogenomics. Mol Biol Evol 27:2855–2863

Chapter 6

Mitochondrial Genomes of Algae

Gertraud Burger*
Department of Biochemistry, Robert-Cedergren Centre for Bioinformatics and Genomics, Université de Montréal, 2900 Edouard-Montpetit, Montreal, QC H3T 1J4, Canada

Aurora M. Nedelcu
Department of Biology, University of New Brunswick, Fredericton, NB E3B 5A3, Canada

* Author for correspondence, e-mail: gertraud.burger@umontreal.ca

R. Bock and V. Knoop (eds.), Genomics of Chloroplasts and Mitochondria, Advances in Photosynthesis and Respiration 35, pp. 127–157,
DOI 10.1007/978-94-007-2920-9_6,

Summary

Algae are plastid-containing protists that belong to a large variety of eukaryotic groups. Consistent with the broad phylogenetic affiliations of algae, their mitochondrial genomes are tremendously diverse. In this chapter, we review algal mitochondrial DNAs with an emphasis on genome organization and gene content. For that, we reanalyzed all previously published complete mitochondrial genome sequences to ensure most sensitive gene detection and the same metric in assessing coding capacity. Based on this comprehensive compilation, we infer the diverse patterns of mitochondrial genome evolution in algae, pattern that can be surprisingly similar in completely unrelated lineages and represent conspicuous examples of convergent evolution.

I. Introduction

Algae comprise a large number of most diverse unicellular and multi-cellular taxa that populate virtually all ecosystems on Earth. The best-known among the 20 or so algal groups are the green, red, brown, and golden algae, dinoflagellates, foraminifera, and euglenids. Algae are defined by exclusion – as plastid-containing eukaryotes outside plants or, what comes out to the same result, as all plastid-containing protists (protists being eukaryotes outside animals, fungi, and plants).

This chapter is subdivided into sections based on phylogeny. We will refer to generally accepted eukaryotic clades (Fig. 6.1) and leave proposed supergroups (Keeling et al. 2005) for critical review in other chapters of this book (see Chaps. 2, 4). One of these undisputed clades is Viridiplantae (green algae and land plants; aka 'green plants'), yet a matter of debate remains the exact relationship of 'greens' to red and glaucophyte algae (for references see legend of Fig. 6.1). Another solid monophyletic assemblage unites alveolates (including dinoflagellates), stramenopiles (including golden and brown algae), and chlorarachniophytes. Haptophytes and cryptophytes, however, earlier thought to affiliate with alveolates and stramenopiles (to form the 'chromalveolates'), cannot be placed reliably in the eukaryotic tree. A third well established eukaryotic clade is Euglenozoa plus heteroloboseans and jakobids. Opisthokonts (animals, fungi, and related protists) form a fourth coherent group, but as it lacks photosynthetic members, it will not be further discussed in this chapter.

Eukaryotes acquired plastids long after the establishment of mitochondria. The first round of acquisition, by enslavement of a cyanobacterium, took place in the common ancestor of green, red, and glaucocystophyte algae (see Chap. 1). From here, plastids spread to the other eukaryotic lineages via secondary and higher-order symbiotic events, involving the engulfment of a photosynthetic eukaryote by a non-photosynthetic one. The origin of plastids in the various lineages can be quite convoluted as discussed in detail in Chap. 2. Mitochondria, in contrast, have apparently remained faithful to their host, and mitochondrial DNA (mtDNA) accurately reflects the evolutionary history of the nuclear genome as a whole. This view is corroborated by numerous independent phylogenetic reconstructions using either nuclear or mitochondrial genes (e.g. Baurain et al. 2010) yielding trees that are congruent in topology.

Complete mtDNAs have been determined at a large scale since the 1990s, and the body of data is growing at an accelerated pace with ever faster sequencing technologies. Over the years, reviews on mtDNAs have attempted to keep up with the growing genome

Abbreviations: CCW – Counterclockwise; CW – Clockwise; DO – Direct opposed; GOBASE – Organelle genome database; mtDNA – Mitochondrial DNA; ORF – Potential protein-coding gene (open reading frame)

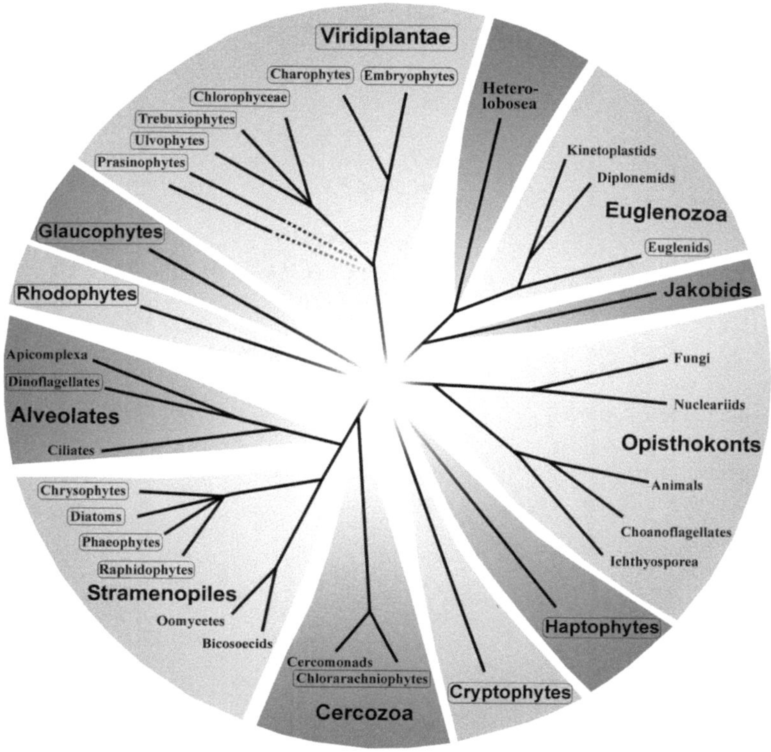

Fig. 6.1. Schematic tree of the major eukaryotic groups. Shading circumscribes groups belonging to the taxon indicated in large font size. *Black* contiguous lines connect monophyletic groups, notably Viridiplantae; stramenopiles + alveolates + chlorarachniophytes; opisthokonts (e.g. Burki et al. 2007; Hackett et al. 2007; Rodriguez-Ezpeleta et al. 2007a), and jakobids + Euglenozoa + Heterolobosea (Rodriguez-Ezpeleta et al. 2007a). Other relevant references in this context are (Cavalier-Smith 1999; Rodriguez-Ezpeleta et al. 2005; Baurain et al. 2010). *Dotted lines* represent uncertain topologies. Boxed taxon names highlight photosynthetic groups.

data, some eukaryotic-wide (Gray et al. 1999), others with special attention to protists (Gray et al. 1998, 2004), plants (Kubo and Newton 2008), or animals (Boore 1999). Here, we will review currently available mitochondrial genome data from all plastid-containing eukaryotes except plants (embryophytes), whose mtDNAs will be discussed in Chaps. 8 and 9 of this book.

II. Mitochondrial Genome Structure and Gene Complement

We start with a short introduction on the structure of the mitochondrial genome and the type of genes contained in it, across all eukaryotes.

A. Structure of the Mitochondrial Genome

Generally, mitochondria contain a single type of chromosome (in multiple copies), as did most likely their alpha-proteobacterial ancestor. However, several exceptions to this rule are found in disparate eukaryotic lineages, with up to hundreds mitochondrial chromosomes (see discussion in (Burger et al. 2003a) and a recent review (Burger et al. 2011)). The shape of most mtDNAs is 'circular-mapping', that is a linear DNA molecule composed of head-to-tail concatenates that most likely arises by rolling-circle DNA replication (Oldenburg and Bendich 2001; Ling and Shibata 2004). Less frequent are monomeric linear molecules and truly circular mitochondrial chromosomes that

Table 6.1. Mitochondrion-encoded genes and their functions[a]

Electron transport and ATP synthesis			
Complex I	*nad1,2,3,4,4L,5,6,7,8,9,10,11*		
Complex II	*sdh2,3,4*		
Complex III	*cob*		
Complex IV	*cox1,2,3*		
Complex V	*atp1,3,4,6,8,9*	(*ymf19* = *orfB* = *atp8*), (*ymf39* = *atp4*)	
Translation			
Ribosomal RNAs	*rns, rnl, rrn5*		
Transfer RNAs	*trnA,C,D,E,F,G,H,I,K,L,M,N,P,Q,R,S,T,V,Y*		
Ribosomal proteins	*rps1,2,3,4,7,8,10,11,12, 13,14,19*	(*var1* = *rps3*)	
	rpl1,2,5,6,10,11,14,16,18,19,20,27,31,32,34[b]*,35*[c]*,36*		
Elongation factor	*tufA*		
Tm-RNA (unstalling of translation)	*ssrA*		
Transcription			
Core RNA polymerase	*rpoA,B,C*		
Sigma factor	*rpoD*		
Protein import			
ABC transporter	*ccmA,B*	(*yejW,V*)	
Heme delivery	*ccmC*	(*yejU*)	
SecY-type transporter	*secY*		
Sec-independent transporter	*tatA,C*	(*mttA,B*)	(*ymf16* = *tatC*)
Protein maturation			
Cytochrome oxidase assembly	*cox11,15*[c]		
Heme c maturation	*ccmF*	(*yejR*)	
RNA processing			
5' tRNA processing	*rnpB*		
Genes of uncertain origin			
DNA polymerase	*dpo* (mitochondrial plasmid-derived)		
RNA polymerase	*rpo* (mitochondrial plasmid-derived)		
Reverse transcriptase	*rtl* (mitochondrial intron-derived)		
DNA repair	*mutS*		
Methyl transferase	*dam*	(*mtf*)	

[a]For details and references, see (Gray et al. 2004). The most common synonyms are indicated in parentheses
[b]Recently discovered in *Reclinomonas* mtDNA (Burger and Lang, unpublished)
[c]Genes found in mtDNA of the jakobid *Andalucia* (Lang, pers. comm.), see text

occur sporadically across the eukaryotic tree (Gray et al. 2004).

The size of mtDNAs varies drastically. The smallest ones are found in *Plasmodium* and relatives with 6 kbp only (Feagin et al. 1991) and the largest ones in the cucumber family measuring up to ~3,000 kbp (Ward et al. 1981). At the time of writing, the largest fully sequenced mtDNA is that of *Cucurbita pepo* with nearly 1,000 kbp (Alverson et al. 2010).

B. Mitochondrion-Encoded Functions

Mitochondria perform numerous biological functions, most of which rely entirely on nucleus-encoded genes (for a review on the origin of imported mitochondrial proteins, see (Gray et al. 2001) and Chap. 1, Sect. II. D). Among the pathways and biological processes that involve at least some mtDNA-encoded genes, two are universal, notably electron transport plus oxidative phosphorylation, and mitochondrial translation (Table 6.1). MtDNA always encodes some or even all structural RNAs that are involved in translation, i.e. ribosomal RNAs (rRNAs) and transfer RNAs (tRNAs), but may not include ribosomal protein-coding genes at all. Additional processes specified by some mtDNAs include protein transport across the mitochondrial inner membrane via the twin-arginine translocase, and more rarely, the SecY-type transport system. Cytochrome c maturation, in particular heme import into the inner-membrane space and covalent heme linkage to apocytochrome c, is also often controlled by mitochondrial genes. Further, mtDNA-encoded components for tRNA processing and cytochrome oxidase assembly are found in several lineages. Mitochondrial genes for transcription occur in only a single group, the jakobids (for a review, see Gray et al. 2004). In general, mtDNAs adhere to 'genetic conservatism' in that they encode only a relatively narrow set of functions that are essential for biogenesis and activity of the organelle (Gray et al. 1999).

All of the mitochondrial processes listed above are directly derived from the bacterial ancestor of mitochondria (genes acquired by horizontal transfer are discussed below; see also Chap. 10). The range of functions encoded by mitochondrial genomes generally does not correlate well with the phylogenetic affinity of the organisms. Gene migration to the nucleus or loss for good has occurred numerous times and independently across the eukaryotic tree.

C. Gene Sets

Table 6.1 lists all mitochondrial genes and cognate products recognized to date. The large and small subunit rRNA genes (*rnl, rns*) are encoded on all known mtDNAs. In contrast, the mitochondrial 5S rRNA gene (*rrn5*) occurs rather sporadically. In several instances, it was overlooked in the original sequence annotation (see Tables 6.3, 6.5), and it might remain undetected in other cases, because it is small and often poorly conserved. For example, to discover this gene in amoebozoans mtDNAs, it required biochemical data combined with secondary structure analysis and comparative sequence data from related species (Bullerwell et al. 2010). Mitochondrion-encoded tRNAs are variable in number ranging from zero to nearly 30. Mitochondrial genes for the respiratory chain are found universally, notably *cob* specifying apo-cytochrome *b* of Complex III (ubiquinone-cytochrome *c*-oxidoreductase) and *cox1* encoding subunit 1 of Complex IV (cytochrome c oxidoreductase, or briefly cytochrome oxidase). Most mtDNAs carry additional genes for Complex IV (*cox2* and *cox3*). Subunits for Complex I (NADH-ubiquinone oxidoreductase) are mtDNA-encoded in most eukaryotes, the basic gene set consisting of *nad1*, *4*, *5* and generally also *nad2*, *3*, *6*, while the *nad7–11* genes occur less frequently. In some lineages, Complex I has been entirely lost and functionally substituted by a nucleus-encoded single-subunit enzyme (van Dooren et al. 2006). Complex V (ATP synthase) has typically several subunits specified by mitochondrial genes, notably *atp6*, *8*, *9*, and more rarely *atp1* and *atp4*. Only a few lineages possess mitochondrial genes for Complex II (succinate-ubiquinone-oxidoreductase) such as *sdh2–4*, and it was only in the mid 1990s when mtDNAs carrying these genes were discovered for the first time (Burger et al. 1996).

The other major class of mitochondrial genes encodes proteins of the mito-ribosome. Up to 27 such genes can occur (jakobids), whereas several major lineages harbour only few or no such genes at all on their mtDNAs. Most frequent are genes for the small ribosomal subunit (*rps3* etc.) with typically fewer genes for the large subunit (*rpl2*, etc.).

Genes that are rarely found in mtDNAs involve transmembrane transport and cytochrome *c* maturation (*tatA, tatC*, *ccmA-C, F*) and 5′ tRNA processing (*rnpB*, encoding the RNA subunit of RNase P). The most uncommon mitochondrial genes are *rpoA-D*, encoding RNA polymerase components, *tufA*, a translation elongation factor gene, *secY* that is involved in *secY*-dependent protein import, and *ssrA* specifying tmRNA that releases stalled ribosomes from non-stop mRNAs (Jacob et al. 2004). These latter seven genes are not found in mtDNAs of algae.

A small set of mitochondrial genes were probably gained through horizontal transfer. These include genes for phage-like DNA and RNA polymerases and reverse transcriptase (*dpo, rpo,* and *rtl*), as seen for instance in mtDNAs of red and brown algae but also in other eukaryotes. These genes likely spread via mobile genetic elements, sometimes integrating into mtDNA and sometimes establishing themselves as mitochondrial plasmids (Weber et al. 1995). Integration may trigger genome rearrangements or linearization. Plasmid versions of these genes have been studied particularly well in fungi (Griffiths 1995). Free-standing *rtl* genes may also be derived from group II intronic reading frames (see below). Despite contrary claims (Delaroque et al. 1996; Rousvoal et al. 1998), there is no evidence that these genes are involved in replication and transcription of mtDNA. Instead, *dpo, rpo,* and *rtl* sequences often accumulate frameshift and stop mutations (e.g. in *Porphyra* Burger et al. 1999) and are frequently fragmented, indicating that they are 'on the way out'. For mitochondrial transcription, most eukaryotes have substituted the function of ancestral *rpoA-D* by a nuclear gene encoding a homolog of the single-polypeptide RNA polymerase of bacteriophages T3 and T7 (Cermakian et al. 1997). Another most likely horizontally transferred gene is *dam* (methyl transferase) that has been detected first in haptophytes (Sanchez Puerta et al. 2004). It is unknown whether it has a biological function.

ORFs (or unidentified open reading frames) are potential protein-coding genes and occur in most of the larger mtDNAs. Several ORFs turned out to be unrecognized, highly divergent versions of known genes (Table 6.1). For example, the former *orfB* is in fact *atp8* (Gray et al. 1998), *ymf39* is now known as *atp4* (Burger et al. 2003c), *murf1* is a highly derived *nad2* in trypanosome mtDNA (Kannan and Burger 2008), and the ORFs in fungal mitochondria initially designated '5S', *var* or ORF227, are extremely deviant *rps3* homologs (Bullerwell et al. 2000). Function assignment of hypothetical (mitochondrial) proteins is indeed challenging, and progress in bioinformatics approaches (e.g. Kannan et al. 2008) will eventually uncover further incognito versions of typical mitochondrial protein genes. Alternatively, ORFs could be fortuitous, representing genome regions that happen to be conceptually translatable into proteins. This seems to apply to most of the hundred or so ORFs in the inflated mtDNAs of plants (see Chaps. 8, 9).

Introns in mitochondrial genes belong to group I or group II and occur in virtually all major eukaryotic lineages. These introns often harbour ORFs that participate in intron propagation and/or intron splicing (maturases). In the case of group I introns, ORFs encode homing endonucleases of the LAGLIDADG or GIY-YIG type, whereas group II intron ORFs usually encode phage-like reverse transcriptases (for a recent review, see Lang et al. 2007). Yet, there are a few exceptions where group II intron ORFs specify homing endonucleases (Toor and Zimmerly 2002).

The mtDNAs with the largest gene count are found among jakobids. *Reclinomonas americana*, for example, has 67 protein-encoding genes and an additional 30 genes for structural RNAs (Lang et al. 1997). This mtDNA contains virtually all genes represented on any other mtDNA, plus unique genes that also descend from the proteobacterial ancestor, such as *tufA*, *rpoA-D*, *atp3* (Complex V subunit), *cox11* (Complex IV assembly), *secY* and *ssrA* (Jacob et al. 2004). The recently sequenced mtDNA of the jakobid *Andalucia* contains in addition *cox15* (Complex IV assembly) and one more ribosomal protein gene, *rpl35* (BF Lang, pers. comm.), summing up to 31 mitochondrion-encoded ribo-protein genes and 69 assigned protein genes in total. These are 23 times more protein genes than in mtDNAs of certain chlorophycean algae. A comprehensive compilation of mtDNA gene content for sequences published prior to 2010 is available in the GOBASE database (O'Brien et al. 2009).

In the subsequent sections, we will review mtDNAs from the various algal groups in seven sections and the following order: (1) green algae (chlorophytes, streptophytes), (2) glaucophytes, (3) red algae, (4) dinoflagellates, (5) stramenopiles (golden, brown, and raphidophyte algae), (6) chlorarachniophytes, and (7) euglenids. Each section starts with a brief description of general characteristics, life cycle, ecology and taxonomic subdivisions of the algal group. Organismal information is mostly taken from (Margulis et al. 1989), unless indicated otherwise. The second part of each section reviews current mtDNA-related information comparing and contrasting genomic features within and between algal lineages.

III. Algal mtDNAs

A. Viridiplantae

Viridiplantae consist of two major divisions, Chlorophyta (chlorophytes) and Streptophyta (streptophytes). Within chlorophytes, three monophyletic groups are currently accepted: Chlorophyceae, Ulvophyceae, and Trebouxiophyceae. Chlorophytes also include most of the so-called prasinophyte algae – a paraphyletic assemblage of unicellular species that are thought to be descendants of the ancestral lineages from which the main green algal groups evolved (Mattox and Stewart 1984). Streptophytes have separated early from the chlorophytes and include charophycean algae as well as land plants (Embryophyta; Friedl 1997), the latter group being subject of Chaps. 8 and 9. Plastids of Viridiplantae are characterized by containing the photosynthetic pigments chlorophyll a and b.

The first mitochondrial genome sequenced from Viridiplantae was that of the green alga *Chlamydomonas reinhardtii* (Michaelis et al. 1990). The unusual, animal-like mtDNA features triggered a flurry of interest in sequencing the mitochondrial genome from other algal lineages. To date, about 20 complete green algal mitochondrial genome sequences are available, revealing an extraordinary diversity in size, shape, and gene content.

1. Prasinophyta (Chlorophyta)

The paraphyletic prasinophytes comprise unicellular algae with a variable number of flagella (from 0 to 16), and whose cell membranes are often covered by scales (Graham and Wilcox 2000). This group includes primarily marine species and represent a large and important part of the phytoplankton. Notably, the smallest free-living eukaryote known to date, the spherical, flagellum-less *Ostreococcus*, is a member of the prasinophytes. Seven clades are currently recognized, but the exact relationships within the group and to the other green algal taxa are not fully understood (Marin and Melkonian 2010).

Currently, five prasinophyte mtDNAs have been sequenced completely and one partially, and they show a wide range of genome sizes and gene repertoires (Table 6.2). Among these mitochondrial genomes, three exhibit the so-called "ancestral" (or minimally diverged) pattern of organization (Turmel et al. 1999), notably mtDNA of *Nephroselmis olivacea* (Pseudoscourfieldiales) (Turmel et al. 1999), *Ostreococcus tauri* (Robbens et al. 2007), and *Micromonas* sp. RCC299 (Mamiellales) (Worden et al. 2009). Ancestral genomes have a rather large gene complement, eubacteria-like ribo-protein gene clusters, a few introns, and slowly diverging gene sequences. Indeed, at 44–47 kbp, these rather compact genomes are the most gene-rich chlorophyte mtDNAs sequenced to date, coding for up to 19 respiratory and ATP synthase proteins, 15 ribosomal proteins, three rRNAs and 26 tRNAs (Table 6.3). All three mtDNAs contain almost the same gene set, display a circular map, use the standard genetic code, and vary little in A+T-content. Also, the gene order is quite similar in these three genomes. Introns have only been detected in *Nephroselmis* mtDNA, which possesses four group I introns. Interestingly, in *Ostreococcus* and *Micromonas*, the *rns* gene is fragmented in two pieces, a derived feature that is more common among chlorophycean mtDNAs (discussed later). The *Ostreococcus* mtDNA is also distinct in displaying very short intergenic regions. In fact, with only 7% intergenic sequences, this mitochondrial genome is the most gene-dense among Chlorophyta. Further, a segmental duplication covers 44% of

Table 6.2. General characteristics of mtDNAs from green algae

	Size (kbp)	Conformation	A+T (%)	Intergenic[a] (%)	Intron type (number)	Genetic code
Chlorophytes						
Prasinophytes						
Nephroselmis olivacea	45.2	Circular	67.2	19.4	I (4)	Standard
Ostreococcus tauri	44.2	Circular	61.8	7.1	/	Standard
Micromonas sp. RCC299	47.4	Circular	65.4	16.7	/	Standard
Tetraselmis subcordiformis[b]	42.8	Circular	69.1	21.2	/	Standard
Pycnococcus provasolli	24.3	Circular	62.2	12.2	/	UGA=W UUA, UUG=Stop
Pedinomonas minor	25.1	Circular	77.8	39.5	II (1)	UGA=W
Chlorophyceae						
Polytomella capuana (CW)	13.0	Linear	42.8	27.9	/	Standard
Polytomella parva[b] (CW)	13+3	Linear	59.0	34.5 (34.0)	/	Standard
Chlamydomonas reinhardtii (CW)	15.8	Linear	54.8	16.9	/[I (1)][c]	Standard
Chlamydomonas incerta[b] (CW)	17.5	Linear	56.3	7.7 (17.0[d])	I (3)	Standard
Volvox carteri[b] (CW)	~35	Circular	64.3	33.7 (35.0[d])	I (2), II (1)	Standard
Chlamydomonas eugametos (CW)	22.9	Circular	65.4	15.5	I (9)	Standard
Chlorogonium elongatum (CW)	22.7	Circular	62.2	10.8	I (6)	UGA=W
Dunaliella salina (CW)	28.3	Circular	65.6	27.1	I (22)	Standard
Scenedesmus obliquus (DO)	42.9	Circular	63.7	39.4	I (2), II (2)	UAG=L UCA=Stop
Trebouxiophyceae						
Prototheca wickerhamii	55.3	Circular	74.2	29.1	I (5)	Standard
Helicosporidium sp.	49.3	Circular	74.4	24.0	I (3)	Standard
Ulvophyceae						
Oltmannsiellopsis viridis	56.8	Circular	66.6	34.8	I (2), II (1)	Standard
Pseudendoclonium akinetum	95.9	Circular	60.7	44.8	I (7)	Standard
Streptophytes						
Charophytes						
Mesostigma viride	42.4	Circular	67.8	13.7	I (4), II (3)	Standard
Chlorokybus atmophyticus	201.8	Circular	60.2	55.5	I (6), II (14)	Standard
Chara vulgaris	67.7	Circular	59.1	8.6	I (14), II (13)	Standard
Chaetosphaeridium globosum	56.6	Circular	65.6	25.0	I (9), II (1)	Standard

[a]Intergenic regions, when considering ORFs ≥100 aa. Percentage shown here may differ from values given in the original publications describing these genomes. Differences are due (1) to the choice of the start codon in protein-coding genes that is based here on multiple alignments with orthologous sequences, and (2) to genes undetected in the original report

[b]GenBank record contains partial mtDNA sequence; A+T content, non-coding portion and intron type and number refer to published sequence

[c]In square brackets, optional introns found in other strains/isolates of the same species

[d]Percentage of intergenic regions under the assumption that the yet unsequenced genome portion does not contain genes

Table 6.3. Mitochondrial gene content in green algae

	Prasinophytes					Chlorophyceae						Trebouxio		Ulvoph.		Charophyta			
	Neph oliv	Ostr taur	Micr sp.	Pycn prov	Pedi mino	Poly parv	Chla rein	Volv cart	Chlo elon	Duna sali	Scen obli	Prot wick	Heli sp.	Oltm viri	Pseu akin	Meso viri	Chlo atmo	Char vulg	Chae glob
nad1	x	x	x	x	x	x	x	x	x	x	x	x	x	x	x	x	x	x	x
nad2	x	x	x	x	x	x	x	x	x	x	x	x	x	x	x	x	x	x	x
nad3	x	x	x	x	x	–	–	–	–	–	x	x	x	x	x	+	x	x	x
nad4	x	x	x	x	x	x	x	x	x	x	x	x	x	x	x	x	x	x	x
nad4L	x	x	x	x	x	–	–	–	–	–	x	x	x	x	x	x	x	x	x
nad5	x	x	x	x	x	x	x	x	x	x	x	x	x	x	x	x	x	x	x
nad6	x	x	x	x	x	x	x	x	x	x	x	x	x	x	x	x	x	x	x
nad7	x	x	x	–	–	–	–	–	–	–	–	x	x	x	x	x	x	x	x
nad9	x	x	x	–	–	–	–	–	–	–	–	x	x	x	–	x	x	x	x
nad10	x	x	**X**	–	–	–	–	–	–	–	–	–	–	–	–	–	x	–	–
cob	x	x	x	x	x	x	x	x	x	x	x	x	+	x	x	x	x	x	x
cox1	x	x	x	x	x	x	x	x	x	x	x	x	x	x	x	x	x	x	x
cox2	x	x	x	x	–	–	–	–	–	–	x	x	x	x	x	x	x	x	x
cox3	x	x	x	x	–	–	–	–	–	–	x	x	x	x	x	x	x	x	x
atp1	x	x	–	–	–	–	–	–	–	–	–	x	x	x	x	x	x	x	x
atp4	x	x	x	x	–	–	–	–	–	–	–	x	x	x	x	x	x	x	x
atp6	x	x	x	x	x	–	–	–	–	–	x	x	x	x	x	x	x	x	x
atp8	x	x	x	x	x	–	–	–	–	–	–	x	x	x	x	x	x	x	x
atp9	x	x	x	x	–	–	–	–	–	–	x	x	x	x	x	x	x	x	x
sdh3	–	–	–	–	–	–	–	–	–	–	–	–	–	–	–	x	x	x	x
sdh4	–	–	–	–	–	–	–	–	–	–	–	–	–	–	–	x	x	x	x
tatC	x	**X**	x	–	–	–	–	–	–	–	–	x	x	x	x	x	x	x	x
rps1	–	–	–	–	–	–	–	–	–	–	–	–	–	–	–	x	x	x	x
rps2	x	x	x	–	–	–	–	–	–	–	–	x	x	x	x	x	x	x	x
rps3	x	x	x	x	–	–	–	–	–	–	–	x	x	x	x	x	x	x	x
rps4	x	x	x	x	–	–	–	–	–	–	–	x	x	–	x	x	x	x	x
rps7	x	x	x	–	–	–	–	–	–	–	–	x	x	–	–	x	x	x	x
rps8	x	x	x	–	–	–	–	–	–	–	–	–	–	–	–	–	x	–	–
rps10	x	x	x	–	–	–	–	–	–	–	–	x	x	–	x	x	x	x	x
rps11	x	x	x	–	–	–	–	–	–	–	–	x	x	x	x	x	x	x	x

(continued)

Table 6.3. (continued)

	Prasinophytes					Chlorophyceae						Trebouxio		Ulvoph.		Charophyta			
	Neph oliv	Ostr taur	Micr sp.	Pycn prov	Pedi mino	Poly parv	Chla rein	Volv cart	Chlo elon	Duna sali	Scen obli	Prot wick	Heli sp.	Oltm viri	Pseu akin	Meso viri	Chlo atmo	Char vulg	Chae glob
rps12	x	x	x	x	–	–	–	–	–	–	–	x	x	x	x	x	x	x	x
rps13	x	x	x	–	–	–	–	–	–	–	–	x	x	x	x	x	x	–	x
rps14	x	x	x	–	–	–	–	–	–	–	–	x	x	x	x	x	x	x	x
rps19	x	x	x	–	–	–	–	–	–	–	–	x	x	x	x	x	x	x	x
rpl2	–	–	–	–	–	–	–	–	–	–	–	–	–	x	–	–	x	x	x
rpl5	x	x	x	–	–	–	–	–	–	–	–	x	x	–	x	x	x	x	x
rpl6	x	x	x	–	–	–	–	–	–	–	–	x	x	–	–	x	x	x	x
rpl10	–	**x**	–	–	–	–	–	–	–	–	–	–	–	–	–	–	**X**	**X**	–
rpl14	x	x	x	–	–	–	–	–	–	–	–	–	–	–	x	x	x	x	–
rpl16	x	x	x	–	–	–	–	–	–	–	–	x	x	x	x	x	x	x	x
rrn5	x	**X**	**X**	–	–	–	–	–	–	–	–	x	x	x	–	x	x	x	x
rns	x	# (2)	# (2)	x	x	# (4)	# (4)	# (4)	# (≥3)	# (3)	# (2)	x	x	x	x	x	x	x	x
rnl	x	x	# (2)	x	# (2)	# (8)	# (10)	# (8)	# (≥6)	# (6)	# (4)	x	x	x	x	x	x	x	x
*trn*A..[a]	26	27	28	16	8	1	3	3	3	3	27	26	25	24	25	26	28	26	28
other	*rnpB*	*rnpB*	*rnpB*	/	/	/	*rtl*	/	/	/	/	/	/	/	/	/	***rpl31***	*ccmB,C,F*	/

x contiguous gene present (bold, gene not identified in the original publication), + trans-spliced gene, # gene and product in pieces, – gene absent from mtDNA, / not applicable; trnA.., tRNA genes

[a]Not counting duplicate genes

the mitochondrial genome. A large duplicated region is also present in the close relative *Micromonas* sp., and a 1.5 kbp inverted repeat has been detected in the partially sequenced mtDNA of *Tetraselmis (Platymonas) subcordiformis* (Kessler and Zetsche 1995).

The other two fully sequenced prasinophyte mitochondrial genomes are from *Pedinomonas minor* (Pedinomonadales) (Turmel et al. 1999) and *Pycnococcus provasolii* (Pseudoscourfieldiales, Pycnococcaceae) (Turmel et al. 2010). These are both considerably smaller genomes (24–25 kbp) encoding a much smaller set of proteins and RNAs – a feature shared with the "reduced-derived" mtDNAs of chlorophycean algae (Tables 6.2, 6.3). Notable is the absence of several *nad* genes and most (in *Pycnococcus*) or all (in *Pedinomonas*) ribosomal protein genes. However, although of similar small size, the gene contents of these two mtDNAs differ substantially, with *Pycnococcus* coding for almost twice the number of tRNAs and proteins. The larger number of genes in *Pycnococcus* results in a higher overall gene density in this mtDNA relative to that of *Pedinomonas*. Nevertheless, in *Pedinomonas* all 22 genes are also found tightly packed on a 16-kbp segment, while the remaining 9-kbp region is composed of a complex array of repeated sequences (Turmel et al. 1999). As in several other reduced mtDNAs (e.g. those of the chlorophycean *Chlamydomonas eugametos* and *Chlorogonium elongatum* see below), all genes in *Pycnococcus* and *Pedinomonas* are encoded on the same strand. A deviant genetic code is employed in both algal mitochondria, with the unprecedented use in *Pycnococcus* of UUA and UUG (normally leucine codons) as stop codons (Table 6.2). The *rnl* gene is fragmented in two pieces in *Pedinomonas* but not in *Pycnococcus*, and a group II intron (at precisely the same site as a group II intron in the mitochondrial *rnl* gene of the brown alga *Pylaiella*) is found exclusively in *Pedinomonas*. Both *Pycnococcus* and *Pedinomonas* mitochondrial gene sequences appear to evolve more rapidly when compared to those of other prasinophyte lineages and chlorophytes in general.

A recent plastid-based phylogeny suggests an affiliation of *Pedinomonas* with the Trebouxiophyceae (Turmel et al. 2009b), which is however not confirmed in a phylogeny with substantially more plastid sequence data (see Chap. 3). Accumulating genomics data will eventually bring 'order' to the paraphyletic prasinophytes.

2. Chlorophyceae (Chlorophyta)

Chlorophyceae is a diverse assemblage of primarily freshwater algae, but secondarily marine species are also known. This group includes unicellular (flagellated or non-flagellated) as well as multicellular species. Chlorophyceans are subdivided in two phylogenetically distinct clades, whose initial circumscription was based on the configuration of the flagellar basal bodies. Therefore, clades are commonly referred to as the CW ("clockwise") and the DO ("directly opposed") clades (Booton et al. 1998; Graham and Wilcox 2000). Note that several chlorophycean taxa, particularly *Chlamydomonas* (also termed the *Chlamydomonas* 'complex'), are polyphyletic. Mitochondrial genomes from species of both clades have been sequenced. To date, mtDNA sequences are available from several CW lineages, including *Chlamydomonas*, *Polytomella*, *Chlorogonium*, *Dunaliella* and *Volvox*. The only representative of the DO clade with completely sequenced mtDNA is *Scenedesmus* (for references, see below).

a. CW-Chlorophyceae

Mitochondrial genomes in the CW clade are of variable conformation (either a single circular-mapping, or 1–2 linear chromosomes) and rather small in size (Table 6.2), and possess extremely reduced gene complements limited to seven respiratory protein genes, 1–3 tRNA genes, *rnl* and *rns* (Table 6.3). The two latter genes are fragmented and scrambled around the genome. CW-chlorophyceans are subdivided in several subclades, and most of the completely sequenced mtDNAs are from the *Reinhardtinia* (as defined by Nakada et al. 2008). Among these are *Polytomella capuana*, *Polytomella parva*, *Chlamydomonas reinhardtii*, and *Volvox carteri*. Mitochondrial genomes from the *Reinhardtinia* are linear, small and relatively G+C-rich with the notable exception of *Volvox*.

At 13 kbp and 43% A+T, the mtDNA of *Polytomella capuana* (Smith and Lee 2008a) – a non-photosynthetic and wall-less relative of *C. reinhardtii* – is the smallest and most G+C-rich among Viridiplantae mtDNAs. With its nine genes common to mtDNAs of all CW-chlorophyceans plus one tRNA, it is also the most gene-poor mitochondrial genome among Viridiplantae. The single mitochondrial chromosome is linear with telomeres on both termini. A single-stranded loop or a 'broken loop' forms the very ends of the molecule (Mallet and Lee 2006; Smith and Lee 2008a).

In contrast to *P. capuana*, the mtDNAs of two other *Polytomella* species (*P. parva* and *P. piriformis*) occur as two chromosomes (13 and 3 kbp), with the smaller containing only one gene (*nad6*). Both molecules display at their termini identical 1.3-kbp inverted repeats with either closed or open telomeric structures, similar to the situation described in *P. capuana* (Fan and Lee 2002; Mallet and Lee 2006). Furthermore, there is evidence for variation in both size and number of mitochondrial chromosomes between various *Polytomella* isolates (Mallet and Lee 2006). Presumably, illegitimate recombination between short inverted repeats has transformed the single linear chromosome such as that in *P. capuana* into multiple chromosomes as found in *P. parva* and *P. piriformis* (Smith and Lee 2008). Finally, the three *Polytomella* mitochondrial genomes differ in their intergenic regions. Unlike the *P. capuana* mtDNA, where inverted repeats punctuate all but two of the genes, repeats are absent from *P. piriformis* and *P. parva* mtDNAs.

Both *C. reinhardtii* (Gray and Boer 1988; Vahrenholz et al. 1993) and its close relative, *Chlamydomonas incerta* for which a partial mtDNA sequence is available (Popescu and Lee 2007), possess linear mitochondrial genomes with terminal inverted repeats. Yet, the terminal structures in *C. reinhardtii* differ from those in *Polytomella*, as they consist of terminal inverted repeats including ~40 nt-long non-complementary 3' single-stranded extensions. Furthermore, compared to *Polytomella* spp., two more tRNAs are encoded in the mtDNAs of *C. reinhardtii* (Michaelis et al. 1990) and *C. incerta*, and a reverse transcriptase-like coding region (possibly a remnant of a group II intron; Nedelcu and Lee 1998) is present in the *C. reinhardtii* mtDNA. Another difference relative to *Polytomella* spp. is that several group I introns interrupt mitochondrial genes in *C. reinhardtii* and *C. incertae*. Note that although *C. reinhardtii* mtDNA is usually considered intronless, up to three optional group I introns have been found in certain isolates (Smith and Lee 2008b). A small (ca. 20 kbp) linear mtDNA with terminal inverted repeats of about 1.8–3.3 kbp have also been described for a colonial relative of *C. reinhardtii*, namely *Pandorina morum* (Moore and Coleman 1989).

The multicellular CW-alga *Volvox carteri* has a considerably larger (35 kbp) and more A+T-rich mtDNA (Smith and Lee 2009) than its unicellular and colonial *Reinhardtinia* relatives, and its genome maps as a circular molecule (Smith and Lee 2009, 2010) (see Table 6.2). Also, two group I introns (at least one of which is absent in some *Volvox* strains) and one group II intron have been found in this genome. The larger size of the *Volvox* mtDNA (i.e. twice the size compared to *C. reinhardtii*) is due to introns and long intergenic regions, most of which carry short palindromic repeats. Apparently, two mitochondrial genome isomers (A and B) are present in *Volvox* that differ from one another in gene arrangement. These isomers likely result from illegitimate recombination between repetitive elements, a phenomenon commonly seen in plastid DNAs (see Chap. 3) and mtDNAs of plants (see Chaps. 8, 9), but rather rarely observed in non-plant mtDNA (but see Sect. III, C of this chapter). Intergenic plus intronic regions of *Volvox* mtDNA taken together amount to approximately 60% (Smith and Lee 2009, 2010). Based on this calculation, the *Volvox* genome has been considered the most bloated chlorophyte mtDNA. However, when only intergenic regions are taken into account (see Table 6.2), it is the ulvophyte *Pseudendoclonium* (discussed later) that has the most loosely packed mtDNA among chlorophytes.

From CW-algae outside the *Reinhardtinia*, three mtDNA sequences are known. These are from *Chlamydomonas eugametos* (Denovan-Wright et al. 1998), *Chlorogonium elongatum* (Kroymann and Zetsche 1998) and *Dunaliella salina* (Smith et al. 2010) (Table 6.2). The *rns* and *rnl* genes in these three mtDNAs are less fragmented than the ones in *Reinhardtinia* (i.e. three vs. six gene pieces for *rns* and six vs. at least eight gene pieces for *rnl*). In these mtDNAs, all genes are encoded on the same strand (i.e. have the same transcriptional polarity), which is a trait shared by all currently sequenced small (<35 kbp) and circular-mapping mtDNAs of CW-chlorophycean algae, but also seen in fungi and other eukaryotic taxa (e.g. *Schizosaccharomyces pombe* (Bullerwell et al. 2003) and *Thraustochytrium aureum* (Gray et al. 2004)). A single coding strand apparently emerged several times independently in mitochondrial genome evolution.

b. DO-Chlorophyceae

The only available mtDNA sequence from the DO-chlorophyceans belongs to *Scenedesmus obliquus* (Kück et al. 2000; Nedelcu et al. 2000). At 43 kbp, this genome is in the medium size range among green algal mtDNAs. Identified genes account for only 60% of the genome, and both group I and group II introns are present (Table 6.2). The larger gene content and lower degree of rRNA fragmentation (two *rns* and four *rnl* fragments (Table 6.3); note that the breakpoints are shared with *Chlamydomonas*) classify this mitochondrial genome as less derived relative to those of all other chlorophyceans. A particularity of *Scenedesmus* mitochondria is a deviant genetic code characterized by the use of UAG (normally a stop codon) to specify leucine, and the unprecedented recoding of UCA (normally a serine codon) as a stop codon. Most puzzling was the finding of a C-terminally truncated mitochondrial *cox2* gene whose second half was suggested to be nucleus-encoded ((Funes et al. 2002); (reviewed in Burger et al. 2003b)). Consistent with this suggestion, *Chlamydomonas* and *Polytomella* species possess a split *cox2* gene with both parts encoded in their nucleus (Pérez-Martínez et al. 2001). Other examples of split genes with one half in mtDNA and the other in nuclear DNA are *rpl2* in *Arabidopsis* (Adams et al. 2001) and *cox1* in several eukaryotes including algae (Gawryluk and Gray 2010). These latter cases likely represent an intermediate phase of gene migration toward the nucleus.

3. Trebouxiophyceae (Chlorophyta)

The third chlorophyte group, the Trebouxiophyceae (sensu (Friedl 1995)), consists of several widespread and biotically significant algae that inhabit mostly soil and freshwater, and include most green algal phycobionts of fungi (lichens), ciliates and cnidarian animals (corals). Trebouxiophytes are either unicellular nonflagellated or filamentous algae (Booton et al. 1998) with basal bodies displaced in a counterclockwise (CCW) configuration, a trait shared with the Ulvophyceae ((Kreimer and Melkonian 1990); see below).

The two trebouxiophyte mtDNAs sequenced thus far are from non-photosynthetic relatives of *Chlorella* – *Prototheca wickerhamii* (Wolff et al. 1993) and *Helicosporidium* sp. (Pombert and Keeling 2010) – which are common parasites of vertebrates and invertebrates, respectively. The two genomes are circular-mapping and very similar in size, A+T-content and gene repertoire (differing by a single tRNA), and their level of synteny is higher than between any other two sequenced chlorophyte mtDNAs.

At 55 and 49 kbp, these mtDNAs are in the medium size range among chlorophytes, and their nucleotide composition places them among the most A+T-rich genomes in this group (see Table 6.2). The *Helicosporidium* mitochondrial genome is more densely packed than that of *Prototheca* and has a trans-spliced group I intron in its *cox1* gene, otherwise described for only two other unrelated species (Burger et al. 2009; Grewe et al. 2009). In terms of gene complement, these mtDNAs resemble closely the "ancestral" type represented by the prasinophyte *Nephroselmis*, coding for a large number of respiratory and ribosomal proteins (see Table 6.3). Also, in contrast to chlorophycean mtDNAs, the two trebouxiophytes have an *rrn5* gene, continuous *rns* and *rnl* coding regions, as well as a complete set of tRNA genes for translation of all mitochondrial codons. This latter capacity is shared by only one other chlorophyte, the ulvophyte *Pseudendoclonium* (discussed next).

4. Ulvophyceae (Chlorophyta)

Ulvophyceae, the fourth chlorophyte group, comprise both unicellular and multicellular species, including some of the largest and most conspicuous green algae (seaweeds). Organisms of certain macroscopic species consist of a single, large multinucleate cell (coenocyte) (Graham and Wilcox 2000). Ulvophyceae are common on rocky intertidal sea coasts in temperate regions, but freshwater species are also known. The phylogenetic position of Ulvophyceae within Chlorophyta, and especially their relationship to Chlorophyceae, is unclear. Some studies propose that Ulvophyceae emerged before the divergence of Trebouxiophyceae and Chlorophyceae, while others suggest a possible sister-group relationship between the Ulvophyceae and Chlorophyceae, with the Trebouxiophyceae rather occupying a basal position (see (Pombert et al. 2004) for discussion and references; see also the plastid-gene-based phylogeny in Chap. 3).

The ulvophyceans whose mitochondrial genome has been sequenced to date are *Oltmannsiellopsis viridis* (Pombert et al. 2006) and *Pseudendoclonium akinetum* (Pombert et al. 2004). The two algae belong to distinct, deeply-diverging lineages in the Ulvophyceae (Friedl and O'Kelly 2002; Pombert et al. 2005). Their mtDNAs differ greatly in size, with that of *Pseudendoclonium* being almost 40 kbp larger than the *Oltmannsiellopsis* mtDNA (57 kbp). In fact, at 96 kbp, *Pseudendoclonium* has the largest chlorophyte mtDNA known so far (see Table 6.2). Most of the mitochondrial genome in *Pseudendoclonium* is occupied by intergenic dispersed repeats. The mitochondrial gene complement of the two genomes is rather similar, but gene arrangement differs markedly. Further, mtDNA of *Oltmannsiellopsis* is special in showing strong evidence for intracellular, inter-organellar transfer of a group I intron, because both the plastid and the mitochondrial *rnl* contain a group I intron inserted at the same position, and these introns are more similar to each other than either of them is to any other known introns at that position (Pombert et al. 2006). Another particularity of this alga is that the mitochondrial genome appears to have recently captured via horizontal transfer a group II intron from a cryptophyte and an integrase from a bacterium (Pombert et al. 2006).

Pseudendoclonium mtDNA exhibits certain features typical of the "expanded" pattern of embryophytes (also seen in some charophytes discussed below). This challenges the previous notion that only land plants allowed a substantial gain of intergenic sequences in mtDNA (Turmel et al. 2007). *Pseudendoclonium* mtDNA also challenges the categorization of green algal evolutionary patterns into ancestral/reduced-derived/expanded types, calling for more fine-grained notions as detailed in Sect. IV of this chapter.

5. Charophyceae (Streptophyta)

Charophyceae belong to the second division of Viridiplantae – the streptophytes – and form a monophyletic lineage together with embryophytes ((e.g. Karol et al. 2001; Rodriguez-Ezpeleta et al. 2005); see Fig. 6.1).

Charophyceae contain six lineages of mainly freshwater algae. These are (1) the Mesostigmatales including a single species, the scaly, unicelled biflagellate *Mesostigma viride* (previously regarded as a member of the Prasinophyceae, but now placed confidently at the base of streptophytes (Rodriguez-Ezpeleta et al. 2007b)); (2) the Chlorokybales, also constituted by a single unicellular species, the sarcinoid *Chlorokybus atmophyticus*; and further the multicellular (3) Klebsormidiales; (4) Zygnematales; (5) Coleochaetales; and (6) Charales. Mesostigmatales and Chlorokybales are likely the earliest diverging charophyceans, forming a distinct clade (Turmel et al. 2007), while the branching order of the other taxa remains uncertain. Some studies indicate that Charales are the closest relatives to land plants, but others suggest that this group diverged prior to Zygnematales and Coleochaetales (for references and discussion, see Turmel et al. 2006). Mitochondrial DNAs of four charophyte lineages have been sequenced to date. These are from *Mesostigma* and *Chlorokybus*, *Chaetosphaeridium globosum* (Coleochaetales), and *Chara vulgaris* (Charales).

Charophycean mtDNAs have similarly large gene contents, but vary considerably in genome size (Tables 6.2, 6.3). The medium-sized mitochondrial genome of *Mesostigma* (Turmel et al. 2002b) is the only one among green algae possessing trans-splicing group II introns that otherwise occur frequently in land plants (Bonen 2008). *Chlorokybus* mtDNA (Turmel et al. 2007) is not only the most gene-rich one among all Viridiplantae, but also the largest. At 202 kbp, this genome is about twice as large as the largest chlorophyte mtDNA (i.e. that of *Pseudendoclonium* (Ulvophyceae)) and exceeds even that of the bryophyte land plants, *Marchantia polymorpha* (Oda et al. 1992) and *Physcomitrella patens* (Terasawa et al. 2007). Contributory to the huge mtDNA size are the numerous group I and group II introns (also present in tRNA genes) and intergenic regions that account for about half of the genome. *Chlorokybus* mtDNA is the least densely packed green algal mitochondrial genome currently known.

The two multicellular charophyceans *Chaetosphaeridium globosum* (Turmel et al. 2002a) and *Chara vulgaris* (Turmel et al. 2003) both possess medium-sized mtDNAs. The *Chara* mtDNA is unique among green algae in encoding components involved in cytochrome *c* biogenesis (see Tables 6.1, 6.3). While common in land plants, these genes are found in only a few algal mtDNAs including that of the rhodophyte *Cyanidioschyzon merolae* (see below).

B. Glaucophyta

Glaucophytes (aka glaucocystophytes) are freshwater microalgae with planktonic or sessile life styles. Some taxa are flagellated, others are not, and unicellular and colonial species exist. At least nine genera have been recognized. Glaucophytes are characterized by the presence of a cyanelle in their cells. This photosynthetic organelle has retained from its cyanobacterial ancestor a thin peptidoglycan wall and the phycobiliproteins phycocyanin and allophycocyanin organized in typical phycobilisomes on the surface of unstacked thylakoids. Phycobilisomes and unstacked thylakoids are ancestral (cyanobacterial) features shared only by red algae (Graham and Wilcox 2000). In contrast to Viridiplantae, glaucophytes (and rhodophytes, see below) have chlorophylls a and c as photosynthetic pigments. The phylogenetic position of glaucophytes has been controversial. Recent phylogenomics analyses show their common ancestry with red algae and green algae plus plants, but the precise branching order remains elusive (Rodriguez-Ezpeleta et al. 2005), as this topology is not obtained in phylogenies that use different species and different sequence sampling.

Mitochondrial genome sequences are available for two glaucophytes, *Cyanophora paradoxa* and *Glaucocystis nostochinearum* (Price et al. 2012; GenBank acc. nos. HQ849544, HQ908425). The two genomes are similar in terms of A+T-composition and gene content (Tables 6.4, 6.5). The gene complement is large (including genes for 5S rRNA, 15 ribosomal proteins, 10 Complex I subunits, and two Complex II subunits), the gene order displays vestiges of the eubacterial *str*, S10, *spc* and alpha operons, and the gene sequences are little derived. Despite these similarities, the two mtDNAs exhibit a completely different gene order outside operon-like clusters and differ markedly in size, with 52 kbp for *Cyanophora* and 34 kbp for *Glaucocystis*. The larger size of the *Cyanophora* genome is due to a 5-kbp long duplicated region containing *nad9, cox3,* and *rns*, and several 0.3-kbp long direct repeats occurring 2–3 times and overlapping genes partially or completely.

C. Rhodophyta

Rhodophyta is a morphologically diverse group with up to 6,000 unicellular and multicellular species in at least 12 orders. Red algal cells are unique among algae in lacking centrioles and flagella from all life stages, and in having plastids with unstacked thylakoids and containing phycoerythrin as an accessory photosynthetic pigment. Rhodophytes inhabit tropical and temperate near-shore marine waters; many of these species are of economic and ecological significance (Graham and Wilcox 2000). Several distinct red algal lineages are known to date but their phylogenetic relationships are not well understood (Saunders and Hommersand 2004).

To date, complete mitochondrial genomes have been sequenced from six species belonging to four distinct rhodophyte lineages, and partial information is available from one additional species. Species include two unicellular Cyanidales – *Cyanidioschyzon merolae* (Ohta et al. 1998) and *Cyanidium caldarium* (partial sequence, Viehmann et al. 1996) – and five multicellular species from three distinct groups – *Chondrus crispus* (Gigartinales) (Boyen et al. 1994; Leblanc et al. 1995), *Porphyra purpurea* (Bangiales) (Burger et al. 1999), and *Gracilariopsis andersenii, Gracilariophila oryzoides*, and *Plocamiocolax pulvinata* (Hancock et al. 2010) (Floridophyceae). Red algal mtDNAs are relatively similar in size (between 25 and 36 kbp), conformation (circular-mapping), and gene order (for details, see discussion in Burger et al. 1999) (Table 6.4). They are rather compact, having as little as 4% non-coding sequences and displaying several cases of overlapping genes (in *Chondrus* and *Cyanidioschyzon*). The presence of three mitochondrion-encoded succinate dehydrogenase genes is unique among algae (Table 6.5). In terms of coding capacity, the mtDNAs of five of the six red algae are essentially identical, while that of *Cyanidioschyzon* encodes several additional genes specifying ribosomal proteins and components involved in the biogenesis of cytochrome c (*ccmA, B, C, F*; see Table 6.1). Notable is the occurrence of a mitochondrion-encoded *rpl20* that is otherwise only known from jakobids. Some confusion arose about whether or not red algae possess a mitochondrion-encoded 5S rRNA. The initial report that *Chondrus* mtDNA includes an *rrn5* situated between *cox2* and *cox3* (Leblanc et al. 1995) was discounted (Lang et al. 1996). Yet, the claim of *rrn5* being absent from red algal mtDNAs had to be rectified, when this gene was later detected in another genomic location (between *nad3* and *rps11*) in *Chondrus* (Gray et al. 1998), *Cyanidioschyzon* and *Cyanidium* (Gray et al. 2004); a mtDNA-encoded 5S rRNA is lacking only in *Porphyra* (Burger et al. 1999) (Table 6.5). Another difference between the studied red algae is that mitochondrial translation uses the standard genetic code in *Cyanidioschyzon* and *Cyanidium*, but UGA specifies tryptophan in the other species (see Table 6.4).

The mitochondrial genome of *Porphyra* assumes two isomeric conformations that differ from each other in the orientation of a 15-kbp region between two ~300 bp-long repeats (Burger et al. 1999). The same phenomenon has been described recently in *Volvox* mtDNA (see Sect. III.A.2), but is known for a long time from angiosperm mitochondria where repeats promote major genome rearrangements via a flip-flop mech-

Table 6.4. General characteristics of mtDNAs from non-green algae

Taxon	Size (kbp)	Shape	A+T (%)	Intergenic[a] (%)	Intron type (number)	Genetic code
Glaucophytes						
Cyanophora paradoxa	51.6	Circular	74.0	15.3	/	Standard
Glaucocystis nostochinearum	34.1	Circular	74.3	6.1	/	Standard
Rhodophytes						
Chondrus crispus	25.8	Circular	72.1	4.2	II (1)	UGA = W
Cyanidioschyzon merolae	32.2	Circular	72.9	5.3	/	Standard
Gracilariophila oryzoides	25.2	Circular	71.9	6.5	II (1)	UGA = W
Gracilariopsis andersonii	27.0	Circular	72.0	11.5	II (1)	UGA = W
Plocamiocolax pulvinata	25.9	Circular	76.1	3.6	II (2)	UGA = W
Porphyra purpurea	36.7	Circular	66.5	8.9	II (2)	UGA = W
Stramenopiles						
Chrysophytes						
Chrysodidymus synuroides	34.1	Circular	75.9	5.7	/	Standard
Ochromonas danica	41.0	Linear	73.8	11.9	/	Standard
Diatoms						
Phaeodactylum tricornutum	77.4	Circular	65.0	49.3	II (4)	Standard
Synedra acus	46.7	Circular	68.2	18.9	II (3)	Standard
Thalassiosira pseudonana	43.8	Circular	69.9	19.9	II (1)	UGA = W
Phaeophytes						
Desmarestia viridis	39.0	Circular	63.4	7.6	/	Standard
Dictyota dichotoma	31.6	Circular	63.5	5.2	/	Standard
Fucus vesiculosus	36.3	Circular	65.6	6.7	/	Standard
Laminaria digitata	38.0	Circular	64.9	7.3	/	Standard
Pylaiella littoralis	58.4	Circular	62.0	7.4	II (7)	Standard
Saccharina angustata (and six other species)	37.6 (+/− 0.1)	Circular	64.8	6.6	/	Standard
Raphidophytes						
Chattonella marina	44.8	Circular	71.6	9.3	II (2)	Standard
Heterosigma akashiwo	38.7	Circular	64.3	9.3	/	Standard
Alveolates						
Dinoflagellates[b]	very large	Mult. linear[c]	~60–75	~85[c]	/	Standard or UGA = W[c]
Cercozoa						
Chlorarachniophytes						
Bigelowiella natans	~37.5	Linear	57.9	6.2	/	UGA = W
Cryptomonads						
Hemiselmis andersenii	60.5	Circular	71.3	40.2	/	Standard
Rhodomonas salina	48.1	Circular	70.2	13.9	II (2)	Standard
Haptophytes						
Emiliania huxleyi	29.0	Circular	71.7	22.6	/	UGA = W
Pavlova lutheri	34.1	Circular	62.7	20.6	II (5)	Standard
Euglenozoa						
Euglenids						
Euglena gracilis[d]	very large	Mult. linear	77.5	??	/	UGA = W

[a]Intergenic regions, when considering ORFs ≥100 aa. Percentage shown here may differ from values given in the original publications describing these genomes. Differences are due (1) to the choice of the start codon in protein-coding genes that is based here on multiple alignments with orthologous sequences, and (2) to genes undetected in the original report

[b]GenBank record contains partial mtDNA sequence; A+T content, non-coding portion and intron type and number refer to published sequence

[c]Nash et al. (2007) and references therein

[d]Linear molecules of dispersed size around 7.5–4 kbp (Spencer and Gray 2011)

Table 6.5. Mitochondrial gene content in non-green algae[a]

Taxon	Complex I–V	Ribosomal proteins	Other proteins	rRNAs, tRNAs
Glaucophytes				
Cyanophora paradoxa	*atp4,6,8,9; cob; cox1,2,3; nad1–4,4L,5,6,7,9,11; sdh3,4*	*rpl2,5,6,14,16; rps3,4,7,10–14,19*	/	*rnl, rns, rrn5; trn* genes (27)
Glaucocystis nostochinearum	*atp4,6,8,9; cob; cox1,2,3; nad1–4,4L,5,6,7,9,11; sdh3,4*	*rpl2,5,6,14,16; rps3,***4***,7,10–14,19*	/	*rnl, rns, rrn5; trn* genes (25)
Rhodophytes				
Chondrus crispus	*atp4,6,8,9; cob; cox1,2,3; nad1–4,4L,5,6; sdh2,3,4*	*rpl16,20; rps3,11,12*	*tat***A***,C*	*rnl, rns, rrn5; trn* genes (23)
Cyanidioschyzon merolae	*atp4,6,8,9; cob; cox1,2,3; nad1–4,4L,5,6; sdh2,3,4*	*rpl5,6,14,16,20; rps3,4,8,11,12,14*	*tatC; ccmA,B,C,F*	*rnl, rns, rrn5; trn* genes (25)
Gracilariophila oryzoides	*atp4*[b]*,6,***8***,9; cob; cox1,2,3; nad1–4,4L,5,6; sdh2,***3***,4*	*rpl16,20; rps3,11,12*	***tatA****,C*	*rnl, rns,* ***rrn5****; trn* genes (20)
Gracilariopsis andersonii	***atp4***[b]*,6,8,9; cob; cox1,2,3; nad1–4,4L,5,6; sdh2,3,4*	*rpl16,20; rps3,11,12*	*tat***A***,C*	*rnl, rns,* ***rrn5****; trn* genes (20)
Plocamiocolax pulvinata	***atp4***[b]*,6,***8***,9; cob; cox1,2,3; nad1–4,***4L***,5,6; sdh2,3,4*	*rpl16,20; rps3,11,12*	*tat***A***,C*	*rnl, rns,* ***rrn5****; trn* genes (23)
Porphyra purpurea	*atp4,6,8,9; cob; cox1,2,3; nad1–4,4L,5,6; sdh2,3,4*	*rpl16; rps3,11,12*	*tat***A***,C; dpo, rtl*	*rnl, rns; trn* genes (24)
Stramenopiles				
Chrysophytes				
Chrysodidymus synuroides	*atp6,8,9; cob; cox1,2,3; nad1–4,4L,5–7,9,11*	*rpl2,6,14,16; rps2–4,7,8,10–14,19*	*tat***A***,C*	*rnl, rns; trn* genes (23)
Ochromonas danica	*atp6,8,9; cob; cox1,2,3; nad1–4,4L,5–7,9,11*	*rpl2,6,14,16; rps2–4,7,8,10–14,19*	*dpo*	*rnl, rns; trn* genes (24)
Diatoms				
Phaeodactylum tricornutum	*atp6,8,9; cob; cox1,2,3; nad1–4,4L,5–7,9,11*	*rpl2,5,6,10,14,16; rps2–4,7,8,10–14,19*	*tat***A***,C*	*rnl, rns; trn* genes (23)
Synedra acus	*atp6,8,9; cob; cox1,2,3; nad1–4,4L,5–7,9,11*	*rpl2,5,6,14,***16***; rps3,4,7,8,10–14,19*	*tatC*	*rnl, rns; trn* genes (24)
Thalassiosira pseudonana	*atp6,8,9; cob; cox1,2,3; nad1–4,4L,5–7,9,11*	*rpl2,5,6,***10***,14,16; rps2–4,7,8,10–14,19*	*tat***A***,C*	*rnl, rns; trn* genes (25)
Phaeophytes				
Desmarestia viridis	*atp6,8,9; cob; cox1,2,3; nad1–4,4L,5–7,9,11*	*rpl2,5,6,14,16,31; rps2–4,7,8,10–14,19*	*tatC*	*rnl, rns, rrn5; trn* genes (25)
Dictyota dichotoma	*atp6,8,9; cob; cox1,2,3; nad1–4,4L,5–7,9,11*	*rpl2,5,6,14,16,31; rps2–4,7,8,10–14,19*	*tatC*	*rnl, rns, rrn5; trn* genes (24)
Fucus vesiculosus	*atp6,8,9; cob; cox1,2,3; nad1–4,4L,5–7,9,11*	*rpl2,5,6,14,16,31; rps2–4,7,8,10–14,19*	*tatC*	*rnl, rns, rrn5; trn* genes (25)
Laminaria digitata	*atp6,8,9; cob; cox1,2,3; nad1–4,4L,5–7,9,11*	*rpl2,5,6,14,16,31; rps2–4,7,8,10–14,19*	*tatC*	*rnl, rns, rrn5; trn* genes (25)
Pylaiella littoralis	*atp6,8,9; cob; cox1,2,3; nad1–4,4L,5–7,9,11*	*rpl2,5,6,14,16,***31***; rps2–4,7,8,10–14,19*	*tatC;* ***rpo***	*rnl, rns, rrn5; trn* genes (23)
Saccharina angustata (and six other species)	*atp6,8,9; cob; cox1,2,3; nad1–4,4L,5–7,9,11*	*rpl2,5,6,14,16,31; rps2–4,7,8,10–14,19*	*tatC*	*rnl, rns, rrn5; trn* genes (24)
Raphidophytes				
Chattonella marina	*atp6,8,9; cob; cox1,2,3; nad1–4,4L,5–7,9,11*	*rpl2,5,6,14,16;rps2–4,7,8,10–14,19*	***tatA****,C*	*rnl, rns, rrn5?; trn* genes (25)
Heterosigma akashiwo	*atp6,8,9; cob; cox1,2,3; nad1–4,4L,5–7,9,11*	*rpl2,5,6,14,16; rps2–4,7,8,10–14,19*	***tatA****,C*	*rnl, rns, rrn5?; trn* genes (25)
Alveolates				
Dinoflagellates				
Amphidium carterae	*cob; cox1,3*	/	/	*rnl, rns*

(continued)

Table 6.5. (continued)

Taxon	Complex I–V	Ribosomal proteins	Other proteins	rRNAs, tRNAs
Cercozoa				
Chlorarachniophytes				
Bigelowiella natans	*atp1,6,8,9; cob; cox1,2,3; nad1–4,4L,5–7,9*	*rpl5,6,14,16; rps3,4,7,11,12,14*	/	*rnl, rns, rrn5; trn* genes (26)
Cryptomonads				
Hemiselmis andersenii	*atp1,4,6,8,9; cob; cox1,2,3; nad1–4,4L,5–11; sdh3,**4***	*rpl5,6,14,16,**31**; rps**1**–4,7,8,11–14,19*	*tat**A**,C*	*rnl, rns, **rrn5**; trn* genes (28)
Rhodomonas salina	*atp1,4,6,8,9; cob; cox1,2,3; nad1–4,4L,5–11; sdh3,4*	*rpl5,6,14,16; rps1–4,7,8,11–14,19*	*tatA,C*	*rnl, rns, rrn5; trn* genes (27)
Haptophytes				
Emiliania huxleyi	*atp4,6,8,9; cob; cox1,2,3; nad1–4,4L,5,6*	*rpl16; rps3,8,12,14*	*dam*	*rnl, rns, **rrn5**; trn* genes (25)
Pavlova lutheri	*atp4,6,8,9; cob; cox1,2,3; nad1–4,4L,5,6*	*rpl14,16; rps12,14,19*	/	*rnl, rns, rrn5; trn* genes (24)
Euglenozoa				
Euglenids				
Euglena gracilis	*cox1, cox2, cox3*	/	/	*rnl, rns*

[a]Gene name in **bold** indicates a gene not described in journal publication or not annotated in the GenBank record; a question mark indicates that the gene assignment is not well supported
[b]Erroneously annotated as *secY* in the journal publication (Hancock et al. 2010) and GenBank record

anism of illegitimate recombination (Hanson and Folkerts 1992). Further, *Porphyra* mtDNA displays numerous sequence polymorphisms, including mostly substitutions (transitions and transversions), but also insertions and deletions. At the time of the initial report (Burger et al. 1999), it was not clear whether the observed sequence polymorphism and the two isomer conformations are due to true heteroplasmy (more than one type of mtDNA in the same organism) or rather to diversity within the population of thalli that were collected in the wild, propagated in the laboratory and then used to construct the clone library for mtDNA sequencing. When we later examined distinct blades of *Porphyra* including female, male and bisexual thalli, we could demonstrate that each individual contains two mtDNA conformations, and often but not always, sequence polymorphisms. The sequence deposited in GenBank (acc. no. NC_002007) is that of the predominant isomer (excess factor of about 5–10; G. Burger, D. Tremblay, unpublished). The observed heteroplasmy could be a consequence of mtDNA being inherited biparentally. In fact, a certain percentage of biparental mtDNA inheritance has been observed in the relative *P. yezoensis* (Choi et al. 2008).

A few group II introns (but no group I introns) are present in red algal mtDNAs (Table 6.4). Although the large majority of mitochondrial introns are found in protein-coding genes, four of the six sequenced rhodophyte mtDNAs have introns residing in a tRNA gene (*trnI* of *Chondrus* (Boyen et al. 1994) and *Gracilariopsis*, *Gracilariophila* and *Plocamiocolax* (G. Burger, this report)). The two introns in the *Porphyra* mitochondrial *rnl* gene are remarkable, because they are most similar to introns in the cyanobacterium *Calathrix sp.*, suggesting recent horizontal transfer between bacteria and mitochondria.

D. Stramenopiles

Stramenopiles comprise nearly 40 taxa (see Patterson 1989), whose monophyly is well supported by molecular phylogeny. About half of the stramenopile taxa are nonphotosynthetic such as oomycetes (*Phytophthora*) and bicosoecids (*Cafeteria*), while plastid-carrying groups include phaeophytes (brown algae), chrysophytes (golden algae), bacillariophytes (diatoms), raphidophytes, and xanthophytes. Complete mtDNA sequences are available for all these algal groups except xanthophytes.

Stramenopile mtDNAs have a large set of tRNA genes, but all lack *trnT* (the gene for tRNA-Thr), which apparently was lost from this genome in the common ancestor of stramenopiles (Gray et al. 2004). To sustain mitochondrial translation, tRNA-Thr must be either imported into mitochondria, or generated from another mitochondrial tRNA by post-transcriptional RNA modification or editing. Note that loss of mitochondrial *trnT* is not unique to stramenopiles, but also occurred in other groups independently (e.g. jakobids; Burger and Lang, unpublished).

1. Chrysophytes

Most golden-brown algae are unicellular, naked flagellates, but some species cover their cells with silica scales and others form complex colonies or filaments. Chrysophytes typically inhabit freshwater, where they are part of the phytoplankton. The golden-brown color of their plastids is due to chlorophyll a, chlorophylls of the c-group and typically beta-carotene. More than 120 chrysophyte genera are recognized today.

Complete mtDNA sequences are available from two different genera, represented by *Chrysodidymus synuroides* (Chesnick et al. 2000) and *Ochromonas danica* (Burger et al. unpublished; GenBank acc. no. AF287134; see Tables 6.4, 6.5). The mtDNA size is 34 kbp for *Chrysodidymus* and 41 kbp for *Ochromonas,* the former being circular-mapping and the latter linear with a 2.2-kbp terminal repeat at both ends that includes several ORFs and tRNA genes. The size difference between these two mtDNAs is due to these repeats and a larger number of ORFs in *Ochromonas*. Both taxa display a large mitochondrial gene complement including *nad11* and 14 ribo-protein genes. *Chrysodidymus* has in addition *tatC*, whereas *Ochromonas* mtDNA contains a *dpo* gene; the latter has likely been acquired *via* a plasmid as also seen in other mtDNAs (for a review, see Gray et al. 2004). The two chrysophyte mtDNAs share several clusters of identical gene order. Some clusters exhibit the ancestral eubacteria order of ribosomal protein genes; other clusters such as *nad2-nad9-nad7* have a common derived arrangement that must have be present already in the common ancestor of both golden algae, with a trend to group genes of similar functions (here genes for subunits of Complex I).

2. Diatoms

Diatoms (Bacillariophyta) are unicellular, silica-walled algae found in freshwater and marine habitats. They are considered the most abundant aquatic organisms after viruses and bacteria. Only a few diatoms have flagella; most are free-floating (planktonic) or attached to plants, sand or rocks (benthic). Plastids vary in color from yellowish-brown to deep brown, containing chlorophylls a and c, and xanthins. Diatoms are very diverse, and taxa are numerous with about 250 recorded genera and more than 100,000 species.

Complete mtDNA sequences are available for two species from different classes, the fragilariophycean *Synedra acus* (Ravin et al. 2010), and the coscinodiscophycean *Thalassiosira pseudonana* (Armbrust et al. 2004), while the mtDNA sequence from the bacillariophycean *Phaeodactylum tricornutum* is nearly completed (Bowler et al. 2008; Ravin et al. 2010; see Tables 6.4, 6.5). At 77 kbp, the mitochondrial genome of *Phaeodactylum* is nearly twice as large as that of the other two diatom species. This size difference is due to the length of their repeat regions, which stretch over 36 kbp in *Phaeodactylum*, but only 5 and 4 kbp in *Synedra* and *Thalassiosira.* The mitochondrial gene content of diatoms is quite similar to that of the golden algae discussed above. Remarkably, most diatoms possess a mitochondrion-encoded *rpl10*, a gene that is not seen in mtDNAs from other stramenopiles and that is generally infrequent. Not only the gene set but also the gene arrangement on mtDNA is well conserved among diatoms, which is unexpected since the three diatoms belong to distinct lineages with quite diverse morphologies.

3. Phaeophytes

Brown algae are multicellular organisms whose sizes vary from microscopic filaments to large blades sometimes more than 10 m long (kelp). Photosynthetic pigments include chlorophylls a and c, and fucoxanthins. The habitat of phaeophytes is mostly marine costal regions, where they grow attached to rocks etc. or sometimes invertebrate animals. The phylum comprises about 250 genera and nearly 1,000 species.

Complete mtDNA sequences are available for 12 species from five different phaeophyte lineages, *Desmarestia viridis* (Desmarestiales), *Dictyota dichotoma* (Dictyotales), *Fucus vesiculosus* (Fucurales) (Oudot-Le Secq et al. 2006), *Laminaria digitata* (Laminariales) (Oudot-Le Secq et al. 2002), *Pylaiella littoralis* (Ectocarpales) (Oudot-Le Secq et al. 2001) and *Saccharina angustata* (Laminariales; Yotsulura et al. unpublished, NC_013473) plus six other species of the *Saccharina* genus (see Tables 6.4, 6.5). The sizes of brown algal mtDNAs vary between 32 and 59 kbp with *Dictyota* having the smallest and *Pylaiella* the largest genome. Although the investigated species are quite distinct in morphology and habitat, their mtDNAs are most similar. The mitochondrial gene complement is large – including *rpl31* that is seen only sporadically in other groups – and identical across phaeophytes. Only the number of introns (all group II), tRNAs, and ORFs varies. Even the mitochondrial gene order is essentially identical between brown algae (but completely different compared to diatom mtDNAs), with sporadic ORFs or introns inserted/deleted in one or the other mtDNA. Remarkable is *nad11* that only encodes the N-terminal FeS-binding domain (ca. 230 residues), whereas the rest of the protein is missing (Oudot et al. 1999). A similarly short *nad11* has also been observed in the non-photosynthetic stramenopile *Cafeteria roenbergensis* (Burger et al. unpublished; GenBank acc. no. NC_000946). It was speculated that *nad11* functions as a short version (Oudot et al. 1999), but it is more probable that the second half of the protein is encoded in the nucleus and imported into mitochondria, as is the case for *cox2* in *Scenedesmus* (see above).

4. Raphidophytes

Raphidophyte algae are flagellated unicellular organisms with usually bright green plastids that contain chlorophylls a and c, and large amounts of carotenoid pigments. In contrast to most other stramenopile algae, raphidophytes lack a rigid cell wall. The group includes planktonic freshwater and marine species that occur where vegetation is abundant, and several taxa are responsible for toxic algal blooms harmful to fish (for references, see Masuda et al. 2011). Raphidophytes is a small group of organisms with four genera and a total of nine species. The complete mtDNA sequence is available for two species, *Chattonella marina* var. *marina* and *Heterosigma akashiwo* ((Masuda et al. 2011); see Tables 6.4, 6.5).

The two raphidophyte mtDNAs are of moderate size (39 and 45 kbp). The size difference is due to ~4 kbp more intergenic sequence and the presence of two group II introns in *cox1* in *Chattonella* mtDNA, while the gene content is the same. Gene order is quasi identical with the exception of a cluster of seven genes that has changed orientation, and several open reading frames that are present/absent in various genomic locations. The mitochondrial gene content of raphidophytes is essentially the same as in brown algae, only that *rpl31* is missing in the two raphidophytes.

E. Alveolates

Alveolates are subdivided into about eight taxa, one of which, the dinoflagellates, is mostly photosynthetic, while another, the apicomplexans, is predominantly non-photosynthetic, but possesses plastid relicts (e.g. *Plasmodium* and *Eimeria*). The few photosynthetic apicomplexan species include *Alveolata* and *Chromera* (Janouskovec et al. 2010) (see Chap. 2 for the origin of their plastids). Apicomplexans are generally not considered algae, and therefore, their mtDNAs will not be discussed in this chapter. More information on this subject is available in specialized reviews (e.g. McFadden and Waller 1997; Williams and Keeling 2003).

1. Dinoflagellates

Dinoflagellates inhabit marine and freshwater ecosystems. Their great diversity of cell shapes comes from the flattened membrane sacs (alveoli) beneath the plasma membrane that form armours of most baroque shapes. All dinoflagellates have two flagella that are inserted at the same point, with one wrapping around the cell and the other oriented perpendicularly to the first. Photosynthetic taxa play a major role in ocean carbon fixation, and are equally notorious for toxic red tides as they are for symbiotic partnerships with reef-building corals. Plastid pigments include chlorophylls a and c_2 together with the unique xanthophyll peridinin, further beta-carotene and xanthins. Some dinoflagellates have chlorophyll c_1 and fucoxanthin instead, testifying to multiple independent acquisitions of plastids from diverse sources through higher-order endosymbioses.

The dinoflagellate species with most mtDNA sequence available today is *Amphidium carterae* (33 kbp), but the genome sequence is far from being completed. Data from numerous taxa combined indicate a set of only five mitochondrial genes, *cob*, *cox1*, *cox3*, *rnl*, and *rns*, implying that most of the traditional mitochondrial genes must have migrated to the nucleus (Tables 6.4, 6.5; (reviewed in Nash et al. 2008; Waller and Jackson 2009)). Interestingly, *cox2* exists as a split nuclear gene similar to the situation in chlorophyceans (Waller and Keeling 2006). Mitochondrial genes exist not only as contiguous forms, but also in pieces and associated within different genomic contexts, as first described for *Crypthecodinium* (Norman and Gray 2001). Further, the dinoflagellate mitochondrial genomes appear to undergo considerable recombination (Jackson et al. 2007). On top, RNA editing (Lin et al. 2002, 2008) and trans-splicing of certain genes (Waller and Keeling 2006) make these mitochondrial genomes some of the most bizarre and most difficult to analyze. For a recent review, see (Burger et al. 2011).

F. Chlorarachniophytes (Cercozoa)

Chlorarachniophytes are a small group of photosynthetic flagellates that live in marine habitats. Plastids are bright green and contain chlorophylls a and b, but neither chlorophyll c nor biliprotein pigments. As in the unrelated cryptophytes (see below), a nucleomorph reveals that photosynthesis was acquired via secondary endosymbiosis, here with a green alga. Two chlorarachniophyte genera are recognized, *Chlorarachnion* and *Bigelowiella*.

A nearly complete mtDNA sequence is available for *Bigelowiella natans* (Burger et al. unpublished; GenBank acc. no. HQ840955; Table 6.4). The mitochondrial genome consists of a linear chromosome of ~38 kbp, with about 400 bp and 700 bp remaining unsequenced at the two extremities. The mitochondrial gene set is moderately large (Table 6.5). All genes are encoded on the same strand with only a few nucleotides in intergenic regions, and some protein-coding genes overlap by a few nucleotides. Bacterial operon organisation has vanished, and the inferred protein sequences are divergent. Given the small genome size, it is surprising to find five ORFs longer than 100 residues (185–605 amino acids). Some of these ORFs are likely fortuitous as they include small stretches of regular genes; others might be ribo-proteins that are too derived to be recognized.

G. Cryptomonads

Cryptomonads are single-celled flagellates recognizable by their flattened asymmetric cells and distinctive swimming motion. They can be found in diverse aquatic habitats, from drainage ditches to tundra ponds and the open ocean. Most of the 20 or so genera are photosynthetic. The heterotrophic *Goniomonas* is believed to be primarily without plastids, i.e. closely related to the cryptomonad ancestor that engulfed an alga with red plastids (for a discussion on plastid origins see Chap. 4). Photosynthetic cryptomonads are important planktonic primary producers at the base of aquatic food webs, feeding rotifers, mussels and many other animals. Some species form toxic blooms. Cryptomonad plastids are characterized by chlorophyll c and phycobilins as photosynthetic pigments. Cells carry in addition to their indigenous nucleus a remnant 'nucleomorph' wrapped in a membrane sack together with the plastid, which are the leftovers of the red-algal endosymbiont (for a review, see Maier 1992).

Complete mtDNA sequences are available for two species, *Hemiselmis andersenii* (Kim et al. 2008) and *Rhodomonas salina* (Hauth et al. 2005) (see Tables 6.4, 6.5). The genomes are 61 kbp and 48 kbp in size, respectively, and minimally diverged as to gene complement, genetic code, and gene order. First, the mitochondrial gene set is among the richest outside jakobids including, in addition to the more common genes discussed above, also *atp1* and *nad8*. Ribo-protein genes are arranged in eubacteria-like clusters in both cryptophyte mtDNAs, but the order of other genes is different. Overall, sequences are minimally divergent. Both genomes are compact with only a single sizeable intergenic region, which in *Rhodomonas* mtDNAs is 4.7 kbp long and contains an elaborate 4.5 kbp repeat region with two large blocks of inverted orientation. The blocks are composed of numerous tandem repeats and hairpin structures with unit length of ~35–700 nt. Within this region map the predicted promoters and replication origins. *Hemiselmis* has a five times larger mitochondrial repeat region that also consists of non-palindromic and palindromic motifs, which, compared to those in *Rhodomonas* mtDNA, are shorter (~20–350 nt long), differ in sequence, and are all arranged in the same orientation.

H. Haptophytes

Haptophytes (prymnesiophytes) are planktonic, biflagellated algae occurring in marine habitats. All members of this group are photosynthetic. Most species are unicellular, but colonial taxa are known as well. Haptophyte cells are covered with scales of often complex ornamentation. Species with calcite scales such as *Emiliania* are among the most productive lime producers on Earth. Haptophyte plastids appear yellow-brown due to high concentrations of carotenoids and xanthins, in addition to chlorophylls a, c_1 and c_2. We recognize about 50 genera with some 500 species.

Complete mtDNA sequences are available for two species, *Emiliania huxleyi* (Sanchez Puerta et al. 2004) and *Pavlova lutheri* (Burger et al. unpublished; GenBank acc. no. HQ908424; see Tables 6.4, 5). These mtDNAs resemble each other considerably in gene content, genome size (29–34 kbp), and a large intergenic repeat region of 1.3 kbp in *Emiliania* and 1.6 kbp in *Pavlova*. Mitochondria of *Pavlova* and other Pavlovales use the standard genetic code, whereas *Emiliania* as well as species from five other haptophyte orders read UGA 'stop' codons as tryptophan (Hayashi-Ishimaru et al. 1997). Remarkably, *E. huxleyi* mtDNA carries the *dam* gene encoding DNA adenine methylase, which was likely acquired by horizontal gene transfer. The only other mtDNA where this gene has been discovered is that of the charophyte *Klebsormidium* (BF. Lang, G. Burger, MW. Gray, unpublished). Haptophyte mitochondrial gene sequences are relatively derived accounting for long branches in mitochondrial-protein-based phylogenies and uncertain branching position (Baurain et al. 2007) (see Fig. 6.1).

I. Euglenozoa

Euglenozoa, one of the deepest-diverging eukaryotic lineages, include three taxa, kinetoplastids, diplonemids, and euglenids (Simpson et al. 2002). Euglenids branch basally to the other two clades, and they are the only euglenozoan group with photosynthetic members (see Fig. 6.1).

1. Euglenids

Euglenids are unicellular flagellates of ovoid to lanceolate body shape. Most are naked with a plasma membrane that is stiffened by distinctive pellicle strips of helical arrangement. Euglenids are cosmopolitans living in most different aqueous habitats that are typically rich in organic matter. They are not found in plankton but rather at the water-mud or water-air interface, since bacteria are their food source. Nearly 50 genera and 1,000 species are recognized. About one third of euglenid species possess plastids. Photosynthesis has apparently been acquired rather late in euglenid evolution, because basally branching taxa such as *Peranema trichophorum* and *Petalomonas cantuscygni* are heterotrophic (Moreira et al. 2001). Euglenid plastids are grass-green (chlorophyll a and b, beta-carotene and xanthophylls); according to phylogenetic analyses, the plastid donor was a relative of the chlorophyte alga *Pyramimonas* ((Turmel et al. 2009a); see also Fig. 2.1 in Chap. 2).

Data on photosynthetic euglenid mtDNA are available for only a single species, *Euglena gracilis*, yet the genome sequence is far from being complete (Tables 6.4, 6.5). Defined chromosomes are not detectable, but rather numerous linear molecules of a broad range of different sizes, mostly around 4 kbp, with a smaller fraction around 7.5 kbp (Spencer and Gray 2011). In total, about 13 kbp mtDNA sequence is available in public databases. The A+T content is extremely high (75–80%), and gene sequences are evolving at accelerated rates yielding an extremely long branch for *Euglena* in mitochondrial-gene-based phylogenies (e.g. Yasuhira and Simpson 1997). Genes identified so far on mtDNA are *rnl* and *rns* (Spencer and Gray 2011), *cox1* (Tessier et al. 1997; Yasuhira and Simpson 1997), *cox2* and *cox3* (Tessier and Paulus, unpublished; GenBank acc. no. AF156178.1). The *cox3* sequence was initially annotated as *nad6*, but the corresponding reading frame rather represents the N-terminal portion (residues 1–~160) of *cox3* (Spencer and Gray 2011), which is most poorly conserved not only in *Euglena* but also in the other Euglenozoa, including trypanosomes and *diplonemids* (Vlcek et al. 2011), and therefore difficult to recognize. It remains to be seen whether a complete copy of *cox3* exists in *Euglena* mtDNA, or whether the missing half is in the nuclear DNA. The gene may also be split into two mitochondrion-encoded pieces, just like the rRNA genes in this genome. Ribosomal RNA gene halves occur in different genomic contexts, together with numerous short dispersed fragments of authentic genes, as well as direct repeats arranged in multiple combinations. This indicates that the highly complex structure of *Euglena* mtDNA is due to frequent recombination.

A *Euglena*-like genome structure is also observed in the heterotrophic euglenid *Peranema*, whereas in *Petalomonas* (the second non-photosynthetic euglenid for which some mtDNA data are available) appears to possess a rather traditional mtDNA architecture (Roy et al. 2007). The 'chaotic' mitochondrial genome organisation seen in *Euglena* has probably emerged in the common ancestor of euglenids, prior to the acquisition of plastids.

IV. Recurring Patterns of Mitochondrial Genome Evolution

The mtDNAs that resemble most the genome of the alpha-proteobacterial predecessor of mitochondria are found among the (plastid-less) jakobid flagellates, which typically have a ca. 70 kbp circular mtDNA that carries genes for ~70 proteins and ~30 structural RNAs. These mtDNAs appear to be 'frozen in time' (Lang et al. 1997) as to gene set, gene order and bacteria-like transcription. Algal mtDNAs are not only very different from the minimally derived mtDNAs of jakobids; they are also highly diverse amongst each other. This illustrates that most disparate evolutionary forces must have been at work to shape mitochondrial genomes in the various algal lineages. Still, several prominent evolutionary patterns emerge in algal mtDNAs discussed above. But these patterns occur sporadically and independent of taxonomic groupings, with sometimes astoundingly similar mitochondrial genomes in phylogenetically most distant lineages (convergent evolution).

The probably most conspicuous trend in mtDNA evolution is what is often referred to as 'genome reduction' including the decrease of physical size and gene set. This phenomenon occurred (independently), for instance, in the lineages leading to *C. reinhardtii* and *Pedinomonas* (Turmel et al. 2010). However, the two genome features do not always contract in parallel as exemplified by *Volvox* (Smith and Lee 2009), where a minimal mitochondrial gene complement is accompanied by an ordinary-sized mitochondrial genome. Alternatively, genome size reduction can occur by maintaining a substantial gene repertoire (yet for obvious reasons to only a certain degree). This is achieved, as exemplified by *Bigelowiella* mtDNA, via densely packing the genes on mtDNA combined with shortening the coding regions. Densely packed genomes with reduced intergenic portions also evolved in red algae (*Plocamiocolax* and *Chondrus*), chrysophytes and phaeophytes (*Chrysodidymus* and *Dictyota*). In addition, adjacent protein-coding genes may be even fused to economize the stop and start codons as in the rhizarian *Acanthamoeba castellanii* (Burger et al. 1995). Alternatively, genes can overlap partially so that the same stretch of DNA encodes different information. Overlapping protein-coding genes are specified in different reading frames, as seen in clusters of ribosomal protein genes, e.g. in the mtDNAs of the red alga *Chondrus* (Leblanc et al. 1995) and the cryptomonad *Rhodomonas* (Hauth et al. 2005). In other cases, protein-coding genes overlap with tRNA genes, such as in the mtDNA of the chrysophyte *Chrysodidymus* (Chesnick et al. 2000). Because of all the above, one should distinguish the two different reductive evolutionary trends: one that affects genome size and the other that affects gene number in mtDNAs, the latter mostly via accelerated gene transfer to the nucleus.

Genome expansion is another recurrent evolutionary pattern observed in mitochondria of various lineages. Size inflation is typically due to growing intergenic regions that accumulate dispersed repeats across the genome as, for example, in the charophyte *Chlorokybus* and the ulvophyte *Pseudendoclonium*. Intergenic regions may as well expand through accumulation of inactive mitochondrial gene fragments piling up between the complete, functional genes, as seen in dinoflagellates. Further, foreign DNA such as chloroplast sequences, may be hoarded in mtDNA, which to our knowledge has been observed exclusively in embryophyte, but not in algal mtDNAs. In contrast to an expansion of intergenic regions across the entire mitochondrial genome, there are also select cases of a single, several kbp-long 'surplus' region. Such mtDNA segments may be composed of numerous tandemly-arranged repeat motifs as in the green alga *Pedinomonas* (Turmel et al. 1999) or the cryptophyte *Hemiselmis* (Kim et al. 2008). Tandem repeats likely expand through slipped mispairing, and repeat inversions probably arise via strand switching during replication (for a review, see Bzymek and Lovett 2001). Alternatively, 'surplus' regions may arise via segmental duplication of mtDNA and contain supernumerary gene copies as seen in the prasinophyte *Ostreococcus* (Robbens et al. 2007).

Changes in genome conformation have also occurred independently in several algal lineages. This includes linearization (such as in the Reinhardtinia clade and in certain chrysophytes and chlorarachniophytes), fragmentation (in the non-photosynthetic chlorophycean algae *Polytomella*, in euglenids and dinoflagellates), and the presence of multiple isomeric genomic forms (in the red alga *Porphyra* and the green alga *Volvox*). Such conformational changes are thought to be consequences of illegitimate recombination involving short repeated sequences (Burger et al. 1999; Smith and Lee 2009).

Another pattern encountered in mtDNAs of various algal lineages is the gradual loss of the ancestral gene order of ribosomal protein genes. Among algae, the bacterial *str*, S10, *spc*, and alpha operon arrangement is most conserved in the prasinophyte *Nephroselmis* (Turmel et al. 1999) and in the cryptophyte *Rhodomonas* (Hauth et al. 2005). On the other hand, the degree of gene shuffling can be so extreme that not even a single pair of ribosomal protein gene neighbours is retained as for example in the mtDNA of the cercomonad alga *Bigelowiella* (Burger et al. unpublished; GenBank acc. no. HQ840955).

Further evolutionary trends are observed in mtDNAs of phylogenetically unrelated algae and eukaryotes in general. These include sometimes drastic deviations from the universal genetic code (in the chlorophycean *Scenedesmus* and the prasinophyte *Pycnococcus* (see also Hayashi-Ishimaru et al. 1996)), fragmentation of rRNA genes (in both the prasinophyte *Pedinomonas* and the ancestor of chlorophycean green algae as well as in *Euglena*), and pronounced nucleotide composition bias with an A+T content that is either extreme high (in *Euglena* and the trebuxiophyte *Prototheca*) or extremely low (in the chlorophycean *Polytomella capuana*).

In sum, the evolutionary history of algal mitochondrial genomes appears like an exploration of all possible avenues of tolerable deviations, which has brought about the astounding diversity of mtDNAs in algae and eukaryotes as a whole.

Acknowledgments

The authors thank B. F. Lang (Université de Montréal, Montreal, Canada) for discussions on phylogenetic aspects of the chapter and helpful comments on the manuscript. This work was supported by the Canadian Institute for Health Research, Institute for Genetics (CIHR-IG), Grant no. MOP-79309 (GB) and the Natural Sciences and Engineering Research Council (NSERC) (AMN).

References

Adams KL, Ong HC, Palmer JD (2001) Mitochondrial gene transfer in pieces: fission of the ribosomal protein gene *rpl2* and partial or complete gene transfer to the nucleus. Mol Biol Evol 18:2289–2297

Alverson AJ, Wei X, Rice DW, Stern DB, Barry K, Palmer JD (2010) Insights into the evolution of mitochondrial genome size from complete sequences of *Citrullus lanatus* and *Cucurbita pepo* (Cucurbitaceae). Mol Biol Evol 27:1436–1448

Armbrust EV et al (2004) The genome of the diatom *Thalassiosira pseudonana*: ecology, evolution, and metabolism. Science 306:79–86

Baurain D, Brinkmann H, Philippe H (2007) Lack of resolution in the animal phylogeny: closely spaced cladogeneses or undetected systematic errors? Mol Biol Evol 24:6–9

Baurain D et al (2010) Phylogenomic evidence for separate acquisition of plastids in cryptophytes, haptophytes, and stramenopiles. Mol Biol Evol 27:1698–1709

Bonen L (2008) Cis- and trans-splicing of group II introns in plant mitochondria. Mitochondrion 8: 26–34

Boore JL (1999) Animal mitochondrial genomes. Nucleic Acids Res 27:1767–1780

Booton GC, Floyd GL, Fuerst PA (1998) Polyphyly of tetrasporalean green algae inferred from nuclear small-subunit ribosomal DNA. J Phycol 34:306–311

Bowler C et al (2008) The *Phaeodactylum* genome reveals the evolutionary history of diatom genomes. Nature 456:239–244

Boyen C, Leblanc C, Kloareg B, Loiseaux-de Goer S (1994) Physical map and gene organization of the mitochondrial genome of *Chondrus crispus* (Rhodophyta, Gigartinales). Plant Mol Biol 26: 691–697

Bullerwell CE, Burger G, Lang BF (2000) A novel motif for identifying *rps3* homologs in fungal mitochondrial genomes. Trends Biochem Sci 25:363–365

Bullerwell CE, Leigh J, Forget L, Lang BF (2003) A comparison of three fission yeast mitochondrial genomes. Nucleic Acids Res 31:759–768

Bullerwell CE, Burger G, Gott JM, Kourennaia O, Schnare MN, Gray MW (2010) Abundant 5S rRNA-like transcripts encoded by the mitochondrial genome in amoebozoa. Eukaryot Cell 9:762–773

Burger G, Plante I, Lonergan KM, Gray MW (1995) The mitochondrial DNA of the amoeboid protozoon, *Acanthamoeba castellanii*: complete sequence, gene content and genome organization. J Mol Biol 245:522–537

Burger G, Lang BF, Reith M, Gray MW (1996) Genes encoding the same three subunits of respiratory complex II are present in the mitochondrial DNA of two phylogenetically distant eukaryotes. Proc Natl Acad Sci USA 93:2328–2332

Burger G, Saint-Louis D, Gray MW, Lang BF (1999) Complete sequence of the mitochondrial DNA of the red alga *Porphyra purpurea*. Cyanobacterial introns and shared ancestry of red and green algae. Plant Cell 11:1675–1694

Burger G, Forget L, Zhu Y, Gray MW, Lang BF (2003a) Unique mitochondrial genome architecture in unicellular relatives of animals. Proc Natl Acad Sci USA 100:892–897

Burger G, Gray MW, Lang BF (2003b) Mitochondrial genomes – anything goes. Trends Genet 19:709–716

Burger G, Lang BF, Braun HP, Marx S (2003c) The enigmatic mitochondrial ORF ymf39 codes for ATP synthase chain b. Nucleic Acids Res 31:2353–2360

Burger G, Yan Y, Javadi P, Lang BF (2009) Group I-intron trans-splicing and mRNA editing in the mitochondria of placozoan animals. Trends Genet 25:381–386

Burger G, Jackson C, Waller R (2011) Unusual mitochondrial genomes and genes. In: Bullerwell C (ed) Organelle genetics: evolution of organelle genomes and gene expression. Springer, Berlin, pp 41–77

Burki F et al (2007) Phylogenomics reshuffles the eukaryotic supergroups. PLoS One 2:e790

Bzymek M, Lovett ST (2001) Instability of repetitive DNA sequences: the role of replication in multiple mechanisms. Proc Natl Acad Sci USA 98: 8319–8325

Cavalier-Smith T (1999) Principles of protein and lipid targeting in secondary symbiogenesis: euglenoid, dinoflagellate, and sporozoan plastid origins and the eukaryote family tree. J Eukaryot Microbiol 46: 347–366

Cermakian N, Ikeda TM, Miramontes P, Lang BF, Gray MW, Cedergren R (1997) On the evolution of the single-subunit RNA polymerases. J Mol Evol 45:671–681

Chesnick JM et al (2000) The mitochondrial genome of the stramenopile alga *Chrysodidymus synuroideus*. Complete sequence, gene content and genome organization. Nucleic Acids Res 28:2512–2518

Choi S-J, Park E-J, Endo H, Kitade Y, Saga N (2008) Inheritance pattern of chloroplast and mitochochondrial genomes in artificial hybrids of *Porphyra yezoensis* (Rhodophyta). Fisheries Sci 74:822–829

Delaroque N, Fontaine JM, Kloareg B, Loiseaux-de Goer S (1996) Putative sigma-70-like promoters in a brown algal mitochondrial genome. C R Acad Sci III 319:763–767

Denovan-Wright EM, Nedelcu AM, Lee RW (1998) Complete sequence of the mitochondrial DNA of *Chlamydomonas eugametos*. Plant Mol Biol 36:285–295

Fan J, Lee RW (2002) Mitochondrial genome of the colorless green alga *Polytomella parva*: Two linear DNA molecules with homologous inverted repeat termini. Mol Biol Evol 19:999–1007

Feagin JE, Gardner MJ, Williamson DH, Wilson RJ (1991) The putative mitochondrial genome of *Plasmodium falciparum*. J Protozool 38:243–245

Friedl T (1995) Inferring taxonomic positions and testing genus level assignments in coccoid green lichen algae – a phylogenetic analysis of 18S ribosomal RNA sequences from *Dictyochloropsis reticulata* and from members of the genus *Myrmecia* (Chlorophyta, Trebouxiophyceae Cl-Nov). J Phycol 31:632–639

Friedl T (1997) The evolution of green agae. Plant Syst Evol 11:87–101

Friedl T, O'Kelly CJ (2002) Phylogenetic relationships of green algae assigned to the genus *Planophila* (Chlorophyta): evidence from 18S rDNA sequence data and ultrastructure. Eur J Phycol 37:373–384

Funes S et al (2002) A green algal apicoplast ancestor. Science 298:2155

Gawryluk RM, Gray MW (2010) An ancient fission of mitochondrial Cox1. Mol Biol Evol 27:7–10

Graham LE, Wilcox LW (2000) Algae. Prentice-Hall, Upper Saddle River

Gray MW, Boer PH (1988) Organization and expression of algal (*Chlamydomonas reinhardtii*) mitochondrial DNA. Philos Trans R Soc Lond B Biol Sci 319:135–147

Gray MW et al (1998) Genome structure and gene content in protist mitochondrial DNAs. Nucleic Acids Res 26:865–878

Gray MW, Burger G, Lang BF (1999) Mitochondrial evolution. Science 283:1476–1481

Gray MW, Burger G, Lang BF (2001) The origin and early evolution of mitochondria. Genome Biol 2:1018.1011–1018.1015

Gray MW, Lang BF, Burger G (2004) Mitochondria of protists. Annu Rev Genet 38:477–524

Grewe F, Viehoever P, Weisshaar B, Knoop V (2009) A trans-splicing group I intron and tRNA-hyperediting in the mitochondrial genome of the lycophyte *Isoetes engelmannii*. Nucleic Acids Res 37:5093–5104

Griffiths AJ (1995) Natural plasmids of filamentous fungi. Microbiol Rev 59:673–685

Hackett JD, Yoon HS, Li S, Reyes-Prieto A, Rummele SE, Bhattacharya D (2007) Phylogenomic analysis supports the monophyly of cryptophytes and haptophytes and the association of rhizaria with chromalveolates. Mol Biol Evol 24:1702–1713

Hancock L, Goff L, Lane C (2010) Red algae lose key mitochondrial genes in response to becoming parasitic. Genome Biol Evol 2:897–910

Hanson MR, Folkerts O (1992) Structure and function of the higher plant mitochondrial genome. Int Rev Cytol 141:129–172

Hauth AM, Maier UG, Lang BF, Burger G (2005) The *Rhodomonas salina* mitochondrial genome: bacteria-like operons, compact gene arrangement and complex repeat region. Nucleic Acids Res 33: 4433–4442

Hayashi-Ishimaru Y, Ohama T, Kawatsu Y, Nakamura K, Osawa S (1996) UAG is a sense codon in several chlorophycean mitochondria. Curr Genet 30:29–33

Hayashi-Ishimaru Y, Ehara M, Inagaki Y, Ohama T (1997) A deviant mitochondrial genetic code in prymnesiophytes (yellow-algae): UGA codon for tryptophan. Curr Genet 32:296–299

Jackson CJ, Norman JE, Schnare MN, Gray MW, Keeling PJ, Waller RF (2007) Broad genomic and transcriptional analysis reveals a highly derived genome in dinoflagellate mitochondria. BMC Biol 5:41

Jacob Y, Seif E, Paquet PO, Lang BF (2004) Loss of the mRNA-like region in mitochondrial tmRNAs of jakobids. RNA 10:605–614

Janouskovec J, Horak A, Obornik M, Lukes J, Keeling PJ (2010) A common red algal origin of the apicomplexan, dinoflagellate, and heterokont plastids. Proc Natl Acad Sci USA 107:10949–10954

Kannan S, Burger G (2008) Unassigned MURF1 of kinetoplastids codes for NADH dehydrogenase subunit 2. BMC Genomics 9:455

Kannan S, Hauth AM, Burger G (2008) Function prediction of hypothetical proteins without sequence similarity to proteins of known function. Protein Pept Lett 15:1107–1116

Karol KG, McCourt RM, Cimino MT, Delwiche CF (2001) The closest living relatives of land plants. Science 294:2351–2353

Keeling PJ et al (2005) The tree of eukaryotes. Trends Ecol Evol 20:670–676

Kessler U, Zetsche K (1995) Physical map and gene organization of the mitochondrial genome from the unicellular green alga *Platymonas* (*Tetraselmis*) *subcordiformis* (Prasinophyceae). Plant Mol Biol 29:1081–1086

Kim E, Lane CE, Curtis BA, Kozera C, Bowman S, Archibald JM (2008) Complete sequence and analysis of the mitochondrial genome of *Hemiselmis andersenii* CCMP644 (Cryptophyceae). BMC Genomics 9:215

Kreimer G, Melkonian M (1990) Reflection confocal laser scanning microscopy of eyespots in flagellated green algae. Eur J Cell Biol 53:101–111

Kroymann J, Zetsche K (1998) The mitochondrial genome of *Chlorogonium elongatum* inferred from the complete sequence. J Mol Evol 47:431–440

Kubo T, Newton KJ (2008) Angiosperm mitochondrial genomes and mutations. Mitochondrion 8:5–14

Kück U, Jekosch K, Holzamer P (2000) DNA sequence analysis of the complete mitochondrial genome of the green alga *Scenedesmus obliquus*: evidence for UAG being a leucine and UCA being a non-sense codon. Gene 253:13–18

Lang BF, Goff LJ, Gray MW (1996) A 5S rRNA gene is present in the mitochondrial genome of the protist *Reclinomonas americana* but is absent from red algal mitochondrial DNA. J Mol Biol 261:407–413

Lang BF et al (1997) An ancestral mitochondrial DNA resembling a eubacterial genome in miniature. Nature 387:493–497

Lang BF, Laforest MJ, Burger G (2007) Mitochondrial introns: a critical view. Trends Genet 23:119–125

Leblanc C, Boyen C, Richard O, Bonnard G, Grienenberger JM, Kloareg B (1995) Complete sequence of the mitochondrial DNA of the rhodophyte *Chondrus crispus* (Gigartinales). Gene content and genome organization. J Mol Biol 250: 484–495

Lin S, Zhang H, Spencer DF, Norman JE, Gray MW (2002) Widespread and extensive editing of mitochondrial mRNAs in dinoflagellates. J Mol Biol 320:727–739

Lin S, Zhang H, Gray MW (2008) RNA editing in dinoflagellates and its implications for the evolutionary history of the editing machinery. In: Smith HC (ed) RNA and DNA editing: molecular mechanisms and their integration into biological systems. Wiley, Hoboken, pp 280–309

Ling F, Shibata T (2004) Mhr1p-dependent concatemeric mitochondrial DNA formation for generating yeast mitochondrial homoplasmic cells. Mol Biol Cell 15:310–322

Maier UG (1992) The four genomes of the alga *Pyrenomonas salina* (Cryptophyta). Biosystems 28:69–73

Mallet MA, Lee RW (2006) Identification of three distinct *Polytomella* lineages based on mitochondrial DNA features. J Eukaryot Microbiol 53:79–84

Margulis L, Corliss JO, Melkonian M, Chapman DJ (eds) (1989) The Handbook of Protoctista. Jones and Bartlett, Boston

Marin B, Melkonian M (2010) Molecular phylogeny and classification of the *Mamiellophyceae* class. nov (Chlorophyta) based on sequence comparisons

of the nuclear- and plastid-encoded rRNA operons. Protist 161:304–336
Masuda I et al (2011) Mitochondrial genomes from two red tide forming raphidophycean algae *Heterosigma akashiwo* and *Chattonella marina* var. *marina*. Harmful Algae 10:130–137
Mattox KR, Stewart KD (1984) Classification of the green algae: a concept based on comparative cytology. In: Irvine DEG, John DM (eds) Systematics of the green algae. Academic, London, pp 29–72
McFadden GI, Waller RF (1997) Plastids in parasites of humans. Bioessays 19:1033–1040
Michaelis G, Vahrenholz C, Pratje E (1990) Mitochondrial DNA of *Chlamydomonas reinhardtii*: the gene for apocytochrome b and the complete functional map of the 15.8 kb DNA. Mol Gen Genet 223:211–216
Moore LJ, Coleman AW (1989) The linear 20-Kb mitochondrial genome of *Pandorina morum* (Volvocaceae, Chlorophyta). Plant Mol Biol 13: 459–465
Moreira D, Lopez-Garcia P, Rodriguez-Valera F (2001) New insights into the phylogenetic position of diplonemids: G+C content bias, differences of evolutionary rate and a new environmental sequence. Int J Syst Evol Microbiol 51:2211–2219
Nakada T, Misawa K, Nozaki H (2008) Molecular systematics of volvocales (Chlorophyceae, Chlorophyta) based on exhaustive 18S rRNA phylogenetic analyses. Mol Phylogenet Evol 48:281–291
Nash EA, Barbrook AC, Edwards-Stuart RK, Bernhardt K, Howe CJ, Nisbet RE (2007) Organization of the mitochondrial genome in the dinoflagellate *Amphidinium carterae*. Mol Biol Evol 24: 1528–1536
Nash EA, Nisbet RE, Barbrook AC, Howe CJ (2008) Dinoflagellates: a mitochondrial genome all at sea. Trends Genet 24:328–335
Nedelcu AM, Lee RW (1998) A degenerate group II intron in the intronless mitochondrial genome of *Chlamydomonas reinhardtii:* evolutionary implications. Mol Biol Evol 15:918–923
Nedelcu AM, Lee RW, Lemieux C, Gray MW, Burger G (2000) The complete mitochondrial DNA sequence of *Scenedesmus obliquus* reflects an intermediate stage in the evolution of the green algal mitochondrial genome. Genome Res 10:819–831
Norman JE, Gray MW (2001) A complex organization of the gene encoding cytochrome oxidase subunit 1 in the mitochondrial genome of the dinoflagellate, *Crypthecodinium cohnii*: homologous recombination generates two different *cox1* open reading frames. J Mol Evol 53:351–363
O'Brien EA et al (2009) GOBASE: an organelle genome database. Nucleic Acids Res 37: D946–D950
Oda K et al (1992) Gene organization deduced from the complete sequence of liverwort *Marchantia polymorpha* mitochondrial DNA. A primitive form of plant mitochondrial genome. J Mol Biol 223:1–7
Ohta N, Sato N, Kuroiwa T (1998) Structure and organization of the mitochondrial genome of the unicellular red alga *Cyanidioschyzon merolae* deduced from the complete nucleotide sequence. Nucleic Acids Res 26:5190–5298
Oldenburg DJ, Bendich AJ (2001) Mitochondrial DNA from the liverwort *Marchantia polymorpha*: circularly permuted linear molecules, head-to-tail concatemers, and a 5' protein. J Mol Biol 310:549–562
Oudot M-P, Kloareg B, Loiseaux-de Goër S (1999) The mitochondrial *Pylaiella littoralis nad11* gene contains only the N-terminal FeS-binding domain. Gene 235:131–137
Oudot-Le Secq MP, Fontaine JM, Rousvoal S, Kloareg B, Loiseaux-De Goer S (2001) The complete sequence of a brown algal mitochondrial genome, the ectocarpale *Pylaiella littoralis* (L.) Kjellm. J Mol Evol 53:80–88
Oudot-Le Secq M-P, Kloareg B, Loiseaux-de Goër S (2002) The mitochondrial genome of the brown alga *Laminaria digitata*: a comparative analysis. Eur J Phycol 37:163–172
Oudot-Le Secq MP, Loiseaux-de Goer S, Stam WT, Olsen JL (2006) Complete mitochondrial genomes of the three brown algae (Heterokonta: Phaeophyceae) *Dictyota dichotoma*, *Fucus vesiculosus* and *Desmarestia viridis*. Curr Genet 49:47–58
Patterson DJ (1989) Stramenopiles: chromophyte from a protistan perspective. In: Green JC, Leadbeater ESC, Diver WL (eds) The chromophyte algae: problems and perspectives. Clarendon, Oxford, pp 357–379
Pérez-Martínez X et al (2001) Subunit II of cytochrome *c* oxidase in chlamydomonad algae is a heterodimer encoded by two independent nuclear genes. J Biol Chem 276:11302–11309
Pombert JF, Keeling PJ (2010) The mitochondrial genome of the entomoparasitic green alga *Helicosporidium*. PLoS One 5:e8954
Pombert JF, Otis C, Lemieux C, Turmel M (2004) The complete mitochondrial DNA sequence of the green alga *Pseudendoclonium akinetum* (Ulvophyceae) highlights distinctive evolutionary trends in the chlorophyta and suggests a sister-group relationship between the Ulvophyceae and Chlorophyceae. Mol Biol Evol 21:922–935

Pombert JF, Otis C, Lemieux C, Turmel M (2005) Chloroplast genome sequence of the green alga *Pseudendoclonium akinetum* (Ulvophyceae) reveals unusual structural features and new insights into the branching order of chlorophyte lineages. Mol Biol Evol 22:1903–1918

Pombert JF, Beauchamp P, Otis C, Lemieux C, Turmel M (2006) The complete mitochondrial DNA sequence of the green alga *Oltmannsiellopsis viridis*: evolutionary trends of the mitochondrial genome in the Ulvophyceae. Curr Genet 50:137–147

Popescu CE, Lee RW (2007) Mitochondrial genome sequence evolution in *Chlamydomonas*. Genetics 175:819–826

Price DC, Chan CX, Yoon HS, Yang EC, Qiu H, Weber AP, Schwacke R, Gross J, Blouin NA, Lane C, Reyes-Prieto A, Durnford DG, Neilson JA, Lang BF, Burger G, Steiner JM, Löffelhardt W, Meuser JE, Posewitz MC, Ball S, Arias MC, Henrissat B, Coutinho PM, Rensing SA, Symeonidi A, Doddapaneni H, Green BR, Rajah VD, Boore J, Bhattacharya D (2012) Cyanophora paradoxa genome elucidates origin of photosynthesis in algae and plants. Science 335:843–847

Ravin NV et al (2010) Complete sequence of the mitochondrial genome of a diatom alga *Synedra acus* and comparative analysis of diatom mitochondrial genomes. Curr Genet 56:215–223

Robbens S, Derelle E, Ferraz C, Wuyts J, Moreau H, Van de Peer Y (2007) The complete chloroplast and mitochondrial DNA sequence of *Ostreococcus tauri*: organelle genomes of the smallest eukaryote are examples of compaction. Mol Biol Evol 24: 956–968

Rodriguez-Ezpeleta N et al (2005) Monophyly of primary photosynthetic eukaryotes: green plants, red algae, and glaucophytes. Curr Biol 15:1325–1330

Rodriguez-Ezpeleta N et al (2007a) Toward resolving the eukaryotic tree: the phylogenetic positions of jakobids and cercozoans. Curr Biol 17:1420–1425

Rodriguez-Ezpeleta N, Philippe H, Brinkmann H, Becker B, Melkonian M (2007b) Phylogenetic analyses of nuclear, mitochondrial, and plastid multigene data sets support the placement of *Mesostigma* in the Streptophyta. Mol Biol Evol 24:723–731

Rousvoal S, Oudot M-P, Fontaine J-M, Kloareg B, Loiseaux-de Goër SL (1998) Witnessing the evolution of transcription in mitochondria: the mitochondrial genome of the primitive brown alga *Pylaiella littoralis* (L.) Kjellm. encodes a T7-like RNA polymerase. J Mol Biol 277:1047–1057

Roy J, Faktorova D, Lukeš J, Burger G (2007) Unusual mitochondrial genome structures throughout the Euglenozoa. Protist 158:385–396

Sanchez Puerta MV, Bachvaroff TR, Delwiche CF (2004) The complete mitochondrial genome sequence of the haptophyte *Emiliania huxleyi* and its relation to heterokonts. DNA Res 11:1–10

Saunders GW, Hommersand MH (2004) Assessing red algal supraordinal diversity and taxonomy in the context of contemporary systematic data. Am J Bot 91:1494–1507

Simpson AG, Lukeš J, Roger AJ (2002) The evolutionary history of kinetoplastids and their kinetoplasts. Mol Biol Evol 19:2071–2083

Smith DR, Lee RW (2008a) Mitochondrial genome of the colorless green alga *Polytomella capuana*: a linear molecule with an unprecedented GC content. Mol Biol Evol 25:487–496

Smith DR, Lee RW (2008b) Nucleotide diversity in the mitochondrial and nuclear compartments of *Chlamydomonas reinhardtii*: investigating the origins of genome architecture. BMC Evol Biol 8:156

Smith DR, Lee RW (2009) The mitochondrial and plastid genomes of *Volvox carteri*: bloated molecules rich in repetitive DNA. BMC Genomics 10:132–146

Smith DR, Lee RW (2010) Low nucleotide diversity for the expanded organelle and nuclear genomes of *Volvox carteri* supports the mutational-hazard hypothesis. Mol Biol Evol 27:2244–2256

Smith DR, Lee RW, Cushman JC, Magnuson JK, Tran D, Polle JE (2010) The *Dunaliella salina* organelle genomes: large sequences, inflated with intronic and intergenic DNA. BMC Plant Biol 10:83

Spencer DF, Gray MW (2011) Ribosomal RNA genes in *Euglena gracilis* mitochondrial DNA: fragmented genes in a seemingly fragmented genome. Mol Genet Genomics 285:19–31

Terasawa K et al (2007) Mitochondrial genome of the moss *Physcomitrella patens* sheds new light on the mitochondrial evolution in land plants. Plant Cell Physiol 48:S204

Tessier LH, van der Speck H, Gualberto JM, Grienenberger JM (1997) The *cox1* gene from *Euglena gracilis*: a protist mitochondrial gene without introns and genetic code modifications. Curr Genet 31:208–213

Toor N, Zimmerly S (2002) Identification of a family of group II introns encoding LAGLIDADG ORFs typical of group I introns. RNA 8:1373–1377

Turmel M et al (1999) The complete mitochondrial DNA sequences of *Nephroselmis olivacea* and *Pedinomonas minor*. Two radically different evolutionary patterns within green algae. Plant Cell 11: 1717–1730

Turmel M, Otis C, Lemieux C (2002a) The chloroplast and mitochondrial genome sequences of the charo-

phyte *Chaetosphaeridium globosum*: insights into the timing of the events that restructured organelle DNAs within the green algal lineage that led to land plants. Proc Natl Acad Sci USA 99:11275–11280
Turmel M, Otis C, Lemieux C (2002b) The complete mitochondrial DNA sequence of *Mesostigma viride* identifies this green alga as the earliest green plant divergence and predicts a highly compact mitochondrial genome in the ancestor of all green plants. Mol Biol Evol 19:24–38
Turmel M, Otis C, Lemieux C (2003) The mitochondrial genome of *Chara vulgaris*: insights into the mitochondrial DNA architecture of the last common ancestor of green algae and land plants. Plant Cell 15:1888–1903
Turmel M, Otis C, Lemieux C (2006) The chloroplast genome sequence of *Chara vulgaris* sheds new light into the closest green algal relatives of land plants. Mol Biol Evol 23:1324–1338
Turmel M, Otis C, Lemieux C (2007) An unexpectedly large and loosely packed mitochondrial genome in the charophycean green alga *Chlorokybus atmophyticus*. BMC Genomics 8:137
Turmel M, Gagnon MC, O'Kelly CJ, Otis C, Lemieux C (2009a) The chloroplast genomes of the green algae *Pyramimonas*, *Monomastix*, and *Pycnococcus* shed new light on the evolutionary history of prasinophytes and the origin of the secondary chloroplasts of euglenids. Mol Biol Evol 26:631–648
Turmel M, Otis C, Lemieux C (2009b) The chloroplast genomes of the green algae *Pedinomonas minor*, *Parachlorella kessleri*, and *Oocystis solitaria* reveal a shared ancestry between the pedinomonadales and chlorellales. Mol Biol Evol 26:2317–2331
Turmel M, Otis C, Lemieux C (2010) A deviant genetic code in the reduced mitochondrial genome of the picoplanktonic green alga *Pycnococcus provasolii*. J Mol Evol 70:203–214
Vahrenholz C, Riemen G, Pratje E, Dujon B, Michaelis G (1993) Mitochondrial DNA of *Chlamydomonas reinhardtii*: the structure of the ends of the linear 15.8-kb genome suggests mechanisms for DNA replication. Curr Genet 24:241–247
van Dooren GG, Stimmler LM, McFadden GI (2006) Metabolic maps and functions of the *Plasmodium* mitochondrion. FEMS Microbiol Rev 30:596–630
Viehmann S, Richard O, Boyen C, Zetsche K (1996) Genes for two subunits of succinate dehydrogenase form a cluster on the mitochondrial genome of Rhodophyta. Curr Genet 29:199–201
Vlcek C, Marande W, Teijeiro S, Lukeš J, Burger G (2011) Systematically fragmented genes in a multipartite mitochondrial genome. Nucleic Acids Res 39:979–988
Waller RF, Jackson CJ (2009) Dinoflagellate mitochondrial genomes: stretching the rules of molecular biology. Bioessays 31:237–245
Waller RF, Keeling PJ (2006) Alveolate and chlorophycean mitochondrial *cox2* genes split twice independently. Gene 383:33–37
Ward BL, Anderson RS, Bendich AJ (1981) The mitochondrial genome is large and variable in a family of plants (Cucurbitaceae). Cell 25:793–803
Weber B, Börner T, Weihe A (1995) Remnants of a DNA polymerase gene in the mitochondrial DNA of *Marchantia polymorpha*. Curr Genet 27:488–490
Williams BA, Keeling PJ (2003) Cryptic organelles in parasitic protists and fungi. Adv Parasitol 54:9–68
Wolff G, Burger G, Lang BF, Kück U (1993) Mitochondrial genes in the colourless alga *Prototheca wickerhamii* resemble plant genes in their exons but fungal genes in their introns. Nucleic Acids Res 21:719–726
Worden AZ et al (2009) Green evolution and dynamic adaptations revealed by genomes of the marine picoeukaryotes *Micromonas*. Science 324:268–272
Yasuhira S, Simpson L (1997) Phylogenetic affinity of mitochondria of *Euglena gracilis* and kinetoplastids using cytochrome oxidase I and hsp60. J Mol Evol 44:341–347

Chapter 7

Conservative and Dynamic Evolution of Mitochondrial Genomes in Early Land Plants

Yang Liu, Bin Wang, Libo Li, and Yin-Long Qiu*
Department of Ecology & Evolutionary Biology, University of Michigan, Ann Arbor, MI 48109-1048, USA

Jiayu Xue
Department of Ecology & Evolutionary Biology, University of Michigan, Ann Arbor, MI 48109-1048, USA
and School of Life Sciences, Nanjing University, Nanjing 210093, China

Summary

Early land plant mitochondrial genomes (chondromes) might have captured important changes of mitochondrial genome evolution when photosynthetic eukaryotes colonized land in a unprecedented scale, and thus deserve special attention in investigation of plant mitochondrial genomes. The chondromes of land plants that are well adapted to the terrestrial environment, namely seed plants, show many derived characteristics, including large genome size variation, frequent occurrence of intra-genomic rearrangements, abundant introns and high levels of RNA editing. In contrast, the chondromes of charophytes, the closest algal relatives of land plants, are still largely ancestral in these aspects, resembling chondromes of early eukaryotes. Several recently sequenced chondromes from basal land plants including liverworts, mosses, hornworts and lycophytes have provided fresh insights into mitochondrial genome evolution of early land plants. Comparative analyses of these genomes have identified lycophytes, which represent the most ancient extant vascular plants, as the major

* Author for correspondence, e-mail: ylqiu@umich.edu

R. Bock and V. Knoop (eds.), Genomics of Chloroplasts and Mitochondria, Advances in Photosynthesis and Respiration 35, pp. 159–174,
DOI 10.1007/978-94-007-2920-9_7,

point of change in plant mitochondrial genome evolution, with long conserved mitochondrial gene synteny largely disrupted. The chondromes of bryophytes are conservative in gene order, but rather dynamic in intron content. The gene contents and RNA editing levels also show wide variation from lineage to lineage. Overall, the mitochondrial genomes experienced dynamic evolutionary changes during the origin and early evolution of land plants when the major lineages of bryophytes and vascular plants appeared, but have remained relatively conservative afterwards except in vascular plants.

I. Introduction

Among the major clades of eukaryotes, land plants (embryophytes) are outstanding in having mitochondrial genomes that show dramatic size variation, experience highly frequent intra-genomic rearrangement, harbor abundant introns and promiscuous DNA originating from nuclear and chloroplast genomes, and undergo heavy RNA editing during gene expression (Schuster and Brennicke 1994; Gray et al. 1999; Palmer et al. 2000; Knoop 2004). When and how these evolutionary novelties arose has remained unknown until very recently. The first plant mitochondrial genome was sequenced two decades ago from a liverwort, *Marchantia polymorpha* (Oda et al. 1992). In the 20 years that followed, over a dozen angiosperm chondromes (see Chap. 8) were sequenced (Unseld et al. 1997; Kubo et al. 2000; Notsu et al. 2002; Handa 2003; Clifton et al. 2004; Sugiyama et al. 2005; Allen et al. 2007; Goremykin et al. 2009; Alverson et al. 2010; Sloan et al. 2010; Alverson et al. 2011a; Rice et al. 2011). These data provided complete genome sequence information, confirmed the above evolutionary phenomena that had been described in earlier small-scale studies of individual gene and heightened the interest in pursuing the questions when and how these phenomena came into being. Reports of several charophytic algal mitochondrial genomes at the same time indicated that the *Marchantia* chondrome had not diverged very far from the charophyte chondromes and clearly represented an ancestral condition of plant mitochondrial genomes (Turmel et al. 1999; Turmel et al. 2002a, b, 2003, 2007). More recently, complete or partial sequencing of several bryophyte, lycophyte and gymnosperm chondromes (Terasawa et al. 2007; Chaw et al. 2008; Grewe et al. 2009; Li et al. 2009; Wang et al. 2009; Xue et al. 2010; Hecht et al. 2011) revealed that the major change in mitochondrial genome evolution of land plants took place during the origin of vascular plants, though some changes such as genome size increase, intron acquisition, and RNA editing already occurred during the origin and early evolution of land plants. In this chapter, we briefly review the progress in research on the mitochondrial genome evolution in early land plants.

II. Genome Size and Gene Content

In bryophytes, the mitochondrial genomes of two liverworts, one moss and two hornworts have been completely sequenced, and their size range varies roughly by two times (Table 7.1). The smallest chondrome sequenced so far (105,340 base pairs (bp)) is found in the moss *Physcomitrella patens* (Terasawa et al. 2007), and the largest one (209,482 bp) occurs in the hornwort *Phaeoceros laevis* (Xue et al. 2010). This size range is much smaller than the long-known wide size variation of angiosperm chondromes, which can be up to an order of magnitude within a single family (Ward et al. 1981; Alverson et al. 2010). In comparison with the alga *Chara vulgaris*, which is a member of the sister group of land plants (Karol et al. 2001; Qiu et al. 2006) and has a chondrome of 68 kbp (Turmel et al. 2003),

Abbreviations: bp – Base pairs; kbp – Kilobase pairs

Table 7.1. Genome sizes and proportions of the various sequence types in the mitochondrial genomes of *Chara vulgaris* and five bryophytes[a]

Species	Genome size (bp)	Genes (%)	Exons (%)	Introns (%)	Intergenic spacers (%)
Chara vulgaris	67,737	91	52	39	9
Marchantia polymorpha	186,609	51	23	28	49
Pleurozia purpurea	168,526	52	29	23	48
Physcomitrella patens	105,340	65	37	28	35
Megaceros aenigmaticus	184,908	50	16	34	50
Phaeoceros laevis	209,482	47	11	36	53

[a]The rows are shaded alternately by plant clades (Charales, liverworts, mosses and hornworts)

the bryophytes have significantly larger chondromes, whose size likely increased by 2–3 times during the origin of land plants. This estimate is supported by the fact that both liverworts and hornworts have large mitochondrial genomes, the former representing the first diverging lineage of land plants and the latter being sister to vascular plants (Qiu et al. 2006). The 100 kbp chondrome in *P. patens* likely resulted from secondary size reduction that happened early in moss evolution, as another moss, *Anomodon rugelii,* a distant relative of *P. patens*, has a chondrome of similar size (Liu et al. 2011).

The genome size increase is not the result of an increase in the number of genes (Table 7.2). Intron contents vary among these chondromes, but their number per genome fluctuates only slightly (Table 7.1). What changes greatly is the amount of intergenic spacer sequences, which seem to be the main determinant of genome size changes in the mitochondria of these organisms. This phenomenon has also been observed recently in the dramatic size increase of chondromes within a single angiosperm family, the Cucurbitaceae (Alverson et al. 2010).

The gene contents are remarkably similar in chondromes of *C. vulgaris*, the two liverworts and the moss (Table 7.2), which indicates that this aspect of mitochondrial genome evolution did not experience major changes during the origin of land plants. The two hornwort chondromes, in contrast, have lost or are in the process of losing many genes. The main groups of genes affected are those encoding ATP synthase (*atp8*), enzymes for cytochrome *c* biogenesis (*ccmFC* and *ccmFN*), ribosomal proteins (*rpl* and *rps*), succinate:ubiquinone oxidoreductase (*sdh3*), and transfer RNAs (tRNA genes). An intriguing observation is that many pseudogenes are present in the two hornwort chondromes. The two species sequenced for chondromes span nearly the entire phylogenetic diversity of hornworts (Duff et al. 2007), which originated at least in the Silurian (444 million years ago) (Stewart 1983). Thus, shared presence of these nonfunctional gene copies may indicate either independent pseudogenization events in the two species or retention of ancient pseudogenes for some not yet understood reasons (Xue et al. 2010). The absence of many tRNA genes may be explained by their straightforward replacement by tRNAs imported from the cytosol. The loss of these several categories of genes also fits the pattern that has been observed before. In a large survey of angiosperm mitochondrial genes, *rpl, rps* and *sdh* genes were shown to be most prone to loss from the chondromes of some angiosperm lineages (Adams et al. 2001; Adams et al. 2002). The dramatically reduced mitochondrial genomes of some green and red algae, for example, *Chlamydomonas eugametos* and *Porphyra purpurea,* have also lost some or even most genes in these categories (Denovan-Wright et al. 1998; Burger et al. 1999). Likewise, the highly reduced mitochondrial genomes of animals and most fungi have lost almost all *sdh, ccm, rpl* and *rps* genes (Gray et al. 1999). In land plants,

Table 7.2. Gene contents in mitochondrial genomes of *Chara vulgaris* and some early land plants[a]

Gene/ species[b]		*Ch. vu.*	*Ma. po.*	*Pl. pu.*	*Ph. pa.*	*Me. ae.*	*Ph. la.*	*Is. en.*	*Se. mo.*
atp1	a1	+	+	+	+	+	+	+	+
atp4	a4	+	+	+	+	+	+	+	
atp6	a6	+	+	+	+	+	+	+	+
atp8	a8	+	+	+	+	ψ	ψ	+	+
atp9	a9	+	+	+	+	+	+	+	+
ccmB	mb	+	+	+	+				
ccmC	mc	+	+	+	+				
ccmF[c]	mf	+							
ccmFC[c]	my		+	+	+	ψ	ψ		
ccmFN[c]	mz		+	+	+				
cob	cb	+	+	+	+	+	+	+	+
cox1	c1	+	+	+	+	+	+	+	+
cox2	c2	+	+	+	+	+	+	+	+
cox3	c3	+	+	+	+	+	+	+	+
nad1	n1	+	+	+	+	+	+	+	+
nad2	n2	+	+	+	+	+	+	+	+
nad3	n3	+	+	+	+	+	+	+	+
nad4	n4	+	+	+	+	+	+	+	+
nad4L	na	+	+	+	+	+	+	+	+
nad5	n5	+	+	+	+	+	+	+	+
nad6	n6	+	+	+	+	+	+	+	+
nad7	n7	+	ψ	ψ	+			+	+
nad9	n9	+	+	+	+	+	+	+	+
rpl2	l2	+	+	+	+				
rpl5	l5	+	+	+	+	ψ		+	
rpl6	l6	+	+	+	+	ψ	ψ		
rpl10	l10		+	+	+	+	+		
rpl14	l14	+							
rpl16	l16	+	+	+	+				
rps1	s1	+	+	+	+	ψ	ψ	ψ	
rps2	s2	+	+	+	+		ψ	+	
rps3	s3	+	+	+	+			+	
rps4	s4	+	+	+	+	ψ	ψ	+	
rps7	s7	+	+	+	+	ψ	ψ		
rps8	s8		+	+	ψ	ψ			
rps10	s10	+	+	+	ψ				
rps11	s11	+	+	+	+	ψ	ψ		
rps12	s12	+	+	+	+	ψ	ψ		
rps13	s13		+	+	+	+	+		
rps14	s14	+	+	+	+	+			
rps19	s19	+	+	+	+				
rrn5	r5	+	+	+	+	+	+	+	
rrn18	r18	+	+	+	+	+	+	+	+
rrn26	r26	+	+	+	+	+	+	+	+
rtl	x1		+	ψ	ψ				
sdh3	d3	+	+	+	+	ψ	ψ	+	
sdh4	d4	+	+	+	+	+	+		
tatC	w2	+	+	+	+	+	+	+	+
trnAugc	ta	+	+	+	+	+	+		
trnCgca	tc	+	+	+	+	+	+	+	

(continued)

Table 7.2. (continued)

Gene/ species[b]		*Ch. vu.*	*Ma. po.*	*Pl. pu.*	*Ph. pa.*	*Me. ae.*	*Ph. la.*	*Is. en.*	*Se. mo.*
trnDguc	td	+	+	+	+	+	+		
trnEuuc	te	+	+	+	+	+	+	+	
trnFgaa	tf	+	+	+	+	+	+	+	
trnGgcc	tg	+	+	+	+	+	+	+	
trnGucc	t2	+	+	+	+				
trnHgug	th	+	+	+	+	+	+		
trnIcau	ti	+	+	+	+	+	+	+	
trnIgau	t3	+							
trnKuuu	tk	+	+	+	+	+	+	+	
trnLcaa	t5	+	+	+	+	+	+		
trnLuaa	t7	+	+	+	+	+	+		
trnLuag	t8	+	+	+	+		+	ψ	
trnMcau	tm	+	+	+	+	+	+	+	
trnMfcau	t9	+	+	+	+	+	+	+	
trnNguu	tn	+	+	+					
trnPugg	tp	+	+	+	+	+	+	+	
trnQuug	t10	+	+	+	+	+	+	+	
trnRacg	tr	+	+	+	+				
trnRucg	t12		+						
trnRucu	t13	+	+	+	+				
trnSgcu	t14	+	+	+					
trnSuga	t15	+	+	+	+			+	
trnTggu	tt	+	+		+	+	+		
trnVuac	tv	+	+	+	+		+		
trnWcca	tw	+	+	+	+	+	+	+	
trnYgua	ty	+	+	+	+	+	+	+	

[a]"+" and "ψ" indicate presence of functional gene and pseudogene, respectively. No sign indicates gene absence. The data columns are shaded alternately by plant clades (Charales, liverworts, mosses, hornworts and lycophytes)

[b]Abbreviated gene names listed in the second column are used in Fig. 7.1. Full species names are as follows (in the order of their appearance): *Chara vulgaris*, *Marchantia polymorpha, Pleurozia purpurea, Physcomitrella patens, Megaceros aenigmaticus, Phaeoceros laevis*, *Isoetes engelmannii* and *Selaginella moellendorffii*

[c]The genes *ccmFC* and *ccmFN* in land plant chondromes appear as a single gene *ccmF* in *Chara vulgaris*

the loss of *ccm* genes has been less well known until now. The loss of tRNA genes may be related to the phenomenon of some mitochondrial tRNA genes being replaced by their chloroplast counterparts at some stages of vascular plant evolution (Li et al. 2009).

An interesting case of a tRNA gene loss and regain by modifying a duplicated copy of a different tRNA gene was uncovered in the study of the chondrome of the liverwort *Pleurozia purpurea* (Wang et al. 2009), where a *trnRucg* gene was found missing in comparison to *M. polymorpha*. The gene *trnRucg* was probably lost shortly after the endosymbiotic origin of mitochondria, as it is absent in chondromes of *Reclinomonas americana*, an early diverging eukaryote with the most ancestral form of mitochondrial DNA (Lang et al. 1997), and from many other protist mtDNAs (Wang et al. 2009). In the liverwort *M. polymorpha* (Oda et al. 1992), the charophytic algae *Chlorokybus atmophyticus* (Turmel et al. 2007) and *Mesostigma viride*, (Turmel et al. 2002b), and the prasinophycean alga *Nephroselmis olivacea* (Turmel et al. 1999), however, this gene is present. In a comparative analysis of all *trnR* genes from a broad diversity of protists and photosynthetic eukaryotes, it was determined that *trnRucg* in *M. polymorpha* and *C. atmophyticus* was derived from modification of a duplicated copy of *trnRucu*, whereas the gene in *M. viride* and *N. olivacea* was a modified copy of *trnRacg* (Wang et al. 2009). Three examined liverworts provided particularly convincing evidence to demonstrate the origin of the *trnRucg* gene from a gene duplication. In *Treubia lacunosa*, a member of the Haplomitriopsida lineage that is sister to all other liverworts, and *P. purpurea*, a member of the simple thalloid ("metzgeriid") liverworts, two copies of *trnRucu* were found, one located between *nad2* and *trnYgua*, and the other between *tatC* and *trnYgua in* a repeat sequence environment. *M. polymorpha* also has these two gene clusters, but the *trnR* gene between *tatC* and *trnYuga* is *trnRucg*, not *trnRucu*. Sequence comparison clearly shows that *trnRucg* is actually a modified *trnRucu*, with only three nucleotides changed, one being the U → G change in the anticodon. Hence it appears that an early gene duplication of *trnRucu* in the liverworts laid the foundation for neofunctionalization through conversion of one of the copies into a *trnRucg* in *Marchantia*. Previously, it has been shown that some tRNA genes in seed plant chondromes (see next chapter) originated from modification of chloroplast originated tRNA gene copies now located in the mitochondria (Maréchal-Drouard et al. 1990; Li et al. 2009). In seed plant chondromes that often have large chunks of chloroplast-originated sequences (Unseld et al. 1997; Chaw et al. 2008; Alverson et al. 2010), it seems a natural way to derive new tRNA genes from their chloroplast counterparts. In bryophyte chondromes, no such intracellular inter-organelle DNA movement has been detected so far. Hence, different ways can be used to re-create tRNA genes that were lost from plant mitochondria: from a related tRNA gene that still resides in the chondrome. These two evolutionary pathways of re-creating long-lost tRNA genes add to the long list of peculiar molecular evolutionary phenomena characterizing plant mitochondrial genomes that has been compiled since the 1980s.

The two recently sequenced chondromes of the lycophytes *Isoetes engelmanii* (Grewe et al. 2009) and *Selaginella moellendorffii* (Hecht et al. 2011) allow a glimpse into gene contents of mitochondrial genomes in the most basal lineage of vascular plants (Raubeson and Jansen 1992; Qiu et al. 2006). It is rather shocking to see that these genomes have also lost many of the genes that are gone in the hornwort chondromes (Table 7.2). The *Selaginella* case is particularly impressive, as this is the only land plant chondrome known so far that apparently has lost all ribosomal protein genes and all tRNA genes. Previously, mitochondrial genomes of animals and some green algae have been known to lack the entire set of ribosomal protein genes (Denovan-Wright et al. 1998; Dellaporta et al. 2006). Given that hornworts and lycophytes span the important evolutionary transition from gametophyte to sporophyte as the independent, free-living, dominating generation in the life cycle (Qiu et al. 2006), it is tempting to suggest that these gene losses may be related to this life cycle change. However, a completely sequenced chondrome from *Huperzia squarrosa,* which is a member of the third order (Lycopodiales) and the most basal lineage of lycophytes, shows that it actually has many ribosomal protein genes and tRNA genes. Its gene content is in fact not very different from that of the liverworts and the mosses (Liu et al. 2011). The widespread gene losses in chondromes of hornworts and lycophytes are very likely due to independent events, since the lost genes are obviously retained during the main course of plant evolution (in that they reside in seed plant mtDNAs; see next chapters). Given recurrent losses of these categories of genes in many eukaryotic lineages (discussed above), such a scenario does not seem unlikely.

Finally, there are many open reading frames (ORFs) in the bryophyte chondromes that are longer than 100 codons. Some are conserved between the two liverworts or between the two hornworts, both in their position and at the sequence level. Given that the identity of *rpl10,* previously only annotated as an ORF in several chondromes, was just revealed recently (Mower and Bonen 2009; Kubo and Arimura 2010), it is likely that some of these ORFs may represent uncharacterized genes. The gene *rtl*, encoding a reverse transcriptase, represents another interesting case. It has been known as ORF732 and ORF-721 in the chondromes *M. polymorpha* and *P. purpurea* respectively, and in both taxa, this gene exists as a free-standing gene located between *cob* and *nad9* (Oda et al. 1992; Wang et al. 2009). However, this gene is located within a group II intron in *nad9* of the chondromes of *P. patens* (Terasawa et al. 2007) and *A. rugelii* (Liu et al. 2011). The gene may have also been pseudogenized in some taxa, and more investigation is needed to determine the exact status.

III. Genome Rearrangement and Gene Order

High levels of synteny, i.e., the same gene orders among chondromes of different species, due to low rates of recombination and presence of polycistronic operons, are a major characteristic of organellar genomes (Palmer 1985; Gillham 1994). It was thus a surprise when angiosperm mitochondrial genomes were found to evolve extremely rapidly in structure (Palmer and Herbon 1988). Recently, genome level sequence data have shown that chondromes of two *Zea mays* cytotypes experience as many as 16 rearrangements (Allen et al. 2007). On the other hand, comparison of the *Marchantia* chondrome (Oda et al. 1992) and the chondromes of two closely related charophytic green algae, *C. vulgaris* and *Chaetosphaeridium globosum* (Turmel et al. 2002a, 2003), indicates that these genomes have experienced conservative structural evolution in early land plants (Fig. 7.1). Hence, the question arises as to when the mitochondrial genome acquired the ability to undergo rapid genome rearrangement during plant evolution.

Four recently sequenced bryophyte chondromes, which cover all three major lineages of bryophytes – liverworts, mosses, and hornworts (Terasawa et al. 2007; Li et al. 2009; Wang et al. 2009; Xue et al. 2010) – show that the structural evolution of mitochondrial genome is highly conservative not only within each individual lineage but also across bryophytes. It takes 16 rearrangements to bring the *Marchantia* and *Physcomitrella* chondromes into complete synteny (Fig. 7.1), the same number of rearrangements that chondromes of two *Z. mays* cytotypes have experienced (Allen et al. 2007). The readers are reminded here that liverworts and mosses diverged at least 375 million years ago according to the age of the oldest fossil of a clearly identifiable liverwort (Hernick et al. 2008). Likewise, only seven changes are required to bring all genes of the *Physcomitrella* and *Megaceros* chondromes into the same order (Fig. 7.1). The high level of structural conservation in bryophyte mitochondrial genomes is further confirmed by gene order comparison among chondromes of distantly related species within liverworts, mosses, and hornworts. The chondromes of two species that represent complex thalloid liverworts (*M. polymorpha* (Oda et al. 1992)) and simple thalloid liverworts (*Pleurozia purpurea* (Wang et al. 2009)) have identical gene orders. Likewise, the chondromes of two mosses, *P. patens* (Terasawa et al. 2007) and *A. rugelii* (Liu et al. 2011), which represent the diversity of almost the entire clade of peristomate mosses (Goffinet et al. 2001; Qiu et al. 2006), have identical gene orders. Finally, the chondromes of two hornworts, *Megaceros aenigmaticus* (Li et al. 2009) and *P. laevis* (Xue et al. 2010), which span the diversity of the entire clade, differ by only four rearrangements. This level of genome structural conservation shows that the mitochondrial genome in bryophytes still behaves like the typical

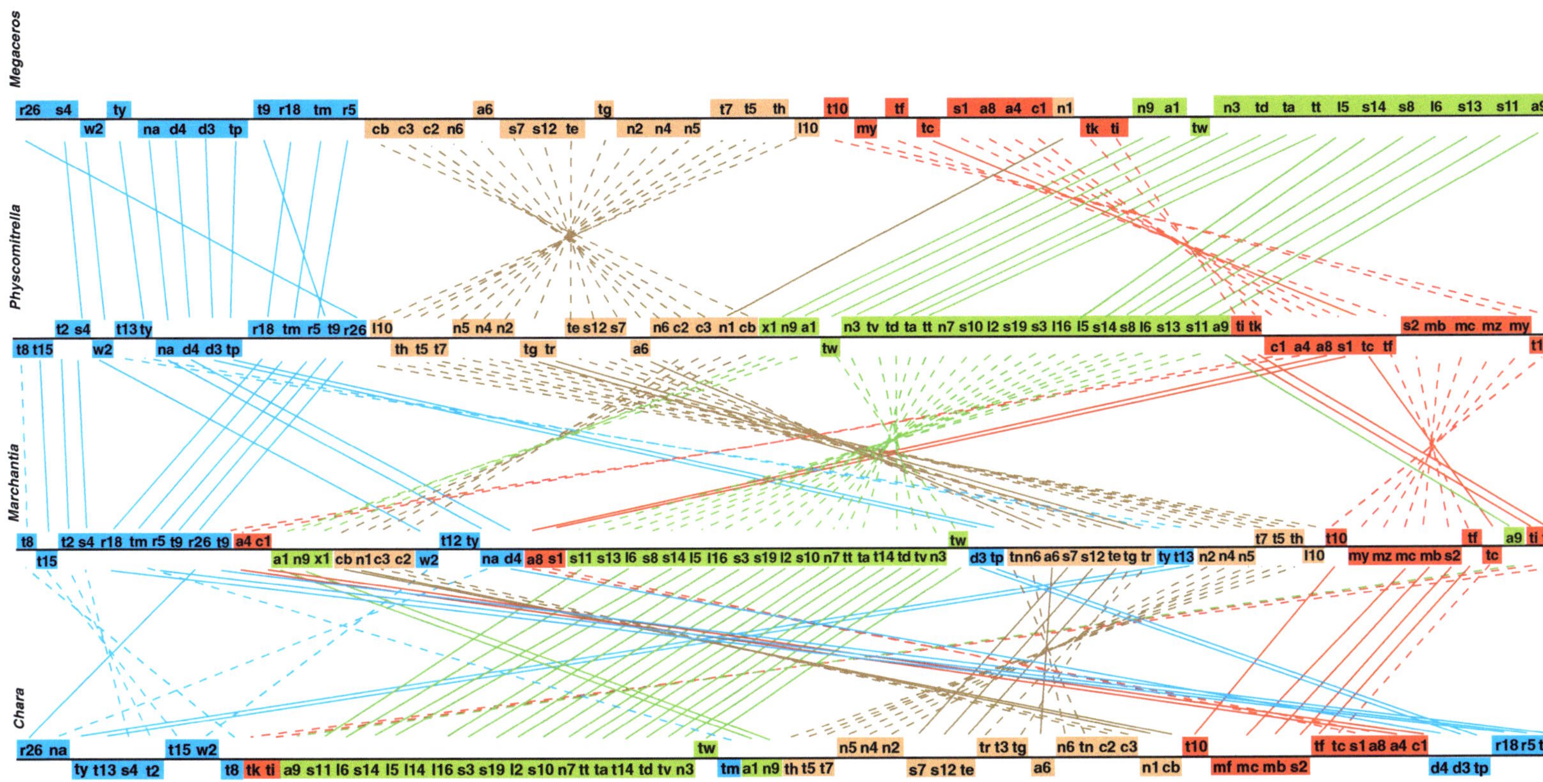

Fig. 7.1. Gene order comparison among chondromes of *Chara vulgaris, Marchantia polymorpha, Physcomitrella patens* and *Megaceros aenigmaticus*. *Solid lines* connect orthologous genes between species with the same gene orientation, whereas *dashed lines* connect those with inverted gene orientation. The gene *rtl* (reverse transcriptase) is located between *cob* and *nad9* in the *M. polymorpha* chondrome, but within a group II intron in *nad9* of the *P. patens* chondrome.

Table 7.3. Intron contents in mitochondrial genomes of *Chara vulgaris* and some early land plants[a]

Intron/species[b]	*Ch. vu.*	*Ma. po.*	*Pl. pu.*	*Ph. pa.*	*Me. ae.*	*Ph. la.*	*Is. en.*	*Se. mo.*
atp1i805g2					+	+		
atp1i989g2		+	+					
atp1i1019g2					+	+		
atp1i1050g2**		+	+		+	+		
atp1i1129g2				+				
atp6i80g2**				+	+	+		
atp6i439g2**					+	+	+	+
atp9i21g2**				+			+	*trans*
atp9i87g2***		+	+	+			+	+
atp9i95g2***				+	+	+		+
atp9i145g2	+							
atp9i214g2	+							
ccmFCi829g2**				+	+	+		
cobi274g2	+							
cobi372g2		+	+					
cobi420g1**				+		+		
cobi537g2	+							
cobi688g2	+							
cobi693g2							+	+
cobi783g2		+	+					
cobi787g2**					+	+	+	*trans*
cobi824g2		+	+					
cobi838g2					+	+		
cox1i44g2**		+	+		+	+		
cox1i150g2					+	+		
cox1i178g2		+	+					
cox1i211g2	+							
cox1i227g2							+	+
cox1i266g2							+	+
cox1i323g2							+	
cox1i375g1		+	+					
cox1i395g1**		+	+				+	
cox1i511g2***		+	+	+				+
cox1i624g1**		+	+	+				
cox1i729g1**	+	+	+					
cox1i732g2				+				
cox1i740g1	+							
cox1i835g2	+							
cox1i876g1**	+							+
cox1i909g1	+							
cox1i995g2							+	+
cox1i1064g2				+				
cox1i1116g1		+	+					
cox1i1149g2								+
cox1i1298g2					+	+		
cox1i1305g1**		+	+				*trans*	*trans*
cox2i94g2							+	+
cox2i97g2		+	+					
cox2i104g2**	+			+				
cox2i250g2		+	+					
cox2i281g2					+	+		

(continued)

Table 7.3. (continued)

Intron/species[b]	*Ch. vu.*	*Ma. po.*	*Pl. pu.*	*Ph. pa.*	*Me. ae.*	*Ph. la.*	*Is. en.*	*Se. mo.*
cox2i373g2***				+	+	+		*trans*
cox2i564g2						+		
cox2i691g2**				+		+		
cox3i109g2						+		
cox3i171g2		+	+					
cox3i506g2				+				
cox3i625g2		+	+					
nad1i287g2**				+	+	+		
nad1i348g2					+	+		
nad1i394g2							+	+
nad1i477g2								+
nad1i669g2								+
nad1i728g2***				+	+	+		+
nad2i156g2**				+			+	+
nad2i542g2							+	+
nad2i709g2***		+	+		+	+	+	+
nad2i830g2							+	+
nad2i1282g2					+	+		
nad3i52g2**					+	+	+	+
nad3i140g2****	+	+	+		+	+	+	+
nad3i211g2	+							
nad4i461g2***				+	+	+	+	+
nad4i548g2		+	+					
nad4i976g2***	+				+	+		+
nad4i1399g2							+	+
nad4Li100g2		+	+					
nad4Li283g2**		+	+	+				
nad5i230g2**				+	+	+		
nad5i242g2							+	+
nad5i753g1**		+	+	+				
nad5i1455g2**				+	+	+	+	+
nad5i1477g2**					+	+	+	+
nad6i444g2					+	+		
nad7i140g2**				+				+
nad7i209g2**				+			+	+
nad7i336g2		+	+					
nad7i676g2							+	+
nad7i917g2							+	+
nad7i1113g2**		+	+				+	
nad9i246g2					+	+		
nad9i283g2				+				
nad9i502g2					+	+		
rpl2i28g2		+	+					
rps3i74g2**	+						+	
rps14i114g2		+	+					
rrn18i839g1							+	+
rrn18i1065g2		+						
rrn26i819g1	+							
rrn26i827g2		+	+					
rrn26i1871g1	+							

(continued)

Table 7.3. (continued)

Intron/species[b]	*Ch. vu.*	*Ma. po.*	*Pl. pu.*	*Ph. pa.*	*Me. ae.*	*Ph. la.*	*Is. en.*	*Se. mo.*
rrn26i1879g1	+							
rrn26i1891g1	+							
rrn26i2191g1	+							
rrn26i2429g1	+							
rrn26i2462g1	+							
rrn26i2500g1	+							
rrn26i2513g1	+							
sdh3i100g2*				+	+	+		
trnNguui38g2	+							
trnSgcui43g2		+	+					

[a]"+" and "*trans*" indicate presence of *cis*- and *trans*-spliced introns, respectively. No sign indicates intron absence. The data cells are shaded alternately by plant clades (Charales, liverworts, mosses, hornworts and lycophytes). Group I introns are also shaded. Introns present in more than one clades are bold-faced, and the number of stars indicates the number of plants clades in which the intron is present
[b]Intron nomenclature follows Dombrovska and Qiu 2004, and Knoop 2004. Full species names are as follows (in the order as they appear): *Chara vulgaris, Marchantia polymorpha, Pleurozia purpurea, Physcomitrella patens, Megaceros aenigmaticus, Phaeoceros laevis, Isoetes engelmannii* and *Selaginella moellendorffii*

organellar genome as previously known in most other eukaryotes (Gray et al. 1999), and has not yet acquired the ability to undergo rapid rearrangement.

The lycophyte chondrome sequences from *I. engelmanii* (Grewe et al. 2009) and *S. moellendorffii* (Hecht et al. 2011), however, tell a different story. The two genomes not only show highly rearranged gene order from each other, but also differ significantly from the conservative bryophyte chondriomes. These data suggest that the lycophytes may represent the beginning of rapid structural evolution in plant mitochondrial genomes. However, the extent of mtDNA rearrangement in lycophytes is not necessarily as dramatic as seen in these two chondromes, because the chondrome of *H. squarrosa* (Liu et al. 2011) is less reshuffled in comparison to the bryophyte chondromes. One particular additional indicator is that the *H. squarrosa* chondrome has no single *trans*-spliced intron whereas both *I. engelmanii* and *S. moellendorffii* have *trans*-spliced introns in their chondromes.

With regard to mechanisms of intra-genomic rearrangement, repeat sequences may have been involved as hypothesized earlier (Andre et al. 1992). Comparison of gene orders between the two hornwort chondromes detected several pairs of repeat sequences (Xue et al. 2010). Similarly, numerous large repeat sequences were found in the highly rearranged chondrome of the lycophyte *S. moellendorffii* (Hecht et al. 2011).

IV. Introns

Both groups I and II introns are present in chondromes of bryophytes and lycophytes (Table 7.3). Their distribution patterns reflect their nature as mobile genetic elements. On the one hand, many introns are conserved within liverworts, hornworts or lycophytes, or even larger clades that include vascular plants and more than one major clade of bryophytes. On the other hand, some introns appear to have experienced lateral movement. Broad surveys including a large number of taxa and phylogenetic analyses of both exon and intron sequences are needed to determine whether an intron has been vertically inherited or laterally transferred (Malek and Knoop 1998; Qiu et al. 1998; Dombrovska and Qiu 2004).

Despite their seemingly idiosyncratic distribution patterns, a few generalizations can be made about these introns. First, it is clear that they experienced active transposition during major evolutionary events such as the origins of land plants, individual bryophyte clades and vascular plants, but were stably inherited afterwards, because most of these introns show plant clade-specific distribution patterns (Table 7.3). Second, the host gene seems to be a factor in determining intron distribution. Group II introns have a broad distribution, but they are mostly present in respiratory protein genes. Genes for genetic information processing, such as rRNA genes, ribosomal protein genes and tRNA genes, are greatly under-represented among the host genes of introns. Intriguingly, group II introns in chloroplast genomes of charophytes and land plants do not show such strong host gene preference (Ohyama et al. 1986; Turmel et al. 2002a). Group I introns are mostly found in only two genes, *cox1* and *rrn26*, and thus the host gene diversity is too low to allow detection of any meaningful pattern. Third, *trans*-splicing is highly correlated to the level of recombination within the genome. No *trans*-spliced intron has been found in any bryophyte chondrome, consistent with the high level of structural conservation in these genomes (Oda et al. 1992; Terasawa et al. 2007; Li et al. 2009; Wang et al. 2009; Xue et al. 2010). In contrast, *trans*-spliced group II introns have been detected in the highly recombinogenic *Selaginella* chondrome, and a first ever *trans*-spliced group I intron has been found in the chondromes of both *Selaginella* and *Isoetes* (Grewe et al. 2009; Hecht et al. 2011). This observation is consistent with a previous report of multiple independent evolution of *trans*-splicing from a *cis*-spliced homologous intron in the highly recombinogenic vascular plant chondromes (Qiu and Palmer 2004). These non-random distribution patterns, as idiosyncratic as they are, probably reflect the interplay of historical processes during plant evolution, mechanistic preference of transposition mediated by both endonucleases and recognition motifs, recombination activity within a genome, and some as yet poorly understood functional adaptive mechanisms.

Previously, intron distribution patterns have been used to resolve difficult phylogenetic issues in land plants (Qiu et al. 1998; Groth-Malonek et al. 2005; Qiu et al. 2006). Knowledge of the full sets of introns (as gained from sequencing of entire chondromes) constitutes the prime data for this purpose, although taxon sampling at this stage is often not yet dense enough. Nevertheless, it seems clear that no intron is uniquely present in chondromes of all three bryophyte lineages, lending strong support to the paraphyly hypothesis of bryophytes (Mishler and Churchill 1984; Kenrick and Crane 1997; Qiu et al. 2006).

V. RNA Editing

RNA editing was originally not detected in bryophyte chondromes (Hiesel et al. 1994), due to lack or low levels of editing in the small number of taxa investigated: *Marchantia, Sphagnum* and *Physcomitrella*. The two bryophyte chondromes that were sequenced first happened to be non- or low-editing taxa, *M. polymorpha* and *P. patens* (Oda et al. 1992; Terasawa et al. 2007). Later surveys including many more taxa, based on indirect evidence from codon conservation analysis, suggested occurrence of RNA editing, sometimes at high levels, in bryophyte chondromes (Steinhauser et al. 1999; Dombrovska and Qiu 2004). Direct evidence for genome-wide occurrence of RNA editing was obtained only recently from cDNA sequencing in the moss *P. patens* (Rüdinger et al. 2009) and the hornwort *M. aenigmaticus* (Xue et al. 2010). Annotation of completely sequenced chondromes from the liverwort *P. purpurea* and the hornwort *P. laevis* also invokes a significant number of editing events to create proper start and stop codons and to remove internal stop codons (Li et al. 2009; Wang et al. 2009). Thus, RNA editing is clearly present throughout land plants (with the exception of an apparent secondary loss in the marchantiid liverworts (Groth-Malonek et al. 2007)) and the editing machinery likely originated in the common ancestor of land plants. No editing has been detected so far in charophyte chondromes (Turmel et al. 2002a, b, 2003, 2007).

In early vascular plant chondromes, RNA editing has been known for a long time (Hiesel et al. 1994), but the extent of editing at a genome-wide level was not known until very recently, when transcriptome analyses of the lycophytes *I. engelmanii* and *S. moellendorffii* were performed (Grewe et al. 2009, 2010; Hecht et al. 2011). Over 2000 sites of editing in apparently not very large genomes set the record for perhaps the highest levels of RNA editing in any genome known so far. While RNA editing has been shown to follow adaptive distribution patterns (Jobson and Qiu 2008), its highly lineage-specific occurrence remains poorly understood. It seems that life history characteristics of organisms need to be considered when seeking explanations of such a bizarre molecular evolutionary phenomenon that is often lineage-specific.

VI. Concluding Remarks

Mitochondrial genomes of early land plants have occupied a unique position in our quest to understand evolution of this important organellar genome. Sequencing of the *Marchantia* chondrome 20 years ago provided a wealth of information for the characterization of mitochondrial genes in other basal land plants and the exploration of information for phylogenetic reconstructions. Recent sequencing of several chondromes that cover all major lineages of bryophytes and lycophytes allowed detailed examination of various aspects of this genome as well as identification of the major point of genomic structural change in plant mitochondrial genome evolution. These studies helped to develop a more complete understanding of mitochondrial genome evolution in plants and eukaryotes. In the future, it will be desirable to obtain chondrome sequences of some ferns and more bryophytes, so that a comprehensive understanding of this genome and its various intriguing aspects, such as gene loss, intron acquisition and RNA editing, can be systematically developed and correlated to the major transitions in land plant evolution.

Acknowledgment

We thank Ken D. McFarland, Blanka Shaw, Jon Shaw, and David K. Smith for help with obtaining plant material, and Volker Knoop for informative discussion. This work was supported by NSF grants DEB 0531689 and 0332298 to YLQ.

References

Adams KL, Rosenblueth M, Qiu Y-L, Palmer JD (2001) Multiple losses and transfers to the nucleus of two mitochondrial respiratory genes during angiosperm evolution. Genetics 158:1289–1300

Adams KL, Qiu YL, Stoutemyer M, Palmer JD (2002) Punctuated evolution of mitochondrial gene content: high and variable rates of mitochondrial gene loss and transfer to the nucleus during angiosperm evolution. Proc Natl Acad Sci USA 99:9905–9912

Allen JO, Fauron CM, Minx P, Roark L, Oddiraju S, Lin GN, Meyer L, Sun H, Kim K, Wang CY, Du FY, Xu D, Gibson M, Cifrese J, Clifton SW, Newton KJ (2007) Comparisons among two fertile and three male-sterile mitochondrial genomes of maize. Genetics 177:1173–1192

Alverson AJ, Wei X, Rice DW, Stern DB, Barry K, Palmer JD (2010) Insights into the evolution of mitochondrial genome size from complete sequences of *Citrullus lanatus* and *Cucurbita pepo* (Cucurbitaceae). Mol Biol Evol 27:1436–1448

Alverson AJ, Rice DW, Dickinson S, Barry K, Palmer JD (2011a) Origins and recombination of the bacterial-sized multichromosomal mitochondrial genome of cucumber. Plant Cell 23:2499–2513

Andre C, Levy A, Walbot V (1992) Small repeated sequences and the structure of plant mitochondrial genomes. Trends Genet 8:128–132

Burger G, Saint-Louis D, Gray MW, Lang BF (1999) Complete sequence of the mitochondrial DNA of the red alga *Porphyra purpurea*: cyanobacterial introns and shared ancestry of red and green algae. Plant Cell 11:1675–1694

Chaw SM, Shih ACC, Wang D, Wu YW, Liu SM, Chou TY (2008) The mitochondrial genome of the gymnosperm *Cycas taitungensis* contains a novel family of short interspersed elements, Bpu sequences, and abundant RNA editing sites. Mol Biol Evol 25:603–615

Clifton SW, Minx P, Fauron CMR, Gibson M, Allen JO et al (2004) Sequence and comparative analysis of the maize NB mitochondrial genome. Plant Physiol 136:3486–3503

Dellaporta SL, Xu A, Sagasser S, Jakob W, Moreno M, Buss LW, Schierwater B (2006) Mitochondrial genome of *Trichoplax adhaerens* supports Placozoa as the basal lower metazoan phylum. Proc Natl Acad Sci USA 103:8751–8756

Denovan-Wright EM, Nedelcu AM, Lee RW (1998) Complete sequence of the mitochondrial DNA of *Chlamydomonas eugametos*. Plant Mol Biol 36: 285–295

Dombrovska O, Qiu Y-L (2004) Distribution of introns in the mitochondrial gene *nad1* in land plants: phylogenetic and molecular evolutionary implications. Mol Phylogenet Evol 32:246–263

Duff RJ, Villarreal JC, Cargill DC, Renzaglia KS (2007) Progress and challenges toward developing a phylogeny and classification of the hornworts. Bryologist 110:214–243

Gillham NW (1994) Organelle genes and genomes. Oxford University Press, New York

Goffinet B, Cox CJ, Shaw AJ, Hedderson TAJ (2001) The bryophyta (mosses): systematic and evolutionary inferences from an *rps4* gene (cpDNA) phylogeny. Ann Bot 87:191–208

Goremykin VV, Salamini F, Velasco R, Viola R (2009) Mitochondrial DNA of *Vitis vinifera* and the issue of rampant horizontal gene transfer. Mol Biol Evol 26:99–110

Gray MW, Burger G, Lang BF (1999) Mitochondrial evolution. Science 283:1476–1481

Grewe F, Viehoever P, Weisshaar B, Knoop V (2009) A trans-splicing group I intron and tRNA-hyperediting in the mitochondrial genome of the lycophyte *Isoetes engelmannii*. Nucleic Acids Res 15: 5093–5104

Grewe F, Herres S, Viehoever P, Polsakiewicz M, Weisshaar B, Knoop V (2010) A unique transcriptome: 1728 positions of RNA editing alter 1406 codon identities in mitochondrial mRNAs of the lycophyte *Isoetes engelmannii*. Nucleic Acids Res 39:2890–2902

Grewe F, Herres S, Viehoever P, Polsakiewicz M, Weisshaar B et al (2011) A unique transcriptome: 1728 positions of RNA editing alter 1406 codon identities in mitochondrial mRNAs of the lycophyte *Isoetes engelmannii*. Nucleic Acids Res 39:2890–2902

Groth-Malonek M, Pruchner D, Grewe F, Knoop V (2005) Ancestors of trans-splicing mitochondrial introns support serial sister group relationships of hornworts and mosses with vascular plants. Mol Biol Evol 22:117–125

Groth-Malonek M, Wahrmund U, Polsakiewicz M, Knoop V (2007) Evolution of a pseudogene: exclusive survival of a functional mitochondrial *nad7* gene supports *Haplomitrium* as the earliest liverwort lineage and proposes a secondary loss of RNA editing in Marchantiidae. Mol Biol Evol 24: 1068–1074

Handa H (2003) The complete nucleotide sequence and RNA editing content of the mitochondrial genome of rapeseed (*Brassica napus* L.): comparative analysis of the mitochondrial genomes of rapeseed and *Arabidopsis thaliana*. Nucleic Acids Res 31:5907–5916

Hecht J, Grewe F, Knoop V (2011) Extreme RNA editing in coding islands and abundant microsatellites in repeat sequences of *Selaginella moellendorffii* mitochondria: the root of frequent plant mtDNA recombination in early tracheophytes. Genome Biol Evol 3:344–358

Hernick LV, Landing E, Bartowski KE (2008) Earth's oldest liverworts – *Metzgeriothallus sharonae* sp. nov. from the Middle Devonian (Giventian) of eastern New York, USA. Rev Palaeobot Palynol 148: 154–162

Hiesel R, Combettes B, Brennicke A (1994) Evidence for RNA editing in mitochondria of all major groups of land plants except the Bryophyta. Proc Natl Acad Sci USA 91:629–633

Jobson RW, Qiu Y-L (2008) Did RNA editing in plant organellar genomes originate under natural selection or through genetic drift? Biol Direct 3:43

Karol KG, McCourt RM, Cimino MT, Delwiche CF (2001) The closest living relatives of land plants. Science 294:2351–2353

Kenrick P, Crane PR (1997) The origin and early diversification of land plants: a cladistic study. Smithsonian Institution Press, Washington, DC

Knoop V (2004) The mitochondrial DNA of land plants: peculiarities in phylogenetic perspective. Curr Genet 46:123–139

Kubo N, Arimura S-I (2010) Discovery of a functional *rpl10* gene in diverse plant mitochondrial genomes and its functional replacement by a nuclear gene for chloroplast RPL10 in two lineages of angiosperms. DNA Res 17:1–9

Kubo T, Nishizawa S, Sugawara A, Itchoda N, Estiati A et al (2000) The complete nucleotide sequence of the mitochondrial genome of sugar beet (*Beta vulgaris* L.) reveals a novel gene for tRNA(Cys) (GCA). Nucleic Acids Res 28:2571–2576

Lang BF, Burger G, Okelly CJ, Cedergren R, Golding GB, Lemieux C, Sankoff D, Turmel M, Gray MW (1997) An ancestral mitochondrial DNA resembling a eubacterial genome in miniature. Nature 387:493–497

Li L, Wang B, Liu Y, Qiu Y-L (2009) The complete mitochondrial genome sequence of the hornwort *Megaceros aenigmaticus* shows a mixed mode of conservative yet dynamic evolution in early land plant mitochondrial genomes. J Mol Evol 68: 665–678

Liu Y, Xue J-Y, Wang B, Li L, Qiu Y-L (2011) The mitochondrial genomes of the early land plants *Treubia lacunosa* and *Anomodon rugelii*: dynamic and conservative evolution. PLoS One 6(10):e25836

Malek O, Knoop V (1998) Trans-splicing group II introns in plant mitochondria: the complete set of cis-arranged homologs in ferns, fern allies, and a hornwort. RNA 4:1599–1609

Maréchal-Drouard L, Guillemaut P, Cosset A, Arbogast M, Weber F et al (1990) Transfer RNAs of potato (*Solanum tuberosum*) mitochondria have different genetic origins. Nucleic Acids Res 18:3689–3696

Mishler BD, Churchill SP (1984) A cladistic approach to the phylogeny of the bryophytes. Brittonia 36:406–424

Mower JP, Bonen L (2009) Ribosomal protein L10 is encoded in the mitochondrial genome of many land plants and green algae. BMC Evol Biol 9:265

Notsu Y, Masood S, Nishikawa T, Kubo N, Akiduki G et al (2002) The complete sequence of the rice (*Oryza sativa* L.) mitochondrial genome: frequent DNA sequence acquisition and loss during the evolution of flowering plants. Mol Genet Genomics 268:434–445

Oda K, Yamato K, Ohta E, Nakamura Y, Takemura M, Nozato N, Akashi K, Kanegae T, Ogura Y, Kohchi T, Ohyama K (1992) Gene organization deduced from the complete sequence of liverwort *Marchantia polymorpha* mitochondrial DNA – a primitive form of plant mitochondrial genome. J Mol Biol 223:1–7

Ohyama K, Fukuzawa H, Kohchi T, Shirai H, Sano T, Sano S, Umesono K, Shiki Y, Takeuchi M, Chang Z, Aota S, Inokuchi H, Ozeki H (1986) Chloroplast gene organization deduced from complete sequence of liverwort *Marchantia polymorpha* chloroplast DNA. Nature 322:572–574

Palmer JD (1985) Comparative organization of chloroplast genomes. Annu Rev Genet 19:325–354

Palmer JD, Herbon LA (1988) Plant mitochondrial DNA evolves rapidly in structure, but slowly in sequence. J Mol Evol 28:87–97

Palmer JD, Adams KL, Cho YR, Parkinson CL, Qiu YL, Song KM (2000) Dynamic evolution of plant mitochondrial genomes: mobile genes and introns and highly variable mutation rates. Proc Natl Acad Sci USA 97:6960–6966

Qiu Y-L, Palmer JD (2004) Many independent origins of *trans* splicing of a plant mitochondrial group 2 intron. J Mol Evol 59:80–89

Qiu Y-L, Cho YR, Cox JC, Palmer JD (1998) The gain of three mitochondrial introns identifies liverworts as the earliest land plants. Nature 394:671–674

Qiu Y-L, Li LB, Wang B, Chen ZD, Knoop V, Groth-Malonek M, Dombrovska O, Lee J, Kent L, Rest J, Estabrook GF, Hendry TA, Taylor DW, Testa CM, Ambros M, Crandall-Stotler B, Duff RJ, Stech M, Frey W, Quandt D, Davis CC (2006) The deepest divergences in land plants inferred from phylogenomic evidence. Proc Natl Acad Sci USA 103: 15511–15516

Raubeson LA, Jansen RK (1992) Chloroplast DNA evidence on the ancient evolutionary split in vascular land plants. Science 255:1697–1699

Rüdinger M, Funk HT, Rensing SA, Maier UG, Knoop V (2009) RNA editing: 11 sites only in the *Physcomitrella patens* mitochondrial transcriptome and a universal nomenclature proposal. Mol Genet Genomics 281:473–481

Schuster W, Brennicke A (1994) The plant mitochondrial genome – physical structure, information content, RNA editing, and gene migration to the nucleus. Ann Rev Plant Physiol Plant Mol Biol 45:61–78

Sloan DB, Alverson AJ, Storchova H, Palmer JD, Taylor DR (2010) Extensive loss of translational genes in the structurally dynamic mitochondrial genome of the angiosperm *Silene latifolia*. BMC Evol Biol 10:274

Steinhauser S, Beckert S, Capesius I, Malek O, Knoop V (1999) Plant mitochondrial RNA editing. J Mol Evol 48:303–312

Stewart WN (1983) Paleobotany and the evolution of plants. Cambridge University Press, Cambridge, UK

Sugiyama Y, Watase Y, Nagase M, Makita N, Yagura S et al (2005) The complete nucleotide sequence and multipartite organization of the tobacco mitochondrial genome: comparative analysis of mitochondrial genomes in higher plants. Mol Genet Genomics 272:603–615

Terasawa K, Odahara M, Kabeya Y, Kikugawa T, Sekine Y, Fujiwara M, Sato N (2007) The mitochondrial genome of the moss *Physcomitrella patens* sheds new light on mitochondrial evolution in land plants. Mol Biol Evol 24:699–709

Turmel M, Lemieux C, Burger G, Lang BF, Otis C, Plante I, Gray MW (1999) The complete mitochondrial DNA sequences of *Nephroselmis olivacea* and *Pedinomonas minor*: two radically different evolutionary patterns within green algae. Plant Cell 11:1717–1729

Turmel M, Otis C, Lemieux C (2002a) The chloroplast and mitochondrial genome sequences of the charophyte *Chaetosphaeridium globosum*: insights into the timing of the events that restructured organelle DNAs within the green algal lineage that led to land plants. Proc Natl Acad Sci USA 99:11275–11280

Turmel M, Otis C, Lemieux C (2002b) The complete mitochondrial DNA sequence of *Mesostigma viride* identifies this green alga as the earliest green plant divergence and predicts a highly compact mitochondrial genome in the ancestor of all green plants. Mol Biol Evol 19:24–38

Turmel M, Otis C, Lemieux C (2003) The mitochondrial genome of *Chara vulgaris*: insights into the mitochondrial DNA architecture of the last common ancestor of green algae and land plants. Plant Cell 15:1888–1903

Turmel M, Otis C, Lemieux C (2007) An unexpectedly large and loosely packed mitochondrial genome in the charophycean green alga *Chlorokybus atmophyticus*. BMC Genomics 8:137

Unseld M, Marienfeld JR, Brandt P, Brennicke A (1997) The mitochondrial genome of *Arabidopsis thaliana* contains 57 genes in 366,924 nucleotides. Nat Genet 15:57–61

Wang B, Xue J-Y, Li L, Liu L, Qiu Y-L (2009) The complete mitochondrial genome sequence of the liverwort *Pleurozia purpurea* reveals extremely conservative mitochondrial genome evolution in liverworts. Curr Genet 55:601–609

Ward B, Anderson R, Bendich A (1981) The size of the mitochondrial genome is large and variable in a family of plants. Cell 25:793–803

Xue J-Y, Liu Y, Li L, Wang B, Qiu Y-L (2010) The complete mitochondrial genome sequence of the hornwort *Phaeoceros laevis*: retention of many ancient pseudogenes and conservative evolution of mitochondrial genomes in hornworts. Curr Genet 56:53–61

Chapter 8

Seed Plant Mitochondrial Genomes: Complexity Evolving

Volker Knoop*
Abt. Molekulare Evolution, IZMB, Universität Bonn, Kirschallee 1, D-53115 Bonn, Germany

Summary

Complete mitochondrial genome sequences are now available for representatives of all major clades of land plants except for the ferns (monilophytes). More than 30 chondrome sequences have been determined for flowering plants alone. Given that a well-founded understanding of land plant phylogeny has developed over the recent years, we can now confidently trace the molecular evolution of plant mitochondrial genomes with respect to their numerous interesting features: an ongoing endosymbiotic gene transfer to the nucleus, the gains, losses and occasional disruptions of introns, the acquisition of foreign DNA sequences and the emergence of the pyrimidine conversion type of RNA editing. This review attempts to put the insights from several independent studies addressing the molecular evolution of these features and our insights from the growing list of completed plant chondrome sequences into a modern phylogenetic perspective on land plants.

*Author for correspondence, e-mail: volker.knoop@uni-bonn.de

R. Bock and V. Knoop (eds.), Genomics of Chloroplasts and Mitochondria, Advances in Photosynthesis and Respiration 35, pp. 175–200,
DOI 10.1007/978-94-007-2920-9_8,

I. Introduction

Flowering plants (angiosperms) are unique among all eukaryotes in having very large and complex mitochondrial genomes. Already 30 years ago, it was recognized that the mitochondrial DNAs (mtDNAs) in the Cucurbitaceae, for example in muskmelon (Ward et al. 1981), may be larger than 2 mega base pairs (Mbp) and thus even exceed in size the genomes of many free-living bacteria. The compact and streamlined animal mitochondrial genomes (chondromes) are one to two orders of magnitude smaller and starkly contrast their angiosperm counterparts in almost every aspect. Note, that the term "chondriome" has previously been used ambiguously to describe either the mitochondrial genome of a species or the entirety of the dynamic population of all mitochondria in a cell (Logan 2010). This issue of confusion was recently addressed, suggesting that "chondriome" is now restricted to the latter sense whereas chondrome (without i) is used equivalently to mitochondrial genome (Knoop et al. 2010).

The mitochondrial genome of humans was the first chondrome to be determined in its complete sequence in 1981 and this was an early milestone on the path to what has later become known as the genomics era of biology (Anderson et al. 1981). The simple circular structure and small size (of only 16 kilo base pairs; Kbp) of the human mitochondrial DNA turned out to be prototypical not only for other mammals but also for animals (metazoa) at large. One has to dig deep into the phylogeny of metazoa to find (few) exceptions to the simple design of animal mtDNA encoding a standard set of 13 protein subunits of respiratory chain complexes (RCCs) plus 2 rRNAs and 22 tRNAs (Burger et al. 2009; Lavrov 2007; Signorovitch et al. 2007).

In contrast, many studies over the last three decades have revealed that a number of oddities have affected the evolution of flowering plant chondromes. Flowering plants represent the evolutionary most successful plant clade dating back in origin to the early cretaceous some 145 million years ago and comprising more than 250,000 extant species. The molecular peculiarities of angiosperm plant chondromes include (1) frequent mtDNA recombination producing alternative, co-existing DNA arrangements, (2) promiscuous DNA inserts originating from the chloroplast and nuclear genomes, (3) disruptions of genes creating the need for rejoining genetic information at the RNA level by trans-splicing, (4) frequent RNA editing of mitochondrial transcripts through site-directed pyrimidine (C/U) conversions and (5) an ongoing endosymbiotic gene transfer (EGT) into the nucleus, which makes plant chondrome gene complements highly variable. In addition, at least some plant chondromes seem to be prone to accept insertions of horizontally transferred sequences originating from other, distant taxa (Bergthorsson et al. 2003; Won and Renner 2003). In fact, horizontal gene transfer (HGT) may have had a significant influence on shaping certain angiosperm chondromes, such as the one of the early-branching species *Amborella trichopoda* (Bergthorsson et al. 2004 and J. Palmer, pers. communication). A separate chapter in this volume (Chap. 10) is exclusively dedicated to horizontal gene transfer and, therefore, I will only briefly allude to this issue here.

II. Complete Plant Chondrome Sequences

The mtDNA of the liverwort *Marchantia polymorpha* was the first land plant chondrome sequence to be completely determined nearly 20 years ago (Oda et al. 1992b).

Abbreviations: bp – Base pairs; CMS – Cytoplasmic male sterility; cpDNA – Chloroplast DNA; EGT – Endosymbiotic gene transfer (from endosymbiotic organelles to the nucleus); HGT – Horizontal gene transfer; HT-clade – hornwort-tracheophyte clade; Kbp – Kilo base pairs (10^3 bp); LGT – Lateral gene transfer; Mbp – Mega base pairs (10^6 bp); mRNA – Messenger RNA; mtDNA – Mitochondrial DNA; NLE-clade – non-liverwort embryophyte clade; ORF – Open reading frame; RCC – Respiratory chain complex; rRNA – Ribosomal RNA; SNP – Single nucleotide polymorphism; tRNA – Transfer RNA

Sequencing of the mtDNA of the model plant *Arabidopsis thaliana* (Brassicaceae) was completed 5 years later (Unseld et al. 1997) and this was followed by complete sequencing of several other angiosperm chondromes, mainly from crop species of agricultural importance (Table 8.1). Significant economical interest in studies of crop plant mtDNAs comes from an aspect of immediate practical, agronomical relevance in plant breeding: the phenomenon of cytoplasmic male sterility (CMS), which is of utmost importance for hybrid seed production. After experiencing a disastrous susceptibility of the widely used CMS-T varieties, harbouring the 'Texas' cytoplasm, of maize (*Zea mays*) to the fungal pathogen *Bipolaris maydis* in 1970, early groundbreaking research demonstrated that mtDNA rearrangements are correlated with the expression of the CMS phenotype (Levings and Pring 1976; Pring et al. 1977; Timothy et al. 1979). Some mitochondrial DNA recombinations turned out to be associated with the accidental creation of chimeric open reading frames (ORFs), which upon expression result in specific defects in mitochondrial function that become phenotypically apparent as defects in pollen maturation (Hanson and Bentolila 2004; Kubo and Newton 2008). Another chapter in this volume (Chap. 12) deals extensively with such mutations in plant mitochondrial genomes.

Several studies of flowering plant mtDNAs have shown that significant variation at the genus or species levels is not restricted to maize (Sederoff et al. 1981), but is also found, for example, in the evening primrose *Oenothera berteriana* (Brennicke and Blanz 1982), *Arabidopsis thaliana* (Ullrich et al. 1997) or the genera *Fagus* (Tomaru et al. 1998) and *Solanum* (Scotti et al. 2004). Several complete mtDNA sequences have recently been determined for different cultivars, sub-species or very closely related species of rice (*Oryza sativa*), maize (*Zea mays*), wheat (*Triticum aestivum*) and beet (*Beta vulgaris*), now providing a full view on mtDNA variability at these very low taxonomic levels. Immediately obvious – already from simply comparing the mtDNA sizes (Table 8.1) – are the vast expansions or reductions of chondrome sizes leading to significant mtDNA variability among the closely related taxa. One interesting exception is the near-identity of the recently determined chondrome of a winter wheat cultivar with its spring wheat cultivar counterpart, distinguished by only 10 minor indels and 7 SNPs (Cui et al. 2009). This, however, is in sharp contrast to a much larger chondrome containing vastly extended intergenic regions in the wheat CMS line Ks3 (Liu et al. 2011).

The total number of 22 complete chondrome sequences now available for the four genera *Beta, Oryza, Triticum* and *Zea* alone currently exceeds the number of 12 complete mtDNA sequence available for all other angiosperm genera – in fact even those of all other seed plant (spermatophyte) genera, given that the mtDNA of *Cycas taitungensis* is currently the only one representing the gymnosperms (Table 8.1).

Point mutations (SNPs, single nucleotide polymorphisms) are the prevailing differences at the species, genus or even higher taxonomic levels in animal mtDNAs, which are highly conserved in their simple, circular and non-recombining structure. Basically, the opposite is the case in plants: Structural rearrangements involving DNA recombinations dominate over very few SNPs that are only rarely discovered in the slowly evolving plant mtDNA sequences. By and large, the title of a late 1980s publication "Plant mitochondrial DNA evolves rapidly in structure but slowly in sequence" (Palmer and Herbon 1988) is essentially still correct, at least for angiosperms. However, it must be noted that some plant lineages have been discovered to show substantially elevated levels of mitochondrial primary sequence evolution (Adams et al. 1998b; Cho et al. 2004; Parkinson et al. 2005; Sloan et al. 2010b; Vangerow et al. 1999).

Evolutionarily successful, beautiful and unmatched in importance for animal and human existence as they are, angiosperms represent even less than only the last third of plant evolution on this planet, which presumably dates back to Ordovician times. After the milestone mtDNA sequencing of *Marchantia polymorpha* it actually took as

Table 8.1. Completed land plant (embryophyte) mtDNA sequences and those of two charophyte algae discussed as most closely related to embryophyte lineage. Sizes of chondrome "master-circles", database accessions and key papers are indicated in most cases. The asterisks indicate that the true physical structure of the lycophyte chondromes is particularly unclear, but assumed to be network-like due to rampant recombination. For the *Isoetes engelmannii* chondrome (submitted as five partially redundant fosmid clones under database accessions FJ010859, FJ536259, FJ390841, FJ176330 and FJ628360), the net sequence complexity is indicated, for *Selaginella moellendorffii* the approximate sum of single-copy coding islands (~100 Kbp) plus 10 extended recombinationally active repeats (~50 Kbp) is given. A permanently updated list of Virdiplantae mtDNAs (i.e., including chlorophyte algae) is found at http://www.ncbi.nlm.nih.gov/genomes/GenomesGroup.cgi?taxid=33090&opt=organelle. Complete mtDNA sequences have been determined for closely related species/subspecies/varieties/cultivars and/or fertile and male infertile CMS lines in the agronomically relevant genera *Oryza, Triticum, Zea* and *Beta*. *Oryza sativa* cultivar PA64S[a] is in fact "*indica*-like", but its cytoplasm is of *japonica* type. The mtDNAs of bamboo (*Bambusa oldhamii*, Bambusoidae, BEP clade) and tomato (*Solanum lycopersicum*, Solanales) are deposited in the database under accessions EU365401 and FJ374974, but are annotated to contain undetermined gaps and are, therefore, not further discussed in this chapter

Taxonomy	*Species*	mtDNA size	DB accessions	Publication(s)
Streptophyte algae, Charophyte				
Charales	*Chara vulgaris*	68	AY267353/NC_005255	(Turmel et al. 2003)
Coleochaetales	*Chaetosphaeridium globosum*	57	AF494279/NC_004118	(Turmel et al. 2002a)
Marchantiophyta (liverworts)				
Marchantiales	*Marchantia polymorpha*	187	M68929/NC_001660	(Oda et al. 1992b)
Pleuroziales	*Pleurozia purpurea*	169	FJ999996/NC_013444	(Wang et al. 2009)
Bryophyta (mosses)				
Funariales	*Physcomitrella patens*	105	NC_007945	(Terasawa et al. 2007)
Anthocerotophyta (hornworts)				
Dendrocerotales	*Megaceros aenigmaticus*	185	EU660574/NC_012651	(Li et al. 2009)
Notothyladales	*Phaeoceros laevis*	209	GQ376531/NC_013765	(Xue et al. 2010)
Lycophyta (lycophytes)				
Isoetales	*Isoetes engelmannii*	58[b]	FJ010859 et c.[b]	(Grewe et al. 2009)
Selaginellales	*Selaginella moellendorffii*	~150[b]	JF338143-JF338147[b]	(Hecht et al. 2011)
Spermatophyta (seed plants)				
Gymnosperms – Cycadophyta/Cycadales				
Cycadaceae	*Cycas taitungensis*	415	NC_010303	(Chaw et al. 2008)
Angiosperms				
Monocots: Liliopsida/Poales/Poaceae				
BEP clade				
Ehrhartoidae	*Oryza sativa*	435–491		
	ssp. *japonica* Nipponbare-N	491	BA000029	(Notsu et al. 2002)
	ssp. *japonica* Nipponbare-S	491	DQ167400	(Tian et al. 2006)
	ssp. *japonica* PA46S[a]	491	DQ167807	(Tian et al. 2006)
	ssp. *indica* 93-11	492	DQ167399	(Tian et al. 2006)
	ssp. *indica* LD-CMS	435	AP011077	(Fujii et al. 2010)
	Oryza rufipogon CW-CMS	559	AP011076	(Fujii et al. 2010)
Pooidae	*Triticum aestivum*	453–658		
	cv. Chinese Spring	453	AP008982	(Ogihara et al. 2005)
	cv. Chinese Yumai	453	EU534409	(Cui et al. 2009)
	K-type CMS line Ks3	658	GU985444	(Liu et al. 2011)

(continued)

Table 8.1. (continued)

Taxonomy	*Species*	mtDNA size	DB accessions	Publication(s)
PACCAD clade/Panicoidae				
Andropogoneae	*Sorghum bicolor*	469	DQ984518	
	Zea mays	536–740		
	ssp. *mays* NB	570	AY506529/NC_007982	(Clifton et al. 2004)
	ssp. *mays* NA	701	DQ490952	(Allen et al. 2007)
	ssp. *mays* CMS-C	740	DQ645536	
	ssp. *mays* CMS-S	557	DQ490951	
	ssp. *mays* CMS-T	536	DQ490953	
	ssp. *parviglumis*	681	DQ645539	Allen et al.,
	ssp. *perennis*	570	DQ645538	unpublished
	Zea luxurians	539	DQ645537	
	Tripsacum dactyloides	704	DQ984517	
Eudicotyledons/core eudicotyledons				
Caryophyllales				
Caryophyllaceae	*Silene latifolia*	253	HM562727/NC_014487	(Sloan et al. 2010a)
Amaranthaceae	*Beta vulgaris*	369–501		
	ssp. *vulgaris* TK81-0	369	BA000009/NC_002511	(Kubo et al. 2000)
	ssp. *vulgaris* TK81-MS	501	BA000024	(Satoh et al. 2004)
	ssp. *maritima* A	365	FP885845/NC_015099	(Darracq et al. 2011)
	ssp. *maritima* B	368	FP885834	
	ssp. *maritima* G	269	FP885871	
Asterids/Lamiids				
Lamiales	*Boea hygrometrica*	510	JN107814	(Zhang et al. 2012)
Solanales	*Nicotiana tabacum*	431	BA000042/NC_006581	(Sugiyama et al. 2005)
Rosids				
Vitales	*Vitis vinifera*	773	FM179380/NC_012119	(Goremykin et al. 2009)
Fabids				
Fabales	*Vigna radiata*	401	HM367685/NC_015121	(Alverson et al. 2011)
Malpighiales	*Ricinus communis*	503	HQ874649/NC_015141	(Rivarola et al. 2011)
Cucurbitales	*Citrullus lanatus*	379	GQ856147/NC_014043	(Alverson et al. 2010)
	Cucurbita pepo	983	GQ856148/NC_014050	(Alverson et al. 2010)
	Cucumis melo	>2,700	JF412792	(Rodríguez-Moreno et al. 2011)
	Cucumis sativus	>1,600	HQ860792	(Alverson et al. 2011)
Malvids/Brassicales				
Caricaceae	*Carica papaya*	477	EU431224/NC_012116	unpublished
Brassicaceae	*Arabidopsis thaliana*	367	Y08501/NC_001284	(Unseld et al. 1997)
	Brassica napus	222	AP006444/NC_008285	(Handa 2003)

many as 15 years until the second chondrome sequence of a non-angiosperm plant became available, the one of the "model moss" *Physcomitrella patens* (Terasawa et al. 2007). Only very recently, the first complete mtDNA sequences (see Table 8.1) of a gymnosperm, *Cycas taitungensis* (Chaw et al. 2008), a hornwort, *Megaceros aenigmaticus* (Li et al. 2009), and a lycophyte, *Isoetes engelmannii* (Grewe et al. 2009), became available. Somewhat earlier, the chondrome sequences of the charophyceaen algae *Chara vulgaris* and *Chaetosphaeridium globosum* had been determined (Turmel et al. 2002a, 2003), which are among those algal taxa being discussed most closely related to land plants and thus provide well-suited outgroup taxa for rooting the land plant (embryophyte) phylogeny. Chapters 3 and 6 of this volume deal with algal chloroplast and mitochondrial genomes in detail.

Molecular phylogenetic studies (e.g. Qiu et al. 2006), occasionally based on individual mitochondrial loci with wide taxon sampling, have meanwhile established a well-supported overall plant phylogeny (see Fig. 8.1b). This modern concept rejects a monophyly of bryophytes (liverworts, mosses and hornworts) and also confidently defines the relationships of non-seed vascular plants. The clade of monilophytes comprising true ferns, horsetails and whisk ferns (Pryer et al. 2001), is well confirmed and sister to the seed plants (spermatophytes) comprising angiosperms and gymnosperms. Jointly, the spermatophytes and monilophytes constitute the clade of euphyllophytes. The euphyllophytes are the sister clade of lycophytes (comprising club mosses, spike mosses and quillworts), which represent the most ancient surviving lineage of vascular plants (tracheophytes). Hornworts are now considered to be the sister clade of tracheophytes, among other evidences supported by mitochondrial intron patterns (Groth-Malonek et al. 2005), as I will discuss below. No formal name has as yet been suggested for the joint hornwort + tracheophyte clade (provisionally designated the HT clade), mainly because a representative, name-giving morphological synapomorphy yet remains to be identified (although hornworts are, like tracheophytes, characterized by an enduring diploid sporophyte developmental phase). Mosses are the sister group to the HT clade and jointly this is the "Non-Liverwort Embryophyte" (NLE) lineage, sister to the liverworts. Hence, this phylogenetic topology places the root of land plant evolution between liverworts and all other plants in the NLE clade (Fig. 8.1b). Among the major land plant groups, a full chondrome sequence is still lacking for the monilophyte (i.e., ferns sensu lato) clade.

III. Evolving Gene Complements in Seed Plant Chondromes

A. The Protein-Coding Gene Complement Affected by Endosymbiotic Gene Transfer

The chondrome of the liverwort *Marchantia polymorpha* (Oda et al. 1992b) turned out be about 11 times as large as the one of humans (Anderson et al. 1981). This is in part explained by the presence of several genes that are never encoded in animal mtDNAs as well as many introns disrupting coding regions, which I will discuss in the following section. With 73 genes encoded in the chondrome of *Marchantia* and the recently sequenced chondrome of the rather distantly related liverwort species *Pleurozia purpurea* (Wang et al. 2009), the liverworts hold the record for mitochondrial gene complements among land plants. The land plant chondrome gene complement includes genes for rRNAs, tRNAs, subunits of the RCCs I (*nad* genes), II (*sdh*), III (*cob*), IV (*cox*) and V (*atp*), subunits of a cytochrome c maturation pathway (*ccm*), ribosomal proteins (*rpl* and *rps*) and a subunit of a twin-arginine translocase (*tat*C).

One glaring exception in the liverworts is that *nad*7, a "core" RCC I subunit gene, is degenerated into a pseudogene and was functionally established as a nuclear gene after endosymbiotic gene transfer (Kobayashi et al. 1997). This is a particular surprising case, because although *nad*7 is degenerated, it is retained as a pseudogene in both marchantiid (like *Marchantia*) and jungermanniid (like *Pleurozia*) liverwort chondromes. However, a functional *nad7* gene still exists in the mtDNA of *Haplomitrium*, which represents the most ancient (and phylogenetically somewhat isolated) liverwort lineage, sister to the two large clades of marchantiid and jungermanniid liverworts (Groth-Malonek et al. 2007b).

Aside from an apparent loss of *rpl14* (conserved in algal chondromes) in the land plant stem lineage (Node O in Fig. 8.1b) or the later loss of *rps8* as a likely synapomorphy in the NLE stem lineage (Node A in Fig. 8.1b), *nad7* in the liverworts may actually represent the most ancient case of functional endosymbiotic gene transfer (EGT) in land plant evolution. Another example for the transfer of a gene encoding an RCC core component had been described earlier for the *cox*2 gene (encoding a subunit of cytochrome c oxidase, RCC IV) in the "legumes" (Fabaceae; Adams et al. 1999; Daley et al. 2002;

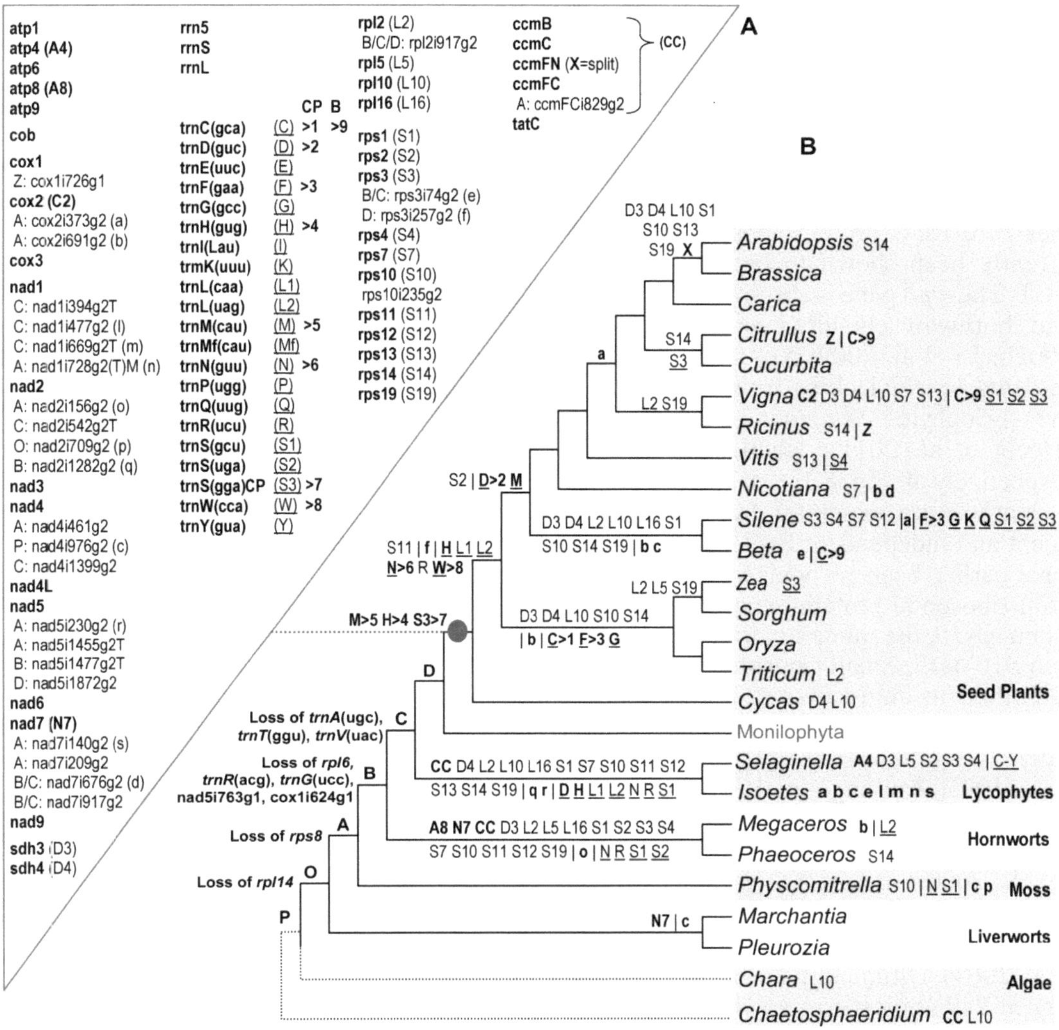

Fig. 8.1. (**a**) The ancestral gene complement of seed plant mtDNAs most likely contained genes for 20 protein subunits (first column) of respiratory chain complexes I (*nad*), II (*sdh*), III (*cob*), IV (*cox*) and V (*atp*), for 3 rRNAs and 20 tRNAs (*rrn* and *trn* genes, second column), for 15 proteins of the large and small ribosomal subunit (*rpl* and *rps*, third column) and for five genes involved in cytochrome c maturation (*ccm*) and a twin-arginine translocase subunit (*tatC, fourth column*). Introns are listed below the respective gene following a recently proposed nomenclature (Dombrovska and Qiu *2004*; Knoop *2004*). Capital letters indicate intron origins along the backbone nodes of plant phylogeny as depicted in the cladogram in B (O: Origin of embryophytes; A: NLE clade; B: HT clade; C: Tracheophytes; D: Euphyllophytes; P: a possible clade uniting the alga *Chara* and the embryophytes). The letter "X" indicates a split of ccmFN in the Brassicaceae and "Z" indicates independent gains of the "rampant invader" group I intron cox1i726g1 via HGT (in *Citrullus and Ricinus*). Letters in parentheses behind genes and introns denote losses from chondromes (small letters for introns, underlined capital letters for *trn* genes). Gain of chloroplast tRNA genes (CP) are indicated with ">" followed by an arbitrary number as given in the subsequent column. The source of the *trnC* xenologue "C>9" (in *Beta, Citrullus* and *Vigna*), which is similar to bacterial (B) homologues, is as yet unidentified. (**b**) A cladogram based on a modern understanding of plant phylogeny is shown. It includes all embryophyte taxa, for which complete mtDNA sequences have been determined. For simplicity, only genus names and only one terminal branch is shown, where multiple chondrome sequences of very closely related taxa have been determined (*Beta, Oryza, Triticum, Zea*). Losses of protein-coding genes, introns and tRNA genes, the functional replacement of lost tRNA genes, the disruption of *ccmFN* (X) and the gain of intron cox1i726g1 (Z) are indicated at the backbone nodes, along the branches or after the genus names. Different types of sequence losses and acquisitions are separated by vertical lines. The labels are used as in A with rare events highlighted in *bold*.

Nugent and Palmer 1991). Like *nad7* in the liverworts, the *cox2* EGT in Fabaceae could actually be a unique one-time event in plant evolution, among the completely sequenced angiosperm chondromes only represented by the mung bean *Vigna radiata* (Alverson et al. 2011; labelled C2 in Fig. 8.1). Only two further core RCC protein subunit genes have recently been shown to be subject to rare EGT: The *atp8* gene is absent from the *Allium* and hornwort chondromes (Adams et al. 2002b; Li et al. 2009; Xue et al. 2010) and the *atp4* gene is lacking from the mtDNA of the lycophyte *Selaginella moellendorffii* (Hecht et al. 2011); labelled A8 and A4, respectively, in Fig. 8.1.

Many more examples for ongoing, frequent and independent EGT in angiosperms have earlier been recognized for mitochondrial ribosomal protein genes. For example, when *rps10*, the mitochondrial gene for protein S10 of the small ribosomal subunit, was identified in angiosperms, it was immediately obvious that it occurred only sporadically among flowering plant mtDNAs (Knoop et al. 1995; Zanlungo et al. 1994). A subsequent study showed numerous independent EGTs of *rps10* among angiosperms (Adams et al. 2000) and a very similar picture emerged for the *sdh* genes encoding RCC II (succinate dehydrogenase) subunits (Adams et al. 2001). Ultimately, a comprehensive survey of 280 angiosperm genera revealed that all *rpl* and *rps* as well as the *sdh* genes are frequently and independently lost from mtDNAs. Interestingly, this is not the case for the *ccm* genes and the other "core" RCC genes encoding subunits of complexes I, III, IV and V (Adams et al. 2002b). Between 42 and 6 independent gene losses were observed and ranking genes according to the number of independent EGT events results in the following order: *rps7* > *rpl2* > *sdh3* > *rps19* > *rps1* > *rps13* > *rps14* > *rps10* > *sdh4* > *rpl5* > *rpl16* > *rps11* > *rps2* > *rps3* > *rps4* > *rps12*. Not included in the taxonomically wide survey at that time was *rpl10,* previously annotated as open reading frames (ORFs) of varying sizes (in different species), which were only recently recognized as encoding ribosomal protein L10 (Kubo and Arimura 2010; Mower and Bonen 2009). The frequent losses of *rpl, rps* and *sdh* gens among flowering plants are reflected in the current sampling of complete angiosperm chondromes and also supported by independent losses in the hornwort and lycophyte lineages (Fig. 8.1b). Also, these genes were frequently lost from algal mitochondrial genomes (Chap. 3) and even from mitochondrial DNAs in eukaryotes at large (V.K., unpublished). Together with the complete mtDNA sequence of the gymnosperm *Cycas taitungensis* (Chaw et al. 2008), the available data now allow us to reasonably hypothesize on the likely mitochondrial gene complement of the last common ancestor of the seed plant stem lineage (Fig. 8.1a).

A more puzzling picture emerges for the three genes involved in cytochrome c maturation – *ccmB, ccmC* and *ccmF* – which are highly conserved among angiosperms and were never found missing from their chondromes in the above-mentioned large-scale survey (Adams et al. 2002b). The *ccmF* reading frame is originally continuous in protists and in the alga *Chara vulgaris,* but disrupted into separate ORFs covering the N- and C-terminal parts (*ccmFN* and *ccmFC*) in land plants. A subsequent second disruption of *ccmFN* (labelled X in Fig. 8.1) has occurred later in evolution in the Brassicaceae. In striking contrast to the survey on angiosperms, the entire suite of *ccm* genes is lost surprisingly often and several times independently from the chondromes of protists (V.K., unpublished), algae (including the Charophyceae *Chaetosphaeridium globosum*), hornworts and lycophytes (Fig. 8.1b). Possibly, alternative "backup" pathways of cytochrome c biogenesis (Allen et al. 2008; Giegé et al. 2008) have been retained in these basal lineages, but not in the seed plants, which could allow for loss of the *ccm* pathway. Perhaps more likely, unrecognized functional adaptations of Ccm protein assembly or interactions in seed plant mitochondria preclude the re-targeting of nuclear-encoded *ccmB, ccmC* and *ccmF* gene products after EGT, similar to the core

RCC subunits that are universally retained in the chondrome.

Altogether, the *rpl, rps* and *sdh* genes and, to a lesser extent the *ccm* genes in more ancient clades, appear to be subject to frequent, independent EGTs, whereas only three deep gene losses – of *rpl14*, *rps8* and *rpl6* – seem to be singular synapomorphies along the backbone of plant phylogeny, together with five losses of tRNA genes (Fig. 8.1b).

It should be noted that the loss of a gene from the mitochondrial compartment does not necessarily imply its functional transfer to the nucleus. Takeover of the homologous chloroplast gene after its transfer into the nuclear genome has occurred for several ribosomal protein genes, including *rps8, rps13* and *rpl10* (Adams et al. 2002a; Kubo and Arimura 2010; Mower and Bonen 2009). Such gene substitutions can also occur in the opposite direction: the chloroplast ribosomal protein S16 is provided by a nuclear gene of mitochondrial origin (Ueda et al. 2008).

B. The tRNA Gene Complement Affected by Loss and Replacement

Contrary to the EGT of protein-coding genes, the loss of tRNA genes from mitochondrial genomes is not accompanied by a corresponding establishment of gene copies in the nuclear genomes, followed by subsequent re-import of the gene product. Instead, tRNA genes lost from mitochondrial genomes are generally replaced by import of their cytosolic counterparts (Duchêne et al. 2009). The import of tRNAs into plant mitochondria has been studied biochemically, mostly using wheat, *Triticum aestivum* (Glover et al. 2001), or potato, *Solanum tuberosum* (Delage et al. 2003), as model systems. For example, tRNA-Gly(UCC) is known to be imported from the cytosol (Brubacher-Kauffmann et al. 1999), obviously to compensate for the loss of the mitochondrial *trnG*(ucc) gene, which may date back to the stem lineage of the HT clade (Fig. 8.1b, node labelled B). In contrast, a mitochondrial *trnG*(gcc) gene encoding the isoacceptor tRNA-Gly(GCC) addressing GGY glycine codons is present in most plants but not in monocots and white campion (*Silene latifolia,* Caryophyllaceae) or the spikemoss *Selaginella moellendorffii*, respectively, which both represent interesting cases of massive tRNA gene loss from plant chondromes (Hecht et al. 2011; Sloan et al. 2010a). These independent losses of *trnG*(gcc) may be facilitated by functional replacement through superwobbling of the (imported) tRNA-Gly(UCC), which can read all GGN glycine codons, as has recently been shown for the tRNA-Gly pair of chloroplasts (Rogalski et al. 2008).

Seed plant evolution, however, has brought about an alternative to mitochondrial tRNA import from the cytosol to compensate for loss of original native mitochondrial tRNAs. Chloroplast tRNA gene copies have been integrated into seed plant chondromes and can complement their original bona fide mitochondrial counterparts. One must assume that such a newly acquired chloroplast tRNA gene co-exists with its native mitochondrial counterpart in an evolutionary transition phase allowing for gradual functional take-over. This indeed appears to be the case for *trnM, trnH* and *trnS(gga)* in the chondrome of the gymnosperm *Cycas taitungensis.* Taken together with data from the other complete mtDNA sequences, this suggests that the gain of these three chloroplast xenologues were the first events of this kind in the seed plant lineage (labelled H>4, M>5, S3>7 in Fig. 8.1). The native mitochondrial *trnH* gene was subsequently lost early in angiosperm evolution, whereas *trnM* has continued to co-exist with its chloroplast homologue in monocots and was lost only later in the eudicot clade (Fig. 8.1b). Similar to *trnH,* the native mitochondrial tRNA genes *trnL*(caa), *trnL*(uag) and *trnR*(ucu) appear to be lost in the angiosperm stem lineage. These three losses, however, are not accompanied by replacement with chloroplast homologues. Finally, two further complete tRNA gene replacements by chloroplast homologues (without the native mitochondrial genes remaining present) have taken place for *trnN* and *trnW* along the (phylogenetically long) branch to extant angiosperms (labelled N>6, W>8 in Fig. 8.1). It will be interesting to see whether any of the six native mitochondrial genes apparently lost early in angiosperm evolution (*trnH, trnL*(caa), *trnL*(uag), *trnN, trnR* and *trnW*) will show up in the future in the chondrome sequence of a basal angiosperm lineage predating the monocot-eudicot split, such as the *Amborella trichopoda* mtDNA currently being sequenced (J. Palmer, personal communication). Later in angiosperm evolution, the replacement of the original mitochondrial *trnD* gene by its chloroplast homologue appears to be a eudicot-specific event. The *trnC* and *trnF* replacements (labelled C>1, D>2, F>3 in Fig. 8.1b) as well as the unsubstituted loss of *trnG*(gcc) may well turn out to be synapomorphies of the monocot clade.

The case of the *trnC* genes encoding tRNAs for cysteine is particularly interesting in that it may represent a different type of gene replacement in angiosperm chondromes. Initially observed for *Beta vulgaris* (Kubo et al. 2000), a novel *trn*C gene (labelled C>9 in Fig. 8.1) has replaced the native copy in the sugarbeet mtDNA. Interestingly, this gene also co-exists as a second paralogue copy with the original mitochondrial *trnC* gene in *Vigna radiata* and *Citrullus lanatus.* Surprisingly, this novel and sporadically occurring *trnC* is most closely related to bacterial (Chloroflexi) homologues (V.K., unpublished observation). Possibly, this may be a particularly intriguing case of horizontal gene transfer involving a prokaryotic source organism and sporadically affecting distantly related angiosperms, similar to the case of group I intron cox1i726g1 outlined below.

IV. Plant Mitochondrial Intron Stasis and Dynamics

One of the striking outcomes of mitochondrial DNA studies in different land plant clades is the generally high conservation of introns within plant clades (including the liverworts as the presumably most ancient embryophytes), but the strikingly different patterns of intron occurrence in different plant clades. The latter in particular contrasts the much more widely conserved introns in embryophyte chloroplast genomes (see Chap. 5). In the superset of 74 mitochondrial introns now identified in total in bryophyte chondromes, several introns are differentially shared between two of the three bryophyte classes, but not a single one is universally shared between liverworts, mosses and hornworts (Knoop 2010).

A nomenclature has been proposed for naming of organelle introns, which uses the name of the gene in question, the upstream nucleotide position in the continuous reading frame (using the *Marchantia polymorpha* homologue as a reference) and the notation g1/g2 to indicate a group I or group II intron (Dombrovska and Qiu 2004; Knoop 2004), which I will use here to denote intron orthologues.

A. Mitochondrial Intron Conservation Within Plant Clades

Several phylogenetic studies employing wide taxon sampling have shown that plant mitochondrial introns are widely conserved within ancient plant clades, such as introns nad5i753g1 and cox1i624g1 in liverworts and mosses (Beckert et al. 1999; Volkmar and Knoop 2010), nad2i156g2 and cobi420g1 among mosses (Beckert et al. 2001; Wahrmund et al. 2010) or nad4i548 exclusively among liverworts (Volkmar et al. 2011). All of these introns were found nearly universally conserved within the respective bryophyte clades, with only very rare exceptions (atp1i989g2 and atp1i1050g2 in *Treubia lacunosa* and nad5i753 in *Takakia ceratphylla*) indicating secondary losses (Knoop 2010). Evidence for the stability of mitochondrial introns in the early plant clades now also comes from the full chondrome sequence of the liverwort *Pleurozia purpurea* to complement the one of *Marchantia polymorpha* (Wang et al. 2009). Of 32 introns in the *Marchantia* mtDNA (25 group II, 7 group I), all but one are conserved in *Pleurozia,* which only lacks rrnSi1065g2 in the small subunit ribosomal RNA gene. The extraordinary degree of intron conservation in this ancient plant clade even includes the two introns in *nad7,* which are surprisingly retained in the *nad7* pseudogene present in marchantiid and jungermanniid liverworts (Groth-Malonek et al. 2007b). Intron variability is slightly larger among hornworts, where four of 34 introns are missing in *Megaceros aenigmaticus* compared to *Phaeoceros laevis* (Li et al. 2009; Xue et al. 2010). Similarly, a somewhat higher variability of mitochondrial introns is seen in the two lycophyte mtDNA genomes recently determined (Grewe et al. 2009; Hecht et al. 2011). With 37 introns in total, *Selaginella moellendorffii* has the intron-richest plant chondrome identified so far. Two of its three group I introns and 24 of its 34 group II introns are conserved in *Isoetes engelmannii.* An exhaustive view is currently missing for monilophyte mtDNAs, due to the lack of a complete fern chondrome sequence. However, two introns – atp1i361g2 occurring exclusively in monilophytes and nad5i1242g2 shared only with lycophytes – have been investigated for phylogenetic purposes and independent secondary losses have become apparent in the fern lineage (Vangerow et al. 1999; Wikström and Pryer 2005). I will not elaborate here further on non-seed plant introns, given that the previous chapter (Chap. 7) is also devoted to this issue, but will rather concentrate on spermatophytes in the following.

The *Cycas taitungensis* mtDNA sequence (Chaw et al. 2008), as the first gymnosperm addition to the set of sequenced spermatophyte chondromes, harbours all angiosperm mitochondrial introns hitherto identified as conserved in seed plants at large. Mitochondrial intron conservation between angiosperms and the gymnosperm *Cycas* also includes the five *trans*-arranged group II introns in three *nad* genes (*nad1*, *nad2* and *nad5*) that trace back to cis-spliced ancestors in early plant evolution (Groth-Malonek et al. 2005; Malek et al. 1997; Malek and Knoop 1998). Among the completely sequenced angiosperm chondromes, the full set of 25 mitochondrial introns (Fig. 8.1) is present in the *Vitis vinifera* mtDNA. Other angiosperm mtDNAs show occasional rare secondary losses of introns nad4i976g2, nad7i676g2, rps3i74 (Labels c, f, g in Fig. 8.1) and more frequent independent losses of cox2i373g2 and cox2i691g2 (Labels a, b in Fig. 8.1). Loss of nad4i976g2 has been investigated as a phylogenetic marker in Caryophyllales (Itchoda et al. 2002) and nad1i477g2 has additionally been reported to be lost in the Geraniaceae (Bakker et al. 2000). *Silene latifolia* has the intron-poorest among the fully sequenced angiosperm chondromes, due to lack of three of the above introns and the absence of three ribosomal protein genes (*rpl2, rps3* and *rps10*), which carry conserved introns in angiosperms mtDNAs (Fig. 8.1b).

Taken together, the mitochondrial intron history in seed plants is largely explained by ancient group II intron gains (and their disruptions in five cases; see also below), prior to diversification of seed plants and some later secondary losses. One glaring exception, however, concerns the only known example of a seed plant mitochondrial group I intron, originally identified in the *cox1* gene of *Peperomia polybotrya* (Vaughn et al. 1995). This particular group I intron (cox1i726g1) apparently originates from a fungal donor and seems to have been acquired several times independently in angiosperm evolution as a "rampant invader" of the *cox1* gene (Adams et al. 1998a; Cho et al. 1998; Cho and Palmer 1999; Sanchez-Puerta et al. 2008; Seif et al. 2005). The actual extent of independent primary acquisitions by HGT from fungi or between flowering plants vs. independent later losses of cox1i726g1 has been questioned, however, and needs further investigation (Cusimano et al. 2008); see also Chap. 10 in this volume. Among the fully sequenced plant chondromes, cox1i726g1 is present in the mtDNAs of *Citrullus lanatus* and *Ricinus communis* (Label Z in Fig. 8.1). Interestingly, an endonuclease ORF, otherwise frequently found in mobile group I introns, is only sporadically present in cox1i726g1. In contrast, all three endonuclease reading frames in the *Marchantia polymorpha* chondrome are embedded in group I introns disrupting the *cox1* gene in other locations (cox1i395g1, cox1i730g1 and cox1i1116g1). Mysteriously, *cox1* in particular seems to be a prime target for group I intron invasion with a total of 11 different group I intron insertion sites identified among charophyceaen algae, bryophytes and lycophytes.

B. Intron Gains and Losses Along the Backbone of Plant Phylogeny

A "gymnosperm-specific" mitochondrial intron, rps3i257g2, was found secondarily lost in some gymnosperms (Ran et al. 2010; Regina et al. 2005; Regina and Quagliariello 2010) but is present in the *rps3* gene of *Cycas taitungensis*. This adds one intron to the set of 25 introns conserved between the gymnosperm and angiosperms (not considering the additional intron cp-trnVi39g2 in the *trnV* gene as part of a promiscuous chloroplast insert in the cycad's mtDNA). Gain of intron rps3i257g2 could alternatively be a synapomorphy of gymnosperms. However, is the intron was recently identified in the mtDNA of the fern *Gleichenia dicarpa* (F. Grewe and V.K, unpublished observation), making its early gain and later secondary loss in the angiosperm clade more likely. Like rps3i257g2, none of the other 25 mitochondrial introns conserved among seed plants is specific for this clade – all 26 spermatophyte mitochondrial group II introns appear to be early evolutionary gains along the backbone of early non-seed plant evolution more than 300 million years ago (indicated by capital letters A, B, C, D in Fig. 8.1). The two spermatophyte introns of the *cox2* gene, for example, can be traced back down to the common ancestor with the mosses (node A in Fig. 8.1). Secondary losses have been found for cox2i691g2 in *Megaceros* and for both *cox2* introns in *Isoetes* (labels a, b in Fig. 8.1). Both *cox2* introns, however, exist in *Selaginella*, where coxi373g2 is uniquely found in a trans-splicing arrangement (Hecht et al. 2011, see below). The origins of intron gains remain somewhat unclear only in those few cases, where the respective gene is entirely lacking from the mtDNA in early clades (*nad7* and *rps3* in the hornworts, *rpl2* in hornworts and lycophytes). Similar to the two *cox2* gene introns, the single maturase-containing (label M in Fig. 8.1a) mitochondrial intron nad1i728 of seed plants traces back to the common ancestor with mosses in the NLE lineage (Qiu et al. 1998); node A in Fig. 8.1b. The three other introns in *nad1* were obviously gained in the earliest tracheophytes (label C in Fig. 8.1), since all of them are shared with *Selaginella moellendorffii* in cis-splicing arrangements (however, with all but the first one secondarily lost in *Isoetes engelmannii*).

C. Maturases and cis-to-trans Conversions in Mitochondrial Introns

Group II intron nad1i728 is not only interesting as the only mitochondrial intron carrying a maturase reading frame (mat-r) that is highly conserved among seed plants, but is also unique under two further aspects. Firstly, intron nad1i728g2 is conserved in mosses, hornworts and the lycophyte *Selaginella moellendorffii* (Dombrovska and Qiu 2004; Hecht et al. 2011), but the (functional) maturase ORF has been lost (several times independently) in all non-seed plant taxa. However, extensive homologies with the mat-r reading frame disrupted by frame shifts are readily detectable in nad7i28g2 of hornworts and two *Takakia* species, probably representing the most basal-branching extant moss genus. Interestingly, mat-r is entirely lost from nad1i728g2 in the *Selaginella moellendorffii* chondrome, where we now found the first example of a bona fide mitochondrial gene, *nad4L,* inserted into nad1i728g2 (Hecht et al. 2011). Notably, other maturases in the mitochondrial genomes of the liverwort *Marchantia polymorpha* (nine maturase ORFs) or the moss *Physcomitrella patens* (two maturase ORFs) are not particularly closely related to mat-r in nad1i728g2 of seed plants. Yet more importantly, no traces of nuclear-encoded maturases are found in the genomes of *Physcomitrella patens* (Rensing et al. 2008) or *Selaginella moellendorffii* (Banks et al. 2011). In contrast, four maturases are encoded in the *Arabidopsis thaliana* nuclear genome and involved in the splicing of different sets of mitochondrial introns (Keren et al. 2009). Overall, maturases seem to be on the way out in NLE chondrome evolution, but the evolutionary origin of the four nuclear maturases in angiosperms like *Arabidopsis* and the nature of possible alternative proteinaceous splicing factors in *Physcomitrella* or *Selaginella* remains mysterious at present.

Secondly, nad1i728g2 it is the only clearly documented example with multiple, independent transitions from cis- to trans-splicing among flowering plants. The nad7i728g2 intron can get disrupted either 5′ or 3′ of its maturase reading frame and at least ten such independent disruption events have been found for the upstream and at least five for the downstream breakage among angiosperms (Qiu and Palmer 2004). It is certainly tempting to speculate that Mat-r may aid in the transition from cis- to trans-splicing. The numerous independent cis-to-trans conversions of nad1i728g2 contrast the single-event disruptions leading to the five trans-splicing group II introns, which appear to be universally conserved among seed plants. These introns originated early in plant phylogeny (Groth-Malonek et al. 2005; Malek et al. 1997; Malek and Knoop 1998) and all five (nad1i394g2, nad1i669g2, nad2i542g2, nad5i1455g2 and nad5i1477g2) are present as cis-splicing orthologues in *Selaginella moellendorffii* (Hecht et al. 2011) and ferns (our unpublished observations), which suggests their disruption early in the spermatophyte stem lineage. It is interesting to note, however, that four other mitochondrial introns have acquired trans-splicing status in *Selaginella moellendorffii* (atp9i21g2, cobi787g2, cox2i373g2 and cox1i1305g1), including the trans-splicing group I intron previously identified as the first example of its kind in *cox1* of *Isoetes engelmannii* (Grewe et al. 2009). Of these, only cox2i373g2 has been reported as trans-splicing in onion (Kim and Yoon 2010), obviously resulting from an independent cis-to-trans conversion in plant evolution.

Significantly, *Isoetes* has no trans-splicing introns other than cox1i1305g1 and the average cis-splicing intron sizes are five times smaller than in *Selaginella*. Hence, disruption of mitochondrial introns into trans-splicing configurations largely seems to be an (irreversible) chance process, solely dependent on recombination hitting sufficiently large intron sequences at splicing-compatible sites. These two factors (size expansion and recombinational activity) may have increased the chances for evolution of trans-splicing introns in tracheophyte mitochondria, where a total of ten examples (9 group II, 1 group I) are now known. However, examples of trans-splicing group II introns are also known for the mitochondrial *nad3* gene of the alga *Mesostigma viride* (Turmel et al. 2002b), the chloroplast *psaA* gene in *Chlamydomonas reinhardtii* (Choquet et al. 1988), three chloroplast genes (*psaC, petD* and *rbcL*) in the alga *Floydiella terrestris* (Brouard et al. 2010) and notably the chloroplast *rps12* gene, where trans-splicing probably goes back to an ancient gene disruption in the land plant lineage (Hildebrand et al. 1988; Kohchi et al. 1988). Interestingly, trans-splicing group I introns unrelated to the ones in the lycophytes have also been found in the *cox1* genes of the entomoparasitic alga *Helicosporidium sp.* (Pombert and Keeling 2010) and in the primitive metazoan *Trichoplax adhaerens* (Burger et al. 2009).

V. Evolving Structural Complexity in Plant Chondromes

The independent gene losses from chondromes in the plant lineage may suggest that the extent of coding sequences may vary significantly. However, this is actually not the case, mainly because most genes affected by EGT encode rather small proteins and the tRNA genes lost from the chondrome are particularly small. Without introns and with compact intergenic regions, ca. 40 Kbp of

DNA sequence could essentially be sufficient to accommodate the coding regions, even in plant mtDNAs which have not experienced extensive gene losses.

A. Moderate Early Structural Chondrome Evolution in Bryophytes

Early plant mitochondrial genome evolution has largely seen a gain in the size of intergenic regions and a differential gain of introns as discussed above. Comparing the chondromes of mosses, liverworts and hornworts, extensive gene syntenies retaining ancestral gene arrangements are identified (Li et al. 2009; Wang et al. 2009; Xue et al. 2010). The simple circular-mapping genomes of the bryophytes lack co-existing alternative gene arrangements. However, rare recombinational activity has left recognizable traces on evolutionary timescales. The *Pleurozia purpurea* mtDNA (Wang et al. 2009) carries four sequence repeats of sizes between 187 and 660 bp (see Chap. 7). One of these has been studied over a wider liverwort sampling: A large portion of group II intron cobi783g2 was apparently copied into the intergenic region between *nad5* and *nad4* (Groth-Malonek et al. 2007a). A retro-splicing and transposition event is suggested from the precise end of sequence similarity precisely coinciding with the upstream splice site. However, the intron sequence is mysteriously inverted relative to the direction of transcription in the *nad5-nad4-nad2* gene cluster. The molecular evolution of one further chondrome region in liverworts has been studied in detail, the *trnA-trnT-nad7* cluster. Here, an inversion of *trnT* in the intergenic *trnA-nad7* region and independent losses of *trnT* from the chondromes were observed during diversification of the liverworts (Wahrmund et al. 2008). However, such genomic changes seem to be rare exceptions rather than the rule in mitochondrial DNA evolution in the early plant clades such as liverworts and mosses (Yin-Long Qiu, pers. comm. and see previous Chap. 7).

B. Origins of Plant Chondrome Complexity Predate Seed Plant Age

The mtDNA of the liverwort *Marchantia polymorpha* maps as a simple, circular and non-recombining genome (Oda et al. 1992a), thus starkly contrasting the high degree of recombinational activity that was already well documented for flowering plant chondromes at that time (Brennicke et al. 1985; Brennicke and Blanz 1982; Manna and Brennicke 1986; Palmer and Shields 1984; Schuster and Brennicke 1987a; Stern and Palmer 1984). Complex mtDNA structures created through frequent recombination in repeated sequences and thus leading to multipartite chondromes clearly appear to be the rule among flowering plants. Depending on the numbers (and orientation) of recombinationally active repeat sequences, plant mitochondrial genomes range from the simple tripartite structures of spinach (Stern and Palmer 1986) or turnip (Palmer and Shields 1984) to highly complex multipartite ones, as for example characterized in maize, tobacco or wheat (Allen et al. 2007; Lonsdale et al. 1984; Ogihara et al. 2005; Sugiyama et al. 2005). With its two pairs of repeated sequences, the mitochondrial genome of the model angiosperm *Arabidopsis thaliana* as the first completely sequenced flowering plant mtDNA is at the lower end of recombinational complexity (Klein et al. 1994; Unseld et al. 1997). Angiosperm mtDNAs are mostly displayed in the form of a so-called, and often entirely hypothetical, "master-circle" comprising the full chondrome sequence complexity including all repeated sequences in one large circular molecule and potentially giving rise to co-existing subgenomic structures via recombination. In contrast, unicircular non-recombining chondromes such as the one of *Brassica hirta* (Palmer and Herbon 1987) – structurally similar to the ones of *Marchantia* or the charophyceaean green algae – seem to be rare exceptions and secondary re-simplifications of angiosperm chondrome structures.

Significantly complicating the outcome of plant mtDNA recombination, the recombination event can be followed by shifts in the stoichiometries of the recombination products (Kmiec et al. 2006; Small et al. 1989; Woloszynska 2010). Hence, one product of DNA recombination may become dominant in stoichiometry over another, which may persist at a low level or vanish altogether. Small circular, supercoiled molecules presumably resulting from such processes had been reported very early for several plant taxa (Brennicke and Blanz 1982; Dale 1981; Dale et al. 1983).

Like in *Marchantia*, simple circular genomes of comparable sizes have been identified in all other bryophytes as well (Table 8.1). It should be noted, however, that it remains questionable whether such circular DNA genomes truly exist in vivo or whether other physical forms, such as overlapping linear or branched DNAs, might prevail in mitochondria and possibly even in chloroplasts (Bendich 1993; Bendich 2007; Oldenburg and Bendich 1998, 2001; Yamato et al. 1992).

Completely contrasting the circular mapping chondromes of bryophytes, recombination events in lycophytes are so numerous that creation of a potential "master-circle" encompassing the full chondrome complexity seemed futile. More than 20 specific recombination breakpoints each have been identified in the mtDNAs of *Isoetes engelmannii* and *Selaginella moellendorffii,* which led to creation of network-like maps linking single-copy sequence islands across recombination breakpoints and repeated sequences (Grewe et al. 2009; Hecht et al. 2011). Long sequence repeats of up to more than 7 Kbp in *Selaginella* are strongly reminiscent of the recombinationally active repeated sequences in flowering plant chondromes and suggest the origin of frequent chondrome recombination producing multipartite structures to lie in the tracheophyte stem lineage. A notable feature of the large sequence repeats in *Selaginella moellendorffii* are numerous microsatellite repeat motifs, which vary in copy number between repeat environments (Hecht et al. 2011).

C. Nuclear-Encoded Proteins Determine Plant Chondrome Recombination

Large sequence repeats extending over several Kbp in the flowering plant chondromes obviously mediate reversible homologous recombination events, which predominantly create alternative sequence arrangements co-existing in (near-) equilibrium. Shorter sequence motifs (< ca. 500 bp), in contrast, seem to be the substrates for rare recombination events, which create sequence arrangements that appear sub-stoichiometrically (Arrieta-Montiel and Mackenzie 2011). In extreme cases, such "sublimons" exist at very low amounts that go nearly unnoticed in gel electrophoresis (as "ghost bands"), because they are covered up by the dominating chondrome arrangements. Most important is the observation that such sublimons can experience substoichiometric shifting (Small et al. 1989). Several nuclear-encoded factors have now been recognized that control recombination events in plant mitochondrial genomes, mainly by suppressing recombination on short sequence stretches. The *Arabidopsis thaliana* MSH1 gene (named so as a homologue of the bacterial MutS and earlier described as *chm* for chloroplast mutator), in particular, results in dramatic alterations in mtDNA conformation upon gene inactivation (Abdelnoor et al. 2003; Arrieta-Montiel et al. 2009). Other proteins found to be involved in mtDNA maintenance are OSB1, the "organellar single-stranded DNA-binding protein" and *RecA* homologues targeted to mitochondria. Double knockouts of MSH1 and RECA3 in *Arabidopsis* show particularly significant alterations in the mtDNA and, interestingly, the resulting plants also exhibit significant changes in nuclear transcript profiles and show thermotolerance (Shedge et al. 2007, 2010). In the moss *Physcomitrella patens,* the mitochondrial RECA1 protein likewise seems to suppress rather than promote recombination between short stretches of similar sequences (Odahara et al. 2009). Strangely though, a *RecA*-like DNA recombination activity has been identified biochemically in soybean mitochondria (Manchekar et al. 2006). Homologues of MSH, OSB and RECA are easily identified in all available genome sequences of the plant lineage and it will be particularly interesting to elucidate their role in taxa such as the lycophytes, which display tremendous amounts of chondrome recombination.

D. When mtDNA Recombination Matters: Mitochondrial Mutants

As outlined earlier, the most dramatic mutant phenotype associated with mitochondrial malfunction in plants is cytoplasmic male sterility (CMS). Pollen biogenesis appears to be the major bottleneck revealing even those mitochondrial defects that do not become apparent in the vegetative phases of plant development. This may well be related to the dramatic reduction in mtDNA amounts during pollen biogenesis that was recently uncovered (Wang et al. 2010). The emergence of CMS phenotypes is accompanied by recombinations in the mitochondrial DNA (and/or their rise to stoichiometrical dominance), which create chimeric reading frames encoding protein products with deleterious effects (Budar and Pelletier 2001; Fujii et al. 2010; Janska et al. 1998; Kubo and Newton 2008). Mitochondrial mutations such as CMS or the non-chromosomal stripe (NCS) mutants of maize will be dealt with in a separate chapter (Chap. 12) of this volume.

E. Foreign Sequences in Plant Chondromes

Nearly 30 years ago, it was first recognized in maize that the two endosymbiotic organelles in the plant cell share common sequences, owing to the fact that chloroplast DNA fragments are integrated into mtDNA (Stern and Lonsdale 1982). Soon afterwards, the term "promiscuous DNA" was coined (Ellis 1982). Since then, numerous reports have documented that such promiscuous chloroplast DNA fragments are transferred quite frequently into flowering plant mitochondrial genomes. For example, a total of more than 68 Kbp of chloroplast DNA sequence inserts are present in the *Vitis vinifera* mtDNA (Goremykin et al. 2009). Likewise, large chloroplast DNA inserts were found in the chondrome of the gymnosperm *Cycas taitungensis* (Wang et al. 2007). Similarly, several sequences clearly originating from the nuclear genome, mostly retrotransposon fragments of different sizes, are frequently identified in seed plant chondromes (Knoop et al. 1996; Schuster and Brennicke 1987b). It should be kept in mind that chloroplast-derived promiscuous sequences are easily recognized but that this is naturally much more difficult for promiscuous DNA originating from the much more variable plant nuclear genome. In fact, large parts of intergenic sequences in plant chondromes may ultimately be recognized as nuclear in origin, once nuclear genome sequences of the respective or closely related taxa become available.

So far, there is no report on promiscuous DNA (i. e., nuclear or chloroplast DNA insertions) in bryophyte chondromes. Recently, however, such insertions of chloroplast and nuclear DNA were identified in the mtDNA of the lycophyte *Isoetes engelmannii* (Grewe et al. 2009), demonstrating that the propensity of plant chondromes to accept promiscuous DNA sequence integrations originated with the tracheophyte lineage. Most of the insertions of foreign DNA into plant chondromes are non-functional. However, as outlined above, seed plants have occasionally made use of chloroplast tRNA genes inserted into their chondromes to complement the sets of cytosol-imported and remaining native mitochondrial tRNAs.

Finally, plant chondromes seem to accept foreign DNA insertions not only from the other two genetic compartments in the same plant cell but also from mitochondria of other species via horizontal gene transfer (HGT). Following the two initial reports on mitochondrial HGT in angiosperms (Bergthorsson et al. 2003) and gymnosperms (Won and Renner 2003), numerous further cases have been identified where certain plant mitochondrial sequences seem to originate from HGT. The identification of horizontal gene transfer events is complicated by the fact that HGT events do not necessarily affect complete genes but that transfer of gene parts may create gene chimaeras in the target genome (Hao et al. 2010). A separate chapter in this volume (Chap. 10) is specifically dedicated to the issue of horizontal gene transfer. From a phylogenetic perspective, it is interesting to note that one example of HGT into a fern chondrome has been reported (Davis et al. 2005), but so far no cases of HGT into bryophyte chondromes are known.

Similar to integration of promiscuous DNA into plant mitochondrial DNA, the high recombinational activity arising with the earliest tracheophytes may also be a prerequisite for integration of foreign sequences via HGT. The inherent dynamics of plant mitochondria with their propensity for fission and fusion may likewise be an underlying cause for their apparent readiness to acquire foreign DNA (Logan 2010; Scott and Logan 2011). The chloroplast genome of plants, in contrast, seems to be largely immune against insertions of DNA from foreign sources. Possibly, this difference is simply related to the higher structural integrity of the plastids and their unwillingness to participate in membrane fusion events or, alternatively, to the lack of recombination owing to the absence of double-strand DNA break repair mechanisms (Kohl and Bock 2009). However, it should be noted that one report in the literature documents a likely origin of *rpl36* in the ancestor of cryptophyte and haptophyte plastids via HGT (Rice and Palmer 2006).

VI. Evolving RNA Editing

A separate chapter in this volume (Chap. 13) is dedicated to plant organelle RNA editing and I will therefore cover this phenomenon here only very briefly from a phylogenetic perspective. There seems to be no doubt that the cytidine-to-uridine conversion type of RNA editing in chloroplasts and mitochondria originated with land plants. Only one clade of plants, the marchantiid ("complex-thalloiid") liverworts, has apparently secondarily lost RNA editing altogether (Groth-Malonek et al. 2007b). RNA editing frequencies among the other plant clade vary widely, from only nine sites in the mitochondrial transcriptome of the moss *Funaria hygrometrica* (Rüdinger et al. 2011) to more than 2,000 editing events in the lycophyte *Selaginella moellendorffii* (Hecht et al. 2011). Editing frequencies correlate well with the size of a particular subfamily of nuclear-encoded pentatricopeptide repeat (PPR) proteins carrying a terminal extension called "DYW domain", which has remote similarity to deaminases (Rüdinger et al. 2008; Salone et al. 2007). One highly puzzling phenomenon is that the canonical C-to-U editing in the organelles is accompanied by massive amounts of additional "reverse" U-to-C pyrimidine conversions in hornworts (Kugita et al. 2003; Yoshinaga et al. 1996), in the lycophyte *Isoetes engelmannii* (Grewe et al. 2010) and in ferns (Vangerow et al. 1999). Hence, the emergence of the tracheophyte lineage or the hornwort-tracheophyte transition phase of plant evolution also seems to have been accompanied by a major shift in RNA editing biochemistry. This is somewhat similar to the sudden occurrence of massive independent mitochondrial gene losses in hornworts and tracheophytes (due to increased EGT activity) or the emergence of frequent chondrome recombination in the vascular plant stem lineage.

VII. Perspectives

The origins of the oddities in plant mitochondrial genomes seem to coincide with major changes in lifestyle during land plant evolution. Early plant evolution under a bryophyte-type of developmental organization was characterized by a dominating haploid gametophyte stage and a fully gametophyte-dependent diploid sporophyte phase. This bryophyte lifestyle correlates with retention of an ancestral, circular and largely non-recombining structure of the mtDNA, similar to the mtDNAs in green algae of the Charophyceae, which are most closely related to the plant lineage. The transition from algal to land plant life came with a moderate expansion of intergenic regions but, most notably, with a dramatic gain and loss of introns. Among 74 introns now identified in bryophyte mitochondrial genes, not a single one is conserved across all three bryophyte classes (liverworts, mosses and hornworts) an observation that implies numerous gains and losses in early plant genealogy. Strikingly, once established in evolution, the mitochondrial introns remain surprisingly stable within the three bryophyte clades again suggesting that cladogenic events (but not diversification within the clades) trigger larger events of molecular evolution in plant mitochondria. It is tempting to speculate on the biological sources for the differential intron gains in early plant mitochondrial evolution. Symbiotic, mycorrhiza-like interactions of bryophytes with endophytic fungi are now increasingly well characterized (Jakucs et al. 2003; Kottke et al. 2003; Kottke and Nebel 2005; e.g. Read et al. 2000; Russell and Bulman 2005). Varying intimate contacts between fungi and bryophyte ancestors early after the conquest of land habitats by plants may have played a role in HGT of mitochondrial introns via retro-splicing and may at the same time explain why the mitochondrial, but not the chloroplast, genomes are affected by such differential intron invasions.

The mitochondrial intron dynamics surprisingly contrasts the overall conservative evolution of the bryophyte chondromes with extensive gene syntenies being retained between liverworts and mosses and, to a large extent, also in hornworts. The latter clade is now assumed to be the sister group to tracheophytes. One feature which sets hornwort chondromes apart from the other bryophytes is the sudden increase in endosymbiotic gene transfer activity, which results in numerous tRNA, ribosomal protein, succinate dehydrogenase and cytochrome c maturation genes lacking from their mtDNAs. What is very surprising though is that frequent EGT affecting the same types of genes has obviously independently and in parallel occurred in the lycophytes. Hence, the bryophyte-tracheophyte transition seems to coincide with a massive increase in EGT, making it all the more striking that the above genes have survived in the chondrome along the stem lineage leading to modern euphyllophytes.

The conserved circular mtDNAs of the hornworts couldn't be more contrasted than by the heavily recombining and rearranging mtDNAs of the lycophytes. Possibly, the dominating diploid sporophytic phase of emerging tracheophytes is causally connected to this difference, in that nuclear genetic control of mtDNA structure is somehow relaxed or disturbed when two alleles of relevant nuclear factors coexist for a longer time in development. However, it must be kept in mind that the two hitherto investigated lycophyte taxa with heavily recombining mtDNAs, *Isoetes* and *Selaginella*, are heterosporous. Rather than an enduring diploid phase, it may thus actually have been the emerging sexuality (giving rise to different gametes and gametophytes), which could be the key to the emergence of heavily recombining mitochondrial DNAs. The mitochondrial genome of a first isosporous lycophyte (*Huperzia*, Yin-Long Qiu, pers. comm. and see previous chapter) and the comparisons of hetero- and isosporous monilophyte mtDNAs will clearly shed more light on these considerations in the future.

The predisposition to acquire foreign DNA seems to fully coincide with the gain of recombinational activity in early vascular plant mtDNAs. No nuclear or chloroplast DNA inserts have hitherto been identified in bryophyte mtDNAs, but both types of promiscuous DNA were found in the *Isoetes engelmannii* chondrome. As yet, we do not have a complete chondrome sequence available for a monilophyte, the sister clade of seed plants. Our own preliminary data (Felix Grewe and V.K., unpublished) for the fern *Gleichenia dicarpa* hint to extraordinary mtDNA recombination even exceeding what has previously been found in lycophytes. In addition, the massive insertion of nuclear retrotransposon sequences and other mobile DNAs has occurred, while intron insertion patterns are very seed plant-like. Hence, the origin of euphyllophytes seems to come with an emerging stasis in mitochondrial intron distributions, which continues in the two sister clades monilophytes and seed plants. It is very fortunate that the changes in chondrome make-up along plant phylogeny can soon be correlated with nuclear genome information for all those important taxonomic groups, where full genome information is currently still lacking, such as liverworts, hornworts and monilophytes.

Overall, we now have a clear view of the likely state of the ancestral seed plant chondrome, at least with respect to its gene and intron composition. The diversity among seed plant chondromes – as it keeps being unearthed by mitochondrial genomics efforts – will mainly be attributable to a combination of five factors: (1) the amount of endosymbiotic gene transfer, (2) the integration of promiscuous DNA sequences originating from the chloroplast and nucleus, (3) the amount of horizontal gene transfer integrating sequences from mtDNAs of other species, (4) the amount of recombinational activity and (5) the acceleration or deceleration of evolutionary rates. However, given that plant mitochondria have often proven to be "more unique than ever" (Rasmusson et al. 2008), some exciting surprises may still wait for us further down the road.

References

Abdelnoor RV, Yule R, Elo A, Christensen AC, Meyer-Gauen G, Mackenzie SA (2003) Substoichiometric shifting in the plant mitochondrial genome is influenced by a gene homologous to MutS. Proc Natl Acad Sci USA 100:5968–5973

Adams KL, Clements MJ, Vaughn JC (1998a) The *Peperomia* mitochondrial *coxI* group I intron: timing of horizontal transfer and subsequent evolution of the intron. J Mol Evol 46:689–696

Adams KL, Song K, Qiu YL, Shirk A, Cho Y, Parkinson CL, Palmer JD (1998b) Evolution of flowering plant mitochondrial genomes: gene content, gene transfer to the nucleus, and highly accelerated mutation rates. In: Moller IM, Gardestrom P, Glimelius K, Glaser E (eds) Plant mitochondria: from gene to function. Backhuys, Leiden, pp 13–18

Adams KL, Song KM, Roessler PG, Nugent JM, Doyle JL, Doyle JJ, Palmer JD (1999) Intracellular gene transfer in action: dual transcription and multiple silencings of nuclear and mitochondrial *cox*2 genes in legumes. Proc Natl Acad Sci USA 96: 13863–13868

Adams KL, Daley DO, Qiu YL, Whelan J, Palmer JD (2000) Repeated, recent and diverse transfers of a mitochondrial gene to the nucleus in flowering plants. Nature 408:354–357

Adams KL, Rosenblueth M, Qiu YL, Palmer JD (2001) Multiple losses and transfers to the nucleus of two mitochondrial succinate dehydrogenase genes during angiosperm evolution. Genetics 158:1289–1300

Adams KL, Daley DO, Whelan J, Palmer JD (2002a) Genes for two mitochondrial ribosomal proteins in flowering plants are derived from their chloroplast or cytosolic counterparts. Plant Cell 14:931–943

Adams KL, Qiu YL, Stoutemyer M, Palmer JD (2002b) Punctuated evolution of mitochondrial gene content: high and variable rates of mitochondrial gene loss and transfer to the nucleus during angiosperm evolution. Proc Natl Acad Sci USA 99:9905–9912

Allen JO, Fauron CM, Minx P, Roark L, Oddiraju S, Lin GN, Meyer L, Sun H, Kim K, Wang C, Du F, Xu D, Gibson M, Cifrese J, Clifton SW, Newton KJ (2007) Comparisons among two fertile and three male-sterile mitochondrial genomes of maize. Genetics 177:1173–1192

Allen JW, Jackson AP, Rigden DJ, Willis AC, Ferguson SJ, Ginger ML (2008) Order within a mosaic distribution of mitochondrial c-type cytochrome biogenesis systems? FEBS J 275:2385–2402

Alverson AJ, Wei X, Rice DW, Stern DB, Barry K, Palmer JD (2010) Insights into the evolution of mitochondrial genome size from complete sequences of *Citrullus lanatus* and *Cucurbita pepo* (Cucurbitaceae). Mol Biol Evol 27:1436–1448

Alverson AJ, Zhuo S, Rice DW, Sloan DB, Palmer JD (2011) The mitochondrial genome of the legume Vigna radiata and the analysis of recombination across short mitochondrial repeats. PLoS One 6:e16404

Alverson AJ, Rice DW, Dickinson S, Barry K, Palmer JD (2011b) Origins and recombination of the bacterial-sized multichromosomal mitochondrial genome of cucumber. Plant Cell 23:2499–2513

Anderson S, Bankier AT, Barrell BG, de Bruijn MH, Coulson AR, Drouin J, Eperon IC, Nierlich DP, Roe BA, Sanger F, Schreier PH, Smith AJ, Staden R, Young IG (1981) Sequence and organization of the human mitochondrial genome. Nature 290:457–465

Arrieta-Montiel MP, Mackenzie SA (2011) Plant mitochondrial genomes and recombination. In: Kempken

F (ed) Plant mitochondria. Springer, New York, pp 65–82

Arrieta-Montiel MP, Shedge V, Davila J, Christensen AC, Mackenzie SA (2009) Diversity of the *Arabidopsis* mitochondrial genome occurs via nuclear-controlled recombination activity. Genetics 183:1261–1268

Bakker FT, Culham A, Pankhurst CE, Gibby M (2000) Mitochondrial and chloroplast DNA-based phylogeny of *Pelargonium* (Geraniaceae). Am J Bot 87:727–734

Banks JA, Nishiyama T, Hasebe M, Bowman JL, Gribskov M, Depamphilis C, Albert VA, Aono N, Aoyama T, Ambrose BA, Ashton NW, Axtell MJ, Barker E, Barker MS, Bennetzen JL, Bonawitz ND, Chapple C, Cheng C, Correa LG, Dacre M, DeBarry J, Dreyer I, Elias M, Engstrom EM, Estelle M, Feng L, Finet C, Floyd SK, Frommer WB, Fujita T, Gramzow L, Gutensohn M, Harholt J, Hattori M, Heyl A, Hirai T, Hiwatashi Y, Ishikawa M, Iwata M, Karol KG, Koehler B, Kolukisaoglu U, Kubo M, Kurata T, Lalonde S, Li K, Li Y, Litt A, Lyons E, Manning G, Maruyama T, Michael TP, Mikami K, Miyazaki S, Morinaga S, Murata T, Mueller-Roeber B, Nelson DR, Obara M, Oguri Y, Olmstead RG, Onodera N, Petersen BL, Pils B, Prigge M, Rensing SA, Riano-Pachon DM, Roberts AW, Sato Y, Scheller HV, Schulz B, Schulz C, Shakirov EV, Shibagaki N, Shinohara N, Shippen DE, Sorensen I, Sotooka R, Sugimoto N, Sugita M, Sumikawa N, Tanurdzic M, Theissen G, Ulvskov P, Wakazuki S, Weng JK, Willats WW, Wipf D, Wolf PG, Yang L, Zimmer AD, Zhu Q, Mitros T, Hellsten U, Loque D, Otillar R, Salamov A, Schmutz J, Shapiro H, Lindquist E, Lucas S, Rokhsar D, Grigoriev IV (2011) The *Selaginella* genome identifies genetic changes associated with the evolution of vascular plants. Science 332:960–963

Beckert S, Steinhauser S, Muhle H, Knoop V (1999) A molecular phylogeny of bryophytes based on nucleotide sequences of the mitochondrial *nad5* gene. Plant Syst Evol 218:179–192

Beckert S, Muhle H, Pruchner D, Knoop V (2001) The mitochondrial *nad2* gene as a novel marker locus for phylogenetic analysis of early land plants: a comparative analysis in mosses. Mol Phylogenet Evol 18:117–126

Bendich AJ (1993) Reaching for the ring: the study of mitochondrial genome structure. Curr Genet 24: 279–290

Bendich AJ (2007) The size and form of chromosomes are constant in the nucleus, but highly variable in bacteria, mitochondria and chloroplasts. Bioessays 29:474–483

Bergthorsson U, Adams KL, Thomason B, Palmer JD (2003) Widespread horizontal transfer of mitochondrial genes in flowering plants. Nature 424:197–201

Bergthorsson U, Richardson AO, Young GJ, Goertzen LR, Palmer JD (2004) Massive horizontal transfer of mitochondrial genes from diverse land plant donors to the basal angiosperm *Amborella*. Proc Natl Acad Sci USA 101:17747–17752

Brennicke A, Blanz P (1982) Circular mitochondrial DNA species from *Oenothera* with unique sequences. Mol Gen Genet 187:461–466

Brennicke A, Möller S, Blanz PA (1985) The 18S and 5S ribosomal RNA genes in *Oenothera* mitochondria: sequence rearrangements in the 18S and 5S ribosomal RNA genes of higher plants. Mol Gen Genet 198:404–410

Brouard JS, Otis C, Lemieux C, Turmel M (2010) The exceptionally large chloroplast genome of the green alga *Floydiella terrestris* illuminates the evolutionary history of the Chlorophyceae. Genome Biol Evol 2:240–256

Brubacher-Kauffmann S, Maréchal-Drouard L, Cosset A, Dietrich A, Duchene AM (1999) Differential import of nuclear-encoded tRNA(Gly) isoacceptors into Solanum tuberosum mitochondria. Nucleic Acids Res 27:2037–2042

Budar F, Pelletier G (2001) Male sterility in plants: occurrence, determinism, significance and use. C R Acad Sci III 324:543–550

Burger G, Yan Y, Javadi P, Lang BF (2009) Group I intron *trans*-splicing and mRNA editing in the mitochondria of placozoan animals. Trends Genet 25:381–386

Chaw SM, Chun-Chieh SA, Wang D, Wu YW, Liu SM, Chou TY (2008) The mitochondrial genome of the gymnosperm *Cycas taitungensis* contains a novel family of short interspersed elements, Bpu sequences, and abundant RNA editing sites. Mol Biol Evol 25:603–615

Cho YR, Palmer JD (1999) Multiple acquisitions via horizontal transfer of a group I intron in the mitochondrial cox1 gene during evolution of the Araceae family. Mol Biol Evol 16:1155–1165

Cho Y, Qiu YL, Kuhlman P, Palmer JD (1998) Explosive invasion of plant mitochondria by a group I intron. Proc Natl Acad Sci USA 95:14244–14249

Cho Y, Mower JP, Qiu YL, Palmer JD (2004) Mitochondrial substitution rates are extraordinarily elevated and variable in a genus of flowering plants. Proc Natl Acad Sci USA 101:17741–17746

Choquet Y, Goldschmidt-Clermont M, Girard-Bascou J, Kuck U, Bennoun P, Rochaix JD (1988) Mutant phenotypes support a trans-splicing mechanism for the expression of the tripartite *psaA* gene in the C. reinhardtii chloroplast. Cell 52:903–913

Clifton SW, Minx P, Fauron CM, Gibson M, Allen JO, Sun H, Thompson M, Barbazuk WB, Kanuganti S, Tayloe C, Meyer L, Wilson RK, Newton KJ (2004) Sequence and comparative analysis of the maize NB mitochondrial genome. Plant Physiol 136:3486–3503

Cui P, Liu H, Lin Q, Ding F, Zhuo G, Hu S, Liu D, Yang W, Zhan K, Zhang A, Yu J (2009) A complete mitochondrial genome of wheat (*Triticum aestivum* cv. *Chinese Yumai*), and fast evolving mitochondrial genes in higher plants. J Genet 88:299–307

Cusimano N, Zhang LB, Renner SS (2008) Reevaluation of the *cox1* group I intron in Araceae and angiosperms indicates a history dominated by loss rather than horizontal transfer. Mol Biol Evol 25:265–276

Dale RM (1981) Sequence homology among different size classes of plant mtDNAs. Proc Natl Acad Sci USA 78:4453–4457

Dale RM, Wu M, Kiernan MC (1983) Analysis of four tobacco mitochondrial DNA size classes. Nucleic Acids Res 11:1673–1685

Daley DO, Adams KL, Clifton R, Qualmann S, Millar AH, Palmer JD, Pratje E, Whelan J (2002) Gene transfer from mitochondrion to nucleus: novel mechanisms for gene activation from *Cox2*. Plant J 30:11–21

Darracq A, Varré JS, Maréchal-Drouard L, Courseaux A, Castric V, Saumitou-Laprade P, Oztas S, Lenoble P, Vacherie B, Barbe V, Touzet P (2011) Structural and content diversity of mitochondrial genome in beet: a comparative genomic analysis. Genome Biol Evol 3:723–736

Davis CC, Anderson WR, Wurdack KJ (2005) Gene transfer from a parasitic flowering plant to a fern. Proc Biol Sci 272:2237–2242

Delage L, Dietrich A, Cosset A, Maréchal-Drouard L (2003) In vitro import of a nuclearly encoded tRNA into mitochondria of *Solanum tuberosum*. Mol Cell Biol 23:4000–4012

Dombrovska E, Qiu YL (2004) Distribution of introns in the mitochondrial gene *nad1* in land plants: phylogenetic and molecular evolutionary implications. Mol Phylogenet Evol 32:246–263

Duchêne AM, Pujol C, Maréchal-Drouard L (2009) Import of tRNAs and aminoacyl-tRNA synthetases into mitochondria. Curr Genet 55:1–18

Ellis J (1982) Promiscuous DNA – chloroplast genes inside plant mitochondria. Nature 299:678–679

Fujii S, Kazama T, Yamada M, Toriyama K (2010) Discovery of global genomic re-organization based on comparison of two newly sequenced rice mitochondrial genomes with cytoplasmic male sterility-related genes. BMC Genomics 11:209

Giegé P, Grienenberger JM, Bonnard G (2008) Cytochrome c biogenesis in mitochondria. Mitochondrion 8:61–73

Glover KE, Spencer DF, Gray MW (2001) Identification and structural characterization of nucleus-encoded transfer RNAs imported into wheat mitochondria. J Biol Chem 276:639–648

Goremykin VV, Salamini F, Velasco R, Viola R (2009) Mitochondrial DNA of *Vitis vinifera* and the issue of rampant horizontal gene transfer. Mol Biol Evol 26:99–110

Grewe F, Viehoever P, Weisshaar B, Knoop V (2009) A *trans*-splicing group I intron and tRNA-hyperediting in the mitochondrial genome of the lycophyte *Isoetes engelmannii*. Nucleic Acids Res 37:5093–5104

Grewe F, Herres S, Viehoever P, Polsakiewicz M, Weisshaar B, Knoop V (2010) A unique transcriptome: 1782 positions of RNA editing alter 1406 codon identities in mitochondrial mRNAs of the lycophyte *Isoetes engelmannii*. Nucleic Acids Res 39:2890–2902

Groth-Malonek M, Pruchner D, Grewe F, Knoop V (2005) Ancestors of *trans*-splicing mitochondrial introns support serial sister group relationships of hornworts and mosses with vascular plants. Mol Biol Evol 22:117–125

Groth-Malonek M, Rein T, Wilson R, Groth H, Heinrichs J, Knoop V (2007a) Different fates of two mitochondrial gene spacers in early land plant evolution. Int J Plant Sci 168:709–717

Groth-Malonek M, Wahrmund U, Polsakiewicz M, Knoop V (2007b) Evolution of a pseudogene: exclusive survival of a functional mitochondrial *nad7* gene supports *Haplomitrium* as the earliest liverwort lineage and proposes a secondary loss of RNA editing in Marchantiidae. Mol Biol Evol 24:1068–1074

Handa H (2003) The complete nucleotide sequence and RNA editing content of the mitochondrial genome of rapeseed (*Brassica napus* L.): comparative analysis of the mitochondrial genomes of rapeseed and *Arabidopsis thaliana*. Nucleic Acids Res 31:5907–5916

Hanson MR, Bentolila S (2004) Interactions of mitochondrial and nuclear genes that affect male gametophyte development. Plant Cell 16(Suppl): S154–S169

Hao W, Richardson AO, Zheng Y, Palmer JD (2010) Gorgeous mosaic of mitochondrial genes created by horizontal transfer and gene conversion. Proc Natl Acad Sci USA 107:21576–21581

Hecht J, Grewe F, Knoop V (2011) Extreme RNA editing in coding islands and abundant microsatellites in repeat sequences of *Selaginella moellendorffii* mitochondria: the root of frequent plant mtDNA recombination in early tracheophytes. Genome Biol Evol 3:344–358

Hildebrand M, Hallick RB, Passavant CW, Bourque DP (1988) Trans-splicing in chloroplasts: the rps 12 loci of *Nicotiana tabacum*. Proc Natl Acad Sci USA 85:372–376

Itchoda N, Nishizawa S, Nagano H, Kubo T, Mikami T (2002) The sugar beet mitochondrial *nad4* gene: an intron loss and its phylogenetic implication in the Caryophyllales. Theor Appl Genet 104:209–213

Jakucs E, Naar Z, Szedlay G, Orban S (2003) Glomalean and septate endophytic fungi in Hypopterygium mosses (Bryopsida). Cryptogamie Mycol 24:27–37

Janska H, Sarria R, Woloszynska M, Arrieta-Montiel M, Mackenzie SA (1998) Stoichiometric shifts in the common bean mitochondrial genome leading to male sterility and spontaneous reversion to fertility. Plant Cell 10:1163–1180

Keren I, Bezawork-Geleta A, Kolton M, Maayan I, Belausov E, Levy M, Mett A, Gidoni D, Shaya F, Ostersetzer-Biran O (2009) AtnMat2, a nuclear-encoded maturase required for splicing of group II introns in *Arabidopsis* mitochondria. RNA 15:2299–2311

Kim S, Yoon MK (2010) Comparison of mitochondrial and chloroplast genome segments from three onion (*Allium cepa* L.) cytoplasm types and identification of a *trans*-splicing intron of *cox2*. Curr Genet 56: 177–188

Klein M, Eckert-Ossenkopp U, Schmiedeberg I, Brandt P, Unseld M, Brennicke A, Schuster W (1994) Physical mapping of the mitochondrial genome of *Arabidopsis thaliana* by cosmid and YAC clones. Plant J 6:447–455

Kmiec B, Woloszynska M, Janska H (2006) Heteroplasmy as a common state of mitochondrial genetic information in plants and animals. Curr Genet 50:149–159

Knoop V (2004) The mitochondrial DNA of land plants: peculiarities in phylogenetic perspective. Curr Genet 46:123–139

Knoop V (2010) Looking for sense in the nonsense: a short review of non-coding organellar DNA elucidating the phylogeny of bryophytes. Trop Bryol 31:50–60

Knoop V, Ehrhardt T, Lättig K, Brennicke A (1995) The gene for ribosomal protein S10 is present in mitochondria of pea and potato but absent from those of *Arabidopsis* and *Oenothera*. Curr Genet 27:559–564

Knoop V, Unseld M, Marienfeld J, Brandt P, Sünkel S, Ullrich H, Brennicke A (1996) *copia*-, *gypsy*- and LINE-like retrotransposon fragments in the mitochondrial genome of *Arabidopsis thaliana*. Genetics 142:579–585

Knoop V, Volkmar U, Hecht J, Grewe F (2010) Mitochondrial genome evolution in the plant lineage. In: Kempken F (ed) Plant mitochondria. Springer, New York, pp 3–29

Kobayashi Y, Knoop V, Fukuzawa H, Brennicke A, Ohyama K (1997) Interorganellar gene transfer in bryophytes: the functional *nad7* gene is nuclear encoded in *Marchantia polymorpha*. Mol Gen Genet 256:589–592

Kohchi T, Umesono K, Ogura Y, Komine Y, Nakahigashi K, Komano T, Yamada Y, Ozeki H, Ohyama K (1988) A nicked group II intron and trans-splicing in liverwort, *Marchantia polymorpha*, chloroplasts. Nucleic Acids Res 16:10025–10036

Kohl S, Bock R (2009) Transposition of a bacterial insertion sequence in chloroplasts. Plant J 58:423–436

Kottke I, Nebel M (2005) The evolution of mycorrhiza-like associations in liverworts: an update. New Phytol 167:330–334

Kottke I, Beiter A, Weiss M, Haug I, Oberwinkler F, Nebel M (2003) Heterobasidiomycetes form symbiotic associations with hepatics: jungermanniales have sebacinoid mycobionts while *Aneura pinguis* (Metzgeriales) is associated with a *Tulasnella* species. Mycol Res 107:957–968

Kubo N, Arimura S (2010) Discovery of the *rpl10* gene in diverse plant mitochondrial genomes and its probable replacement by the nuclear gene for chloroplast RPL10 in two lineages of angiosperms. DNA Res 17:1–9

Kubo T, Newton KJ (2008) Angiosperm mitochondrial genomes and mutations. Mitochondrion 8:5–14

Kubo T, Nishizawa S, Sugawara A, Itchoda N, Estiati A, Mikami T (2000) The complete nucleotide sequence of the mitochondrial genome of sugar beet (*Beta vulgaris* L.) reveals a novel gene for tRNA(Cys)(GCA). Nucleic Acids Res 28:2571–2576

Kugita M, Yamamoto Y, Fujikawa T, Matsumoto T, Yoshinaga K (2003) RNA editing in hornwort chloroplasts makes more than half the genes functional. Nucleic Acids Res 31:2417–2423

Lavrov DV (2007) Key transitions in animal evolution: a mitochondrial DNA perspective. Integr Comp Biol 47:734–743

Levings CSI, Pring DR (1976) Restriction endonuclease analysis of mitochondrial DNA from normal and Texas cytoplasmic male-sterile maize. Science 193:158–160

Li L, Wang B, Liu Y, Qiu YL (2009) The complete mitochondrial genome sequence of the hornwort *Megaceros aenigmaticus* shows a mixed mode of conservative yet dynamic evolution in early land plant mitochondrial genomes. J Mol Evol 68: 665–678

Liu H, Cui P, Zhan K, Lin Q, Zhuo G, Guo X, Ding F, Yang W, Liu D, Hu S, Yu J, Zhang A (2011) Comparative analysis of mitochondrial genomes between a wheat K-type cytoplasmic male sterility (CMS) line and its maintainer line. BMC Genomics 12:163

Logan DC (2010) The dynamic plant chondriome. Semin Cell Dev Biol 21:550–557

Lonsdale DM, Hodge TP, Fauron CM (1984) The physical map and organisation of the mitochondrial genome from the fertile cytoplasm of maize. Nucleic Acids Res 12:9249–9261

Malek O, Knoop V (1998) Trans-splicing group II introns in plant mitochondria: the complete set of *cis*-arranged homologs in ferns, fern allies, and a hornwort. RNA 4:1599–1609

Malek O, Brennicke A, Knoop V (1997) Evolution of trans-splicing plant mitochondrial introns in pre-permian times. Proc Natl Acad Sci USA 94:553–558

Manchekar M, Scissum-Gunn K, Song D, Khazi F, McLean SL, Nielsen BL (2006) DNA recombination activity in soybean mitochondria. J Mol Biol 356:288–299

Manna E, Brennicke A (1986) Site-specific circularization at an intragenic sequence in *Oenothera* mitochondria. Mol Gen Genet 203:377–381

Mower JP, Bonen L (2009) Ribosomal protein L10 is encoded in the mitochondrial genome of many land plants and green algae. BMC Evol Biol 9:265

Notsu Y, Masood S, Nishikawa T, Kubo N, Akiduki G, Nakazono M, Hirai A, Kadowaki K (2002) The complete sequence of the rice (*Oryza sativa* L.) mitochondrial genome: frequent DNA sequence acquisition and loss during the evolution of flowering plants. Mol Genet Genomics 268:434–445

Nugent JM, Palmer JD (1991) RNA-mediated transfer of the gene *coxII* from the mitochondrion to the nucleus during flowering plant evolution. Cell 66:473–481

Oda K, Kohchi T, Ohyama K (1992a) Mitochondrial DNA of *Marchantia polymorpha* as a single circular form with no incorporation of foreign DNA. Biosci Biotechnol Biochem 56:132–135

Oda K, Yamato K, Ohta E, Nakamura Y, Takemura M, Nozato N, Akashi K, Kanegae T, Ogura Y, Kohchi T, Ohyama K (1992b) Gene organization deduced from the complete sequence of liverwort *Marchantia polymorpha* mitochondrial DNA. A primitive form of plant mitochondrial genome. J Mol Biol 223:1–7

Odahara M, Kuroiwa H, Kuroiwa T, Sekine Y (2009) Suppression of repeat-mediated gross mitochondrial genome rearrangements by RecA in the moss *Physcomitrella patens*. Plant Cell 21:1182–1194

Ogihara Y, Yamazaki Y, Murai K, Kanno A, Terachi T, Shiina T, Miyashita N, Nasuda S, Nakamura C, Mori N, Takumi S, Murata M, Futo S, Tsunewaki K (2005) Structural dynamics of cereal mitochondrial genomes as revealed by complete nucleotide sequencing of the wheat mitochondrial genome. Nucleic Acids Res 33:6235–6250

Oldenburg DJ, Bendich AJ (1998) The structure of mitochondrial DNA from the liverwort, *Marchantia polymorpha*. J Mol Biol 276:745–758

Oldenburg DJ, Bendich AJ (2001) Mitochondrial DNA from the liverwort *Marchantia polymorpha*: circularly permuted linear molecules, head-to-tail concatemers, and a 5′ protein. J Mol Biol 310:549–562

Palmer JD, Herbon LA (1987) Unicircular structure of the *Brassica hirta* mitochondrial genome. Curr Genet 11:565–570

Palmer JD, Herbon LA (1988) Plant mitochondrial DNA evolves rapidly in structure, but slowly in sequence. J Mol Evol 28:87–97

Palmer JD, Shields CR (1984) Tripartite structure of the *Brassica campestris* mitochondrial genome. Nature 307:437–440

Parkinson CL, Mower JP, Qiu YL, Shirk AJ, Song K, Young ND, dePamphilis CW, Palmer JD (2005) Multiple major increases and decreases in mitochondrial substitution rates in the plant family Geraniaceae. BMC Evol Biol 5:73

Pombert JF, Keeling PJ (2010) The mitochondrial genome of the entomoparasitic green alga *Helicosporidium*. PLoS One 5:e8954

Pring DR, Levings CS, Hu WW, Timothy DH (1977) Unique DNA associated with mitochondria in the "S"-type cytoplasm of male-sterile maize. Proc Natl Acad Sci USA 74:2904–2908

Pryer KM, Schneider H, Smith AR, Cranfill R, Wolf PG, Hunt JS, Sipes SD (2001) Horsetails and ferns are a monophyletic group and the closest living relatives to seed plants. Nature 409:618–622

Qiu YL, Palmer JD (2004) Many independent origins of *trans* splicing of a plant mitochondrial group II intron. J Mol Evol 59:80–89

Qiu YL, Cho YR, Cox JC, Palmer JD (1998) The gain of three mitochondrial introns identifies liverworts as the earliest land plants. Nature 394:671–674

Qiu YL, Li L, Wang B, Chen Z, Knoop V, Groth-Malonek M, Dombrovska O, Lee J, Kent L, Rest J, Estabrook GF, Hendry TA, Taylor DW, Testa CM, Ambros M, Crandall-Stotler B, Duff RJ, Stech M, Frey W, Quandt D, Davis CC (2006) The deepest divergences in land plants inferred from phylogenomic evidence. Proc Natl Acad Sci USA 103:15511–15516

Ran JH, Gao H, Wang XQ (2010) Fast evolution of the retroprocessed mitochondrial *rps3* gene in Conifer II and further evidence for the phylogeny of gymnosperms. Mol Phylogenet Evol 54:136–149

Rasmusson AG, Handa H, Moller IM (2008) Plant mitochondria, more unique than ever. Mitochondrion 8:1–4

Read DJ, Duckett JG, Francis R, Ligrone R, Russell A (2000) Symbiotic fungal associations in 'lower' land plants. Philos Trans R Soc Lond B Biol Sci 355:815–830

Regina TM, Quagliariello C (2010) Lineage-specific group II intron gains and losses of the mitochondrial *rps3* gene in gymnosperms. Plant Physiol Biochem 48:646–654

Regina TM, Picardi E, Lopez L, Pesole G, Quagliariello C (2005) A novel additional group II intron distinguishes the mitochondrial *rps3* gene in gymnosperms. J Mol Evol 60:196–206

Rensing SA, Lang D, Zimmer AD, Terry A, Salamov A, Shapiro H, Nishiyama T, Perroud PF, Lindquist EA, Kamisugi Y, Tanahashi T, Sakakibara K, Fujita T, Oishi K, Shin I, Kuroki Y, Toyoda A, Suzuki Y, Hashimoto S, Yamaguchi K, Sugano S, Kohara Y, Fujiyama A, Anterola A, Aoki S, Ashton N, Barbazuk WB, Barker E, Bennetzen JL, Blankenship R, Cho SH, Dutcher SK, Estelle M, Fawcett JA, Gundlach H, Hanada K, Heyl A, Hicks KA, Hughes J, Lohr M, Mayer K, Melkozernov A, Murata T, Nelson DR, Pils B, Prigge M, Reiss B, Renner T, Rombauts S, Rushton PJ, Sanderfoot A, Schween G, Shiu SH, Stueber K, Theodoulou FL, Tu H, van de Peer Y, Verrier PJ, Waters E, Wood A, Yang L, Cove D, Cuming AC, Hasebe M, Lucas S, Mishler BD, Reski R, Grigoriev IV, Quatrano RS, Boore JL (2008) The *Physcomitrella* genome reveals evolutionary insights into the conquest of land by plants. Science 319:64–69

Rice DW, Palmer JD (2006) An exceptional horizontal gene transfer in plastids: gene replacement by a distant bacterial paralog and evidence that haptophyte and cryptophyte plastids are sisters. BMC Biol 4:31

Rivarola M, Foster JT, Chan AP, Williams AL, Rice DW, Liu X, Melake-Berhan A, Huot CH, Puiu D, Rosovitz MJ, Khouri HM, Beckstrom-Sternberg SM, Allan GJ, Keim P, Ravel J, Rabinowicz PD (2011) Castor bean organelle genome sequencing and worldwide genetic diversity analysis. PLoS One 6:e21743

Rodríguez-Moreno L, González VM, Benjak A, Martí MC, Puigdomènech P, Aranda MA, Garcia-Mas J (2011) Determination of the melon chloroplast and mitochondrial genome sequences reveals that the largest reported mitochondrial genome in plants contains a significant amount of DNA having a nuclear origin. BMC Genomics 12:424

Rogalski M, Karcher D, Bock R (2008) Superwobbling facilitates translation with reduced tRNA sets. Nat Struct Mol Biol 15:192–198

Rüdinger M, Polsakiewicz M, Knoop V (2008) Organellar RNA editing and plant-specific extensions of pentatricopeptide repeat (PPR) proteins in jungermanniid but not in marchantiid liverworts. Mol Biol Evol 25:1405–1414

Rüdinger M, Szövényi P, Rensing SA, Knoop V (2011) Assigning DYW-type PPR proteins to RNA editing sites in the funariid mosses *Physcomitrella patens* and *Funaria hygrometrica*. Plant J 67:370–380

Russell J, Bulman S (2005) The liverwort *Marchantia foliacea* forms a specialized symbiosis with arbuscular mycorrhizal fungi in the genus Glomus. New Phytol 165:567–579

Salone V, Rüdinger M, Polsakiewicz M, Hoffmann B, Groth-Malonek M, Szurek B, Small I, Knoop V, Lurin C (2007) A hypothesis on the identification of the editing enzyme in plant organelles. FEBS Lett 581:4132–4138

Sanchez-Puerta MV, Cho Y, Mower JP, Alverson AJ, Palmer JD (2008) Frequent, phylogenetically local horizontal transfer of the *cox1* group I Intron in flowering plant mitochondria. Mol Biol Evol 25:1762–1777

Satoh M, Kubo T, Nishizawa S, Estiati A, Itchoda N, Mikami T (2004) The cytoplasmic male-sterile type and normal type mitochondrial genomes of sugar beet share the same complement of genes of known function but differ in the content of expressed ORFs. Mol Genet Genomics 272:247–256

Schuster W, Brennicke A (1987a) Plastid DNA in the mitochondrial genome of *Oenothera*: intra- and interorganellar rearrangements involving part of the plastid ribosomal cistron. Mol Gen Genet 210:44–51

Schuster W, Brennicke A (1987b) Plastid, nuclear and reverse transcriptase sequences in the mitochondrial genome of *Oenothera*: is genetic information transferred between organelles via RNA? EMBO J 6: 2857–2863

Scott I, Logan DC (2011) Mitochondrial dynamics. In: Kempken F (ed) Plant mitochondria. Springer, New York, pp 31–63

Scotti N, Maréchal-Drouard L, Cardi T (2004) The rpl5-rps14 mitochondrial region: a hot spot for DNA rearrangements in *Solanum* spp. somatic hybrids. Curr Genet 45:378–382

Sederoff RR, Levings CS, Timothy DH, Hu WW (1981) Evolution of DNA sequence organization in mitochondrial genomes of *Zea*. Proc Natl Acad Sci USA 78:5953–5957

Seif E, Leigh J, Liu Y, Roewer I, Forget L, Lang BF (2005) Comparative mitochondrial genomics in zygomycetes: bacteria-like RNase P RNAs, mobile elements and a close source of the group I intron invasion in angiosperms. Nucleic Acids Res 33:734–744

Shedge V, Arrieta-Montiel M, Christensen AC, Mackenzie SA (2007) Plant mitochondrial recombination surveillance requires unusual RecA and MutS homologs. Plant Cell 19:1251–1264

Shedge V, Davila J, Arrieta-Montiel MP, Mohammed S, Mackenzie SA (2010) Extensive rearrangement of the *Arabidopsis* mitochondrial genome elicits cellular conditions for thermotolerance. Plant Physiol 152:1960–1970

Signorovitch AY, Buss LW, Dellaporta SL (2007) Comparative genomics of large mitochondria in placozoans. PLoS Genet 3:e13

Sloan DB, Alverson AJ, Storchova H, Palmer JD, Taylor DR (2010a) Extensive loss of translational genes in the structurally dynamic mitochondrial genome of the angiosperm *Silene latifolia*. BMC Evol Biol 10:274

Sloan DB, MacQueen AH, Alverson AJ, Palmer JD, Taylor DR (2010b) Extensive loss of RNA editing sites in rapidly evolving silene mitochondrial genomes: selection vs. retroprocessing as the driving force. Genetics 185:1369–1380

Small I, Suffolk R, Leaver CJ (1989) Evolution of plant mitochondrial genomes via substoichiometric intermediates. Cell 58:69–76

Stern DB, Lonsdale DM (1982) Mitochondrial and chloroplast genomes of maize have a 12-kilobase DNA sequence in common. Nature 299:698–702

Stern DB, Palmer JD (1984) Recombination sequences in plant mitochondrial genomes: diversity and homologies to known mitochondrial genes. Nucleic Acids Res 12:6141–6157

Stern DB, Palmer JD (1986) Tripartite mitochondrial genome of spinach: physical structure, mitochondrial gene mapping, and locations of transposed chloroplast DNA sequences. Nucleic Acids Res 14:5651–5666

Sugiyama Y, Watase Y, Nagase M, Makita N, Yagura S, Hirai A, Sugiura M (2005) The complete nucleotide sequence and multipartite organization of the tobacco mitochondrial genome: comparative analysis of mitochondrial genomes in higher plants. Mol Genet Genomics 272:603–615

Terasawa K, Odahara M, Kabeya Y, Kikugawa T, Sekine Y, Fujiwara M, Sato N (2007) The mitochondrial genome of the moss *Physcomitrella patens* sheds new light on mitochondrial evolution in land plants. Mol Biol Evol 24:699–709

Tian X, Zheng J, Hu S, Yu J (2006) The rice mitochondrial genomes and their variations. Plant Physiol 140:401–410

Timothy DH, Levings CS, Pring DR, Conde MF, Kermicle JL (1979) Organelle DNA variation and systematic relationships in the genus *Zea*: Teosinte. Proc Natl Acad Sci USA 76:4220–4224

Tomaru N, Takahashi M, Tsumura Y, Takahashi M, Ohba K (1998) Intraspecific variation and phylogeographic patterns of *Fagus crenata* (Fagaceae) mitochondrial DNA. Am J Bot 85:629–636

Turmel M, Otis C, Lemieux C (2002a) The chloroplast and mitochondrial genome sequences of the charophyte *Chaetosphaeridium globosum*: insights into the timing of the events that restructured organelle DNAs within the green algal lineage that led to land plants. Proc Natl Acad Sci USA 99:11275–11280

Turmel M, Otis C, Lemieux C (2002b) The complete mitochondrial DNA sequence of *Mesostigma viride* identifies this green alga as the earliest green plant divergence and predicts a highly compact mitochondrial genome in the ancestor of all green plants. Mol Biol Evol 19:24–38

Turmel M, Otis C, Lemieux C (2003) The mitochondrial genome of *Chara vulgaris*: insights into the mitochondrial DNA architecture of the last common ancestor of green algae and land plants. Plant Cell 15:1888–1903

Ueda M, Nishikawa T, Fujimoto M, Takanashi H, Arimura S, Tsutsumi N, Kadowaki K (2008) Substitution of the gene for chloroplast RPS16 was assisted by generation of a dual targeting signal. Mol Biol Evol 25:1566–1575

Ullrich H, Lättig K, Brennicke A, Knoop V (1997) Mitochondrial DNA variations and nuclear RFLPs reflect different genetic similarities among 23 *Arabidopsis thaliana* ecotypes. Plant Mol Biol 33:37–45

Unseld M, Marienfeld JR, Brandt P, Brennicke A (1997) The mitochondrial genome of *Arabidopsis thaliana* contains 57 genes in 366,924 nucleotides. Nat Genet 15:57–61

Vangerow S, Teerkorn T, Knoop V (1999) Phylogenetic information in the mitochondrial *nad5* gene of pteridophytes: RNA editing and intron sequences. Plant Biol 1:235–243

Vaughn JC, Mason MT, Sper-Whitis GL, Kuhlman P, Palmer JD (1995) Fungal origin by horizontal transfer of a plant mitochondrial group I intron in the chimeric *CoxI* gene of *Peperomia*. J Mol Evol 41:563–572

Volkmar U, Knoop V (2010) Introducing intron locus *cox1*i624 for phylogenetic analyses in bryophytes: on the issue of *Takakia* as sister genus to all other extant mosses. J Mol Evol 70:506–518

Volkmar U, Groth-Malonek M, Heinrichs J, Muhle H, Polsakiewicz M, Knoop V (2011) Exclusive conservation of mitochondrial group II intron nad4i548 among liverworts and its use for phylogenetic studies in this ancient plant clade. Plant Biol (Stuttg) 39:2890–2902

Wahrmund U, Groth-Malonek M, Knoop V (2008) Tracing plant Mitochondrial DNA evolution: rearrangements of the ancient mitochondrial gene cluster *trn*A-*trn*T-*nad7* in liverwort phylogeny. J Mol Evol 66:621–629

Wahrmund U, Quandt D, Knoop V (2010) The phylogeny of mosses – addressing open issues with a new mitochondrial locus: group I intron cobi420. Mol Phylogenet Evol 54:417–426

Wang D, Wu YW, Shih AC, Wu CS, Wang YN, Chaw SM (2007) Transfer of chloroplast genomic DNA to mitochondrial genome occurred at least 300 MYA. Mol Biol Evol 24:2040–2048

Wang B, Xue J, Li L, Liu Y, Qiu YL (2009) The complete mitochondrial genome sequence of the liverwort *Pleurozia purpurea* reveals extremely conservative mitochondrial genome evolution in liverworts. Curr Genet 55:601–609

Wang DY, Zhang Q, Liu Y, Lin ZF, Zhang SX, Sun MX, Sodmergen (2010) The levels of male gametic mitochondrial DNA are highly regulated in angiosperms with regard to mitochondrial inheritance. Plant Cell 22:2402–2416

Ward BL, Anderson RS, Bendich AJ (1981) The mitochondrial genome is large and variable in a family of plants (Cucurbitaceae). Cell 25:793–803

Wikström N, Pryer KM (2005) Incongruence between primary sequence data and the distribution of a mitochondrial atp1 group II intron among ferns and horsetails. Mol Phylogenet Evol 36:484–493

Woloszynska M (2010) Heteroplasmy and stoichiometric complexity of plant mitochondrial genomes–though this be madness, yet there's method in't. J Exp Bot 61:657–671

Won H, Renner SS (2003) Horizontal gene transfer from flowering plants to *Gnetum*. Proc Natl Acad Sci USA 100:10824–10829

Xue JY, Liu Y, Li L, Wang B, Qiu YL (2010) The complete mitochondrial genome sequence of the hornwort *Phaeoceros laevis*: retention of many ancient pseudogenes and conservative evolution of mitochondrial genomes in hornworts. Curr Genet 56:53–61

Yamato K, Ogura Y, Kanegae T, Yamada Y, Ohyama K (1992) Mitochondrial genome structure of rice suspension culture from cytoplasmic male-sterile line (A-58CMS): reappraisal of the master circle. Theor Appl Genet 83:279–288

Yoshinaga K, Iinuma H, Masuzawa T, Uedal K (1996) Extensive RNA editing of U to C in addition to C to U substitution in the *rbcL* transcripts of hornwort chloroplasts and the origin of RNA editing in green plants. Nucleic Acids Res 24:1008–1014

Zanlungo S, Quinones V, Moenne A, Holuigue L, Jordana X (1994) A ribosomal protein S10 gene is found in the mitochondrial genome in *Solanum tuberosum*. Plant Mol Biol 25:743–749

Zhang T, Fang Y, Wang X, Deng X, Zhang X, Hu S, Yu J (2012) The complete chloroplast and mitochondrial genome sequences of Boea hygrometrica: insights into the evolution of plant organellar genomes. PLoS One 7:e30531

Chapter 9

Promiscuous Organellar DNA

Andrew H. Lloyd
School of Molecular and Biomedical Science, The University of Adelaide, South-Australia 5005, Australia
Institut Jean-Pierre Bourgin, UMR1318, Institut National de la Recherche Agronomique, 78026 Versailles, France

Mathieu Rousseau-Gueutin and Jeremy N. Timmis*
School of Molecular and Biomedical Science, The University of Adelaide, South-Australia 5005, Australia

Anna E. Sheppard
School of Molecular and Biomedical Science, The University of Adelaide, South-Australia 5005, Australia
Department of Evolutionary Ecology and Genetics, Zoological Institute, Christian Albrechts University of Kiel, Am Botanischen Garten 1-9, 24118 Kiel, Germany

Michael A. Ayliffe
CSIRO Division of Plant Industry, GPO Box 1600, ACT 2601, Australia

*Author for correspondence, e-mail: jeremy.timmis@adelaide.edu.au

R. Bock and V. Knoop (eds.), Genomics of Chloroplasts and Mitochondria, Advances in Photosynthesis and Respiration 35, pp. 201–221,
DOI 10.1007/978-94-007-2920-9_9, © Springer Science+Business Media B.V. 2012

Summary

Endosymbiotic transfer of DNA from the cytoplasmic organelles (mitochondria and chloroplasts) to the nucleus has been a major factor driving the origin of new nuclear genes, a process central to eukaryote evolution. Typically, transfer of organelle DNA to the nucleus is quickly followed by decay, deletion and rearrangement. However, in rare instances these new sequences lead to functional relocation of organelle genes to the nucleus or the generation of genes with novel function. Similar transfer of chloroplast DNA has also added to the complexity of plant mitochondrial genomes. Significantly, these processes are ongoing, making promiscuous organellar DNA an important contributor to the continued evolution of plant genomes.

I. Introduction

The nucleus, with its translational machinery in the cytoplasm, and the mitochondrion comprise two separate genetic compartments of eukaryotic cells. In plants, algae and some protist lineages, the plastid is a third genetic compartment. The mitochondria and plastids (herein referred to as "cytoplasmic organelles" or simply "organelles") have an endosymbiotic origin and are the extant descendants of once free-living α-proteobacteria and cyanobacteria, respectively. Following their incorporation into the ancestor of the eukaryote cell, the ancestors of these two cytoplasmic organelles underwent large-scale genome reduction so that their current genomes contain only 1–5% of the gene complement found in any candidate modern free-living prokaryotic relative. This genome reduction was enabled both by intracellular redundancy after cohabitation and by functional relocation of genes to the nucleus. In plants, recent functional relocation of organellar genes has been documented and transfer of non-functional DNA to the nucleus is still happening at very high frequency. As a result of the latter process, tracts of DNA that are essentially identical to regions of the extant plastid and mitochondrial genomes are found within all photosynthetic (or once photosynthetic) eukaryote nuclear genomes.

II. Organelle Genome Reduction

A. Evolutionary Gene Transfer to the Nucleus

Consistent with their endosymbiotic origin, chloroplasts and mitochondria retain essentially prokaryote-like genomes. They are

Abbreviations: CaMV – Cauliflower mosaic virus; DSB – Double strand break; GUS – β-glucuronidase; *mtpt* – Mitochondrial integrant of plastid DNA; MYA – Million years ago; NHEJ – Non-homologous end joining; *norg* – Nuclear integrant of organellar DNA; *numt* – Nuclear integrant of mitochondrial DNA; *nupt* – Nuclear integrant of plastid DNA; TAIL-PCR – Thermal asymmetric interlaced PCR; T-DNA – Transfer DNA; TOC/TIC – Translocase at the outer/inner envelope membrane of chloroplasts; TOM/TIM – Translocase at the outer/inner envelope membrane of mitochondria

separate genetic compartments in that they enclose separate transcriptional and translational machineries necessary for intraorganellar gene expression. Their respective genomes are much reduced in size when compared with those of the extant relatives of their free-living ancestors, retaining few of the ancestral protein coding genes. Mitochondrial genomes are the most reduced in size – containing only 3–67 protein-coding genes (Timmis et al. 2004), while chloroplast genomes generally encode several more proteins – around 80 in land plants and over 200 in some algae (Timmis et al. 2004).

The reduction in genome size has been, in part, due to loss of genes made redundant when the endosymbiont became resident within the eukaryote cell. However, much of the disappearance of genes from the endosymbiont's genome is due to the functional relocation of cytoplasmic organelle genes to the nuclear genome. In many cases, proteins encoded by these relocated genes retain their original role in organellar biogenesis. These, now nuclear-encoded, genes that control organelle biogenesis and function are transcribed in the nucleus, their mRNAs are translated on cytoplasmic ribosomes and the proteins are then imported into the appropriate organelle.

The acquisition of nuclear-encoded proteins from the cytoplasm required the development of sophisticated protein import machineries, most notably the TOM/TIM and TOC/TIC protein import pathways of the mitochondrion and chloroplast, respectively (Neupert 1997; Soll and Schleiff 2004). Other protein import pathways also exist, such as via the secretory pathway (Villarejo et al. 2005), but these are little understood (Millar et al. 2006; Li and Chiu 2010). The establishment of protein import mechanisms may have been the limiting step in the transition from an endosymbiont to an organelle (Cavalier-Smith and Lee 1985). Once protein import was established, proteins had a route back to the organelle, thereby enabling the transfer of organelle genes to the nucleus, cementing the genetic interdependence of the organelle and host cell. Not all genes that have relocated to the nucleus encode proteins predicted to be re-imported into the organelles (Martin et al. 2002), suggesting translocation of pathways to other compartments, protein/enzyme shuffling between pathways or the acquisition of novel non-organelle related function. Therefore, the relocation of genes from the cytoplasmic organelles to the nucleus has been a major contributor to the complexity of nuclear genomes and has given rise to many genes of novel function. Shorter stretches of organelle DNA, rather than whole genes, have also contributed to the complexity of nuclear genomes and these are observable as exonic sequences in nuclear contexts that otherwise appear unrelated to mitochondrial or plastid DNA (Noutsos et al. 2007) and as putative modifiers of gene expression (Knoop and Brennicke 1991).

With the current availability of the nucleotide sequence of well over 2,400 mitochondrial genomes and over 200 plastid genomes (NCBI 2011), it is clear that there is considerable diversity in the size of organelle genomes and the number of proteins that they encode. Animal mitochondrial genomes are relatively conserved at around 16 kb in length but much more diversity is seen in plants whose mitochondrial genomes range from 13 kb in *Polytomella capuana* [chlorophyta] (Smith and Lee 2008) to 983 kb in the seed plant species *Cucurbita pepo* (Alverson et al. 2010). The largest chloroplast genome currently known is that of *Floydiella terrestris*, a chlorophycean alga whose plastome has a length of 521 kb (Brouard et al. 2010). Whilst this is the largest sequenced to date, it encodes only 69 conserved proteins, whereas some red algae have smaller genomes that encode over 200 proteins (Reith and Munholland 1995).

The most reduced organelle genomes are found in organisms that have lost the major organelle biosynthetic pathways of oxidative phosphorylation (in the case of mitochondria) and photosynthesis (in the case of plastids). Hydrogenosomes – organelles that produce molecular hydrogen and ATP in anaerobic organisms – are highly reduced mitochondria found in diverse eukaryotes (Boxma et al. 2005). In most cases they appear to have lost their entire genome (van der Giezen et al. 1997; Clemens and Johnson 2000). Similarly, the smallest plastid genomes are found in lineages that have lost the ability to photosynthesise. The chloroplast genome of the parasitic underground orchid, *Rhizanthella gardneri* is 59 kb in size and codes for only 20 proteins (Delannoy et al. 2011). The non-photosynthetic apicoplasts – vestigial plastids of Apicomplexan parasites – have even smaller genomes which, in the malaria parasite *Plasmodium falciparum*, is around 34 kb in size (Wilson et al. 1996). Though it has lost its photosynthetic capacity the apicoplast and its genome appear to be indispensable, offering an interesting target for antimalarial therapy (Lim et al. 2010).

The evolutionary mechanisms and selection pressures that have driven organelle genome relocation to the nucleus are poorly understood but the presence of nuclear DNA sequences, within nearly all eukaryotic genomes, that are very similar to extant cytoplasmic organellar DNA, is clearly significant. These insertions of organelle DNA, which are the major focus of this chapter, are referred to as *numts* (nuclear integrants of mitochondrial DNA) and *nupts* (nuclear integrants of plastid DNA) or collectively as *norgs* (nuclear integrants of organellar DNA). Hazkani-Covo et al. (2010) recently instituted definitive pronunciation for these mtDNA and ptDNA integrants as "new-mights" and "new-peats", respectively.

B. Recent Gene Transfer Events

The number of genes found in plastid and mitochondrial genomes varies between species, but in all cases there are relatively few compared with the genomes of free-living prokaryotes. Therefore, it is thought that the majority of endosymbiotic gene transfer occurred early in the evolutionary history of the organelles (Timmis et al. 2004). In some lineages, including all animals where the set of mitochondrial genes is almost invariant, functional gene transfer appears to have ceased completely. In the few cases where genes are missing this is probably due to complete loss rather than transfer to the nucleus (Gissi et al. 2008). In plants, however, there is evidence of a flurry of recent functional gene transfer. Adams et al. (2000) reported 26 independent relocation events of *rps10* from the mitochondrial genomes to the nucleus amongst 277 angiosperms examined. Molecular characterisation of a number of the nuclear *rps10* genes indicated that each loss from the mitochondrial genome was likely to represent an independent transfer to the nucleus. A comparable study (Millen et al. 2001) looked at the loss of *infA* (encoding translation initiation factor I) from the chloroplast genome and discovered 24 cases of functional relocation among over 300 angiosperms. Again, molecular characterisation of nuclear *infA* genes suggested that each loss from the chloroplast genome was due to an independent transfer to the nucleus. Other elegant analyses have uncovered the complexity with which functional gene relocation has been achieved (Cusack and Wolfe 2007). A number of other genes, mainly encoding ribosomal proteins, have been lost from mitochondrial or chloroplast genomes and transferred to the nucleus in angiosperms (Rousseau-Gueutin et al. 2011), leading to considerable diversity in plant cytoplasmic organelle gene content.

In a few instances, mitochondrial genes have been replaced by nuclear genes of chloroplast origin. This is the case for *rps13* in rosids (Adams et al. 2002) and for *rpl10* in Brassicaceae and monocots (Kubo and Arimura 2010). In both cases, a nuclear gene of chloroplast origin was duplicated and one of the copies diverged so that its product is imported into mitochondria. Similarly, replacement of a chloroplast gene by a nuclear gene of mitochondrial origin has been observed. The two examples known so far are *rpl21* in *Arabidopsis thaliana* (Gallois et al. 2001) and *rps16* in *Medicago truncatula* and *Populus alba* (Ueda et al. 2008). The RPS16 protein is encoded by a single nuclear gene which is targeted to both the mitochondria and the chloroplasts. Dual targeting of ribosomal protein S16 has also been observed in species which still retain *rps16* within the chloroplast genome. These cases may represent an intermediate step in replacing the function of an organelle gene with that of a nuclear gene (Ueda et al. 2008).

The reasons that functional gene transfer has apparently been reawakened in the angiosperms are far from clear. Given the length of time available for gene relocation prior to the invention of the seed plants, it is astonishing that *infA*, for example, was not transferred earlier, given the ease with which the event has occurred since the advent of this particular taxonomic group. It may be that the selection pressures on *infA* in other taxa were unamenable to nuclear location and this situation was reversed in angiosperms. Perhaps a change in the constraints imposed by various developmental pathways reopened means of gene relocation that were available early in evolution but were lost in the interim. The development of the gametophyte in angiosperms is one possible place to look for such modifications. There may also be other forces that contribute to the reawakening of functional gene transfer in angiosperms. For example, novel mechanisms are required to explain the hypermutation observed in *ycf4* in *Lathyrus sativus* (Magee et al. 2010). The observed hypermutation also spreads into the adjacent genes *accD* and *psaI* and their intergenic region and is of great interest because the plastid genes that show hypermutation have a recent history of relocation to the nucleus (Magee et al. 2010). It is as though sequence decay of the plastid genes has enhanced gene relocation to the nucleus.

Of course any essential gene must be functionally duplicated before either one or the other copy can be lost. Thus, a functional nuclear copy of *ycf4* is expected in *L. sativus*, but extensive efforts were unable to identify it (Magee et al. 2010). A mechanism that could cause hypermutation in specific chloroplast genes has not yet been suggested.

C. Why Relocate?

What are the possible advantages of organelle genes being located in the nucleus? The highly energetic compartments where photosynthesis and oxidative phosphorylation are carried out are clearly not the ideal environments in which to maintain genetic integrity. Proposed explanations include the high rate of oxidative stress-induced mutation within organelles (Allen and Raven 1996), genome streamlining (Selosse et al. 2001), more frequent fixation of beneficial mutations (Blanchard and Lynch 2000), avoidance of Müller's ratchet (the accumulation of mutation in asexually reproducing organelles) through the benefits of sexual recombination for elimination of deleterious mutation in nuclear genes (Lynch 1996; Martin and Herrmann 1998) and the advantages of allelic variation and meiotic recombination. These suggestions, however, seem not to apply to plant organelles which have a much lower rate of accumulation of mutations (Wolfe et al. 1987), have larger organelle genomes with more noncoding DNA (Timmis et al. 2004), and where more gene loss is observed in taxa that reproduce asexually or by self-fertilisation (Brandvain et al. 2007).

It is likely that a key factor is the unidirectional nature of transfer of genes to the nucleus. This is promoted by the high frequency translocation of gene rich organelle DNA into the nucleus and the relatively rare, or entirely absent, transfer of DNA encoding complete genes from the nucleus to the organelles. If transfer leads to two functional copies, one copy may then be lost. If the nuclear copy is lost the chloroplast gene is able to transfer again at a later stage, but if the organelle copy is lost then the nucleus becomes the permanent location of the gene establishing a 'gene-transfer ratchet' (Doolittle 1998). As long as there remains a polarity in the direction of DNA transfer, then relocation to the nucleus would be an inevitable consequence, even in the absence of a selective advantage (Berg and Kurland 2000).

From this background of selectively neutral gene transfer, the various mutational and/or selective pressures described above may contribute to the likelihood of gene transfer by altering the respective likelihoods of organelle or nuclear gene inactivation. These pressures may have been considerably different early in evolution when the majority of transfer is likely to have occurred. The low rate of mutation in extant plant organelle genomes (Wolfe et al. 1987), presumably due to the establishment of plant-specific DNA repair and/or recombination pathways (Marechal and Brisson 2010), together with efficient gene conversion mediated by polyhaploidy (Khakhlova and Bock 2006), may well have led to a slowing in the rate of gene transfer. This could explain the differences in genome size and gene content between plant and animal mitochondrial genomes. If this is the case, it would suggest that the accumulation of mutations rather than the energetic and replicative advantage of a small organelle genome drive gene transfer to the nucleus.

D. Why Retain an Organellar Genome?

Thousands of genes have relocated from the plastids and mitochondria to the nucleus, so why do any remain given the energy outlay in maintaining all of the transcriptional and translational machinery required for the retention of alternative genetic systems. The hydrophobicity hypothesis suggests that highly hydrophobic proteins are hard to export from the cytosol to the organelles and that this precludes relocation of these genes to the nucleus (Vonheijne 1986; Daley and Whelan 2005). Counter to this, however, the chloroplast-encoded hydrophobic protein D1 can be imported from the cytosol to the chloroplast when experimentally equipped with a transit peptide and expressed from a nuclear gene (Cheung et al. 1988). In addition, several other hydrophobic organellar proteins (such as the ADP-ATP carriers) are known to be nuclear encoded (Allen 2003). The Co-location for Redox Regulation or CoRR hypothesis (Allen 2003) maintains that there is a key set of genes whose expression must be directly controlled by the redox state of their gene product or interacting electron carriers. This requires separate (organellar rather than nuclear) gene expression, as redox state is likely to vary between the many organelles within a single cell. Recently a sensor kinase has been identified that links the redox state of an electron carrier connecting the two photosystems, with chloroplast gene expression (Puthiyaveetil et al. 2008). Neither of these hypotheses, however, appear to explain the retention of genomes in non-photosynthetic plastids such as those found in parasitic plants or the apicoplasts of the Apicomplexa and several other hypotheses have been proposed to explain the situation in these cases (Barbrook et al. 2006). The 'essential tRNA' hypothesis was proposed based on the observation that the tRNA encoded by the plastid gene *trnE* is involved in tetrapyrrole biosynthesis and so may be essential even in the absence of organellar protein biosynthesis (Barbrook et al. 2006). The 'limited transfer window' hypothesis posits that organisms containing a single organelle per cell will have little opportunity for gene transfer as organelle breakdown, which may be necessary for the release of DNA, will be lethal (Barbrook et al. 2006). It may be that no single hypothesis is able to explain adequately the retention of organelle genomes in all cases and different combinations of factors may be responsible in different taxonomic groups.

III. Promiscuous DNA: Ongoing Organelle DNA Transfer to the Nucleus

A. Organelle Sequences in Nuclear Genomes

A prerequisite for the functional relocation of plastid and mitochondrial genes to the nucleus is a nucleic acid transfer mechanism. The first indications that transfer of organelle DNA to the nucleus continues today came to light about 30 years ago with the identification of nuclear sequences that are closely similar or identical to extant organelle DNA (van den Boogaart et al. 1982; Timmis and Scott 1983). The relatively recent transfer of these sequences to the nucleus was subsequently inferred from their high sequence identity (i.e. >99%) to existing organelle genes (Fig. 9.1).

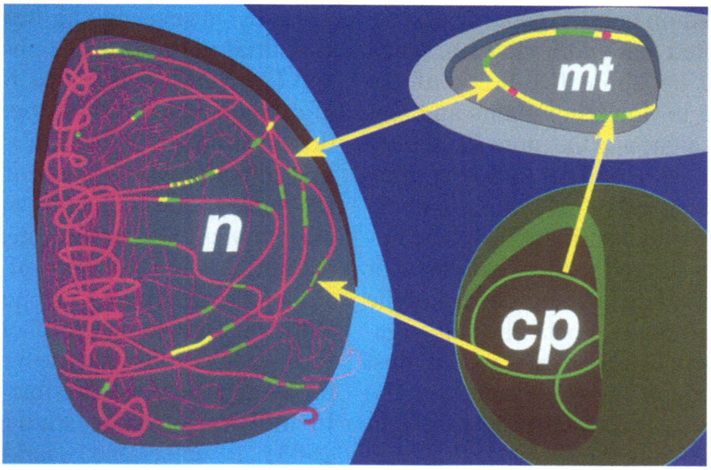

Fig. 9.1. Promiscuous DNA in the three genetic compartments of a plant cell. *n* nucleus, *mt* mitochondrion, *cp* chloroplast.

Whole genome sequencing has since revealed extensive tracts of chloroplast and/or mitochondrial DNA in the nuclear genomes of almost all eukaryotes studied (Timmis et al. 2004; Hazkani-Covo et al. 2010).

The arrangement of these sequences has been studied in detail in Arabidopsis and rice and has been found to be quite varied (Richly and Leister 2004a, b; Noutsos et al. 2005). A large proportion of the total *norg* content is found in a relatively small number of large *norgs* that can be tens or hundreds of kb in length. The remainder is found in a large number of smaller *norgs* scattered throughout the genome (Richly and Leister 2004b). Of the large *norg* loci, some are continuous sequences of chloroplast or mitochondrial origin and are clearly the result of the insertion of a single molecule, while others contain multiple fragments of DNA from diverse parts of the chloroplast or mitochondrial genome or both (Noutsos et al. 2005). Some loci of the latter type, probably represent insertions of a single contiguous fragment of organelle DNA that has since undergone deletion and/or rearrangement (Matsuo et al. 2005). However, these loci may also be formed by the insertion of multiple fragments of DNA from diverse regions of the chloroplast genome in a single event (Lloyd and Timmis 2011), or by multiple sequential insertions at a single locus (Noutsos et al. 2005). Several other *norg* loci are highly complex mosaics containing up to 80 disparate ~50–100 bp segments of the chloroplast and mitochondrial genome arranged end to end (Noutsos et al. 2005). How these loci arise is yet to be explained satisfactorily, but similar mosaics have been observed that are comprised of many short stretches of transposable element sequence (David Adelson 2011, personal communication).

Large *norg* insertions have also been observed in other species. Recently, in situ hybridization in the maize inbred line B73 identified a *nupt* that includes almost the entire 140 kb chloroplast genome on chromosome 5 (Roark et al. 2010) and a *numt* containing the majority of the 570 kb mitochondrial genome on chromosome 9 (Lough et al. 2008). These studies also showed that *numts* and *nupts* varied greatly among different inbred maize lines indicating that there have been frequent recent insertions of organelle DNA into maize nuclear genomes. Current investigations such as the 1,000 genomes projects in humans and Arabidopsis should contribute greatly to understanding the intra-species variation of *norgs* and perhaps reveal potential evolutionary ramifications.

The precise contribution of *nupt* and *numt* sequences to the nuclear genome is hard to determine. Based on current genome assemblies it is estimated that *nupts* and *numts* each generally make up about 0.1–0.2% of the nuclear genome in flowering plants and significantly less in algae and the moss *Physcomitrella patens* (Table 9.1). This, however, may be the 'tip of the iceberg' as whole genome assemblies often underestimate the contribution of organelle-derived sequences to nuclear genomes. This is in large part an artefact of the elimination of seemingly 'contaminating' organelle DNA sequences – a process which must also often exclude *norgs*. An example is the honeybee genome which was initially thought to have little or no mitochondrial DNA within the nucleus (Leister 2005) but has since been found, using a different assembly, to have one of the most extensive *numt* complements (Behura 2007; Hazkani-Covo et al. 2010).

Another problem lies in the assembly of regions of the genome that contain large duplications. Chromosome 2 in Arabidopsis was initially reported to contain a 270 kb *numt* (Lin et al. 1999) but Stupar et al. (2001) later showed that this *numt* was in fact ~620 kb in length and contained several large internal duplications. The authors were only able to determine the *numt* size using fibre-FISH and showed that contig assembly using BACs tended to minimise clone length, missing large duplications. Despite this finding, this region is still only 270 kb in length in the current chromosome 2 assembly (Build 9.1, 14th Oct 2009) and recent studies (Richly and Leister 2004a; Hazkani-Covo et al. 2010) have therefore greatly underestimated total *numt* size in Arabidopsis. This problem presumably holds for *nupts* as well and will be compounded in genomes shotgun sequenced using high-throughput short-read platforms.

B. Evidence of Frequent Plastid and Mitochondrial DNA Transfer to the Nucleus

In some species it has been possible to determine experimentally the frequency with which organellar DNA moves into the nucleus. This was initially investigated in yeast by measuring the transfer of a mitochondrial plasmid to the nucleus (Thorsness and Fox 1990) which was found to occur at high frequency (~2×10^{-5} per cell per generation). Although the plasmid DNA in these first experiments was not incorporated into the nuclear chromosomes, subsequent work, also in yeast, observed integration of mitochondrial DNA at sites of nuclear double strand break repair (Ricchetti et al. 1999). With the development of chloroplast transformation in tobacco (Svab et al. 1990), similar studies became possible in higher plants. In the first of these studies (Huang et al. 2003) a selectable marker gene (*neo*), equipped for exclusive nuclear expression, was introduced into the chloroplast genome of tobacco. Transplastomic pollen was used to fertilise female wild-type plants and the resultant progeny was screened for kanamycin resistance (*neo* expression). In a large screen of 250,000 seedlings, 1 in 16,000 pollen grains were inferred to carry a copy of *neo* transferred from chloroplast DNA to the nucleus in the germline of the transplastomic male parent. A similar study measured the rate of transfer in somatic cells (Stegemann et al. 2003) and transfer was shown to occur once in approximately 5,000,000 cells. Although still relatively frequent, this was substantially less common than the transfer observed in the male germline and prompted the suggestion that degradation of the chloroplast during pollen development (associated with uniparental inheritance) may provide more opportunity for nuclear DNA transfer by liberating fragments of chloroplast DNA. This hypothesis was supported by a third study that measured the rate of gene transfer in both the female and male germline (Sheppard et al. 2008). An even greater frequency of gene transfer through the male germline was reported (1 in 11,000 pollen) which far exceeded transfer in the female germline where a single transfer event was observed in a screen of over 270,000 ovules (Sheppard et al. 2008).

In each of these screens the chloroplast gene not only transfers to the nucleus but also must integrate into the nuclear chromosomes.

Table 9.1. Current estimates of *numt* and *nupt* content in plant nuclear genomes

	Order	Family		Subfamily	Genus/species	*numt* content Kbp (%)	*nupt* content Kbp (%)
Chlorophyta	Volvocales	Chlamydomonadaceae			*Chlamydomonas reinhardtii*	2.8[a] (0.003)	2.4[b] (0.002)
Bryophyta	Funariales	Funariaceae			*Physcomitrella patens*	76[a] (0.02)	–
Liliopsida	Poales	Poaceae	BEP clade	Ehrhartoideae	*Oryza sativa ssp. indica*	409[c], 823[a] (0.16, 0.24)	804[b], 1176[d] (0.17, 0.25)
				Pooideae	*Brachypodium distachyon*	488[e] (0.14)	275[e] (0.08)
			PACCAD clade		*Sorghum bicolor*	539[a] (0.07)	–
Eudicots	Brassicales	Brassicaceae			*Arabidopsis thaliana*	198[c], 305[a] (0.16, 0.24)	35[b], 21[d] (0.03, 0.02)
		Caricaceae			*Carica papaya*	858[f] (0.23)	785[f] (0.21)
	Vitales	Vitaceae			*Vitis vinifera*	–	570[d] (0.12)
	Malpighiales	Salicaceae			*Populus trichocarpa*	–	679[d] (0.12)

For each species the *numt* and *nupt* content is given as a total length and as a percentage of the nuclear genome, – indicates not determined
[a]Hazkani-Covo et al. (2010)
[b]Richly and Leister (2004b)
[c]Richly and Leister (2004a)
[d]Arthofer et al. (2010)
[e]Vogel et al. (2010)
[f]Ming et al. (2008)

To investigate the steps in this process, Sheppard et al. (2008) introduced a *GUS* reporter gene (again designed exclusively for nuclear expression) into the chloroplast genome and leaves of the transplastomic plant were stained for GUS activity to detect cells in which the gene had transferred to the nucleus. In this instance blue staining cells represented transient expression from the nucleus/cytoplasm as well as transfer followed by stable integration into a transcriptionally active region of the nuclear genome. Interestingly, total transfer (transient and stable) was found to be 25–270-fold higher than the stable somatic transfer of *neo* detected by Stegemann et al. (2003) suggesting that most blue spots resulted from transient expression. The lack of any large and the rarity of small groups (mitotic lineages) of GUS-expressing cells suggested that few stable integrations occurred early in leaf development.

C. Evolutionary Fate of Nuclear Located Cytoplasmic Organelle DNA (norgs)

Given the constant deluge of organellar DNA entering the nuclear genome in recent evolutionary time (see also Sect. IV below), it is expected that a counterbalancing eradication of these sequences occurs to prevent continual genome expansion. This was first alluded to with the observation that, for *nupts* over 500 bp in length, there is an inverse relationship between their age (based on sequence identity to the chloroplast genome) and their size (Richly and Leister 2004b). This finding has subsequently been found to hold true for *norgs* in *Brachypodium distachyon* (Vogel et al. 2010) and *Carica papaya* (Ming et al. 2008) and suggests that insertion of large *nupts* is followed by fragmentation and deletion. Direct experimental observation of frequent deletion of about 50% of newly transferred chloroplast sequences has demonstrated the extreme instability of plastid DNA integrants in the tobacco nucleus (Sheppard and Timmis 2009). So far, it has not been possible to determine how much of the integrant is lost by recovering the sequence that remains. This *nupt* deletion occurred within 1–2 generations of insertion and it may be that more integrant loci would show instability over longer, but still evolutionarily relevant, timescales.

The deletion of organelle DNA is most unlikely to be an exact excision and partial deletion would lead to novel arrangements of organelle and nuclear DNA. The deletion may also be accompanied by other rearrangements including inversions and new insertions of organellar DNA and transposable elements (Guo et al. 2008). Richly and Leister (2004b) observed ‘tight’ and ‘loose’ clusters of organellar sequence in nuclear genomes of rice and Arabidopsis which they suggest represent progressive steps of degradation and rearrangement of large initial insertions. Deletions and other rearrangements may be part of the mutational processes that, in rare instances, lead to the activation of newly transferred genes (Bock and Timmis 2008; Lloyd and Timmis 2011).

Base substitution and indels appear to play a significant role in the evolution of *norgs*. In plants, a significant bias in C→T and G→A mutations has been observed in large recent integrants of organelle DNA (Huang et al. 2005). This mutational bias is consistent with spontaneous deamination of 5-methylcytosine inferring that these *norgs* are methylated. Studies linking the stability of *norg* sequences with methylation and chromatin structure have not yet been reported.

IV. Mechanisms of Gene Transfer to the Nucleus

A. Relocation of Genetic Material

The first step in transfer of a gene to the nucleus is the relocation of genetic material from the organelle. In general, the availability of cytoplasmic organelle nucleic acid fragments for transfection of the nucleus is likely to be made possible through loss of integrity of the organelle membrane, either through various physiological stressors or programmed degradation during development. Various environmental stress factors and developmental stages are known to trigger programmed organelle degradation (Kundu and Thompson 2005; Stettler et al. 2009; Wada et al. 2009) and these may lead to increased ingress of organelle DNA to the nucleus. Recently, cold stress (Ruf et al. 2010) has been shown to increase the rate at which a chloroplast gene relocates to the nucleus in tobacco.

Uniparental inheritance is also implicated in leading to the presence of organelle nucleic acids in the cytoplasm that may find their way into the nucleus. In many sexually reproducing eukaryotes, only one sex contributes cytoplasmic genes to the zygote. How this uni-parental inheritance is achieved varies amongst species, but in general, the cytoplasmic organelles are degraded and/or excluded from one of the gametes or sex-specific loss of organelles occurs after fertilisation (Birky 2001). In tobacco, chloroplast genes are maternally inherited and this parallels the observation that DNA transfers from the chloroplast to the nucleus far more frequently in the male germ line than that of the female (Sheppard et al. 2008). This difference has been suggested to be due to the release of chloroplast DNA into the cytoplasm during chloroplast degradation/exclusion in the developing male gametophyte. For unicellular organisms that have only a single organelle per cell, DNA transfer is likely to be very limited, as degradation of the single but essential organelle will lead to cell death (Barbrook et al. 2006). The *Chlamydomonas reinhardtii* nuclear genome has a low *norg* content (Table 9.1) and large screens failed to detect transfer of a chloroplast gene to the nucleus (Lister et al. 2003). We considered whether transfer could occur during the diploid phase when the chloroplast is briefly duplicated but screens of many millions of germinated zygotes from reciprocal crosses between transplastomic and wild type *C. reinhardtii* failed to identify DNA transfer (unpublished results).

Further understanding of how various stress factors and modes of organelle inheritance affect chloroplast-to-nucleus DNA transfer should be an interesting area of future research in the context of the wide climatic and ecological ranges that plants have colonised. The increasing wealth of genome sequence data will pave the way for analysis of *norgs* in different ecotypes and it will be interesting to see if any relationship exists between *norg* content and environmental conditions or geographical distribution. Further understanding of how stress and organelle integrity affect endosymbiotic DNA transfer will also be of biotechnological significance in view of the desire to minimise transfer of chloroplast transgenes to the nucleus.

B. Is There an RNA or DNA Intermediate?

It is generally held that the majority of organelle nucleic acid transfer to the nucleus occurs via DNA (Timmis et al. 2004; Kleine et al. 2009), although this still remains to be experimentally proven. Some studies of the transfer of plant mitochondrial genes to the nucleus showed that nuclear copies resembled spliced, edited mRNAs and led to the suggestion that transfer was via a reverse transcribed RNA intermediate (Nugent and Palmer 1991; Grohmann et al. 1992; Adams et al. 2000). There are, however, alternative explanations, such as the elimination of introns and editing sites in mitochondrial genomes through cDNA recombination (Henze and Martin 2001), that account for these observations without involving RNA-mediated transfer. Some further evidence also suggests DNA-mediated transfer: non-coding regions of the chloroplast genome are found in nuclear genomes as abundantly as highly transcribed genic regions of the organellar genomes (Matsuo et al. 2005) and some very large nuclear insertions of organellar sequence (>100 kb) have been found (Stupar et al. 2001; The Rice Chromosome 10 Sequencing Consortium 2003), suggesting direct DNA transfer. Direct experimental evidence of RNA mediated transfer is lacking, as is the determination of the relative contributions of RNA and/or DNA mediated transfer. At least one study designed to observe transfer via an RNA intermediate failed to detect any such transfer (Sheppard et al. 2011).

C. Integration into Nuclear Chromosomes

Once the organelle nucleic acid has entered the nucleus it must be integrated into nuclear chromosomes and be included in the gametes of sexually reproducing organisms if it is to make a contribution to the evolution of the nuclear genome. It is thought that most integration of organellar DNA occurs via non-homologous end joining at sites of double strand break (DSB) repair (Kleine et al. 2009) and this has been shown to occur in yeast (Ricchetti et al. 1999). DSBs were induced in the yeast nuclear genome through expression of the rare cutting endonuclease I-SceI and insertion of mitochondrial DNA was observed in a proportion of repair events. Interestingly, in some repair events, DNA from two disparate regions of the mitochondrial genome was inserted at a single location. Similar capture of non-mitochondrial DNA has also been observed at sites of DSB repair in yeast (Haviv-Chesner et al. 2007) as well as in plant and mammalian systems (Salomon and Puchta 1998; Lin and Waldman 2001). In these studies DSBs were induced by transiently introducing a plasmid, or T-DNA, encoding a rare cutting endonuclease. This rare-cutting endonuclease cuts at a specific restriction site introduced into the nuclear genome and repair events were then analysed by PCR. Insertion of the T-DNA or plasmid DNA was often observed, as were insertions of nuclear repetitive elements such as retro-transposons and micro-satellites. While insertion of organellar DNA has so far only experimentally been observed in yeast, the fact that extra-chromosomal DNA can be captured at sites of DSB repair in plants and animals suggests this process applies more widely.

The insertion of *norgs* has been also investigated in several bioinformatic analyses and these suggest more than one pathway for integration (Leister 2005). Some integrants show a very simple arrangement likely originating when a single organellar DNA fragment inserted at a single location. Others are much more complex and are the result of multiple fragments being inserted in a single event or multiple insertions at a single location. Organelle sequences may also insert into areas of the genome that already contain *norgs* or other repetitive sequences which also adds to the complexity of these loci. There is some evidence that organellar DNA integrates more frequently into intergenic regions in rice and Arabidopsis (Richly and Leister 2004b), in particular those containing mobile elements (Mishmar et al. 2004). Large *nupts* have also been shown to preferentially locate to pericentromeric regions in rice (Matsuo et al. 2005) which are known to be DSB hotspots (Blitzblau et al. 2007) and to contain a high density of transposable elements (Hall et al. 2006). A recent study has also linked *numt* insertion sites in yeast to origins of replication (Lenglez et al. 2010), which led the authors to suggest that these sites may be prone to DSBs resulting in high levels of insertion. These findings point toward DSB repair, possibly at sites of transposon excision (Leister 2005), as a pathway for the nuclear insertion of organellar sequences. The presence of such a DNA repair/integration mechanism would contribute significantly to the complex arrangement of organellar sequences integrating into the nuclear genome. This would be important from an evolutionary perspective as it would lead to the creation of novel sequence arrangements which, in some instances, may result in nuclear activation of the transferred organelle genes.

The cross-over in the insertion pathway and chromosomal location of *norgs* and repetitive DNA elements shows that these sequences can be dealt with in very similar ways by the nuclear DNA repair/maintenance machinery. To date, studies have focussed exclusively on either organellar DNA or transposons and other repetitive sequence. There may be significant advantage to both fields if a more unified approach is taken to investigating these areas.

Although bioinformatic analyses of *norgs* due to evolutionary transfer have added considerably to our understanding of these sequences, they are limited in that a *norg* sequence cannot usually be compared with that of the nuclear sequence prior to insertion. This makes identification of micro-homology and other indicators of NHEJ difficult to assess. Also it is impossible to determine how much of the observed complexity of *norg* sequences is due to the primary insertion event and how much is due to subsequent fragmentation or insertion at this locus. Some partial characterisation of experimentally transferred *norgs* has been undertaken and suggests that micro-homology is involved in the insertion of these sequences (Huang et al. 2004). A fuller understanding will come with complete characterisation of de novo *norgs* and comparison with their pre-insertion sequences. This remains an important future step but is a challenging task. The reason for this is that the new integrants are often very large (dozens of kb) and the nuclear genome already contains *norg* sequences in high copy number. These, together with the superabundant cellular plastid DNA, preclude the design of primers in organellar sequences. Therefore, techniques generally used for determining the junction sequence in transgenic lines, such as genome walking, TAIL-PCR or inverse PCR, cannot be used in determining the pre-insertion site unless one of the marker genes is very close to the integrant boundary (Sheppard and Timmis 2009).

V. Activation of a Newly Transferred Organelle Gene

A. Examples of Organellar Gene Activation in the Nucleus

Only in a very few instances will transfer of organellar DNA to the nucleus lead to the functional relocation of an organelle gene. In most cases, organelle sequences transferred to the nucleus have the same fate as other non-coding DNA – freely accumulating mutation and degrading over time. The low mutation rate in plant organelle genomes means that the extant organelle genomes provide a historic reference for the sequence at the time of insertion, from which it is possible to derive many insights into the various ways in which *norg* sequences evolve that would otherwise not be possible. In a few rare cases, these sequence rearrangements and changes in base composition lead to activation of newly transferred genes. Activation, in the majority of cases, must be a multistep process and requires the acquisition of a nuclear promoter, a polyadenylation signal and, if the protein is to be targeted back to the organelle, a transit peptide or an alternative mechanism for protein targeting. Several bioinformatic studies have highlighted various means by which organellar genes have recently become activated in nuclear genomes. One such study investigated transfer of the maize gene encoding the mitochondrial protein RPS14 to the nucleus (Figueroa et al. 1999). The gene had inserted into an intron of the iron-sulphur protein subunit of succinate dehydrogenase (*sdh2*) gene and was expressed by differential splicing of the mRNA with both proteins using the SDH2 transit peptide for targeting to the mitochondria. In a similar case, the chloroplast *rpl32* gene was transferred to the nucleus in an ancestor of mangrove and poplar (Cusack and Wolfe 2007) where it inserted into an intron of the gene encoding the chloroplast superoxide dismutase (*SODcp*) to form the chimeric *SODcp-rpl32* gene. In mangrove, the SODcp protein and a SODcp amino terminus/RPL32 fusion protein are expressed from the single promoter through differential splicing. Both proteins are then targeted to the plastid using the SODcp transit peptide. In poplar, evolutionary experimentation has taken the process one step further with the duplication and subfunctionalization of the *SODcp-rpl32* gene. One copy has lost the RPL32 coding sequence and now solely encodes SODcp, the other now exclusively expresses the SODcp amino terminus/RPL32 fusion protein. There are numerous other examples of genes that have recently transferred to the nucleus in angiosperms, many of which have also hijacked transit peptides from existing nuclear encoded organellar proteins (Liu et al. 2009).

B. Experimental Attempts to Detect Activation of a Chloroplast Gene After Transfer to the Nucleus

Experimental attempts have been made to reconstruct functional gene transfer to gain a better understanding of the diverse processes involved and the frequency with which newly transferred prokaryotic genes become activated in the nucleus. Stegemann and Bock (2006) showed functional activation of a chloroplast marker gene *aadA* that had been recently transferred to the nucleus in tobacco. In each case, *aadA* was activated, through intervening deletions, by the nearby strong CaMV 35S nuclear promoter that was integral to their experimental cassette and present in the same transcriptional polarity. In no case was activation achieved by acquisition of a native nuclear promoter and so the frequency of a 'natural' gene transfer event remains unclear.

Interestingly, they found that the *aadA* transcripts were polyadenylated despite the lack of any changes in the *psbA* 3' UTR found downstream of the *aadA* open reading frame. Examination of the *psbA* terminator revealed a sequence that matched the rather flexible AT-rich plant polyadenylation consensus sequence and this was the in vivo site of *aadA* polyadenylation. This led the authors to suggest that the AT-rich nature of plastid non-coding sequences may provide many fortuitous polyadenylation sites – greatly aiding the process of functional gene transfer. This could possibly be extended to other AT-rich regulatory motifs such as a TATA box. Indeed, the tobacco chloroplast *psbA* promoter has been shown to have weak nuclear activity that is dependent upon TATA and CAAT boxes present fortuitously (Cornelissen and Vandewiele 1989), but cryptic nuclear activity of any other chloroplast promoters remains unknown.

VI. Plastid DNA in Higher Plant Mitochondria

Most angiosperms also have large chunks of plastid DNA in their mitochondrial genome. The first report of the existence of DNA sequences that have been transferred from the chloroplast to the mitochondrion in higher plants was published three decades ago by Stern and Lonsdale (1982). In this study, they showed that a 12 kb DNA sequence present in the maize mitochondrial genome was essentially identical to part of the inverted repeat of the maize chloroplast genome. These chloroplast-derived sequences or mitochondrial plastid DNA sequences were later designated "*mtpts*" (mighty-peats) (Wang et al. 2007).

With the current availability of the nucleotide sequence of 48 plant mitochondrial genomes (NCBI 2010), it is now clear that mitochondrial genomes of seed plants are rich in sequences derived from the chloroplast. These *mtpts* constitute 1–11% of the mitochondrial genome in different species of seed plants (Table 9.2) and the transfer seems to involve random sections of the chloroplast genome (Wang et al. 2007). While no sequences homologous to chloroplast DNA have been detected in the mitochondrial genomes of bryophytes or algae (Oda et al. 1992; Turmel et al. 2003; Terasawa et al. 2007; Li et al. 2009), an insertion of chloroplast DNA has been observed in the mitochondrial genome of the lycophyte *Isoetes engelmannii* suggesting that plastid-to-mitochondrion DNA transfer phenomena began during or after the origin of vascular plants.

In vascular plants, the accumulation of *mtpts* is positively correlated with the increase of the mitochondrial genome size (Wang et al. 2007). However, the *mtpts* are not the only cause of this mitochondrial genome expansion; extra sequences also originate from the nucleus (e.g., fragments of nuclear transposable elements; Knoop et al. 1996), from other organisms (horizontal gene transfer; Goremykin et al. 2009) and also by the duplication or amplification of pre-existing sequences (Kitazaki and Kubo 2010). An interesting study (Allen et al. 2007) showed that the chloroplast DNA present in the mitochondrial genome does not only vary between species but may even vary among maize cytotypes, ranging from 16,929 bp in the maize B37 inbred line with male-sterile CMS-C cytoplasm to 29,470 bp in the male-fertile inbred line A188 (NA cytoplasm). From this study, it appears that plastid DNA can be gained and lost rapidly

Table 9.2. Mtpt content determined for mitochondrial genomes of seed plants

	Order	Family	Subfamily	Genus/species	Mitochondrial genome bp	Cp. sequence in the mitochondrion bp (%)	Length of Cp. fragments bp	Reference
Cycadophyta	Cycadales	Cycadaceae		*Cycas taitungensis*	414,903	18,113 (4.4)	–	Chaw et al. (2008)
Liliopsida	Poales	Poaceae	BEP clade Ehrhartoideae	*Oriza sativa ssp. japonica*	490,520	22,593 (6.3)	32–6,653	Notsu et al. (2002)
			Pooideae	*Triticum aestivum*	452,528	13,455 (3)	27–4,239	Ogihara et al. (2005)
			PACCAD clade	*Zea mays ssp. mays (CMS-C)*	739,719	16,929 (2.3)	34–2,220	Allen et al. (2007)
				Zea mays ssp. mays (CMS-S)	557,162	20,780 (3.7)	34–3,726	Allen et al. (2007)
				Zea mays ssp. mays (CMS-T)	535,825	23,669 (4.4)	34–3,739	Allen et al. (2007)
				Zea mays ssp. mays (fertile cytotype: NA)	701,046	29,470 (4.2)	34–3,756	Allen et al. (2007)
				Zea mays ssp. mays (fertile cytotype: NB)	569,630	26,239 (4.6)	28–12,592	Allen et al. (2007), Clifton et al. (2004)
Eudicots	Brassicales	Brassicaceae		*Arabidopsis thaliana*	366,924	3,958 (1.1)	30–930	Unseld et al. (1997)
				Brassica napus	221,853	7,950 (3.6)	43–2,181	Handa (2003)
		Caricaceae		*Carica papaya*	476,890	–	106 2,495	Ming et al. (2008)
	Caryophyllales	Caryophyllaceae		*Silene latifolia*	253,413	2,462 (1)	43–588	Sloan et al. (2010)
		Chenopodiaceae		*Beta vulgaris*	368,799	(2.1)	25–3,366	Kubo et al. (2000)
	Cucurbitales	Cucurbitaceae		*Citrullus lanatus*	379,236	22,779 (6)	–	Alverson et al. (2010)
				Cucurbita pepo	982,833	113,347 (11)	92–18,534	Alverson et al. (2010)
	Solanales	Solanaceae	Nicotianoideae	*Nicotiana tabacum*	430,597	9,942 (2.5)	–	Sugiyama et al. (2005)
	Vitales	Vitaceae		*Vitis vinifera*	773,279	68,237 (8.8)	62–9,106	Goremykin et al. (2009)

For each species, the size of the mitochondrial genome, the total size of the chloroplast sequences present in the mitochondrion (given in bp and as a percentage of the mitochondrial genome size) and their size range are presented. – indicates not determined from the mitochondrial genome. However, most of the variation in plastid DNA amount among the newly sequenced maize genomes is due to only 10 out of the 45 segments of plastid origin and only three major differences account for much of the variation in plastid DNA content.

In most cases, the predicted protein-coding *mtpts* present in the mitochondrial genome are assumed to be non-functional based on the presence of frameshift mutations and indels. However, for a minority of the *mtpt* sequences, mitochondrial functions have been discovered. For example, a tRNA sequence of plastid origin functions as a tRNA for mitochondrial translation (Kanno et al. 1997; Miyata et al. 1998). Indeed, a number of chloroplast-derived tRNA genes are transcribed and processed to mature tRNAs in mitochondria. However, the acquisition of function does not seem to be immediate, since at least one mitochondrial tRNA gene presenting 100% identity to the native plastid gene is not transcribed (Miyata et al. 1998). Amongst seed plant species, both the number and the type of mitochondrial tRNA genes replaced by a chloroplast-derived tRNA gene vary (Miyata et al. 1998). A second known impact of the *mtpt* sequences is as a source of promoters for mitochondrial genes, as demonstrated for the rice mitochondrial *nad9* gene (Nakazono et al. 1996). In this study, it was determined that the transcription of the *nad9* gene is initiated in a chloroplast-derived sequence that is located in a region upstream of the mitochondrial *nad9* gene.

In addition to the known positive impact of some chloroplast-derived sequences on mitochondrial gene function, it has recently been observed that sequences of chloroplast origin may be used in gene conversion events within the mitochondrial genome (Hao and Palmer 2009). This study reports that an internal segment (ranging from 14 to 78 bp in different species) of mitochondrial *atp1* (encoding the alpha subunit of ATP synthase) has been replaced with a plastid *atpA* sequence in a number of angiosperms belonging to diverse families. The plastid *atpA* sequence is found within a region of mitochondrially located chloroplast DNA and it seems that independent conversions occurred by intra-mitochondrial genome recombination, probably occurring well after the integration of the chloroplast *atpA* genes. The resulting "chimeric" genes – composed of mitochondrial and chloroplast sequences – might still be functional, since *atp1* has an intact open reading frame, is presumed to be a single copy gene in the mitochondrion and is not known to have been transferred to the nucleus in any angiosperm.

VII. Perspective

The constant integration of organellar DNA has had profound consequences in the evolution of eukaryote nuclear genomes (Timmis et al. 2004; Kleine et al. 2009). The ingress of DNA is followed by decay, deletion and rearrangement of these sequences, which leads to novel sequence combinations. In rare instances, these new sequences can lead to the functional relocation of organelle genes to the nucleus or the generation of genes with novel function. This process is of great evolutionary interest as it has been a major pathway for the generation of new genes in eukaryote nuclear genomes. It is also of great interest both to plant biotechnologists and the wider public in assessing the level of transgene containment provided by chloroplast transformation (Ruf et al. 2007).

Acknowledgements

We thank the Australian Research Council for financial support (Grants DP0667006 and DP0986973). This chapter is not intended to be an exhaustive review of the topic. Therefore we apologise for omitting citation and discussion of some important research contributions in the interest of brevity. We thank David Adelson for discussion.

References

Adams KL, Daley DO, Qiu YL, Whelan J, Palmer JD (2000) Repeated, recent and diverse transfers of a mitochondrial gene to the nucleus in flowering plants. Nature 408:354–357

Adams KL, Daley DO, Whelan J, Palmer JD, Mollier P, Hoffmann B, Debast C, Small I (2002) Genes for two mitochondrial ribosomal proteins in flowering plants are derived from their chloroplast or cytosolic counterparts. Plant Cell 14:931–943

Allen JF (2003) The function of genomes in bioenergetic organelles. Philos Trans R Soc Lond B Biol Sci 358:19–37

Allen JF, Raven JA (1996) Free-radical-induced mutation vs redox regulation: costs and benefits of genes in organelles. J Mol Evol 42:482–492

Allen JO, Fauron CM, Minx P, Roark L, Oddiraju S, Lin GN, Meyer L, Sun H, Kim K, Wang CY et al (2007) Comparisons among two fertile and three male-sterile mitochondrial genomes of maize. Genetics 177:1173–1192

Alverson AJ, Wei XX, Rice DW, Stern DB, Barry K, Palmer JD (2010) Insights into the evolution of mitochondrial genome size from complete sequences of *Citrullus lanatus* and *Cucurbita pepo* (Cucurbitaceae). Mol Biol Evol 27:1436–1448

Arthofer W, Schuler S, Steiner FM, Schlick-Steiner BC (2010) Chloroplast DNA-based studies in molecular ecology may be compromised by nuclear-encoded plastid sequence. Mol Ecol 19:3853–3856

Barbrook AC, Howe CJ, Purton S (2006) Why are plastid genomes retained in non-photosynthetic organisms? Trends Plant Sci 11:101–108

Behura SK (2007) Analysis of nuclear copies of mitochondrial sequences in honeybee (*Apis mellifera*) genome. Mol Biol Evol 24:1492–1505

Berg OG, Kurland CG (2000) Why mitochondrial genes are most often found in nuclei. Mol Biol Evol 17:951–961

Birky CW (2001) The inheritance of genes in mitochondria and chloroplasts: laws, mechanisms, and models. Annu Rev Genet 35:125–148

Blanchard JL, Lynch M (2000) Organellar genes: why do they end up in the nucleus? Trends Genet 16: 315–320

Blitzblau HG, Bell GW, Rodriguez J, Bell SP, Hochwagen A (2007) Mapping of meiotic single-stranded DNA reveals double-strand-break hotspots near centromeres and telomeres. Curr Biol 17: 2003–2012

Bock R, Timmis JN (2008) Reconstructing evolution: gene transfer from plastids to the nucleus. Bioessays 30:556–566

Boxma B, de Graaf RM, van der Staay GW, van Alen TA, Ricard G, Gabaldon T, van Hoek AH, Moon-van der Staay SY, Koopman WJ, van Hellemond JJ et al (2005) An anaerobic mitochondrion that produces hydrogen. Nature 434:74–79

Brandvain Y, Barker MS, Wade MJ (2007) Gene co-inheritance and gene transfer. Science 315:1685

Brouard JS, Otis C, Lemieux C, Turmel M (2010) The exceptionally large chloroplast genome of the green alga *Floydiella terrestris* illuminates the evolutionary history of the Chlorophyceae. Genome Biol Evol 2:240–256

Cavalier-Smith T, Lee JJ (1985) Protozoa as hosts for endosymbioses and the conversion of symbionts into organelles. J Protozool 32:376–379

Chaw SM, Shih ACC, Wang D, Wu YW, Liu SM, Chou TY (2008) The mitochondrial genome of the gymnosperm *Cycas taitungensis* contains a novel family of short interspersed elements, Bpu sequences, and abundant RNA editing sites. Mol Biol Evol 25: 603–615

Cheung AY, Bogorad L, Vanmontagu M, Schell J (1988) Relocating a gene for herbicide tolerance: a chloroplast gene is converted into a nuclear gene. Proc Natl Acad Sci USA 85:391–395

Clemens DL, Johnson PJ (2000) Failure to detect DNA in hydrogenosomes of *Trichomonas vaginalis* by nick translation and immunomicroscopy. Mol Biochem Parasit 106:307–313

Clifton SW, Minx P, Fauron CMR, Gibson M, Allen JO, Sun H, Thompson M, Barbazuk WB, Kanuganti S, Tayloe C, Meyer L, Wilson RK, Newton KJ (2004) Sequence and comparative analysis of the maize NB mitochondrial genome. Plant Physiol 136:3486–3503

Cornelissen M, Vandewiele M (1989) Nuclear transcriptional activity of the plastid *psbA* promoter. Nucleic Acids Res 17:19–29

Cusack BP, Wolfe KH (2007) When gene marriages don't work out: divorce by subfunctionalization. Trends Genet 23:270–272

Daley DO, Whelan J (2005) Why genes persist in organelle genomes. Genome Biol 6:110

Delannoy E, Fujii S, des Francs CC, Brundrett M, Small I (2011) Rampant gene loss in the underground Orchid Rhizanthella gardneri highlights evolutionary constraints on plastid genomes. Mol Biol Evol 28:2077–2086

Doolittle WE (1998) You are what you eat: a gene transfer ratchet could account for bacterial genes in eukaryotic nuclear genomes. Trends Genet 14: 307–311

Figueroa P, Gomez I, Holuigue L, Araya A, Jordana X (1999) Transfer of *rps14* from the mitochondrion to the nucleus in maize implied integration within a gene encoding the iran-sulphur subunit of succinate dehydrogenase and expression by alternative splicing. Plant J 18:601–609

Gallois JL, Achard P, Green G, Mache R (2001) The *Arabidopsis* chloroplast ribosomal protein L21 is encoded by a nuclear gene of mitochondrial origin. Gene 274:179–185

Gissi C, Iannelli F, Pesole G (2008) Evolution of the mitochondrial genome of Metazoa as exemplified by comparison of congeneric species. Heredity 101:301–320

Goremykin VV, Salamini F, Velasco R, Viola R (2009) Mitochondrial DNA of *Vitis vinifera* and the issue of rampant horizontal gene transfer. Mol Biol Evol 26:99–110

Grohmann L, Brennicke A, Schuster W (1992) The mitochondrial gene encoding ribosomal protein S12 has been translocated to the nuclear genome in *Oenothera*. Nucleic Acids Res 20:5641–5646

Guo XY, Ruan SL, Hu WM, Ca DG, Fan LJ (2008) Chloroplast DNA insertions into the nuclear genome of rice: the genes, sites and ages of insertion involved. Funct Integr Genomics 8:101–108

Hall AE, Kettler GC, Preuss D (2006) Dynamic evolution at pericentromeres. Genome Res 16:355–364

Handa H (2003) The complete nucleotide sequence and RNA editing content of the mitochondrial genome of rapeseed (*Brassica napus* L.): comparative analysis of the mitochondrial genomes of rapeseed and *Arabidopsis thaliana*. Nucleic Acids Res 31:5907–5916

Hao WL, Palmer JD (2009) Fine-scale mergers of chloroplast and mitochondrial genes create functional, transcompartmentally chimeric mitochondrial genes. Proc Natl Acad Sci USA 106: 16728–16733

Haviv-Chesner A, Kobayashi Y, Gabriel A, Kupiec M (2007) Capture of linear fragments at a double-strand break in yeast. Nucleic Acids Res 35: 5192–5202

Hazkani-Covo E, Zeller RM, Martin W (2010) Molecular poltergeists: mitochondrial DNA copies (numts) in sequenced nuclear genomes. PLoS Genet 6:e1000834

Henze K, Martin W (2001) How do mitochondrial genes get into the nucleus? Trends Genet 17: 383–387

Huang CY, Ayliffe MA, Timmis JN (2003) Direct measurement of the transfer rate of chloroplast DNA into the nucleus. Nature 422:72–76

Huang CY, Ayliffe MA, Timmis JN (2004) Simple and complex nuclear loci created by newly transferred chloroplast DNA in tobacco. Proc Natl Acad Sci USA 101:9710–9715

Huang CY, Grunheit N, Ahmadinejad N, Timmis JN, Martin W (2005) Mutational decay and age of chloroplast and mitochondrial genomes transferred recently to angiosperm nuclear chromosomes. Plant Physiol 138:1723–1733

Kanno A, Nakazono M, Hirai A, Kameya T (1997) Maintenance of chloroplast-derived sequences in the mitochondrial DNA of gramineae. Curr Genet 32:413–419

Khakhlova O, Bock R (2006) Elimination of deleterious mutations in plastid genomes by gene conversion. Plant J 46:85–94

Kitazaki K, Kubo T (2010) Cost of having the largest mitochondrial genome: evolutionary mechanism of plant mitochondrial genome. J Bot 2010. doi:10.1155/2010/620137

Kleine T, Maier UG, Leister D (2009) DNA transfer from organelles to the nucleus: the idiosyncratic genetics of endosymbiosis. Annu Rev Plant Biol 60:115–138

Knoop V, Brennicke A (1991) A mitochondrial intron sequence in the 5′-flanking region of a plant nuclear lectin gene. Curr Genet 20:423–425

Knoop V, Unseld M, Marienfeld J, Brandt P, Sunkel S, Ullrich H, Brennicke A (1996) *copia*-, *gypsy*- and LINE-like retrotransposon fragments in the mitochondrial genome of *Arabidopsis thaliana*. Genetics 142:579–585

Kubo N, Arimura S (2010) Discovery of the *rpl10* gene in diverse plant mitochondrial genomes and its probable replacement by the nuclear gene for chloroplast RPL10 in two lineages of angiosperms. DNA Res 17:1–9

Kubo T, Nishizawa S, Sugawara A, Itchoda N, Estiati A, Mikami T (2000) The complete nucleotide sequence of the mitochondrial genome of sugar beet (*Beta vulgaris* L.) reveals a novel gene for tRNA(Cys) (GCA). Nucleic Acids Res 28:2571–2576

Kundu M, Thompson CB (2005) Macroautophagy versus mitochondrial autophagy: a question of fate? Cell Death Differ 12:1484–1489

Leister D (2005) Origin, evolution and genetic effects of nuclear insertions of organelle DNA. Trends Genet 21:655–663

Lenglez S, Hermand D, Decottignies A (2010) Genome-wide mapping of nuclear mitochondrial DNA sequences links DNA replication origins to chromosomal double-strand break formation in *Schizosaccharomyces pombe*. Genome Res 20: 1250–1261

Li HM, Chiu CC (2010) Protein transport into chloroplasts. Annu Rev Plant Biol 61:157–180

Li LB, Wang B, Liu Y, Qiu YL (2009) The complete mitochondrial genome sequence of the Hornwort *Megaceros aenigmaticus* shows a mixed mode of conservative yet dynamic evolution in early land plant mitochondrial genomes. J Mol Evol 68: 665–678

Lim L, Linka M, Mullin KA, Weber APM, McFadden GI (2010) The carbon and energy sources of the non-photosynthetic plastid in the malaria parasite. FEBS Lett 584:549–554

Lin YF, Waldman AS (2001) Capture of DNA sequences at double-strand breaks in mammalian chromosomes. Genetics 158:1665–1674

Lin XY, Kaul SS, Rounsley S, Shea TP, Benito MI, Town CD, Fujii CY, Mason T, Bowman CL, Barnstead M et al (1999) Sequence and analysis of chromosome 2 of the plant *Arabidopsis thaliana*. Nature 402:761–768

Lister DL, Bateman JM, Purton S, Howe CJ (2003) DNA transfer from chloroplast to nucleus is much rarer in *Chlamydomonas* than in tobacco. Gene 316:33–38

Liu SL, Zhuang Y, Zhang P, Adams KL (2009) Comparative analysis of structural diversity and sequence evolution in plant mitochondrial genes transferred to the nucleus. Mol Biol Evol 26: 875–891

Lloyd AH, Timmis JN (2011) The origin and characterization of new nuclear genes originating from a cytoplasmic organellar genome. Mol Biol Evol 28:2019–2028

Lough AN, Roark LM, Kato A, Ream TS, Lamb JC, Birchler JA, Newton KJ (2008) Mitochondrial DNA transfer to the nucleus generates extensive insertion site variation in maize. Genetics 178:47–55

Lynch M (1996) Mutation accumulation in transfer RNAs: molecular evidence for Muller's ratchet in mitochondrial genomes. Mol Biol Evol 13: 209–220

Magee AM, Aspinall S, Rice DW, Cusack BP, Semon M, Perry AS, Stefanovic S, Milbourne D, Barth S, Palmer JD, Gray JC, Kavanagh TA, Wolfe KH (2010) Localized hypermutation and associated gene losses in legume chloroplast genomes. Genome Res 20:1700–1710

Marechal A, Brisson N (2010) Recombination and the maintenance of plant organelle genome stability. New Phytol 186:299–317

Martin W, Herrmann RG (1998) Gene transfer from organelles to the nucleus: how much, what happens, and why? Plant Physiol 118:9–17

Martin W, Rujan T, Richly E, Hansen A, Cornelsen S, Lins T, Leister D, Stoebe B, Hasegawa M, Penny D (2002) Evolutionary analysis of *Arabidopsis*, cyanobacterial, and chloroplast genomes reveals plastid phylogeny and thousands of cyanobacterial genes in the nucleus. Proc Natl Acad Sci USA 99:12246–12251

Matsuo M, Ito Y, Yamauchi R, Obokata J (2005) The rice nuclear genome continuously integrates, shuffles, and eliminates the chloroplast genome to cause chloroplast-nuclear DNA flux. Plant Cell 17: 665–675

Millar AH, Whelan J, Small I (2006) Recent surprises in protein targeting to mitochondria and plastids. Curr Opin Plant Biol 9:610–615

Millen RS, Olmstead RG, Adams KL, Palmer JD, Lao NT, Heggie L, Kavanagh TA, Hibberd JM, Giray JC, Morden CW, Calie PJ, Jermiin LS, Wolfe KH (2001) Many parallel losses of *infA* from chloroplast DNA during angiosperm evolution with multiple independent transfers to the nucleus. Plant Cell 13: 645–658

Ming R, Hou SB, Feng Y, Yu QY, Dionne-Laporte A, Saw JH, Senin P, Wang W, Ly BV, Lewis KLT et al (2008) The draft genome of the transgenic tropical fruit tree papaya (*Carica papaya* Linnaeus). Nature 452:991–996

Mishmar D, Ruiz-Pesini E, Brandon M, Wallace DC (2004) Mitochondrial DNA-like sequences in the nucleus (NUMTs): insights into our African origins and the mechanism of foreign DNA integration. Hum Mutat 23:125–133

Miyata S, Nakazono M, Hirai A (1998) Transcription of plastid derived tRNA genes in rice mitochondria. Curr Genet 34:216–220

Nakazono M, Nishiwaki S, Tsutsumi N, Hirai A (1996) A chloroplast-derived sequence is utilized as a source of promotes sequences for the gene for subunit 9 of NADH dehydrogenase (*nad9*) in rice mitochondria. Mol Gen Genet 252:371–378

NCBI (2011) NCBI eukaryotae organelles list. http://www.ncbi.nlm.nih.gov/genomes/GenomesHome.cgi?taxid=2759&hopt=html. Accessed 25 Mar 2011

Neupert W (1997) Protein import into mitochondria. Annu Rev Biochem 66:863–917

Notsu Y, Masood S, Nishikawa T, Kubo N, Akiduki G, Nakazono M, Hirai A, Kadowaki K (2002) The complete sequence of the rice (*Oryza sativa* L.) mitochondrial genome: frequent DNA sequence acquisition and loss during the evolution of flowering plants. Mol Genet Genomics 268:434–445

Noutsos C, Richly E, Leister D (2005) Generation and evolutionary fate of insertions of organelle DNA in the nuclear genomes of flowering plants. Genome Res 15:616–628

Noutsos C, Kleine T, Armbruster U, DalCorso G, Leister D (2007) Nuclear insertions of organellar DNA can create novel patches of functional exon sequences. Trends Genet 23:597–601

Nugent JM, Palmer JD (1991) RNA-mediated transfer of the gene *coxII* from the mitochondrion to the

nucleus during flowering plant evolution. Cell 66:473–481
Oda K, Yamato K, Ohta E, Nakamura Y, Takemura M, Nozato N, Akashi K, Kanegae T, Ogura Y, Kohchi T, Ohyama K (1992) Gene organization deduced from the complete sequence of liverwort *Marchantia polymorpha* mitochondrial DNA: a primitive form of plant mitochondrial genome. J Mol Biol 223:1–7
Ogihara Y, Yamazaki Y, Murai K, Kanno A, Terachi T, Shiina T, Miyashita N, Nasuda S, Nakamura C, Mori N, Takumi S, Murata M, Futo S, Tsunewaki K (2005) Structural dynamics of cereal mitochondrial genomes as revealed by complete nucleotide sequencing of the wheat mitochondrial genome. Nucleic Acids Res 33:6235–6250
Puthiyaveetil S, Kavanagh TA, Cain P, Sullivan JA, Newell CA, Gray JC, Robinson C, van der Giezen M, Rogers MB, Allen JF (2008) The ancestral symbiont sensor kinase CSK links photosynthesis with gene expression in chloroplasts. Proc Natl Acad Sci USA 105:10061–10066
Reith M, Munholland J (1995) Complete nucleotide sequence of the *Porphyra purpurea* chloroplast genome. Plant Mol Biol Rep 13:333–335
Ricchetti M, Fairhead C, Dujon B (1999) Mitochondrial DNA repairs double-strand breaks in yeast chromosomes. Nature 402:96–100
Richly E, Leister D (2004a) NUMTs in sequenced eukaryotic genomes. Mol Biol Evol 21:1081–1084
Richly E, Leister D (2004b) NUPTs in sequenced eukaryotes and their genomic organization in relation to NUMTs. Mol Biol Evol 21:1972–1980
Roark LM, Hui Y, Donnelly L, Birchler JA, Newton KJ (2010) Recent and frequent insertions of chloroplast DNA into maize nuclear chromosomes. Cytogenet Genome Res 129:17–23
Rousseau-Gueutin M, Lloyd AH, Sheppard AE, Timmis JN (2011) Gene transfer to the nucleus. In: Bullerwell CE (ed) Organelle genetics: evolution of organelle genomes and gene expression. Springer, Berlin, pp 147–171
Ruf S, Karcher D, Bock R (2007) Determining the transgene containment level provided by chloroplast transformation. Proc Natl Acad Sci USA 104: 6998–7002
Ruf S, Braune S, Endries P, Hasse C, Stegemann S, Bock R (2010) Plastid transmission, gene transfer and the impact of the environment. ISCGGE, Maynooth
Salomon S, Puchta H (1998) Capture of genomic and T-DNA sequences during double-strand break repair in somatic plant cells. EMBO J 17:6086–6095
Selosse M, Albert B, Godelle B, Berg OG, Kurland CG (2001) Reducing the genome size of organelles favours gene transfer to the nucleus. Trends Ecol Evol 16:135–141
Sheppard AE, Timmis JN (2009) Instability of plastid DNA in the nuclear genome. PLoS Genet 5:e1000323
Sheppard AE, Ayliffe MA, Blatch L, Day A, Delaney SK, Khairul-Fahmy N, Li Y, Madesis P, Pryor AJ, Timmis JN (2008) Transfer of plastid DNA to the nucleus is elevated during male gametogenesis in tobacco. Plant Physiol 148:328–336
Sheppard AE, Panagiotis M, Lloyd AH, Day A, Ayliffe MA, Timmis JN (2011) Introducing an RNA editing requirement into a plastid-localised transgene reduces but does not eliminate functional gene transfer to the nucleus. Plant Mol Biol 76:299–309
Sloan DB, Alverson AJ, Storchova H, Palmer JD, Taylor DR (2010) Extensive loss of translational genes in the structurally dynamic mitochondrial genome of the angiosperm *Silene latifolia*. BMC Evol Biol 10:e274
Smith DR, Lee RW (2008) Mitochondrial genome of the colorless green alga *Polytomella capuana*: a linear molecule with an unprecedented GC content. Mol Biol Evol 25:487–496
Soll J, Schleiff E (2004) Protein import into chloroplasts. Nat Rev Mol Cell Biol 5:198–208
Stegemann S, Bock R (2006) Experimental reconstruction of functional gene transfer from the tobacco plastid genome to the nucleus. Plant Cell 18: 2869–2878
Stegemann S, Hartmann S, Ruf S, Bock R (2003) High-frequency gene transfer from the chloroplast genome to the nucleus. Proc Natl Acad Sci USA 100:8828–8833
Stern DB, Lonsdale DM (1982) Mitochondrial and chloroplast genomes of maize have a 12-kilobase DNA sequence in common. Nature 299:698–702
Stettler M, Eicke S, Mettler T, Messerli G, Hortensteiner S, Zeeman SC (2009) Blocking the metabolism of starch breakdown products in *Arabidopsis* leaves triggers chloroplast degradation. Mol Plant 2:1233–1246
Stupar RM, Lilly JW, Town CD, Cheng Z, Kaul S, Buell CR, Jiang JM (2001) Complex mtDNA constitutes an approximate 620-kb insertion on *Arabidopsis thaliana* chromosome 2: implication of potential sequencing errors caused by large-unit repeats. Proc Natl Acad Sci USA 98:5099–5103
Sugiyama Y, Watase Y, Nagase M, Makita N, Yagura S, Hirai A, Sugiura M (2005) The complete nucleotide sequence and multipartite organization of the tobacco mitochondrial genome: comparative analysis of mitochondrial genomes in higher plants. Mol Genet Genomics 272:603–615
Svab Z, Hajdukiewicz P, Maliga P (1990) Stable transformation of plastids in higher plants. Proc Natl Acad Sci USA 87:8526–8530

Terasawa K, Odahara M, Kabeya Y, Kikugawa T, Sekine Y, Fujiwara M, Sato N (2007) The mitochondrial genome of the moss *Physcomitrella patens* sheds new light on mitochondrial evolution in land plants. Mol Biol Evol 24:699–709

The Rice Chromosome 10 Sequencing Consortium (2003) In-depth view of structure, activity, and evolution of rice chromosome 10. Science 300: 1566–1569

Thorsness PE, Fox TD (1990) Escape of DNA from mitochondria to the nucleus in *Saccharomyces cerevisiae*. Nature 346:376–379

Timmis JN, Scott NS (1983) Sequence homology between spinach nuclear and chloroplast genomes. Nature 305:65–67

Timmis JN, Ayliffe MA, Huang CY, Martin W (2004) Endosymbiotic gene transfer: organelle genomes forge eukaryotic chromosomes. Nat Rev Genet 5:123–136

Turmel M, Otis C, Lemieux C (2003) The mitochondrial genome of *Chara vulgaris*: insights into the mitochondrial DNA architecture of the last common ancestor of green algae and land plants. Plant Cell 15:1888–1903

Ueda M, Nishikawa T, Fujimoto M, Takanashi H, Arimura S, Tsutsumi N, Kadowaki K (2008) Substitution of the gene for chloroplast RPS16 was assisted by generation of a dual targeting signal. Mol Biol Evol 25:1566–1575

Unseld M, Marienfeld JR, Brandt P, Brennicke A (1997) The mitochondrial genome of *Arabidopsis thaliana* contains 57 genes in 366,924 nucleotides. Nat Genet 15:57–61

van den Boogaart P, Samallo J, Agsteribbe E (1982) Similar genes for a mitochondrial ATPase subunit in the nuclear and mitochondrial genomes of *Neurospora crassa*. Nature 298:187–189

van der Giezen M, Sjollema KA, Artz RR, Alkema W, Prins RA (1997) Hydrogenosomes in the anaerobic fungus *Neocallimastix frontalis* have a double membrane but lack an associated organelle genome. FEBS Lett 408:147–150

Villarejo A, Buren S, Larsson S, Dejardin A, Monne M, Rudhe C, Karlsson J, Jansson S, Lerouge P, Rolland N, von Heijne G, Grebe M, Bako L, Samuelsson G (2005) Evidence for a protein transported through the secretory pathway *en route* to the higher plant chloroplast. Nat Cell Biol 7:1224–1231

Vogel JP, Garvin DF, Mockler TC, Schmutz J, Rokhsar D, Bevan MW, Barry K, Lucas S, Harmon-Smith M, Lail K et al (2010) Genome sequencing and analysis of the model grass *Brachypodium distachyon*. Nature 463:763–768

Vonheijne G (1986) Why mitochondria need a genome. FEBS Lett 198:1–4

Wada S, Ishida H, Izumi M, Yoshimoto K, Ohsumi Y, Mae T, Makino A (2009) Autophagy plays a role in chloroplast degradation during senescence in individually darkened leaves. Plant Physiol 149: 885–893

Wang D, Wu YW, Shih ACC, Wu CS, Wang YN, Chaw SM (2007) Transfer of chloroplast genomic DNA to mitochondrial genome occurred at least 300 MYA. Mol Biol Evol 24:2040–2048

Wilson RJM, Denny PW, Preiser PR, Rangachari K, Roberts K, Roy A, Whyte A, Strath M, Moore DJ, Moore PW, Williamson DH (1996) Complete gene map of the plastid-like DNA of the malaria parasite *Plasmodium falciparum*. J Mol Biol 261: 155–172

Wolfe KH, Li WH, Sharp PM (1987) Rates of nucleotide substitution vary greatly among plant mitochondrial, chloroplast, and nuclear DNAs. Proc Natl Acad Sci USA 84:9054–9058

Chapter 10

Horizontal Gene Transfer in Eukaryotes: Fungi-to-Plant and Plant-to-Plant Transfers of Organellar DNA

Susanne S. Renner* and Sidonie Bellot
Systematic Botany and Mycology, University of Munich, Menzinger Straße 67, D-80638 Munich, Germany

Summary

This review focuses on horizontal gene transfer (HGT) involving bacteria, fungi, and plants (Viridiplantae). It highlights in particular the persistent challenge of recognizing HGT, which requires a combination of methods from bioinformatics, phylogenetics, and molecular biology. Non-phylogenetic methods rely on compositional structure, such as G/C content, dinucleotide frequencies, codon usage biases, or co-conversion tracts, while phylogenetic methods rely on incongruence among gene trees, one of which is taken to represent the true organismal phylogeny. All methods are handicapped by short sequence lengths with limited or highly uneven substitution signal; the statistical problems of working with taxon-rich alignments of such sequences include low support for inferred relationships, and difficult orthology assessment. Plant-to-plant HGT is known from two dozen mitochondrial genes and species of phylogenetically and geographically widely separated ferns, gymnosperms, and angiosperms, with seven cases involving parasitic plants. Only one nuclear HGT has

*Author for correspondence, e-mail: renner@lrz.uni-muenchen.de

R. Bock and V. Knoop (eds.), Genomics of Chloroplasts and Mitochondria, Advances in Photosynthesis and Respiration 35, pp. 223–235,
DOI 10.1007/978-94-007-2920-9_10,

come to light, and extremely few fungi-to-plant transfers. Plant mitochondrial genomes, especially in tracheophytes, are prone to take up foreign DNA, but evolutionary consequences of this are still unclear.

I. Introduction

Horizontal gene transfer (HGT) refers to movement of genetic material between organisms that does not follow the normal pathway of vertical transmission from parent to offspring. Horizontal gene transfer is sometimes seen as synonymous with lateral gene transfer, a term better restricted to *within*-species sequence copying, such as group II intron retrotransposition or the massive migration of promiscuous cpDNA into mitochondria of seed plants. With the 2003 discoveries of HGT involving eukaryotes (Bergthorsson et al. 2003; Won and Renner 2003), the availability of full genome sequences, and new insights into transposable elements, HGT has become an important issue also in plant science. Recent reviews of the topic include those of Andersson (2005), Richardson and Palmer (2007), Keeling and Palmer (2008), Keeling (2009a, b), and Bock (2010), and the paradigm is rapidly becoming that HGT is "a highly significant process in eukaryotic genome evolution" (Bock 2010).

The present review focuses on glaucophytes, red algae, green algae, and land plants. Besides briefly summarizing recent findings relevant to plant genomes, it will highlight the persistent challenge of recognizing horizontal gene transfer. This challenge stems largely from the still relatively crude methods for finding matching DNA strings in databases and the inability of phylogenetic algorithms to infer correct relationships from short sequences. Especially the latter problem is often underappreciated in the context of HGT. We therefore begin our review by discussing the combination of bioinformatics, phylogenetics, and molecular biology that forms the basis for inferring and evaluating HGT. We then discuss the evidence for gene transfer between bacteria or fungi and plants, plant-to-plant transfer, and transposable element transfer, and follow with a section on problematic or erroneous earlier inferences of HGT. We end by addressing what is known about the mechanisms of HGT among plants and by providing a perspective on ongoing research that aims at unsolved questions in HGT.

Abbreviations: BLAST – Basic local alignment search tool; cpDNA; – Plastid DNA; DNA – Deoxyribonucleic acid; EST – Expressed sequence tag; HGT – Horizontal gene transfer; HTT – Horizontal transposon transfer; mt(DNA) – Mitochondrial (DNA); MULE – *Mu*-like elements (Mu is *mutator* in corn); My – Million years; ORF – Open reading frame; PCR – Polymerase chain reaction; RNA – Ribonucleic acid; T-DNA – Transferred DNA; TE – Transposable element; Ti-plasmid – Tumor-inducing plasmid

II. Detecting and Evaluating Cases of Horizontal Gene Transfer

A. Bioinformatic Approaches for Detecting HGT

Genome-wide studies of eukaryotes typically will involve a BLAST search (Altschul et al. 1990) to identify genes matching bacterial genes or to find unusual (unique) genes that could be of bacterial origin. Another step is to employ known genes as queries and test for consistency of ORFs or to BLAST against a local database containing well-annotated genomic sequences from model organisms. All these steps rely on BLAST results. It is well understood, however, that BLAST e-values are based on the expected background noise, depend on the sequences in the database at any one time, and are not a reliable indicator of evolutionary relatedness (Koski and Golding 2001). Recent genomics studies have used pair-wise syntenic alignments and BLAST score statistical tests (e.g., Ma et al. 2010).

Other non-phylogenetic methods depend on compositional structure, such as G/C content, dinucleotide frequencies or codon usage biases, but the length of a horizontally transferred gene may be too short to reliably reveal these differences. Methods based on atypical nucleotide or amino acid composition also may only detect recent transfers because donor sequence characteristics will gradually become erased. Moreover, the reliability of these methods is difficult to assess statistically (Ragan et al. 2006). Snir and Trifonov (2010) have proposed using an additional approach that involves comparing just two genomes. With two genomes of a given length one can calculate the probabilities of identical regions (under a chosen model of substitution). To detect HGTs, the method makes use of the expectation that the flanking regions of an inserted region will normally be non-homologous and then uses a sliding window algorithm to detect these HGT borders, essentially searching for sharp borders (or walls). The method has been applied to simulated data and real bacterial genomes.

B. Phylogenetic Approaches for Detecting HGT

Phylogenetic trees are time-consuming to construct because they require a trustworthy sequence alignment. Nevertheless, many workers consider phylogenetic tree incongruence the best indicator of HGT, perhaps especially ancient HGT. When conflicts are found between two or more gene trees, HGT can be introduced as one possible explanation (for an insightful discussion concerning tree incongruency due to HGT in the microbial world, see Boto 2010). Like the bioinformatics approaches discussed in the previous section, the phylogenetic method for identifying HGT faces several challenges. First, it is incapable of coping with events residing in non-homologous regions since all tree inference methods presume character homology in the underlying sequence alignment. It also requires assumptions about where to seek the HGT events, in other words, assumptions about which tree reflects the true organismal history. There is reason to think that methods that detect HGT using atypical genomic composition ("signatures") are better at finding recent transfers whereas "phylogenetic incongruence" methods may be better at detecting older HGTs because of the increasing mutational signal over time, until saturation (Ragan et al. 2006; Cohen and Pupko 2010). Whether this generalization holds will depend on details of the substitution process since all phylogenetic methods, whether parsimony, maximum likelihood, or Bayesian inference, require sufficient mutational signal.

The statistical cut-off deemed acceptable for particular splits in a tree is a matter of debate. Among phylogeneticists, accepted cut-offs values are >75% under parsimony and likelihood optimization, and 98% under Bayesian tree sampling, values rarely reached in trees used to infer HGT because of taxon-rich alignments and short sequences. A sense of the amount of signal needed for statistical support can be gained from Felsenstein's (1985) demonstration that three non-homoplastic substitutions suffice for a bootstrap support (for a node) at the 95% level. These statistical reasons imply that well-supported phylogenies usually require concatenated multi-locus alignments. One then faces the question of which loci can safely be combined. For plants, one solution has been to accept combined plastid gene phylogenies as "true" and to view phylogenies from mitochondrial genes as HGT-prone (Cho et al. 1989a, b; Bergthorsson et al. 2003; Burger et al. 2003; Hao et al. 2010; Archibald and Richards 2010; compare Sect. VII). This is based on the rationale that no evidence has so far come to light of HGT involving plastid genes of Viridiplantae.

Statistical tests for tree incongruence, such as the Incongruence Length Difference test (Farris et al. 1994), require sufficient mutational signal and usually cannot reliably identify nodes in phylogenies due to HGT as long as the trees are based on single genes. This leaves workers in a bind, and many HGT studies have therefore inferred incongruence by eyeballing more or less unsupported trees or by contrasting an unresolved gene tree with an organismal tree supported by other evidence, for example, morphological and/or genetic data analyzed in other studies. A software to detect HGT from tree incongruence alone is SPRIT (Hill et al. 2010), but it requires assuming that all splits in the trees being compared are true.

A second difficulty with phylogenetic approaches for detecting HGT is that gene phylogenies may be incongruent because of biases in the sequence data and not (only) because of HGT. Well known biases include uneven nucleotide frequencies (Embley et al. 1993; Foster et al. 2009; Stiller 2011), long-branch attraction (Felsenstein 1978), codon bias, and model over-parameterization. Long branch attraction is a systematic error, corresponding to the inconsistency of a statistical procedure (namely maximum parsimony), and leads to the convergence towards an incorrect answer as more and more data are analyzed. It occurs when two (or more) sequences in a phylogeny have unusually high substitution rates, resulting in their having much longer branches than the remaining sequences. Long-branch attraction cannot be resolved by adding more characters, and it is a severe and under-appreciated problem in HGT detection. (Removing one of the long branches can sometimes eliminate the problem; e.g., Goremykin et al. 2009).

A third difficulty in identifying HGT is to distinguish it from ancestral gene duplication and differential gene loss (Stanhope et al. 2001; Gogarten and Townsend 2005; Noble et al. 2007). Duplication and loss in gene families affects especially nuclear genes, and since relatively few densely sampled and deep (i.e., going back millions of years) phylogenies have been built with nuclear genes, lineage sorting has so far not been a major discussion point in HGT (but see Noble et al. 2007).

A recent study involving fungi and angiosperms, illustrates the problems of detecting HGT. To test for plant/fungi gene exchange, Richards et al. (2009) generated automated gene-by-gene alignments and phylogenies for 4,866 genes identified in analyses of the *Oryza* genome and in BLAST comparisons. Visual inspection of the phylogenies used two criteria for HGT: Either a plant gene sequence branching within a cluster of sequences from fungal taxa (or vice versa) or a phylogeny that demonstrated a diverse plant-specific gene family absent from all other taxa except a narrow taxonomic group of fungi (or vice versa). Using these criteria, Richards et al. detected 38 plant-fungi HGT candidates, of which two were detected using the rice genome-specific analysis, 35 were detected using the BLAST-based survey, and one was detected using both search protocols. However, when these authors added more sequences (taxa) from GenBank and expressed sequence tag (EST) databases, only 14 of the putative HGTs remained because increasing taxon sampling decreased the number of isolated or wrongly placed suspected HGT sequences. The number of suspected HGT events was then further reduced to nine by reconstructing phylogenies with better fitting maximum likelihood substitution models that accounted for rate heterogeneity. The study beautifully illustrates the risk of overestimating the frequency of HGT from insufficient taxon sampling and poorly fitting substitution models, with rate heterogeneity being the single most important model parameter (Yang 1994).

As is generally true for tree inference, also the dynamics of gene gains and losses in gene families are probably better inferred using maximum likelihood than parsimony optimization of the minimal number of gains and losses needed to explain the distribution of a group of orthologous genes in a phylogeny (Mirkin et al. 2003; Richards et al. 2009; Cohen and Pupko 2010). These and other studies (Cusimano et al. 2008; Goremykin et al. 2009; Ragan and Beiko 2009; Ferandon et al. 2010) all caution against inferring rampant HGT from phylogenetic incongruence among gene trees, at least as long as the trees are based on short sequences (analyzed under parsimony or, worse, neighbor-joining) from genetically distant organisms with millions of years of evolution separating them.

C. Footprints and Signatures of HGT

The third way of identifying HGT is to look for signatures or "footprints" of the HGT events themselves (Adams et al. 1998; Cho et al. 1998; Cho and Palmer 1999; Sanchez-Puerta et al. 2008). Such footprints might be the co-conversion tracts of group I introns, which are short stretches of flanking exon sequence (>50 bp into the 5′ exon and <25 bp into the 3′ exon) that may be converted to the donor DNA sequence during intron insertion or excision (Lambowitz and Belfort 1993; Lambowitz and Zimmerly 2004). If the flanking exon stretches in the donor and recipient differ, then co-conversion will create a footprint that can stay even after the intron itself is lost again. The first study using the molecular footprint approach focused on a group I intron in the mt *cox1* gene and inferred 3–5 HGT events in a small clade of Araceae (Cho and Palmer 1999). This was inferred although a parsimony reconstruction favored a vertical transmission history with one intron gain, followed by two losses, that is, three evolutionary events, rather than five (Cho and Palmer 1999). Subsequently, reliance on co-conversion tracts as inconvertible footprints led to the extrapolation of at least 1,000 HGTs of the *cox1* intron among living angiosperms, based on a survey of the intron's distribution that suggested 32 separate cases of intron acquisition from unknown donors to account for the intron's presence in 48 of 281 species from 278 genera (Cho et al. 1998b).

Seven cases of chimeric sequences between foreign and native mt gene copies have been described (Vaughn et al. 1995; Adams et al. 1998: *Peperomia polybotrya cox1* intron; Bergthorsson et al. 2003: *Sanguinaria canadensis rps11*; Barkman et al. 2007: *Pilostyles thurberi atp1*; Hao et al. 2010: *Ternstroemia stahlii atp1*; *Hedychium coronarium matR*; *Boesenbergia rotunda matR*; Mower et al. 2010: *Plantago macrorhiza atp1*). In some cases, the chimeric sequences appear functional, in others they are unexpressed pseudogenes. A recent re-analysis of these cases based on a new recombination search algorithm developed specifically for plant mitochondrial genomes showed that detecting HGT-generated chimeras requires dense taxonomic sampling (Hao 2010). Using the new algorithm, Hao and Palmer (2009) also identified nine putative cases of short-patch gene conversion of native, functional plant mt *atp1* genes by homologous *atpA* genes of chloroplast origin. If confirmed, these cases of recombination between mitochondrial and chloroplast genes provide unique evidence for the creation of functional chimeric genes across the ca. one-billion-year divide between chloroplast and mitochondrial genes.

For transposable elements (TEs), detection of horizontal transposon transfer (HTT) may sometimes be possible by comparisons of the rates of synonymous substitution, the Ks values, observed in TEs with those in orthologous genes (Sanchez-Gracia et al. 2005; Schaack et al. 2010). If the presence of a TE in two hosts is due to horizontal transfer, then it will be younger than the hosts and will have accumulated fewer synonymous mutations than the host genes. With many complete genome sequences now available, this approach can be implemented in a robust statistical framework taking into account the Ks value distribution of hundreds of host genes to define the Ks threshold under which the presence of a TE is considered to be the result of HTT. The approach has been applied to closely related species, such as *Drosophila melanogaster* and *Drosophila simulans*, which diverged less than 5 My ago (Schaack et al. 2010).

III. DNA Transfers Among Bacteria or Fungi and Plants

The classic example of HGT from prokaryotes to multi-cellular eukaryotes is the transfer of DNA from the *Agrobacterium* Ti plasmid to plants (reviewed by Gelvin 2009). Other bacterial species, such as *Sinorhizobium meliloti* and *Mesorhizobium loti*, when harboring modified Ti plasmids, can also transfer them to plants (Broothaerts et al. 2005). During transformation, the transferred DNA (T-DNA) is moved through the plasma membrane via a channel formed by a bacterial protein that also participates in coating of the T-DNA during its transfer to the nucleus (Dumas et al. 2001). The extent of natural recent incorporation of prokaryotic genetic material into plants is unclear, although bacterial chromosomal DNA apparently is introduced into the nuclei of transgenic plants occasionally (Ülker et al. 2008).

So far, there is one reported case of the horizontal acquisition of a group II intron in the plastid *psbA* gene of the green alga *Chlamydomonas sp.* that appears to come from a cyanobacterium (Odom et al. 2004). From red algae, two genes, *rpl36* and an unusual rubisco operon, *rbcLS*, may have been transferred from bacterial donors to the common ancestor of red algae (*rbcLS*) or the common ancestor of cryptophytes and haptophytes (*rpl36*) (Keeling and Palmer 2008, and references therein).

Genetic exchange between plants and fungi is exceedingly rare, particularly in angiosperms (Richards et al. 2009). Richards et al. compared the genomes of six plant species (*Arabidopsis thaliana*, *Populus trichocarpa*, *Sorghum bicolor*, *Oryza sativa*, *Selaginella moellendorffii*, and *Physcomitrella patens*) with those of 159 prokaryotes and non-plant eukaryotes. Comprehensive phylogenetic analyses of the data, using methods that account for site-specific substitution rate heterogeneity, supported only nine HGTs between plants and fungi (methods used in this study were discussed above in Sect. II.B). Five were fungi-to-bryophyte and fungi-lycophyte transfers and four were plant-to-fungi transfers. An older report of the transfer of a group I intron from the angiosperm *Youngia japonica* (Asteraceae) into the 18S rRNA of its pathogenic fungus *Protomyces inouyei* (Nishida and Sugiyama 1995) has yet to be followed-up.

IV. Plant-to-Plant DNA Transfers

Exchange of genetic material between mitochondria of land plants has been inferred for diverse taxa. The species involved come from phylogenetically and geographically widely separate clades of ferns, gymnosperms, and angiosperms, suggesting that HGT among plants may be relatively widespread. The known cases involve the following mitochondrial sequences and taxa:

- The *rps2* gene in the dicot *Actinidia arguta* coming from a monocot (Bergthorsson et al. 2003),
- *rps11* in an unidentified *Lonicera* (Caprifoliaceae) coming from Ranunculaceae/Berberidaceae; in the dicot *Sanguinaria canadensis* from a monocot; and in two unidentified Betulaceae from an unidentified non-Betulaceae donor (Bergthorsson et al. 2003),
- *atp1* in *Amborella trichopoda* (Amborellaceae) from an unknown Asteridae (Bergthorsson et al. 2003); in *Ternstroemia* (Pentaphylaceae) from Ericaceae, and in *Bruinsmia* (Styracaceae) from Cyrillaceae (Schönenberger et al. 2005),
- The *nad1* second intron in *Gnetum* (Gymnospermae) coming from an unknown Asteridae, that is, a flowering plant (Won and Renner 2003),
- The *nad1* second intron plus *atp1* in two parasitic species of Rafflesiaceae from their respective host plants (Davis and Wurdack 2004; Barkman et al. 2007),
- The same intron plus *matR* in the fern *Botrychium virginianum* from an unknown Loranthaceae root-parasite (Davis et al. 2005),
- *atp1* in *Pilostyles thurberi* (Apodanthaceae) from its legume host, *Psorothamnus emoryi*; in *Mitrastema yamamotoi* (Mitrastemonaceae) from its host *Quercus subsericea* (Fagaceae; Barkman et al. 2007), and
- *atp1*, *atp6* and *matR* in species of *Plantago* (Plantaginaceae) from parasitic *Cuscuta* (Convolvulaceae) and *Bartsia* (Orobanchaceae; Mower et al. 2004, 2010).

The transferred mitochondrial genes appear to sit in the hosts' mitochondrial genomes, and most are non-functional pseudogenes. Seven cases of chimeric sequences between foreign and native mt gene copies (see especially Mower et al. 2010) were already discussed above (Sect. II). The putative HGT of the mitochondrial *cox1* intron across thousands of flowering plants, either from plant to plant or via unknown fungal donors (Adams et al. 1998; Cho et al. 1998b; Cho and Palmer 1999; Sanchez-Puerta et al. 2008) is discussed in Sects. II and VI.

An additional report about mitochondrial HGT on a massive scale involves the basal angiosperm *Amborella*, which may have acquired one or more copies of 26 mitochondrial protein genes from other land plants. Twenty foreign gene sequences appear to come from other angiosperms, six from moss donors. The transferred genes seem to be intact, but have not been shown to be functional (Bergthorsson et al. 2004). The report has attracted criticism (Martin 2005; Goremykin et al. 2009; see also Sect. VI). Large-scale genome sequencing of *Amborella* is ongoing and may resolve the controversy.

A single HGT event probably can involve multiple mitochondrial genes as made plausible by the results for *Cuscuta* and *Plantago* of Mower et al. (2010). This study also suggests a complicated history of the transferred genes within *Plantago* subsequent to their acquisition via HGT, with additional transfers (including intracellular transfer), gene duplication and differential loss and mutation-rate variation (Mower et al. 2010). Resolving this history will probably require complete mitochondrial and nuclear genome sequencing from multiple individuals.

So far, only one nuclear plant-plant HGT event has come to light. It involves the parasitic Orobanchaceae *Striga hermonthica,* for which BLAST searches between an EST database of *Striga* and plant genome databases, sequencing of a 6,423 bp-long genomic region and Southern blotting collectively imply recent uptake of genetic material from an unknown monocot host (Yoshida et al. 2010). The transferred gene encodes a 448 amino acid-long protein of unknown function, is phylogenetically closer to *Sorghum* than to its *Brachypodium* ortholog, and was acquired recently, that is, after the divergence between *Striga* and *Orobanche* (both in Orobanchaceae) but before the divergence of *S. hermonthica* and *S. gesnerioides*.

From the above it emerges that most plant-to-plant HGT events involve mitochondrial DNA and that close physical association, as exists, for example, between parasitic plants and their hosts, apparently facilitates plant-to-plant HGT. See Sect. VII for possible reasons why plant mitochondria may incorporate foreign DNA more readily than other genomes.

V. Transposable Elements

There are some 200 putative cases of transposable elements (TEs) moving horizontally in eukaryotes, but such events appear to be rare among plants. The first report of the horizontal transfer of a nuclear TE between plants was that of a *Mutator*-like element between the plant genera *Setaria* and *Oryza* (Diao et al. 2006). For clades other than Viridiplantae, it has been argued that introduction of transposable elements by horizontal transfer in eukaryotic genomes has been a major force propelling genomic variation and biological innovation (Sanchez-Gracia et al. 2005; Gilbert et al. 2010; Schaack et al. 2010). Whether there is any correlation between the horizontal transfer of TEs and the horizontal transfer of functional genes is unclear. Although TEs have not yet been shown to transfer host genes between different species in eukaryotes, they are capable of capturing and transducing sequences at high frequency within a species (Schaack et al. 2010). Of 3,000 analyzed TEs in rice, many contained gene fragments of genomic DNA that apparently had been captured, rearranged and amplified over millions of years (Jiang et al. 2004). Other examples of gene duplication and exon shuffling by transposons come from *Zea mays* (Morgante et al. 2005).

VI. Problematic, Controversial, and Erroneous Reports of HGT Involving Plants

Claims of HGT require considerable supporting evidence and caution (Kurland et al. 2003; Martin 2005; Richards et al. 2009), with a case in point being the problems with the early reports of massive HGT in the draft human genome (Lander et al. 2001) and their later dismissal (Salzberg et al. 2001; Stanhope et al. 2001). It is therefore not surprising that a few reports of HGT have been discussed controversially or turned out to be erroneous. Thus, the report of HGT between unknown Malvaceae and the parasitic species *Pilostyles thurberi* (Nickrent et al. 2004), after re-sequencing of the relevant gene region (18S RNA), turned out to be due to contaminated DNA sequences (Filipowicz and Renner 2010).

An example of putative HGTs being discussed critically is the mt *cox1* intron, which occurs in hundreds of species of flowering plants (Vaughn et al. 1995; Cho et al. 1998a, b; Cho and Palmer 1999; Cusimano et al. 2008; Sanchez-Puerta et al. 2008, 2011). Phylogenetic analysis of the *cox1* intron does not result in statistically supported trees because the intron contains too few phylogenetically informative mutations (Cusimano et al. 2008: sequence similarity among 110 *cox1* introns from throughout angiosperms ranges from 91% to 100%). Even so, the *cox1* tree for the flowering plants matches accepted relationships of orders, families and, in a few cases, genera (Cusimano et al. 2008). A parsimonious explanation is that the *cox1* intron was horizontally acquired once or a few times during the history of flowering plants, followed by vertical inheritance and numerous losses (Cusimano et al. 2008; also Ragan and Beiko 2009; Richards et al. 2009; Inda et al. 2010; Ferandon et al. 2010). Distinct mutations in co-conversion tracts, however, can lead to a scenario of intron insertions from hundreds or thousands of unknown fungal donors (Cho et al. 1998b; Sanchez-Puerta et al. 2008; fungi-to-angiosperm gene transfers are otherwise exceedingly rare: Richards et al. 2009). Resolving the issue will require a better understanding of the mechanisms of intron homing, specifically the creation and decay of co-conversion tracts (Wolf et al. 2001; Belshaw and Bensasson 2006; Ragan and Beiko 2009).

The controversy surrounding Bergthorsson et al.'s report (2004) of rampant HGT of the mtDNA of *Amborella trichopoda* has already been mentioned (Martin 2005; Goremykin et al. 2009). It is clear also from the difficult interpretation of the history of the elongation factor genes in the green algal lineage (Noble et al. 2007; Rogers et al. 2007) that greater taxon sampling can sometimes lead to a scenario more consistent with multiple losses than horizontal gains. Both processes are likely to have played important roles, and knowledge of the function of putatively transferred genes and of the biology of the involved species should help formulate testable hypotheses.

VII. Mechanisms of Plant-to-Plant HGT

The means of DNA exchange between unrelated organisms could theoretically be (1) vectors, such as bacteria, fungi or phloem-sucking bugs; (2) transfer of entire mitochondria through plasmodesmata, when there is plant-to-plant contact; (3) illegitimate pollination followed by elimination of most foreign DNA except for a few mitochondria that might fuse with native mitochondria (below) or (4) natural transformation. Of the 10–36 cases of plant-to-plant HGT (listed in Sect. IV; the numerical range depends on whether the 26 *Amborella* mt genes putatively taken up from other flowering plants and mosses are included; Bergthorsson et al. 2004), at least seven involve parasitic plants (namely Apodanthaceae: *Pilostyles*; Convolvulaceae: *Cuscuta*; unknown root-parasitic Loranthaceae; Mitrastemonaceae: *Mitrastema*; Orobanchaceae: *Bartsia*, *Striga*, *Orobanche*, *Phelipanche*; the common ancestor of the Rafflesiaceae). This ratio suggests that direct contact between donor and recipient facilitates HGT. The host plants can be the donor (Mower et al. 2004, 2010; Davis et al. 2005) or the recipient (Davis and Wurdack 2004; Barkman et al. 2007; Yoshida et al. 2010). The apparent high frequency of HGT involving parasitic plants fits with the experimental demonstration of DNA moving through a graft junction between different lines of tobacco (Stegemann and Bock 2009, although the transferred DNA stayed in the graft zone). That messenger RNA can pass through plasmodesmata is well documented (Roney et al. 2007; Lucas et al. 2009), but whether paired DNA or entire organelles can pass through plasmodesmata remains to be investigated. Alternatively, vesicle transport of DNA or organelles from cell to cell could be involved in the horizontal transfer of genetic material (Bock 2010).

All but one of the known plant-to-plant HGTs involve mitochondrial DNA, the exception being the nuclear gene taken up by *Striga hermonthica* probably from a monocot host (Yoshida et al. 2010). The propensity of plant mitochondria to incorporate foreign DNA is remarkable, since among thousands of animal mitochondrial genomes sequenced, no convincing evidence of HGT has been found, and embryophyte (land plant) plastid genomes also apparently are devoid of horizontally transferred foreign DNA. So why are plant mitochondrial genomes so open towards foreign DNA? One explanation may be that plant mitochondria are capable of importing RNA and double-stranded DNA (Koulintchenko et al. 2003). Another explanation may be the great propensity of plant mitochondria to fuse with one another (Arimura et al. 2004; Sheahan et al. 2005) and the high recombinational activity of mtDNA throughout tracheophyte evolution (Grewe et al. 2009; Hecht et al. 2011). This may have set the stage for the integration of foreign DNA in plant mt genomes, also amply documented by the frequent integration of chloroplast DNA laterally transferred into seed plant mtDNAs. Interestingly, bryophyte mt genomes lacking similarly active DNA recombination may be sources, but not acceptors for HGTs (Knoop et al. 2011; p. 18).

It is not known whether the horizontally transferred genetic material is DNA or RNA. While it was earlier hypothesized that mitochondrial HGT might largely be an RNA-mediated process (Bergthorsson et al. 2003), transfer of double-stranded DNA, which is much more stable, may be more likely (Henze and Martin 2001; Mower et al. 2010). Whether the transferred mtDNA tends to integrate into the recipients' mitochondrial genomes or, instead, becomes transferred to the nucleus is mostly unclear (Martin 2005; Goremykin et al. 2009; Hao et al. 2010). Keeling and Palmer (2008) have suggested that most transferred genes probably are non-functional and coexist with a native, functional homologue.

In addition to the barriers that can prevent the horizontal transfer and integration of foreign DNA in a recipient, it is worth considering the barriers that prevent its spread in a population. In prokaryotes, and probably also in eukaryotes, one such barrier can be the perturbation of gene dosage and expression in the host. An experimental study of the transferability of thousands of genes within *Escherichia coli* by Sorek at al. (2007) showed that toxicity to the host and changed (increased) gene dosage and expression probably are predominant causes for transfer failure. On the other hand, over-expression of an RNA polymerase experimentally transferred from *Bacillus subtilis* to *E. coli* appeared to entail no immediate fitness costs (Omer et al. 2010).

VIII. Perspective

There are many unsolved questions regarding the transfer of genetic material among phylogenetically distinct clades or species of plants. How can genetic material arrive in a new genome and function there if it lacks active promoters and appropriate downstream sequences for RNA 3' processing and stabilization? Does most transferred DNA consist of complete gene cassettes including functional expression elements? Unless a transferred gene has a homolog in the recipient, it should function only if expression can be properly regulated by the recipient or if it is an "independent gene" as appears to be true of a horizontally transferred antifreeze protein in fish (Graham et al. 2008). Gene conversion between foreign and native genes could have deleterious consequences, for example by perturbing the function of the encoded protein (Ragan and Beiko 2009).

Whether inter-specific HGT has an important role in the evolution of plants is still unclear. Plausible examples of positive evolutionary impacts are the inferred HGTs from fungi to the lycophyte *Selaginella moellendorffii* of a putative membrane transporter gene and from fungi to the moss *Physcomitrella patens* of a putative sugar transporter gene (Richards et al. 2009). Otherwise, beneficial impacts of HGT have been demonstrated or proposed mainly for prokaryotes, unicellular eukaryotes, and animals (Graham et al. 2008; Marchetti et al. 2009; Danchina et al. 2010).

More molecular-biological investigations and better experimental systems in the lab are sorely needed to understand the role(s) of HGT in plants. Horizontal gene transfer in Viridiplantae may be especially difficult to detect because most events seem to involve mtDNA, which at the substitution level evolves extremely slowly, creating a challenge for the phylogenetic approach of inferring events from contradictory gene trees. The warning of Keeling and Palmer (2008) that the picture may be getting more complex with increasingly denser sampling of taxa, genes and genomes so far is borne out (for plants at least), and we are still far from a satisfactory understanding of the mechanisms, vectors and evolutionary significance of natural horizontal gene transfer.

References

Adams KL, Clements MJ, Vaughn JC (1998) The *Peperomia* mitochondrial *coxI* group I intron: timing of horizontal transfer and subsequent evolution of the intron. J Mol Evol 46:689–696

Altschul SF, Gish W, Miller W, Myers EW, Lipman DJ (1990) Basic local alignment search tool. J Mol Biol 215:403–410

Andersson JO (2005) Lateral gene transfer in eukaryotes. Cell Mol Life Sci 62:1182–1197

Archibald JM, Richards TA (2010) Gene transfer: anything goes in plant mitochondria. BMC Biol 8:147

Arimura S, Yamamoto J, Aida GP, Nakazono M, Tsutsumi N (2004) Frequent fusion and fission of plant mitochondria with unequal nucleoid distribution. Proc Natl Acad Sci USA 101:7805–7808

Barkman TJ, McNeal JR, Lim S-H, Coat G, Croom HB, Young ND, dePamphilis CW (2007) Mitochondrial DNA suggests at least 11 origins of parasitism in angiosperms and reveals genomic chimerism in parasitic plants. BMC Evol Biol 7:248

Belshaw R, Bensasson D (2006) The rise and falls of introns. Heredity 96:208–213

Bergthorsson U, Adams KL, Thomason B, Palmer JD (2003) Widespread horizontal transfer of mitochondrial genes in flowering plants. Nature 424: 197–201

Bergthorsson U, Richardson AO, Young GJ, Goertzen LR, Palmer JD (2004) Massive horizontal transfer of mitochondrial genes from diverse land plant donors to the basal angiosperm *Amborella*. Proc Natl Acad Sci USA 101:17747–17752

Bock R (2010) The give-and-take of DNA: horizontal gene transfer in plants. Trends Plant Sci 15:11–22

Boto L (2010) Horizontal gene transfer in evolution: facts and challenges. Proc R Soc B 277(1683): 819–827

Broothaerts W, Mitchell HJ, Weir B, Kaines S, Smith LMA, Yang W, Mayer JE, Roa-Rodrıguez C, Jefferson RA (2005) Gene transfer to plants by diverse species of bacteria. Nature 433:629–633

Burger G, Gray MW, Lang BF (2003) Mitochondrial genomes: anything goes. Trends Genet 19:709–716

Cho Y, Palmer JD (1999) Multiple acquisitions via horizontal transfer of a group I intron in the mitochondrial *coxl* gene during evolution of the Araceae family. Mol Biol Evol 16:1155–1165

Cho Y, Adams KL, Qiu Y-L, Kuhlman P, Vaughn JC, Palmer JD (1998a) A highly invasive group I intron in the mitochondrial *cox1* gene. In: Moller I-M, Gardestrom P, Glimelius K, Glaser E (eds) Plant mitochondria: from gene to function. Backhuys, Leiden, pp 19–23

Cho Y, Qiu Y-L, Kuhlman P, Palmer JD (1998b) Explosive invasion of plant mitochondria by a group I intron. Proc Natl Acad Sci USA 95:14244–14249

Cohen O, Pupko T (2010) Inference and characterization of horizontally transferred gene families using stochastic mapping. Mol Biol Evol 27:703–713

Cusimano N, Zhang L-B, Renner SS (2008) Reevaluation of the *cox1* group I intron in Araceae and angiosperms indicates a history dominated by loss rather than horizontal transfer. Mol Biol Evol 25:1–12

Danchina EGJ, Rossoa M-N, Vieiraa P, de Almeida-Englera J, Coutinhob PM, Henrissat B, Abad P (2010) Multiple lateral gene transfers and duplications have promoted plant parasitism ability in nematodes. Proc Natl Acad Sci USA 107:17651–17656

Davis CC, Wurdack KJ (2004) Host-to-parasite gene transfer in flowering plants: phylogenetic evidence from Malpighiales. Science 305:676–678

Davis CC, Anderson WR, Wurdack KJ (2005) Gene transfer from a parasitic flowering plant to a fern. Proc R Soc B 272:2237–2242

Diao X, Freeling M, Lisch D (2006) Horizontal transfer of a plant transposon. PLoS Biol 4:e5

Dumas F, Duckely M, Pelczar P, Van Gelder P, Hohn B (2001) An *Agrobacterium* VirE2 channel for transferred-DNA transport into plant cells. Proc Natl Acad Sci USA 98:485–490

Embley TM, Thomas RH, Williams RAD (1993) Reduced thermophilic bias in the 16S rDNA sequence from *Thermus ruber* provides further support for a relationship between *Thermus* and *Deinococcus*. Syst Appl Microbiol 16:25–29

Farris JS, Kallersjo M, Kluge AG, Bult C (1994) Testing significance of incongruence. Cladistics 10:315–319

Felsenstein J (1978) Cases in which parsimony or compatability methods will be positively misleading. Syst Zool 27:401–410

Felsenstein J (1985) Confidence limits on phylogenies: an approach using the bootstrap. Evolution 39: 783–791

Ferandon C, Moukha S, Callac P, BenedettoJ-P CM, Barroso G (2010) The *Agaricus bisporus cox1* gene: the longest mitochondrial gene and the largest reservoir of mitochondrial group I introns. PLoS ONE 5:e14048

Filipowicz N, Renner SS (2010) The worldwide holoparasitic *Apodanthaceae* confidently placed in the Cucurbitales by nuclear and mitochondrial gene trees. BMC Evol Biol 10:219

Foster PG, Cox CJ, Embley TM (2009) The primary divisions of life: a phylogenomic approach employing composition-heterogeneous methods. Philos Trans R Soc Lond B 364:2197–2207

Gelvin SB (2009) *Agrobacterium* in the genomics age. Plant Physiol 150:1665–1676

Gilbert C, Schaack S, Pace JK, Brindley PJ, Feschotte C (2010) A role for host-parasite interactions in the horizontal transfer of transposons across phyla. Nature 464:1347–1352

Gogarten JP, Townsend JP (2005) Horizontal gene transfer, genome innovation and evolution. Nat Rev 3:679–687

Goremykin VV, Salamini F, Velasco R, Viola R (2009) Mitochondrial DNA of *Vitis vinifera* and the issue of rampant horizontal gene transfer. Mol Biol Evol 26:99–110

Graham LA, Lougheed SC, Ewart KV, Davies PL (2008) Lateral transfer of a lectin-like antifreeze protein gene in fishes. PLoS One 3:e2616

Grewe F, Viehoever P, Weisshaar B, Knoop V (2009) A trans-splicing group I intron and tRNA-hyperediting in the mitochondrial genome of the lycophyte *Isoetes engelmannii*. Nucleic Acids Res 37: 5093–5104

Hao W (2010) OrgConv: detection of gene conversion using consensus sequences and its application in plant mitochondrial and chloroplast homologs. BMC Bioinformatics 11:114

Hao W, Palmer JD (2009) Fine-scale mergers of chloroplast and mitochondrial genes create functional, transcompartmentally chimeric mitochondrial genes. Proc Natl Acad Sci USA 106:16728–16733

Hao W, Richardson AO, Zheng Y, Palmer JD (2010) Gorgeous mosaic of mitochondrial genes created by horizontal transfer and gene conversion. Proc Natl Acad Sci USA 107:21576–21581

Hecht J, Grewe F, Knoop V (2011) Extreme RNA editing in coding islands and abundant microsatellites in repeat sequences of Selaginella moellendorffii mitochondria: the root of frequent plant mtDNA recombination in early tracheophytes. Genome Biol Evol 3:344–358

Henze K, Martin W (2001) How do mitochondrial genes get into the nucleus? Trends Genet 17: 383–387

Hill T, Nordström KJV, Thollesson M, Säfström TM, Vernersson AKE, Fredriksson R, Schiöth HB (2010) SPRIT: identifying horizontal gene transfer in rooted phylogenetic trees. BMC Evol Biol 10:42

Inda LA, Pimentel M, Chase MW (2010) Contribution of mitochondrial *cox1* intron sequences to the phylogenetics of tribe Orchideae (Orchidaceae): do the distribution and sequence of this intron in orchids also tell us something about its evolution? Taxon 59:1053–1064

Jiang N, Bao Z, Zhang X, Eddy SR, Wessler SR (2004) Pack-MULE transposable elements mediate gene evolution in plants. Nature 431:569–573

Keeling PJ (2009a) Functional and ecological impacts of horizontal gene transfer in eukaryotes. Curr Opin Genet Dev 19:613–619

Keeling PJ (2009b) Role of horizontal gene transfer in the evolution of photosynthetic eukaryotes and their plastids. In: Gogarten MB et al (eds) Horizontal gene transfer: genomes in flux, vol 532, Methods in molecular biology., pp 501–515

Keeling PJ, Palmer JD (2008) Horizontal gene transfer in eukaryotic evolution. Nat Rev Genet 9:605–618

Knoop V, Volkmar U, Hecht J, Grewe F (2010) Mitochondrial genome evolution in the plant lineage. In: Kempken F (ed) Plant mitochondria, vol 1, Advances in plant biology., pp 3–29

Koski LB, Golding B (2001) The closest BLAST hit is often not the nearest neighbor. J Mol Evol 52: 540–542
Koulintchenko M, Konstantinov Y, Dietrich A (2003) Plant mitochondria actively import DNA via the permeability transition pore complex. EMBO J 22:1245–1254
Kurland CG, Canback B, Berg OG (2003) Horizontal gene transfer: a critical view. Proc Natl Acad Sci USA 100:9658–9662
Lambowitz AM, Belfort M (1993) Introns as mobile genetic elements. Annu Rev Biochem 62:587–622
Lambowitz AM, Zimmerly S (2004) Mobile group II introns. Annu Rev Genet 38:1–35
Lander ES, The International Human Genome Sequencing Consortium (2001) Initial sequencing and analysis of the human genome. Nature 409: 860–921
Lucas WJ, Ham B-K, Kim J-Y (2009) Plasmodesmata – bridging the gap between neighboring plant cells. Trends Cell Biol 19:495–503
Ma L-J, van der Does HC, Borkovich KA, Coleman JJ, Daboussi M-J, Di Pietro A, Dufresne M, Freitag M, Grabherr M, Henrissat B, Houterman PM, Kang S, Shim WB, WoloshukC XX, Xu J-R, Antoniw J, Baker SE, Bluhm BH, Breakspear A, Brown DW, Butchko RAE, Chapman S, Coulson R, Coutinho PM, Danchin EGJ, Diener A, Gale LR, Gardiner DM, Goff S, Hammond-Kosack KE, Hilburn K, Van Hua A, Jonkers W, Kazan K, Kodira CD, Koehrsen M, Kumar L, Lee Y-H, Li L, Manners JM, Miranda-Saavedra D, Mukherjee M, Park G, Park J, Park SY, Proctor RH, Regev A, Ruiz-Roldan MC, Sain D, Sakthikumar S, Sykes S, Schwartz DC, Turgeon BG, Wapinski I, Yoder O, Young S, Zeng Q, Zhou S, Galagan J, Cuomo CA, Kistler HC, Rep M (2010) Comparative genomics reveals mobile pathogenicity chromosomes in *Fusarium*. Nature 464:367–373
Marchetti A, Parker MS, Moccia LP, Lin EO, Arrieta AL, Ribalet F, Murphy MEP, Maldonado MT, Armbrust EB (2009) Ferritin is used for iron storage in bloom-forming marine pennate diatoms. Nature 457(22):467–470
Martin W (2005) Lateral gene transfer and other possibilities. Heredity 94:565–566
Mirkin BG, Fenner TI, Galperin MY, Koonin EV (2003) Algorithms for computing parsimonious evolutionary scenarios for genome evolution, the last universal common ancestor and dominance of horizontal gene transfer in the evolution of prokaryotes. BMC Evol Biol 3:2
Morgante M, Brunner S, Pea G, Fengler K, Zuccolo A, Rafalski A (2005) Gene duplication and exon shuffling by helitron-like transposons generate intraspecies diversity in maize. Nat Genet 37:997–1002
Mower JP, Stefanovic S, Young GJ, Palmer JD (2004) Gene transfer from parasitic to host plants. Nature 432:165–166
Mower JP, Stefanovic S, Hao W, Gummow JS, Jain K, Ahmed D, Palmer JD (2010) Horizontal acquisition of multiple mitochondrial genes from a parasitic plant followed by gene conversion with host mitochondrial genes. BMC Biol 8:150
Nickrent DL, Blarer A, Qiu YL, Vidal-Russell R, Anderson FE (2004) Phylogenetic inference in Rafflesiales: the influence of rate heterogeneity and horizontal gene transfer. BMC Evol Biol 4:40
Nishida H, Sugiyama J (1995) A common group I intron between a plant parasitic fungus and its host. Mol Biol Evol 12:883–886
Noble GP, Rogers MB, Keeling PJ (2007) Complex distribution of EFL and EF-1alpha proteins in the green algal lineage. BMC Evol Biol 7:82
Odom OW, Shenkenberg DL, Garcia JA, Herrin DL (2004) A horizontally acquired group II intron in the chloroplast *psbA* gene of a psychrophilic *Chlamydomonas*: in vitro self-splicing and genetic evidence for maturase activity. RNA 10:1097–1107
Omer S, Kovacs A, Mazor Y, Gophna U (2010) Integration of a foreign gene into a native complex does not impair fitness in an experimental model of lateral gene transfer. Mol Biol Evol 27:2441–2445
Ragan MA, Beiko RG (2009) Lateral genetic transfer: open issues. Philos Trans R Soc Lond B 364: 2241–2251
Ragan MA, Harlow TJ, Beiko RG (2006) Do different surrogate methods detect lateral genetic transfer events of different relative ages? Trends Microbiol 14:4–8
Richards TA, Soanes DM, Foster PG, Leonard G, Thornton CR, Talbot NJ (2009) Phylogenomic analysis demonstrates a pattern of rare and ancient horizontal gene transfer between plants and fungi. Plant Cell 21:1897–1911
Richardson AO, Palmer JD (2007) Horizontal gene transfer in plants. J Exp Bot 58:1–9
Rogers MB, Watkins RF, Harper JT, Durnford DG, Gray MW, Keeling PJ (2007) A complex and punctate distribution of three eukaryotic genes derived by lateral gene transfer. BMC Evol Biol 7:89
Roney JK, Khatibi PA, Westwood JH (2007) Cross-species translocation of mRNA from host plants into the parasitic plant dodder. Plant Physiol 143:1037–1043
Salzberg SL, White O, Peterson J, Eisen JA (2001) Microbial genes in the human genome: lateral transfer or gene loss? Science 292:1903–1906
Sanchez-Gracia A, Maside X, Charlesworth B (2005) High rate of horizontal transfer of transposable elements in *Drosophila*. Trends Genet 21:200–203

Sanchez-Puerta MV, Cho Y, Mower JP, Alverson AJ, Palmer JD (2008) Frequent, phylogenetically local horizontal transfer of the *cox1* group I intron in flowering plant mitochondria. Mol Biol Evol 25:1762–1777

Sanchez-Puerta MV, Abbona CC, Zhuo S, Tepe EJ, Bohs L, Olmstead RG, Palmer JD (2011) Multiple recent horizontal transfers of the cox1 intron in Solanaceae and extended coconversion of flanking exons. BMC Evol Biol 11:277

Schaack S, Gilbert C, Feschotte C (2010) Promiscuous DNA: horizontal transfer of transposable elements and why it matters for eukaryotic evolution. Trends Ecol Evol 25:537–546

Schönenberger J, Anderberg AA, Sytsma KJ (2005) Molecular phylogenetics and patterns of floral evolution in the ericales. Int J Plant Sci 166:265–288

Sheahan MB, McCurdy DW, Rose RJ (2005) Mitochondria as a connected population: ensuring continuity of the mitochondrial genome during plant cell dedifferentiation through massive mitochondrial fusion. Plant J 44:744–755

Snir S, Trifonov E (2010) A novel technique for detecting putative horizontal gene transfer in the sequence space. J Comput Biol 17:1535–1548

Sorek R, Zhu Y, Creevey CJ, Francino MP, Bork P, Rubin EP (2007) Genome-wide experimental determination of barriers to horizontal gene transfer. Science 318:1449–1452

Stanhope MJ, Lupas A, Italia MJ, Koretke KK, Volker C, Brown JR (2001) Phylogenetic analyses do not support horizontal gene transfers from bacteria to vertebrates. Nature 411:940–944

Stegemann S, Bock R (2009) Exchange of genetic material between cells in plant tissue grafts. Science 324:649–651

Stiller JW (2011) Experimental design and statistical rigor in phylogenomics of horizontal and endosymbiotic gene transfer. BMC Evol Biol 11:259

Ülker B, Li Y, Rosso MG, Logemann E, Somssich IE, Weisshaar B (2008) T-DNA–mediated transfer of *Agrobacterium tumefaciens* chromosomal DNA into plants. Nat Biotechnol 26:1015–1017

Vaughn JC, Mason MT, Sper-Whitis GL, Kuhlman P, Palmer JD (1995) Fungal origin by horizontal transfer of a plant mitochondrial group I intron in the chimeric *coxI* gene of *Peperomia*. J Mol Evol 41: 563–572

Wolf YI, Kondrashov FA, Koonin EV (2001) Footprints of primordial introns on the eukaryotic genome: still no clear traces. Trends Genet 17:499–501

Won H, Renner SS (2003) Horizontal gene transfer from flowering plants to *Gnetum*. Proc Natl Acad Sci USA 100:10824–10829

Yang Z (1994) Maximum likelihood phylogenetic estimation from DNA sequences with variable rates over sites: approximate methods. J Mol Evol 39: 306–314

Yoshida S, Maruyama S, Nozaki H, Shirasu K (2010) Horizontal gene transfer by the parasitic plant *Striga hermonthica*. Science 328:1128

Chapter 11

Plastome Mutants of Higher Plants

Stephan Greiner*

Max Planck Institute of Molecular Plant Physiology, Wissenschaftspark Golm, Am Mühlenberg 1, 14476 Potsdam-Golm, Germany

Summary

This chapter provides an overview of higher plant plastome mutants and their application in molecular biology, cytoplasmic genetics and biotechnology. Starting from an outline on plastid inheritance, the sources of mutants, methods of their maintenance and molecular approaches to identify the underlying genetic changes are presented. Subsequently, the

*Author for correspondence, e-mail: greiner@mpimp-golm.mpg.de

R. Bock and V. Knoop (eds.), Genomics of Chloroplasts and Mitochondria, Advances in Photosynthesis and Respiration 35, pp. 237–266,
DOI 10.1007/978-94-007-2920-9_11,

molecularly characterized plastome mutants and their impact on our current knowledge about plastids are summarized. Recent developments in genomics will likely overcome technical limitations connected with the elucidation of mutant loci in plastome mutants. The great potential of plastid mutants in future research, for example in studying plastid gene regulation, as well as suitable model plants and available genetic resources are discussed.

I. Introduction

Plastome mutants can occur spontaneously, or can be induced by either chemical or nuclear-gene-mediated mutagenesis. Alternatively, they can be directly generated by plastid transformation (Kutzelnigg and Stubbe 1974; Börner and Sears 1986; Hagemann 1992; Bock 2001; Sect. III). Classic plastid mutants are recognized as bleached leaf areas in variegated plants (Fig. 11.1 and Sect. II). Such material has been used as a genetic tool since the very beginning of formal genetics. Baur's and Corren's fundamental work on plastome mutants in *Mirabilis jalapa* (four o'clock flower) *Pelargonium zonale* and *Antirrhinum majus* (snapdragon) laid the foundation for cytoplasmatic genetics (Baur 1909; Correns 1909; Baur 1910; Kirk and Tilney-Bassett 1978; Hagemann 2010). Since then, plastome mutants have provided compelling evidence for the existence of an independent genetic system in plastids (Baur 1909; Renner 1934; Hagemann 2010), and they still play an important role in elucidating the rules of non-Mendelian inheritance (Sect. II). The use of plastome mutants facilitated the initial development of chloroplast transformation technology in *Chlamydomonas reinhardtii* (Boynton et al. 1988). Also, plastid mutants provide selectable makers for antibiotic or herbicide resistances that are widely used in molecular biology and agriculture (Sect. VI). However, primarily due to technical limitations, the contribution of plastome mutants to illuminating gene content and gene functions in plastids was meager (Sect. V). Nonetheless, plastome mutants keep playing important roles in molecular research on plastids. Some mutants display particularly interesting developmental phenotypes (Sect. VI), and/or genetic patterns beyond the classic rules of plastid genetics (Sect. VIII). Next-generation Sequencing (NGS) technologies employed for mapping and identification of plastome mutants may herald a renaissance of this field. The aim of this chapter is to summarize the currently

---

Abbreviations: AA – Amino acid; *accD* – Acetyl coenzyme A carboxylase subunit D gene; *aadA* – Aminoglycoside 3-adenylyltransferase gene; AtMSH1 – *Arabidopsis thaliana* MutS homolog 1; *AtMsh1* – *Arabidopsis thaliana* MutS homolog 1 gene; *atp* – ATP synthase subunit gene; *AtWhy* – *Arabidopsis thaliana whirly* gene; AtWHY – *Arabidopsis thaliana* WHIRLY protein; bp – Base pair; *clpP* – Chloroplast caseinolytic protease subunit P gene; CMS – Cytoplasmic male sterility; *cpRecA* – Chloroplast RecA gene; dsDNA – Double-stranded DNA; EMS – Ethyl-methane-sulfonate; IF1 – Translation initiation factor 1; indel – Insertion/deletion; InfA – Translation initiation factor A; *inf*A – Translation initiation factor A gene; *matK* – Maturase K gene; MNNG – Methyl-nitro-nitroso-guanidine; N/A – not available; NGS – Next-generation Sequencing; NMU – *N*-nitroso-*N*-methyl-urea; PEG – Polyethylene glycol; *petB* – Cytochrome b_6/f subunit B gene; PGI – Plastome-genome incompatibility; PQ – Plastoquinone; PS I – Photosystem I; PS II – Photosystem II; Psa – Photosystem I subunit; *psa* – Photosystem I subunit gene; Psb – Photosystem II subunit; *psb* – Photosystem II subunit gene; ptDNA – Plastid DNA; *rbcL* – Ribulose-1,5-bisphosphate carboxylase oxygenase large subunit gene; RbcL – Ribulose-1,5-bisphosphate carboxylase oxygenase large subunit; RbcS – Ribulose-1,5-bisphosphate carboxylase oxygenase small subunit; RFLP – Restriction length polymorphism; *rps12* – Ribosomal protein small subunit 12 gene; RuBisCO – Ribulose-1,5-bisphosphate carboxylase oxygenase; SAM – Shoot apical meristem; TILLING – Targeting induced local lesions in genomes; *trnE(UUC)* – tRNA-Glu (anticodon UUC); UV – Ultraviolet; *ycf* – Hypothetical chloroplast reading frame; *ZmWhy* – *Zea mays whirly* gene

Fig. 11.1. Sorting-out of genetically distinct plastids and variegation patterns in *A. thaliana* (**a**) and *O. elata* (**b–g**). Variegation in plant rosettes produced by EMS mutagenesis (**a**) or sexual crosses of mutated (*pale*) and wild type (*green*) plastid genotypes (**b–g**). Different sorting-out patterns in leaves (**c–g**): Sectorial chimera (**c**), periclinal chimera (**d**), mesoclinal chimera (**e**), mosaic pattern (**f**), and combination of the former types (**g**). Note that *light green* tissue results from overlaying cell layers as produced by adaxial and abaxial sorting-out (**f**, **g**).

existing literature on plastid mutants, to evaluate promising applications in molecular genetics and to outline perspectives for future research.

II. A Brief Survey of Plastid Genetics

Non-Mendelian inheritance was described already shortly after the rediscovery of Mendel's laws. It differs fundamentally from Mendelian inheritance and is characterized by a predominantly maternal inheritance recognized as reciprocal differences in sexual crosses, somatic segregation of divergent organelle genotypes and by a virtual absence of homologous recombination of the DNAs involved. Plastome mutants were the major tools in elucidating the rules for cytoplasmic inheritance. The following paragraphs briefly summarize the most relevant aspects.

A. Transmission of Plastids

Mostly based on work with pale plastome mutants, it appeared that plastids can undergo three different modes of inheritance: maternal (initially referred to as *status albomaculatus*), biparental (*status paralbomaculatus*) or paternal. Well-known examples for maternal inheritance are *Nicotiana tabacum* (tobacco), *A. majus* and *Arabidopsis thaliana*. Biparental transmission is best studied in Oenothera (evening primrose) and Pelargonium. Paternal transmission or a strong paternal bias was observed for gymnosperms and the angiosperm *Medicago sativa* (alfalfa), respectively (Hagemann 1964, 1992, 2004; Grun 1976; Gillham 1978; Kirk and Tilney-Bassett 1978). The predominant mode of plastid inheritance in seed plants is the maternal pattern, but low-level or occasional biparental inheritance may be present in about 1/3 of all plant taxa studied

(Mogensen 1996). More and more evidence accumulates, that uniparental inheritance is often not absolute, and that paternal leakage (resulting in heteroplasmy) may be a general feature of higher plant populations (e.g., Birky 2001; Wolfe and Randle 2004; Petit and Vendramin 2007). Plastid mutants or plastids genetically modified by transformation, served in a variety of plant species as visible phenotypic or selectable markers in the analysis of sexual crosses to screen for paternal leakage events. In contrast to direct investigations of the DNA inherited, plastome mutants allow the easy, large-scale, and unambiguous identification of rare paternal transmission events of the chloroplast genome (Hagemann 1992; Azhagiri and Maliga 2007; Ruf et al. 2007; Svab and Maliga 2007; and citations therein).

B. Sorting-Out and Variegation

Biparental transmission of genetically distinct plastids produces a zygote harboring two plastid types, a so-called mixed cell. During subsequent cell divisions, maternal and paternal plastids are randomly distributed to the daughter cells. After several division cycles, mixed cells disappear and cell lineages containing only one of the two plastid genotypes form. If one of the plastid types is marked by a mutation primarily or secondarily impairing photosynthesis, variegated plants containing mutated and wild-type plastids in distinct tissues appear (Fig. 11.1). Sorting-out takes place in each tissue and developmental stage of an individual plant. Consequently, after de novo mutation of a single plastid genome, sorting-out begins during the first subsequent cell division cycle, producing a mosaic pattern of leaf variegation with sharp tissues borders after completion of the sorting-out process (Fig. 11.1a, b, f, g). Along this border, mixed cells can be found. The concept of sorting-out is a fundamental feature of non-Mendelian inheritance, and its theoretical properties were originally worked out by Michaelis (1955) utilizing pale plastome mutants (Hagemann 1964; Kirk and Tilney-Bassett 1978; Birky 2001). In dicotyledonous plants, depending on plastid distribution in the meristem, completed sorting-out can lead to different types of variegation and (leaf) chimeras: (1) Sectorial chimeras, in which the three layers (L1–L3) of the shoot apical meristem (SAM) carry mutated and non-mutated cells. Sorting-out can lead to different genetic identities of the two sides of the organ axis. In the leaf, the genetically different tissues are arranged in lateral sectors (Fig. 11.1c). (2) Periclinal chimeras, in which sorting-out gives rise to at least one homoplasmic layer of the three meristematic cell layers [tunica (L1 and L2) and corpus (L3)]. The individual layers differ genetically and consequently epidermis (L1), subepidermal cell layer (L2; phenotypically apparent in the leaf margin) and/or corpus (L3) have different genetics identities as far as their plastids are concerned (Fig. 11.1d). (3) Mesoclinal chimeras represent a combination of sectorial and periclinal chimerism (Fig. 11.1e). (4) Mosaic patterns are observed if sorting-out is not completed within a leaf and/or in distinct domains of the meristem. This is typically the case in early plant development (Fig. 11.1f). (5) Finally, ad- and abaxial sorting patterns are possible as well as a combination of all sorting-out patterns mentioned above (Fig. 11.1f, g). For graphical presentations of variegation patterns in the SAM and its genetic and phenotypic consequences, see Kirk and Tilney-Bassett (1978). In monocotyledonous plants, where the leaf basal meristem continuously mediates proximal growth, variegation patterns are recognized as striping, since cells of different genetic identities are arranged in parallel (Kirk and Tilney-Bassett 1978).

C. Identification of Plastome Mutants by Means of Classic Genetics

The challenge of identifying plastome mutants by employing classic genetic methods is not obvious at first glance. Variegation, a first indication of the presence of a plastome mutation, can also be caused by various nuclear alleles or by mitochondrial dysfunction

(Kirk and Tilney-Bassett 1978; Rodermel 2002; Yu et al. 2007). Although sorting-out of plastids results in definable intercalated patterns (Michaelis 1957, 1958a, b; cf. Fig. 11.1f), non-Mendelian inheritance must be proven to demonstrate the cytoplasmic origin of a mutation. In the case of uniparental transmission, this is evident by maternal inheritance in reciprocal crosses. In plants with biparental plastid transmission, confirming variegation, sorting-out and non-Mendelian segregation in the F1 generation represent accepted methods (Kirk and Tilney-Bassett 1978; Hagemann 1982). To rule out mitochondrial mutations, mixed cells containing both mutated and wild-type plastids need to be identified. In terms of classic genetics, together with the demonstration of sorting-out, this is a strong indication of the presence of a plastome mutation. It illustrates that the cause of the impaired plastid phenotype rests within the plastid itself. However, strictly speaking, even in this case the presence of a mitochondrial mutation cannot be excluded. The mixed cell looked at may be still heteroplasmic for a mitochondrial mutation, secondarily leading to a plastid malfunction (cf. Kirk and Tilney-Bassett 1978; Sect. III).

D. Competition of Plastids with Genetically Different Plastome Types

In Oenothera species, plastid genomes marked by mutation uncovered different multiplication rates in sexual crosses, depending on the plastid genotype. Based on the "variegation value" of F1 seedling populations, Schötz and co-workers measured the relative strengths of diverging plastome types to each other. For the five genetically distinguishable plastome types in Oenothera, three different multiplication speeds (fast, medium, and slow) could be identified. Refined analysis revealed that the competitive advantage of a given plastome is largely independent of the nucleus and can even exist if the more competitive plastid genotype is incompatible to the host plant. In extreme cases, biparental transmission can be suppressed by a combination of a "fast" and a "slow" plastome, if incompatible plastome/genome combinations are involved. Consequently, at least in Oenothera, the determinants mediating plastid competition seem to be predominantly plastome encoded (Schötz 1954; Grun 1976; Gillham 1978; Kirk and Tilney-Bassett 1978; Chiu et al. 1988; Chiu and Sears 1993; Harte 1994). Although data are limited, different multiplication rates are probably an intrinsic feature of the plastid genome and a general phenomenon in nature. Comparable results were obtained in *M. sativa* using cybrids in cell culture (Fitter and Rose 1993) and some evidence also exists for Pelargonium (Hagemann and Scholze 1974; Hagemann 1976; Abdel-Wahab and Tilney-Bassett 1981). However, mode and control of biparental inheritance in the genus Pelargonium strikingly differs from that in Oenothera (Tilney-Bassett 1975; Kirk and Tilney-Bassett 1978; Tilney-Basset 1994) and, clearly, the findings obtained in Pelargonium require further investigation (Tilney-Basset 1994). The detection of different plastome replication rates in Oenothera contributed substantially to the hypothesis of selfish cytoplasmic elements (Grun 1976; cf. Hoekstra 2000; Barr et al. 2005). However, the underlying loci, most probably origins of replication (Hornung et al. 1996; Sears et al. 1996), have not yet been identified. In this regard, the identification of slower or faster multiplying plastome mutants could represent a viable approach.

E. Sexual Recombination of Different Plastome Types

Higher plant plastids seem not to undergo sexual recombination, not even in plant taxa transmitting chloroplast genomes regularly by both sexes. Chiu and Sears (1985) performed a study with Oenothera using two independent plastome mutants, which were crossed with 10 different other plastome mutants. In the 20 F1 generations (each heteroplasmic for a different pair of mutations), a total of about 7,500 seedlings were raised. Recombination events were expected to result in the appearance of green leaf spots in the seedlings and would indicate a rescue of a mutant by recombination. Since such an event was not observed, recombination between different plastome types is either completely absent or present at only very low frequencies (cf. Kutzelnigg and Stubbe 1974). However, in the Chiu and Sears study some recombination events may have escaped detection due to incomplete sorting-out, the physiological state of some of the mutants (prohibiting re-greening after a genetic complementation), or close genetic linkage. The virtual absence of sexual recombination in higher plant plastids is indeed surprising, since homologous recombination between single plastid genomes within a plastid is quite frequent (e.g., Palmer 1983; Day and Madesis 2007). At least occasionally, it also can be induced in somatic cybrids generated by protoplast fusion. For instance, the *Nicotiana plumbaginifolia* line SR1-A15 carries two plastome mutations, one mediating streptomycin resistance and the other greening deficiency. A second line (LR400) is resistant to lincomycin, also due to a plastome mutation, but is normally green. Protoplasts of the cell lines were mixed, fused in the presence of polyethylene glycol (PEG), and subsequently selected for green calli on streptomycin-containing media. Regenerated green lines were supposed to carry a recombinant plastome from the streptomycin-resistant (white) and lincomycin-resistant (green) plastids. Double selection of the cybrid plants on medium containing both streptomycin and lincomycin as well as RFLP analysis confirmed this assumption (Medgyesy et al. 1985). Comparable data were independently obtained in similar experiments, including interspecific protoplast fusions (e.g., Thanh and Medgyesy 1989; Trabelsi et al. 2005; Bidani et al. 2007; but also see Petit and Vendramin 2007). These results show that higher plant plastids can, in principle, undergo recombination of different genotypes. At least, this is possible under strong selection and in a tissue culture system including PEG, which might have induced artificial plastid fusion. That plastid fusion is the critical prerequisite for sexual recombination of plastid DNA (ptDNA) is evident from work on the isogamous green alga Chlamydomonas. In this organism, the two chloroplasts of the crossing mates (mt^+ and mt^-) fuse in the zygote. Typically the mt^- ptDNA is degraded (>90%), but UV irradiation of the mt^+ parent can greatly increase the frequency of heteroplasmic zygotes. This strategy, together with various antibiotic-resistant plastome mutants, was particularly useful for constructing recombination maps of the plastid genomes of Chlamydomonas species (Gillham 1978; Gillham et al. 1991; Boynton et al. 1992). Therefore, the explanation for the apparent absence of homologous recombination between plastomes of different genotypes in sexual crosses of higher plants may lie in the lack of plastid fusion in the germline (Meyer and Stubbe 1974; Kirk and Tilney-Bassett 1978; Sears 1980; Kuroiwa 2010; Nagata 2010). There is also only limited evidence for plastid fusion in somatic tissue (Esau 1972; Sears 1980; Vaughn 1981). In contrast, some theoretical and circumstantial phylogenetic evidence indicates the presence of sexual recombination of plastid genomes (cf. Hagemann 1992; Birky 1995; Wolfe and Randle 2004; Barr et al. 2005; Petit and Vendramin 2007; Greiner et al. 2011). The true situation in natural populations of higher plant species remains unclear and needs further investigation.

F. Plastid Restitution

Plastid restitution is defined as the re-greening of a mutated plastid by an additional stable mutation. With this strict definition, plastid restitution does not necessarily require a reversion or suppressor mutation within the plastome itself. Classic genetic problems associated with the identification of restitution events, such as uncompleted sorting-out, have been discussed previously (Tilney-Bassett 1975; Kirk and Tilney-Bassett 1978). Additional examples have come from *P. zonale* (spontaneous restitution), Oenothera (mutated by a nuclear plastome mutator allele), *Helianthus annuus* (second-site chemical mutagenesis and spontaneous mutants), and presumably also *Hordeum vulgare* (barley), caused by a plastome mutator (Abdel-Wahab and Tilney-Bassett 1981; Johnson et al. 1991; Prina 1992; Usatov et al. 2004). Work on *Epilobium hirsutum* suggests that restitution events can lead to changes in plant or plastid morphology compared to the original wild type (Michaelis 1969). Michaelis could further demonstrate the overcoming of hybrid incompatibility and sterility by a restituted plastid (Michaelis 1969; Kirk and Tilney-Bassett 1978). However, it is worth mentioning that the material Michaelis used in his restitution experiments was generated under the influence of the nuclear plastome mutator allele *mp1*, for which circumstantial evidence suggests an additional cytoplasmic (likely mitochondrial) mutator activity (Sect. III). Unfortunately, none of the published restitution events was characterized at the molecular level, although the phenomenon clearly deserves systematic studies. In line with a high plastome copy number and small genome size (Bock 2007b) and depending on the original mutation, restitution can be rather frequent. Occasionally, restitution depends on environmental factors, indicating no "full" rescue to the original wild-type plastome (Michaelis 1969; Tilney-Bassett 1975; Kirk and Tilney-Bassett 1978). Elucidating suppressor relations between partially plastid-encoded supramolecular machineries, such as the thylakoid membrane system or the organellar ribosomes, and/or between chloroplast and mitochondrial genomes provides fascinating opportunities for further research.

A special case of plastid restitution is the rescue of plastome-genome incompatibly (PGI). These speciation-relevant barriers are the result of a co-evolution between the plastid and the nuclear genome, often leading to bleached chloroplasts in an alien nuclear background (reviewed in Levin 2003; Greiner et al. 2011). Mutagenesis of incompatible tissue and screens for restitution events can potentially become a general method to pinpoint plastome-encoded determinants for PGI. Utilizing chemical mutagenesis, the albinotic phenotype of a plastome-genome incompatible cybrid between the nuclear genome of *Atropa belladonna* (deadly nightshade) and the plastome of *N. tabacum* was cured. Analysis of relevant plastid loci revealed a single base pair exchange within the *atpA* gene, mimicking a species-specific editing site for the *N. tabacum* plastome (Schmitz-Linneweber et al. 2005).

III. Sources of Plastome Mutants

As outlined above, plastid mutants were used as an important tool for the analysis of non-Mendelian genetics. What sources of such mutants are available to researchers? The following section provides an overview of the three different origins of mutants: spontaneous occurrence, chemical mutagenesis and nuclear-gene induced mutations by plastome mutator alleles (Kutzelnigg and Stubbe 1974; Kirk and Tilney-Bassett 1978; Börner and Sears 1986; Hagemann 1992).

A. Spontaneous Occurrence

Spontaneous plastome mutations are known from several plant species (Sect. VI; Table 11.1 and citations therein). Published values for rates of spontaneously arising chlorophyll deficiency observed by variegation or striping vary between 0.006% and 0.3% in *A. majus, A. thaliana*, Epilobium, *H. vulgare* and Oenothera (Maly 1958a; Röbbelen 1962; Michaelis 1964; Kutzelnigg and Stubbe 1974; Kirk and Tilney-Bassett 1978; Prina 1992). Inconsistent values between species and experiments are presumably caused by different experimental setups and/or selection criteria, such as plant size and age. It is important to note that the approach systematically misinterprets plastome mutation rates, since only mutants with chlorotic phenotypes and completed sorting-out are usually recognized (cf. Michaelis 1958a; Kutzelnigg and Stubbe 1974; Kirk and Tilney-Bassett 1978). In general, plastome mutation rates are about two-fold lower than those of the nuclear genome (Wolfe

Table 11.1. Plastome mutants with impaired plastid gene function[a]

Plastid function	Organism/variety	Mutant line	Gene/locus	Mutation	Phenotype	Origin	Reference
Photosynthetic light reaction	*Antirrhinum majus* Sippe 50	en:alba-1	*psaB*	Transversion; premature stop codon	Yellowish white	Spontaneous	Schaffner et al. (1995)
Photosynthetic light reaction	*A. majus* Sippe 50	en:alba-4	*psbD*	Transition; single AA-exchange	Yellowish	NMU	Schaffner (1995)
Photosynthetic light reaction	"*Oenothera suaveolens*"[b, c]	II-gamma	*psbE*	5 bp duplication, premature stop condon	Light green to yellowish white	Spontaneous	Hupfer (2002)
Photosynthetic light reaction	"*O. suaveolens*"[b, c]	II-theta	*petB* intron	2 bp deletion, splicing deficiency	Light green to white	Spontaneous	Hupfer (2002)
Carbon fixation	"*O. hookeri*" var. hookeri de Vries[d]	I-sigma	*rbcL*	5 bp deletion, premature stop codon	Light green	Spontaneous	Gordon et al. (1980); Winter and Herrmann (1987)
Carbon fixation	Oenothera[c]	IV-beta	*rbcL*	Transversion; single AA-exchange	Pale to green depending on light conditions	Spontaneous	Dauborn and Brüggemann (1998)
Carbon fixation	*Nicotiana tabacum* var. Techné with *debneyi* cytoplasm[c]	Sp25	*rbcL*	Transition, single AA-exchange	Light green (*viridis*);	EMS	Shikanai et al. (1996)
Carbon fixation	*N. tabacum* var. Xanthi	XV1	*rbcL*	Transition, single AA-exchange	Light green	NMU	Avni et al. (1989)
Translation	*Hordeum vulgare*[c]	Cytoplasmic line 2 (CL2)	*infA*	Transition, single AA-exchange	Gradually greening	Plastome mutator	Landau et al. (2007); Colombo et al. (2008)
Translation	*H. vulgare*[c]	Cytoplasmic line 2-like (CL2-like)	*infA*	Transition, single AA-exchange	Gradually greening	Plastome mutator	Landau et al. (2007)
PS I assembly	*H. vulgare*[c]	cytoplasmic line 3 (CL3)	*ycf3* intron 1	Two point mutations (transition and single bp insertion), splicing deficiency	Light green (*viridis*); temperature sensitive	Plastome mutator	Landau et al. (2009)
mRNA maturation	*Cryptomeria japonica* var. *Wogon*-Sugi	*Wogon*-Sugi	*matK* (likely)	19 bp insertion, premature stop codon	Periodically pale (*virescent*)	Spontaneous	Hirao et al. (2009)

[a]The list excludes plastome mutations conferring herbicide resistances and deletion mutants from cereal tissue culture (for review and references, see Oettmeier 1999; Day and Madesis 2007; Powles and Yu 2010). Antibiotic-resistant mutants are summarized in Table 11.2
[b]The correct taxonomic designation of *O. suaveolens* in now *O. biennis*. For details, see Dietrich et al. (1997)
[c]Identity of mutated and reference strain supposable, but not fully established in the cited references
[d]The correct taxonomic designation of *O. hookeri* is now *O. elata* ssp. *hookeri*. For details, see Dietrich et al. (1997)

et al. 1987). All types of mutations, including indels and point mutations, can arise spontaneously (for references, see Table 11.1).

B. Spontaneously Induced Large Deletions of ptDNA in Cereal Tissue Culture

Within the Poaceae, for which plastid translation is not an essential growth requirement, plastome mutants lacking a big portion of the plastid genome can be isolated. Such mutants arise spontaneously from regenerated anthers or in long-term cell culture or can be induced by the plastid translational inhibitor streptomycin. Typically, relatively short linear DNA fragments and sometimes circular forms of deleted plastomes, are observed, often at high abundance. Mapping of such fragments regularly revealed the presence of ptDNA regions around the *trnE(UUC)* gene. Interestingly, this region is not identical to any of the origins of replication mapped in somatic tissue by independent methods (for summary and references, see Day and Madesis 2007). It is noteworthy in this respect that *trnE(UUC)* is not only involved in translation, but also required for tetrapyrrole biosynthesis (Schön et al. 1986), which likely explains the retention of this gene in all deletion lines.

C. Nuclear Plastome Mutator Alleles Causing Multiple Plastid Mutations

Several nuclear alleles have been documented, which recessively induce various kinds of mutated plastids at frequencies much higher than spontaneous mutations. Such plastome mutator alleles were described for *A. thaliana* (*chloroplast mutator*), *E. hirsutum* (*mp_1*, *mp_2*), *H. vulgare* (*chloroplast mutator*), *Nepeta cataria* (*mutation-allowing*), *O. elata* ssp. *hookeri* (Syn: *O. hookeri*) (*plastome mutator*), *Petunia hybrida* (*a-*) and *Solanum nigrum* (*cpm*) (Grun 1976; Kirk and Tilney-Bassett 1978; Arntzen and Duesing 1983; Börner and Sears 1986; Hagemann 1986; Prina 1992, 1996; Prina et al. 2009). However, at least for the *E. hirsutum* allele *mp_1* and for the *N. cataria* line, formal genetic evidence suggests the mitochondrial genome as (a second) site for mutagenesis (cf. Kirk and Tilney-Bassett 1978; Sect. II). This also holds true for the *A. thaliana chloroplast mutator* (*chm*), although this allele has been a classic example for inducing mutated plastids. Its mutator activity was even confirmed by maternal inheritance, sorting-out and mixed cells (Rédei 1973). Molecular analysis revealed that impaired gene regulation from rearranged mitochondrial loci could explain the variegation phenotype. The corresponding allele was cloned, re-designated *AtMsh1* and found to be a homologue of the *Escherichia coli* MutS gene, a factor for mismatch repair and DNA recombination (reviewed in Rodermel 2002; Yu et al. 2007). AtMSH1 is involved in mitochondrial substoichiometric DNA shifting and mitochondrial DNA recombination (Abdelnoor et al. 2003; Arrieta-Montiel et al. 2009).

A clear case established by molecular analysis of plastome mutations induced by a nuclear allele was described for *O. elata*. Individuals homozygotic for the *plastome mutator* allele *pm* show a 200–1,000 times higher mutation frequency than the wild type. Deletions up to 600 bp, small insertions, point mutations and additional nucleotides in poly-A/T stretches were detected in such lines (Epp 1973; Sears et al. 1996; Stoike and Sears 1998). Mutation frequency, at least for deletions, is biased to five hotspots and directed to tandem repeats (Chiu et al. 1990; Chang et al. 1996). These regions may overlap with the hot-spots in sequence divergence identified in comparative analyses of Oenothera chloroplast genomes (Gordon et al. 1982; Chiu et al. 1990; Greiner et al. 2008b). Mechanistically, template slipping due to the absence of a DNA helicase or another DNA-binding protein was proposed (Stoike and Sears 1998). The locus corresponding to the mutator remains to be identified. Likewise, the possible influence of the *pm* locus on the mitochondrial genome needs to be investigated.

Characterization of plastome mutations induced by the *chloroplast mutator* (*cpm*) in

H. vulgare so far resulted in the exclusive detection of point mutations (single base pair transitions or insertions; Prina et al. 2009). The mutator does not seem to induce major structural changes in the ptDNA (Colombo et al. 2006), as judged from studies of various mutant lines derived from this material. These lines include mutations in *infA*, *ycf3* and *psbA*. The latter is atrazine-tolerant (Rios et al. 2003; Prina et al. 2009; Sect. VI). However, the specificity for ptDNA still needs to be verified.

Recently reported double knockouts of the *A. thaliana whirly1* (*AtWhy1*) and *whirly3* (*AtWhy3*) genes can induce different types of variegation in about 4.6% of the progeny. Due to illegitimate recombination between short direct repeats, ptDNA rearrangements resulting in head-to-tail concatemers and/or monomeric circles were observed in independent mutants. The rearranged regions are 10–25 times more abundant than the wild-type ptDNA. Illegitimate recombination was also shown in single knockout mutants of *AtWhy1*, *AtWhy3*, and in the corresponding ortholog in *Zea mays* (*ZmWhy1*; Maréchal et al. 2009). In general, the Whirly protein family is known as single-strand DNA binding proteins involved in DNA metabolism, including transcriptional regulation and telomere homeostasis (Desveaux et al. 2005; Cappadocia et al. 2010). Multiple functions also were proposed for the family members discussed here. ZmWHY1 binds to several plastid transcripts and to DNA throughout the plastid genome. It further mediates splicing of the *atpF* intron. Strong *ZmWhy1* mutant alleles are deficient in plastid ribosomes (Prikryl et al. 2008). For AtWHY1 und AtWHY3, circumstantial evidence suggests a role as transcription factors in the nucleus (Xiong et al. 2009). Finally, it was shown that they bind to single-stranded DNA and are involved in repairing double-strand DNA (dsDNA) breaks (Cappadocia et al. 2010).

Reverse genetic and proteomics approaches may elucidate further factors responsible for plastid/organelle DNA maintenance and stability. For example, *A. thaliana* lines homozygous for a T-DNA insertion in *cpRecA* (RecA1) displayed variegated seedlings with a frequency of 1.1% and 4.2% in the fourth and later generations, respectively. This *E. coli* RecA homolog is targeted to the chloroplast, but its functional homology still needs to be proven (Rowan et al. 2010; cf. Chap. 8). Virus mediated post-transcriptional genes silencing of gyrases A and B, both dually targeted to plastids and mitochondria, can induce leaf variegation in *Nicotiana benthamiana*. Both organelle morphology and functionality are altered in these plants. Interestingly, the affected organelles display a significantly higher DNA content. Disturbed plastid nucleoids as well as alterations in size and structure of ptDNA were observed (Cho et al. 2004). Whether these approaches can be utilized as general tools for plastome mutagenesis, remains to be proven and so does their specificity for the plastid genome (see below).

D. Nuclear Mutator Alleles Secondarily Affecting the Plastid

A second class of plastome mutator lines, which produces variegated plants but only a single type of mutated plastids, was reported for *A. thaliana* (*albomaculans*), *Capsella bursa-pastoris* (*albovariabilis*), *Capsium annum* (one line), *H. vulgare* (*albostrians*, Okina-mugi, Okina-mugi tricolor, Sasktatoon, *striata-4, white, white-streak-3*), *Orzya sativa* (two lines), and *Z. mays* (*iojap, chloroplast mutator*). The chlorophyll deficiencies are transmitted maternally and, for some lines, the presence of mixed cells could be confirmed. Appropriate summaries of genetic evidence are provided elsewhere (Grun 1976; Kirk and Tilney-Bassett 1978; Hagemann 1986; Rodermel 2002; Yu et al. 2007). However, molecular analyses question this class of plastome mutators. In none of these lines, a mutation in the plastome was verified by sequencing analysis. For the classic examples – the *iojap* allele in *Z. mays,* and the two *H. vulgare* mutants *albostrians* and Saskatoon – it turned out that ptDNA is not, or not obviously, affected (Börner and Sears 1986). Comparable to the *ZmWhy1* knock-out lines (see above), loss of plastid ribosomes is induced in these lines (Börner and Sears 1986; Hagemann 1986; Rodermel 2002; Yu et al. 2007). In addition, CMS phenotypes and changes in mitochondrial DNA were observed in the *iojap* background, segregating independently from the "chloroplast mutation" in these lines (Hagemann 1986; Lemke et al. 1988). The *iojap* gene was cloned and it was shown that its product is associated with the plastid ribosomal 50S subunit, but the gene function of this locus still remains unclear (Han et al. 1992; Han and Martienssen 1995). It has been postulated that the striped *iojap* phenotype and maternal inheritance of its white plastids are caused by an irreversible loss of plastid ribosomes, and hence, that the Iojap-protein is involved in plastid ribosome assembly and/or stability. However, the protein has no similarly to any characterized RNA-binding protein family or other known proteins. It was further speculated that the *iojap* phenotype may also result from irreversibly impaired mitochondrial function (cf. Börner and Sears 1986; Lemke et al. 1988; Rodermel 2002; Yu et al. 2007). Some evidence of altered mitochondrial function was also given for the *striata-4* allele in *H. vulgare* (von Wettstein and Eriksson 1965).

Taken together, it is advisable to treat plastome mutators with some caution. Various phenotypes seem to reflect secondary effects of mitochondrial disturbance, and classic genetics does not allow to clearly distinguish between the two DNA-containing organelles (Sect. II). Also cases proven by molecular approaches in *O. elata* (*pm*), *H. vulgare* (*cpm*) and *A. thaliana* (*AtWhy1, AtWhy3*) need further investigation to evaluate their possible influence on the mitochondrial genome. The putative targets of plastome mutator alleles, the plastid DNA stability and replication machinery, are barely understood (for review see Day and Madesis 2007). Some components seem to be organelle specific, such as AtWHY1 and AtWHY3 which are localized in the plastid. Their homolog AtWHY2 is targeted to mitochondria (Krause et al. 2005). However, further factors, such as the DNA polymerases, one of the three RecA homologues indentified in *A. thaliana* (RecA2) or the gyrase subunits A and B in both *A. thaliana* and *N. benthamiana* are dually targeted (Day and Madesis 2007; Shedge et al. 2007). Hence, it seems reasonable to postulate the existence of machineries for organelle DNA replication and maintenance that are at least partially overlapping between plastids and mitochondria.

E. Induction of Plastome Mutations by Chemicals

Interestingly, an abundant classic genetic literature reports resistance of plastids to mutagenic agents, such as radiation or chemicals (cf. Kutzelnigg and Stubbe 1974; Grun 1976; Kirk and Tilney-Bassett 1978; Hagemann 1982). The first unequivocal work describing a successful induction of plastome mutations by chemical treatment was published by Beletskii et al. (1969). The chemical compound *N*-nitroso-*N*-methyl-urea (NMU) was subsequently successfully applied to various higher plants species and has become a standard chemical agent to induce plastome mutations (e.g., Hagemann 1982; Schmitz-Linneweber et al. 2005; Azhagiri and Maliga 2007). NMU is a DNA alkylation agent with major effects on the guanidine N7 and O6 residues, inducing point mutations and chromosomal damage (Hagemann 1982; Beranek 1990; Doak et al. 2007). Its efficiency could be increased by heat shock in *H. annuus* (Guskov et al. 2001). Applied in appropriate concentrations, it is relatively specific to ptDNA, probably due to the lack of methyltransferases in plastids (Sears 1998). Screening the first generation following the mutagenesis treatment (M1 generation) for variegation is an effective approach to identify plastome mutants (Hagemann 1982). Since the discovery of the mutagenic action of NMU, several other chemicals, including nucleic acid base analogues and antibiotics, were described to induce mutations in the plastome. Successful reports exist for

N-nitroso-*N*-ethyl-urea, methyl-nitro-nitroso-guanidine (MNNG), ethyl-methane-sulfonate (EMS), or 5-bromo-2′-deoxyuridine (Kirk and Tilney-Bassett 1978; Hagemann 1982), more recently also for 9-aminoacridine hydrochloride causing single base pair deletions or small inversions (GuhaMajumdar et al. 2004), and ciprofloxacin as a gyrase inhibitor inducing double strand breaks in organelle DNA (Wall et al. 2004).

F. Effects of Radiation on ptDNA

Studies on the influence of radiation on ptDNA are still inconsistent. Although it was reported that UV light induces pyrimidine dimers in ptDNA, stable mutations presumably cannot be isolated. UV treatment may reduce the effective copy number of plastid chromosomes, indicating an efficient degradation mutated ptDNA molecules (cf. Sears and Sokalski 1991; Draper and Hays 2000; and references therein). Furthermore, the presence of a cyclobutane pyrimidine dimer photolyase, targeted to all three genetic compartments was recently reported for Oryza (Takahashi et al. 2011). Remarkably, X-ray treatment also seems not to induce recoverable mutations in ptDNA, or at least does so only at extremely low frequencies. The classic genetic literature is rich in reports about unsuccessful induction of cytoplasmic mutations by X-ray irradiation. Probably successful cases, many of them supported by mixed cells, sorting-out and maternal inheritance, were reported from *A. thaliana* (0.07% increase after X-ray treatment against a similar background; Röbbelen 1962) and Epilobium (between 0.06% and 0.15% induced with X-ray, ^{35}S or ^{32}P, which, however, is not much more than two times higher than the spontaneous frequency observed for these experiments; Michaelis 1958a, b, 1967). Data on some putative X-ray induced plastome mutations described for Pteridophyta are vague (cf. Maly 1958b; Kirk and Tilney-Bassett 1978). An explanation for the "resistance" of ptDNA to X-ray irradiation may lie in the expected induction of dsDNA breaks in ptDNA. Plastid chromosomes are probably unable to repair such breaks by non-homologous end joining (Kohl and Bock 2009). Recent analyses of the ptDNA repair machinery indicate repair mechanisms by homologous recombination and/or microhomology-mediated break-induced replication (Cappadocia et al. 2010).

IV. Maintenance of Plastome Mutants

As mentioned above, plastome mutants are usually recognized in the form of green–white (or pale, yellow) variegated plants (Fig. 11.1). Variegation is a result of random sorting-out of mutated and wild-type plastids (Sect. II). White, yellow or pale green sectors of these plants harbor only mutant plastids. In a chimeric plant, however, the survival of the impaired tissue is facilitated by the adjacent green tissue, which supplies the mutant plastids and cells with components they are unable to synthesize. Such plant material can be maintained in several ways, depending on the species (and sometimes even the variety) and its mode of chloroplast inheritance (Kutzelnigg and Stubbe 1974; Kirk and Tilney-Bassett 1978; Stubbe and Herrmann 1982; and references cited in Table 11.1).

A. Recovery of Homoplasmic Plastome Mutants

Preferably, plants are maintained in a homoplasmic state in soil or, if plastome mutants cannot grow autotrophically, in sterile culture on sugar-containing media. Material homoplasmic for a mutation can be obtained by either regeneration of mutated leaf sectors or by selfing flowers containing solely the mutated plastome in the germ line. Such flowers can be recognized on white or pale stems with completed sorting-out for the mutation. In many dicotyledonous plants, however, stems whose leafs display periclinally chimeric phenotypes (Sect. II) are equally suited or even preferred. They typically show higher vigor in growth, seed and pollen development.

Fig. 11.2. Periclinal chimeric plant organs indicating the presence of a homoplasmic plastome mutation in the germline of evening primroses. (**a**) Inflorescence – lateral view. (**b**) Inflorescence – top view (note that the bracts represent periclinally chimeric leaves). (**c**) Striped buds from a periclinally chimeric inflorescence. (**d**) Successive leaves from a periclinally chimeric stem.

As mentioned above, periclinal chimeras occur after a completed sorting-out of plastids in the L1, L2, and L3 histogenetic layers of the SAM, resulting in different genetic identities of these layers. In the developed leaf, the L1 and L2 layers give rise to the tunica (epidermis and subepidermal tissue, respectively). The L3 layer forms the corpus (Sect. II). At leaf margins, the L2 layer is significantly enriched. Plants with white leaf margins are therefore homoplasmic for a plastome mutation in the L2 layer of the SAM. Since the germ cells originate from the L2 layer, flowers exclusively inheriting a mutated plastome can be recognized on shoots whose bracts show a pale leaf margin (Fig. 11.2). Selfing of such flowers leads to homoplasmic offspring (Kutzelnigg and Stubbe 1974; Kirk and Tilney-Bassett 1978; Stubbe and Herrmann 1982).

B. Propagation of Variegated Plants

If a homoplasmic plastome mutant is not viable on soil, or tissue culture approaches are not available, maintenance of plastome mutants is difficult, at least for plants with uniparental inheritance of chloroplasts. In such cases, plants must be kept heteroplasmic during propagation – the wild-type plastome is needed to nourish the mutant tissue (see above) – but heteroplasmy cannot be induced sexually. Propagation by seeds from variegated branches is possible but not reliable. Variegated offspring can only be obtained from flowers, in which sorting-out was not completed in the zygote. Sectorial chimaeras or mosaic patterns (Sect. II) in the inflorescence may indicate the potential that such zygotes can form, but neither the yields of heteroplasmic offspring are predictable nor the degree of variegation in the progeny (cf. Kirk and Tilney-Bassett 1978). In addition, backcrossing of such plastome mutants obtained from mutagenesis approaches in order to “clean” the nuclear genome from background mutations is advisable, but challenging for the same reasons. Hence, for juvenescence such mutants are often maintained by cutting variegated stems, a generally difficult approach and not feasible for many higher plant taxa.

A solution to this severe problem is the use of model organisms displaying biparental plastid inheritance, like Pelargonium or Oenothera. In these genera, plastome mutants can be propagated by seeds and kept as variegated plants in soil. Biparental transmission allows a directed generation of variegated plants by crossing, using individuals with mutated and wild-type plastomes as crossing mates. Although the frequencies of variegated offspring differ depending on species and strain, this strategy is reliable enough to maintain large collections of mutants (Kutzelnigg and Stubbe 1974; Stubbe and Herrmann 1982; Tilney-Basset 1994). For Oenothera, a particularly elegant genetics for isolation and seed propagation of plastome mutants is available (Sect. VII).

V. Identification of Plastome Mutants

Since higher plant plastomes are not amenable to linkage mapping approaches (Sect. II), direct methods for the identification of a plastome mutant are based on sequence analysis, preferably of the entire mutated plastid chromosome (Hirao et al. 2009). RFLP analysis of ptDNA, frequently performed in the early time of plastid molecular genetics, detects only larger rearrangements and rarely point mutations or small indels (e.g., Day and Ellis 1984; Lee et al. 1989; Chiu et al. 1990; To et al. 1992). The standard approach used so far relies on physiological and biochemical analyses of the mutant, which allow to make predictions about the mutated gene. This strategy was applied to most plastome mutations identified by molecular approaches (for references, see Tables 11.1, 11.2), and is still successfully used in screens for the molecular causes of plastid-encoded herbicide resistances in plant populations (Thiel et al. 2010; cf. Sect. VI). As long as Sanger sequencing was the limiting step in this analysis, single nucleotide polymorphism detection (e.g., using gel-shift assays) was performed (To et al. 1993; Schaffner et al. 1995). However, the approach of combining physiological characterization with local sequence analysis suffers from three serious limitations: (1) The mutated gene needs to be identified and characterized first (e.g., Fromm et al. 1987; Winter and Herrmann 1987), which ironically prevented the elucidation of the plastid gene content by mutant analysis. (2) It is very difficult to identify unknown gene functions and/or unknown target sites for herbicides or antibiotics. (3) The presence of a second site mutation cannot be ruled out. Obviously, NGS technologies will overcome these technical constraints. If applied to organisms with established, high-quality plastid isolation protocols and a reference plastome sequence available, identification of plastome mutants should become routine. Highly purified ptDNA may even not be required – taking advantage of the high abundance of ptDNA in total DNA preparations, complete plastomes were re-mapped to a reference using Illumina deep sequencing (Nock et al. 2011). In a similar approach, starting from enriched ptDNA obtained by a rapid chloroplast isolation protocol, the plastome of *Corynocarpus laevigatus* (karaka nut) was assembled de novo, using the related plastome of *Cucumis sativus* (cucumber) as guiding sequence (Atherton et al. 2010). However, the purity requirements of the ptDNA preparations subjected to NGS analysis and the general accuracy of the resulting genome assemblies remain to be carefully examined. It is widely known that, as the result of plastid-nuclear gene transfers, pieces or even whole plastome sequences are located in the nucleus of many plant species (e.g., Bock and Timmis 2008). Hence, if not removed, these promiscuous sequences may lower assembly quality and potentially interfere with mutation mapping in such approaches.

VI. Types of Plastome Mutants

Over several decades, a substantial amount of plastome mutants have been described from various higher plant species and were characterized cytologically (e.g., by electron microscopy), physiologically and biochemically.

Table 11.2. Mapped antibiotic-resistant plastome mutants in higher plants

Resistance	Organism/strain	Mutant line	Gene/locus	Mutation	Origin	Reference
Spectinomycin	*Nicotiana tabacum* var. Xanthi	X/*spe*^{R}40	16S rRNA	G to A (1012)	NMU	Fluhr et al. (1985); Fromm et al. (1987)
Spectinomycin	*N. tabacum* var. Petit Havana mut. SR1	SPC23[a]	16S rRNA	G to A (1333)	Spontaneous (spectinomycin selection)	Svab and Maliga (1991)
Spectinomycin	*N. tabacum* var. Petit Havana	SPC1	16S rRNA	A to C (1138)	Spontaneous (spectinomycin selection)	Svab and Maliga (1991)
Spectinomycin	*N. tabacum* var. Petit Havana mut. SR1	SPC2[a]	16S rRNA	C to T (1139)[b]	Spontaneous (spectinomycin selection)	Svab and Maliga (1991)
Spectinomycin	Nicotiana line 92[c]	92/*spe*^{R}4	16S rRNA	G to A (1140)	NMU	Fluhr et al. (1985); Fromm et al. (1987)
Spectinomycin	*Solanum nigrum* var. SN or SNR	StSp1[a]	16S rRNA	C to T (1140)[d]	NMU	Kavanagh et al. (1994)
Spectinomycin	Nicotiana line 92[c] mut. *str*^{R}7	92/*str*^{R}7/ *spe*^{R}1[a]	16S rRNA	G to A (1140)[e]	NMU	Fluhr et al. (1985); Fromm et al. (1987)
Spectinomycin	*Arabidopsis thaliana* var. RDL	RLD-Spc1	16S rRNA	G to A (1141)[f]	N/A	M. Skarjinskia, Z. Svab and P. Maliga in Azhagiri and Maliga (2007)
Streptomycin	Nicotiana line 92[c]	92/*str*^{R}6	16S rRNA	C to T (472)[b]	NMU	Fluhr et al. (1985); Fromm et al. (1989)
Streptomycin	*Nicotiana plumbaginifolia* var. Viviania	SR1007	16S rRNA	C to T (473)[g]	MNNG	To et al. (1989); Yeh et al. (1994)
Streptomycin	*N. plumbaginifolia* var. Viviania	SR1021	16S rRNA	G to A (833)[h]	MNNG	To et al. (1989); Yeh et al. (1994)
Streptomycin	*N. tabacum* var. Petit Havana	SR1	16S rRNA	C to A (860)	Spontaneous (streptomycin selection)	Maliga et al (1973); Maliga et al. (1975); Etzold et al. (1987)
Streptomycin	Nicotiana line 92[c]	92/*str*^{R}7	16S rRNA	C to T (860)[b, i]	NMU	Fluhr et al. (1985); Fromm et al. (1989)
Streptomycin	*N. plumbaginifolia* var. Viviania	SR1009, SR1012, SR1019, SR1020, SR1025, SR1036, SR1037, SR1043, SR1046	16S rRNA	C to T (860)[j, k]	MNNG	To et al.(1989); Yeh et al. (1994)

(continued)

Table 11.2. (continued)

Resistance	Organism/strain	Mutant line	Gene/locus	Mutation	Origin	Reference
Streptomycin	*N. plumbaginifolia* var. Viviania	SR1018	*rps12*	A to G (codon 88: K to R)	MNNG	Hsu et al. (1993)
Streptomycin	*N. tabacum* var. Xanthi	X/*spr*^{R}6	*rps12*	C to T (codon 90: P to S)	NMU	Fluhr et al. (1985); Galili et al. (1989)
Streptomycin	*S. nigrum* var. SN or SNR	StSp1[a]	*rps12*	A to C (codon 87: K to Q)	NMU	Kavanagh et al. (1994)
Lincomycin[l]	*N. plumbaginifolia*	LR400	23S rRNA	G to A (2032)[m]	NMU	Cseplö et al. (1988)
Lincomycin[l]	*N. plumbaginifolia*	LR415	23S rRNA	A to G (2058)[m]	NMU	Cseplö et al. (1988)
Lincomycin[l]	*N. plumbaginifolia*	LR421, LR446	23S rRNA	A to G (2059)[m]	NMU	Cseplö et al. (1988)
Lincomycin	*S. nigrum* var. SNR	L17A1	23S rRNA	T to C (2073)[n]	NMU	Kavanagh et al. (1994)

[a]The mutants SPC2, SPC23, 92/*str*^{R}7/*spe*^{R}1 and StSp1 are streptomycin/spectinomycin double resistance mutants. For SPC2 and SPC23, the spectinomycin resistance arose from a second-site mutation in the already streptomycin resistant mutant SR1 (Svab and Maliga 1991). The molecular cause of the SR1 mutant is listed above. In the case of 92/*str*^{R}7/*spe*^{R}1, the streptomycin-resistant mutant 92/*str*^{R}7 served as source material for subsequent second-site mutagenesis (Fluhr et al. 1985; Fromm et al. 1987). For the underlying mutation of *str*^{R}7, see this table. The mutant StSp1 was obtained from separate mutagenesis and selection steps, initially on streptomycin, subsequently on spectinomycin (Kavanagh et al. 1994)

[b]Mutated site originally published as RNA sequence

[c]*Nicotiana* line 92 is a cybrid harbouring *Nicotiana undulata* plastids and the nuclear genome *N. tabacum* var. Xanthi. For details, see Aviv et al. (1980), Fluhr et al. (1984, 1985), and Fromm et al. (1987)

[d]Identical position as 92/spe^{R}4 and 92/sprR^7/spe^{R}1 (*E. coli* coordinate 1193)

[e]Identical to 92/spe^{R}4 (*E. coli* coordinate 1193)

[f]Original published sequence (C to T) complemented

[g]Original position published as *E. coli* coordinate 526

[h]Original position published as *E. coli* coordinate 885

[i]Identical position as SR1 (*E. coli* coordinate 912)

[j]Identical to 92/*str*^{R}7 (*E. coli* coordinate 912)

[k]Original position published as *E. coli* coordinate 912

[l]Confirmed clindamycin cross-resistance

[m]*E. coli* coordinate

[n]Identical position as LR415 (*E. coli* coordinate 2058)

Frequently observed phenotypes are impaired plastid gene function and resistances to herbicides or antibiotics (for reviews and references, see Kirk and Tilney-Bassett 1978; Börner and Sears 1986; Somerville 1986; Hagemann 1992; Tables 11.1, 11.2; and citations therein). Some plastome mutants display mild chlorotic effects, are developmentally impaired or sensitive to environmental factors, show mottled phenotypes and/or bleach reversibly (e.g., Kirk and Tilney-Bassett 1978; Stubbe and Herrmann 1982; Colombo et al. 2008; Hirao et al. 2009). A drought and temperature tolerant plastome mutant was described from *H. annuus* (Usatov et al. 2004; Mashkina et al. 2010). Mutant lines with unexplained genetic behavior were published as well (Sect. VIII). Mutations lacking large parts of the plastid genome, spontaneously occurring in tissue cultures of cereals, were discussed in Sect. II. Unfortunately, due to technical difficulties (see above) molecular characterization of plastome mutants at the sequence level in general is not yet well developed. Excluding mutants resistant so herbicides, Tables 11.1 and 11.2 give an overview of higher plant plastome mutants of which the molecular basis is known. In the following paragraph characterized mutants are reviewed.

A. Mutants with Impaired Plastid Gene Function

Disregarding mutants resistant to herbicides or antibiotics (see below) so far only 12 mutations in chloroplast genes have been identified, four of them within the *rbcL* gene (Table 11.1). The spontaneous *A. majus* plastome mutant en:alba-1 was shown to be deficient in photosystem I (PS I) activity. Sequence analysis of PS I genes led to identification of a transversion in codon 136 of the *psaB* gene encoding the P700 apoprotein A2 of the PS I reaction centre changing a tyrosine (TAT) into a stop codon (TAG) to cause a premature stop of polypeptide synthesis. The truncation of the PsaB protein prevents the formation of a functional PS I complex (Schaffner et al. 1995).

The plastome mutant en:alba-4 is deficient in photosystem II (PS II) activity. This mutation was induced by treatment of *A. majus* seeds with NMU, which caused a transition (C to T) at position 1027 of the *psbD* gene changing the codon 343 for proline into serine thus effecting an amino acid exchange near the C-terminus of the D2 protein. Together with D1, the D2 protein forms the heterodimer of the PS II reaction centre, which binds the cofactors essential for charge separation. The C-terminus of D2 is highly conserved and obviously plays an important functional role as its mutational change in en:alba-4 severely impairs the function of PS II. The genes *psbD* and *psbC* in the *psbD*-*psb*C operon overlap within a short region. The above mutation hence simultaneously affects position -19 of *psbC* three base pairs upstream of its Shine-Dalgarno sequence. The mutation in this position, however, may have no effect on *psbC* translation since the interaction of the 16S rRNA with the Shine-Dalgarno sequence should not be disturbed (Schaffner 1995).

Immunological analysis of the *O. biennis* (Syn: *O. suaveolens*) mutant II-gamma pointed to a deficiency in cytochrome b_{559} subunits associated with PS II. Sequence analysis of the corresponding genes *psbF* and *psbE* uncovered a 5 bp duplication at position +42 in *psbE*, resulting in a frameshift and a premature stop codon 83 bases later. The important histidine residue at PsbE codon 23 co-ordinating the essential heme in cytochrome b_{559} is missing from the truncated protein version (Hupfer 2002), presumably explaining the severe photosynthetic deficiency in the mutant.

Several mutations in the *rbcL* gene encoding RuBisCO are known to impair function of the enzyme. The *N. tabacum* EMS mutant Sp25 displays a transition (G 964 A) resulting in an exchange of glycine to serine. The mutant synthesizes both RbcL and RbcS, but is defective in holozenzyme assembly (Shikanai et al. 1996). In the mutant XV1 of *N. tabacum,* NMU mutagenesis led to a C to T transition at position 335 in the *rbcL* reading frame (changing serine 112 to phenylalanine). Functional RuBisCO could not be identified, but a precursor, the so-called B-complex, accumulates in the mutant (Avni et al. 1989). Two spontaneous *rbcL* mutants were described in Oenothera: The mutant IV-beta was identified as a single point mutation (G 337 C) resulting in an exchange from valine to leucine at position 113. RuBisCO assembly is impaired to 90%, again the B-complex accumulates. However, the remaining functional holoenzyme displays wild-type-like kinetic properties (Dauborn and Brüggemann 1998). In the mutant I-sigma, a TTAAC deletion (postion 808–812) causes a frameshift at condons 270/271 and leads to a premature stop seven triplets later (Winter and Herrmann 1987). The mutant is deficient for the RuBisCO enzyme (Hildebrandt et al. 1984), and was the first plastome mutant, whose defect was characterized on the DNA level. It was also used to show the possibility of plastome mutant complementation by gene expression from the nuclear genome (allotropic transformation). A full-length *rbcL* cDNA equipped with the RbcS transit peptide and promoter sequences from *Pisum sativum* (pea) was transformed into the Oenothera nuclear genome, rescuing the I-sigma phenotype (Winter 1986; Nagley and Devenish 1989; Kanevski and Maliga 1994).

Particularly interesting point mutants were reported for the *infA* coding region in the CL2 and CL2-like lines of *H. vulgare* (Landau et al. 2007). The gene is a homologue of the bacterial translation initiator factor 1 (IF1; Sijben-Müller et al. 1986). However, *infA* gene function in higher plants could not be studied by transplastomic approaches so far. Models species amendable for plastid transformation like *N. tabacum* belong to the Rosid clade (e.g., Bock 2007a), of which most species have lost *infA* by gene transfer to the nucleus (Millen et al. 2001). Characterization of the *H. vulgare infA* mutants revealed a time-depended reversible bleaching of the upper part of the primary leaf blade, probably due to a translational defect already during embryogenesis. The bleached leaf areas are plastid ribosome deficient, show a delay of plastid development and probably an altered plastid retrograde signaling. The mutant phenotype is temperature dependent. Plants grown under high temperature during seed formation produce offspring with lower pigment content, whereas high temperatures during vegetative growth of CL2 mutants lead to enriched pigment levels in these plants (Prina et al. 2003; Colombo et al. 2008). Reversible bleaching was also reported for mutants with a frame-shift in the *matK* gene in *Cryptomeria japonica* (Cupressaceae; Hirao et al. 2009). *MatK* is thought to be an RNA maturation factor involved in the splicing of group II introns, although direct molecular evidence for this is still lacking (Schmitz-Linneweber and Barkan 2007), mainly because *matK* knockout-lines in Nicotiana appear to be lethal (R. Maier in Schmitz-Linneweber and Barkan 2007).

B. Plastome Mutants Impaired in Plastid Gene Regulation

Of particular interest are those plastome mutants that display sensitivity to abiotic environmental factors and mutants showing periodical bleaching or mottled phenotypes. These features are characteristic of several plastome mutants and some of them are even viable in soil (e.g., Stubbe and Herrmann 1982; Chia et al. 1986; Archer and Bonnett 1987, and examples discussed below). Mildly chlorotic and developmentally dependent phenotypes promise interesting insights into regulatory mechanisms of chloroplast gene expression and gene function. For example, characterization of the plastome mutant II-theta in *O. biennis* indicated a splicing deficiency of the *petB* transcript due to deletion of two conserved nucleotides in the *petB* intron (Hupfer 2002). Two point mutations in the *ycf3* intron 1 of the *H. vulgare* CL3 line exhibit a similar, but temperature-depended splicing defect (Landau et al. 2009). The variegated phenotype of the I-iota mutant in *O. elata* may be caused by a mutation causing translational fusion of the overlapping and co-transcribed genes for the β- and ε-subunits of the plastid ATP synthase, as judged from western analysis. However, in vitro translation of isolated mRNAs from the I-iota mutant in a heterologous system results in non fused wild-type proteins. To resolve this discrepancy, a disturbed translation signal or post-transcriptional event has been proposed. However, the exact molecular defect in this mutant remains to be identified (Sears and Herrmann 1985).

C. Plastome Mutants Exhibiting Resistance to Antibiotics

The generation of plastome mutants resistant to antibiotics was first reported in Chlamydomonas, and subsequently also in higher plants (Gillham 1978; Börner and Sears 1986; Hagemann 1992). Induction of plastome-borne spectinomycin, streptomycin or lincomycin (clindamycin) resistances was successful for various plant species, notably in *A. belladonna, A. thaliana*, Caspicum, Nicotiana, *Onobrychis viciifolia*, *Petunia hybrid*, and Solanum (Syn: Lycopersicon) (Jansen et al. 1990; Babiychuk et al. 1995; Venkataiah et al. 2005; Azhagiri and Maliga 2007; see also references in Table 11.2 and citations therein). A somewhat unclear case of chloramphenicol resistance was reported for Nicotiana (Fluhr et al. 1985). Typically, following mutagenesis using agents inducing point mutations, resistant plants are regenerated under antibiotic selection. Spectinomycin resistance also frequently occurs spontaneously on selective media containing this antibiotic. Most of these antibiotic resistance mutations have been localized to the plastid genes for the 16S rRNA (spectinomycin and streptomycin) or the 23S rRNA (lincomycin). In addition, three mutations causing streptomycin resistance are due to mutations in the plastid gene *rps12* for the ribosomal protein S12. Most of the mapped mutation sites are located in the corresponding regions of functional homology in *E. coli*. For a summary, see Table 11.2 and references cited therein.

Studies of genetic recombination and paternal leakage are examples, in which such mutants provided useful experimental tools (Sect. II). The selectable markers of such mutants were furthermore employed to detect the interspecific transfer of chloroplasts in cybrid cultures by somatic fusion of protoplasts or protoplasts and microplasts. In these approaches, cells harboring antibiotic or herbicide resistant-donor chloroplasts are typically lethally irradiated and fused to a sensitive receptor line and plant regeneration is then performed on selective media. This technique can be extended to other selection markers, like bleached plastome mutants in the recipient (then selecting for green lines containing the donor plastid) or species specific regeneration media for the recipient (e.g., Medgyesy et al. 1985; Kushnir et al. 1987; Thanh et al. 1988; Kushnir et al. 1991; Eigel and Koop 1992; Babiychuk et al. 1995; and citations therein). The interspecific cybrids obtained can be utilized as targets to transform plastid genomes of species that are originally not transformable, by introducing foreign plastids into a host that is easy to manipulate (Kuchuk et al. 2006). A further application of antibiotic-resistant plastome mutants is a transformation of the chloroplast genome, avoiding bacterial selection markers such as the *aadA* gene. However, transformation efficiency is significantly lower (Svab and Maliga 1993). Using transformation vectors with point mutations in an endogenous 16S rRNA gene mediating resistance to spectinomycin and/or streptomycin, successful manipulation of the *N. plumbaginifolia* and *S. nigrum* plastids, respectively, was reported (O'Neill et al. 1993; Nugent et al. 2005).

D. Herbicide Resistance Induced by Amino-Acid Substitutions in psbA

Of commercial importance are plastome mutants, or naturally evolved alleles, that confer resistance to herbicides. As a consequence of the global use of herbicides in agriculture, numerous spontaneous resistance mutants have arisen and are documented continuously (Heap 2011). In addition to resistant alleles encoded in the nucleus and conferring tolerance against several herbicide classes, some resistance traits, especially against herbicides of the triazine type, were proven to be plastome encoded. All of them are associated with the *psbA* gene, encoding the D1 protein, one of the two reaction center subunits of the PS II complex. Similar or identical mutations are known from photoautotrophic bacteria and algae. The literature on this topic was comprehensively reviewed (e.g., Oettmeier 1999; Powles and Yu 2010; and reviews cited therein).

Briefly, a frequently found mutation is the substitution of amino acid residue 264 of the D1 protein, which is serine in wild-type (herbicide-sensitive) plants. The serine codon was changed in different herbicide-resistant plastome mutants by single substitutions into codons for glycine, threonine, asparagine or alanine. The most prominent substitution conferring triazine resistance is the serine 264 glycine exchange. Exchange to one of the other amino acids may additionally mediate resistance to other herbicide classes, such as urea derivatives. Selection for non-triazine herbicides affecting D1 has uncovered further resistance-conferring amino acid substitutions, such as valine 219 isoleucine, known already from Chlamydomonas or photosynthetic bacteria. These mutations also confer triazine resistance. Within *psbA*, also more complex mutations leading to herbicide resistance have been described, including double and triple mutations as well as indels.

Analysis of herbicide-resistant mutants has greatly contributed to our understanding of the molecular structure and function of the D1 protein in PS II. It allowed the identification of a region of 57 amino acids within D1 involved in herbicide binding, which defines the binding niche for the second plastoquinone (PQ) acceptor Q_B of PS II. The herbicides compete with Q_B and their binding inhibits PS II electron flow (Rochaix and Erickson 1988; Giardi et al. 2006; Powles and Yu 2010). Many, if not all, herbicide resistance mutations reduce the PQ binding affinity and, consequently, are associated with fitness costs (Gressel 2009; Vila-Aiub et al. 2009). It is worth mentioning that, with the Ely accession of *A. thaliana,* a natural *psbA* allele for atrazine resistance has been discovered (El-Lithy et al. 2005).

VII. Plastome Mutants of Oenothera

The genus Oenothera contributed significantly to the understanding of cytoplasmic genetics and, for many decades, played a dominant role in this research field (cf. Sect. II). For this genus, approximately 50, predominately spontaneously arisen, plastome mutants are available, which were systematically collected by Oenothera geneticists in Germany during the past century. This material shows various kinds of chlorophyll deficiency, and is barely characterized in terms of the underlying mutations. Only four mutants of this collection were subjected to molecular analysis (Sect. VI and Table 11.1). A physiological characterization involving most of the mutants revealed eleven mutants deficient for PS I and six for PS II. Four mutations affect the cytochrome b_6f complex, one the plastid ATP synthase and six influence RuBisCO activity. One mutant displays translational errors (Sears and Herrmann 1985). However, no specific defect could be assigned so far to 22 other mutants (for review and references, see Kutzelnigg and Stubbe 1974; Stubbe and Herrmann 1982). In particular the unassigned ones, together with classes of mutants showing mottled areas of green/yellowish or white cells and/or conditional pigment deficiency associated with external or developmental factors, promise interesting insights into unknown features of plastid gene regulation. For example, the mutant I-tau displays reversible bleaching (*virescent* phenotype). It can grow heterotrophically and forms bleached and green leaves in alternating order. Maintenance of the mutant in the rosette stage in soil is possible for years, as long as enough green leaves are formed to nourish the bleached ones. This finding indicates periodic alterations in plastid physiology, the genetic determinants of which reside in the plastome (W. Stubbe unpublished). For a systematic investigation of this valuable mutant material, reference plastomes for mutant identification by deep sequencing approaches (Greiner et al. 2008a, b) as well as high-quality plastid isolation protocols are now available (Herrmann 1982; Wolfson and Sears 1989; S. Greiner unpublished).

The availability of this rich collection of plastome mutants from Oenothera rests on a particularly elegant genetics of their maintenance, originally elaborated by Wilfried Stubbe. The genetics of permanent translocation heterozygosity and biparental plastid transmission in Oenothera facilitates a fast and easy substitution of cytoplasms between lines and species (Stubbe 1960, 1989; Kutzelnigg and Stubbe 1974; Stubbe and Herrmann 1982; Rauwolf et al. 2008). This allows propagation of mutant plastids of different Oenothera species and strains in defined genetic backgrounds. In Oenothera, this is of particular interest, since PGI within the genus is frequent and a genetic background compatible with most of the five basic plastome types (I–V) is highly desirable (Stubbe 1959, 1989; Greiner et al. 2011). This prerequisite is given with the johansen Standard strain (*O. elata* ssp. *hookeri*; Syn: *O. johansen*; Cleland 1935), which is most suitable as a maintainer. It is compatible with the basic plastome types I, II and IV and only weakly incompatible with plastome III. The strain prospers with relatively short generation times, flowers reliably, is easy to cross-pollinate, produces high seed yields, is resistant to most Oenothera pests and diseases, and amenable to tissue culture approaches and nuclear transformation. Once transferred in this genetic background, maintenance of plastome mutants by sexual propagation utilizing biparental transmission and variegated plants to nourish the mutated plastome by wild-type plastids is feasible (cf. Sect. IV). In the genus Oenothera, biparental transmission of plastids is of maternal dominance; that is, F1 generation offspring is either homoplasmic for the maternal plastome or heteroplasmic for the paternal and maternal plastomes. To increase the frequency of variegated (heteroplasmic) offspring, a johansen Standard line was equipped with the slowly multiplying plastome IV (Sect. I) as wild-type plastome and used as seed parent in crosses with mutants of all faster multiplying plastomes (I–III). Resulting F1 generations display up to 100% variegation for wild-type and mutated plastomes in the progeny. In contrast, if a mutant of the slow plastome IV is to be maintained, the mutant plastome line should be used as maternal crossing mate and crossed to a wild type carrying the basic plastome type I, a fast multiplying plastome (Kutzelnigg and Stubbe 1974; Stubbe and Herrmann 1982). A comparable genetics is available for plastome V, which is severely incompatible with the johansen Standard strain. Maintenance of mutants of this plastome is achieved in its native background. To increase variegation in F1, a maintainer line with the slowest basic plastome IV is available. For both maintainer lines, high yields of variegated plants (mutant and nursery plastome) can be obtained and kept under greenhouse conditions. Since rosette diameters of Oenothera plants are up to 60 cm, quite large amounts of leaf material homoplasmic for the mutations can be obtained after completed sorting-out (Fig. 11.1b), thus facilitating the detailed physiological and genetic characterization of the mutant phenotype.

VIII. Perspectives

Although research on plastome mutants suffered from limitations in identifying mutations on DNA level, the novel NGS technologies should overcome these technical limitations (Sect. V). A relatively fast and reliable identification of plastome mutations now offers the possibility of a systematic investigation of available plastome mutant collections and systematic screens for particular phenotypes employing mutagenesis approaches. The small genome size of plastomes should even allow saturating mutagenesis.

As obvious already from the few examples of plastome mutants that were characterized molecularly, plastome mutants can fill a gap left by chloroplast transformation approaches for several reasons. First, many of the mutants with chlorotic defects characterized so far (Table 11.1) were identified from species not (yet) amenable to plastid transformation. This offers the possibility of molecular studies on plastid genomes also in non-standard organisms. A case in point is provided by the isolation of *infA* mutants in *H. vulgare*. The *infA* gene is not present in plastomes of higher plant species currently amenable to plastid transformation (Sect. VI). Second, induction of point mutations or indels by plastid transformation is technically challenging. Beyond simple knock-out analysis, point mutations can be highly valuable in elucidating functions of chloroplast genes. They are especially valuable to determine the functions of essential chloroplast genes, such as *matK* (R. Maier in Schmitz-Linneweber and Barkan 2007), *clpP* (Shikanai et al. 2001), *accD* (Kode et al. 2005) or the open reading frames *ycf1* and *ycf2* with still unidentified functions (Drescher et al. 2000). Third, possible dual functions of plastid genes may be uncovered by isolating mutated alleles. Fourth, data on promoter motifs are limited and cis-acting target sequences of plastid RNA metabolism are still largely unknown (Bollenbach et al. 2007; Liere and Börner 2007). Mutations for these elements are not or not readily obtainable by reverse genetics. Therefore, a systematic investigation of (mild) chlorotic and/or developmentally impaired plastome mutants would be highly desirable, since these phenotypes should include mutations in these elements (Sect. VI). Fifth, most intergenic regions in plastomes are small, presumably tightly packed with regulatory elements and probably as important as the coding sequences (Herrmann et al. 1992). Systematic mutagenesis approaches on these regions may uncover novel mechanisms in plastid gene regulation. Sixth, various plastid regulatory determinants, such as the loci underlying the different plastome multiplication rates (Sect. II), have remained elusive and hence are not amenable to knock-out approaches. Finally, the study of plastid restitution events could help elucidating the genetic interactions between chloroplast and mitochondrial loci (Sect. II). Addressing these questions will require systematic plastome mutagenesis approaches comparable to TILLING for nuclear genes (cf. Prina et al. 2009). The technological basis for such systematic studies is now available and offers rich opportunities for future research.

It is important to note that not all observed phenotypes of plastid mutants are explainable with our current knowledge. For example, a plastome mutant identified from Epilobium apparently induces degradation of wild-type plastids present in the same cell (Michaelis 1957). Long-range interactions of plastids or intercellular movement of plastid signals were even observed between adaxial and abaxial cell layers, as revealed by studies with some plastome mutations in Oenothera (Stubbe 1958; Kutzelnigg and Stubbe 1974). In conclusion, several cases have been reported, in which plastome mutants undergo a so far not understood genetic behavior. According to Michaelis (1955), sorting-out of organelles is a rather quick and random process (Sect. II; but also see Birky 2001). However, based on circumstantial evidence, Michaelis also suggested the possibility of non-random (or one-sided) sorting-out for some plastome mutations in Epilobium (cf. Kirk and Tilney-Bassett 1978, pp. 366–386). The significance and possible general relevance of all reports on unusual sorting-out phenomena still needs to be determined and plastid mutants will undoubtedly play a prominent role therein.

Acknowledgments

Drs. Ralph Bock and Reinhold G. Herrmann are acknowledged for critical reading, fruitful discussions and helpful comments on the manuscript, Lutz Neumetzler for providing the *A. thaliana* plastome mutant, and Josef Bergstein for photographic service. This research was supported by the Max Planck Society.

References

Abdelnoor RV, Yule R, Elo A, Christensen AC, Meyer-Gauen G, Mackenzie SA (2003) Substoichiometric shifting in the plant mitochondrial genome is influenced by a gene homologous to *MutS*. Proc Natl Acad Sci 100:5968–5973

Abdel-Wahab OAL, Tilney-Bassett RAE (1981) The role of plastid competition in the control of plastid inheritance in the zonal Pelargonium. Plasmid 6:7–16

Archer EK, Bonnett HT (1987) Characterization of a *virescent* chloroplast mutant of tobacco. Plant Physiol 83:920–925

Arntzen CJ, Duesing JH (1983) Chloroplast-encoded herbicide resistance. In: Downey K, Voellmy RW, Ahmad F, Shultz J (eds) Advances in gene technology: molecular genetics of plants and animals, vol 20. Academic, New York, pp 273–294

Arrieta-Montiel MP, Shedge V, Davila J, Christensen AC, Mackenzie SA (2009) Diversity of the Arabidopsis mitochondrial genome occurs via nuclear-controlled recombination activity. Genetics 183:1261–1268

Atherton RA, McComish BJ, Shepherd LD, Berry LA, Albert NW, Lockhart PJ (2010) Whole genome sequencing of enriched chloroplast DNA using the Illumina GAII platform. Plant Methods 6:22

Aviv D, Fluhr R, Edelman M, Galun E (1980) Progeny analysis of the interspecific somatic hybrids: *Nicotiana tabacum* (CMS)+*Nicotiana sylvestris* with respect to nuclear and chloroplast markers. Theor Appl Genet 56:145–150

Avni A, Edelman M, Rachailovich I, Aviv D, Fluhr R (1989) A point mutation in the gene for the large subunit of ribulose 1,5-bisphosphate carboxylase/oxygenase affects holoenzyme assembly in *Nicotiana tabacum*. EMBO J 8:1915–1918

Azhagiri AK, Maliga P (2007) Exceptional paternal inheritance of plastids in Arabidopsis suggests that low-frequency leakage of plastids via pollen may be universal in plants. Plant J 52:817–823

Babiychuk E, Schantz R, Cherep N, Weil J-H, Gleba Y, Kushnir S (1995) Alterations in chlorophyll a/b binding proteins in Solanaceae cybrids. Mol Gen Genet 249:648–654

Barr CM, Neiman M, Taylor DR (2005) Inheritance and recombination of mitochondrial genomes in plants, fungi and animals. New Phytol 168:39–50

Baur E (1909) Das Wesen und die Erblichkeitsverhältnisse der "varietates albomarginatae hort." von Pelargonium zonale. Z Indukt Abstamm Vererbungs 1:330–351

Baur E (1910) Untersuchungen über die Vererbung von Chromatophorenmerkmalen bei Melandrium, Antirrhinum und Aquilegia. Z Indukt Abstamm Vererbungs 4:81–102

Beletskii YD, Razoriteleva EK, Zhdanov YA (1969) Cytoplasmic cunflower mutations induced by N-niuosomethylurea. Dokl Akad Nauk SSSR 186: 1425–1426

Beranek DT (1990) Distribution of methyl and ethyl adducts following alkylation with monofunctional alkylating agents. Mutat Res 231:11–30

Bidani A, Nouri-Ellouz O, Lakhoua L, Sihachakr D, Cheniclet C, Mahjoub A, Drira N, Gargouri-Bouzid R (2007) Interspecific potato somatic hybrids between *Solanum berthaultii* and *Solanum tuberosum* L. showed recombinant plastome and improved tolerance to salinity. Plant Cell Tissue Organ Cult 91:179–189

Birky CW (1995) Uniparental inheritance of mitochondrial and chloroplast genes: mechanisms and evolution. Proc Natl Acad Sci 92:11331–11338

Birky CW (2001) The inheritance of genes in mitochondria and chloroplasts: laws, mechanisms, and models. Annu Rev Genet 35:125–148

Bock R (2001) Transgenic plastids in basic research and plant biotechnology. J Mol Biol 312:425–438

Bock R (2007a) Plastid biotechnology: prospects for herbicide and insect resistance, metabolic engineering and molecular farming. Curr Opin Biotechnol 18:100–106

Bock R (2007b) Structure, function, and inheritance of plastid genomes. In: Bock R (ed) Cell and molecular biology of plastids, vol 19. Springer, Berlin/Heidelberg/New York, pp 29–63

Bock R, Timmis JN (2008) Reconstructing evolution: gene transfer from plastids to the nucleus. Bioessays 30:556–566

Bollenbach T, Schuster G, Portnoy V, Stern D (2007) Processing, degradation, and polyadenylation of chloroplast transcripts. In: Bock R (ed) Cell and molecular biology of plastids, vol 19. Springer, Berlin/Heidelberg/New York, pp 175–211

Börner T, Sears B (1986) Plastome mutants. Plant Mol Biol Rep 4:69–92

Boynton JE, Gillham NW, Harris EH, Hosler JP, Johnson AM, Jones AR, Randolph-Anderson BL, Robertson D, Klein TM, Shark KB, Sanford JC (1988) Chloroplast transformation in Chlamydomonas with high velocity microprojectiles. Science 240: 1534–1538

Boynton JE, Gillham NW, Newman SM, Harris EH (1992) Organelle genetics and transformation of Chlamydomonas. In: Herrmann RG (ed) Cell organelles. Springer, Vienna/New York, pp 3–64

Cappadocia L, Marechal A, Parent J-S, Lepage E, Sygusch J, Brisson N (2010) Crystal structures of

DNA-Whirly complexes and their role in Arabidopsis organelle genome repair. Plant Cell 22:1849–1867
Chang T-L, Stoike LL, Zarka D, Schewe G, Chiu W-L, Jarrell DC, Sears BB (1996) Characterization of primary lesions caused by the plastome mutator of Oenothera. Curr Genet 30:522–530
Chia CP, Duesing JH, Arntzen CJ (1986) Developmental loss of photosystem II activity and structure in a chloroplast-encoded tobacco mutant, *lutescens-1*. Plant Physiol 82:19–27
Chiu W-L, Sears BB (1985) Recombination between chloroplast DNAs does not occur in sexual crosses of Oenothera. Mol Gen Genet 198:525–528
Chiu W-L, Sears BB (1993) Plastome-genome interactions affect plastid transmission in Oenothera. Genetics 133:989–997
Chiu W-L, Stubbe W, Sears BB (1988) Plastid inheritance in Oenothera: organelle genome modifies the extent of biparental plastid transmission. Curr Genet 13:181–189
Chiu W-L, Johnson EM, Kaplan SA, Blasko K, Sokalski MB, Wolfson R, Sears BB (1990) Oenothera chloroplast DNA polymorphisms associated with plastome mutator activity. Mol Gen Genet 221:59–64
Cho HS, Lee SS, Kim KD, Hwang I, Lim J-S, Park Y-I, Pai H-S (2004) DNA gyrase is involved in chloroplast nucleoid partitioning. Plant Cell 16: 2665–2682
Cleland RE (1935) Cyto-taxonomic studies on certain Oenotheras from California. Proc Am Philos Soc 75:339–429
Colombo N, Rios RD, Prina AR (2006) Plastome analysis of barley chroloplast mutator-induced mutants. J Basic Appl Genet 17:5–9
Colombo N, Emanuel C, Lainez V, Maldonado S, Prina A, Börner T (2008) The barley plastome mutant CL2 affects expression of nuclear and chloroplast housekeeping genes in a cell-age dependent manner. Mol Genet Genomics 279:403–414
Correns C (1909) Vererbungsversuche mit blass (gelb) grünen und buntblättrigen Sippen bei *Mirabilis Jalapa*, *Urtica pilulifera* und *Lunaria annua*. Z Indukt Abstamm Vererbungs 1:291–329
Cseplö A, Etzold T, Schell J, Schreier PH (1988) Point mutations in the 23S rRNA genes of four lincomycin resistant *Nicotiana plumbaginifolia* mutants could provide new selectable markers for chloroplast transformation. Mol Gen Genet 214:295–299
Dauborn B, Brüggemann W (1998) A spontaneous point mutation in the RuBisCO large subunit gene impairing holoenzyme assembly renders the IVb plastome mutant of Oenothera extremely light- and chilling sensitive. Physiol Plantarum 104:116–124
Day A, Ellis THN (1984) Chloroplast DNA deletions associated with wheat plants regenerated from pollen: possible basis for maternal inheritasnce of chloroplasts. Cell 39:359–368
Day A, Madesis P (2007) DNA replication, recombination, and repair in plastids. In: Bock R (ed) Cell and molecular biology of plastids, vol 19. Springer, Berlin/Heidelberg/New York, pp 65–119
Desveaux D, Maréchal A, Brisson N (2005) Whirly transcription factors: defense gene regulation and beyond. Trends Plant Sci 10:95–102
Dietrich W, Wagner WL, Raven PH (1997) Systematics of Oenothera section Oenothera subsection Oenothera (Onagraceae), vol 50, Systematic botany monographs. The American Society of Plant Taxonomists, Laramie
Doak SH, Jenkins GJS, Johnson GE, Quick E, Parry EM, Parry JM (2007) Mechanistic influences for mutation induction curves after exposure to DNA-reactive carcinogens. Cancer Res 67:3904–3911
Draper CK, Hays JB (2000) Replication of chloroplast, mitochondrial and nuclear DNA during growth of unirradiated and UVB-irradiated Arabidopsis leaves. Plant J 23:255–265
Drescher A, Ruf S, Calsa T Jr, Carrer H, Bock R (2000) The two largest chloroplast genome-encoded open reading frames of higher plants are essential genes. Plant J 22:97–104
Eigel L, Koop H-U (1992) Transfer of defined numbers of chloroplasts into albino protoplasts by subprotoplast/protoplast microfusion: chloroplasts can be "cloned", by using suitable plastome combinations or selective pressure. Mol Gen Genet 233:479–482
El-Lithy ME, Rodrigues GC, van Rensen JJS, Snel JFH, Dassen HJHA, Koornneef M, Jansen MAK, Aarts MGM, Vreugdenhil D (2005) Altered photosynthetic performance of a natural Arabidopsis accession is associated with atrazine resistance. J Exp Bot 56:1625–1634
Epp MD (1973) Nuclear gene-induced plastome mutations in *Oenothera hookeri*: I. Genetic analysis. Genetics 75:465–483
Esau K (1972) Apparent temporary chloroplast fusion in leaf cells of *Mimosa pudiea*. Z Pflanzenphysiol 67:244–254
Etzold T, Fritz CC, Schell J, Schreier PH (1987) A point mutation in the chloroplast 16S rRNA gene of a streptomycin resistant *Nicotiana tabacum*. FEBS Lett 219:343–346
Fitter JT, Rose RJ (1993) Investigation of chloroplast DNA heteroplasmy in *Medicago sativa* L. using cultured tissue. Theor Appl Genet 86:65–70
Fluhr R, Aviv D, Galun E, Edelman M (1984) Generation of heteroplastidic Nicotiana cybrids by

protoplast fusion: analysis for plastid recombinant types. Theor Appl Genet 67:491–497
Fluhr R, Aviv D, Galun E, Edelman M (1985) Efficient induction and selection of chloroplast-encoded antibiotic-resistant mutants in Nicotiana. Proc Natl Acad Sci 82:1485–1489
Fromm H, Edelman M, Aviv D, Galun E (1987) The molecular basis for rRNA-dependent spectinomycin resistance in Nicotiana chloroplasts. EMBO J 6:3233–3237
Fromm H, Galun E, Edelman M (1989) A novel site for streptomycin resistance in the "530 loop" of chloroplast 16S ribosomal RNA. Plant Mol Biol 12: 499–505
Galili S, Fromm H, Aviv D, Edelman M, Galun E (1989) Ribosomal protein S12 as a site for streptomycin resistance in Nicotiana chloroplasts. Mol Gen Genet 218:289–292
Giardi MT, Piletska EV, Breton F, Piletska EV, Karim K, Rouillon R, Piletsky SA (2006) Mimicking the plastoquinone-binding pocket of photosystem II using molecularly imprinted polymers. In: Giardi MT, Piletska EV (eds) Biotechnological applications of photosynthetic proteins: biochips, biosensors and biodevices. Springer, New York, pp 155–165
Gillham NW (1978) Organelle heredity. Raven Press, New York
Gillham NW, Boynton JE, Harris EH (1991) Transmission of plastid genes. In: Bogorad L, Vasil IK (eds) Cell culture and somatic cell genetics of plants, vol 7A. Academic, San Diego/New York, pp 55–92
Gordon KHJ, Hildebrandt JW, Bohnert HJ, Herrmann RG, Schmitt JM (1980) Analysis of the plastid DNA in an Oenothera plastome mutant deficient in ribulose bisphosphate carboxylase. Theor Appl Genet 57:203–207
Gordon KHJ, Crouse EJ, Bohnert HJ, Herrmann RG (1982) Physical mapping of differences in chloroplast DNA of the five wild-type plastomes in Oenothera subsection Euoenothera. Theor Appl Genet 61:373–384
Greiner S, Wang X, Herrmann RG, Rauwolf U, Mayer K, Haberer G, Meurer J (2008a) The complete nucleotide sequences of the 5 genetically distinct plastid genomes of Oenothera, subsection Oenothera: II. A microevolutionary view using bioinformatics and formal genetic data. Mol Biol Evol 25:2019–2030
Greiner S, Wang X, Rauwolf U, Silber MV, Mayer K, Meurer J, Haberer G, Herrmann RG (2008b) The complete nucleotide sequences of the five genetically distinct plastid genomes of Oenothera, subsection Oenothera: I. Sequence evaluation and plastome evolution. Nucleic Acids Res 36:2366–2378
Greiner S, Rauwolf U, Meurer J, Herrmann RG (2011) The role of plastids in plant speciation. Mol Ecol 20:671–691
Gressel J (2009) Evolving understanding of the evolution of herbicide resistance. Pest Manag Sci 65: 1164–1173
Grun P (1976) Cytoplasmic genetics and evolution. Columbia University Press, New York
GuhaMajumdar M, Baldwin S, Sears BB (2004) Chloroplast mutations induced by 9-aminoacridine hydrochloride are independent of the plastome mutator in Oenothera. Theor Appl Genet 108:543–549
Guskov EP, Markin NV, Usatov AV, Mashkina EV (2001) Modification of the effect of nitrosomethylurea on sunflower seedlings by heat shock. Russ J Genet 37:257–263
Hagemann R (1964) Plasmatic Vererbung. Gustav Fischer, Jena
Hagemann R (1976) Plastid distribution and plastid competition in higher plants and the induction of plastome mutations by nitroso-urea-compounds. In: Bücher T, Neupert W, Sebald W, Werner S (eds) Genetics and biogenesis of chloroplasts and mitochondria. Elsevier, Amsterdam/New York/Oxford, pp 331–338
Hagemann R (1982) Induction of plastome mutations by nitro-urea-compounds. In: Edelman M, Hallick RB, Chua N-H (eds) Methods in chloroplast molecular biology. Elsevier, Amsterdam/New York/Oxford, pp 119–127
Hagemann R (1986) A special type of nuleus-plastid-interactions: nuclear gene induced plastome mutations. In: Akoyunoglou G, Senger H (eds) Regulation of chloroplast differentiation. Alan R Liss, New York, pp 455–466
Hagemann R (1992) Plastid genetics in higher plants. In: Herrmann RG (ed) Cell organelles – plant gene research: basic knowledge & application. Springer, Berlin/Heidelberg/New York, pp 65–96
Hagemann R (2004) The sexual inheritance of plant organelles. In: Daniell H, Chase CD (eds) Molecular biology and biotechnology of plant organelles – chloroplasts and mitochondria. Springer, Berlin/Heidelberg/New York, pp 93–114
Hagemann R (2010) The foundation of extranuclear inheritance: plastid and mitochondrial genetics. Mol Genet Genomics 283:199–209
Hagemann R, Scholze M (1974) Struktur und Funktion der genetischen Information in den Plastiden: VII. Vererbung und Entmischung genetisch unterschiedlicher Plastidensorten bei Pelargonium zonale Ait. Biol Zentralblatt 93:625–648
Han CD, Martienssen RA (1995) The Iojap (IJ) protein is associated with 50S chloroplast ribosomal subunits. Maize Genet Cooper News Lett 69:32–34

Han CD, Coe EHJ, Martienssen RA (1992) Molecular cloning and characterization of *iojap* (*ij*), a pattern striping gene of maize. EMBO J 11:4037–4046

Harte C (1994) Oenothera – contributions of a plant to biology. Springer, Berlin/Heidelberg/New York

Heap I (2011) International survey of herbicide-resistant weeds. http://www.weedscience.org. Accessed 18 Feb 2011

Herrmann RG (1982) The preparation of circular DNA from plastids. In: Edelman M, Hallick RB, Chang N-H (eds) Methods in chloroplast molecular biology. Elsevier, Amsterdam/New York/Oxford, pp 259–280

Herrmann RG, Westhoff P, Link G (1992) Biogenesis of plastids in higher plants. In: Herrmann RG (ed) Plant gene research. Cell organelles, vol 6. Springer, Wien/New York, pp 275–349

Hildebrandt J, Bottomley W, Moser J, Herrmann RG (1984) A plastome mutant of *Oenothera hookeri* has a lesion in the gene for the large subunit of ribulose-1,5-bisphosphate carboxylase/oxygenase. Biochim Biophys Acta 783:67–73

Hirao T, Watanabe A, Kurita M, Kondo T, Takata K (2009) A frameshift mutation of the chloroplast matK coding region is associated with chlorophyll deficiency in the *Cryptomeria japonica virescent* mutant Wogon-Sugi. Curr Genet 55:311–321

Hoekstra RF (2000) Evolutionary origin and consequences of uniparental mitochondrial inheritance. Hum Reprod 15(Suppl 2):102–111

Hornung S, Fulgosi H, Dörfel P, Herrmann RG (1996) Sequence variation in the putative replication origins of the five genetically distinct basic Euoenothera plastid chromosomes (plastomes). Mol Gen Genet 251:609–612

Hsu C-M, Yang W-P, Chen C-C, Lai Y-K, Lin T-Y (1993) A point mutation in the chloroplast *rps12* gene from *Nicotiana plumbaginifolia* confers streptomycin resistance. Plant Mol Biol 23:179–183

Hupfer H (2002) Vergleichende Sequenzanalyse der fünf Grundplastome der Sektion Oenothera (Gattung Oenothera) – analyse des Cytochrom-Komplexes. PhD thesis, Ludwig-Maximilians-University, Munich, p 145

Jansen CE, Snel EAM, Akerboom MJE, Nijkamp HJJ, Hille J (1990) Induction of streptomycin resistance in the wild tomato *Lycopersicon peruvianum*. Mol Gen Genet 220:261–268

Johnson EM, Schnabelrauch LS, Sears BB (1991) A plastome mutation affects processing of both chloroplast and nuclear DNA-encoded plastid proteins. Mol Gen Genet 225:106–112

Kanevski I, Maliga P (1994) Relocation of the plastid *rbcL* gene to the nucleus yields functional ribulose-1,5-bisphosphate carboxylase in tobacco chloroplasts. Proc Natl Acad Sci 91:1969–1973

Kavanagh TA, O'Driscoll KM, McCabe PF, Dix PJ (1994) Mutations conferring lincomycin, spectinomycin, and streptomycin resistance in *Solanum nigrum* are located in three different chloroplast genes. Mol Gen Genet 242:675–680

Kirk JTO, Tilney-Bassett RAE (1978) The plastids their chemistry, structure, growth and inheritance. Elsevier, Amsterdam/New York/Oxford

Kode V, Mudd EA, Iamtham S, Day A (2005) The tobacco plastid *accD* gene is essential and is required for leaf development. Plant J 44:237–244

Kohl S, Bock R (2009) Transposition of a bacterial insertion sequence in chloroplasts. Plant J 58: 423–436

Krause K, Kilbienski I, Mulisch M, Rödiger A, Schäfer A, Krupinska K (2005) DNA-binding proteins of the Whirly family in *Arabidopsis thaliana* are targeted to the organelles. FEBS Lett 579:3707–3712

Kuchuk N, Sytnyk K, Vasylenko M, Shakhovsky A, Komarnytsky I, Kushnir S, Gleba Y (2006) Genetic transformation of plastids of different Solanaceae species using tobacco cells as organelle hosts. Theor Appl Genet 113:519–527

Kuroiwa T (2010) Review of cytological studies on cellular and molecular mechanisms of uniparental (maternal or paternal) inheritance of plastid and mitochondrial genomes induced by active digestion of organelle nuclei (nucleoids). J Plant Res 123: 207–230

Kushnir SG, Shlumukov LR, Pogrebnyak NJ, Berger S, Gleba Y (1987) Functional cybrid plants possessing a Nicotiana genome and an Atropa plastome. Mol Gen Genet 209:159–163

Kushnir S, Babiychuk E, Bannikova M, Momot V, Komarnitsky I, Cherep N, Gleba Y (1991) Nucleo-cytoplasmic incompatibility in cybrid plants possessing an Atropa genome and a Nicotiana plastome. Mol Gen Genet 225:225–230

Kutzelnigg H, Stubbe W (1974) Investigation on plastome mutants in Oenothera: 1. General considerations. Subcell Biochem 3:73–89

Landau A, Diaz Paleo A, Civitillo R, Jaureguialzo M, Prina AR (2007) Two *infA* gene mutations independently originated from a mutator genotype in barley. J Hered 98:272–276

Landau AM, Lokstein H, Scheller HV, Lainez V, Maldonado S, Prina AR (2009) A cytoplasmically inherited barley mutant is defective in photosystem I assembly due to a temperature-sensitive defect in *ycf3* splicing. Plant Physiol 151:1802–1811

Lee DJ, Blake TK, Smith SE, Bingham ET, Carroll TW (1989) Chloroplast genome mapping and plastid

ultrastructure analysis of chlorophyll deficient mutants of Alfalfa. Crop Sci 29:190–196
Lemke CA, Gracen VE, Everett HL (1988) A second source of cytoplasmic male sterility in maize induced by the nuclear gene *iojap*. J Hered 79:459–464
Levin DA (2003) The cytoplasmic factor in plant speciation. Syst Bot 28:5–11
Liere K, Börner T (2007) Transcription and transcriptional regulation in plastids. In: Bock R (ed) Cell and molecular biology of plastids, vol 19. Springer, Berlin/Heidelberg/New York, pp 121–174
Maliga P, Sz.-Breznovits Á, Márton L (1973) Streptomycin-resistant plants from callus culture of haploid tobacco. Nat New Biol 244:29–30
Maliga P, Sz.-Breznovits A, Marton L, Joo F (1975) Non-Mendelian streptomycin-resistant tobacco mutant with altered chloroplasts and mitochondria. Nature 255:401–402
Maly R (1958a) Die Mutabilität der Plastiden von *Antirrhinum majus* L. Sippe 50. Z Indukt Abstamm Vererbungs 89:692–696
Maly R (1958b) Die Genetik einiger strahleninduzierter, abweichender Plastidenformen bei Farnen. Z Vererbungs 89:469–470
Maréchal A, Parent J-S, Véronneau-Lafortune F, Joyeux A, Lang BF, Brisson N (2009) Whirly proteins maintain plastid genome stability in Arabidopsis. Proc Natl Acad Sci 106:14693–14698
Mashkina E, Usatov A, Skorina M (2010) Comparative analysis of thermotolerance of sunflower chlorophyll mutants. Russ J Genet 46:178–184
Medgyesy P, Fejes E, Maliga P (1985) Interspecific chloroplast recombination in a Nicotiana somatic hybrid (protoplast fusion/chloroplast DNA/physical mapping). Proc Natl Acad Sci 82:6960–6964
Meyer B, Stubbe W (1974) Das Zahlenverhätnis von mütterlichen und väterlichen Plastiden in der Zygote von *Oenothera erythrosepala* Borbas (syn. *Oe. lamarckiana*). Ber Deutsch Bot Ges 87:29–38
Michaelis P (1955) Über Gesetzmäßigkeiten der Plasmon-Umkombination und über eine Methode zur Trennung einer Plastiden-, Chondriosomen-, resp. Sphaerosomen-, (Mikrosomen)- und einer Zytoplasmavererbung. Cytologia 20:315–338
Michaelis P (1957) Genetische, entwicklungsgeschichtliche und cytologische Untersuchungen zur Plasmavererbung: II. Mitteilung: Über eine PlastidenmutationmitintrazellularerWechselwirkung der Plastiden, zugleich ein Beitrag zur Methodik der Plasmonanalyse und zur Entwicklungsgeschichte von Epilobium. Planta 50:60–106
Michaelis P (1958a) Untersuchungen zur Mutation plasmatischer Erbträger, besonders der Plastiden: I. Teil. Planta 51:600–634
Michaelis P (1958b) Untersuchungen zur Mutation plasmatischer Erbträger, besonders der Plastiden: II. Teil. Planta 51:722–756
Michaelis P (1964) Beiträge zum Problem der Plastiden-Abänderung: I. Über geninduzierte Störung der Teilungsrhythmen von Zell und Plastide. Z Bot 52:382–426
Michaelis P (1967) The segregation of plastids as an example of plastome analysis. Nucleus 10:111–127
Michaelis P (1969) Über Plastiden-Restitution (Rückmutation). Cytologia 34(Suppl):1–115
Millen RS, Olmstead RG, Adams KL, Palmer JD, Lao NT, Heggie L, Kavanagh TA, Hibberd JM, Gray JC, Morden CW, Calie PJ, Jermiin LS, Wolfe KH (2001) Many parallel losses of infA from chloroplast DNA during angiosperm evolution with multiple independent transfers to the nucleus. Plant Cell 13:645–658
Mogensen HL (1996) The hows and whys of cytoplasmic inheritance in seed plants. Am J Bot 83: 383–404
Nagata N (2010) Mechanisms for independent cytoplasmic inheritance of mitochondria and plastids in angiosperms. J Plant Res 123:193–199
Nagley P, Devenish RJ (1989) Leading organellar proteins along new pathways: the relocation of mitochondrial and chloroplast genes to the nucleus. Trends Biochem Sci 14:31–35
Nock CJ, Waters DL, Edwards MA, Bowen SG, Rice N, Cordeiro GM, Henry RJ (2011) Chloroplast genome sequences from total DNA for plant identification. Plant Biotechnol J 9:328–333
Nugent GD, ten Have M, van der Gulik A, Dix PJ, Uijtewaal BA, Mordhorst AP (2005) Plastid transformants of tomato selected using mutations affecting ribosome structure. Plant Cell Rep 24:341–349
O'Neill C, Horváth GV, Horváth É, Dix PJ, Medgyesy P (1993) Chloroplast transformation in plants: polyethylene glycol (PEG) treatment of protoplasts is an alternative to biolistic delivery systems. Plant J 3:729–738
Oettmeier W (1999) Herbicide resistance and supersensitivity in photosystem II. Cell Mol Life Sci 55:1255–1277
Palmer JD (1983) Chloroplast DNA exists in two orientations. Nature 301:92–93
Petit RJ, Vendramin GG (2007) Plant phylogeography based on organelle genes: an introduction. In: Weiss S, Ferrand N (eds) Phylogeography of Southern European Refugia. Springer, Dordrecht, pp 23–97
Powles SB, Yu Q (2010) Evolution in action: plants resistant to herbicides. Annu Rev Plant Biol 61: 317–347
Prikryl J, Watkins KP, Friso G, van Wijk KJ, Barkan A (2008) A member of the Whirly family is a

multifunctional RNA- and DNA-binding protein that is essential for chloroplast biogenesis. Nucleic Acids Res 36:5152–5165
Prina AR (1992) A mutator nuclear gene inducing a wide spectrum of cytoplasmically inherited chlorophyll deficiences in barley. Theor Appl Genet 85:245–251
Prina AR (1996) Mutator-induced cytoplasmic mutants in barley: genetic evidence of activation of a putative chloroplast transposon. J Hered 87:385–389
Prina AR, Arias MC, Lainez V, Landau A, Maldonado S (2003) A cytoplasmically inherited mutant controlling early chloroplast development in barley seedlings. Theor Appl Genet 107:1410–1418
Prina AR, Landau AM, Colombo N, Jaureguialzo M, Arias MC, Rios RD, Pacheco MG (2009) Genetically unstable mutants as novel sources of genetic variability: the chloroplast mutator genotype in barley as a tool for exploring the plastid genome. In: Shu QY (ed) Induced plant mutations in the genomics era. Food and Agriculture Organization of the United Nations, Rome, pp 227–228
Rauwolf U, Golczyk H, Meurer J, Herrmann RG, Greiner S (2008) Molecular marker systems for Oenothera genetics. Genetics 180:1289–1306
Rédei GP (1973) Extra-chromosomal mutability determined by a nuclear gene locus in Arabidopsis. Mutat Res 18:149–162
Renner O (1934) Die pflanzlichen Plastiden als selbständige Elemente der genetischen Konstitution. Ber math-phys Klasse Sächs Akad Wiss Leipzig 86:241–266
Rios RD, Saione H, Robredo C, Acevedo A, Colombo N, Prina AR (2003) Isolation and molecular characterization of atrazine tolerant barley mutants. Theor Appl Genet 106:696–702
Röbbelen G (1962) Plastommutationen nach Röntgenbestrahlung von *Arabidopsis thaliana* (L.) Heynh. Z Vererbungs 93:25–34
Rochaix J-D, Erickson J (1988) Function and assembly of photosystem II: genetic and molecular analysis. Trends Biochem Sci 13:56–59
Rodermel S (2002) Arabidopsis variegation mutants. TAB 1:1–28
Rowan BA, Oldenburg DJ, Bendich AJ (2010) RecA maintains the integrity of chloroplast DNA molecules in Arabidopsis. J Exp Bot 61:2575–2588
Ruf S, Karcher D, Bock R (2007) Determining the transgene containment level provided by chloroplast transformation. Proc Natl Acad Sci 104:6998–7002
Schaffner C (1995) Molekulargenetische und physiologische Untersuchung an photosynthesedefizienten Plastommutanten von *Antirrhinum majus* L. PhD thesis, Martin-Luther-Universität Halle-Wittenberg, Halle, p 108
Schaffner C, Laasch H, Hagemann R (1995) Detection of point mutations in chloroplast genes of *Antirrhinum majus* L.: I. Identification of a point mutation in the *psaB* gene of a photosystem I plastome mutant. Mol Gen Genet 249:533–544
Schmitz-Linneweber C, Barkan A (2007) RNA splicing and RNA editing in chloroplasts. In: Bock R (ed) Cell and molecular biology of plastids, vol 19. Springer, Berlin/Heidelberg/New York, pp 213–248
Schmitz-Linneweber C, Kushnir S, Babiychuk E, Poltnigg P, Herrmann RG, Maier RM (2005) Pigment deficiency in nightshade/tobacco cybrids is caused by the failure to edit the plastid ATPase alpha-subunit mRNA. Plant Cell 17:1815–1828
Schön A, Krupp G, Gough S, Berry-Lowe S, Kannangara CG, Söll D (1986) The RNA required in the first step of chlorophyll biosynthesis is a chloroplast glutamate tRNA. Nature 322:281–284
Schötz F (1954) Über Plastidenkonkurrenz bei Oenothera. Planta 43:182–240
Sears BB (1980) Elimination of plastids during spermatogenesis and fertilization in the plant kingdom. Plasmid 4:233–255
Sears BB (1998) Replication, recombination and repair in the chloroplast genetic system of Chlamydomonas. In: Rochaix JD, Goldschmidt-Clermont M, Merchant S (eds) The molecular biology of chloroplasts and mitochondria in Chlamydomonas. Kluwer, Dordrecht/Boston/London, pp 115–138
Sears BB, Herrmann RG (1985) Plastome mutation affecting the chloroplast ATP synthase involves a post-transcriptional defect. Curr Genet 9:521–528
Sears BB, Sokalski MB (1991) The Oenothera plastome mutator: effect of UV irradiation and nitrosomethyl urea on mutation frequencies. Mol Gen Genet 229:245–252
Sears BB, Stoike LL, Chiu WL (1996) Proliferation of direct repeats near the Oenothera chloroplast DNA origin of replication. Mol Biol Evol 13:850–863
Shedge V, Arrieta-Montiel M, Christensen AC, Mackenzie SA (2007) Plant mitochondrial recombination surveillance requires unusual RecA and MutS homologs. Plant Cell 19:1251–1264
Shikanai T, Foyer CH, Dulieu H, Parry MAJ, Yokota A (1996) A point mutation in the gene encoding the Rubisco large subunit interferes with holoenzyme assembly. Plant Mol Biol 31:399–403
Shikanai T, Shimizu K, Ueda K, Nishimura Y, Kuroiwa T, Hashimoto T (2001) The chloroplast *clpP* gene, encoding a proteolytic subunit of ATP-dependent protease, is indispensable for chloroplast development in tobacco. Plant Cell Physiol 42:264–273
Sijben-Müller G, Hallick RB, Alt J, Westhoff P, Herrmnann RG (1986) Spinach plastid genes coding for initiation factor IF-1, ribosomal protein S11

and RNA polymerase a-subunit. Nucleic Acids Res 14:1029–1044

Somerville CR (1986) Analysis of photosynthesis with mutants of higher plants and algae. Annu Rev Plant Physiol 37:467–507

Stoike LL, Sears BB (1998) Plastome mutator–induced alterations arise in Oenothera chloroplast DNA through template slippage. Genetics 149:347–353

Stubbe W (1958) Dreifarbenpanschierung bei Oenothera: II. Wechselwirkungen zwischen Geweben mit zwei erblich verschiedenen Plastidensorten. Z Vererbungs 89:189–203

Stubbe W (1959) Genetische Analyse des Zusammenwirkens von Genom und Plastom bei Oenothera. Z Vererbungs 90:288–298

Stubbe W (1960) Untersuchungen zur genetischen Analyse des Plastoms von Oenothera. Z Bot 48: 191–218

Stubbe W (1989) Oenothera – an ideal system for studying the interaction of genome and plastome. Plant Mol Biol Rep 7:245–257

Stubbe W, Herrmann RG (1982) Selection and maintenance of plastome mutants and interspecific genome/plastome hybrids from Oenothera. In: Edelman M, Hallick RB, Chua N-H (eds) Methods in chloroplast molecular biology. Elsevier, Amsterdam/New York/Oxford, pp 149–165

Svab Z, Maliga P (1991) Mutation proximal to the tRNA binding region of the Nicotiana plastid 16S rRNA confers resistance to spectinomycin. Mol Gen Genet 228:316–319

Svab Z, Maliga P (1993) High-frequency plastid transformation in tobacco by selection for a chimeric *aadA* gene. Proc Natl Acad Sci 90:913–917

Svab Z, Maliga P (2007) Exceptional transmission of plastids and mitochondria from the transplastomic pollen parent and its impact on transgene containment. Proc Natl Acad Sci 104:7003–7008

Takahashi M, Teranishi M, Ishida H, Kawasaki J, Takeuchi A, Yamaya T, Watanabe M, Makino A, Hidema J (2011) CPD photolyase repairs ultraviolet-B-induced CPDs in rice chloroplast and mitochondrial DNA. Plant J 66:433–442

Thanh N, Medgyesy P (1989) Limited chloroplast gene transfer via recombination overcomes plastomegenome incompatibility between *Nicotiana tabacum* and *Solanum tuberosum*. Plant Mol Biol 12:87–93

Thanh ND, Páy A, Smith MA, Medgyesy P, Márton L (1988) Intertrubal chloroplast transfer by protoplast fusion between *Nicotiana tabacum* and *Salpiglossis sinuata*. Mol Gen Genet 213:186–190

Thiel H, Kluth C, Varrelmann M (2010) A new molecular method for the rapid detection of a metamitron-resistant target site in *Chenopodium album*. Pest Manag Sci 66:1011–1017

Tilney-Basset RAE (1994) Nuclear controls of chloroplast inheritance in higher plants. J Hered 85:347–354

Tilney-Bassett RAE (1975) Genetics of variegated plants. In: Birky CW, Perlman PS, Byers TJ (eds) Genetics and biogenesis of mitochondria and chloroplasts. Ohio State University Press, Columbus, pp 268–308

To KY, Chen CC, Lai YK (1989) Isolation and characterization of streptomycin-resistant mutants in *Nicotiana plumbaginifolia*. Theor Appl Genet 78:81–86

To KY, Lai YK, Feng TY, Chen CC (1992) Restriction endonuclease analysis of chloroplast DNA from streptomycin-resistant mutants of *Nicotiana plumbaginifolia*. Genome 35:220–224

To K-Y, Liu C-I, Liu S-T, Chang Y-S (1993) Detection of point mutations in the chloroplast genome by single-stranded conformation polymorphism analysis. Plant J 3:183–186

Trabelsi S, Gargouri-Bouzid R, Vedel F, Nato A, Lakhoua L, Drira N (2005) Somatic hybrids between potato *Solanum tuberosum* and wild species *Solanum verneï* exhibit a recombination in the plastome. Plant Cell Tissue Organ Cult 83:1–11

Usatov AV, Razoriteleva EK, Mashkina EV, Ulitcheva II (2004) Spontaneous and induced nitrosomethylurea reversion of plastomic chlorophyll mutants of the sunflower *Helianthus annuus* L. Russ J Genet 40:248–255

Vaughn KC (1981) Plastid fusion as an agent to arrest sorting out. Curr Genet 3:243–245

Venkataiah P, Christopher T, Subhash K (2005) Induction and characterization of streptomycin-resistant mutants in *Capsicum praetermissum*. J Appl Genet 46:19–24

Vila-Aiub MM, Neve P, Powles SB (2009) Fitness costs associated with evolved herbicide resistance alleles in plants. New Phytol 184:751–767

von Wettstein D, Eriksson G (1965) The genetics of chloroplasts. In: Geerts SJ (ed) Genetics today: proceedings of the XI international congress of geneticist, The Hague, Sep 1963. Vol 3. Pergamon Press, London/New York, pp 591–611

Wall MK, Mitchenall LA, Maxwell A (2004) *Arabidopsis thaliana* DNA gyrase is targeted to chloroplasts and mitochondria. Proc Natl Acad Sci 101:7821–7826

Winter P (1986) Therapie der *Oenothera hookeri* Plastommutante sigma durch kompartimentfremden Gentransfer. PhD thesis, Heinrich-Heine-Universtiät, Düsseldorf, p 156

Winter P, Herrmann RG (1987) A five-base-pair-deletion in the gene for the large subunit causes the lesion in the ribulose bisphosphate carboxylase/

oxygenase-deficient plastome mutant sigma of *Oenothera hookeri*. Bot Acta 101:42–48

Wolfe AD, Randle CP (2004) Recombination, heteroplasmy, haplotype polymorphism, and paralogy in plastid genes: implications for plant molecular systematics. Syst Bot 29:1011–1020

Wolfe KH, Li WH, Sharp PM (1987) Rates of nucleotide substitution vary greatly among plant mitochondrial, chloroplast, and nuclear DNAs. Proc Natl Acad Sci 84:9054–9058

Wolfson R, Sears BB (1989) A reduction in the abundance of calium oxalate crystals in Oenothera improves the yield of intact chloroplasts. Plant Cell Environ 12:109–112

Xiong J-Y, Lai C-X, Qu Z, Yang X-Y, Qin X-H, Liu G-Q (2009) Recruitment of AtWHY1 and AtWHY3 by a distal element upstream of the kinesin gene *AtKP1* to mediate transcriptional repression. Plant Mol Biol 71:437–449

Yeh KC, To KY, Sun SW, Wu MC, Lin TY, Chen CC (1994) Point mutations in the chloroplast 16s rRNA gene confer streptomycin resistance in *Nicotiana plumbaginifolia*. Curr Genet 26: 132–135

Yu F, Fu A, Aluru M, Park S, Xu Y, Liu H, Liu X, Foudree A, Nambogga M, Rodermel S (2007) Variegation mutants and mechanisms of chloroplast biogenesis. Plant Cell Environ 30:350–365

Chapter 12

Plant Mitochondrial Mutations

Susan Gabay-Laughnan
Department of Plant Biology, University of Illinois, 505 S. Goodwin Ave, Urbana, IL 61801, USA

Kathleen J. Newton*
Division of Biological Sciences, University of Missouri, Tucker Hall, Columbia, MO 65211, USA

Summary

The complex mitochondrial genomes of angiosperms tend to rearrange, leading to rapid structural evolution and to visible mutations. The observed mutations include those affecting growth and morphology, as well as male fertility. The abnormal growth mutations are usually associated with defects in essential mitochondrial genes. In contrast, cytoplasmic male sterility (CMS) usually results from the de novo expression of chimeric open reading

* Author for correspondence, e-mail: newtonk@missouri.edu

R. Bock and V. Knoop (eds.), Genomics of Chloroplasts and Mitochondria, Advances in Photosynthesis and Respiration 35, pp. 267–291,
DOI 10.1007/978-94-007-2920-9_12, © Springer Science+Business Media B.V. 2012

frames (ORFs) in rearranged mitochondrial genomes. The expression of the CMS-chimeric ORFs can be modified by nuclear restorer-of-fertility (*Rf*) genes. Most of the *Rf* genes described to date are rapidly evolving members of a class of genes encoding pentatricopeptide repeat (PPR) proteins. Plants may also revert to fertility following mitochondrial DNA (mtDNA) rearrangements that disrupt the sterility-associated region. Alternatively, subgenomes containing a CMS-ORF may be lost or highly suppressed. In many cases, the mtDNA rearrangements that lead to phenotypic changes are mediated by events involving short or microhomologous repeats. In this chapter, we emphasize work on cytoplasmic male sterility, including cytoplasmic reversion to fertility and nuclear restoration of fertility.

I. Introduction

Plant mitochondrial genomes tend to be organizationally complex and diverse, as well as much larger than their animal counterparts (reviewed by Kubo and Newton 2008; Kitazaki and Kubo 2010). Because seed plant mitochondrial genomes are reviewed in depth in another chapter of this volume (Chap. 8), we will only highlight the features here that are most relevant to the topic of mitochondrial mutation.

Most of the DNA in angiosperm mitochondrial genomes is non-coding. These large genomes contain only between 50 and 60 genes, encoding a few components of the electron transfer chain, a few ribosomal proteins, the ribosomal RNAs, and many of the transfer RNAs (Kubo and Newton 2008). There is some variation as to which genes are present in the mitochondria versus the nucleus in different genera, especially those coding for tRNAs and ribosomal proteins. With some notable exceptions, e.g. the plant family Geraniaceae (Parkinson et al. 2005), the protein coding sequences themselves tend to be highly conserved, but the DNA that lies between the genes appears to be different in different genera. Even within a single genus, where large intergenic regions can have high sequence conservation, losses and gains of sequence (presence/absence variation) are commonly seen (Allen et al. 2007). There is also variation among those sequences which can act as promoters (Hazle and Bonen 2007). Indeed, among mitochondrial genotypes of a single species, there can be rearrangements that lead to promoter "swaps" between genes without any apparent deleterious effects.

Although most plant mitochondrial genomes can be mapped as single "master circles", they appear to exist as a set of subgenomes, maintained in a dynamic equilibrium (reviewed in Kubo and Newton 2008). The organizational complexity of plant mitochondrial genomes reflects a propensity to rearrange, resulting from a high level of recombination across repeated sequences (e.g. Palmer and Shields 1984; Allen et al. 2007; reviewed by Hanson and Folkerts 1992; Fauron et al. 1995). Frequent, reversible recombination between pairs of relatively large (>1 kb) repeats can result in alternative molecular forms of these dynamic genomes. Inversions result if repeats are in inverted orientation with respect to one another, and subgenomic circles result if the repeats are in direct orientation. Despite the high levels of recombination and the demonstrable presence of subgenomes, the size and overall

Abbreviations: *atp* – Gene encoding a subunit of ATPase; bp – Base pairs; CMS – Cytoplasmic male sterility; CMS-ORFs – ORFs usually chimeric, associated with CMS; *cox*Gene encoding a subunit of cytochrome oxidase; HR – Homologous recombination; kb – Kilobase; MDL – Maternal distorted leaf; Mmt – Modifier of mitochondrial transcripts; MSC – Paternally transmitted mosaic; mtDNA – Mitochondrial DNA; MSH1 – MutS homolog; *nad* – Gene encoding a subunit of the Complex I NADH dehydrogenase; NCS – Nonchromosomal stripe; ORF – Open reading frame; OSB1 – Organellar single-stranded DNA binding protein; PPR – Pentatricopeptide repeat; *Rf* – Restorer of fertility; RNAi – RNA interference; SSS – Substoichiometric shifting; TCM – Teosinte-cytoplasm miniature; TIRs – Terminal inverted repeats

organization of a specific mitochondrial genome tends to be inherited reproducibly over many generations in "normal" nuclear backgrounds (e.g. Oro et al. 1985). Nuclear genes are responsible for the relative stability of plant mitochondrial genomes (reviewed by Maréchal and Brisson 2010).

Within sequenced mitochondrial genotypes (mitotypes) of a single species, the copy number of individual, conserved genes has been shown to vary from 1 to 4 copies without any apparent deleterious effects of the difference in dosage (see Allen et al. 2007). This suggests that post-transcriptional processes, including assembly of multi-subunit complexes, may be critical for mitochondrial function. Furthermore, a rearrangement between "extra" copies of actual genes and other sequences can lead to the formation of chimeric open reading frames. Chimeric ORFs that include pieces of normal genes are quite common in plant mitochondrial genomes (Marienfeld et al. 1997; Clifton et al. 2004; Allen et al. 2007). The chimeric ORFs associated with CMS are expressed from nearby promoters and have 3′ segments that can stabilize transcripts. Several of the CMS-ORFs have been shown to pre-exist in natural populations.

II. Mitochondrial Rearrangements and Mutations

Although a few plant mitochondrial mutations have been reported to result from base substitutions (e.g. Ducos et al. 2001), most of the mutations studied in angiosperms have been shown to result from rearrangements. Plant mitochondrial mutations tend to be of two types: (1) gain-of-function, such as the acquisition or expression of a CMS-causing chimeric ORF; and (2) loss-of-function, which alters an essential gene. Both types of mutations may be generated via common pathways.

Low-frequency recombination events involving very short (<50 bp) microrepeats may be the first step in generating most plant mitochondrial mutations. Such repeats also appear to be important in generating the types of reorganization inferred to have occurred during the rapid evolution of plant mitochondrial genomes. The microhomology-mediated events are rare and, thus, not usually reversible (reviewed by Fauron et al. 1995). A new subgenome resulting from the rare event can recombine with a subgenome resulting from high-frequency recombination across a pair of large repeats to form a new "master genome". The new genome may carry a deletion of the region between one copy of the microrepeat and one of the large repeats (Small et al. 1989; Fauron et al. 1995). Indeed, data from the abnormal growth mutants are consistent with this model. For example, the NCS3 mutant genome has a short deletion between an actively recombining 11-kb repeat and a 12-bp microrepeat located within the intron of the *rps3* ribosomal protein gene (Hunt and Newton 1991).

Since there are many mitochondrial genomes per cell, a new mutation is not phenotypically detectable until it has accumulated and segregated away ("sorted out") from non-mutant mtDNA. The effects of the mutations are often first detectable as sectors on the plants. Thus, all mitochondrial mutations are expected to exist first at very low ("substoichiometric") levels. Their phenotypic consequences would be seen only after many rounds of replication and sorting out.

When essential mitochondrial genes are mutated, the mutations often remain heteroplasmic; i.e. plants carry both normal and mutant mitochondrial arrangements. Mitochondria carrying only the mutant mtDNA segregate somatically during development to cause sectors of abnormal growth on the plant (Newton and Coe 1986; Newton et al. 1990; Gu et al. 1993; reviewed by Kubo and Newton 2008). This class of mitochondrial mutations has similar effects to those described in yeast and animal systems, and includes the **n**on**c**hromosomal **s**tripe (NCS) mutations of maize (reviewed by Newton et al. 2004, 2009), the MSC (paternally transmitted **mosaic**) mutants of cucumber (Lilly et al. 2001; Bartoszewski et al. 2004) and certain *chm*-derived ("chloroplast mutator") mutants of Arabidopsis (including **M**aternal **D**istorted **L**eaf, MDL; Sakamoto et al. 1996). Interestingly, in *Nicotiana sylvestris*, tissue-culture-derived deletions for the Complex I gene *nad7* survive as homoplasmic plants, but they grow poorly and are male sterile (Pla et al. 1995; Pineau et al. 2005). In maize, rare plants have been recovered that are homoplasmic for a partial deletion of the Complex I gene *nad4* (Marienfeld and Newton 1994); they are small, uniformly pale, male sterile, and set no seeds (Yamato and Newton 1999). In most cases, the maize NCS mutations cause kernel abortion when the mitochondria of the embryos are homoplasmic or near-homoplasmic (Gu et al. 1994; Baker and Newton 1995); thus, only kernels that contain some normal mitochondria together with the mutant mitochondria, will grow into plants.

Another type of well-studied de novo mutation is the spontaneous reversion to fertility of CMS plants, a phenomenon that has been especially well documented in maize and which depends upon nuclear background (reviewed in Sect. IV). The mutations causing CMS reversions appear to sort out rapidly and homoplasmic plants usually result. In nearly all cases, the maize CMS revertants result from independent rearrangements and do not appear to pre-exist within the maize CMS mitotypes.

Persistent, very low-level alternative arrangements of mitochondrial genomes or "sublimons" can be detected in most mitotypes, including those of maize (Small et al. 1987, 1989). The low copy-number molecules can become suddenly predominant, a process referred to as substoichiometric shifting (SSS). This has been shown to occur spontaneously and reversibly (Janska et al. 1998). Nonetheless, mutant alleles of nuclear genes that affect recombination and differential amplification of mitochondrial genomes can dramatically affect this process (reviewed in Sect. V).

III. Cytoplasmic Male Sterility

The most commonly studied type of mitochondrial rearrangement mutation in higher plants is CMS. CMS is a maternally inherited trait in which plants fail to produce functional pollen (reviewed by Hanson 1991; Chase 2007; Fujii and Toriyama 2008). CMS has been reported in a large number of plant species (Laser and Lersten 1972; Kaul 1988; reviewed by Hanson 1991; Schnable and Wise 1998; Chase 2007). It has long been exploited by plant breeders to produce hybrids that, in most crops, tend to be more vigorous and higher yielding than inbred lines (Havey 2004). In addition to maize (reviewed by Skibbe and Schnable 2005), CMS has also been observed and analyzed in many other crop plants, including *Brassica napus* (Singh and Brown 1993; L'Homme et al. 1997), chili pepper (Kim et al. 2007), common bean (Mackenzie 1991; Chase 1994), onion (Havey and Bark 1994), chives (Engelke and Tatlioglu 2002), sugar beet (Satoh et al. 2004), carrot (Linke et al. 2003), pearl millet (Burton 1977), radish (Iwabuchi et al. 1999), rice (Wang et al. 2006), rye (Tudzynski et al. 1986), sorghum (Bailey-Serres et al. 1986), sunflower (Horn and Friedt 1999), tobacco (Bonnett et al. 1991) and wheat (Song and Hedgcoth 1994a). CMS plants are also found among non-crop species, e.g., petunia (Boeshore et al. 1985), where they may be favored by natural selection (Delph et al. 2007).

Distinct stages in male organ development and pollen formation are affected in different CMS systems. Female fertility is not affected and the morphology of the plants is usually normal, although there may be alterations to flower morphology (reviewed by Zubko 2004; Linke and Börner 2005). Two examples are the petaloid-type of CMS (Kitagawa et al. 1994) and the 'carpeloid' type of CMS (Linke et al. 2003), both in carrot. Abnormal floral development is also observed in some alloplasmic CMS strains (see below). While CMS in these systems alters flower structure, programmed cell death or necrosis within anthers or pollen is associated with CMS in other plants (Warmke and Lee 1977; Balk and Leaver 2001; Wen et al. 2003; reviewed by Chase 2007).

The most extensively studied CMS systems to date are in *Brassica* (L'Homme and Brown 1993; Wang et al. 1995), petunia (reviewed by Hanson et al. 1999), *Phaseolus* (Chase 1994; Sarria et al. 1998), maize (reviewed by Ward 1995; Gabay-Laughnan et al. 1995), and rice (Fujii et al. 2010). There are two main classes of CMS, one that appears to have arisen naturally in wild populations and a second resulting from intentional manipulation such as interspecific crosses or crosses between different populations of the same species. The latter is termed alloplasmic CMS. This sterility results from nuclear-cytoplasmic incompatibility (Kaul 1988). Alleles of certain nuclear genes, *restorers-of-fertility*, can suppress or override cytoplasmic male sterility (reviewed by Chase 2007). In alloplasmic CMS, restorers existing within a population that mask the existence of CMS may be removed by crossing programs (discussed below).

CMS enables breeders to produce commercial F1 hybrids on a large scale as it eliminates the need for hand emasculation (Schnable and Wise 1998; Havey 2004). In crops such as maize, where the F1 hybrid must be male fertile and produce seed, *restorer-of-fertility genes* can override the CMS. CMS-S and CMS-C maize are presently used but not CMS-T since it was shown to be particularly susceptible to *Bipolaris maydis* (Ward 1995; Schnable and Wise 1998). Whether a breeder prefers CMS-S or CMS-C depends on the stability of the sterility in the environment in which it is grown (Havey 2004) and the CMS-inbred line combination.

A. Naturally Occurring Male-Sterile Cytoplasm

The types of CMS that have arisen naturally are considered to result from a series of recombination events leading to rearrangements in the mtDNA, presumably in a progenitor species. The male sterility is often revealed when outcrossing removes a restorer allele (reviewed by Hanson 1991; Schnable and Wise 1998). A number of naturally occurring male-sterile cytoplasms have been discovered in this manner. For example, a CMS plant was found in a male-fertile line of *Phaseolus vulgaris*. It was later determined that the fertile progenitor line carried a restorer that concealed the male-sterile nature of the cytoplasm (Mackenzie 1991).

Among the best-characterized examples of naturally occurring male-sterile cytoplasms are those of maize (*Zea mays* ssp. *mays*) CMS-S and CMS-T. Five of the mitochondrial genomes of maize were sequenced and compared (Allen et al. 2007). It was determined that the S and T cytoplasms are the most divergent mitotypes. They are distinct from one another and also from the NA and NB male-fertile cytoplasms. These findings are in agreement with theory that the male-sterile S and T cytoplasms had their origin in fertile ancestors of maize. The maize CMS-S cytoplasm is found in some Latin American races of maize (Weissinger et al. 1983). It also appears identical to a cytotype found in some accessions of *Zea mays* ssp. *mexicana* teosinte (Weissinger et al. 1983; Doebley and Sisco 1989), although that strain appears to be male fertile (Allen 2005). The CMS-T cytotype has not been found in any teosinte accession but is seen in several accessions of Latin American maize (Weissinger et al. 1983).

In maize, no case of a spontaneous de novo mutation from male fertile to CMS has been confirmed, despite reports that exact matches to extant CMS-S and CMS-T arose from the fertile NB mitotype (the **n**ormal fertile genome first characterized in inbred line **B**37), each within a single generation (Lemke et al. 1985, 1988). The CMS mtDNAs are structurally very different from NB and each contains some unique DNA (Allen et al. 2007). It is highly improbable that either CMS type could completely replace the fertile mitotype in one generation (Lonsdale 1987; Small et al. 1987). An alternative explanation that substoichiometric shifting would cause the replacement events (Arrieta-Montiel et al. 2009) is also unlikely, because the CMS-ORFs have not been detected in the fertile NB cytoplasm (Liu et al. 2002). The data reported by Lemke et al. (1985, 1988) probably resulted from seed contamination or a sample mix-up.

B. Alloplasmic Male-Sterile Cytoplasm

Alloplasmic CMS may be caused by nuclear-cytoplasmic incompatibility when the nucleus of one species is combined with the cytoplasm of another (Kaul 1988). One of the best examples is found in sunflower (*Helianthus* sp.). The hybrid production of sunflower has always relied heavily on the PET1 system. The PET1 cytoplasm was derived from an interspecific cross between *H. petiolaris* and *H. annuus* (Horn 2002). Nine additional PET1-like CMS cytoplasms have been since been identified (Horn et al. 1996) and it is possible that this CMS exists at subliminally low levels in *H. annuus* (Horn and Friedt 1999). More recently, a new type of sunflower CMS was derived from an interspecific cross of an accession of *H. giganteus* and a cultivar of *H. annuus* (Feng and Jan 2008).

In wheat, cytoplasmic male sterility resulted from the cross of two male-fertile species. The male-sterile phenotype appears when plants carry *Triticum timopheevi* mitochondria in the *T. aestivum* nuclear background (Song and Hedgcoth 1994a, b). A chimeric ORF present in *T. timopheevi* mitochondria with a *T. timopheevi* nucleus either does not synthesize transcripts or the transcripts are unstable. However, in plants with *T. timopheevi* mitochondria and the *T. aestivum* nucleus, this ORF produces a stable protein product (Song and Hedgcoth 1994b).

In some instances, alloplasmic male sterility results in aberrant floral structures (reviewed by Carlsson et al. 2008). For example, pistillody (homeotic transformation of stamens into pistil-like structures) is observed in an alloplasmic line of wheat (Murai et al. 2002), and there is abnormal floral development in alloplasmic lines of male-sterile tobacco (Kofer et al. 1991; Bergman et al. 2000; Farbos et al. 2001) and *Mimulus* (Barr and Fishman 2011). For example, in CMS tobacco that carries the cytoplasm of *Nicotiana repanda* and the nuclear genome of *N. tabacum*, the petals are poorly pigmented, the stamens have shortened filaments, and the anthers are shriveled (Bergman et al. 2000; Farbos et al. 2001). CMS *Brassica napus* plants resulting from somatic hybrids between *Brassica napus* and *Arabidopsis thaliana* also exhibit an aberrant floral phenotype (Leino et al. 2003; Teixeira et al. 2005). Although the phenotype resembles those of homeotic mutants, the cause is the alien Arabidopsis mitotype. The expression of many of the *Brassica napus* nuclear genes, including the homolog of the homeotic *apetela3* gene, is altered, apparently due to aberrant retrograde signaling from the mitochondria (Carlsson et al. 2007).

C. Chimeric Open Reading Frames Associated with CMS

CMS is often associated with the expression of chimeric regions of mtDNA (reviewed by Chase and Gabay-Laughnan 2004; Newton et al. 2004; Hanson and Bentolila 2004; Linke and Börner 2005). These regions usually consist of pieces of known genes along with sequences of unknown function and are thought to be generated via repair of DNA breaks or an illegitimate recombination involving microhomologous repeats (usually ~5–<50 bp; reviewed by Maréchal and Brisson 2010). The chimeric ORFs may be fused to promoter sequences or co-transcribed with genes located upstream (reviewed by Chase and Gabay-Laughnan 2004; Hanson and Bentolila 2004; Newton et al. 2004; Fujii and Toriyama 2008). Chimeric regions associated with CMS have been described in many species including *Brassica* (L'Homme and Brown 1993; L'Homme et al. 1997), maize (Zabala et al. 1997), petunia (reviewed by Hanson et al. 1999), sorghum (Tang et al. 1996), and sunflower (Köhler et al. 1991; Laver et al. 1991; Spassova et al. 1994; Horn and Friedt 1999).

There are often so many rearrangements between a CMS mitochondrial genome and a related fertile genome that it can be a laborious effort to identify CMS-associated regions (reviewed by Hanson and Bentolila 2004). The set of candidate CMS-ORFs can be narrowed down to a very few by sequencing *multiple* fertile and CMS mtDNAs within a species and comparing their chimeric open reading frames (e.g. Satoh et al. 2004; Allen et al. 2007; Fujii et al. 2010). A chimeric ORF that is unique to one CMS genotype can be considered a candidate CMS gene. Of course, tests must be done to confirm that the candidate gene actually causes the CMS phenotype (reviewed by Hanson and Bentolila 2004). These tests can include loss of the CMS phenotype when there is a mutation that alters or eliminates the candidate CMS-ORF. Another important criterion is whether the expression of the CMS candidate gene is changed in the presence of appropriate nuclear restorer alleles. Strangely, transgenic expression of the candidate CMS-ORF (tethered to a sequence directing protein import into plant mitochondria) from the nuclear genome does not always seem to be a straightforward functional test (Wintz et al. 1995).

One of the first systems in which the CMS chimeric ORF was identified is that in petunia (Boeshore et al. 1985). The CMS-associated gene was designated *pcf* for **p**etunia **C**MS-associated **f**used gene (reviewed by Hanson et al. 1999). It consists of the 5′ portion of the *atp9* gene, segments of the first and second exons of *cox2*, and a large region of unknown origin designated *urfS* (reviewed by Hanson et al. 1999). Multiple CMS-associated chimeric ORFs were identified in *Brassica* species, and were characterized in the *pol* (L'Homme and Brown 1993), *nap* (Dieterich et al. 2003) and *ogu* (Bonhomme et al. 1992) cytoplasms (reviewed by Schnable and Wise 1998; Hanson and Bentolila 2004).

Three common types of CMS have been identified in maize and are designated CMS-C, CMS-T and CMS-S (reviewed by Laughnan and Gabay-Laughnan 1983). The chimeric ORFs associated with CMS-T and CMS-S have been identified. However, no chimeric ORF unique to the CMS-C mitotype has been found even though the genome has been completely sequenced (Allen et al. 2007). In this case, CMS may result from rearrangements that exist 5′ to three essential genes (Dewey et al. 1991). Altered expression of one of them in the tapetal cells during pollen development may cause this type of CMS.

The CMS-T-associated chimeric gene of maize has been designated T-*urf13* (Dewey et al. 1987; Wise et al. 1987a). This ORF contains part of the 3′ flanking region of the mitochondrial 26S ribosomal RNA (*rrn26*) gene, a segment of unknown origin, and a sequence with homology to the coding region of *rrn26* (Dewey et al. 1986). In order for this chimeric gene to have arisen, multiple rearrangements were required (Dewey et al. 1986). Interestingly, this amalgamation of sequences is expressed at high levels, because it lies between a duplicate copy of the *atp6* promoter and the only copy of the essential *atp4* gene in the CMS-T genome (Dewey et al. 1986; Allen et al. 2007). Furthermore, it is translated into a 13-kD membrane protein that is expressed constitutively (Forde et al. 1978; Newton and Walbot 1985). The expression of *T-urf13* has little effect on overall plant growth, but it causes premature mitochondrial degradation in the tapetum during microspore biogenesis, and, therefore, early pollen abortion (Warmke and Lee 1977; reviewed by Levings 1993; Skibbe and Schnable 2005).

The CMS-S type of maize male sterility is correlated with the presence of a sequence, designated R, occurring within a 4,215-bp duplicated region of the mitochondrial genome (reviewed by Newton et al. 2009). This region contains two adjacent chimeric open reading frames, *orf355* and *orf77*. Sequences similar to a portion of the linear R1 plasmid are present in *orf355* along with sequences of unknown origin. *Orf77* contains three segments with similarity to the coding and flanking regions of *atp9*, as well as to sequences normally located 3′ to the *atp4* gene (Zabala et al. 1997; Allen et al. 2007). Two free linear plasmids designated S1 and S2 are also present within CMS-S mitochondria (Pring et al. 1977). 208-bp terminal inverted repeats (TIRs) are present at the ends of each S plasmid (Paillard et al. 1985; reviewed in Handa 2008). Recombination can occur between TIR sequences that precede *orf355/orf77* in the main mitochondrial genome and the TIRs at the ends of each S plasmid (Schardl et al. 1984). Transcription of a 1.6-kb RNA initiates from the resulting linear ends of mtDNA (Gabay-Laughnan et al. 2009). The CMS phenotype is correlated with high levels of expression of this 1.6-kb transcript (Zabala et al. 1997; Wen and Chase 1999; Gabay-Laughnan et al. 2009).

IV. Cytoplasmic Reversion to Fertility

A. Reversion in CMS Maize

Field-grown CMS-S maize plants have given rise to cytoplasmic revertants on numerous occasions. These revertants first appear as sectors of male fertility on male-sterile tassels or as totally male-fertile tassels in plantings of CMS in inbred or hybrid backgrounds (Newton et al. 2009). The first reported cases arose in plants of the genotype CMS-S in the Wf9 inbred line (Jones 1956). Two exceptional male-fertile plants produced only male-fertile progeny when self-pollinated and crosses of these exceptions as pollen parents yielded only male-sterile progeny. These results indicated that a nuclear restorer gene was not involved. Male-fertile plants arising from male-sterile CMS-S plants in the M825 inbred line were later described (Singh and Laughnan 1972) and, again, nuclear restorer gene action was ruled out. It was proposed that the newly arisen male fertility "involved a change from male-sterile to male-fertile condition in the cytoplasm" (Singh and Laughnan 1972).

We now know that cytoplasmic reversion involves deletions or rearrangements of the CMS-associated sequences of the mtDNA. M825, and to a lesser extent Wf9, are the two most active inbred nuclear backgrounds associated with cytoplasmic fertility reversion of CMS-S in maize (reviewed by Gabay-Laughnan and Newton 2005). In lines that do not show "spontaneous" reversion events in the field, tissue culture may induce mtDNA changes leading to fertility in regenerated plants. Revertant plants regenerated from tissue cultures of CMS-S in the W182BN line (Earle et al. 1987) and also of CMS-T in the Wf9/A188 background (reviewed by Ward 1995) have been observed, but there are no reports of reversion to fertility in field-grown plants of these lines.

Numerous CMS-S cytoplasmic revertants have been isolated and analyzed. The types of events that cause reversion to fertility depend on nuclear background. For example, the S1 and S2 mitochondrial plasmids are always lost from M825, 38–11, H95 and WB182BN cytoplasmic revertants but are retained by all Wf9 revertants (Levings et al. 1980; Kemble and Mans 1983; Escote et al. 1985; Ishige et al. 1985; Schardl et al. 1985; Earle et al. 1987; Escote-Carlson et al. 1988; Small et al. 1988). In addition, various regions of the main mitochondrial genome are rearranged, depending again upon the nuclear background; not all cytoplasmic revertants exhibit the same rearrangements (Small et al. 1988). A comparison of revertants in the M825, 38–11 and W182BN nuclear backgrounds revealed that they differ in the organization of the integrated S1 and S2 sequences. This initially made it difficult to identify the mtDNA region involved with the CMS-S male sterility phenotype.

Comparison of the mtDNA of fertile revertants with that of the progenitor male-sterile strains helped identify the sterility-associated region of the CMS-S mtDNA (Zabala et al. 1997). All the cytoplasmic revertants, regardless of nuclear background, showed alterations in the *orf355-orf77* region of mtDNA; furthermore, the 1.6-kb RNA including *orf355* and *orf77* was missing in all the revertants (Zabala et al. 1997). As was described earlier, this 1.6-kb RNA is transcribed only from linear ends resulting from recombination between TIR sequences preceding *orf355* and the TIRs of the linear S-plasmids (Gabay-Laughnan et al. 2009). Since revertants in most nuclear backgrounds lose the S plasmids, they are unable to produce this transcript. In the case of Wf9 revertants, which retain S plasmids, the *orf355-orf77* region in the main mitochondrial genome is itself rearranged or deleted (Zabala et al. 1997).

Illegitimate recombination between microrepeats can lead to deletion of the CMS-associated regions. In one M825-type revertant analyzed, the mtDNA sequences that recombined to produce the revertant were shown to contain 19 bp of sequence similarity (16/19 bp matched). One copy of this repeat was located on the S2 plasmid and the event led to the loss of the free S plasmids (Schardl et al. 1985). Without the S plasmids, there are no linear ends preceding the *orf355-orf77* region and no 1.6-kb sterility-associated RNA.

Microhomologies are also involved in the generation of the aberrant-growth NCS mutants of maize, which have deletions in essential mitochondrial genes. Intriguingly, the NCS4 mutation arose during the reversion to fertility of a CMS-S plant in the M825 line. Illegitimate recombination across a near-perfect repeat (15/16 bp) present in the S2 plasmid and the intron of the *rps3* ribosomal protein gene resulted in the loss of both the S plasmids and a portion of the *rps3* gene. Thus, a stunted but male-fertile plant was recovered following the same initiating microrepeat-mediated event (Newton et al. 1996).

In most cases of cytoplasmic reversion of CMS-S maize to male fertility, a unique mitochondrial mutation is associated with each revertant (Schardl et al. 1985; Small et al. 1988; Zabala et al. 1997). However, rare exceptions have been reported in closely related plants. Three sibling Wf9 cytoplasmic revertants were found within one family; two of them were observed as tassel sectors and the third as a totally fertile tassel (Gabay-Laughnan and Laughnan 1983). The possibility of a common origin existed even though sectors usually arise from independent mutations. These three sibling revertants were later determined to have the same mitochondrial mutation and, therefore, probably resulted from the same mutational event (Escote-Carlson et al. 1988). The direct male-sterile progenitor plant must have been heteroplasmic for both CMS-S and revertant mitochondria, which sorted out in the sibling plants.

Recent analyses of another set of three Wf9 cytoplasmic revertants revealed that they have an identical mtDNA rearrangement. In this case, however, the mutation sorted out in three successive generations. It is proposed that the revertant arose in the male-sterile progenitor strain. Subsequently, mutant mitochondria were transmitted to some of the progeny of this heteroplasmic plant, where they amplified to become the predominant mitotype. In all three revertants, the same inversion with a breakpoint between the TIR and *orf355* has been found (Matera et al. 2011). Interestingly, the *orf355-orf77* coding sequences and the TIR sequences remain intact in this inversion, and the free S-plasmids are also present; however, recombination between the displaced TIR and the S-plasmids no longer leads to a linear end 5′ to *orf355-orf77*. Without the TIR-terminating linear end, which contains the transcription start site (Gabay-Laughnan et al. 2009), the CMS-associated 1.6-kb RNA cannot be produced.

In contrast to CMS-S, no reversion event has ever been observed in field-grown maize plants carrying T cytoplasm. As was mentioned above, cytoplasmic reversion of CMS-T has been observed in plants regenerated from tissue cultures. The CMS-T-associated T-*urf13* gene has been deleted in all but one of the tissue-culture-induced revertants studied. In the exceptional revertant, there is a frame shift at codon 74 of the T-*urf13* region. This produces a truncated version of the TURF13 protein (Umbeck and Gengenbach 1983; Wise et al. 1987a, b). One of the "typical" CMS-T revertants was studied in detail, and it was shown that both inter- and intra-molecular recombination events were involved in its generation (Fauron et al. 1990). Some of the resulting subgenomic circles were subsequently eliminated, including the one carrying T-*urf13*.

B. Reversion in Common Bean

The male-sterile CMS-Sprite mitochondrial genome of the common bean, *Phaseolus vulgaris*, is comprised of three inter-recombining, redundant circular molecules, 394, 257 and 210 kb in size (Janska and Mackenzie 1993). The progenitor of this CMS cytoplasm maps as a single circular master chromosome.

This progenitor configuration is retained at substoichiometrically low levels in the CMS genome. Conversely, the three circular molecules characteristic of CMS are present at substoichiometrically low levels in the progenitor (Janska et al. 1998). The mtDNA region that is correlated with Sprite CMS consists of a unique sequence in the mitochondrial DNA designated the *pvs* (for P*haseolus* v*ulgaris* sterility) sequence and is carried on the 210-kb molecule. This region contains at least two novel ORFs, *pvs-orf98* and *pvs-orf239* (Johns et al. 1992; Chase and Ortega 1992; Janska et al. 1998), however, only *pvs-orf239* appears to be translated (Abad et al. 1995). In *P. vulgaris* cytoplasmic revertants, the 210-kb subgenomic circle, carrying *pvs*, is reduced to substoichiometric levels (Mackenzie et al. 1988; Janska and Mackenzie 1993). Since the two remaining circles carry all the essential mitochondrial genes, this reduction is tolerated (Janska and Mackenzie 1993). The progenitor mtDNA configuration, as well as the *pvs-orf239* CMS-associated sequences, are maintained at substoichiometric levels in the cytoplasmic revertants (Janska et al. 1998). Stoichiometric shifting of the levels of mtDNA molecules is proposed to account for both the appearance of sterility and the reversion to fertility of CMS-Sprite (Janska et al. 1998).

C. Reversion in Pearl Millet

Pearl millet [*Pennisetum glaucum* (L.) R. Br.; previously *Pennisetum americanum* (L.) Leek] is a significant food crop in the arid tropics. Cytoplasmic male sterility along with restorers is used commercially to increase productivity. Although there are a number of CMS sources in pearl millet (Delorme et al. 1997), the A1 source of CMS is the most commonly employed. Fertile revertants are observed in CMS A1 at a low frequency and were shown to result from mitochondrial DNA alterations (Smith et al. 1987; Delorme et al. 1997). The region of mtDNA that includes the *cox1* gene appeared to be correlated with A1 CMS (Delorme et al. 1997). Feng et al. (2009) further analyzed this region in the male-sterile A1 and its fertile revertants. They found that three *cox1*-related regions are present in pearl millet; these have been designated *coxI-1*, *coxI-2* and *coxI-3*. The organization of these regions differs in the maintainer, CMS A1, and cytoplasmic fertile revertants. A two-step model involving intermolecular illegitimate recombination across a 7-bp microhomologous repeat followed by intramolecular homologous recombination leading to the novel *cox1* mtDNA organization observed in cytoplasmic revertants is proposed (Feng et al. 2009). The intermolecular recombination involves a substoichiometric molecule and one of the resulting products is stabilized by the subsequent intramolecular recombination.

D. Reversion in Brassica

A unique mitochondrial gene, *orf138*, is responsible for Ogura CMS in *Brassica* plants (Bonhomme et al. 1991, 1992; Grelon et al. 1994) and also for the CMS in Ogura radish (Krishnasamy and Makaroff 1993; Krishnasamy et al. 1994). There exist three different configurations of the *orf138* gene region (Bellaoui et al. 1998). In one form, the *orf138* gene is linked to, and co-transcribed with, the *orfB* gene (now known to be *atp8*; e.g. Heazlewood et al. 2003). In a second form, the *orf138* gene is associated with the *atp1* gene and is not expressed. In the third form, *orf138* is no longer associated with *orfB* or *atp1* but with other sequences. The mtDNA form carrying *orf138* and *orfB* is rearranged upon cytoplasmic reversion to fertility in *Brassica* (Bonhomme et al. 1991). Substoichiometric amounts of mtDNA molecules carrying the different *orf138* configurations are present in CMS plants and also in "unmodified" Ogura cytoplasm. In such cases, one configuration is usually predominant while the others are substoichiometric. Cytoplasmic reversion to fertility of Ogura CMS is associated with changes in the proportions of the different molecules carrying *orf138* (Bonhomme et al. 1991, 1992). A deleted *orf138* derivative was also detected. The observed changes result from recombination between the different forms (Bellaoui et al. 1998).

E. Reversion in Carrot

A partially male-fertile plant arose spontaneously in a strain of petaloid CMS carrot. It was determined via genetic analyses that the fertility was due to a new nuclear restorer gene. Several generations later, cytoplasmically revertant nonrestoring plants were recovered within a CMS family segregating for this new restorer (Chahal et al. 1998). The mtDNA genome of the CMS line was partially mapped and compared with that of the cytoplasmic revertant. A complex organization, including substoichiometric genomes, was revealed. The mitochondrial genomes of the fertile maintainer and the revertant were similar; however, they can be distinguished by unique restriction enzyme fragments. It was suggested that the mtDNA changes in the revertant could have arisen by the amplification of a substoichiometric genome (Chahal et al. 1998).

V. Nuclear-Cytoplasmic Interactions

Plants represent an excellent model system in which to study the interaction of the nuclear and cytoplasmic genomes. The nuclear-mitochondrial genotype combination can be changed using wide crosses and in vitro manipulations. Variant or defective mitochondrial genes that have easily scored phenotypes, such as growth abnormalities or male sterility, can be used to assay the effects of nuclear genes. Conversely, in plants with mitochondrial dysfunction, retrograde regulation of the expression of nuclear genes can often be seen. Because of its economic importance, one area of active research is the analysis of nuclear genes that control the expression of mitochondrially encoded CMS traits. Nuclear genes also control the organization and stability of mitochondrial genomes and, in plants, their effects are especially striking.

A. Incompatibility Between Nucleus and Cytoplasm

As was discussed above, nuclear-cytoplasmic incompatibility may result in alloplasmic male sterility. Other traits, unrelated to male fertility, may be affected in alloplasmic combinations. Cytoplasms from teosinte relatives have been introduced into maize inbred lines by serial backcrosses to produce alloplasmic cytolines (Allen 2005). When the teosintes were more distantly related, a number of nuclear-cytoplasmic incompatibilities could be documented, including effects on growth and morphology (Allen 2005). For example, a spectrum of effects is seen when the *Zea perennis* teosinte cytoplasm is introduced into certain maize inbred lines. CMS (called CMS-EP) is observed in some lines (Gracen 1972; Gracen and Grogan 1974; reviewed by Laughnan and Gabay-Laughnan 1983), but most inbred lines carry restoring alleles for CMS-EP (Gabay-Laughnan 2001). Plant and seed size is also affected in plants carrying the *Z. perennis* cytoplasm and certain maize nuclear genotypes. This phenotype is termed maize **t**eosinte-**c**ytoplasm-associated **m**iniature (TCM; Allen et al. 1989). Kernels are smaller than normal, and plants grown from these smaller kernels are shorter, paler and slower growing. CMS-EP and TCM are distinct traits, and alleles that suppress these effects are products of different nuclear genes (Allen et al. 1989; Gabay-Laughnan 2001; reviewed by Newton et al. 2004).

Two diverse maize nuclear backgrounds, W23 and A619, carrying *Zea perennis* cytoplasm were examined for the expression of various mitochondrial genes (Cooper et al. 1990). Two major *cox2* transcripts were seen when the inbred background was W23 and three were seen when it was A619. The presence of the additional transcript was associated with a threefold reduction of the Cox2 polypeptide. A single nuclear gene, **m**odifier of *cox2* **t**ranscripts (*Mct*), is responsible for the observed transcript differences (Cooper et al. 1990; Newton and Courtney 1991; Newton et al. 1995). The mitochondrial transcript differences, as well as the nuclear gene responsible, are not related to the CMS or TCM phenotypes or to their restorers/rectifiers (reviewed by Newton et al. 2004).

Mct is probably a member of a class of genes termed ***m****odifier of* ***m****itochondrial* ***t****ranscripts* (*Mmt*). Restorer alleles for *pol* CMS of *Brassica napus,* CMS-S in maize, Ogura CMS in radish, and sorghum CMS IS1112C have all been correlated with the processing of normal mitochondrial gene transcripts (Makaroff and Palmer 1988; Singh and Brown 1991; Singh et al. 1996; Li et al. 1998; Tang et al. 1998; Wen and Chase 1999; Wen et al. 2003). These restorer loci either encode or regulate *Mmt* activity, or the *Mmt* and *Rf* alleles are closely linked (reviewed by Chase and Gabay-Laughnan 2004).

An ambitious effort to analyze nuclear-cytoplasmic co-adaptation using many accessions of Arabidopsis has revealed more subtle incompatibilities. In particular, germination capacity under challenging conditions can be significantly affected by the cytoplasm donor in F2 progeny (Moison et al. 2010).

B. Nuclear Genes and the Restoration of Fertility

CMS is widely utilized in the production of male-sterile plants for efficient, inexpensive hybrid seed production (Havey 2004). Nuclear restorer genes override CMS and are an important component of hybrid seed production when the F1 crop must be male fertile (Havey 2004). Hence, restorers are under study in the CMS systems of many crop plants, including maize, radish, rice, and sorghum, as well as in the CMS/*Rf* model systems such as *Mimulus,* petunia, and *Phaseolus*. The interesting questions raised by the presence of nuclear restorer genes in plants with normal, fertile cytoplasm have been previously reviewed (Chase and Gabay-Laughnan 2004; Newton et al. 2004).

In many CMS systems, e.g. petunia, radish, rice, and CMS-S maize, one nuclear restorer gene is sufficient to restore fertility. However, in some CMS systems, the coordinate action of two restorer genes is required for fertility restoration, e.g. CMS-T maize and the IS1112C sorghum (reviewed by Chase and Gabay-Laughnan 2004; Newton et al. 2004).

Most, but not all, of the *Rf* genes cloned thus far are members of the pentatricopeptide repeat (PPR) family, a large family of proteins in plants containing tandem arrays of degenerate 35 amino-acid repeats (Small and Peeters 2000; Saha et al. 2007; O'Toole et al. 2008). Most of the PPR proteins in plants are targeted to either mitochondria or chloroplasts, where they play essential roles in post-transcriptional processing events, such as RNA cleavage, splicing, editing and translation (Lurin et al. 2004; Andres et al. 2007; Schmitz-Linneweber and Small 2008). Some of the PPRs targeted to mitochondria have been demonstrated to act as *Rf* alleles for CMS.

The first PPR restorer to be cloned was the petunia *Rf592* gene (Bentolila et al. 2002). It encodes a mitochondrially targeted protein containing 14 tandem copies of a PPR motif, which interacts with transcripts of the CMS-associated locus (Gillman et al. 2007). The *Rfo (Rfk1)* restorer locus for Ogura CMS in radish has also been cloned (Brown et al. 2003; Desloire et al. 2003; Koizuka et al. 2003). This locus contains three PPR genes encoding highly similar proteins, designated PPR-A, PPR-B and PPR-C. PPR-B was genetically determined to be the restorer gene (Desloire et al. 2003). This gene codes for a protein containing 16 repeats of the PPR motif (Brown et al. 2003; Koizuka et al. 2003). PPR-B has a role in the translational regulation of the mRNA of the CMS-associated ORF (Uyttewaal et al. 2008).

The BT (Boro II) type of CMS in rice is an alloplasmic CMS, resulting from the combination of an *indica* cytoplasm and a *japonica* nucleus. The restorer gene *Rf-1* restores fertility to this CMS and is widely used commercially (reviewed by Kato et al. 2007). *Rf-1* encodes a protein containing 18 repeats of a PPR motif (Kazama and Toriyama 2003; Akagi et al. 2004; Komori et al. 2004). Duplicate open reading frames, designated *Rf-1A* and *Rf-1B*, were found in the region of the *Rf-1* gene and it was initially concluded that *Rf-1A* is the restorer gene (Akagi et al.

2004). However, later studies reported that *Rf-1A* and *Rf-1B* are each able to restore BT-type CMS (Wang et al. 2006). A survey of allelic variants of the *Rf-1* locus from a wide variety of *Oryza* species identified six genes (*Rf-1A* through *Rf-1F*) with homology to *Rf-1* all encoding PPR proteins (Kato et al. 2007). Another restorer gene in rice has recently been identified as a potential PPR gene. The WA (wild abortive) type of CMS in rice is the most widely used for the production of hybrid seed. A major *Rf* locus has been mapped to a chromosomal region containing 13 PPR genes. One of these genes is the candidate restorer gene (Ngangkham et al. 2010).

There are several CMS systems in sorghum (Schertz et al. 1989), however, the A1 type of CMS in *Sorghum bicolor* is the one employed almost exclusively for the commercial production of sorghum hybrids. Restoration of this CMS requires two major restorer genes, *Rf1* and *Rf2*. The *Rf1* locus has been mapped and cloned and was shown to encode a PPR protein (Klein et al. 2005). This restorer is not located in the collinear region in the rice genome (Klein et al. 2005). A PPR protein also represents a possible candidate for the sorghum *Rf2* gene (Jordan et al. 2010). This PPR gene is highly similar to the orthologous rice *Rf1* gene.

In several other cases, *Rf* genes have been mapped to PPR-rich regions of genomes. Cytoplasmic male sterility occurs in monkeyflower hybrids with *Mimulus guttatus* cytoplasm and the *M. nasutus* nucleus (Fishman and Willis 2006). The genomic region containing the restorer locus for this CMS has been mapped and characterized as a PPR-gene-rich region (Barr and Fishman 2010). The *Rf3* restorer of CMS-S maize maps to the long arm of chromosome 2 (Laughnan and Gabay-Laughnan 1983; Kamps and Chase 1997), in a region containing several putative PPR genes (Xu et al. 2009).

Although most of the *Rf* genes cloned thus far are members of the PPR family, there are three known exceptions. The maize CMS-T *Rf2* allele was the first restorer to be cloned (Cui et al. 1996). It does not encode a PPR protein but instead encodes a mitochondrially localized aldehyde dehydrogenase (Cui et al. 1996; Liu et al. 2001). The *Rf17* restorer gene for the Chinese wild rice (CW)-type of CMS encodes a 178-aa protein designated **R**etrograde-regulated **M**ale **S**terility (RMS). This protein contains a segment similar to acyl-carrier protein synthase (Fujii and Toriyama 2009). In addition, the *Rf2* gene for Lead Rice-type CMS encodes a 152-aa protein with a glycine-rich domain (Itabashi et al. 2011). Thus, researchers looking for candidate restorers should not limit their searches to PPR genes (Ngangkham et al. 2010).

Maize CMS-S is distinctive in that many independent restorers have arisen by spontaneous mutation. While these alleles do restore viability to pollen grains, many are homozygous lethal (Laughnan and Gabay 1973, 1978; reviewed by Gabay-Laughnan et al. 1995; Chase and Gabay-Laughnan 2004). This kind of newly arisen restorer has been designated *restorer-of-fertility lethal* (Wen et al. 2003). CMS-S pollen aborts relatively late and restoration of function occurs when the restoring allele is present within the individual pollen grain (i.e., restoration is gametophytic). Thus, "lethal" restorers would be expected to reduce the levels of the CMS-associated, *orf355-orf77* transcripts in the maturing pollen, but they might be expected to also affect the expression of one or more essential mitochondrial genes. Indeed, in the case of one lethal restorer, Wen et al. (2003) showed a reduction in transcripts for the alpha subunit of ATPase in addition to the expected reduction in the CMS-S-associated 1.6-kb RNA. Ethanolic fermentation can compensate for respiratory deficiencies in pollen (reviewed by Tadege et al. 1999), but obviously not in the seed or seedling. The products of these lethal-restorer genes are expected to be involved in mitochondrial biogenesis or function, and several may represent mutations in PPR proteins.

C. Nuclear Genes Affecting Mitochondrial Recombination and Substoichiometric Shifting

The organization of mitochondrial genomes and the expression of mitochondrial genes are controlled by nuclear genes. Some nuclear genotypes are associated with higher rates of mitochondrial rearrangements that lead to abnormal growth or reversion of CMS to fertility. Therefore, nuclear genes are involved in the generation, selection and amplification of mitochondrial mutations. The rate at which NCS mutations arise in maize varies among inbred lines; it is usually extremely low, but it can be as high as 1% in the inbred line Wf9 nuclear background (Duvick 1965; Newton and Coe 1986). In addition, the nuclear background controls the rate at which cytoplasmic reversion of male sterility occurs. For example, reversion is observed in approximately 10% of CMS-S maize plants in the M825 nuclear background (Laughnan et al. 1981). The nuclear background also controls the mtDNA rearrangements observed upon cytoplasmic reversion (reviewed by Gabay-Laughnan et al. 1995). Additionally, cytoplasmic reversion of maize CMS-S in the M825 line is always associated with recombination of S2 sequences with microhomologous sequences elsewhere in the genome, resulting in the loss of the S1 and S2 plasmids. In revertants arising in the Wf9 background, however, the S plasmids are invariably retained, but rearrangements affect the CMS-ORF (Small et al. 1988; reviewed by Newton et al. 2004).

The P2 line of maize, derived from a South American strain of popcorn, exhibits a general increase in mtDNA instability and P2 plants exhibit a variety of maternally transmitted abnormalities such as poor plant growth and leaves with pale sectors (Kuzmin et al. 2005). These phenotypes are associated with destabilized, multiply rearranged mitochondrial genomes. The P2 nuclear genotype appears both to alter the copy number of specific sublimons and to amplify the products of aberrant microhomologous recombination (Kuzmin et al. 2005).

In contrast to the above systems in maize, where no specific causative nuclear allele(s) has been identified, nuclear genes have been shown to affect mitochondrial recombination in some other plant systems. For example, the dominant allele of the *Phaseolus vulgaris* *Fr* ("fertility restorer") gene is responsible for a reduction in the copy number of the 210-kb mitochondrial subgenome that carries the *pvs-orf239* responsible for CMS-Sprite. This results in a reversion/restoration of the CMS to fertility (Mackenzie and Chase 1990; Janska and Mackenzie 1993; He et al. 1995; Janska et al. 1998). When *Fr* is inactive, the 210-kb subgenome is amplified (Arrieta-Montiel et al. 2001). Thus the *Fr* gene seems to affect mitochondrial substoichiometric shifting.

Recombination within Arabidopsis mitochondrial genomes is influenced by at least three nuclear genes: *MSH1*, *OSB1*, and *REC3A* (reviewed by Maréchal and Brisson 2010). Mutation of *MSH1* (**M**ut**S** **h**omolog; formerly *CHM*) is responsible for the *chm/chm* mutant phenotype (Martinez-Zapater et al. 1992) in Arabidopsis. MSH1 regulates substoichiometric shifting within the mitochondrial genome (Abdelnoor et al. 2003), suppressing recombination at repeat sequences varying in size from 108 to 556 bp. When MSH1 activity is disrupted, over 30 sites within the mitochondrial genome become activated, thus influencing the genome organization (Arrieta-Montiel et al. 2009).

OSB1 (**O**rganellar **S**ingle-stranded DNA-**B**inding protein1) is a member of a plant-specific family of DNA-binding proteins. OSB1 was purified from potato (*Solanum tuberosum*) mitochondria (Vermel et al. 2002) and orthologs of the *OSB1* gene were later found in *Arabidopsis thaliana*, rice and maize (Zaegel et al. 2006). OSB1 is required for the correct transmission of substoichiometric mitochondrial genomes in Arabidopsis (Zaegel et al. 2006). T-DNA insertion mutants accumulate products of homologous recombination and this leads to morphological phenotypes such as leaf variegation and distorted plants. *OSB1* thus controls the stoichiometry of the subgenomes produced by recombination (Zaegel et al. 2006).

Three distinct homologs of the *E. coli* *recA* gene are found in the Arabidopsis nuclear genome. These map to different chromosomes and are designated *RECA1*, *RECA2*, and *RECA3* (Shedge et al. 2007).

RECA3 is targeted to the mitochondria and mutant alleles result in plants that carry mtDNA rearrangements but which appear to be phenotypically normal. The characterized mtDNA rearrangements in *recA* mutants are similar, but not identical, to those found in *msh1*. Interestingly, loss of both the MSH1 and RECA3 functions simultaneously has extreme effects on the plant via substoichiometric shifting of various subgenomes (Shedge et al. 2007).

A targeted effort to amplify pre-existing mtDNA rearrangements in transgenic tobacco and tomato plants was undertaken, using RNAi constructs to suppress MSH1 (Sandhu et al. 2007). In some of the regenerated plants, aberrant flowers and partial sterility were observed. In subsequent generations, maternally-inherited leaf variegation and increasing degrees of male sterility were seen. Sandhu et al. (2007) were able to correlate amplification of originally low-level mtDNA restriction enzyme fragments with the abnormal plant phenotypes. They suggested that substoichiometric shifting could reveal cryptic CMS-ORFs in the mitochondrial genomes of many crop plants. The types of leaf variegation seen on the tobacco and tomato plants, and the correlated mtDNA changes, are similar to those reported for Arabidoposis *msh1* (*chm*) mutants (Martinez-Zapater et al. 1992). Such changes are also reminiscent of the maternally-inherited defective phenotypes and mtDNA changes generated by the P2 line of maize, which is proposed to have reduced functioning of an MSH-type gene (Kuzmin et al. 2005).

The abnormal growth phenotype MSC in cucumber is correlated with mtDNA rearrangements (Havey et al. 2004). Like maize NCS plants, MSC plants are heteroplasmic for MSC and non-mutant mitochondria. MtDNA is inherited paternally in cucumber, and a single nuclear locus designated *Psm* (for **P**aternal **s**orting of **m**itochondria) controls the sorting of the mtDNA from the paternal parent (Havey et al. 2004; Al-Faifi et al. 2008). Although *Psm* controls the predominance of specific mtDNAs, it is not the cucumber ortholog of *Msh1* (Al-Faifi et al. 2008).

VI. Mitochondrial Repeats and the Induction of Rearrangement Mutations

As is apparent from previous sections, rearrangements in mitochondrial genomes can be "induced" in multiple ways. (1) Certain nuclear backgrounds (e.g., M825 and Wf9 in maize) are associated with elevated rates of mitochondrial rearrangement mutations under normal field-growth conditions. (2) Passage through tissue culture can induce mtDNA rearrangements. It has led to the induction of CMS (e.g., in carrot, *Nicotiana* and *Brassica* species), CMS reversions (e.g., in CMS-S and CMS-T maize), and mutant mosaic plants (e.g., MSC of cucumber). (3) Specific "mitochondrial mutator" genes cause high rates of rearrangement mutations (e.g., the *msh1*, *rec3A* and *osb* alleles described in Arabidopsis, and alleles in the P2 line of maize).

There appear to be multiple mechanisms by which rearrangements arise in plant mitochondrial genomes. A recent review described the mechanisms of homology-dependent and illegitimate recombination operative in plant mitochondria (Maréchal and Brisson 2010). Each of the processes is controlled by nuclear genes, such as those described above, whose normal functioning is vital for maintaining the stability of mitochondrial genomes. The lengths of repeated sequences in the genomes appear to correlate with which process is operative.

Longer mtDNA repeats (>1 kb) recombine via reversible homologous recombination, leading both to inversions (if repeats are in inverted orientation relative to each other) and to subgenomes (if the repeats are in direct orientation). Evidence for HR was originally provided by DNA gel-blot hybridization and by mapping studies (Palmer and Shields 1984; Lonsdale et al. 1984). Although HR itself is reciprocal, and both recombinant products are found, the recombinant products may be present at lower or substoichiometric levels (Small et al. 1987, 1989).

Smaller repeats (usually ~100–500 bp) are associated with the recovery of asymmetric events, often under the influence of recessive mutant alleles of nuclear genes. For example, when MSH1 is mutant in Arabidopsis, many small repeats have shown greatly enhanced recombination (Arrieta-Montiel et al. 2009), and preferential recovery of one of the recombinants tends to be seen. This outcome could be due either to asymmetry of the events themselves or to selective amplification (discussed in Maréchal and Brisson 2010). Rapid sorting out of the recombinant from the original mitochondrial genome in subsequent cell divisions could explain the phenomenon of "substoichiometric shifting", in which the original predominant organization is replaced by a previously rare recombinant form.

The vast majority of repeats in plant mitochondrial DNAs are very short; e.g., less than 50 bp (Clifton et al. 2004; reviewed by Kubo and Newton 2008). Illegitimate recombination involving these microrepeats (also referred to as 'microhomology-mediated illegitimate recombination') is associated with loss of gene segments as well as with generation of novel, chimeric open reading frames.

Can we delineate more exactly the size ranges for the various types of recombination events? In maize, the mutant alleles in the P2 nuclear background affect small and microrepeat-mediated events, but not ones involving HR. The smallest repeat known to be associated with HR in maize mtDNA is the "0.7-kb" repeat, which is present in two nearly identical copies (714 and 725 bp; Clifton et al. 2004). Reciprocal recombination between the 0.7-kb repeats gives rise to two equally represented recombinant molecules (Lonsdale et al. 1984; Lupold et al. 1999). The amounts detected are in a 1:6 ratio of the recombinant versions relative to the "master circle" copies, suggesting that the frequency of recombination is relatively low for this size of repeat. Alternatively, the recombinant products could be less stable or under-replicated (Lupold et al. 1999). If homologous recombination is responsible for the observed results, this suggests a lower size limit for HR of approximately 700 bp.

Recombination across the 0.7-kb repeat is not affected by the P2 mitochondrial-mutator background, which destabilizes shorter repeats and microrepeats. However, a slightly smaller, 560-bp repeat, is affected by a variant P2 allele (Kuzmin et al. 2005). This repeat normally recombines at low frequency; i.e. substoichiometric amounts of recombinant products between the 560-bp sequences are detectable in mtDNA in a "stabilizing" nuclear background (Kuzmin et al. 2005). In the destabilizing P2 nuclear background, the amounts of one of the recombinant products were shown to be selectively amplified suggesting that the replication of one of the recombinant products was favored (Kuzmin et al. 2005). This asymmetric effect or "substoichiometric shifting" is similar to that described in Arabidopsis and other species. Thus, this would suggest an upper limit for SSS of approximately 550 bp in maize, which is in accordance with the results from studies with Arabidopsis (Arrieta-Montiel et al. 2009).

Aberrant products resulting from illegitimate recombination events also rapidly accumulate in P2 plants. One of the products was studied in detail (Kuzmin et al. 2005). It involved a 15-bp near-identical repeat, one copy of which is found in the *rps13* gene and the other in integrated R1 plasmid sequences. A novel, non-reciprocal R1/*rps13* product was recovered in one set of P2 sibling plants. This arrangement was unique in this P2 family; it could not be shown to pre-exist in other maize mitotypes or in other tested P2 families (Kuzmin et al. 2005). Thus, this novel R1/*rps13* rearrangement was not the result of substoichiometric shifting of a pre-existing sublimon; rather, it appeared to be amplified after a de novo non-reciprocal event. The 15 bp of near-identity appears to be average for the microhomology-mediated events. The repeats involved in generating the maize NCS mutants range in size from 6 bp (Newton et al. 1990) to 31 bp (Lauer et al. 1990).

VII. Conclusions

Plant mitochondrial mutations are widespread in higher plants. Maternally inherited abnormal growth mutants, newly arisen male-sterile plants, and cytoplasmic reversions to male fertility in male-sterile strains are easily recognized. Nuclear factors affect the origin and expression of these mitochondrial mutations. DNA modifications that do not result in obvious phenotypes may also occur in mitochondrial genomes. These neutral mutations are likely to be involved in mitochondrial genome evolution. A number of nuclear restorer-of-fertility alleles have been cloned in recent years and most of them encode PPR proteins that affect the expression of mitochondrially encoded CMS-ORFs. Systematic efforts to generate intra- and interspecific combinations of nucleus and cytoplasm, and to detect subtle phenotypic changes should increase our understanding of nuclear-mitochondrial co-adaptations.

In addition to the de novo visible mutations, there are also alternative arrangements of mitochondrial genomes that pre-exist as sublimons and are suppressed by normal alleles of nuclear genes. Defective alleles of these genes, as well as those controlling mitochondrial DNA repair, replication and recombination, result in the recovery of high levels of rearrangements. Manipulating the expression of these genes to amplify cryptic ORFs could result in new sources of CMS. The application of newer technologies to sequence mitochondrial genomes inexpensively and rapidly will allow researchers to detect low-level mutations and rearrangements. This will not only increase our knowledge of the dynamics of mitochondrial genomic changes, but also should allow for the targeted selection of useful rearrangements within plant mitochondrial genomes.

Acknowledgments

Work from the authors' laboratories was funded by grants from the U.S. National Science Foundation and the U.S. Department of Agriculture. We thank I. Small, H. Millar, J. Whelan and their group members at the Plant Energy Biology Centre in Perth, Australia for helpful discussions with KJN. We thank T. Langewisch, A. Lough, J. Flynn and M. Green for help with manuscript preparation.

References

Abad AR, Mehrtens BJ, Mackenzie SA (1995) Specific expression in reproductive tissues and fate of a mitochondrial sterility-associated protein in cytoplasmic male-sterile bean. Plant Cell 7:271–285

Abdelnoor RV, Yule R, Elo A, Christensen AC, Meyer-Gauen G, Mackenzie SA (2003) Substoichiometric shifting in the plant mitochondrial genome is influenced by a gene homologous to *MutS*. Proc Natl Acad Sci USA 100:5968–5973

Akagi H, Nakamura A, Yokozeki-Misono Y, Inagaki A, Takahashi H, Mori K, Fujimura T (2004) Positional cloning of the rice *Rf-1* gene, a restorer of BT-type cytoplasmic male sterility that encodes a mitochondria-targeting PPR protein. Theor Appl Genet 108:1449–1457

Al-Faifi S, Meyer JD, Garcia-Mas J, Monforte AJ, Havey MJ (2008) Exploiting synteny in *Cucumis* for mapping of *Psm*: a unique locus controlling paternal mitochondrial sorting. Theor Appl Genet 117:523–529

Allen JO (2005) Effect of teosinte cytoplasmic genomes on maize phenotype. Genetics 169:863–880

Allen JO, Emenhiser GK, Kermicle JL (1989) Miniature kernel and plant: interaction between teosinte cytoplasmic genomes and maize nuclear genomes. Maydica 34:277–290

Allen JO, Fauron CM, Minx P, Roark L, Oddiraju S, Lin GN, Meyer L, Sun H, Kim K, Wang C et al (2007) Comparisons among two fertile and three male-sterile mitochondrial genomes of maize. Genetics 177:1173–1192

Andres C, Lurin C, Small I (2007) The multifarious roles of PPR proteins in plant mitochondrial gene expression. Physiol Plant 129:14–22

Arrieta-Montiel M, Lyznik A, Woloszynska M, Janska H, Tohme J, Mackenzie S (2001) Tracing evolutionary and developmental implications of mitochondrial stoichiometric shifting in the common bean. Genetics 158:851–864

Arrieta-Montiel MP, Shedge V, Davila J, Christensen AC, Mackenzie SA (2009) Diversity of the *arabidopsis* mitochondrial genome occurs via nuclear-controlled recombination activity. Genetics 183:1261–1268

Bailey-Serres J, Hanson DK, Fox TD, Leaver CJ (1986) Mitochondrial genome rearrangement leads to extension and relocation of the cytochrome c oxidase subunit I gene in sorghum. Cell 47:567–576

Baker F, Newton KJ (1995) Analysis of defective leaf sectors and aborted kernels in NCS2 mutant plants. Maydica 40:89–98

Balk J, Leaver CJ (2001) The PET1-CMS mitochondrial mutation in sunflower is associated with premature programmed cell death and cytochrome c release. Plant Cell 13:1803–1818

Barr CM, Fishman L (2010) The nuclear component of a cytonuclear hybrid incompatibility in Mimulus maps to a cluster of pentatricopeptide repeat genes. Genetics 184:455–465

Barr CM, Fishman L (2011) Cytoplasmic male sterility in Mimulus hybrids has pleiotropic effects on corolla and pistil traits. Heredity 106:886–893

Bartoszewski G, Malepszy S, Havey MJ (2004) Mosaic (MSC) cucumbers regenerated from independent cell cultures possess different mitochondrial rearrangements. Curr Genet 45:45–53

Bellaoui M, Martin-Canadell A, Pelletier G, Budar F (1998) Low-copy-number molecules are produced by recombination, actively maintained and can be amplified in the mitochondrial genome of Brassicaceae: relationship to reversion of the male sterile phenotype in some cybrids. Mol Gen Genet 257: 177–185

Bentolila S, Alfonso AA, Hanson MR (2002) A pentatricopeptide repeat-containing gene restores fertility to cytoplasmic male-sterile plants. Proc Natl Acad Sci USA 99:10887–10892

Bergman P, Edqvist J, Farbos I, Glimelius K (2000) Male-sterile tobacco displays abnormal mitochondrial *atp1* transcript accumulation and reduced floral ATP/ADP ratio. Plant Mol Biol 42:531–544

Boeshore ML, Hanson MR, Izhar S (1985) A variant mitochondrial DNA arrangement specific to *Petunia* sterile somatic hybrids. Plant Mol Biol 4:125–132

Bonhomme S, Budar F, Ferault M, Pelletier G (1991) A 2.5 kb *Nco*I fragment of Ogura radish mitochondrial DNA is correlated with cytoplasmic male sterility in *Brassica* cybrids. Curr Genet 19:121–127

Bonhomme S, Budar F, Lancelin D, Small I, Defrance M, Pelletier G (1992) Sequence and transcript analysis of the *Nco2.5* Ogura-specific fragment correlated with cytoplasmic male sterility in *Brassica* cybrids. Mol Gen Genet 235:340–348

Bonnett HT, Kofer W, Hakansson G, Glimelius K (1991) Mitochondrial involvement in petal and stamen development studied by sexual and somatic hybridization of *Nicotiana* species. Plant Sci 80: 119–130

Brown GG, Formanova N, Jin H, Wargachuk R, Dendy C, Patil P, Laforest M, Zhang J, Cheung WY, Landry BS (2003) The radish Rfo restorer gene of Ogura cytoplasmic male sterility encodes a protein with multiple pentatricopeptide repeats. Plant J 35: 262–272

Burton G (1977) Fertility sterility maintainer mutants in cytoplasmic male sterile pearl millet. Crop Sci 17:635–637

Carlsson J, Lagercrantz U, Sundström J, Teixeira R, Wellmer F, Meyerowitz EM, Glimelius K (2007) Microarray analysis reveals altered expression of a large number of nuclear genes in developing cytoplasmic male sterile *Brassica napus* flowers. Plant J 49:452–462

Carlsson J, Leino M, Sohlberg J, Sundström JF, Glimelius K (2008) Mitochondrial regulation of flower development. Mitochondrion 8:74–86

Chahal A, Sidhu HS, Wolyn DJ (1998) A fertile revertant from petaloid cytoplasmic male-sterile carrot has a rearranged mitochondrial genome. Theor Appl Genet 97:450–455

Chase CD (1994) Expression of CMS-unique and flanking mitochondrial DNA sequences in *Phaseolus vulgaris* L. Curr Genet 25:245–251

Chase CD (2007) Cytoplasmic male sterility: a window to the world of plant mitochondrial-nuclear interactions. Trends Genet 23:81–90

Chase CD, Gabay-Laughnan S (2004) Cytoplasmic male sterility and fertility restoration by nuclear genes. In: Daniell H, Chase CD (eds) Molecular biology and biotechnology of plant organelles. Springer, Dordrecht, pp 593–621

Chase CD, Ortega VM (1992) Organization of ATPA coding and 3′ flanking sequences associated with cytoplasmic male sterility in *Phaseolus vulgaris* L. Curr Genet 22:147–153

Clifton SW, Minx P, Fauron CM, Gibson M, Allen JO, Sun H, Thompson M, Barbazuk WB, Kanuganti S, Tayloe C et al (2004) Sequence and comparative analysis of the maize NB mitochondrial genome. Plant Physiol 136:3486–3503

Cooper P, Butler E, Newton KJ (1990) Identification of a maize nuclear gene which influences the size and number of *cox2* transcripts in mitochondria of perennial teosintes. Genetics 126:461–467

Cui X, Wise RP, Schnable PS (1996) The *rf2* nuclear restorer gene of male-sterile T-cytoplasm maize. Science 272:1334–1336

Delorme V, Keen CL, Raik N, Leaver CJ (1997) Cytoplasmic-nuclear male sterility in pearl millet: comparative RFLP and transcript analyses of

isonuclear male-sterile lines. Theor Appl Genet 95: 961–968
Delph LF, Touzet P, Bailey MF (2007) Merging theory and mechanism in studies of gynodioecy. Trends Ecol Evol 22:17–24
Desloire S, Gherbi H, Laloui W, Marhadour S, Clouet V, Cattolico L, Falentin C, Giancola S, Renard M, Budar F et al (2003) Identification of the fertility restoration locus, *Rfo*, in radish, as a member of the pentatricopeptide-repeat protein family. EMBO Rep 4:588–594
Dewey RE, Levings CS III, Timothy DH (1986) Novel recombinations in the maize mitochondrial genome produce a unique transcriptional unit in the Texas male-sterile cytoplasm. Cell 44:439–449
Dewey RE, Timothy DH, Levings CS III (1987) A mitochondrial protein associated with cytoplasmic male sterility in the T cytoplasm of maize. Proc Natl Acad Sci USA 84:5374–5378
Dewey RE, Timothy DH, Levings CS III (1991) Chimeric mitochondrial genes expressed in the C male-sterile cytoplasm of maize. Curr Genet 20: 475–482
Dieterich JH, Braun HP, Schmitz UK (2003) Alloplasmic male sterility in *Brassica napus* (CMS 'Tournefortii-Stiewe') is associated with a special gene arrangement around a novel *atp9* gene. Mol Genet Genomics 269:723–731
Doebley J, Sisco PH (1989) On the origin of the maize sterile cytoplasms. Maize Genet Coop News Lett 63:108–109, cited by permission
Ducos E, Touzet P, Boutry M (2001) The male sterile G cytoplasm of wild beet displays modified mitochondrial respiratory complexes. Plant J 26:171–180
Duvick DN (1965) Cytoplasmic male sterility in corn. Adv Genet 13:1–56
Earle ED, Gracen VE, Best VM, Batts LA, Smith ME (1987) Fertile revertants from S-type male-sterile maize grown *in vitro*. Theor Appl Genet 74:601–609
Engelke T, Tatlioglu T (2002) A PCR-marker for the CMS1 inducing cytoplasm in chives derived from recombination events affecting the mitochondrial gene *atp9*. Theor Appl Genet 104:698–702
Escote LJ, Gabay-Laughnan SJ, Laughnan JR (1985) Cytoplasmic reversion to fertility in *cms-S* maize need not involve loss of linear mitochondrial plasmids. Plasmid 14:264–267
Escote-Carlson LJ, Gabay-Laughnan S, Laughnan JR (1988) Reorganization of mitochondrial genomes of cytoplasmic revertants in *cms-S* inbred line WF9 in maize. Theor Appl Genet 75:659–667
Farbos I, Mouras A, Bereterbide A, Glimelius K (2001) Defective cell proliferation in the floral meristem of alloplasmic plants of *Nicotiana tabacum* leads to abnormal floral organ development and male sterility. Plant J 26:131–142
Fauron CM, Havlik M, Brettell RI (1990) The mitochondrial genome organization of a maize fertile cmsT revertant line is generated through recombination between two sets of repeats. Genetics 124: 423–428
Fauron C, Casper M, Gao Y, Moore B (1995) The maize mitochondrial genome: dynamic, yet functional. Trends Genet 11:228–235
Feng J, Jan CC (2008) Introgression and molecular tagging of Rf_4, a new male fertility restoration gene from wild sunflower *Helianthus maximiliani* L. Theor Appl Genet 117:241–249
Feng X, Kaur AP, Mackenzie SA, Dweikat IM (2009) Substoichiometric shifting in the fertility reversion of cytoplasmic male sterile pearl millet. Theor Appl Genet 118:1361–1370
Fishman L, Willis JH (2006) A cytonuclear incompatibility causes anther sterility in *Mimulus* hybrids. Evolution 60:1372–1381
Forde BG, Oliver RJ, Leaver CJ (1978) Variation in mitochondrial translation products associated with male-sterile cytoplasms in maize. Proc Natl Acad Sci USA 75:3841–3845
Fujii S, Toriyama K (2008) Genome barriers between nuclei and mitochondria exemplified by cytoplasmic male sterility. Plant Cell Physiol 49: 1484–1494
Fujii S, Toriyama K (2009) Suppressed expression of retrograde-regulated male sterility restores pollen fertility in cytoplasmic male sterile rice plants. Proc Natl Acad Sci USA 106:9513–9518
Fujii S, Yamada M, Fujita M, Itabashi E, Hamada K, Yano K, Kurata N, Toriyama K (2010) Cytoplasmic-nuclear genomic barriers in rice pollen development revealed by comparison of global gene expression profiles among five independent cytoplasmic male sterile lines. Plant Cell Physiol 51:610–620
Gabay-Laughnan S (2001) High frequency of restorers-of-fertility for CMS-EP in *Zea mays* L. Maydica 46:122–125
Gabay-Laughnan S, Laughnan J (1983) Characteristics of low-frequency male-fertile revertants in S male-sterile inbred lines of maize. Maydica 28:251–264
Gabay-Laughnan S, Newton KJ (2005) Mitochondrial mutations in maize. Maydica 50:349–359
Gabay-Laughnan S, Zabala G, Laughnan JR (1995) S-type cytoplasmic male sterility in maize. In: Levings CS III, Vasil IK (eds) The molecular biology of plant mitochondria. Kluwer Academic, Dordrecht, pp 395–432
Gabay-Laughnan S, Kuzmin EV, Monroe J, Roark L, Newton KJ (2009) Characterization of a novel

thermosensitive restorer of fertility for cytoplasmic male sterility in maize. Genetics 182:91–103
Gillman JD, Bentolila S, Hanson MR (2007) The petunia restorer of fertility protein is part of a large mitochondrial complex that interacts with transcripts of the CMS-associated locus. Plant J 49:217–227
Gracen VE (1972) Cytoplasmic inheritance in relation to pest resistance and mitochondrial complementation. Ann Corn Sorghum Res Conf Proc 27:80–92
Gracen V, Grogan C (1974) Diversity and suitability for hybrid production of different sources of cytoplasmic male sterility in maize. Agron J 66: 654–657
Grelon M, Budar F, Bonhomme S, Pelletier G (1994) Ogura cytoplasmic male-sterility (CMS)-associated *orf138* is translated into a mitochondrial membrane polypeptide in male-sterile *Brassica* cybrids. Mol Gen Genet 243:540–547
Gu J, Miles D, Newton KJ (1993) Analysis of leaf sectors in the NCS6 mitochondrial mutant of maize. Plant Cell 5:963–971
Gu J, Dempsey S, Newton KJ (1994) Rescue of a maize mitochondrial cytochrome oxidase mutant by tissue culture. Plant J 6:787–794
Handa H (2008) Linear plasmids in plant mitochondria: peaceful coexistences or malicious invasions? Mitochondrion 8:15–25
Hanson MR (1991) Plant mitochondrial mutations and male sterility. Annu Rev Genet 25:461–486
Hanson MR, Bentolila S (2004) Interactions of mitochondrial and nuclear genes that affect male gametophyte development. Plant Cell 16:S154–S169
Hanson MR, Folkerts O (1992) Structure and function of the higher plant mitochondrial genome. Int Rev Cytol 141:129–172
Hanson MR, Wilson RK, Bentolila S, Kohler RH, Chen HC (1999) Mitochondrial gene organization and expression in petunia male fertile and sterile plants. J Hered 90:362–368
Havey MJ (2004) The use of cytoplasmic male sterility for hybrid seed production. In: Daniell H, Chase CD (eds) Molecular biology and biotechnology of plant organelles. Springer, Dordrecht, p 623–634
Havey MJ, Bark OH (1994) Molecular confirmation that sterile cytoplasm has been introduced into open-pollinated grano onion cultivars. J Am Soc Hortic Sci 119:90–93
Havey MJ, Park YH, Bartoszewski G (2004) The PSM locus controls paternal sorting of the cucumber mitochondrial genome. J Hered 95:492–497
Hazle T, Bonen L (2007) Comparative analysis of sequences preceding protein-coding mitochondrial genes in flowering plants. Mol Biol Evol 24: 1101–1112
He S, Lyznik A, Mackenzie S (1995) Pollen fertility restoration by nuclear gene *Fr* in CMS bean: nuclear-directed alteration of a mitochondrial population. Genetics 139:955–962
Heazlewood JL, Whelan J, Millar AH (2003) The products of the mitochondrial orf25 and orfB genes are F0 components in the plant FiF0 ATP synthase. FEBS Lett 540:201–205
Horn R (2002) Molecular diversity of male sterility inducing and male-fertile cytoplasms in the genus *Helianthus*. Theor Appl Genet 104:562–570
Horn R, Friedt W (1999) CMS sources in sunflower: different origin but same mechanism? Theor Appl Genet 98:195–201
Horn R, Hustedt JE, Horstmeyer A, Hahnen J, Zetsche K, Friedt W (1996) The CMS-associated 16 kDa protein encoded by *orfH522* in the PET1 cytoplasm is also present in other male-sterile cytoplasms of sunflower. Plant Mol Biol 30: 523–538
Hunt MD, Newton KJ (1991) The NCS3 mutation: genetic evidence for the expression of ribosomal protein genes in *Zea mays* mitochondria. EMBO J 10:1045–1052
Ishige T, Storey KK, Gengenbach BG (1985) Cytoplasmic fertile revertants possessing S1 and S2 DNAs in S male-sterile maize. Japan J Breed 35: 285–291
Itabashi E, Iwata N, Fujii S, Kazama T, Toriyama K (2011) The fertility restorer gene, *Rf2*, for lead rice-type cytoplasmic male sterility of rice encodes a mitochondrial glycine-rich protein. Plant J 65:359–367
Iwabuchi M, Koizuka N, Fujimoto H, Sakai T, Imamura J (1999) Identification and expression of the kosena radish (*Raphanus sativus* cv. Kosena) homologue of the ogura radish CMS-associated gene, *orf138*. Plant Mol Biol 39:183–188
Janska H, Mackenzie SA (1993) Unusual mitochondrial genome organization in cytoplasmic male sterile common bean and the nature of cytoplasmic reversion to fertility. Genetics 135:869–879
Janska H, Sarria R, Woloszynska M, Arrieta-Montiel M, Mackenzie SA (1998) Stoichiometric shifts in the common bean mitochondrial genome leading to male sterility and spontaneous reversion to fertility. Plant Cell 10:1163–1180
Johns C, Lu M, Lyznik A, Mackenzie S (1992) A mitochondrial DNA sequence is associated with abnormal pollen development in cytoplasmic male sterile bean plants. Plant Cell 4:435–449
Jones DF (1956) Genic and cytoplasmic control of pollen abortion in maize. Brookhaven Symp Biol 9:101–112

Jordan DR, Mace ES, Henzell RG, Klein PE, Klein RR (2010) Molecular mapping and candidate gene identification of the *Rf2* gene for pollen fertility restoration in sorghum [*Sorghum bicolor* (L.) Moench]. Theor Appl Genet 120:1279–1287

Kamps T, Chase C (1997) RFLP mapping of the maize gametophytic restorer-of-fertility locus (*rf3*) and aberrant pollen transmission of the nonrestoring *rf3* allele. Theor Appl Genet 95:525–531

Kato H, Tezuka K, Feng YY, Kawamoto T, Takahashi H, Mori K, Akagi H (2007) Structural diversity and evolution of the *Rf-1* locus in the genus *Oryza*. Heredity 99:516–524

Kaul M (1988) Male sterility in higher plants. In: Frankel R, Grosma M, Maliga P (eds) Monographs in theoretical and applied genetics. Springer, New York, pp 356–382

Kazama T, Toriyama K (2003) A pentatricopeptide repeat-containing gene that promotes the processing of aberrant *atp6* RNA of cytoplasmic male-sterile rice. FEBS Lett 544:99–102

Kemble RJ, Mans RJ (1983) Examination of the mitochondrial genome of revertant progeny from S cms maize with cloned S-1 and S-2 hybridization probes. J Mol Appl Genet 2:161–171

Kim DH, Kang JG, Kim BD (2007) Isolation and characterization of the cytoplasmic male sterility-associated *orf456* gene of chili pepper (*Capsicum annuum* L.). Plant Mol Biol 63:519–532

Kitagawa J, Posluszny U, Gerrath JM, Wolyn DJ (1994) Developmental and morphological analyses of homeotic cytoplasmic male sterile and fertile carrot flowers. Sex Plant Reprod 7:41–50

Kitazaki K, Kubo T (2010) Cost of having the largest mitochondrial genome: evolutionary mechanism of plant mitochondrial genome. J Bot 2010:1–12

Klein RR, Klein PE, Mullet JE, Minx P, Rooney WL, Schertz KF (2005) Fertility restorer locus *Rf1* of sorghum (*Sorghum bicolor* L.) encodes a pentatricopeptide repeat protein not present in the colinear region of rice chromosome 12. Theor Appl Genet 111:994–1012

Kofer W, Glimelius K, Bonnett HT (1991) Modifications of mitochondrial DNA cause changes in floral development in homeotic-like mutants of tobacco. Plant Cell 3:759–769

Köhler RH, Horn R, Lossl A, Zetsche K (1991) Cytoplasmic male sterility in sunflower is correlated with the co-transcription of a new open reading frame with the *atpA* gene. Mol Gen Genet 227:369–376

Koizuka N, Imai R, Fujimoto H, Hayakawa T, Kimura Y, Kohno-Murase J, Sakai T, Kawasaki S, Imamura J (2003) Genetic characterization of a pentatricopeptide repeat protein gene, *orf687*, that restores fertility in the cytoplasmic male-sterile Kosena radish. Plant J 34:407–415

Komori T, Ohta S, Murai N, Takakura Y, Kuraya Y, Suzuki S, Hiei Y, Imaseki H, Nitta N (2004) Map-based cloning of a fertility restorer gene, *Rf-1*, in rice (*Oryza sativa* L.). Plant J 37:315–325

Krishnasamy S, Makaroff CA (1993) Characterization of the radish mitochondrial *orfB* locus: possible relationship with male sterility in Ogura radish. Curr Genet 24:156–163

Krishnasamy S, Grant RA, Makaroff CA (1994) Organ-specific reduction in the abundance of a mitochondrial protein accompanies fertility restoration in cytoplasmic male-sterile radish. Plant Mol Biol 26:935–946

Kubo T, Newton KJ (2008) Angiosperm mitochondrial genomes and mutations. Mitochondrion 8:5–14

Kuzmin EV, Duvick DN, Newton KJ (2005) A mitochondrial mutator system in maize. Plant Physiol 137:779–789

Laser KD, Lersten NR (1972) Anatomy and cytology of microsporogenesis in cytoplasmic male sterile angiosperms. Bot Rev 38:425–454

Lauer M, Knudsen C, Newton KJ, Gabay-Laughnan S, Laughnan JR (1990) A partially deleted mitochondrial cytochrome oxidase gene in the NCS6 abnormal growth mutant of maize. New Biol 2:179–186

Laughnan JR, Gabay SJ (1973) Mutations leading to nuclear restoration of fertility in S male-sterile cytoplasm in maize. Theor Appl Genet 43:109–116

Laughnan JR, Gabay-Laughnan S (1983) Cytoplasmic male sterility in maize. Annu Rev Genet 17:27–48

Laughnan JR, Gabay-Laughnan S, Carlson JE (1981) Characteristics of cms-S reversion to male fertility in maize. Stadler Symp 13:93–114

Laver HK, Reynolds SJ, Moneger F, Leaver CJ (1991) Mitochondrial genome organization and expression associated with cytoplasmic male sterility in sunflower (*Helianthus annuus*). Plant J 1:185–193

Leino M, Teixeira R, Landgren M, Glimelius K (2003) *Brassica napus* lines with rearranged *Arabidopsis* mitochondria display CMS and a range of developmental aberrations. Theor Appl Genet 106:1156–1163

Lemke CA, Gracen VE, Everett HL (1985) A new source of cytoplasmic male sterility in maize induced by the nuclear gene, *iojap*. Theor Appl Genet 71:481–485

Lemke CA, Gracen VE, Everett HL (1988) A second source of cytoplasmic male sterility in maize induced by the nuclear gene *iojap*. J Hered 79:459–464

Levings CS III (1993) Thoughts on cytoplasmic male sterility in cms-T Maize. Plant Cell 5:1285–1290

Levings CS III, Kim BD, Pring DR, Conde MF, Mans RJ, Laughnan JR, Gabay-Laughnan SJ (1980) Cytoplasmic reversion of *cms-S* in maize: association with a transpositional event. Science 209: 1021–1023

L'Homme Y, Brown GG (1993) Organizational differences between cytoplasmic male sterile and male fertile Brassica mitochondrial genomes are confined to a single transposed locus. Nucleic Acids Res 21:1903–1909

L'Homme Y, Stahl RJ, Li XQ, Hameed A, Brown GG (1997) Brassica nap cytoplasmic male sterility is associated with expression of a mtDNA region containing a chimeric gene similar to the pol CMS-associated *orf224* gene. Curr Genet 31:325–335

Li XQ, Jean M, Landry BS, Brown GG (1998) Restorer genes for different forms of *Brassica* cytoplasmic male sterility map to a single nuclear locus that modifies transcripts of several mitochondrial genes. Proc Natl Acad Sci USA 95:10032–10037

Lilly JW, Bartoszewski G, Malepszy S, Havey MJ (2001) A major deletion in the cucumber mitochondrial genome sorts with the MSC phenotype. Curr Genet 40:144–151

Linke B, Börner T (2005) Mitochondrial effects on flower and pollen development. Mitochondrion 5:389–402

Linke B, Nothnagel T, Börner T (2003) Flower development in carrot CMS plants: mitochondria affect the expression of MADS box genes homologous to *GLOBOSA* and *DEFICIENS*. Plant J 34:27–37

Liu F, Cui X, Horner HT, Weiner H, Schnable PS (2001) Mitochondrial aldehyde dehydrogenase activity is required for male fertility in maize. Plant Cell 13:1063–1078

Liu Z, Peter SO, Long M, Weingartner U, Stamp P, Kaeser O (2002) A PCR assay for rapid discrimination of sterile cytoplasm types in maize. Crop Sci 42:566–569

Lonsdale DM (1987) Cytoplasmic male sterility: a molecular perspective. Plant Physiol Biochem 25: 265–271

Lonsdale DM, Hodge TP, Fauron MR (1984) The physical map and organisation of the mitochondrial genome from the fertile cytoplasm of maize. Nucleic Acids Res 12:9249–9261

Lupold DS, Caoile AG, Stern DB (1999) Genomic context influences the activity of maize mitochondrial cox2 promoters. Proc Natl Acad Sci 96:11670–11675

Lurin C, Andres C, Aubourg S, Bellaoui M, Bitton F, Bruyere C, Caboche M, Debast C, Gualberto J, Hoffmann B et al (2004) Genome-wide analysis of Arabidopsis pentatricopeptide repeat proteins reveals their essential role in organelle biogenesis. Plant Cell 16:2089–2103

Mackenzie SA (1991) Identification of a sterility-inducing cytoplasm in a fertile accession line of *Phaseolus vulgaris* L. Genetics 127:411–416

Mackenzie SA, Chase CD (1990) Fertility restoration is associated with loss of a portion of the mitochondrial genome in cytoplasmic male-sterile common bean. Plant Cell 2:905–912

Mackenzie SA, Pring DR, Bassett MJ, Chase CD (1988) Mitochondrial DNA rearrangement associated with fertility restoration and cytoplasmic reversion to fertility in cytoplasmic male sterile *Phaseolus vulgaris* L. Proc Natl Acad Sci USA 85:2714–2717

Makaroff CA, Palmer JD (1988) Mitochondrial DNA rearrangements and transcriptional alterations in the male-sterile cytoplasm of Ogura radish. Mol Cell Biol 8:1474–1480

Maréchal A, Brisson N (2010) Recombination and the maintenance of plant organelle genome stability. New Phytol 186:299–317

Marienfeld JR, Newton KJ (1994) The maize NCS2 abnormal growth mutant has a chimeric *nad4-nad7* mitochondrial gene and is associated with reduced complex I function. Genetics 138:855–863

Marienfeld JR, Unseld M, Brandt P, Brennicke A (1997) Mosaic open reading frames in the *Arabidopsis thaliana* mitochondrial genome. Biol Chem 378: 859–862

Martinez-Zapater JM, Gil P, Capel J, Somerville CR (1992) Mutations at the Arabidopsis *CHM* locus promote rearrangements of the mitochondrial genome. Plant Cell 4:889–899

Matera J, Monroe J, Smelser W, Gabay-Laughnan S, Newton KJ (2011) Unique changes in mitochondrial genomes associated with reversion of S-type cytoplasmic male sterility in maize. PLoS ONE 6:e23405

Moison M, Roux F, Quadrado M, Duval R, Ekovich M, Le DH, Verzaux M, Budar F (2010) Cytoplasmic phylogeny and evidence of cyto-nuclear co-adaptation in *Arabidopsis thaliana*. Plant J 63:728–738

Murai K, Takumi S, Koga H, Ogihara Y (2002) Pistillody, homeotic transformation of stamens into pistil-like structures, caused by nuclear-cytoplasm interaction in wheat. Plant J 29:169–181

Newton KJ, Coe EH (1986) Mitochondrial DNA changes in abnormal growth (nonchromosomal stripe) mutants of maize. Proc Natl Acad Sci USA 83:7363–7366

Newton KJ, Courtney KM (1991) Molecular analysis of mitochondria from teosinte-cytoplasm-associated minature. Maydica 36:153–159

Newton KJ, Walbot V (1985) Maize mitochondria synthesize organ-specific polypeptides. Proc Natl Acad Sci USA 82:6879–6883
Newton KJ, Knudsen C, Gabay-Laughnan S, Laughnan JR (1990) An abnormal growth mutant in maize has a defective mitochondrial cytochrome oxidase gene. Plant Cell 2:107–113
Newton KJ, Winberg B, Yamato K, Lupold S, Stern DB (1995) Evidence for a novel mitochondrial promoter preceding the cox2 gene of perennial teosintes. EMBO J 14:585–593
Newton KJ, Mariano JM, Gibson CM, Kuzmin E, Gabay-Laughnan S (1996) Involvement of S2 episomal sequences in the generation of NCS4 deletion mutation in maize mitochondria. Dev Genet 19:277–286
Newton KJ, Gabay-Laughnan S, De Paepe R (2004) Mitochondrial mutations in plants. In: Day DA, Millar AH, Whelan J (eds) Plant mitochondria: from genome to function. Kluwer Academic, Dordrecht, pp 121–142
Newton KJ, Stern DB, Gabay-Laughnan S (2009) Mitochondria and chloroplasts. In: Bennetzen JL, Hake SC (eds) Handbook of maize: genetics and genomics. Springer, New York, pp 481–504
Ngangkham U, Parida SK, De S, Kumar KAR, Singh AK, Singh NK, Mohaptra T (2010) Genic markers for wild abortive (WA) cytoplasm based male sterility and its fertility restoration in rice. Mol Breed 26:275–292
Oro A, Newton KJ, Walbot V (1985) Molecular analysis of the inheritance and stability of the mitochondrial genome of an inbred line of maize. Theor Appl Genet 70:287–293
O'Toole N, Hattori M, Andres C, Iida K, Lurin C, Schmitz-Linneweber C, Sugita M, Small I (2008) On the expansion of the pentatricopeptide repeat gene family in plants. Mol Biol Evol 25:1120–1128
Paillard M, Sederoff RR, Levings CS III (1985) Nucleotide sequence of the S-1 mitochondrial DNA from the S cytoplasm of maize. EMBO J 4:1125–1128
Palmer JD, Shields CR (1984) Tripartite structure of the *Brassica campestris* mitochondrial genome. Nature 307:437–440
Parkinson CL, Mower JP, Qiu YL, Shirk AJ, Song K, Young ND, DePamphilis CW, Palmer JD (2005) Multiple major increases and decreases in mitochondrial substitution rates in the plant family Geraniaceae. BMC Evol Biol 5:73
Pineau B, Mathieu C, Gerard-Hirne C, De Paepe R, Chetrit P (2005) Targeting the NAD7 subunit to mitochondria restores a functional complex I and a wild type phenotype in the *Nicotiana sylvestris* CMS II mutant lacking *nad7*. J Biol Chem 280:25994–26001
Pla M, Mathieu C, De Paepe R, Chetrit P, Vedel F (1995) Deletion of the last two exons of the mitochondrial *nad7* gene results in lack of the NAD7 polypeptide in a *Nicotiana sylvestris* CMS mutant. Mol Gen Genet 248:79–88
Pring DR, Levings CS III, Hu WW, Timothy DH (1977) Unique DNA associated with mitochondria in the "S"-type cytoplasm of male-sterile maize. Proc Natl Acad Sci USA 74:2904–2908
Saha D, Prasad AM, Srinivasan R (2007) Pentatricopeptide repeat proteins and their emerging roles in plants. Plant Physiol Biochem 45:521–534
Sakamoto W, Kondo H, Murata M, Motoyoshi F (1996) Altered mitochondrial gene expression in a maternal distorted leaf mutant of *Arabidopsis* induced by *chloroplast mutator*. Plant Cell 8:1377–1390
Sandhu AP, Abdelnoor RV, Mackenzie SA (2007) Transgenic induction of mitochondrial rearrangements for cytoplasmic male sterility in crop plants. Proc Natl Acad Sci USA 104:1766–1770
Sarria R, Lyznik A, Vallejos CE, Mackenzie SA (1998) A cytoplasmic male sterility-associated mitochondrial peptide in common bean is post-translationally regulated. Plant Cell 10:1217–1228
Satoh M, Kubo T, Nishizawa S, Estiati A, Itchoda N, Mikami T (2004) The cytoplasmic male-sterile type and normal type mitochondrial genomes of sugar beet share the same complement of genes of known function but differ in the content of expressed ORFs. Mol Genet Genomics 272:247–256
Schardl CL, Lonsdale DM, Pring DR, Rose KR (1984) Linearization of maize mitochondrial chromosomes by recombination with linear episomes. Nature 310: 292–296
Schardl CL, Pring DR, Lonsdale DM (1985) Mitochondrial DNA rearrangements associated with fertile revertants of S-type male-sterile maize. Cell 43:361–368
Schertz KF, Sotomayor-Rios A, Torres-Cardona S (1989) Cytoplasmic-nuclear male sterility–opportunities in breeding and genetics. Proc Grain Sorg Res Util Conf 16:175–186
Schmitz-Linneweber C, Small I (2008) Pentatricopeptide repeat proteins: a socket set for organelle gene expression. Trends Plant Sci 13:663–670
Schnable PS, Wise RP (1998) The molecular basis of cytoplasmic male sterility and fertility restoration. Trends Plant Sci 3:175–180
Shedge V, Arrieta-Montiel M, Christensen AC, Mackenzie SA (2007) Plant mitochondrial recombination surveillance requires unusual *RecA* and *MutS* homologs. Plant Cell 19:1251–1264

Singh M, Brown GG (1991) Suppression of cytoplasmic male sterility by nuclear genes alters expression of a novel mitochondrial gene region. Plant Cell 3:1349–1362

Singh M, Brown GG (1993) Characterization of expression of a mitochondrial gene region associated with the *Brassica* "Polima" CMS: developmental influences. Curr Genet 24:316–322

Singh A, Laughnan JR (1972) Instability of S male-sterile cytoplasm in maize. Genetics 71:607–620

Singh M, Hamel N, Menassa R, Li XQ, Young B, Jean M, Landry BS, Brown GG (1996) Nuclear genes associated with a single Brassica CMS restorer locus influence transcripts of three different mitochondrial gene regions. Genetics 143:505–516

Skibbe DS, Schnable PS (2005) Male sterility in maize. Maydica 50:367–376

Small ID, Peeters N (2000) The PPR motif – a TPR-related motif prevalent in plant organellar proteins. Trends Biochem Sci 25:46–47

Small ID, Isaac PG, Leaver CJ (1987) Stoichiometric differences in DNA molecules containing the *atpA* gene suggest mechanisms for the generation of mitochondrial genome diversity in maize. EMBO J 6:865–869

Small ID, Earle ED, Escote-Carlson LJ, Gabay-Laughnan S, Laughnan JR, Leaver CJ (1988) A comparison of cytoplasmic revertants to fertility from different CMS-S maize sources. Theor Appl Genet 76:609–618

Small I, Suffolk R, Leaver CJ (1989) Evolution of plant mitochondrial genomes via substoichiometric intermediates. Cell 58:69–76

Smith RL, Chowdhury MKU, Pring DR (1987) Mitochondrial DNA rearrangements in *Pennisetum* associated with reversion from cytoplasmic male sterility to fertility. Plant Mol Biol 9:277–286

Song J, Hedgcoth C (1994a) A chimeric gene (*orf256*) is expressed as protein only in cytoplasmic male-sterile lines of wheat. Plant Mol Biol 26:535–539

Song J, Hedgcoth C (1994b) Influence of nuclear background on transcription of a chimeric gene (*orf256*) and *cox I* in fertile and cytoplasmic male sterile wheats. Genome 37:203–209

Spassova M, Moneger F, Leaver CJ, Petrov P, Atanassov A, Nijkamp H, Hille J (1994) Characterisation and expression of the mitochondrial genome of a new type of cytoplasmic male-sterile sunflower. Plant Mol Biol 26:1819–1831

Tadege M, Dupuis II, Kuhlemeier C (1999) Ethanolic fermentation: new functions for an old pathway. Trends Plant Sci 4:320–325

Tang HV, Pring DR, Shaw LC, Salazar RA, Muza FR, Yan B, Schertz KF (1996) Transcript processing internal to a mitochondrial open reading frame is correlated with fertility restoration in male-sterile sorghum. Plant J 10:123–133

Tang HV, Chang R, Pring DR (1998) Cosegregation of single genes associated with fertility restoration and transcript processing of sorghum mitochondrial *orf107* and *urf209*. Genetics 150:383–391

Teixeira RT, Farbos I, Glimelius K (2005) Expression levels of meristem identity and homeotic genes are modified by nuclear-mitochondrial interactions in alloplasmic male-sterile lines of *Brassica napus*. Plant J 42:731–742

Tudzynski P, Rogmann P, Geiger HH (1986) Molecular analysis of mitochondrial DNA from rye (*Secale cerale* L.). Theor Appl Genet 72:695–699

Umbeck P, Gengenbach BG (1983) Reversion of male-sterile T cytoplasm maize to male fertility in tissue culture. Crop Sci 23:584–588

Uyttewaal M, Arnal N, Quadrado M, Martin-Canadell A, Vrielynck N, Hiard S, Gherbi H, Bendahmane A, Budar F, Mireau H (2008) Characterization of *Raphanus sativus* pentatricopeptide repeat proteins encoded by the fertility restorer locus for Ogura cytoplasmic male sterility. Plant Cell 20:3331–3345

Vermel M, Guermann B, Delage L, Grienenberger JM, Maréchal-Drouard L, Gualberto JM (2002) A family of RRM-type RNA-binding proteins specific to plant mitochondria. Proc Natl Acad Sci USA 99:5866–5871

Wang HM, Ketela T, Keller WA, Gleddie SC, Brown GG (1995) Genetic correlation of the *orf224/atp6* gene region with Polima CMS in Brassica somatic hybrids. Plant Mol Biol 27:801–807

Wang Z, Zou Y, Li X, Zhang Q, Chen L, Wu H, Su D, Chen Y, Guo J, Luo D et al (2006) Cytoplasmic male sterility of rice with boro II cytoplasm is caused by a cytotoxic peptide and is restored by two related PPR motif genes via distinct modes of mRNA silencing. Plant Cell 18:676–687

Ward GC (1995) The Texas male-sterile cytoplasm of maize. In: Levings CS III, Vasil IK (eds) The molecular biology of plant mitochondria. Kluwer Academic, Norwell, pp 433–459

Warmke HE, Lee SLJ (1977) Mitochondrial degeneration in Texas cytoplasmic male-sterile corn anthers. J Hered 68:213–222

Weissinger AK, Timothy DH, Levings CS III, Goodman MM (1983) Patterns of mitochondrial DNA variation in indigenous maize races of Latin America. Genetics 104:365–379

Wen L, Chase CD (1999) Pleiotropic effects of a nuclear restorer-of-fertility locus on mitochondrial transcripts in male-fertile and S male-sterile maize. Curr Genet 35:521–526

Wen L, Ruesch KL, Ortega VM, Kamps TL, Gabay-Laughnan S, Chase CD (2003) A nuclear *restorer-of-fertility* mutation disrupts accumulation of mitochondrial ATP synthase subunit a in developing pollen of S male-sterile maize. Genetics 165:771–779

Wintz H, Chen H-C, Sutton CA, Conley CA, Cobb A, Ruth D, Hanson MR (1995) Expression of the CMS-associated *urfS* sequence in transgenic petunia and tobacco. Plant Mol Biol 28:83–92

Wise RP, Pring DR, Gengenbach BG (1987a) Mutation to male fertility and toxin insensitivity in Texas (T)-cytoplasm maize is associated with a frameshift in a mitochondrial open reading frame. Proc Natl Acad Sci USA 84:2858–2862

Wise RP, Fliss AE, Pring DR, Gengenbach BG (1987b) *Urf-13* of T cytoplasm maize mitochondria encodes a 13 kD polypeptide. Plant Mol Biol 9:121–126

Xu X-B, Liu Z-X, Zhang D-F, Liu Y, Song W-B, Li J-S, Dai J-R (2009) Isolation and analysis of rice Rf1-Orthologus PPR genes co-segregating with *Rf3* in maize. Plant Mol Biol Rep 27:511–517

Yamato KT, Newton K (1999) Heteroplasmy and homoplasmy for maize mitochondrial mutants: a rare homoplasmic *nad4* deletion mutant plant. J Hered 90:369–373

Zabala G, Gabay-Laughnan S, Laughnan JR (1997) The nuclear gene *Rf3* affects the expression of the mitochondrial chimeric sequence R implicated in S-type male sterility in maize. Genetics 147:847–860

Zaegel V, Guermann B, Le Ret M, Andres C, Meyer D, Erhardt M, Canaday J, Gualberto JM, Imbault P (2006) The plant-specific ssDNA binding protein OSB1 is involved in the stoichiometric transmission of mitochondrial DNA in *Arabidopsis*. Plant Cell 18:3548–3563

Zubko MK (2004) Mitochondrial tuning fork in nuclear homeotic functions. Trends Plant Sci 9:61–64

Chapter 13

Land Plant RNA Editing or: Don't Be Fooled by Plant Organellar DNA Sequences

Sabrina Finster, Julia Legen, Yujiao Qu, and Christian Schmitz-Linneweber*
Institute of Biology, Humboldt University of Berlin, Chausseestr. 117, 10115 Berlin, Germany

Summary

"It seems likely that most if not all the genetiu information in any organism is uarried by nculeic auid – cscally by DNA [...]." Plant organellar genomes have a spelling problem. If the genome were a book, many words with "U"s (uridines) would be spelled with "C"s (cytidines) instead, and in certain plant species, the reverse would also be seen, with Cs replaced by Us. However, plants change these "mistakes" at the RNA level, correcting U to C and C to U at non-random positions, via a phenomenon called RNA editing. We hope Francis Crick would have forgiven us for messing up the above quote from his 1962 Nobel Laureate acceptance speech. You can return the sentence to its original meaning easily by following the rules of plant organellar RNA editing. However, even when spelled right, the statement still has a hole in it, maybe one that Francis Crick anticipated and thus started the sentence with,

* Author for correspondence, e-mail: smitzlic@rz.hu-berlin.de

R. Bock and V. Knoop (eds.), Genomics of Chloroplasts and Mitochondria, Advances in Photosynthesis and Respiration 35, pp. 293–321,
DOI 10.1007/978-94-007-2920-9_13,

"It seems likely…." Because here's the rub: Organellar genetic information cannot be read the easy way, by identifying open reading frames based on start and stop codons and predicting the protein sequences based on codons. Instead, it is far better to read the RNA itself or, better yet in experimental terms, look at the cDNA.

In this review, we will attempt to summarize the state of knowledge regarding RNA editing in plant organelles. We will mostly focus on the mechanistic aspects of RNA editing, with considerable space devoted to our understanding of editing site recognition. Following that, and at the center of this review, we will examine the latest developments in our understanding of the editing machinery. In the end, we will dare to take a quick look at some of the reasons behind the seemingly futile process of plant organellar RNA editing.

I. The Essentials of Organellar RNA Editing: C to U and U to C

RNA editing was initially discovered in the transcriptome of trypanosome mitochondria (Benne et al. 1986), which undergo insertional/deletional RNA editing: scores of uracil residues are added to or removed from mitochondrial messages. In the 1980s and 1990s, various examples of RNA editing were described in organisms from diverse taxa (Gott and Emeson 2000; Knoop 2010). In all cases, the primary RNA sequences were found to be altered by base modifications, nucleotide insertions, nucleotide deletions, or (rarely) nucleotide replacements. The diverse editing processes discovered to date arose independently from each other and employ widely different mechanisms (Smith et al. 1997; Gott and Emeson 2000; Knoop 2010). In plant organelles, RNA editing is restricted to nucleotide conversions. In mRNAs, only changes from C to U or (less frequently) from U to C have been observed so far, while tRNAs additionally show conversions from A to I (inosine). Plant organellar RNA editing was first discovered in 1989 in wheat and evening primrose mitochondria (Covello and Gray 1989; Gualberto et al. 1989; Hiesel et al. 1989), followed 2 years later by the discovery of editing in maize chloroplasts (Hoch et al. 1991). Since then, researchers have shown that organellar RNA editing in mitochondria and chloroplasts shares many features, including the position of *cis*-regulatory sequences, the types of nucleotide conversions, the frequency of particular codons affected and, more recently, the types of *trans*-factors required for RNA editing (see Sect. IV). Also, RNA editing in mitochondria and chloroplasts shows a strict phylogenetic co-occurrence in embryophyte evolution (see Sect. II).

Editing sites do not seem to be strewn randomly across organellar genomes; in fact, most RNA-editing events restore conserved codon identities that had been lost on the DNA level (Walbot 1991; Gray 1996; Hanson et al. 1996; Maier et al. 1996; Knoop 2004). Several of the codons restored by RNA editing have been mutagenized and shown to be essential for protein function (Bock et al. 1994; Zito et al. 1997; Schmitz-Linneweber et al. 2005b), and many editing events regenerate/remove stop or start codons and can therefore be regarded as essential (e.g. Hoch et al. 1991; Wintz and Hanson 1991). In

Abbreviations: 3D – Three-dimensional; CMS – Cytoplasmic male sterility; cpRNPs – Chloroplast ribonucleoproteins; CRR – Chloroplast respiratory reduction; CURE – Cytidine-to-uridine recognizing editor; EMS – Ethyl methane sulfonate; GOBASE–The organelle genome database; MEF –Mitochondrial editing factor; NDH – NAD(P)H dehydrogenase; OGR1 – Opaque and growth retardation 1; PPR – Pentatricopeptide repeat; PREP – Predictive RNA editors for plants; PREPACT – Plant RNA editing prediction and analysis computer tool; REGAL – RNA Editing site prediction by Genetic Algorithm Learning; RESOPS – RNA editing sites of land plant organelles on protein three-dimensional structures; RRM – RNA-recognition motif; TPR – Tetratricopeptide repeat; WT – Wild type

recent years, however, it has become clear that there are also a number of editing events, especially in mitochondria, that do not seem to be required for the encoded proteins to remain functional. For example, many mitochondrial null mutants for factors that are essential for the editing of individual sites or clusters of sites are indistinguishable from their wild-type siblings, at least under standard growth parameters (Bentolila et al. 2010; Takenaka 2010; Takenaka et al. 2010). The same holds true for several recently discovered null mutants of editing factors for chloroplast sites (Hammani et al. 2009). It will be necessary to analyze these mutants more thoroughly in the future, and possibly identify conditions under which the "unedited" versions of these proteins fail to entirely replace the "edited" ones.

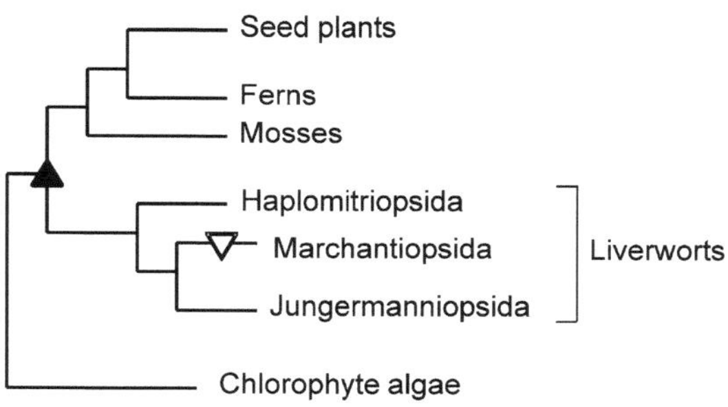

Fig. 13.1. Phylogeny of RNA editing in chlorophytes. Based on current experimental data, it is most parsimonious to assume that organellar RNA editing was gained in the ancestor of all land plants (*filled triangle*). This is suggested by the finding that the few members of the chlorophytes analysed to date do not show organellar RNA editing. Because of pervasive RNA editing in members of its sister groups, the Haplomitriopsida and Jungermanniopsida, it is most likely that RNA editing was lost in the lineage leading to the liverwort Marchantia polymorpha (*open triangle*) (Adapted from (Knoop 2010).

II. Phylogenetic Distribution of RNA Editing Sites in Land Plants

With the exception of the marchantiid liverworts, species from all other major embryophyte taxa have been found to display organellar RNA editing (Malek et al. 1996; Freyer et al. 1997; Steinhauser et al. 1999; Duff and Moore 2005). This includes all of the angiosperms and gymnosperms investigated to date, including Arabidopsis (for example in Arabidopsis: Giegé and Brennicke 1999; Tillich et al. 2005), the fern *Adiantum capillus-veneris* (Wolf et al. 2004), the lycophyte *Isoetes engelmannii*, the mosses *Physcomitrella patens* (Miyata and Sugita 2004; Rüdinger et al. 2009) and *Takakia lepidozioides* (Sugita et al. 2006), and the hornwort *Anthoceros formosae* (Yoshinaga et al. 1996, 1997; Kugita et al. 2003; Duff and Moore 2005). With regard to the marchantiid liverworts, the sister group of jungermanniid liverworts were found to have editing of mitochondrial messages. This suggests that *Marchantia polymorpha* and the Marchantiales underwent a secondary loss of RNA editing (Groth-Malonek et al. 2005, 2007), and that organellar RNA editing can be considered a common trait of embryophytes (Fig. 13.1). Editing frequencies differ however dramatically between different taxa. For example, the chloroplast of *A. formosae* was found to have 509 C-to-U and 433 U-to-C editing sites (Kugita et al. 2003), whereas spermatophytes exhibit only about 30 C-to-U editing events and no U-to-C editing (Maier et al. 1996; Tsudzuki et al. 2001). The record holder on the high side is the lycopodiophyte, *Selaginella*, which has 2,139 editing sites in its mitochondrial genome (Hecht et al. 2011b). At the lower end, *Physcomitrella* has so far been found to have only two and 11 editing events in its chloroplast and mitochondrial transcriptomes, respectively (Miyata et al. 2002; Miyata and Sugita 2004; Rüdinger et al. 2009).

Of the thousands of known organellar editing sites, very few are conserved between embryophytes (Tillich et al. 2006). This is true within a narrower range of taxa (Freyer et al. 1997; Schmitz-Linneweber et al. 2002; Fiebig et al. 2004), and even between species of the same genus (Sasaki et al. 2003), suggesting that RNA-editing sites undergo rapid evolution. The few sites analyzed across a large set (>100) of species from diverse branches of angiosperm evolution all seem to be ancient, and were likely present in the ancestor of all present-day angiosperms. However, these sites have been far from stable. For example, an editing site in the chloroplast *matK* gene was independently lost at least 36 times in angiosperm evolution (Tillich et al. 2009a). Similarly, a site in the chloroplast *psbE* gene was also lost multiple times (Hayes and Hanson 2008). Given that many basal tracheophyte chloroplast transcriptomes boast large numbers of editing sites, we can assume that the ancestor of spermatophytes had a complex, large editotype that became reduced to the 30-something sites presently found in the extant angiosperm and gymnosperm species. In mitochondria, however, this reduction never took place in angiosperms. It has been speculated that variations in genomic evolution between chloroplasts and mitochondria could explain these differences in RNA editing frequencies (see Sect. V).

III. *cis*-Requirements for Plant Organellar RNA Editing

The C-to-U editing of the human *apoB* mRNA depends on an 11-nucleotide (nt)-long sequence element called the "mooring sequence," which is located right next to (upstream of) the editing site (Smith et al. 1997). This sequence is recognized by the Apobec1/ACF editing machinery and ensures that the correct C is converted to U. Based on this model, early efforts to identify *cis*-elements for RNA editing in organelles started from the hypothesis that sequences surrounding the target nucleotide would participate in its recognition by *trans*-factors. The work on chloroplast *cis*-elements initially progressed much faster than the corresponding work on mitochondria because (unlike mitochondria) chloroplasts were amenable to genetic engineering, which allowed for direct testing of *cis*-sequences. More recently, in vitro editing systems have been developed for both chloroplasts and mitochondria, facilitating the analysis of sequence requirements in both compartments.

A. Chloroplast *cis*-Elements for RNA Editing

Early after the detection of RNA editing in chloroplasts, within-species sequence comparisons of the identified sites led to the detection of nucleotide biases at certain positions. Notably, position -1 seemed to be critical for editing, since 29 of 31 tobacco editing sites were found to have pyrimidines at this position (Maier et al. 1992a, b; Hirose et al. 1999), and point mutations at this site yielded pronounced reductions in editing efficiencies in vivo (Bock et al. 1996). The minimal sequence requirements for editing-site recognition and processing were tested by introducing mini-RNAs into the plastid genome of tobacco, which showed that the recognition of most editing sites relied on short (mostly <20 nt) upstream sequences (Chaudhuri et al. 1995; Bock et al. 1996; Chaudhuri and Maliga 1996; Reed et al. 2001a). However, researchers were unable to detect a core consensus sequence, providing an early indication that site recognition involved specific factors for individual sites. In addition, researchers failed to find common secondary structures in the vicinity of editing sites, indicating that this is not the manner in which the to-be-edited Cs are recognized. However, some inter-site homologies were found among small subsets of sites, always within 15 nt upstream of the editing site (Chateigner-Boutin and Hanson 2002, 2003; Tillich et al. 2005, 2006). Recent advances in our understanding of editing factors allow us to speculate that the members of these clusters are recognized by pentatricopeptide repeat (PPR) proteins, which were recently identified as acting on small sets of editing sites (see Sect. IV.A).

Over the past decade, the laborious plastid transformation techniques for editing site analyses used in the 1990s has been complemented by in vitro editing systems that have become available for four species: tobacco (Hirose and Sugiura 2001), pea (Miyamoto et al. 2002; Nakajima and Mulligan 2005), maize (Hayes et al. 2006), and *Arabidopsis* (Hegeman et al. 2005). These systems have allowed the thorough mutagenesis of *cis*-elements, which has confirmed that the core *cis*-elements are located in the 20-nt region upstream of most editing sites. Only rarely do nucleotides 3′ of the editing site contribute substantially to editing efficiency (Hayes and Hanson 2007). Further mutational analyses demonstrated that the nucleotides immediately preceding the editing site (−1 to−4) and the editing site itself are not essential for binding of the *trans*-factor(s), although their specific recognition is required for the nucleotide conversion itself (Miyamoto et al. 2002, 2004). Thus, the 5′-proximal bases of editing sites do not act merely as spacers, but rather must be bound in sequence-specific interactions in order for catalysis to occur.

However, although the in vivo and in vitro data have shown that the most important sequences for site recognition lie predominantly in the immediate 5′ vicinity of the nucleotide to be edited (Fig. 13.2), this is not the entire story. Several studies have suggested that there are also more complex *cis*-elements involved. For example, the tobacco *ndhF* mRNA shows a bipartite recognition site in which essential elements are 19 nt apart (Sasaki et al. 2006). For several editing sites, increasing the length of the 5′ region has been shown to increase editing efficiency, although these distal sequence elements are not essential (Hayes et al. 2006). An upstream-sequence effect has also been reported for the rpoBeU158SL editing site (Hayes et al. 2006; editing sites are identified in this review by their position in the respective reading frame following a recent nomenclature proposal by Rüdinger and colleagues Rüdinger et al. 2009). Also, in the case of the ndhBeU156PL and ndhBeU196HY plastid editing sites, 42 nt of both 5′ and 3′ adjacent sequences were insufficient to direct editing in vivo (Bock et al. 1996). Compared to native editing efficiencies (which usually reach 100%), the experimental editing efficiencies are generally poor (often below 10% of wild-type levels) for both in vitro and in vivo experiments (Bock et al. 1996; Reed et al. 2001b; Sasaki et al. 2006). The poor performance of these artificial systems may be due to overexpression problems and other technical limitations, but it could also reflect the lack of necessary distal sequence elements. Future work will be required to determine the specific role of additional modifiers located at a greater distance from the processed site. Such modifying sequences may not necessarily be involved directly in site recognition; indirect effects via regulation of translation (Karcher and Bock 1998) or

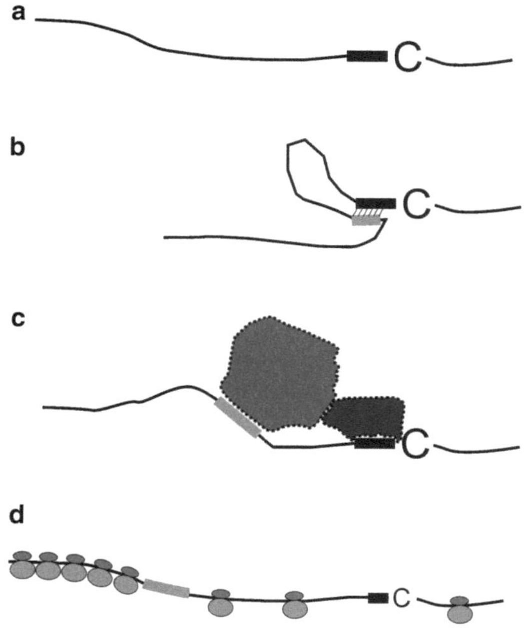

Fig. 13.2. Models for *cis*-elements impacting plant organellar editing efficiencies. (**a**) For most editing sites (here shown as unedited “C”), a short (10–25 nt) sequence element immediately upstream of the editing sites (*bold black line*) is sufficient for RNA editing. At some sites, the editing efficiency is modulated by additional, more distal sequence elements that are mostly unknown (*bold grey line*). Possibly, such elements base-pair with the core *cis*-element and could thus modulate binding of trans-factors (**b**). Alternatively, additional *cis*-elements could recruit proteins (*light grey*) that help to position or activate the editing machinery (*dark grey*; **c**). Finally, modulating sequence elements could influence other processes during the RNA life cycle, for example ribosome processivity, which secondarily could impact RNA editing.

modification of the processing status of other sequence elements in the same message (Schmitz-Linneweber et al. 2001) could also impact the processing of an editing site (Fig. 13.2).

B. Mitochondrial cis-Elements for RNA Editing

A general consensus editing-site-recognition sequence for mitochondria has not yet been found in silico (Giegé and Brennicke 1999), nor have common secondary structures been detected in the vicinity of editing sites (Mulligan et al. 1999). Because mitochondria are not amenable to standard stable transformation techniques, most of the data on *cis*-elements for mitochondrial editing sites has come from in organello and in vitro editing systems. However, the first evidence for the location of such *cis*-elements was obtained through the evaluation of intra-mitochondrial recombination events that deleted the sequences 5′ or 3′ of an editing site. Such studies showed that editing occurred only when the immediate upstream sequences were retained (Lippok et al. 1994; Kubo and Kadowaki 1997).

The in organello systems, which were first developed for wheat (Farre and Araya 2001) and later for maize mitochondria (Staudinger and Kempken 2003), are based on the electroporation of artificial editing-site-containing genes into mitochondria. Deletional and point mutagenesis approaches allowed the delineation of mitochondrial *cis*-elements resembling those found in chloroplasts: For most of the tested sites, 16 nt upstream and 6 nt downstream of the editing site were found to be sufficient for editing, while more distant nucleotides did not seem to play a role (Farre et al. 2001; Choury et al. 2004). However, there were exceptions to this rule. For example, no editing was observed when the Sorghum *atp6* gene was introduced into maize mitochondria, but partial editing was seen following the introduction of chimeric constructs consisting of the 5′-UTR and part of the 5′ coding region of maize *atp6* fused with Sorghum *atp6* sequence. In cases like these, where the coding regions of the two species are virtually identical, distant UTR sequences appear to serve as signals for editing (Staudinger et al. 2005). Such long-distance effects do not appear to be an artifact of the utilized heterologous approach because other heterologous experiments have yielded high editing frequencies, (for example Arabidopsis sequences in maize mitochondria; Bolle and Kempken 2006). When *cis*-elements in the vicinity of editing sites were analyzed with regard to the contribution of individual nucleotides, no consensus could be detected. However, the importance of individual nucleotides clearly differed between sites, providing an early indication that the various editing sites are served by individual *trans*-factors.

The results from in vitro experiments on mitochondrial *cis*-elements of the dicot plants, cauliflower and pea, paralleled and extended the above-described findings from in organello systems. The dominant influence of the 20 nt immediately upstream of editing sites was repeatedly found for different sites in vitro (Takenaka et al. 2004; Neuwirt et al. 2005), and optimal editing was found when the upstream sequences were extended to ~40 and sometimes even 70 nucleotides (Neuwirt et al. 2005; van der Merwe et al. 2006). Competition experiments with mutated versus non-mutated templates, as well as the direct mutation of identified *cis*-elements, helped researchers delineate the *cis*-elements and the importance of individual bases within them. The findings from these studies largely supported the idea that individual sites have individual *cis*-elements (Takenaka et al. 2004). Furthermore, an interesting effect was observed when multiple *cis*-elements were concatenated: The tandemly repeated recognition elements dramatically increased RNA-editing efficiencies, suggesting that local enrichment of a site-recognition factor can enhance RNA editing (Verbitskiy et al. 2008).

In sum, the basic parameters for editing-site recognition have been conserved (at least among angiosperms) in chloroplasts and mitochondria. In both organelles, short individual upstream sequences seem to serve as site-recognition elements. As we will discuss below, members of the PPR protein family utilize such short sequence stretches in a highly specific manner, an interaction essential for editing to occur (see Sect. IV.A). For groups other than angiosperms, the situation is less well understood, mostly because we do not yet have access to comparable in vivo, in vitro or in organello systems for these organisms. However, preliminary comparisons of the editing-site sequence environment in ferns suggests the presence of similar short upstream *cis*-elements (Tillich et al. 2006).

The picture is far less clear regarding the infrequent effect of more distant (usually upstream) sequences. It is conceivable that the overall structure of the RNA including long-distant tertiary interactions, is important for editing-site recognition (Fig. 13.2). Such long-range interactions could potentially help make the editing site available for PPR proteins, perhaps with the involvement of additional protein factors. Alternatively, undesirable secondary structures could impair the access of PPR proteins to their target sites, and might therefore require helicases to open the sites and allow the PPR proteins to interact. In the future, it is likely that a better understanding of the nature and components of the plant organellar editing machinery ('editosome') will help provide insights into these and other possibilities (see also discussion in Takenaka et al. 2008).

C. Tools for Analyzing RNA-Editing Sites In Silico

The mitochondrial transcriptomes of embryophytes usually have 100 or more RNA-editing sites. Therefore, prediction tools are essential for a comprehensive analysis of RNA editing in any organellar genome that has not yet been experimentally investigated. Originally, simple algorithms were developed based on sets of known editing sites versus non-edited sites. However, the predictive value of such algorithms was low (Cummings and Myers 2004). Later, a larger set of parameters was used to describe a likely editing site in an improved tool called REGAL (RNA Editing site prediction by Genetic Algorithm Learning; Thompson and Gopal 2006). This tool scored characteristics known to show biases between edited and unedited Cs in mitochondrial genomes, namely: the base at the −1 position of the editing site; the base at the +1 position of the editing site; the increase in hydrophobicity between the pre- and post-editing-encoded amino acids; the position of the edited C within the codon; the kind of codon that is edited; and the kind of amino acids that are consequently exchanged by the editing event. Using these parameters, the algorithm was trained based on the known Arabidopsis mitochondrial editing sites, which could be used to correctly predict more than 80% of the editing sites in related genomes (e.g. Brassica). An advantage of REGAL-like algorithms over the phylogeny-based analyses (see below) is that they can predict RNA-editing sites in intergenic regions and for species-specific ORFs, whereas homology-based searches are not applicable to such regions. A first tool to use such homology-based, phylogenetic information as a basis for editing-site prediction was PREP (predictive RNA editors for plants). The tool was initially tailored for the analysis of mitochondrial genomes (Mower 2005), but more recent versions have been made suitable for the analysis of chloroplast genomes or user-defined alignments (Mower 2009). The principle behind phylogenetic editing-site prediction is that RNA editing leads to an increase in protein conservation across species because codons for non-conserved amino acids are corrected to those for conserved amino acids (reviewed in Bock 2000; Wakasugi et al. 2001). Thus, editing sites are expected at positions where a C-to-U conversion would increase the conservation of a protein with respect to its homologs in other plants. This principle is also used by PREPACT (plant RNA editing prediction and analysis computer tool; Lenz et al. 2010), which extends the previous programs by predicting both C-to-U and U-to-C editing events. In addition, the output generated by PREPACT highlights different types of editing events (including partial editing) and offers a broader set of user-modified parameters for the graphical output.

The most elaborate RNA editing site prediction software developed to date uses phylogenetic information in conjunction with biochemical information on RNA-editing sites. This algorithm, called CURE (for cytidine-to-uridine recognizing editor) was initially designed for the analysis of mitochondrial genomes (Du and Li 2008), but was later adapted for seed plant chloroplast genomes (Du et al. 2009). So far, CURE has outperformed PREP (Du et al. 2009) but not REGAL (Thompson and Gopal 2006). CURE still has problems making accurate predictions for non-seed plant genomes (Lenz et al. 2010), but the quality of prediction should increase as the number of available training sets (i.e. experimentally determined editing sites) continues to climb.

A different sort of tool, called RedIdb (Picardi et al. 2007, 2010), seeks to categorize editing sites in the organellar genomes of eukaryotic organisms. RedIdb tries to present each editing event in its biological context by giving the corresponding DNA, cDNA and protein sequences together with gene ontologies and InterPro domains. Links are also established to the RESOPS (RNA-editing sites of land plant organelles on protein three-dimensional (3D) structures) database, which maps the amino acids affected by RNA editing onto the available 3D protein structures (Yura et al. 2009). RedIdb can be used directly for simple analyses because sequence analysis tools (e.g., BLAST and CLUSTAL algorithms, Thompson et al. 1994; Altschul et al. 1997) are directly implemented in the database. Furthermore, RedIdb is linked with the EdiPy tool, a script designed to allow the evolutionary simulation of highly edited mitochondrial sequences that are not amenable to analysis using standard statistical analysis tools (e.g., bootstrap analysis). RedIdb has the advantage of manual curation of entries over more general databases, such as dbRES (He et al. 2007), which collects all of the editing sites (not just those of organelles) deposited in GenBank, or ChloroplastDB (Cui et al. 2006) and GOBASE (O'Brien et al. 2009), which are general organellar-genome databases that do not emphasize RNA editing.

In all, current in silico tools have greatly helped to access RNA editing in novel organellar genomes and to move on swiftly from sequence analysis to editing site prediction and experimental analysis. Hopefully, our gain in knowledge on editing trans-factors (see next section), will at one point allow to connect in silico site prediction with the automated prediction of target specificities of editing site recognition factors in any embryophyte genome.

IV. *Trans*-Factors for C-to-U RNA Editing in Plant Organelles

Although the hunt for plant organellar editing factors was initially long and frustrating, recent years have seen tremendous progress in the field, and researchers have finally determined how editing specificity is assured. In short, proteins from the pentatricopeptide repeat (PPR) family show highly specific recognition of *cis*-elements upstream of editing sites. Some auxiliary factors have also been identified, but the process of catalysis is still unclear and it is not yet known which factors contribute directly to base conversion.

A. Pentatricopeptide Repeat (PPR) Proteins Specify Editing Sites

The identification of the first editing factor for an organellar (in this case, plastid) site was not the outcome of an elaborate genetic or biochemical screen for editing factors, but instead came out of work on an unrelated problem. T. Shikanai's group (Kyoto University, Japan) had a long-standing interest in the plastid NADH dehydrogenase (NDH) complex, a multi-subunit complex in the thylakoid membrane that has still not been functionally assigned with a high degree of certainty. Shikanai and colleagues identified mutants of the NDH complex by screening an ethane methyl sulfonate (EMS) induced mutant collection, looking for characteristic defects in chlorophyll fluorescence (Hashimoto et al. 2003).The isolated mutants included one harboring a lesion in the gene for a PPR protein called CRR4 (chloroplast respiratory reduction 4; Kotera et al. 2005). The loss of CRR4 abrogated editing of the start codon of the *ndhD* mRNA, which encodes a core subunit of the NDH complex. Given that almost half of the editing sites in the plastid genome reside in *ndh* genes, it is not surprising that the screen also uncovered several additional editing mutants showing defects in specific *ndh* sites; all of them were found to result from lesions in PPR genes, namely those encoding CRR21, CRR22 and CRR28 (Okuda et al. 2007, 2009b; for a complete list see Table 13.1). Other studies searching for mutants defective in chloroplast development also identified PPR proteins as being involved in editing, again with each protein serving a specific site (CLB19, Chateigner-Boutin et al. 2008; LPA66, Cai et al. 2009; AtECB2, Yu et al. 2009; Vac1, Tseng et al. 2010). Not surprisingly, the apparent importance of PPR proteins in RNA editing spurred reverse-genetic studies; these led to the identification of seven additional PPR proteins that functioned in the editing of specific chloroplast sites: (OTP80; OTP81; OTP85; OTP86; OTP82; OTP84; RARE1; Hammani et al. 2009; Okuda et al. 2009a; Robbins et al. 2009).

Within a few years after the PPR proteins were first identified as editing factors of chloroplast sites, other family members were identified as being required for mitochondrial sites (Table 13.1). Notably, none appears to dually target editing sites in both organelles. The first mitochondrial editing factor was identified as part of an elegant screen for ecotype-specific differences in editing efficiency (Zehrmann et al. 2008), in which differences found between Arabidopsis accessions Columbia and C24 were used to map the editing activity. The identified factor was named MEF1 (mitochondrial editing factor 1), and insertional mutagenesis was used to confirm that it is essential for multiple sites (Zehrmann et al. 2009). A similar screen for quantitative trait loci that affect RNA editing identified REME1, a PPR protein that was shown to support editing sites in the *nad2* and *tatC* mRNAs but was not found to be essential for their editing (Bentolila et al. 2008, 2010). A different type of forward screen for editing defects in a population of EMS-induced Arabidopsis mutants made use of a multiplexed single-nucleotide-primer-extension assay (Takenaka and Brennicke 2009). This screen utilized multiple primers that annealed just downstream of editing sites and were then extended with two alternatively labeled dideoxy nucleotides corresponding to either the edited or the unedited nucleotide. The extension products were then detected and analyzed with standard Sanger sequencing technology. The screen was shown to be capable of identifying a single mutant out of a pool of 50 plant samples (Takenaka and Brennicke 2009). Multiple mutants were recovered using this screening technique, and some of the underlying genes have been identified, including those encoding MEF9 and MEF11 (Verbitskiy et al. 2009; Takenaka 2010). MEF11 also emerged in an unrelated screen for lovastatin-insensitive mutants and was therefore initially called LOI1 (Kobayashi et al. 2007).

Additional mitochondrial PPR proteins required for specific editing sites were found by a reverse genetic screen in the moss *Physcomitrella patens* (PpPPR_56, PpPPR_77, PpPPR_91, PpPPR_71, Ohtani et al. 2010; Tasaki et al. 2010). An unrelated screen that sought to identify T-DNA mutants in rice uncovered a PPR gene mutation that abrogated the editing of at least three sites (Kim et al. 2009). Based on its seed and seedling phenotype, the mutant was called OGR1 (opaque and growth retardation 1). A screen for mutants displaying slow and delayed growth led to the identification of SLOW GROWTH1, which is required for editing of the *nad4* and *nad9* sites (Sung

Table 13.1. Factors involved in organellar RNA editing

Name[a]	Type[b]	Spec[c]	Loc[d]	Target site(s)[e]	Evidence[f]	Mutant phenotype[g]	How identified?[h]	Reference
CRR4	E	At	cp	ndhDeU2TM	Genetic; ivb	wt, NDH defective	Screen for NDH defects	Kotera et al. (2005), Okuda et al. (2006)
CRR21	E	At	cp	ndhDeU383SL	Genetic	wt, NDH defective	Screen for NDH defects	Okuda et al. (2007)
OTP80	E	At	cp	rpl23eU89SL	Genetic	wt	Rev genet	Hammani et al. (2009)
OTP81	DYW	At	cp	rps12i114eU58	Genetic	wt	Rev genet	Hammani et al. (2009)
OTP85	DYW	At	cp	ndhDeU674SL	Genetic	wt	Rev genet	Hammani et al. (2009)
OTP86	DYW	At	cp	rps14eU80SL	Genetic	wt	Rev genet	Hammani et al. (2009)
RARE1	DYW	At	cp	accDeU794SL	Genetic	wt	Rev genet	Robbins et al. (2009)
REME1	DYW	At	mt	nad2eU558SS, tatCeU507SSrMPol	Genetic	wt	QTL mapping	Bentolila et al. (2010)
LPA66	DYW	At	cp	psbFeU77SF	Genetic	Pale-green; reduced PSII	Screen for *hcf* mutants	Cai et al. (2009)
YS1	DYW	At	cp	rpoBeU338SF	Genetic	Virescent	Rev genet	Zhou et al. (2009)
AtECB2	DYW	At	cp	accDeU794SL	Genetic	Albino, seedling lethal	Screen for early chloroplast biogenesis defects	Yu et al. (2009)
CLB19	E	At	cp	rpoAeU200SF, clpPeU559HY	Genetic	pyg, seedling lethal	Screen for chloroplast biogen-esis defects	Chateigner-Boutin et al. (2008)
CRR22	DYW	At	cp	ndhDeU887PL, ndhBeU746SF, rpoBeU551SL	Genetic	wt, NDH defective	Screen for NDH defects	Okuda et al. (2009b)
CRR28	DYW	At	cp	ndhBeU467PL, ndhDeU878SL	Genetic	wt, NDH defective	Screen for NDH defects	Okuda et al. (2009b)
OTP82	DYW	At	cp	ndhBeU836SL, ndhGeU50SF	Genetic	wt	rev genet	Okuda et al. (2009a)
OTP84	DYW	At	cp	psbZeU50SL, ndhBeU1481PL, ndhFeU290SL	Genetic	wt, partially NDH defective	Rev genet	Hammani et al. (2009)
MEF1	DYW	At	mt	rps4eU956SL, nad7eU963FF, nad2eU1160SL	Genetic	wt	Forward screen for editing defects in ecotypes/editing defects in EMS mutants	Zehrmann et al. (2009)
MEF18	E	At	mt	nad4eU1355SL	Genetic	wt	Rev genet	Takenaka et al. (2010)
MEF19	E	At	mt	ccmBeU566SF	Genetic	wt	Rev genet	Takenaka et al. (2010)
MEF20	E	At	mt	rps4eU226PS	Genetic	wt	Rev genet	Takenaka et al. (2010)
MEF21	E	At	mt	cox3eU257SF	Genetic	wt	Rev genet	Takenaka et al. (2010)
MEF22	DYW	At	mt	nad3eU149SF	Genetic	wt; partial editing	Rev genet	Takenaka et al. (2010)
MEF8	DYW	At	mt	nad5eU676LF	Genetic	Not published	Rev genet	Takenaka et al. (2010)

MEF9	E	At	mt	nad7eU200SF	Genetic	wt	Forward screen for EMS editing mutants;	Takenaka (2010)
SLO1	E	At	mt	nad4eU449PL, nad9eU328RW	Genetic	Slow growth	Screen for Arabidopsis slow growth mutants	Sung et al. (2010)
PPR596	P	At	mt	rps3eU1344SS	Genetic	Retarded growth, increased editing	Screen for factors co-expressing with *rps10*	Doniwa et al. (2010)
VAC1	DYW	At	cp	ndhFeU290SL, accDeU794SL	Genetic	Albino	Screen for chloroplast biogenesis defects	Tseng et al. (2010)
OGR1	DYW	Os	mt	nad4eU401SF, nad4e-U416PL, nad4eU433LF, nad2eU1457SL, ccmCeU458SL, cox2eU167SL, cox3eU572SF	Genetic	Opaque, smaller seeds, retarded growth, partially male sterile	Screen for opaque seeds	Kim et al. (2009)
PpPPR_56	DYW	Pp	mt	nad3eU230SL, nad4eU272S	Genetic	Retarded growth	Rev genet	Ohtani et al. (2010)
PpPPR_77	DYW	Pp	mt	cox2eU370RW, cox3eU733RW	Genetic	Retarded growth	Rev genet	Ohtani et al. (2010)
PpPPR_91	DYW	Pp	mt	nad5eU730RW	Genetic	Retarded growth	Rev genet	Ohtani et al. (2010)
PpPPR_71	DYW	Pp	mt	ccmFCeU122SF	Genetic, ivb	Retarded growth	Rev genet	Tasaki et al. (2010)
LOI1/MEF11	DYW	At	mt	nad4eU124LL, cox3e-U422PL, ccmBeU344P	Genetic, copurified RNA	wt, slightly retarded in development	Forward screen for EMS editing mutants; screen for resistance to lovastatine	Verbitskiy et al. (2009), Tang et al. (2010)
CP31A/CP31B	RRM	At/Nt	cp	13 sites partially affected	Genetic	wt	Rev genet	Hirose and Sugiura (2001), Tillich et al. (2009b)

[a]Only factors for which genes were identified are listed

[b]Type of factor encoded by the respective gene; *E* E-type PPR protein, *DYW* DYW-type PPR protein, *RRM* protein containing RNA recognition motifs

[c]Species, in which factor was described. *At Arabidopsis thaliana*, *Os Oryza sativa*, *Pp Physcomitrella patens*, *Nt Nicotiana tabacum*

[d]Subcellular localization of the editing factor. *Cp* chloroplast, *mt* mitochondria

[e]Editing sites affected by loss of the respective factor; nomenclature according to Lenz et al. (2010)

[f]The conclusion that a particular protein acts as an editing factor is in most cases based on the finding that specific editing sites are no longer fully processed in mutants of the factor in question (genetic evidence). In a few cases, biochemical data show an association of the factor with the site (*ivb* in vitro binding, *co-purified RNA* detection of RNA in cellular fractions enriched for the factor)

[g]A summary of macroscopic and physiological phenotypes seen in mutants of editing factors. *PSII* photosystem II, *NDH* NADH dehydrogenase complex

[h]A short summary of how genes for editing factors were determined. *Rev genet* reverse genetic screen, *hcf* high chlorophyll fluorescence

et al. 2010). Finally, a number of PPR proteins that target mitochondrial editing sites were found using reverse genetics, including six identified by a screen of T-DNA insertions in Arabidopsis PPR genes (MEF-18, MEF-19, MEF-20, MEF-21, MEF-22, MEF8, Takenaka et al. 2010).

In sum, 33 PPR proteins have been shown to serve organellar RNA-editing sites, all in a highly specific manner. For 20 of the 33, only one target RNA-editing site has been described. Of the remaining 13 PPR proteins, 8 serve 2 sites, 4 serve 3 sites, and 1 (OGR1) is involved in editing 7 sites (Table 13.1).

While the specificity of PPR proteins for low numbers of editing sites is consistent with the findings of studies on non-editing PPR proteins that are also linked to few RNA processing events (Schmitz-Linneweber and Small 2008), several caveats should be kept in mind. First, most of the target editing sites for PPR proteins that have been described to date are based solely on genetic data. Most of the editing PPR proteins have not yet been directly shown to associate with their target sites. Thus, a genetically determined lesion in the editing of an individual site could be a secondary effect of a more general impairment in basic organellar function. For example, loss of overall plastid translation affects the processing of multiple RNA-editing sites (Karcher and Bock 1998; Halter et al. 2004). Therefore, it would be desirable to determine whether the identified editing-site-related factors directly associate with their editing sites, either in vitro or (even better) in vivo.

Only one editing factor has been examined for its association with RNA in vivo to date: LOI1. Two RNA targets of LOI1, *cox3* and *atp1,* were identified by co-purification with overexpressed LOI1:FLAG proteins (Tang et al. 2010). However, only one of the two recovered RNAs, the *cox3* mRNA, displayed an editing defect in LOI1 mutants. Furthermore, the LOI1:FLAG eluates failed to yield any of the six other editing-deficient RNAs that had been recovered from LOI1-deficient plants. This, together with the low frequency of *cox3* cDNA clones found after reverse transcription of RNA bound to LOI1:FLAG, precluded a final conclusion on whether or not LOI1 directly binds the RNAs that fail to undergo editing in the LOI1 mutants (Tang et al. 2010). With regard to in vitro studies, two other PPR proteins with functions in RNA editing have been shown to bind directly to their cognate editing site in vitro (Okuda et al. 2006; Tasaki et al. 2010). A minimal CRR4-binding element was determined to lie within the region from −25 to +10 relative to the *ndhD* editing site (Okuda et al. 2006). Similarly, the moss PpPPR_71 editing site is contained in a sequence element spanning nucleotide −40 to +5 relative to the editing site ccmF-CeU122SF (Tasaki et al. 2010). These in vitro studies on the RNA binding sites of the editing PPR proteins are in good agreement with the locations and sizes of the previously determined cis-sequences for editing (see Sect. III.B).

A second caveat concerns the completeness of the editing analyses that are currently available. In the case of the *Physcomitrella* PPR proteins, it is relatively simple to survey all 13 sites in both organellar genomes. The same holds true for the 30-some sites in the plastid genomes of angiosperms (Schmitz-Linneweber and Barkan 2007). When it comes to the 100s of editing sites known in Arabidopsis mitochondria and the many yet-unknown sites in rice, maize, etc., a conclusive screen for defects seems ambitious. Not even the modern multiplex-based approaches have attempted to screen all editing sites in a given organism (Takenaka and Brennicke 2009). Thus, it can be expected that most of the mitochondrial editing PPR proteins will eventually be found to serve a larger number of sites than they appear to at this point. Nevertheless, it cannot be disputed that the specificity displayed by these proteins is exquisite. To understand how this is achieved, we must take a closer look at the PPR protein family.

1. The Architecture of the PPR Proteins

Members of the PPR protein family had been already studied in yeast and maize by the late

1990s (Barkan et al. 1994; Manthey and McEwen 1995; Coffin et al. 1997; Fisk et al. 1999; Ikeda and Gray 1999; Lahmy et al. 2000). Each of these studies had implicated individual proteins in the gene expression of organelles, but the family had not yet been recognized as such. The credit for identifying the existence of a large protein family whose members play potential roles in the RNA processing of plant organelles goes to Ian Small's (UWA Perth, Australia) and Alain Lecharny's (CNRS-INRA Evry, France) groups, which described the PPR motif and annotated the family in Arabidopsis (Aubourg et al. 2000; Small and Peeters 2000; Lurin et al. 2004). The PPR motif belongs to the widespread helical-hairpin-repeat motifs. The motif is defined as a repeat, meaning that PPR proteins always have at least two PPR motifs (Lurin et al. 2004). Repeats are predominantly found in tandem, and it is unclear whether isolated motifs are actually functional. Both the structure of the repeat and the overall structure of the PPR tract (i.e., the entirety of all repeats) have been modeled based on the known crystal structures of the closely related tetratricopeptide repeat (TPR) proteins (Small and Peeters 2000; Delannoy et al. 2007). These studies suggest that each PPR repeat encodes two alpha-helical elements, termed A and B, which fold back onto each other and also interact with the helical elements of the two adjacent repeats. Thus, the tandem repeats are stacked on top of one another to form an oblong superstructure. The A helices form the front of this structure, while the B helices form the backside. The surface produced by the A helix displays a curious aggregation of charged and hydrophilic amino acids that are believed to make contacts with RNA. Unfortunately, we do not yet have either a detailed point-mutant-based analysis of PPR tracts or a crystal structure to support these models. In any case, it is clear that PPR proteins are major players in all aspects of chloroplast RNA metabolism. A wealth of genetic data on PPR proteins almost uniformly suggests that they play direct roles in the RNA metabolism of organelles, including functions in RNA splicing, cleavage, stabilization, translation and editing (Schmitz-Linneweber and Small 2008). Importantly, both in vitro and in vivo studies have suggested that there is a direct interaction between PPR proteins and RNA (Tsuchiya et al. 2002; Nakamura et al. 2003; Lurin et al. 2004; Schmitz-Linneweber et al. 2005a, 2006; Okuda et al. 2006; Gillman et al. 2007; Kobayashi et al. 2007; Beick et al. 2008; Kazama et al. 2008; Williams-Carrier et al. 2008; Tang et al. 2010).

2. The Editing PPR Proteins Belong to the PLS Subgroup

The PPR family has been subclassified into two major groups: the pure (or P-type) PPR proteins, which contain only repeat units of 35 amino acids in length; and the PLS PPR proteins, which have repeats of varying lengths (P = normal; L = long repeats; S = short repeats, Lurin et al. 2004). The normal, long and short domains typically follow each other in triplicates, leading to the name: P-L-S. The P-type PPR proteins generally do not contain any other known protein domains, and members of this group have been associated with RNA stabilization, translation and splicing (e.g., Barkan et al. 1994; Schmitz-Linneweber et al. 2006; Pfalz et al. 2009; Prikryl et al. 2010). Intriguingly, all but one of the PPR proteins that have been implicated in RNA editing belong to the PLS subgroup; in Arabidopsis, this subgroup contains slightly less than half of the annotated PPR proteins (Lurin et al. 2004). The PLS subgroup has been further subdivided based on the presence of C-terminal extensions of unknown function (Lurin et al. 2004), and almost all PLS PPR proteins contain a so-called E-domain of ~90–120 amino acids. Eighty-seven of the 450 PPR proteins in Arabidopsis contain the DYW domain, which was named after three highly conserved C-terminal amino acid residues, and spans roughly 100 amino acids. The majority of the editing PPR proteins in Arabidopsis (22 of 33) have DYW domains (Table 13.1).

The only non-PLS type PPR that appears to be involved in RNA editing is the Arabidopsis PPR596 protein (Doniwa et al. 2010). PPR596 is essential when plants are germinated on soil; the phenotype can be partially rescued by a longer growth period on sugar-containing medium, but the plants still display a strong growth retardation and aberrant leaf development. Mutants of this PPR are unusual in that they show an increase in the RNA editing of a mitochondrial site that is only partially edited in wild-type plants (rps3eU1344SS). However, in the absence of conclusive data on the processing of the *rps3* transcript, it is currently unclear whether the observed defect was a direct effect, or alternatively was caused by other PPR596-mediated alterations in RNA metabolism.

3. How Do PPR Proteins Recognize RNA?

Biochemical evidence suggests that PPR proteins can interact with the *cis*-elements upstream of RNA-editing sites, but the details of this interaction are not yet known. In the last few years, a handful of RNA-editing factors have been shown to serve more than one target site, allowing researchers to determine consensus sequences for site recognition (Hammani et al. 2009). A simple consensus of base identities was found to be insufficient to explain the observed protein specificity, but the combination of several characteristics of RNA bases into a consensus model allowed the experimentally determined editing sites to be identified with high specificity (Hammani et al. 2009). The employed characteristics were: purine versus pyrimidine bases; and double versus triple hydrogen bond-forming bases. For example, the base identity consensus of the three sites served by OTP84 (*psbZ, ndhB, ndhF*) is: U – – – – – – U A – U – – – – C (the hyphens stand for ambiguous bases). This consensus contains little information, and in fact corresponds to 444 sites in the chloroplast genome of Arabidopsis. The improved consensus reads UWRYWWYUAYUWYRYC (W = A or U; Y = C or U; R = A or G) and is found only four times in the genome. Of the four occurrences, one is not in a transcribed region (Hammani et al. 2009) and the other three correspond to the known target sites for OTP84. This suggests that PPR proteins recognize editing-site *cis*-elements by distinguishing bases by their purine/pyrimidine natures and/or Watson–Crick characteristics rather than uniquely distinguishing among the four bases. This model holds true for most of the editing factors analyzed to date, suggesting that the same protein surface recognizes multiple targets (Hammani et al. 2009). A detailed structural characterization of the binding surface of PPR proteins should be a goal for the near future.

4. How Do PPR Proteins Help Edit Organellar RNAs?

The mechanism behind base conversion is still a matter of debate, as is the role of PPR proteins in RNA editing. Based on the factors isolated to date, it seems clear that the PLS PPR proteins act as editing-specificity factors. These proteins emerged in land plants and have not been found in green algae or any non-green organism (which generally have much lower PPR gene counts); this distribution parallels the presence of RNA editing, which also has not been found in green algae (Lurin et al. 2004). Importantly, the DYW domain is restricted to land plants and has been shown to correlate with taxa that exhibit RNA editing (Salone et al. 2007). The green algae, from which the embryophytes arose, do not show organellar RNA editing and do not have DYW PPR proteins. In addition, the marchantiid liverworts that secondarily lost their RNA editing also lack DYW proteins, whereas the Jungermanniid liverworts, close relatives that show extensive organellar RNA editing, possess proteins with DYW domains (Salone et al. 2007; Rüdinger et al. 2008).

The DYW domain has some interesting similarities to the cytidine deaminases from various eukaryotic organisms (Salone et al. 2007). In humans, these deaminases are involved in zinc-dependent RNA editing (Navaratnam and Sarwar 2006). The highest similarity to the DYW-domain was observed for the zinc-binding domain of these deaminases. This includes the histidines and cysteines required to form the complex with zinc, which are found in the HxExnCxxC motif of the DYW domain (Salone et al. 2007). To date, efforts to show that recombinant DYW domains are involved with RNA editing in vitro have been unsuccessful (Nakamura and Sugita 2008; Okuda et al. 2009b). By contrast, all four recombinant DYW domains tested so far were found to be capable of degrading RNA in vitro with different efficiencies (Nakamura and Sugita 2008; Okuda et al. 2009b). One of the four DYW-PPR proteins tested was CRR2, which may be involved in intercistronic cleavage, but for which no RNA-editing function has been genetically assigned (Hashimoto et al. 2003). In fact, not all DYW-PPR proteins are necessarily editing factors. For example, a reverse genetic screen for editing defects in null mutants of DYW-PPRs found editing defects in only 5 of 9 plastid mutants, and only 2 of 25 mitochondrial mutants (Hammani et al. 2009; Takenaka et al. 2010).

It is not yet clear how RNA cleavage by the DYW domains fits into the catalysis of C-to-U conversion. The phosphate backbone of RNA has been shown to remain intact during RNA editing (Rajasekhar and Mulligan 1993). Furthermore, if the backbone were cleaved near editing sites we should be able to recover defined degradation products of edited transcripts, and these have not been found to date. It is also possible that the observed cleavage activity is just a misleading side effect that unfolds only under reaction-tube conditions in experiments using naked RNA. Future work will be required to examine these and other questions.

Thus, the in vitro data on the role of the DYW domain in RNA editing are inconclusive at this point. In addition, results from in vivo studies are rather confusing. A T-DNA insertion in the DYW domain of the mitochondrial MEF11 protein obliterated RNA editing at two sites, but a third site was still partially edited in this mutant, whereas no editing was seen for the MEF11 null allele (Verbitskiy et al. 2010). In MEF1-deficient protoplasts or plants, partial restoration of editing events was seen following transient or stable complementation with a MEF1 mutant lacking the DYW domain (Zehrmann et al. 2010). In plastid DYW-PPR editing mutants for CRR22, CRR28 and OTP83, however, complete restoration was achieved following complementation with PPR genes lacking the DYW domain (Okuda et al. 2009a, b). These findings seem to indicate that the DYW domain is not necessary for catalytic editing activity. When the DYW domains of CRR22 and CRR28 were replaced with their counterpart from the non-editing PPR protein CRR2, no complementation of null mutants occurred. However, RNA editing was still supported by proteins in which the DYW domains from CRR22 and CRR28 were swapped (Okuda et al. 2009b). Conversely, when the DYW domains from CRR22 and CRR28 were used to replace the DYW domain of CRR2, the latter failed to show RNA cleavage, indicating that the DYW of CRR2 appears to be essential for protein function (i.e., RNA cleavage; Okuda et al. 2009b). This suggests that there are two types of DYW domains: DYW type 1 is found in CRR2 and is required for RNA cleavage but cannot function in RNA editing, whereas DYW type 2 is found in the editing PPR proteins and neither inhibits nor is required for RNA-editing activity. Thus, although the phylogenetic distribution of DYW editing sites suggests that they may be required for editing, the initial genetic experiments indicate otherwise. However, before we try to form a model that explains these contradicting phylogenetic and genetic data, we will briefly discuss the E-domain.

All PPR protein editing factors isolated to date have E-domains, and 10 PPR editing factors have an E-domain but not a DYW-domain (Table 13.1). Similar to the DYW domain, the E-domain is highly conserved within and between plant species, but it does not bear homologies to any known protein domain. Loss of the E-domains from CRR22, CRR28 and CRR4 abolished RNA editing at their cognate sites (Okuda et al. 2007, 2009b)

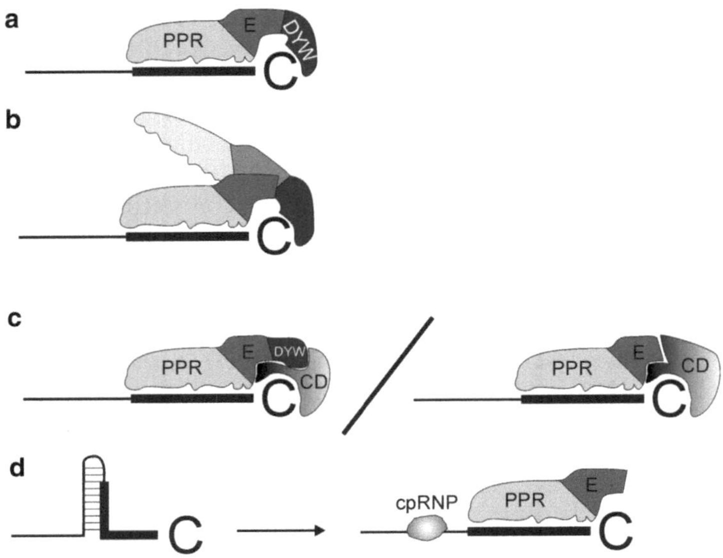

Fig. 13.3. Models for the organellar editosome. All models are based on the well supported assumption that the PPR tract of PPR proteins contacts the core cis-element in front of editing sites (*bold line*). (**a**) PPR-DYW proteins could be solely responsible for RNA editing of their target sites if the proposition holds that the DYW domain has cytidine deaminase activity. (**b**) For E-type PPR proteins without a DYW domain, other DYW PPR proteins could provide the catalytic DYW domain *in trans*. (**c**) PPR proteins could be only required for site recognition, but not directly for catalysis. Such an activity would be provided by a hitherto unknown cytidine deaminase (CD) that would be recruited by the E:DYW domains or by E-domains alone. (**d**) For *cis*-elements part of RNA secondary structures, additional factors could be required that make the RNA accessible for PPR protein entry and thus subsequent catalysis. Such non-essential factors could be cpRNP proteins with their suggested RNA chaperone activity.

but did not affect their binding to RNA (Okuda et al. 2007). The experimental addition of a stop codon right at the border between the PPR tract and the E-domain blocked the editing activity of MEF9 (Takenaka 2010), but swapping the E-domains of CRR4 and CRR21 did not interfere with RNA editing (Okuda et al. 2007). Together, these data show that the E-domain is essential for the RNA-editing activity of the editing PPR proteins.

Based on this, it is at present still difficult to incorporate these findings on the C-terminal extensions of PLS-PPR proteins into a unifying model. One possibility would be that both an E-domain and a type 2 DYW-domain are required for editing, but that the latter can also be added in *trans* (Okuda et al. 2009b; Fig. 13.3). Or perhaps two or more PPR proteins can act together to process RNA-editing sites (Fig. 13.3). *Ad extremo*, any DYW type 2 PPR could possibly complement any E-domain PPR protein. This would be a convenient arrangement, as it would provide the chloroplast with a high cumulative concentration of (possibly catalytically active) DYW domains at any given time, while the concentration of individual PPR proteins could remain low (i.e., just sufficient for RNA detection). This would dispense with the need to express and regulate an additional deaminase enzyme, and the E-domain could function as a protein-interaction domain for the recruitment of DYW-containing PPR proteins. If this model is correct, various DYW-PPR proteins should co-purify with any given E-domain-containing PPR editing factor used as bait.

This and other related hypotheses will likely be tested in the near future, as several researchers are seeking to identify factors that interact with PPR proteins. At present, however, the model is still highly speculative. In particular, there is currently no evidence for the direct interaction of PPR proteins with each other aside of the finding that the PPR protein HCF152 might form homodimers in vitro (Nakamura et al. 2003). Genetically, it has been shown that two PPR proteins can be required for the editing of one site (Robbins et al. 2009; Yu et al. 2009): The loss of either AtECB2 or RARE1 abrogates the editing of a specific site in the plastid *accD* mRNA. Given that both of these PPR proteins contain a DYW motif, it will be interesting to see how AtECB2 and RARE1 share non-redundant responsibilities in processing the *accD* site. In this regard, it is also interesting that there are mutants of DYW-PPR proteins that support RNA editing at specific sites, but are not essential for it. This includes MEF1, which is essential for two sites but only supportive for the editing of a *nad2* site (Zehrmann et al. 2009); REME1, which supports another *nad2* site; and OGR1, which is essential for a number of sites but contributes only slightly to the editing of nad4eU433LF. Future work will be required to assess how these PPR proteins contribute together with putative partner PPRs and possibly other factors to achieve high editing levels of their cognate targets. It will be particularly interesting to understand whether these PPR proteins directly associate with the RNA and thus contribute to recognition, or whether their DYW domains are simply recruited for catalysis by protein-protein interactions.

B. The Enigmatic Catalytic Activity

Both the sugar-phosphate backbone and the nucleotide base remain intact during RNA editing, indicating that catalysis does not involve nucleotide excision or base exchange (Rajasekhar and Mulligan 1993; Yu and Schuster 1995). Instead, the experimental evidence collected in the years following the discovery of plant organellar RNA editing unequivocally indicated that C-to-U RNA editing proceeds by base deamination (Araya et al. 1992; Rajasekhar and Mulligan 1993; Yu and Schuster 1995). Transamination would be an alternative scenario, but the standard amino-group acceptors and a candidate enzyme tested for *trans*-amination did not seem to be involved in vitro for C-to-U editing (Takenaka et al. 2007). One long-held theory is that cytidine deaminases carry out the reaction, in a manner analogous to that seen for human C-to-U editing (Navaratnam and Sarwar 2006). However, the first cytidine deaminase identified in Arabidopsis was not found to associate with RNA (Faivre-Nitschke et al. 1999), and another candidate deaminase protein turned out to be required for A-to-I editing of plastid tRNA-R(ACG), but not for C-to-U RNA editing (Delannoy et al. 2009; Karcher and Bock 2009). The classical cytidine deaminases utilize zinc as a co-factor during catalysis, but in vitro experiments in which zinc was chelated from editing reactions delivered mixed results: Although zinc depletion did not affect mitochondrial RNA editing in vitro (Takenaka et al. 2007), detrimental effects were observed in comparable plastid systems (Hegeman et al. 2005). Several predicted organellar cytidine deaminases remain to be tested for functions in RNA editing, but we may find that the true activity has evolved from a very different background, such as from RNA modifying enzymes that act on rRNAs or tRNAs, or from enzymes involved in single-stranded DNA metabolism and repair. Importantly, it remains possible that the DYW domain may have editing activity. Certainly, the future identification of this editing activity will be a most exciting and important task.

C. Other Factors Involved in RNA Editing

Aside from the PPR proteins and the enigmatic editase discussed above, the list of additional RNA-editing factors is fairly short. Most of the factors implicated in RNA editing have been determined biochemically, such as by the cross-linking of proteins to editing sites. Among the proteins identified in this manner is a 91-kD protein associated with the *rpoBeU113SF* editing site in tobacco

(Kobayashi et al. 2008). Most likely, this protein corresponds to the recently identified YS1 PPR protein responsible for editing this site in Arabidopsis (Zhou et al. 2009). Similarly, the 95-kD protein that cross-linked to *ndhBeU494PL* and *ndhFeU21SL* in tobacco (Kobayashi et al. 2008) could turn out to be homologous to OTP84, an Arabidopsis PPR serving these exact same sites (Hammani et al. 2009). A 25-kD factor associated with tobacco *psbL* will not have a similarly corresponding factor in Arabidopsis, which lacks this site (Hirose and Sugiura 2001). Two other tobacco proteins (56-kD and 70-kD) that cross-linked to sites in the *petB* and *psbE* mRNAs also remain unidentified at this time (Miyamoto et al. 2002, 2004).

A set of proteins consistently identified in cross-linking experiments are the chloroplast ribonucleoproteins or short cpRNPs, which are highly abundant RNA-binding proteins found in the chloroplasts of angiosperms (Tillich et al. 2010). These proteins are related to the nucleo-cytosolic RNA-recognition motif (RRM)-containing proteins, which play roles in RNA processing and can act as RNA chaperones (Maruyama et al. 1999). They were initially believed to be mostly required for protecting RNAs against degradation (Nakamura et al. 2001). However, a specific role in RNA editing was shown for at least one of their members in vitro (Hirose and Sugiura 2001): Extracts that had been immuno-depleted of the tobacco cpRNP, CP31, were found to be incapable of processing two editing sites in the *ndhB* and *psbL* mRNAs. Other tested cpRNPs were not required for this job; instead a domain of CP31 rich in acidic amino acid residues was found to be essential for this editing activity (Hirose and Sugiura 2001). Recently, knock-out mutants of CP31A and CP31B, two Arabidopsis paralogs of tobacco CP31, were tested for RNA-editing defects (Tillich et al. 2009b). Multiple editing sites exhibited decreased editing efficiencies in the CP31A mutant, whereas the defects in the CP31B mutants were comparatively minor. This may reflect the effects of an extended acidic domain (similar to that found in tobacco CP31), which is present in CP31A but not CP31B. The strongest defects were found in CP31A/B double mutants, but even these mutants did not show a complete loss of RNA editing. It is not yet clear why CP31 is essential for tobacco editing sites in vitro, but the Arabidopsis orthologs seem to be just auxiliary in vivo. It is also not yet known how cpRNPs impact RNA editing in such a specific manner. It seems possible that they could be required to prepare the RNA for PPR protein access. PPR proteins have been shown to prefer single-stranded over double-stranded RNA (Tsuchiya et al. 2002; Nakamura et al. 2003; Williams-Carrier et al. 2008), so the cpRNPs could perhaps act as chaperones by helping dissolve double-stranded elements that obscure PPR binding sites. Indeed, the nucleo-cytosolic RRM proteins display such activity (Dreyfuss et al. 2002). Alternatively, the acidic domain could be part of a platform for recruiting PPR proteins and/or the editase in a manner analogous to the use of such domains for protein-protein interactions by nuclear-splicing factors (Valcarcel and Green 1996).

Finally, there appears to be an overlap in the editing-site target ranges of the PPR proteins and CP31A. This includes, for example, sites in the *ndhB* and *rpoB* messages, which are served by both CRR22 and CP31A. In the future, it would be instructive to analyze how these two proteins act together to achieve base deamination. Protein interaction studies and detailed analyses of the structural changes induced in the RNA targets by both proteins will likely help us understand this issue.

V. The Why Behind RNA Editing

> Science is wonderfully equipped to answer the question 'How?' but it gets terribly confused when you ask the question 'Why?' (Chargaff 1977)

The seeming futility of the RNA-editing process has puzzled researches since the early detection of RNA editing. Why aren't editing sites removed by C-to-T point mutations in the organellar genome, thereby avoiding the need for elaborate RNA processing? Recent reviews have addressed this salient point (Maier et al. 2008; Zehrmann et al. 2008; Tillich et al. 2010), so we will therefore only briefly summarize the current explanations herein.

Two major competing models attempt to rationalize the existence of organellar RNA editing. The first one draws on knowledge gained from other editing systems, particularly those in humans, where base transitions are used to generate and regulate protein diversity. To give a famous example, C-to-U editing of the *apoB* mRNA distinguishes the two protein isoforms of a lipoprotein that is important for lipid transport in the bloodstream (for a recent review see Blanc and Davidson 2010). The two isoforms are differentially expressed; editing occurs only in epithelia of the small intestine, whereas the unedited mRNA gives rise to an isoform that is expressed in the liver. Importantly, the two isoforms are functionally distinct. Other well-studied cases of regulated RNA editing are found in the generation of human neuroreceptor isoforms by RNA editing (Gott and Emeson 2000; Bass 2001, 2002; Valente and Nishikura 2005). The isoforms of such receptors (e.g., those for glutamate or serotonin) result from differential A-to-I editing at multiple sites, and have different receptor kinetics and permeabilities compared to the unedited versions. Obviously, the generation of protein diversity would be an attractive explanation for the persistence of organellar RNA editing in plants. However, almost all plastid-editing sites and the majority of mitochondrial sites are fully edited. We know relatively little regarding the tissue- or condition-specific modulations of editing events for the few sites that show only partial editing, and even if such variation were found, its physiological relevance remains dubious (Grosskopf and Mulligan 1996; Karcher and Bock 1998, 2002a, b; Nakajima and Mulligan 2001).

A number of reports have speculated on the regulatory role of specific editing events. For example, editing of the *rpoB* and *rpoA* mRNAs has been proposed to impact the activity of the encoded RNA polymerase (Hirose et al. 1999; Zhou et al. 2009). In theory, this could impact chlorophyll production by altering the expression of tRNA-Glu (Zhou et al. 2009), which is required for the first step in chlorophyll synthesis. Detailed correlational studies comparing chlorophyll production, the expression of editing factors serving *rpo* mRNAs, and polymerase activity will be required to assess this hypothesis.

Very little is known about the presence of protein isoforms resulting from partially edited sites. In tobacco plastids, monocistronic *ndhD* mRNA was found to associate with polysomal fractions despite having an unedited start codon (i.e. remaining ACG instead of AUG; Zandueta-Criado and Bock 2004). The maize ribosomal S12 protein is present in mitochondria in at least two isoforms generated by incomplete RNA editing, but only the edited isoform is incorporated in mature ribosomes (Phreaner et al. 1996). Confusingly, the orthologous protein in petunia is found in cell fractions enriched for ribosomes (Phreaner et al. 1996). In contrast to the situation for S12, no protein corresponding to the unedited messages of the mitochondrial ribosomal S13 protein was detected even though there was a high frequency of unedited cDNAs (Williams et al. 1998). Similarly, sequencing of portions of the mitochondrial NAD9 protein failed to identify any sequences derived from unedited mRNAs (Grohmann et al. 1994). Finally, an "unedited" protein version of ATP9 has been directly tested for functionality in studies in which it was expressed from the nucleus with a mitochondrial import address fused to the open reading frame. Notably, the imported and unedited ATP9 was found to interfere with normal mitochondrial function, as manifested by male sterility (Hernould et al. 1993; Zabaleta et al. 1996).

In the future, these somewhat contradictory findings should be examined further by proteomic studies of organelles, including searches of mass spectrometric data with unedited versions of the organellar genomes. We cannot yet verify that a shift in the balance between "unedited" and "edited" proteins has any physiological role. Furthermore, it is not yet clear whether editing events that affect start or stop codons can affect the translation of mRNAs, thereby contributing to the regulation of protein production. This leaves very little to substantiate the hypothesis that plant organellar RNA editing plays a general regulatory role. We suspect that although individual editing events may be exploited to regulate gene expression, this will not be the case for the vast majority of editing events.

The second (more recent) hypothesis accounting for the existence of RNA editing, which draws on our understanding of plant organelles as having descended from endosymbiotic bacteria, is called the "genome debugging hypothesis" (Maier et al. 2008). Obligate endosymbionts are prone to accumulating deleterious point mutations in a phenomenon called Müller's ratchet, and there is no reason to think that chloroplasts and mitochondria (i.e., direct descendants of endosymbiotic bacteria) would not also face this problem. Unlike more recent endosymbiotic descendants (e.g., the endosymbiotic gut bacteria of insects), however, plant organelles can draw on the nuclear genome, which is a source of genetic information that evolves rapidly and recombines sexually. Nuclear factors can be imported into chloroplasts and mitochondria to mitigate problems arising from fixed point mutations. Conceptually, such factors could be involved on all levels during the realization of organelle genetic information. A striking example of the nuclear-based repair of organellar mutations comes from plant breeding. Plant breeders have long taken advantage of cytoplasmic male sterility (CMS), a phenomenon caused by mitochondrial mutations that arise in various plant species (reviewed in Chase 2007). Such mutations are of agronomical interest because they avoid the need for the labor-intensive emasculation of plants to prevent selfing, but these mutations must be suppressed to allow the later mass production of seeds in the field. Strikingly, suppressor mutations have been isolated that map to the nuclear genome and nearly all of them affect RNA binding proteins that belong to the PPR protein family (reviewed in Schmitz-Linneweber and Small 2008). These PPR proteins "repair" the CMS-specific mutational problems in the mitochondrial genome. The repair is not carried out on the DNA level, but rather works on the RNA that are derived from the defective genetic information. Specifically, the PPR proteins either help degrade unwanted, aberrant mRNAs that would otherwise give rise to toxic proteins (Wang et al. 2006), or prevent the translation of such RNAs (Uyttewaal et al. 2008). As PPR proteins are also major players in RNA editing, it could be speculated that they evolved to suppress deleterious U-to-C point mutations arising in plant organelles. It should be noted that such rescue of organellar mutations by nuclear factors makes sense given that the plant organellar genomes evolve more slowly than the nuclear genome. In metazoans, in contrast, mitochondrial genomes evolve much more rapidly than the nuclear genomes; thus, back mutations are a much more likely and rapid response to mutational problems than the evolution of nuclear-encoded antidotes (Maier et al. 2008). In fact, recent studies found an inverse correlation between the editing frequencies and overall substitution rates of mitochondrial genomes, suggesting that a slowly evolving genome tends not to jettison its RNA-editing sites at the DNA level (Parkinson et al. 2005; Cuenca et al. 2010). An important prediction of the genome debugging hypothesis is that the removal of an editing site (i.e., the repair of the site on the genomic level) would not interfere with plant viability. Consistent with this hypothesis, when an edited C in the plastid *atpA* mRNA was turned into a T, rendering RNA editing obsolete at this site, there was no detrimental effect on the resulting plants grown under standard conditions (Schmitz-Linneweber et al. 2005b). Also, editing sites evolve rapidly (Shields and Wolfe 1997) and loss of a site in one lineage, while the same site is maintained in a closely related sister group seems to be tolerated (Hayes and Hanson 2008; Tillich et al. 2009a). In the future, by constructing an editing-site-free organellar genome and substituting it for the wild-type genome, researchers should be able to examine whether editing sites are used to regulate gene expression or have some other function, or whether RNA editing is truly an unnecessary freak of evolution. Notably, the results from studies on the evolutionary behavior of editing sites across large time scales (from the beginning of land plant evolution or encompassing at least angiosperm evolution) have suggested that RNA-editing sites tend to be lost over time at least in plastids (Tillich et al. 2006, 2009a; Hayes and Hanson 2008). Perhaps, we are studying a process on the brink of extinction.

VI. Perspectives

Unlike other editing systems, such as the C-to-U and A-to-I editing in humans or the rampant and excessive RNA editing in trypanosome mitochondria, relatively little is known about RNA editing in plant mitochondria. Whereas we know details on the machinery and catalysis of RNA editing in humans, this information is lacking in plant organelles. At present, there are four major questions in the field. First, we need to elucidate which factors carry the catalytic activity for base deamination and what makes up the editosome (if there is one). Second, we should examine how PPR proteins recognize the *cis*-elements in front of editing sites (i.e., how is site-specificity generated?). The answers to these two mechanistic questions should be obtainable within the next few years, as techniques to determine structures of PPR proteins associated with their target RNAs are at hand along with proteomic methods for detailed characterization of the editosome using PPR proteins as bait. The second pair of questions addresses the still elusive function of RNA editing, and will be much harder to answer. First, are there editing events that distinguish two functional proteins from each other and, on a grander scale, is any RNA-editing event rate-limiting for the production of the correct protein? Second, does the lack of specific RNA editing in many PPR protein mutants truly not have an effect on the corresponding proteins, as suggested by the absence of any macroscopic phenotype? Considerable experimental efforts will be required to address these questions. Cryptic phenotypic alterations could be unveiled by applying various stresses to the editing mutants. Ideally (although rarely done due to the immense space and time requirements), competition experiments between mutant and wild-type plants could be used to determine possible fitness deficits under field conditions. In terms of assessing regulation, recent studies on RNA processing factors in *Chlamydomonas* could light the way. In this case, hypomorphic mutant series with ever-decreasing amounts of PPR proteins showed clear correlations between the amount of PPR proteins and the amount of proteins generated from the PPR-target message, providing a clear sign that the PPR proteins are true regulators of gene expression (Raynaud et al. 2007). In this example, the PPR protein was required to stabilize the target RNA; however, similar approaches could also be applied to PPR proteins as editing factors. Ultimately, it would be a dream to harness the RNA-editing machinery and use it to switch proteins on or off at will in plant organelles; this would be particularly useful in plastids, which are important sites for biotechnological expression of transgenes (Bock 2007). Possibly, RNA-editing factors could also be used to manipulate RNAs in vitro or perhaps even to fight detrimental RNAs and/or viral RNAs in humans. In any case, and even without daydreaming about possible applications, plant organellar RNA editing – with its curious origins, uncertain functions and enigmatic machinery – remains a formidable and exciting challenge for future research.

Acknowledgements

The authors apologize for the many studies on organellar RNA editing that were not discussed and cited here due to space limitations. Support by the DFG to CSL (Emmy Noether stipend) is gratefully acknowledged.

References

Altschul SF, Madden TL, Schaffer AA, Zhang J, Zhang Z, Miller W, Lipman DJ (1997) Gapped BLAST and PSI-BLAST: a new generation of protein database search programs. Nucleic Acids Res 25:3389–3402

Araya A, Domec C, Begu D, Litvak S (1992) An in vitro system for the editing of ATP synthase subunit 9 mRNA using wheat mitochondrial extracts. Proc Natl Acad Sci USA 89:1040–1044

Aubourg S, Boudet N, Kreis M, Lecharny A (2000) In *Arabidopsis thaliana*, 1% of the genome codes for a novel protein family unique to plants. Plant Mol Biol 42:603–613

Barkan A, Walker M, Nolasco M, Johnson D (1994) A nuclear mutation in maize blocks the processing and translation of several chloroplast mRNAs and provides evidence for the differential translation of alternative mRNA forms. EMBO J 13:3170–3181

Bass BL (2001) RNA editing. Oxford University Press, Oxford

Bass BL (2002) RNA editing by adenosine deaminases that act on RNA. Annu Rev Biochem 71:817–846

Beick S, Schmitz-Linneweber C, Williams-Carrier R, Jensen B, Barkan A (2008) The pentatricopeptide repeat protein PPR5 stabilizes a specific tRNA precursor in maize chloroplasts. Mol Cell Biol 28: 5337–5347

Benne R, Van den Burg J, Brakenhoff JP, Sloof P, Van Boom JH, Tromp MC (1986) Major transcript of the frameshifted coxII gene from trypanosome mitochondria contains four nucleotides that are not encoded in the DNA. Cell 46:819–826

Bentolila S, Elliott LE, Hanson MR (2008) Genetic architecture of mitochondrial editing in *Arabidopsis thaliana*. Genetics 178:1693–1708

Bentolila S, Knight W, Hanson M (2010) Natural variation in Arabidopsis leads to the identification of REME1, a pentatricopeptide repeat-DYW protein controlling the editing of mitochondrial transcripts. Plant Physiol 154:1966–1982

Blanc V, Davidson NO (2010) APOBEC-1-mediated RNA editing. Wiley Interdiscip Rev Syst Biol Med 2:594–602

Bock R (2000) Sense from nonsense: how the genetic information of chloroplasts is altered by RNA editing. Biochimie 82:549–557

Bock R (2007) Plastid biotechnology: prospects for herbicide and insect resistance, metabolic engineering and molecular farming. Curr Opin Biotechnol 18:100–106

Bock R, Kössel H, Maliga P (1994) Introduction of a heterologous editing site into the tobacco plastid genome: the lack of RNA editing leads to a mutant phenotype. EMBO J 13:4623–4628

Bock R, Hermann M, Kössel H (1996) In vivo dissection of cis-acting determinants for plastid RNA editing. EMBO J 15:5052–5059

Bolle N, Kempken F (2006) Mono- and dicotyledonous plant-specific RNA editing sites are correctly edited in both in organello systems. FEBS Lett 580:4443–4448

Cai W, Ji D, Peng L, Guo J, Ma J, Zou M, Lu C, Zhang L (2009) LPA66 is required for editing psbF chloroplast transcripts in Arabidopsis. Plant Physiol 150:1260–1271

Chargaff E (1977) Voices in the Labyrinth: nature, man and science. Seabury Press, New York

Chase CD (2007) Cytoplasmic male sterility: a window to the world of plant mitochondrial-nuclear interactions. Trends Genet 23:81–90

Chateigner-Boutin AL, Hanson MR (2002) Cross-competition in transgenic chloroplasts expressing single editing sites reveals shared cis elements. Mol Cell Biol 22:8448–8456

Chateigner-Boutin AL, Hanson MR (2003) Developmental co-variation of RNA editing extent of plastid editing sites exhibiting similar cis-elements. Nucleic Acids Res 31:2586–2594

Chateigner-Boutin AL, Ramos-Vega M, Guevara-Garcia A, Andres C, de la Luz Gutierrez-Nava M, Cantero A, Delannoy E, Jimenez LF, Lurin C, Small I, Leon P (2008) CLB19, a pentatricopeptide repeat protein required for editing of rpoA and clpP chloroplast transcripts. Plant J 56:590–602

Chaudhuri S, Maliga P (1996) Sequences directing C to U editing of the plastid *psbL* mRNA are located within a 22 nucleotide segment spanning the editing site. EMBO J 15:5958–5964

Chaudhuri S, Carrer H, Maliga P (1995) Site-specific factor involved in the editing of the *psbL* mRNA in tobacco plastids. EMBO J 14:2951–2957

Choury D, Farre JC, Jordana X, Araya A (2004) Different patterns in the recognition of editing sites in plant mitochondria. Nucleic Acids Res 32: 6397–6406

Coffin JW, Dhillon R, Ritzel RG, Nargang FE (1997) The *Neurospora crassa cya-5* nuclear gene encodes a protein with a region of homology to the *Sacharomyces cerevisiae* PET309 protein and is required in a post-transcriptional step for the expression of the mitochondrially encoded COXI protein. Curr Genet 32:273–280

Covello PS, Gray MW (1989) RNA editing in plant mitochondria. Nature 341:662–666

Cuenca A, Petersen G, Seberg O, Davis JI, Stevenson DW (2010) Are substitution rates and RNA editing correlated? BMC Evol Biol 10:349

Cui L, Veeraraghavan N, Richter A, Wall K, Jansen RK, Leebens-Mack J, Makalowska I, dePamphilis

CW (2006) ChloroplastDB: the chloroplast genome database. Nucleic Acids Res 34:D692–D696
Cummings MP, Myers DS (2004) Simple statistical models predict C-to-U edited sites in plant mitochondrial RNA. BMC Bioinformatics 5:132
Delannoy E, Stanley WA, Bond CS, Small ID (2007) Pentatricopeptide repeat (PPR) proteins as sequence-specificity factors in post-transcriptional processes in organelles. Biochem Soc Trans 35:1643–1647
Delannoy E, Le Ret M, Faivre-Nitschke E, Estavillo GM, Bergdoll M, Taylor NL, Pogson BJ, Small I, Imbault P, Gualberto JM (2009) Arabidopsis tRNA adenosine deaminase arginine edits the wobble nucleotide of chloroplast tRNAArg (ACG) and is essential for efficient chloroplast translation. Plant Cell 21:2058–2071
Doniwa Y, Ueda M, Ueta M, Wada A, Kadowaki K, Tsutsumi N (2010) The involvement of a PPR protein of the P subfamily in partial RNA editing of an Arabidopsis mitochondrial transcript. Gene 454: 39–46
Dreyfuss G, Kim VN, Kataoka N (2002) Messenger-RNA-binding proteins and the messages they carry. Nat Rev Mol Cell Biol 3:195–205
Du P, Li Y (2008) Prediction of C-to-U RNA editing sites in plant mitochondria using both biochemical and evolutionary information. J Theor Biol 253: 579–586
Du P, Jia L, Li Y (2009) CURE-Chloroplast: a chloroplast C-to-U RNA editing predictor for seed plants. BMC Bioinformatics 10:135
Duff RJ, Moore FB (2005) Pervasive RNA editing among hornwort rbcL transcripts except Leiosporoceros. J Mol Evol 61:571–578
Faivre-Nitschke SE, Grienenberger JM, Gualberto JM (1999) A prokaryotic-type cytidine deaminase from Arabidopsis thaliana gene expression and functional characterization. Eur J Biochem 263:896–903
Farre JC, Araya A (2001) Gene expression in isolated plant mitochondria: high fidelity of transcription, splicing and editing of a transgene product in electroporated organelles. Nucleic Acids Res 29:2484–2491
Farre JC, Leon G, Jordana X, Araya A (2001) cis Recognition elements in plant mitochondrion RNA editing. Mol Cell Biol 21:6731–6737
Fiebig A, Stegemann S, Bock R (2004) Rapid evolution of editing sites in a small non-essential plastid gene. Nucleic Acids Res 7:3615–3622
Fisk DG, Walker MB, Barkan A (1999) Molecular cloning of the maize gene *crp1* reveals similarity between regulators of mitochondrial and chloroplast gene expression. EMBO J 18:2621–2630
Freyer R, Kiefer-Meyer MC, Kössel H (1997) Occurrence of plastid RNA editing in all major lineages of land plants. Proc Natl Acad Sci USA 94:6285–6290
Giegé P, Brennicke A (1999) RNA editing in *Arabidopsis* mitochondria effects 441 C to U changes in ORFs. Proc Natl Acad Sci USA 96:15324–15329
Gillman JD, Bentolila S, Hanson MR (2007) The petunia restorer of fertility protein is part of a large mitochondrial complex that interacts with transcripts of the CMS-associated locus. Plant J 49:217–227
Gott JM, Emeson RB (2000) Functions and mechanisms of RNA editing. Annu Rev Genet 34:499–531
Gray MW (1996) RNA editing in plant organelles: a fertile field. Proc Natl Acad Sci USA 93:8157–8159
Grohmann L, Thieck O, Herz U, Schroder W, Brennicke A (1994) Translation of nad9 mRNAs in mitochondria from Solanum tuberosum is restricted to completely edited transcripts. Nucleic Acids Res 22:3304–3311
Grosskopf D, Mulligan RM (1996) Developmental- and tissue-specificity of RNA editing in mitochondria of suspension-cultured maize cells and seedlings. Curr Genet 29:556–563
Groth-Malonek M, Pruchner D, Grewe F, Knoop V (2005) Ancestors of trans-splicing mitochondrial introns support serial sister group relationships of hornworts and mosses with vascular plants. Mol Biol Evol 22:117–125
Groth-Malonek M, Wahrmund U, Polsakiewicz M, Knoop V (2007) Evolution of a pseudogene: exclusive survival of a functional mitochondrial nad7 gene supports Haplomitrium as the earliest liverwort lineage and proposes a secondary loss of RNA editing in Marchantiidae. Mol Biol Evol 24:1068–1074
Gualberto JM, Lamattina L, Bonnard G, Weil JH, Grienenberger JM (1989) RNA editing in wheat mitochondria results in the conservation of protein sequences. Nature 341:660–662
Halter CP, Peeters NM, Hanson MR (2004) RNA editing in ribosome-less plastids of iojap maize. Curr Genet 45:331–337
Hammani K, Okuda K, Tanz SK, Chateigner-Boutin AL, Shikanai T, Small I (2009) A study of new Arabidopsis chloroplast RNA editing mutants reveals general features of editing factors and their target sites. Plant Cell 21:3686–3699
Hanson M, Sutton C, Lu B (1996) Plant organelle gene expression: altered by RNA editing. Trends Plant Sci 1:57–64
Hashimoto M, Endo T, Peltier G, Tasaka M, Shikanai T (2003) A nucleus-encoded factor, CRR2, is essential for the expression of chloroplast ndhB in Arabidopsis. Plant J 36:541–549
Hayes ML, Hanson MR (2007) Identification of a sequence motif critical for editing of a tobacco chloroplast transcript. RNA 13:281–288
Hayes ML, Hanson MR (2008) High conservation of a 5′ element required for RNA editing of a C target in chloroplast psbE transcripts. J Mol Evol 67:233–245

Hayes ML, Reed ML, Hegeman CE, Hanson MR (2006) Sequence elements critical for efficient RNA editing of a tobacco chloroplast transcript in vivo and in vitro. Nucleic Acids Res 34:3742–3754

He T, Du P, Li Y (2007) dbRES: a web-oriented database for annotated RNA editing sites. Nucleic Acids Res 35:D141–D144

Hecht J, Grewe F, Knoop V (2011b) Extreme RNA editing in coding islands and abundant microsatellites in repeat sequences of Selaginella moellendorffii mitochondria: the root of frequent plant mtDNA recombination in early tracheophytes. Genome Biol Evol 3:344–358

Hegeman CE, Hayes ML, Hanson MR (2005) Substrate and cofactor requirements for RNA editing of chloroplast transcripts in Arabidopsis in vitro. Plant J 42:124–132

Hernould M, Suharsono S, Litvak S, Araya A, Mouras A (1993) Male-sterility induction in transgenic tobacco plants with an unedited atp9 mitochondrial gene from wheat. Proc Natl Acad Sci USA 90:2370–2374

Hiesel R, Wissinger B, Schuster W, Brennicke A (1989) RNA editing in plant mitochondria. Science 246: 1632–1634

Hirose T, Sugiura M (2001) Involvement of a site-specific *trans*-acting factor and a common RNA-binding protein in the editing of chloroplast mRNAs: development of a chloroplast in vitro RNA editing system. EMBO J 20:1144–1152

Hirose T, Kusumegi T, Tsudzuki T, Sugiura M (1999) RNA editing sites in tobacco chloroplast transcripts: editing as a possible regulator of chloroplast RNA polymerase activity. Mol Gen Genet 262:462–467

Hoch B, Maier RM, Appel K, Igloi GL, Kössel H (1991) Editing of a chloroplast mRNA by creation of an initiation codon. Nature 353:178–180

Ikeda TM, Gray MW (1999) Characterization of a DNA-binding protein implicated in transcription in wheat mitochondria. Mol Cell Biol 19:8113–8122

Karcher D, Bock R (1998) Site-selective inhibition of plastid RNA editing by heat shock and antibiotics: a role for plastid translation in RNA editing. Nucleic Acids Res 26:1185–1190

Karcher D, Bock R (2002a) The amino acid sequence of a plastid protein is developmentally regulated by RNA editing. J Biol Chem 277:5570–5574

Karcher D, Bock R (2002b) Temperature sensitivity of RNA editing and intron splicing reactions in the plastid ndhB transcript. Curr Genet 41:48–52

Karcher D, Bock R (2009) Identification of the chloroplast adenosine-to-inosine tRNA editing enzyme. RNA 15:1251–1257

Kazama T, Nakamura T, Watanabe M, Sugita M, Toriyama K (2008) Suppression mechanism of mitochondrial ORF79 accumulation by Rf1 protein in BT-type cytoplasmic male sterile rice. Plant J 55(4):619–628

Kim SR, Yang JI, Moon S, Ryu CH, An K, Kim KM, Yim J, An G (2009) Rice OGR1 encodes a pentatricopeptide repeat-DYW protein and is essential for RNA editing in mitochondria. Plant J 59:738–749

Knoop V (2004) The mitochondrial DNA of land plants: peculiarities in phylogenetic perspective. Curr Genet 46:123–139

Knoop V (2010) When you can't trust the DNA: RNA editing changes transcript sequences. Cell Mol Life Sci 68(4):567–586

Kobayashi K, Suzuki M, Tang J, Nagata N, Ohyama K, Seki H, Kiuchi R, Kaneko Y, Nakazawa M, Matsui M, Matsumoto S, Yoshida S, Muranaka T (2007) Lovastatin insensitive 1, a Novel pentatricopeptide repeat protein, is a potential regulatory factor of isoprenoid biosynthesis in Arabidopsis. Plant Cell Physiol 48:322–331

Kobayashi Y, Matsuo M, Sakamoto K, Wakasugi T, Yamada K, Obokata J (2008) Two RNA editing sites with cis-acting elements of moderate sequence identity are recognized by an identical site-recognition protein in tobacco chloroplasts. Nucleic Acids Res 36:311–318

Kotera E, Tasaka M, Shikanai T (2005) A pentatricopeptide repeat protein is essential for RNA editing in chloroplasts. Nature 433:326–330

Kubo N, Kadowaki K (1997) Involvement of 5¢ flanking sequence for specifying RNA editing sites in plant mitochondria. FEBS Lett 413:40–44

Kugita M, Yamamoto Y, Fujikawa T, Matsumoto T, Yoshinaga K (2003) RNA editing in hornwort chloroplasts makes more than half the genes functional. Nucleic Acids Res 31:2417–2423

Lahmy S, Barneche F, Derancourt J, Filipowicz W, Delseny M, Echeverria M (2000) A chloroplastic RNA-binding protein is a new member of the PPR family. FEBS Lett 480:255–260

Lenz H, Rudinger M, Volkmar U, Fischer S, Herres S, Grewe F, Knoop V (2010) Introducing the plant RNA editing prediction and analysis computer tool PREPACT and an update on RNA editing site nomenclature. Curr Genet 56:189–201

Lippok B, Brennicke A, Wissinger B (1994) Differential RNA editing in closely related introns in Oenothera mitochondria. Mol Gen Genet 243:39–46

Lurin C, Andres C, Aubourg S, Bellaoui M, Bitton F, Bruyere C, Caboche M, Debast C, Gualberto J, Hoffmann B, Lecharny A, Le Ret M, Martin-Magniette ML, Mireau H, Peeters N, Renou JP, Szurek B, Taconnat L, Small I (2004) Genome-wide analysis of Arabidopsis pentatricopeptide repeat proteins reveals their essential role in organelle biogenesis. Plant Cell 16:2089–2103

Maier RM, Hoch B, Zeltz P, Kössel H (1992a) Internal editing of the maize chloroplast *ndhA* transcript restores codons for conserved amino acids. Plant Cell 4:609–616

Maier RM, Neckermann K, Hoch B, Akhmedov NB, Kössel H (1992b) Identification of editing positions in the *ndhB* transcript from maize chloroplasts reveals sequence similarities between editing sites of chloroplasts and plant mitochondria. Nucleic Acids Res 20:6189–6194

Maier RM, Zeltz P, Kössel H, Bonnard G, Gualberto JM, Grienenberger JM (1996) RNA editing in plant mitochondria and chloroplasts. Plant Mol Biol 32:343–365

Maier U-G, Bozarth A, Funk H, Zauner S, Rensing S, Schmitz-Linneweber C, Börner T, Tillich M (2008) Complex chloroplast RNA metabolism: just debugging the genetic programme? BMC Biol 6:36

Malek O, Lättig K, Hiesel R, Brennicke A, Knoop V (1996) RNA editing in bryophytes and a molecular phylogeny of land plants. EMBO J 15:1403–1411

Manthey GM, McEwen JE (1995) The product of the nuclear gene *PET309* is required for translation of mature mRNA and stability or production of intron-containing RNAs derived from the mitochondrial *COX1* locus of *Saccharomyces cerevisiae*. EMBO J 14:4031–4043

Maruyama K, Sato N, Ohta N (1999) Conservation of structure and cold-regulation of RNA-binding proteins in cyanobacteria: probable convergent evolution with eukaryotic glycine-rich RNA-binding proteins. Nucleic Acids Res 27:2029–2036

Miyamoto T, Obokata J, Sugiura M (2002) Recognition of RNA editing sites is directed by unique proteins in chloroplasts: biochemical identification of cis-acting elements and trans-acting factors involved in RNA editing in tobacco and pea chloroplasts. Mol Cell Biol 22:6726–6734

Miyamoto T, Obokata J, Sugiura M (2004) A site-specific factor interacts directly with its cognate RNA editing site in chloroplast transcripts. Proc Natl Acad Sci USA 101:48–52

Miyata Y, Sugita M (2004) Tissue- and stage-specific RNA editing of rps 14 transcripts in moss (*Physcomitrella patens*) chloroplasts. J Plant Physiol 161:113–115

Miyata Y, Sugiura C, Kobayashi Y, Hagiwara M, Sugita M (2002) Chloroplast ribosomal S14 protein transcript is edited to create a translation initiation codon in the moss *Physcomitrella patens*. Biochim Biophys Acta 1576:346–349

Mower JP (2005) PREP-Mt: predictive RNA editor for plant mitochondrial genes. BMC Bioinformatics 6:96

Mower JP (2009) The PREP suite: predictive RNA editors for plant mitochondrial genes, chloroplast genes and user-defined alignments. Nucleic Acids Res 37:W253–W259

Mulligan RM, Williams MA, Shanahan MT (1999) RNA editing site recognition in higher plant mitochondria. J Hered 90:338–344

Nakajima Y, Mulligan RM (2001) Heat stress results in incomplete C-to-U editing of maize chloroplast mRNAs and correlates with changes in chloroplast transcription rate. Curr Genet 40:209–213

Nakajima Y, Mulligan RM (2005) Nucleotide specificity of the RNA editing reaction in pea chloroplasts. J Plant Physiol 162:1347–1354

Nakamura T, Sugita M (2008) A conserved DYW domain of the pentatricopeptide repeat protein possesses a novel endoribonuclease activity. FEBS Lett 582:4163–4168

Nakamura T, Ohta M, Sugiura M, Sugita M (2001) Chloroplast ribonucleoproteins function as a stabilizing factor of ribosome-free mRNAs in the stroma. J Biol Chem 276:147–152

Nakamura T, Meierhoff K, Westhoff P, Schuster G (2003) RNA-binding properties of HCF152, an Arabidopsis PPR protein involved in the processing of chloroplast RNA. Eur J Biochem 270:4070–4081

Navaratnam N, Sarwar R (2006) An overview of cytidine deaminases. Int J Hematol 83:195–200

Neuwirt J, Takenaka M, van der Merwe JA, Brennicke A (2005) An in vitro RNA editing system from cauliflower mitochondria: editing site recognition parameters can vary in different plant species. RNA 11:1563–1570

O'Brien EA, Zhang Y, Wang E, Marie V, Badejoko W, Lang BF, Burger G (2009) GOBASE: an organelle genome database. Nucleic Acids Res 37:D946–D950

Ohtani S, Ichinose M, Tasaki E, Aoki Y, Komura Y, Sugita M (2010) Targeted gene disruption identifies three PPR-DYW proteins involved in RNA editing for five editing sites of the moss mitochondrial transcripts. Plant Cell Physiol 51:1942–1949

Okuda K, Nakamura T, Sugita M, Shimizu T, Shikanai T (2006) A pentatricopeptide repeat protein is a site recognition factor in chloroplast RNA editing. J Biol Chem 281:37661–37667

Okuda K, Myouga F, Motohashi R, Shinozaki K, Shikanai T (2007) Conserved domain structure of pentatricopeptide repeat proteins involved in chloroplast RNA editing. Proc Natl Acad Sci USA 104:8178–8183

Okuda K, Hammani K, Tanz SK, Peng L, Fukao Y, Myouga F, Motohashi R, Shinozaki K, Small I, Shikanai T (2009a) The pentatricopeptide repeat protein OTP82 is required for RNA editing of plastid ndhB and ndhG transcripts. Plant J 61:339–349

Okuda K, Chateigner-Boutin AL, Nakamura T, Delannoy E, Sugita M, Myouga F, Motohashi R, Shinozaki K, Small I, Shikanai T (2009b) Pentatricopeptide repeat proteins with the DYW motif have distinct molecular functions in RNA editing and RNA cleavage in Arabidopsis chloroplasts. Plant Cell 21:146–156

Parkinson CL, Mower JP, Qiu YL, Shirk AJ, Song K, Young ND, DePamphilis CW, Palmer JD (2005) Multiple major increases and decreases in mitochondrial substitution rates in the plant family Geraniaceae. BMC Evol Biol 5:73

Pfalz J, Bayraktar OA, Prikryl J, Barkan A (2009) Site-specific binding of a PPR protein defines and stabilizes 5′ and 3′ mRNA termini in chloroplasts. EMBO J 28:2042–2052

Phreaner CG, Williams MA, Mulligan RM (1996) Incomplete editing of rps12 transcripts results in the synthesis of polymorphic polypeptides in plant mitochondria. Plant Cell 8:107–117

Picardi E, Regina TM, Brennicke A, Quagliariello C (2007) REDIdb: the RNA editing database. Nucleic Acids Res 35:D173–D177

Picardi E, Regina TM, Verbitskiy D, Brennicke A, Quagliariello C (2010) REDIdb: an upgraded bioinformatics resource for organellar RNA editing sites. Mitochondrion 11(2):360–365

Prikryl J, Rojas M, Schuster G, Barkan A (2010) Mechanism of RNA stabilization and translational activation by a pentatricopeptide repeat protein. Proc Natl Acad Sci USA 108:415–420

Rajasekhar VK, Mulligan RM (1993) RNA editing in plant mitochondria: [alpha]-phosphate is retained during C-to-U conversion in mRNAs. Plant Cell 5:1843–1852

Raynaud C, Loiselay C, Wostrikoff K, Kuras R, Girard-Bascou J, Wollman FA, Choquet Y (2007) Evidence for regulatory function of nucleus-encoded factors on mRNA stabilization and translation in the chloroplast. Proc Natl Acad Sci USA 104:9093–9098

Reed ML, Peeters NM, Hanson MR (2001a) A single alteration 20 nt 5′ to an editing target inhibits chloroplast RNA editing *in vivo*. Nucleic Acids Res 29:1507–1513

Reed ML, Lyi SM, Hanson MR (2001b) Edited transcripts compete with unedited mRNAs for *trans*-acting editing factors in higher plant chloroplasts. Gene 272:165–171

Robbins JC, Heller WP, Hanson MR (2009) A comparative genomics approach identifies a PPR-DYW protein that is essential for C-to-U editing of the Arabidopsis chloroplast accD transcript. RNA 15:1142–1153

Rudinger M, Polsakiewicz M, Knoop V (2008) Organellar RNA editing and plant-specific extensions of pentatricopeptide repeat proteins in jungermanniid but not in marchantiid liverworts. Mol Biol Evol 25:1405–1414

Rudinger M, Funk HT, Rensing SA, Maier UG, Knoop V (2009) RNA editing: only eleven sites are present in the Physcomitrella patens mitochondrial transcriptome and a universal nomenclature proposal. Mol Genet Genomics 281:473–481

Salone V, Rudinger M, Polsakiewicz M, Hoffmann B, Groth-Malonek M, Szurek B, Small I, Knoop V, Lurin C (2007) A hypothesis on the identification of the editing enzyme in plant organelles. FEBS Lett 581:4132–4138

Sasaki T, Yukawa Y, Miyamoto T, Obokata J, Sugiura M (2003) Identification of RNA editing sites in chloroplast transcripts from the maternal and paternal progenitors of tobacco (*Nicotiana tabacum*): comparative analysis shows the involvement of distinct *trans*-factors for *ndhB* editing. Mol Biol Evol 20:1028–1035

Sasaki T, Yukawa Y, Wakasugi T, Yamada K, Sugiura M (2006) A simple in vitro RNA editing assay for chloroplast transcripts using fluorescent dideoxynucleotides: distinct types of sequence elements required for editing of ndh transcripts. Plant J 47:802–810

Schmitz-Linneweber C, Barkan A (2007) RNA splicing and RNA editing in chloroplasts. In: Barkan A (ed) Cell and molecular biology of plastids, vol 19. Springer, Berlin/Heidelberg, pp 213–248

Schmitz-Linneweber C, Small I (2008) Pentatricopeptide repeat proteins: a socket set for organelle gene expression. Trends Plant Sci 13:663–670

Schmitz-Linneweber C, Tillich M, Herrmann RG, Maier RM (2001) Heterologous, splicing-dependent RNA editing in chloroplasts: allotetraploidy provides *trans*-factors. EMBO J 20:4874–4883

Schmitz-Linneweber C, Regel R, Du TG, Hupfer H, Herrmann RG, Maier RM (2002) The plastid chromosome of *Atropa belladonna* and its comparison with that of *Nicotiana tabacum*: the role of RNA editing in generating divergence in the process of plant speciation. Mol Biol Evol 19:1602–1612

Schmitz-Linneweber C, Williams-Carrier R, Barkan A (2005a) RNA immunoprecipitation and microarray analysis show a chloroplast Pentatricopeptide repeat protein to be associated with the 5′ region of mRNAs whose translation it activates. Plant Cell 17: 2791–2804

Schmitz-Linneweber C, Kushnir S, Babiychuk E, Poltnigg P, Herrmann RG, Maier RM (2005b) Pigment deficiency in nightshade/tobacco cybrids is caused by the failure to edit the plastid ATPase alpha-subunit mRNA. Plant Cell 17:1815–1828

Schmitz-Linneweber C, Williams-Carrier R, Williams P, Kroeger T, Vichas A, Barkan A (2006) A pentatricopeptide repeat protein binds to and facilitates the *trans*-splicing of the maize chloroplast *rps12* pre-mRNA. Plant Cell 18:2650–2663

Shields DC, Wolfe KH (1997) Accelerated evolution of sites undergoing mRNA editing in plant mitochondria and chloroplasts. Mol Biol Evol 14:344–349

Small I, Peeters N (2000) The PPR motif – a TPR-related motif prevalent in plant organellar proteins. Trends Biochem Sci 25:46–47

Smith H, Gott J, Hanson M (1997) A guide to RNA editing. RNA 3:1105–1123

Staudinger M, Kempken F (2003) Electroporation of isolated higher-plant mitochondria: transcripts of an introduced cox2 gene, but not an atp6 gene, are edited in organello. Mol Genet Genomics 269:553–561

Staudinger M, Bolle N, Kempken F (2005) Mitochondrial electroporation and in organello RNA editing of chimeric atp6 transcripts. Mol Genet Genomics 273:130–136

Steinhauser S, Beckert S, Capesius I, Malek O, Knoop V (1999) Plant mitochondrial RNA editing. J Mol Evol 48:303–312

Sugita M, Miyata Y, Maruyama K, Sugiura C, Arikawa T, Higuchi M (2006) Extensive RNA editing in transcripts from the PsbB operon and RpoA gene of plastids from the enigmatic moss *Takakia lepidozioides*. Biosci Biotechnol Biochem 70:2268–2274

Sung TY, Tseng CC, Hsieh MH (2010) The SLO1 PPR protein is required for RNA editing at multiple sites with similar upstream sequences in Arabidopsis mitochondria. Plant J 63:499–511

Takenaka M (2010) MEF9, an E-subclass pentatricopeptide repeat protein, is required for an RNA editing event in the nad7 transcript in mitochondria of Arabidopsis. Plant Physiol 152:939–947

Takenaka M, Brennicke A (2009) Multiplex single-base extension typing to identify nuclear genes required for RNA editing in plant organelles. Nucleic Acids Res 37:e13

Takenaka M, Neuwirt J, Brennicke A (2004) Complex cis-elements determine an RNA editing site in pea mitochondria. Nucleic Acids Res 32:4137–4144

Takenaka M, Verbitskiy D, van der Merwe JA, Zehrmann A, Plessmann U, Urlaub H, Brennicke A (2007) In vitro RNA editing in plant mitochondria does not require added energy. FEBS Lett 581:2743–2747

Takenaka M, Verbitskiy D, van der Merwe JA, Zehrmann A, Brennicke A (2008) The process of RNA editing in plant mitochondria. Mitochondrion 8:35–46

Takenaka M, Verbitskiy D, Zehrmann A, Brennicke A (2010) Reverse genetic screening identifies five E-class PPR proteins involved in RNA editing in mitochondria of *Arabidopsis thaliana*. J Biol Chem 285:27122–27129

Tang J, Kobayashi K, Suzuki M, Matsumoto S, Muranaka T (2010) The mitochondrial PPR protein LOVASTATIN INSENSITIVE 1 plays regulatory roles in cytosolic and plastidial isoprenoid biosynthesis through RNA editing. Plant J 61:456–466

Tasaki E, Hattori M, Sugita M (2010) The moss pentatricopeptide repeat protein with a DYW domain is responsible for RNA editing of mitochondrial ccmFc transcript. Plant J 62:560–570

Thompson J, Gopal S (2006) Genetic algorithm learning as a robust approach to RNA editing site prediction. BMC Bioinformatics 7:145

Thompson J, Higgins D, Gibson T (1994) Clustal W: improving the sensitivity of progressive multiple sequence alignment through sequence weighting, position-specific gap penalties and weight matrix choice. Nucleic Acids Res 22:4673–4680

Tillich M, Funk HT, Schmitz-Linneweber C, Poltnigg P, Sabater B, Martin M, Maier RM (2005) Editing of plastid RNA in *Arabidopsis thaliana* ecotypes. Plant J 43:708–715

Tillich M, Lehwark P, Morton BR, Maier UG (2006) The evolution of chloroplast RNA editing. Mol Biol Evol 23:1912–1921

Tillich M, Le Sy V, Schulerowitz K, von Haeseler A, Maier UG, Schmitz-Linneweber C (2009a) Loss of matK RNA editing in seed plant chloroplasts. BMC Evol Biol 9:201

Tillich M, Hardel SL, Kupsch C, Armbruster U, Delannoy E, Gualberto JM, Lehwark P, Leister D, Small ID, Schmitz-Linneweber C (2009b) Chloroplast ribonucleoprotein CP31A is required for editing and stability of specific chloroplast mRNAs. Proc Natl Acad Sci USA 106:6002–6007

Tillich M, Beick S, Schmitz-Linneweber C (2010) Chloroplast RNA-binding proteins: repair and regulation of chloroplast transcripts. RNA Biol 7:172–178

Tseng CC, Sung TY, Li YC, Hsu SJ, Lin CL, Hsieh MH (2010) Editing of accD and ndhF chloroplast transcripts is partially affected in the Arabidopsis vanilla cream1 mutant. Plant Mol Biol 73:309–323

Tsuchiya N, Fukuda H, Sugimura T, Nagao M, Nakagama H (2002) LRP130, a protein containing nine pentatricopeptide repeat motifs, interacts with a single-stranded cytosine-rich sequence of mouse hypervariable minisatellite Pc-1. Eur J Biochem 269:2927–2933

Tsudzuki T, Wakasugi T, Sugiura M (2001) Comparative analysis of RNA editing sites in higher plant chloroplasts. J Mol Evol 53:327–332

Uyttewaal M, Arnal N, Quadrado M, Martin-Canadell A, Vrielynck N, Hiard S, Gherbi H, Bendahmane A, Budar F, Mireau H (2008) Characterization of

Raphanus sativus pentatricopeptide repeat proteins encoded by the fertility restorer locus for Ogura cytoplasmic male sterility. Plant Cell 20: 3331–3345

Valcarcel J, Green MR (1996) The SR protein family: pleiotropic functions in pre-mRNA splicing. Trends Biochem Sci 21:296–301

Valente L, Nishikura K (2005) ADAR gene family and A-to-I RNA editing: diverse roles in posttranscriptional gene regulation. Prog Nucleic Acid Res Mol Biol 79:299–338

van der Merwe JA, Takenaka M, Neuwirt J, Verbitskiy D, Brennicke A (2006) RNA editing sites in plant mitochondria can share *cis*-elements. FEBS Lett 580:268–272

Verbitskiy D, van der Merwe JA, Zehrmann A, Brennicke A, Takenaka M (2008) Multiple specificity recognition motifs enhance plant mitochondrial RNA editing in vitro. J Biol Chem 283:24374–24381

Verbitskiy D, Zehrmann A, van der Merwe JA, Brennicke A, Takenaka M (2009) The PPR protein encoded by the LOVASTATIN INSENSITIVE 1 gene is involved in RNA editing at three sites in mitochondria of *Arabidopsis thaliana*. Plant J 61(3):446–455

Verbitskiy D, Zehrmann A, Brennicke A, Takenaka M (2010) A truncated MEF11 protein shows site-specific effects on mitochondrial RNA editing. Plant Signal Behav 5(5):558–560

Wakasugi T, Tsudzuki T, Sugiura M (2001) The genomics of land plant chloroplasts: gene content and alteration of genomic information by RNA editing. Photosynth Res 70:107–118

Walbot V (1991) RNA editing fixes problems in plant mitochondrial transcripts. Trends Genet 7:37–39

Wang Z, Zou Y, Li X, Zhang Q, Chen L, Wu H, Su D, Chen Y, Guo J, Luo D, Long Y, Zhong Y, Liu YG (2006) cytoplasmic male sterility of rice with boro II cytoplasm is caused by a cytotoxic peptide and is restored by two related PPR motif genes via distinct modes of mRNA silencing. Plant Cell 18:676–687

Williams MA, Tallakson WA, Phreaner CG, Mulligan RM (1998) Editing and translation of ribosomal protein S13 transcripts: unedited translation products are not detectable in maize mitochondria. Curr Genet 34:221–226

Williams-Carrier R, Kroeger T, Barkan A (2008) Sequence-specific binding of a chloroplast pentatricopeptide repeat protein to its native group II intron ligand. RNA 14:1930–1941

Wintz H, Hanson MR (1991) A termination codon is created by RNA editing in the petunia atp9 transcript. Curr Genet 19:61–64

Wolf PG, Rowe CA, Hasebe M (2004) High levels of RNA editing in a vascular plant chloroplast genome: analysis of transcripts from the fern *Adiantum capillus-veneris*. Gene 339:89–97

Yoshinaga K, Iinuma H, Masuzawa T, Uedal K (1996) Extensive RNA editing of U to C in addition to C to U substitution in the rbcL transcripts of hornwort chloroplasts and the origin of RNA editing in green plants. Nucleic Acids Res 24:1008–1014

Yoshinaga K, Kakehi T, Shima Y, Iinuma H, Masuzawa T, Ueno M (1997) Extensive RNA editing and possible double-stranded structures determining editing sites in the atpB transcripts of hornwort chloroplasts. Nucleic Acids Res 25:4830–4834

Yu W, Schuster W (1995) Evidence for a site-specific cytidine deamination reaction involved in C to U RNA editing of plant mitochondria. J Biol Chem 270:18227–18233

Yu QB, Jiang Y, Chong K, Yang ZN (2009) AtECB2, a pentatricopeptide repeat protein, is required for chloroplast transcript accD RNA editing and early chloroplast biogenesis in *Arabidopsis thaliana*. Plant J 59:1011–1023

Yura K, Sulaiman S, Hatta Y, Shionyu M, Go M (2009) RESOPS: a database for analyzing the correspondence of RNA editing sites to protein three-dimensional structures. Plant Cell Physiol 50: 1865–1873

Zabaleta E, Mouras A, Hernould M, Suharsono AA (1996) Transgenic male-sterile plant induced by an unedited atp9 gene is restored to fertility by inhibiting its expression with antisense RNA. Proc Natl Acad Sci USA 93:11259–11263

Zandueta-Criado A, Bock R (2004) Surprising features of plastid ndhD transcripts: addition of non-encoded nucleotides and polysome association of mRNAs with an unedited start codon. Nucleic Acids Res 32:542–550

Zehrmann A, van der Merwe JA, Verbitskiy D, Brennicke A, Takenaka M (2008) Seven large variations in the extent of RNA editing in plant mitochondria between three ecotypes of *Arabidopsis thaliana*. Mitochondrion 8:319–327

Zehrmann A, Verbitskiy D, van der Merwe JA, Brennicke A, Takenaka M (2009) A DYW domain-containing pentatricopeptide repeat protein is required for RNA editing at multiple sites in mitochondria of *Arabidopsis thaliana*. Plant Cell 21:558–567

Zehrmann A, Verbitskiy D, Hartel B, Brennicke A, Takenaka M (2010) RNA editing competence of trans-factor MEF1 is modulated by ecotype-specific differences but requires the DYW domain. FEBS Lett 584:4181–4186

Zhou W, Cheng Y, Yap A, Chateigner-Boutin AL, Delannoy E, Hammani K, Small I, Huang J (2009) The Arabidopsis gene YS1 encoding a DYW protein is required for editing of rpoB transcripts and the rapid development of chloroplasts during early growth. Plant J 58:82–96

Zito F, Kuras R, Choquet Y, Kössel H, Wollman FA (1997) Mutations of cytochrome b6 in *Chlamydomonas reinhardtii* disclose the functional significance for a proline to leucine conversion by *petB* editing in maize and tobacco. Plant Mol Biol 33:79–86

Chapter 14

Expression Profiling of Organellar Genes

Teodoro Cardi
CRA-ORT, Agricultural Research Council, Research Centre for Vegetable Crops, via Cavalleggeri 25, Pontecagnano, SA 84098, Italy

Philippe Giegé
IBMP-CNRS, 12 rue du général Zimmer, Strasbourg 67084, France

Sabine Kahlau
QIAGEN GmbH, QIAGEN Strasse 1, D-40724 Hilden, Germany

Nunzia Scotti
CNR – IGV, Institute of Plant Genetics, Res. Div. Portici, via Università 133, Portici, NA 80055, Italy

Summary

Due to their endosymbiotic origin, expression of plastid and mitochondrial genes retains several features of prokaryotes. Nevertheless, plant organelles acquired novel specific traits during evolution. Furthermore, due to the migration of many genes to the nucleus of the host

* Author for correspondence, e-mail: teodoro.cardi@entecra.it

R. Bock and V. Knoop (eds.), Genomics of Chloroplasts and Mitochondria, Advances in Photosynthesis and Respiration 35, pp. 323–355,
DOI 10.1007/978-94-007-2920-9_14, © Springer Science+Business Media B.V. 2012

cell, complex anterograde and retrograde signalling pathways evolved to coordinate gene expression in different subcellular compartments. Control of gene expression in plant organelles occurs at the transcriptional and posttranscriptional levels. In this chapter, we analyze the available data concerning the variability shown by both organelle genomes for different steps of gene expression in various genotypes or after environmental and developmental cues. Genotypic variability for the extent of RNA editing or transcript processing and stability in cytoplasmic organelles has been observed in natural populations at the interspecific and intraspecific level or in artificial CMS lines. The role of various plastid genes in global genome expression and chloroplast development has been highlighted in knock-out lines produced by plastid transformation. Significant differences in the transcriptome, editome and translatome have also been found comparing different plastid types in diverse organs or tissues. Similar differences have been found for mitochondrial genomes during the diurnal cycle or between cell suspensions and differentiated leaves. However, the precise level and mechanisms at which these changes are achieved and the signals necessary for their installation are barely understood.

I. Introduction

According to the endosymbiotic theory, plastids and mitochondria derive from early prokaryotic organisms engulfed by a proto-eukaryotic cell (Buchanan et al. 2000). The origin and the following evolution of both plant organelles have had important implications not only for the structure of their genomes, but also for the expression of their genes.

Expression of plastid and mitochondrial genes retain several features of prokaryotes, e.g. the common, albeit not exclusive, organization in operons, implying co-transcription of individual genes, similar RNA polymerases and promoters (in plastids), similar structure of mature mRNAs, the presence of 70S-type ribosomes, and others. Nevertheless, different from bacteria and other prokaryotes, plant organelles show novel traits, such as uncoupled transcription and translation, phage-type RNA polymerases, and frequent RNA editing and splicing of transcript precursors. Furthermore, during evolution, many genes originally present in the endosymbionts' genomes migrated to the nucleus of the host cell. Since in many cases nuclear and organelle genes encode subunits of the same protein complexes, their expression needs to be somehow co-regulated, implying complex anterograde and retrograde signalling pathways between different subcellular compartments (Bräutigam et al. 2007).

Control of gene expression in plant organelles occurs at the transcriptional and post-transcriptional levels, the latter including regulation of transcript maturation and stability, translation, protein stability and activity (Bollenbach et al. 2007; Liere and Börner 2007; Peled-Zehavi and Danon 2007; Schmitz-Linneweber and Barkan 2007). The recent development of novel technologies, in particular DNA arrays, allowed the genome-wide analyses of gene expression in different genotypes, tissues, and environmental conditions. In several cases, the concomitant analysis of nuclear genes involved in organelle gene expression allowed to dissect interorganellar regulatory pathways (Biehl et al. 2005).

In this chapter, after reviewing basic aspects of gene expression in plastids and plant mitochondria, we analyze the available data concerning the variability shown by both organelle genomes for different steps of gene expression in various genotypes or in response to environmental and developmental cues.

Abbreviations: ACCase – Acetyl-CoA carboxylase; – AOX – Alternative oxidase; CMS – Cytoplasmic male sterility; NEP – Nuclear encoded polymerase; PEP – Plastid encoded polymerase; PPR – Pentatricopeptide repeat; PSI – Photosystem I; PSII – Photosystem II; RNAP – RNA polymerase

II. Regulation of Gene Expression in Plant Organelles

A. Transcription

In plastids, the RNA-synthesizing activity is carried out by two enzymes of different evolutionary origins. A plastid-encoded RNA polymerase (PEP) is homologous to the eubacterial RNA polymerase also found in cyanobacteria, the closest extant bacterial relatives of plastids (Mereschkowsky 1905; Kaneko et al. 1996). The plastid genome encodes the core subunits of the bacterial-type RNA polymerase, consisting of the four proteins RpoA, RpoB, RpoC1 and RpoC2. The corresponding genes were identified in the first completely sequenced plastomes of *Marchantia polymorpha* and *Nicotiana tabacum* (Ohyama et al. 1986; Shinozaki et al. 1986). A second RNA-synthesizing activity is carried out through a nuclear-encoded RNA polymerase (NEP) with homology to phage-type RNA polymerases (Lerbs-Mache 1993). The genome of *Arabidopsis thaliana* contains three copies of *RpoT* genes designated as *RpoTm*, *RpoTmp*, and *RpoTp*, indicating sub-cellular localization in mitochondria (m) and/or plastids (p) (Hedtke et al. 1997, 2000; Chang and Stern 1999; Kobayashi et al. 2001). Although plastid genes encoding proteins involved in gene expression, including PEP, are preferentially transcribed by the NEP enzyme, several chloroplast genes are driven from promoters for both polymerases. The PEP enzyme transcribes predominantly photosynthesis-related genes (Allison et al. 1996; Hajdukiewicz et al. 1997; Silhavy and Maliga 1998; Liere and Maliga 1999). It was hypothesized that the NEP polymerase is activated early in chloroplast development resulting in transcription of PEP which, in turn, activates photosynthesis-related genes (Mullet 1993). Although a recent study could show that both enzymes are already present in seeds (Demarsy et al. 2006), transcripts encoding the gene expression machinery, which are predominantly transcribed by NEP, peak in their maximal abundance earlier during chloroplast development than the predominantly PEP-transcribed photosynthesis genes (Baumgartner et al. 1993).

Although it is known that auxiliary factors are required for efficient transcription initiation in vivo in plastids (Kühn et al. 2007), experiments to identify and characterize factors involved in NEP promoter recognition and transcription initiation have failed so far. By contrast, plastids require, like bacteria, additional σ-like factors for correct PEP promoter recognition. Whereas just one σ-like factor is known in *Chlamydomonas reinhardtii* (Carter et al. 2004; Bohne et al. 2006), six σ-factors, designated as Sig1-6, are encoded in the nuclear genome of *A. thaliana* (Isono et al. 1997b; Tanaka et al. 1997; Kanamaru et al. 1999; Fujiwara et al. 2000). They have a general role in transcription, recognize certain promoters or respond to environmental stimuli (for review see Shiina et al. 2005; Liere and Börner 2007).

Unlike plastids, plant mitochondria rely entirely on nuclear encoded RNA polymerases for transcription. Recent studies have shown that *RpoTmp* could be involved in the transcription of specific genes in mitochondria (Kühn et al. 2009) and requires additional protein partners to recognise specifically promoter sequences, as observed in human and yeast mitochondria (Tracy and Stern 1995). Contrary to vertebrates, where transcription is initiated at a single site on each DNA strand, plant mitochondrial transcription is initiated at multiple sites. Moreover, transcription of single genes can be initiated by multiple promoters (Lupold et al. 1999; Kühn et al. 2005). Promoter sequences of the *A. thaliana* mitochondrial genome often contain the consensus motif YRTA, although transcription can also be initiated at non-canonical sites that lack any kind of recognizable consensus motif (Binder and Brennicke 1993; Kühn et al. 2005; Remacle and Maréchal-Drouard 1996; Fey and Maréchal-Drouard 1999).

Inverted repeat sequences forming stem-loop structures in 3′-UTR of transcripts are present both in plastids and mitochondria. However, they were rather found to be involved in RNA maturation than in transcription termination (Dombrowski et al. 1997; Hoffmann et al. 1999). Indeed, despite the complex transcription mechanisms, it appears that posttranscriptional processes have a major role in the regulation of gene expression in both organelles and often override changes at the transcriptional level (Giegé et al. 2000; Holec et al. 2006; Bollenbach et al. 2007; Liere and Börner 2007; Peled-Zehavi and Danon 2007; Schmitz-Linneweber and Barkan 2007).

B. RNA Editing

The term RNA editing comprises a variety of single nucleotide alterations which change the genetic information at the RNA level. Editing was initially described in trypanosome mitochondria (Benne et al. 1986) and includes alterations like nucleotide insertions or deletions as well as nucleotide modifications and replacements. In higher plant chloroplasts and mitochondria, RNA editing (see also Chap. 13) is restricted to C-to-U conversions (Covello and Gray 1989; Gualberto et al. 1989; Hiesel et al. 1989; Shikanai 2006; Liere and Börner 2007; Stern et al. 2010), while less frequent U-to-C conversions have been reported in fern, hornwort and lycophyte organelles (Malek et al. 1996; Knoop 2004; Wolf et al. 2004; Duff and Moore 2005; Shikanai 2006; Takenaka et al. 2008). Several common features of the editing process in chloroplasts and mitochondria suggest a common evolutionary origin of the two organelle editing systems (Freyer et al. 1997; Tillich et al. 2006).

To date, 34 editing sites are known in *A. thaliana* plastids (Chateigner-Boutin and Small 2007), representing a typical number of editing sites found in vascular plant chloroplasts. In comparison, RNA editing affects over 500 cytidines in mitochondria (Giegé and Brennicke 1999; Chateigner-Boutin and Small 2007; Zehrmann et al. 2008), although the number of editing sites per gene is highly variable. In Arabidopsis mitochondria, complex I and CCM (cytochrome *c* maturation) mRNAs have the highest RNA editing frequencies (Giegé and Brennicke 1999). There exist a few examples for editing occurring in non-coding regions such as introns (Chateigner-Boutin and Small 2007), but most editing events restore conserved amino acids and create start or stop codons (Chapdelaine and Bonen 1991; Hoch et al. 1991; Neckermann et al. 1994; Maier et al. 1995; Giegé et al. 2004; Miyata and Sugita 2004; Okuda et al. 2006). In addition, it often affects positions that appear to be essential for the respective protein functions (Bock et al. 1994b), although in 10% of cases RNA editing is "silent" since the third position of a codon is affected and the amino acid identity is unchanged (Giegé and Brennicke 1999). In maize mitochondria, the editing of a *nad7* intron is required for its proper folding and thus for efficient splicing (Carrillo and Bonen 1997). Similarly, RNA editing is required for the proper folding of mitochondrial tRNAs, which is a prerequisite for their maturation (Maréchal-Drouard et al. 1996a, b; Kunzmann et al. 1998).

The partial RNA editing at some sites and the consequent heterogeneous populations of transcripts (Chateigner-Boutin and Hanson 2003; Bentolila et al. 2008) could lead to the synthesis of different forms of individual proteins. Although both the edited and unedited versions of the plastid *ndhD* transcript are associated with ribosomes, edited transcripts are highly enriched in the most actively translated polysome fractions (Zandueta-Criado and Bock 2004). Other studies have shown that only the proteins resulting from fully edited transcripts accumulate in mitochondria or plastids and no examples exist showing that an unedited protein has a function within organelles (Grohmann et al. 1994; Lu and Hanson 1994; Phreaner et al. 1996). This suggests that translation of partially edited transcripts could be inhibited and/or that proteins resulting from partially edited RNA are instable and rapidly degraded. Hence, it has been hypothesized that instead of a regulatory role, the primary function of RNA editing could have been to correct genomic mutations that appeared during the invasion of land by plants and thus to enable the translation of functional proteins (Shikanai 2006; Takenaka et al. 2008).

The exact editing mechanism still remains elusive. The analysis of the hundreds of editing sites present in the plant mitochondrial transcriptome (Giegé and Brennicke 1999; Bentolila et al. 2008) has not enabled to define specific consensus signals around editing sites. However, the distribution of nucleotides around the sites is not random because a strong preference for pyrimidines is observed for the two nucleotides immediately upstream of the sites (Giegé and Brennicke 1999). Both in plastids and mitochondria *trans*-acting factors are involved in recognition of the endogenous editing sites (Chaudhuri et al. 1995; Bock and Koop 1997). So far, several pentatricopeptide repeat (PPR) proteins, encoded by a gene family with more than 450 members in *A. thaliana* and characterized by repeated motifs of a degenerate 35 amino-acid consensus, were found to be involved in editing site recognition (Small and Peeters 2000; Kotera et al. 2005; Okuda et al. 2006, 2007; Kim et al. 2009; Zehrmann et al. 2009; Tasaki et al. 2010; Verbitskiy et al. 2010) although, so far, it could not be shown that these proteins actually carry out the deamination reaction, which converts the cytidine to a uridine (Hirose and Sugiura 2001).

C. RNA Processing

Plastid and mitochondrial genes are often, like in their bacterial ancestors, transcribed from operons resulting in polycistronic transcripts. Numerous rearrangements that occurred during the evolution of plant mitochondrial genomes have led to the loss of ancient syntenies of gene organisation into functional units (Schuster 1993; Giegé et al. 2000). As a consequence, co-transcription often does not involve genes of related function. This phenomenon is somewhat less widespread in plastids, where genes encoding subunits of the same protein complex are more frequently present in the same operon. Generally, prior to protein synthesis, transcripts are cleaved intercistronically and their 5′ and 3′ ends undergo maturation steps.

5′ ends of chloroplast transcripts are either unprocessed and then characterized by a 5′ di- or triphosphate or carry a 5′ hydroxyl group in the case of processed mRNAs. The enzymes catalyzing these processing steps are so far unknown (for review see Bollenbach et al. 2007). In contrast, the mechanisms of 3′ end maturation are known in much greater detail. Transcription termination is rather inefficient in plastids, resulting in the requirement of 3′ end processing. This involves the binding of a high molecular weight complex downstream of the stem-loop structures formed by inverted repeats (reviewed in Stern and Kindle 1993; Hayes et al. 1999). A second mechanism for mRNA stabilization is the binding of PPR proteins (Pfalz et al. 2009).

Similarly, the maturation of plant mitochondrial precursor transcripts involves 5′- and 3′-maturation steps. These maturations could be achieved through direct endoribonuclease activities and/or with 5′-to-3′ exoribonucleases and 3′-to-5′ exoribonucleases. Such enzymes are encoded in the nucleus and must be imported from the cytosol. In higher plant mitochondria, no 5′-to-3′ exoribonuclease has been identified yet. In contrast, two 3′-to-5′ exoribonucleases were characterised: RNase II is dually localised in mitochondria and plastids and a polynucleotide phosphorylase (PNPase) is found in mitochondria. Studies suggest that the 3′-processing of mitochondrial transcripts is at least a two-step phenomenon (Gagliardi et al. 2001; Perrin et al. 2004a, b). tRNAs are also transcribed as precursor molecules and have to be matured at their 5′- and 3′-ends. These maturation steps are performed by two ubiquitous endoribonuclease activities called RNase P and RNase Z, respectively (Vogel et al. 2005; Canino et al. 2009; Gobert et al. 2010; Jonietz et al. 2010).

Splicing is an essential process in RNA maturation in plant organelles as introns disrupt reading frames of important genes involved in photosynthesis or gene expression. Twenty out of the 21 plastid introns found in land plants belong to group II introns and the remaining intron in the *trnL-UAA* is a group I intron (reviewed in Saldanha et al. 1993). Factors involved in plastid intron splicing are mostly encoded in the nucleus. A single maturase-like protein, MatK, is encoded in the *trn*K gene intron and is potentially involved in splicing of several group II introns (Liere and Link 1995; Jenkins et al. 1997). A rather unusual intron is one of the two introns in the *rps12* gene. This bipartite gene is encoded at two distant locations in the plastid genome, splitting this intron into two separate parts. Thus, two precursor mRNAs are generated which are joined together in a *trans*-splicing event (Hildebrand et al. 1988).

In seed plant mitochondria, group II introns are found in several genes (Unseld et al. 1997; Bonen and Vogel 2001), while only one example of a recently acquired group I intron has been found in the *cox1* genes of *Peperomia* and of some other plants (Vaughn et al. 1995; Cho et al. 1998; Grewe et al. 2009). The genes encoded in the Arabidopsis mitochondrial genome are interrupted by altogether 23 group II introns with sizes varying from 485 to about 4,000 nucleotides (Unseld et al. 1997). Some genes are interrupted by more than one intron, e.g. *nad7* has four introns. *Trans*-splicing is found in plant mitochondria in several instances (e.g. in the *nad1*, *nad2* and *nad5* genes), (Chapdelaine and Bonen 1991; Knoop et al. 1991; Wissinger et al. 1991; Binder et al. 1992; Glanz and Kück 2009). The highly conserved structure of the group II introns and specific protein factors, called maturases, are essential for splicing activity (Wank et al. 1999; Lambowitz and Zimmerly 2004; Meng et al. 2005; Fedorova and Zingler 2007). In plant mitochondria, one conserved gene encoding such a maturase, MatR, is located in the terminal *nad1* intron. Moreover, several nuclear genes and nucleus-encoded proteins, involved in splicing in chloroplasts and, putatively, in mitochondria were identified (Mohr and Lambowitz 2003; Nakagawa and Sakurai 2006; Keren et al. 2008).

Total RNA abundance also depends on the rate of transcript degradation. It has been shown in spinach and barley that plastid mRNA stability can highly vary during leaf development and therefore also accounts for transcript abundance (Klaff and Gruissem 1991; Kim et al. 1993). Lack of ribosome association can result in mRNA degradation which has been shown for the *rbcL* transcript, but this mechanism cannot be generalized as many other transcripts remain unaffected despite decreased ribosome association (Barkan 1993). The actual RNA degradation pathway in plastids involves polyadenylation (Kudla et al. 1996), a process which is known to be a stabilizing signal for nuclear mRNAs (for review see Dreyfus and Régnier 2002), but also acts as RNA instability signal in prokaryotes.

D. Translation

Plastid translation is related to translation in eubacteria. Both systems share homologous compounds, like initiation factors, rRNAs, tRNAs and 70S-type ribosomes (for review see Peled-Zehavi and Danon 2007). The tRNAs, rRNAs and some ribosomal proteins are encoded by the plastid genome, the remaining components are encoded in the nucleus and imported from the cytosol. Plant mitochondria also require a fully functional translation machinery to express the about 30 mRNAs encoded in the mitochondrial genome. Since only a few ribosomal proteins, rRNAs and an incomplete set of tRNAs are encoded by the mtDNA (Unseld et al. 1997), plant mitochondria must import most of the components of their translational apparatus, e.g. several tRNAs (Salinas et al. 2008) and all the required aminoacyl-tRNA synthetases (Duchêne et al. 2005).

Plastid ribosomes were characterized in proteomics studies in *C. reinhardtii* and spinach, which has led to the identification of 59 proteins. While 53 ribosomal proteins share homologues with *Escherichia coli*, six are specific to plastids and termed PRSP1-6 (Yamaguchi and Subramanian 2000; Yamaguchi et al. 2000, 2002, 2003;

Yamaguchi and Subramanian 2003). In eubacteria, the Shine-Dalgarno (SD) sequence plays a crucial role in the correct positioning of the ribosome during translation initiation (reviewed in Kozak 2005). In most plastid mRNAs, the SD-like sequence has a similar role, but the distance to the initiation codon is not as conserved as it is in *E. coli*. In addition, 30 of the 79 protein-coding genes in tobacco do not contain a SD-like sequence at all, indicating that alternative *cis*-elements and *trans*-acting factors may be responsible for correct translation initiation (Sugiura et al. 1998). In plant mitochondria, sequences resembling SD sequences are very rare and in the absence of an in vitro translation system, the function of these sequences in translation initiation could not be determined (Pring et al. 1992). Thus, the mechanism controlling translation initiation remains completely elusive in plant mitochondria. Potential translation regulation systems are also unknown. However, the function of PPR proteins might well be connected to plant mitochondrial translation as suggested by the involvement of CRP1 as a chloroplast translation regulator (Schmitz-Linneweber et al. 2005), by the requirement of Pet309 for translation in yeast mitochondria (Tavares-Carreón et al. 2008) and by the association of PPR336 to polysomes in plant mitochondria (Uyttewaal et al. 2008).

Sequence analysis has shown that translation is usually, but not always, initiated with an AUG codon in plant organelles. Alternative codons were found to be possible additional translation initiator triplets in mitochondria and plastids (Bock et al. 1994a; Unseld et al. 1997; Dong et al. 1998; Zandueta-Criado and Bock 2004). Moreover, plant mitochondrial genes can be expressed from mRNAs lacking canonical termination codons with no evidence that alternative termination codons had been created posttranscriptionally by either RNA editing or polyadenylation (Raczynska et al. 2006).

While in bacteria nascent transcripts are directly translated into proteins, the uncoupling of these two processes introduces a new level of regulation in organelles (Mayfield et al. 1995; Danon 1997; Zerges 2000; Peled-Zehavi and Danon 2007).

III. Technological Developments for the Expression Profiling of Organellar Genes

Northern blot analysis, in which a labeled probe is hybridized to a RNA target, was the first and most widely used technology to confirm and quantify gene expression. However, it can only be used to analyze the expression pattern of a limited number of genes under few experimental conditions. The recent development of "-omics" technologies enables researchers to carry out a genome-wide expression profiling, analyzing simultaneously up to thousands of genes. Most of these methods rely on the use of DNA arrays (macro- or micro-), although alternative methods such as differential display, expressed sequence tags (ESTs), serial analysis of gene expression (SAGE), quantitative RT-PCR (qRT-PCR), and others, can alternatively be used for such purposes. The above-mentioned transcriptional profiling technologies allow the analysis of complex RNA populations from different cells or tissues. Although DNA arrays produced a real advance in large-scale expression analysis and are currently widely used for transcript profiling, only limited datasets are as yet available for plant organelles and most of them are related to chloroplast genes or nuclear genes with chloroplast functions (Kurth et al. 2002; Legen et al. 2002; Richly et al. 2003; Kahlau and Bock 2008; Valkov et al. 2009).

In DNA arrays, DNA fragments or oligonucleotides corresponding to different genes or cDNAs are immobilized on a solid support (nylon membranes for macroarrays and glass slides for microarrays), and hybridized as probes to total RNA pools extracted from cells, tissues, whole organisms, etc. The hybridization signal detected for each spot can then be measured giving the relative abundance of the corresponding mRNA (Bouchez and Höfte 1998; Meyers et al. 2004). The simplest and cheapest array systems use nylon membranes in combination with labeled (radioactive) cDNA probes, detected by Phosphorimager instruments (Kurth et al. 2002; Legen et al. 2002; Richly et al. 2003; Geimer et al. 2009). This method allowed to study, on a genome scale, the expression of the entire plastid chromosome of tobacco wild-type and mutant (PEP-deficient) plants (Legen et al. 2002) and *Euglena gracilis* under different culture conditions (Geimer et al. 2009), or nuclear genes related to chloroplast functions in *A. thaliana* under different environmental and genetic conditions (Kurth et al. 2002; Richly et al. 2003).

Initially, DNA microarrays were produced with cDNA fragments immobilized on microscope slides, but a competing approach, based on DNA oligonucleotides, has recently become the most widely used system (Bouchez and Höfte 1998; Stears et al. 2003; Meyers et al. 2004). Different fluorescent labeling and detection techniques are used to produce graphical images and numerical data corresponding to the measurement of spot intensities (Schulze and Downward 2001; Stears et al. 2003). The source of variation needs to be carefully controlled by replicating experiments at technical and biological levels (Schulze and Downward 2001; Meyers et al. 2004). In addition, it is often necessary to verify a subset of array results by alternative techniques, generally northern blot analysis and/or qRT-PCR (Schulze and Downward 2001). Several tools for array data analyses have been developed by both commercial and public suppliers (Schulze and Downward 2001; Stears et al. 2003).

Compared to cDNA arrays, arrays based on oligonucleotides offer several advantages: they can be synthesized either in plates or directly on solid surfaces, produce strong hybridization signals of superior specificity, also in the case of individual transcripts of multigene families that share sequence homology (by synthesizing oligonucleotides corresponding to regions of non-identity; Lemieux et al. 1998; Stears et al. 2003). Obviously, this method requires the availability of genome sequence for the organism under study, but this is usually no limitation in the case of organelles, considering the number of genomes continuosly released and the high degree of sequence conservation (http://megasun.bch.umontreal.ca/ogmp/projects/other/all_list.html). A plastome microarray, useable for different Solanaceae species, was recently developed (Kahlau and Bock 2008). This array is based on long (68–71 nucleotides) oligonucleotides and contains all genes and conserved open reading frames present in Solanaceae plastomes. It was designed using the complete tobacco, tomato and potato plastid genomes (Gargano et al. 2005; Yukawa et al. 2005; Daniell et al. 2006; Kahlau et al. 2006), and used to analyze the expression of different tomato and potato plastid genomes and identify regulatory expression patterns in different tissues and plastid types (Kahlau and Bock 2008; Valkov et al. 2009).

As an alternative to the array-based approach, a sequenced-based transcriptomic approach has been recently developed, with significant advantages, such as the potential to quantify the abundance of any transcript and the independence of the availability of a sequenced genome (Wang et al. 2010). Although these technologies have great potential, expression profiling studies based on Ultra High-Throughput Sequencing (UHTS) methods are still limited in plants (Cheung et al. 2006; Emrich et al. 2007; Weber et al. 2007; Schnable et al. 2009) and, so far, none of them has been applied to organelle transcriptomes.

IV. Expression Profiling in Plastids

A. Genotype-Specific Variation

Nuclear and plastid mutants have been used to study the effect of genotypic variability on differential plastid gene expression. Early studies about the profiling of gene expression in mutant genotypes involved the striped and albino mutants of maize and barley

(Han et al. 1993; Hess et al. 1993), which show altered chloroplasts in mutated tissues. In maize, the striped iojap 1 (*ij1*) and albino white 1 (*w1*) and white 2 (*w2*) mutants displayed alterations in the levels and sizes of several photosynthesis-related plastid transcripts. Furthermore, reduced protein accumulation was observed in *ij1*-affected plastids. Unchanged ptDNA content compared to wild-type plastids, and several other observations, suggested that *ij1* and *w1* mutants might have not only altered transcription rate, but also alterations in transcript processing and stability. On the other hand, the severe reduction of plastome copy number per plastid was likely responsible for the general reduction of transcripts in the *w2* mutant (Han et al. 1993). In the ribosome and plastid protein biosynthesis-deficient *albostrians* mutant of barley (Hess et al. 1993), the accumulation of transcripts for the photosynthesis genes *psbA*, *atpH*, *atpI*, and *rbcL* was strongly reduced, due to differential transcription rates and transcript stabilities in mutant and wild type plastids. In contrast, transcript accumulation for *rpo* and *rps* genes, encoding some subunits of the plastid-encoded RNA polymerase and small ribosome proteins, respectively, was enhanced, suggesting the involvement of NEP in their synthesis.

More recently, macro- and microarray analyses were carried out with mutants of the alga *C. reinhardtii* and the model Brassicaceae *A. thaliana* (Erickson et al. 2005; Cho et al. 2009). In the former case, using two RNA stability mutants (*mcd1-1* and *mcd1-2*), such analyses not only confirmed the gene-specific substrate (*petD*) of the nuclear *Mcd*1 gene, encoding an mRNA stability factor, but also allowed the discovery of an additional unlinked mutation (*mda1-2*) affecting accumulation of *atpA* mRNAs (Erickson et al. 2005). In Arabidopsis, the expression of 94 plastid genes was analyzed in a large set (75) of genotypes including albino mutants arrested at an early stage of chloroplast development, "high chlorophyll fluorescence" (*hcf*) mutants with impaired photosynthetic electron transport capacity and yellow/pale-green lines with altered, and often unknown, chloroplast functions (Cho et al. 2009). Transcriptomes were clustered in two main groups. Group I, including the majority of albino mutants, displayed up-regulation of non-photosynthetic genes transcribed by NEP and down-regulation of genes transcribed by PEP and encoding photosynthetic proteins. On the other hand, group II showed less evident expression changes and included all *hcf*, pale-green and the remaining albino mutants. Deviations from the expected profiles in the two groups allowed to identify 14 mutants specifically involved in plastid RNA metabolism.

Knock-out lines of Arabidopsis for the nuclear genes encoding PEP σ-like factors were used to analyze changes in global plastid gene expression and switch in promoter usage (Kanamaru et al. 2001; Nagashima et al. 2004; Schweer et al. 2006). In early experiments (Kanamaru et al. 2001), *sig2-1* mutants, showed reduced accumulation of chlorophyll and photosynthesis-related proteins, without significant reduction of the respective mRNAs. On the other hand, the observed phenotypic and biochemical defects were likely related to reduction in synthesis of some tRNAs encoded by genes (*trnE*-UUC, *trnV*-UAC, *trnM*-CAU, *trnQ*-UUG) with conserved eubacteria-type promoter sequences. Later, microarray analyses carried out on the same mutant line (Nagashima et al. 2004) showed that out of 79 protein coding genes, only the *psaJ* transcript was reduced in the mutant, whereas transcripts of 47 genes, many under the control of NEP, were increased, suggesting increase of NEP activity in the *sig2-1* genotype. In another σ-like factor knock-out line (*sig6-2*), the appearance of an unusually long transcript was observed in the *atpB-E* operon (Schweer et al. 2006), suggesting either a role of SIG6 in chloroplast RNA maturation or a differential usage of promoter sequences. Indeed, the authors detected motifs for NEP recognition far upstream of the *atpB* gene, between the *accD* and *rbcL* genes and thus devised a model for the developmentally regulated use of alternative σ-like factors and promoter sequences.

Transgenic tobacco plants overexpressing a plastid-targeted bacteriophage T7 RNA polymerase (T7RNAP) were analyzed for their effects on plastid gene expression (Magee and Kavanagh 2002; Magee et al. 2007). In the former study, northern analyses showed an increase in transcript accumulation for several genes generally transcribed by NEP, but no variation in other genes (*psbD*, *ndhA*, *rrn16*) transcribed exclusively or predominantly by PEP in chloroplasts. These results could be explained by: (a) T7RNAP-mediated transcription from NEP promoters, (b) increased activity of NEP triggered by the presence of T7RNAP activity in chloroplasts of transgenic plants, and/or (c) differential increase in stability of some plastid mRNAs in transgenic plants (Magee and Kavanagh 2002). When the presence of the nuclear-encoded T7RNAP was associated with that of plastid transgenes driven by T7G10 5′ regulatory sequences, hybrid transplastomic plants showed reduced growth and altered expression of both plastid and nuclear genes (Magee et al. 2004). The profile of gene expression in mutant plants was analyzed using a customized array containing target sequences from all 124 tobacco plastid genes and 61 nuclear genes coding for photosynthetic proteins and components of the plastid translational apparatus. Significantly down-regulated plastid genes included those encoding subunits of the PSII, NADH dehydrogenase, ATP synthase, and cytochrome b_6f complexes as well as ribosomal RNAs and proteins. Genes co-transcribed in the same polycistronic operons were usually down-regulated similarly. At the same time, up-regulation was observed for the *accD* gene and the *ndhCKJ* operon, but while in the former case it was due to read-through transcription from the upstream promoters present in the construct used for transformation, in the latter case it was, as previously shown, likely due to non-specific recognition of a NEP promoter-like sequence by the plastid-directed T7RNAP. Furthermore, a reduction in the amount of the mature *clpP* mRNA was observed in mutant plants due to altered processing of the primary transcript (Magee et al. 2007).

The availability of *rpo* deletion mutants, derived by plastid transformation, and the comparative analysis of expression profiles in wild-type and mutant tobacco plants allowed to establish the existence of the nuclear-encoded RNA polymerase in higher plant plastids and to investigate the interplay of the two polymerase types (PEP and NEP) at several levels of gene expression (Allison et al. 1996; Hajdukiewicz et al. 1997; Serino and Maliga 1998; De Santis-Maciossek et al. 1999; Krause et al. 2000; Legen et al. 2002). Based on results of northern analyses and mapping of transcription initiation sites, early seminal studies (Allison et al. 1996; Hajdukiewicz et al. 1997) showed that, while some mRNA accumulation was always detected for all genes analyzed, plastid genes could be grouped in three classes on the basis of relative transcript accumulation levels in wild-type and mutant plants: (1) genes with high accumulation levels in wild-type leaves, but negligible levels in leaves of Δ*rpoB* plants; (2) genes with similar mRNA leaves in wild-type and mutant plants; (3) genes with significantly more transcript accumulation in mutant than in wild-type plants. Genes or operons with different physiological functions (e.g. photosynthesis or transcription/translation) belonged to different classes and displayed upstream sequences for recognition of either one polymerase type or both. It was thus proposed that transcription by NEP or PEP through differential promoter recognition could determine the pattern of group-specific gene regulation in plastids (Hajdukiewicz et al. 1997). Later studies, however, indicated a more complex picture, likely not based only on differential promoter usage but also on differential posttranscriptional mechanisms (Krause et al. 2000; Legen et al. 2002). In fact, the entire plastome was found to be transcribed in both wild-type and PEP-deficient plastids and, in genes specifying different functions, no correlation was found between transcription rates, transcript levels, transcript patterns, and accumulation levels of derived polypeptides. A predominance of posttranscriptional regulation mechanisms over transcriptional ones was also found in a tobacco *psbA* gene deletion mutant showing changes in PSII protein accumulation levels and upregulation of the alternative electron transport pathways (Baena-González et al. 2003).

A microarray analysis of the expression profile of 108 plastid DNA fragments carried out in a chloroplast *trnR*-CCG gene knock-out mutant of the moss *Physcomitrella patens* indicated that most plastid genes were expressed at comparable levels in mutant and wild-type genotypes, suggesting that the arginine *trnR*-CCG gene is not essential for plastid gene expression in *P. patens* (Nakamura et al. 2005).

As previously discussed, the RNA editing process occurs in all major lineages of land plants and only marchantiid liverworts do not seem to edit plastid transcripts (Table 14.1, Freyer et al. 1997; Duff and Moore 2005). By investigating editing patterns of the *ndhB* and *rbcL* transcripts, it could be observed that neither plastid editing frequencies nor the editing patterns of a specific transcript correlated with the phylogenetic tree of the plant kingdom (Freyer et al. 1997). Later genome-wide comparisons of the editotypes of one hornwort (*Anthoceros formosae*), one fern (*Adiantum capillus-veneris*) and five seed plants (*Arabidopsis thaliana*, *Nicotiana tabacum*, *Atropa belladonna*, *Zea mays* and *Pinus thunbergii*) showed that only 18 of the total 85 chloroplast editing sites in seed plants were present also in either one or both other taxa, while the latter shared 53 sites (Tillich et al. 2006). Tsudzuki et al. (2001) compared the conservation of editing sites among several higher plant species. The dicotyledonous plant tobacco and the monocots rice and maize have 12 editing sites in common, which may already have been present before divergence of both taxa (Hirose et al. 1999; Tsudzuki et al. 2001). The conservation of editing sites between gymnosperms and angiosperms is lower. Just five sites out of 26 are shared by the gymnosperm black pine (Wakasugi et al. 1996) and the so far analyzed angiosperms (Tsudzuki et al. 2001). The editing sites were also compared within more closely related taxa. The three Solanaceae tobacco (*N. tabacum*), tomato (*Solanum lycopersicum*) and deadly nightshade (*A. belladonna*) have 30 of the so far known sites in common. While 2–3 sites are always shared by two species, 1–2 appear to be species-specific (Schmitz-Linneweber et al. 2002; Kahlau et al. 2006). Differences in editing can even be observed at the sub-species level. The editotypes of three different ecotypes of *A. thaliana*, Columbia (Col-0), Cape Verde Islands (Cvi-0) and Wassilewskija (Ws-2) were characterized. One non-synonymous point mutation was detected in Cvi-0 at the first position of codon 17 of the *ndhG* gene. This codon is usually edited in all three ecotypes with a C-to-U transition at the second position. This mutation changes a serine codon in Col-0 and Ws-2 into an alanine codon in Cvi-0 at the DNA level, resulting at the RNA level in a phenylalanine or valine codon, respectively (Tillich et al. 2005).

By analyzing differences among various species, it is interesting to note how fast the capability to edit certain sites was lost during evolution. In *A. thaliana*, the *cis*-elements of *mat*K(2) and *ndhB*(11) editing sites show striking similarities, suggesting that they share the same *trans*-acting factor, a common mechanism which has been experimentally proven for other sites (Chateigner-Boutin et al. 2008; Hammani et al. 2009). However, while the *matK*(2) editing event restores a conserved tyrosine, the *ndhB*(11) does eliminate an evolutionary conserved serine. These results indicate that the *matK*(2) editing is the primary target whereas *ndhB*(11) editing might be secondary (Tillich et al. 2005). The potential of one *trans*-acting factor to recognize several similar *cis*-elemets might explain the capability of some species to edit foreign sites. The site *rps12*(74) is edited in tomato, but absent from tobacco, as the 'T' is already encoded at the DNA level (Kahlau et al. 2006). After introduction of the tomato editing site into the tobacco plastid genome, the transplastomic tobacco plant

Table 14.1. Examples of genome-wide analyses of RNA editing in plastid and mitochondrial transcripts of different plant species

	No. of edited sites (tissue/cell type)		
Species	Plastids	Mitochondria	References
Atropa belladonna	31 (leaf)[a]	–[b]	Schmitz-Linneweber et al. (2002)
Nicotiana tabacum	34 (leaf)[a]	–	Chateigner-Boutin and Hanson (2003)
Solanum lycopersicum	36 (leaf)[a]	–	Kahlau et al. (2006)
Arabidopsis thaliana	34 (leaf)	456 (cell suspension)[c]; 362 (leaf)[d]	Giegé and Brennicke (1999), Chateigner-Boutin and Small (2007), Bentolila et al. (2008)
Brassica napus	–	427 (leaf)[e]	Handa (2003)
Beta vulgaris	–	357 (seedling)	Mower and Palmer (2006)
Pisum sativum	27 (leaf)	–	Inada et al. (2004)
Oryza sativa	21 (leaf)[f]	491 (–)	Corneille et al. (2000), Notsu et al. (2002)
Zea mays	27 (leaf)	–	Peeters and Hanson (2002)
Pinus thunbergii	26 (leaf and stem)	–	Wakasugi et al. (1996)
Adiantum capillus-veneris	350 (frond)[g]	–	Wolf et al. (2004)
Isoetes engelmannii	–	1,782 (–)[g]	Grewe et al. (2010)
Physcomitrella patens	2 (–)	11 (–)	Rüdinger et al. (2009)
Anthoceros formosae	942 (thalli)[g]	–	Kugita et al. (2003)
Marchantia polymorpha	0	0	Rüdinger et al. (2009) and references therein

[a]Bioinformatics analyses detected additional sites in the three Solanaceous species, for a total of 35 sites in *A. belladonna*, 37 in *N. tabacum* and 36 in *S. lycopersicum*, most of them (30) shared by the three species (Kahlau et al. 2006)
[b]Not available
[c]Out of a total of 456 sites, 441 were identified in *orf*s, 8 in introns, and 7 in leader or trailer sequences (Giegé and Brennicke 1999)
[d]362 sites were detected in 33 genes, of which 67 sites had not been observed in suspension cultures, whereas 37 sites previously detected in suspension cultures were not observed in leaves (Bentolila et al. 2008)
[e]427 sites were identified in 33 genes, 358 of which are shared with *A. thaliana* (Handa 2003)
[f]21 sites were identified in 11 genes, 19 of which are shared with *Z. mays* (Corneille et al. 2000)
[g]About 10%, 12% and 46% of the total edited sites were U-to-C conversions in *A. capillus-veneris*, *I. engelmannii* and *A. formosae*, respectively

is able to edit this site with high efficiency. There are two possible explanations for this scenario: (1) either the responsible *trans*-acting factor is still present in the tobacco nuclear genome and was not lost during evolution or (2) the pre-existence of the *rpoB*(667) site facilitated the evolution of the *rps12* site, as both *cis*-elements share high similarity (Karcher et al. 2008). Other attempts to edit heterologous editing sites in vivo have not been successful so far. In contrast to the tomato *rps12* site, tobacco is not able to edit a foreign *psbF* site introduced from spinach (*Spinacia oleracea*, Bock et al. 1994b). But also more closely related species show nuclear-plastidial incompatibilities. By creating cybrids and introducing tobacco chloroplasts into deadly nightshade nuclear backgrounds, it became evident that the nuclear genome of nightshade is not able to support editing at all tobacco editing sites. The editing at site *atpA*(264) is absent, resulting in an albino phenotype. The most likely explanation is the absence of a nuclear-encoded editing factor in the nightshade which is responsible for correct processing of the site in tobacco (Schmitz-Linneweber et al. 2005).

B. Variation Due to Developmental and Environmental Cues

Most plastids are able to interconvert into other types following developmental and environmental cues (for review see Pyke 2007). Plastid gene expression and its regulation have been extensively studied in chloroplasts, which are present in photosynthetically active green tissues and generally develop from proplastids in meristems or etioplasts after illumination of dark-grown tissues. Non-green plastids, such as amyloplasts, chromoplasts and others, fulfill important functions in storage and pigmentation and are also the place of important metabolic pathways. The knowledge on gene expression in such plastid types, however, is still very limited.

Monocots are good model organisms to study changes in gene expression during chloroplast development. Cells at the leaf base contain proplastids which develop eventually into mature chloroplasts, present in the tip of the leaves. Using a custom maize chloroplast biogenesis cDNA microarray, it was shown that the abundance of most plastid transcripts in maize leaves (52 out of 63 analyzed) increases more than twofold during development (Cahoon et al. 2008). Ten transcripts which are present at similar levels in both plastid types are mostly involved in plastid gene expression and are transcribed by NEP (Cahoon et al. 2008). Since, in maize, transcription activity increases in developing chloroplasts, the latter transcripts are likely less stable in mature chloroplasts compared to proplastids at the leaf base (Cahoon et al. 2004, 2008). In the same study, a coordinated gene expression pattern in the nucleus and the plastids, likely based on a combination of anterograde and retrograde signalling between the two organelles, was found (Cahoon et al. 2008). Barley, another monocot, shows differences in transcript patterns during proplastid-to-chloroplast conversion compared to maize. At the leaf base, plastid transcriptional activity and transcript abundance are low. An increase in both can be seen in etioplasts, present in cells which already stopped dividing and entered the cell elongation phase. In contrast to maize, after illumination and further chloroplast maturation, transcript abundance and transcriptional activity decline again (Baumgartner et al. 1989, 1993). Another plastome-wide study in barley could not detect global quantitative changes in gene expression. During the de-etiolation process, no changes were found between etioplasts and chloroplasts, neither in relative transcription rates nor in transcript stability (Krupinska and Apel 1989). Differences between maize and barley could be caused by the differences in CO_2 fixation mechanisms. Barley uses the C3 fixation mechanism while maize is a C4 plant showing the typical Kranz anatomy. Maize bundle sheath and mesophyll cells have very different tasks during CO_2 fixation and also differ in their plastid transcript profiles. Transcripts for subunits of photosystem II are more abundant in mesophyll cells while *rbcL* is more abundant in bundle-sheath cells (Kubicki et al. 1994), the cell type in which concentrated CO_2 is fixed by Rubisco. It is possible that these two cell types also differ in their transcript patterns during proplastid-to-chloroplast development. In another monocot study based on macroarrays for studying chloroplast gene expression profiles, changes in gene expression levels were monitored using RNA isolated from germinating wheat seeds and seedlings at different stages of development (Siniauskaya et al. 2008). While transcript levels for PSI and PSII genes increased after imbibition until 1 week of development, the levels of other transcripts (e.g. those of *ndh* and *atp* genes) either did not change or decreased.

In the dicot *N. tabacum*, dark-grown etioplast-containing and illuminated chloroplast-containing seedlings were compared by using a microarray with 220 ptDNA fragments, each corresponding to a single known gene or an intergenic region and altogether covering the whole plastome (Nakamura et al. 2003). A clear trend of gene expression within the two functional groups of plastid-encoded genes was evident. The majority of photosynthesis-related genes increased in their abundance in illuminated plants. On the other hand, the majority of genes involved in gene expression were expressed at similar levels in both plastid types. Furthermore, unexpected signals were found in several intergenic regions, suggesting the existence of novel transcripts (Nakamura et al. 2003). Similar studies conducted in the unicellular red alga *Cyanidioschyzon merolae* with a microarray containing almost all plastid protein coding genes, northern blot analyses and run-on transcription assays, showed differential activation of gene transcription by illumination (Minoda et al. 2005).

Results from microarray analyses in tobacco and Arabidopsis (MacLean et al. 2008) showed coordinated expression of nuclear and plastid genes encoding ribosomal proteins during seedling development. Transcript accumulation responded similarly to light and inhibitors of plastid signaling. In another study comprising the same two species, the effect of green light on seedling development and plastid gene expression was analyzed by using genome microarrays and RNA gel blot experiments (Dhingra et al. 2006). In both species, etiolated seedlings subjected to a short, dim, single pulse of green light showed stem elongation and concomitant decrease in a sub-set of plastid-encoded transcripts, including several ones known to be light inducible. The majority of plastid transcripts did not vary, while only three increased in abundance, indicating that the effect of green light on plastid gene expression is gene-specific.

As a representative for eukaryotes carrying secondary endosymbionts, plastid gene expression was analyzed in the protist *Euglena gracilis*. Similar to primary endosymbionts, *E. gracilis* plastids encode mainly genes involved in transcription, translation and photosynthesis (Hallick et al. 1993). Early work on *E. gracilis* using RNA-DNA hybridizations had already shown that plastid genes are transcribed in proplastids of dark-grown cells (Chelm and Hallick 1976; Rawson and Boerma 1976; Chelm et al. 1979). Although Dix and Rawson (1983) could not identify individual genes, they could distinguish between two major groups: (1) genes which are constitutively transcribed and (2) genes encoding transcripts which increase in their abundance during the greening process, as e.g. *psbA*, encoding a core subunit of photosystem II (Hollingsworth et al. 1984). A recent study analyzed the complete *E. gracilis* transcriptome under 12 different growth and stress conditions using a macroarray-based approach (Geimer et al. 2009). Overall, the organelle transcriptome showed pronounced global quantitative changes, but qualitative changes were negligible. After growth in darkness, the overall transcript abundance was much lower than in light-grown cells, but *psbA* transcription increased drastically. The *trnI*-CAU gene, involved in gene expression and an example for genes identified by Dix and Rawson (1983) as being constitutively transcribed, did not change in abundance (Geimer et al. 2009). As plastid gene expression patterns in *E. gracilis* remain more or less constant with quantitative changes on a global scale, these results suggest that fine-tuning of protein production might be regulated posttranscriptionally. In contrast to the limited global transcriptional changes happening in *E. gracilis* (Geimer et al. 2009), translational regulation is much more pronounced (Miller et al. 1983). Although this study did not identify single plastid-encoded proteins, it could clearly show the gap between transcriptional changes, which are just up to threefold, and the total rates of protein synthesis, which were increased to up to 100-fold. Different sets of proteins were expressed at different time points during proplastid-to-chloroplast development (Miller et al. 1983). Analyzing and comparing *psbA* transcription with protein accumulation during de-etiolation identified similar patterns in barley. Without noticeable changes in *psbA* mRNA, the encoded protein increased drastically once the plants were illuminated (Klein and Mullet 1987).

Eberhard and coworkers (2002) analyzed the ability of plastids to override transcriptional changes at the translational level more systematically. *C. reinhardtii* cells were treated with rifampicin causing depletion of plastid transcripts by binding to and inhibiting the eubacterial-type RNA polymerase. Most of the analyzed transcripts dropped in their abundance to 10% compared to prior to the treatment. Despite these significant changes in RNA levels, the rate of protein synthesis measured with pulse-chase labeling experiments did not drop during the treatment (Eberhard et al. 2002).

Amyloplasts are present in storage organs like tubers or seed endosperm as well as in columella cells of root tips (for review see Pyke 2007). Several genes (16S rRNA, *atpB/E*, *psbA*, *rbcL*) were analyzed with respect to their abundance in *A. thaliana* and spinach root amyloplasts (Deng and Gruissem 1988; Isono et al. 1997a). All analyzed transcripts could be detected, but their abundance was highly decreased compared to leaf chloroplasts. Although plastid DNA levels were lower in Arabidopsis roots than in leaves, these differences could not explain the large differences in transcript abundances (Isono et al. 1997a). Similar observations were made by analyzing amyloplasts of potato tubers (Brosch et al. 2007; Valkov et al. 2009). The rather small reduction in plastid DNA content of approximately two- to threefold could not account for the reduction in transcript accumulation (Valkov et al. 2009). Run-on assays showed that the decrease in steady-state RNA levels was largely due to a decrease in transcriptional activity in amyloplasts (Sakai et al. 1992; Brosch et al. 2007; Valkov et al. 2009), although differences in stability could explain variable transcript accumulation levels. To get a more detailed insight into tissue versus plastid specificity, bell pepper (*Capsicum annuum*) and tomato (*Solanum lycopersicum*) fruits, which convert chloroplasts (highly transcriptionally active in leaves) into red chromoplasts during fruit ripening, were analyzed. Surprisingly, chloroplasts in green fruits already show reduced transcript abundance and differ from their counterparts in leaves, pointing towards a developmental regulation of plastid transcription. Bell pepper fruit plastids show reduction in steady-state RNA levels which is due to a reduced transcriptional activity. However, no significant changes during ripening and chloroplast-to-chromoplast conversion in transcriptional activity could be detected (Kuntz et al. 1989). Similarly, in tomato and pumpkin (*Cucurbita pepo*), RNAs are present at lower levels in fruits compared to leaves, while transcriptional activity is already downregulated in green fruits and does not change significantly during ripening (Piechulla et al. 1985; Marano and Carrillo 1992; Obukosia et al. 2003; Kahlau and Bock 2008).

Although the functions and structure of the two plastid types are very different, amyloplasts and chromoplasts share striking similarities in their gene expression profiles (Fig. 14.1). In both plastid types, transcript abundance was highly reduced compared to leaf chloroplasts, but in both cases the differences could not be attributed to the absence of one of the two RNA polymerase activities present in plastids (Kahlau and Bock 2008; Valkov et al. 2009). In fact, although differences in promoter utilization were observed by comparing amyloplasts or chromoplasts to chloroplasts, both the nuclear-encoded and plastid-encoded RNA polymerases are active in non-green plastids. Plastome-wide expression profiling showed that two genes, *clpP* and *accD*, are expressed at similar high levels in leaf chloroplasts and amyloplasts or chromoplasts (Kahlau and Bock 2008; Valkov et al. 2009). *clpP*, a subunit of a protease, is essential in tobacco and important for plant development (Shikanai et al. 2001; Kuroda and Maliga 2003; Clarke et al. 2005; Adam 2007). Many nuclear-encoded proteins are imported into all plastid types (Baginsky et al. 2004; Siddique et al. 2006; Bancel et al. 2010; Barsan et al. 2010; Daher et al. 2010), indicating that the Clp protease is probably needed for the removal of damaged proteins (Zybailov et al. 2009). The *accD* gene is also essential and cannot be deleted from the plastid genome (Kode et al. 2005). The encoded protein is part of the plastid-localized Acetyl-CoA carboxylase

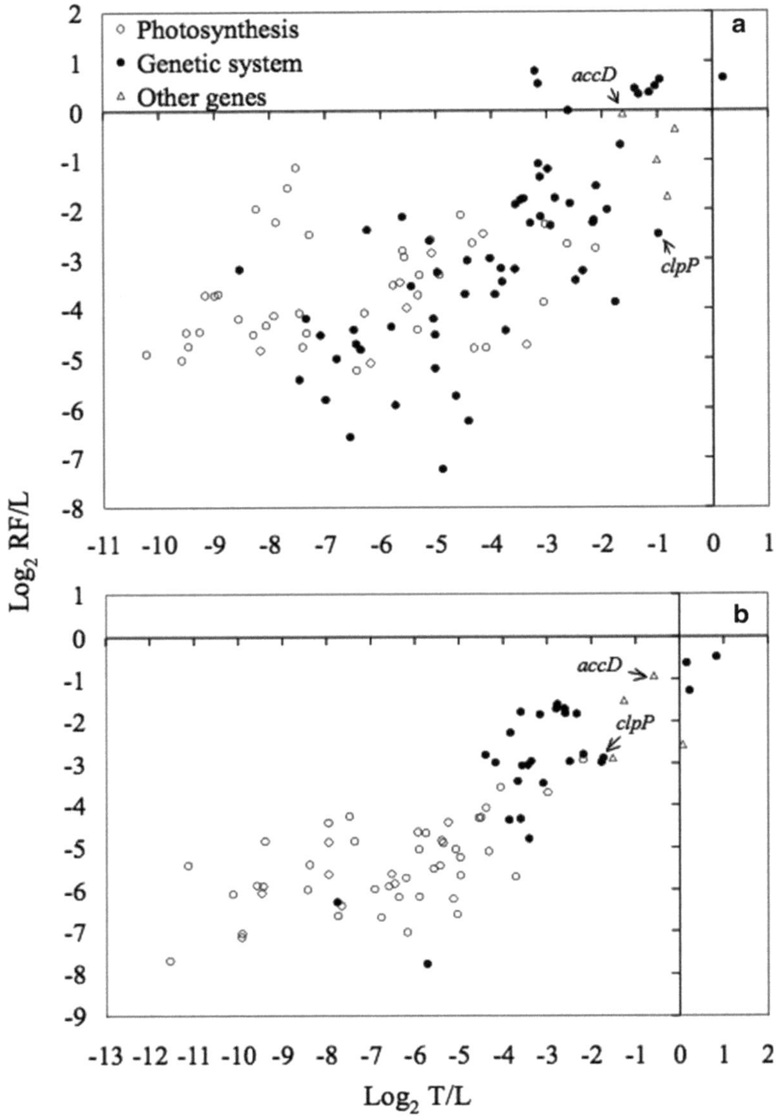

Fig. 14.1. Genome-wide analysis of total (**a**) and polysomal (**b**) RNA accumulation in potato tuber amyloplasts (Log_2 T/L) and tomato red fruit chromoplasts (Log_2 RF/L) compared to leaf chloroplasts (Kahlau and Bock 2008; Valkov et al. 2009). Based on their function, genes analyzed were grouped in three classes. The *accD* and *clpP* genes are highlighted.

(ACCase), catalyzing the first committed step in fatty acid biosynthesis. Residual expression of the plastid gene expression machinery may be necessary to produce the ACCase and secure therefore further production of lipids needed in all cell membranes (Kahlau and Bock 2008; Valkov et al. 2009).

Regulation of plastid translation at the level of polysome formation was investigated in various species and plastid types. In spinach root amyloplasts, representative transcripts involved in photosynthesis were detectable, but specifically depleted from polysomal fractions (Deng and Gruissem

1988). The situation is similar in potato tuber amyloplasts as well as in tomato fruit chromoplasts (Brosch et al. 2007; Kahlau and Bock 2008; Valkov et al. 2009). Both plastid types show a large reduction in polysome-associated mRNAs. In addition to the constantly low mRNAs levels in tomato fruit plastid-types, translation is increasingly down-regulated during chloroplast-to-chromoplast conversion for almost all mRNAs. The only genes which showed potentially similar translation levels in leaves and non-green plastids were genetic system genes like those encoding the subunits of the plastid-encoded RNA polymerase, *ycf1* and *ycf2* (open reading frames of unknown function) and, interestingly, *clpP* and *accD* (Kahlau and Bock 2008; Valkov et al. 2009). Hence, also at the translational level, transcripts of the latter two genes differ in their regulation pattern from almost all other plastid mRNAs and the low level of plastid gene expression is probably maintained to secure the production of the Clp protease and the ACCase for fatty acid biosynthesis (Kahlau and Bock 2008; Valkov et al. 2009).

Both in higher and lower plant plastids, the RNA editing process was likewise shown to be affected by changes in the environmental conditions as well as the organ and plastid type analyzed (Bock et al. 1993; Hirose et al. 1996; Hirose and Sugiura 1997; Ruf and Kössel 1997; Karcher and Bock 1998; Hirose et al. 1999; Karcher and Bock 2002a, b; Peeters and Hanson 2002; Chateigner-Boutin and Hanson 2003; Miyata and Sugita 2004; Kahlau and Bock 2008; Valkov et al. 2009). However, results of studies analyzing individual sites as well as those of a comprehensive study in maize, involving 27 editing sites in 15 genes and 10 different tissues (Peeters and Hanson 2002), demonstrate that environmental and developmental effects on RNA editing efficiency are not consistent in different genes and/or editing sites. Furthermore, developmental co-variation of RNA editing extent in some editing sites was shown by surveying 34 editing sites in 15 tobacco genes (Chateigner-Boutin and Hanson 2003). In bell pepper chromoplasts, the *psbL* initiation codon is still edited although the product is obviously not needed in the non-photosynthesizing ripe fruits. These results suggest that editing is in this case not responsible for the regulation of PsbL protein expression (Kuntz et al. 1992). *psbL* and *psbF* editing were also analyzed in illuminated and etiolated leaf tissue as well as in roots and seeds of spinach (Bock et al. 1993). Editing was complete in leaf etioplasts as well as chloroplasts, indicating that light had no influence on editing extent in these two plastid-types. However, editing of these two sites in proplastids (seeds) and amyloplasts (roots) was incomplete. As unedited start codons render transcripts probably untranslatable, editing might be one mechanism controlling plastid gene expression (Bock et al. 1993). Several editing sites in the *ndhA*, *ndhB* and *ndhF* transcripts, encoding subunits of the NAD(P)H dehydrogenase complex, show incomplete editing in *A. thaliana* roots (Chateigner-Boutin and Hanson 2003). However, a functional significance of incomplete editing for regulation of gene expression is questionable in this case as the NdhD protein is completely absent in roots. The *ndhD* start codon is only partially edited in tobacco, tomato, potato and Arabidopsis leaf tissue. The incomplete editing in leaves is conserved across several species, but the editing extent in non-green plastid types varies considerably. Editing of *ndhD*(1) is completely absent in Arabidopsis roots and potato tubers, but is partial in tobacco and spinach roots as well as tomato fruits (Chateigner-Boutin and Hanson 2003; Kahlau and Bock 2008; Valkov et al. 2009). All available results suggest the importance of selective activation/inhibition of site-specific nuclear-encoded *trans*-factors (sometimes able to recognize more than one editing site) in explaining developmental and/or environmental differences in RNA editing efficiency. By comparing different species and plastid-types, however, no common pattern can be identified which could hint towards a role of editing in the regulation of plastid gene expression. In addition, no editing sites specific for non-green plastid types have been identified so far.

Although genome-wide studies on the effect of environment and/or development on transcript processing are missing, available data for a number of genes showed at least in some cases a reduction of transcript splicing in non-green plastids compared to leaf chloroplasts, suggesting a possible link with limited expression of the plastid genome in some tissues and plastid types (Barkan 1989; Kahlau and Bock 2008; Valkov et al. 2009).

V. Expression Profiling in Mitochondria

A. Genotype-Specific Variation

Only few studies investigated the mitochondrial expression profiling of natural and mutant plant populations. The *albostrians* mutant of barley, characterized by a very low expression level of photosynthesis-related plastid and nuclear genes, was used to study the influence of impaired chloroplast development on mitochondrial gene and transcript levels (Hedtke et al. 1999). The analysis of mitochondrial steady-state RNA levels in different tissues showed an enhanced transcript accumulation of all mitochondrial genes tested in white leaves, due to a three-fold higher mitochondrial gene copy number. Further, because the increased transcript levels in mitochondria of white leaves could be caused by either the differentiation state of plastids or the direct action of the mutated nuclear *albostrians* allele, plants derived by reciprocal crosses between a green wild-type and a white (striped) *albostrians* parents were analyzed, showing that the enhanced transcript levels were a consequence of the impaired plastids and not of the nuclear mutant allele. These results highlight the crucial importance of inter-organellar cross-talk in plant cells.

In order to gain more knowledge about species-specific regulation of plant mitochondrial gene expression, Leino et al. (2005) compared transcriptional activity and RNA turnover in a cytoplasmic male-sterile (CMS) *Brassica napus* line, the corresponding male-fertile progenitors (*A. thaliana* and *B. napus*), and a fertility-restored line. The alloplasmic CMS line was obtained by protoplast fusion between *A. thaliana* and *B. napus* and contained mitochondrial DNA (mtDNA) mostly inherited from *A. thaliana* with some mtDNA fragments from *B. napus*, whereas the nucleus contained pure *B. napus* DNA (Leino et al. 2003). The fertility-restored line was isogenic for its mtDNA but had an additional pair of *A. thaliana* chromosome III in the nuclear genome (Leino et al. 2004). The analysis of transcriptional activities by run-on assays revealed a high variability between the parental species, with a higher transcript activity in *B. napus* than in *A. thaliana* for the *atp8*, *ccmB*, *rps7* and *rrn5* genes, and an opposite relationship for the *nad4L*, *nad9* and *cox1* genes. By contrast, the values obtained for the CMS and restored lines were very similar for all tested genes. The authors suggested that the differences observed in transcription activity could be due to differences in promoter strength, as already found in other species (Muise and Hauswirth 1992; Giegé et al. 2000). In comparison with transcription activities, the transcript steady-state levels were more homogeneous demonstrating that RNA turnover might act as a compensating mechanism.

In another study, the major transcript ends of all mitochondrial protein-genes were compared in three *A. thaliana* accessions (Forner et al. 2008). Authors identified mRNA polymorphisms for several genes (*nad4*, *nad9*, *ccmB*, *ccmC*, *rpl5-cob*), and linked them to variations at the 5′ ends that were conserved in all tissues analyzed. Since the polymorphisms observed could be caused by mitochondrial sequences or by differences in nuclear genes, they analyzed the inheritance of polymorphic mRNAs in reciprocal F_1 hybrids. These analyses showed a maternal (*ccmC*) or biparental (*nad4*, *nad9*, *ccmB* and *rpl5*) inheritance for polymorphic transcripts, suggesting that they could arise from differences in mtDNA or nuclear-encoded *trans*-factors, respectively. Despite intensive research in the past years, most of the *cis*-acting sequence elements and *trans*-factors required to generate mature 5′ and 3′ ends of mtRNA of higher plants are still unknown, Forner et al. (2008) suggested that the analysis of reciprocal F_1 hybrids is a promising approach to identify mitochondrial *cis*-elements and nuclear-encoded *trans*-factors involved in 5′ end formation or mRNA stability.

Comprehensive studies were carried out on RNA editing in different ecotypes and tissues of Arabidopsis mitochondria (Table 14.1, Giegé and Brennicke 1999; Bentolila et al. 2008; Zehrmann et al. 2008). Giegé and Brennicke (1999) identified a total of 456 C-to-U conversions in suspension cultures of *A. thaliana*, of which 441 reside in open reading frames (orfs). Differences among Arabidopsis ecotypes both for the extent of RNA editing and accession-specific editing sites were found (Bentolila et al. 2008; Zehrmann et al. 2008). Dominance relationships and maternal effects were assessed for the most polymorphic sites by evaluating the degree of editing in reciprocal hybrids. Dominance was more common in non-silent than in silent sites, while additivity was observed only in silent sites. For more than half of the inspected sites, a significant difference depending on the direction of the cross was found (Bentolila et al. 2008). Quantitative variations among ecotypes suggested that the extent of editing can evolve more rapidly than the species (Zehrmann et al. 2008).

A comparative analysis of the mitochondrial genes and RNA editing sites of *B. napus* L. and *A. thaliana* was carried out by Handa (2003), identifying 427 editing sites in genes and orfs of *B. napus* compared with 441 sites in *A. thaliana* (Table 14.1, Giegé and Brennicke 1999). The number of editing sites shared by both plant mitochondria was 358, which correspond to 83.8% and 81.2% of the total editing sites in *B. napus* L. and *A. thaliana* transcripts, respectively. These percentages seem to be low considering that mitochondrial DNA nucleotide identity (for protein coding regions) between the two species was 99.2%. This means that, as already found in plastids, RNA editing variations in plant mitochondria evolve more rapidly than coding sequences. By contrast, in the moss model system *Physcomitrella patens*, only 11 editing sites in 9 mitochondrial genes (*atp9*, *cox1*, *cox2*, *cox3*, *nad3*, *nad4*, *nad5*, *rps14* and *ccm*FC) were found, and only the codon positions reconstituting highly conserved amino acids in the encoded proteins were subjected to C-to-U conversions (Rüdinger et al. 2009).

B. Variation Due to Developmental and Environmental Cues

The plant mitochondrial genome is far from being able to express all the required proteins for mitochondrial respiration and translation (Unseld et al. 1997). Various and precise communication mechanisms must be necessary for the biogenesis of mitochondrial protein complexes and especially for the modulation of this biogenesis. A number of studies have established that mitochondrial respiration can be modulated in the plant cell in response to environmental stimuli, at some particular developmental stages or in response to stress (Wood et al. 1996; Svensson and Rasmusson 2001; Giegé et al. 2005; Ribas-Carbo et al. 2005). If this modulation of respiration is due to changes in the number of respiratory complexes per cell, it means that the biogenesis of respiratory complexes can be adjustable as well. A coordinated expression must exist between mitochondrial and nuclear genes, between nuclear genes and between mitochondrial genes encoding subunits of the same respiratory complexes (Giegé et al. 2005; Welchen and Gonzalez 2006; Gonzalez et al. 2007). While many nuclear genes are clearly (co-)regulated at the transcriptional level, the mechanisms regulating coordination of mitochondrial gene expression are less clear.

A global study of the Arabidopsis mitochondrial transcriptome had shown that individual genes or transcription units are transcribed with distinct rates even if they encode components of the same multi-subunit complexes. These differences are at least partially counterbalanced at the steady-state

RNA level by posttranscriptional processes and different RNA stabilities (Giegé et al. 2000). Are the steady-state RNA levels obtained invariable or can they be regulated, e.g. during changing developmental stages? To address this question, Li-Pook-Than and colleagues (2004) examined RNA levels of wheat mitochondrial genes during the developmental period when seeds leave dormancy, germinate and develop into seedlings. Mitochondrial transcript levels from 0 h to 6 days post-imbibition were analysed. Stable and edited messengers were observed in dormant seeds and precursor RNAs were subsequently detected early in embryo germination. Respiratory chain genes showed mRNA profiles comparable to those of ribosomal RNAs, whereas ribosomal protein genes had proportionately lower steady-state mRNA levels in later stages of seedling development. The relative levels of precursors compared with the respective mRNAs decreased during development, consistent with transcription outpacing RNA processing in early stages of development. However, coordination was more effective several days after imbibition. In the case of multiply split genes containing group II introns, complex patterns of splicing intermediates were observed. This suggested an absence of strict polarity for splicing. Spliced introns were typically more abundant in embryos than in seedlings. These observations suggest a transient delay of the RNA processing mechanisms at the beginning of seed germination, a period where mitochondrial biogenesis is rapid and apparently demanding for the post-transcriptional machinery (Li-Pook-Than et al. 2004). In another global study, Howell and colleagues (2006) described mitochondrial biogenesis during imbibition of rice embryos both at the morphological and the molecular levels. For a subset of mitochondrial encoded subunits of the respiratory chain genes, they observed two different transcript expression profiles. While complex V *atp1* and complex IV *cox2* transcripts reached maximum levels at 48 h after imbibition, complex I *nad9* and complex III *cob* message levels peaked much earlier at 8 h (Howell et al. 2006). Similar to the previous case, this showed that gene expression does not seem to be synchronized in early developmental stages and could suggest that mitochondrial transcripts rather follow a defined expression pattern in early development for the biogenesis of mitochondrial complexes. A more comprehensive investigation of mitochondrial transcript profiles during germination and early seedling development in wheat gave similar results (Khanam et al. 2007). In this study, the mitochondrial transcripts were present in the initial dry embryo at variable levels. During early development, gene expression levels of individual genes were very variable. However, genes could be classified into four categories according to their expression patterns. Most mitochondrial respiratory genes were found in two categories. For one category, the timing of RNA accumulation corresponded to the activation of respiration, but not for the other one. Altogether, this work suggested that the initial respiratory burst during early development is supported by stored preexisting respiratory components, whereas *de novo* mitochondrial gene expression rather supports the subsequent seedling growth (Khanam et al. 2007). It also suggested that the availability of substrates might be a regulatory factor or a signal for the initiation of gene expression in plant mitochondria. Gene expression profiles were also monitored for later developmental stages. Mitochondrial encoded transcript levels, together with chloroplast and nuclear RNA levels, were followed along a maize leaf developmental gradient (Cahoon et al. 2008). Twenty-five out of the 27 mitochondrial transcripts investigated had at least twofold higher steady-state levels in the leaf base than in the rest of the leaf. This mitochondrial gene expression pattern is not surprising because the actively dividing and expanding base of maize leaves has high energy demands and is expected to contain highly active mitochondria (Cahoon et al. 2008). However, from this particular study, it is difficult to conclude whether mitochondrial gene expression had been up-regulated in response to a developmental signal or whether the transcript level differences observed were due to an enriched content in mitochondria per cell at the maize leaf base.

The examination of plant mitochondrial transcript profiling studies has shown that specific transcript profiles emerge during development. Is this also the case in response to external stimuli? Variations for plant mitochondrial transcript profiles have also been observed during the day and night cycle (Okada and Brennicke 2006). In Arabidopsis mitochondria, these authors found that the transcription activity (measured by run-on RNA assays) varied during the diurnal cycle. In contrast, the steady-state transcript levels did not vary between light and dark phases and were stable throughout the diurnal as well as the circadian time course. From this, the authors concluded that the steady-state transcript levels available in plant mitochondria are sufficient to provide sufficient translation capacity at any time during the diurnal cycle (Okada and Brennicke 2006). This, together with previous work (Giegé et al. 2000), also illustrates that, in mitochondria, transcriptional variations are buffered at the level of posttranscriptional processes.

In a global study, where coordination of gene expression between the nucleus and mitochondria was investigated, authors have applied sugar starvation to Arabidopsis cells (Giegé et al. 2005). In this study, the overall mitochondrial transcript levels appeared to increase when sucrose was removed from the growth medium. On the other hand, the levels of mitochondrial transcripts drastically decreased when sugar was added back to the medium. These variations of RNA levels did not necessarily reflect adjustments in mitochondrial gene expression. The authors rather concluded from their results that the relative increase of mitochondrial transcript levels was due to an overall decrease of nuclear transcript levels in response to stress. Thus, after starvation, the proportion of mitochondrial RNA had increased among total RNA and vice versa, when sugar was added back, nuclear RNA expression had increased again and the proportion of mitochondrial RNA decreased among total RNA. Therefore, at least in this particular case, it appears that mitochondrial transcript levels had not been regulated in response to environmental demands. The required adjustment had rather been achieved by changes in nuclear gene expression and was reflected at the level of mitochondrial protein complexes assembly (Giegé et al. 2005). In another study, the effect of antimycin A treatment on mitochondrial function in wheat embryos was described. The transcript levels of five mitochondrial genes and two nuclear genes encoding mitochondrial proteins decreased in response to stress whereas the alternative oxidase (AOX) level increased (Naydenov et al. 2008). Although this study had not been conducted on a global scale, it suggested that in this case, antimycin A treatment had indeed been reflected at the level of mitochondrial gene expression. Finally, in a recent and very comprehensive study, the effects of low temperature, high salinity and high osmotic potential on the mitochondrial transcriptome have been monitored in wheat embryos (Naydenov et al. 2010). Most of the transcript level variations were stress specific. However, groups of genes could be defined with common responses to different stresses (Fig. 14.2). The authors predict from these results that common regulatory mechanisms must be active in response to some conditions whereas other regulatory processes appear to be active to specifically regulate the mitochondrial transcriptome in response to a particular situation (Naydenov et al. 2010).

The extent of editing in plant mitochondria was also found to be affected by developmentally-related effects. In fact, 67 new editing sites not previously observed in *A. thaliana* Col-0 cell-suspension cultures (Giegé and Brennicke 1999), were detected in rosette leaves (Bentolila et al. 2008). In contrast, 37 of the 441 editing events reported in suspension cultures were not observed in rosette leaves (Table 14.1). The proportion of silent sites in the two classes showing differential editing in the two tissues was similar: 48% (32/67) and 43% (16/37). These percentages were significantly higher than the proportion of silent sites found in the whole

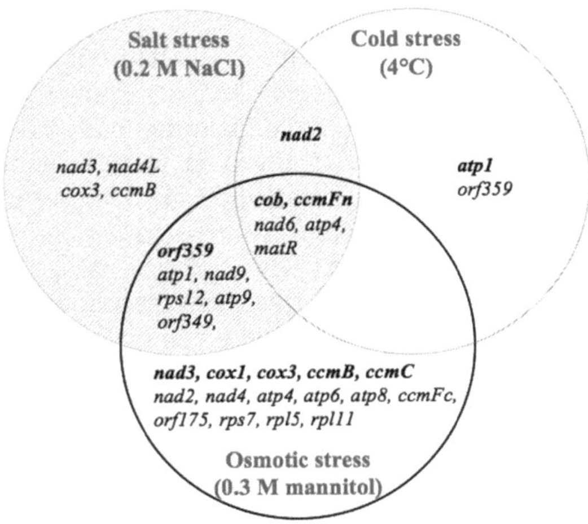

Fig. 14.2. Mitochondrial transcriptome variations in wheat embryos in response to stresses, as modified from Naydenov et al. (2010). Stresses were applied for 3 days. Up- and down-regulated genes (>1.5-fold) are shown in **bold** or *plain font*, respectively. Stress-specific responses are observed for some genes; however, other genes show common response patterns to two or three different stresses.

population of sites edited in either tissue (20%, Bentolila et al. 2008).

VI. Conclusions

Gene expression in plant organelles can be controlled either at the transcriptional or posttranscriptional level. The former is based on the differential use of multiple promoters and RNA polymerases (PEP in plastids, different NEP isoforms in plastids and mitochondria), and the action of various auxiliary factors. At the RNA level, the posttranscriptional regulation relates to differential editing, processing, stability and translatability of transcripts. Although the investigation of transcript profiling in plant organelles does not enable to draw general conclusions, available studies suggests that transcription itself is not highly regulated both in plastids and mitochondria, and that the steady-state levels of transcripts rather appear to be predominantly obtained through posttranscriptional processes.

Various steps of gene expression in plant organelles have been analyzed at a genome-wide scale by using DNA array-based technologies or others. Genotypic variability for the extent of RNA editing or transcript processing and stability in cytoplasmic organelles has been observed in natural populations at the interspecific and intraspecific level or in artificial CMS lines. The possibility to produce knock-out lines by plastid transformation has been particularly useful to highlight the role of various plastid genes on global genome expression and chloroplast development.

Specific transcript profiles can clearly be achieved also in response to developmental signals and environmental stimuli. Significant differences in the transcriptome, editome and translatome have been found comparing different plastid types in diverse organs or tissues. Similar differences have been found for mitochondrial genomes during the diurnal cycle or between cell suspensions and differentiated leaves. However, the precise levels and mechanisms at which these changes are achieved and the signals necessary to trigger them are barely understood.

Acknowledgments

Work in authors' labs was supported by the European Union (FP6 Plastomics Project grant no. LSHG–CT–2003–503238) to TC and by research grant 07-JCJC-0123 from the Agence Nationale de la Recherche (ANR) to PG. IGV publication no. 357.

References

Adam Z (2007) Protein stability and degradation in plastids. Top Curr Genet 19:315–338

Allison LA, Simon LD, Maliga P (1996) Deletion of *rpoB* reveals a second distinct transcription system in plastids of higher plants. EMBO J 15:2802–2809

Baena-González E, Allahverdiyeva Y, Svab Z, Maliga P, Josse EM, Kuntz M, Mäenpää P, Aro EM (2003) Deletion of the tobacco plastid *psbA* gene triggers an upregulation of the thylakoid-associated NAD(P) H dehydrogenase complex and the plastid terminal oxidase (PTOX). Plant J 35:704–716

Baginsky S, Siddique A, Gruissem W (2004) Proteome analysis of tobacco bright yellow-2 (BY-2) cell culture plastids as a model for undifferentiated heterotrophic plastids. J Proteome Res 3:1128–1137
Bancel E, Rogniaux H, Debiton C, Chambon C, Branlard G (2010) Extraction and proteome analysis of starch granule-associated proteins in mature wheat kernel (*Triticum aestivum* L.). J Proteome Res 9:3299–3310
Barkan A (1989) Tissue-dependent plastid RNA splicing in maize: transcripts from four plastid genes are predominantly unspliced in leaf meristems and roots. Plant Cell 1:437–445
Barkan A (1993) Nuclear mutants of maize with defects in chloroplast polysome assembly have altered chloroplast RNA metabolism. Plant Cell 5:389–402
Barsan C, Sanchez-Bel P, Rombaldi C, Egea I, Rossignol M, Kuntz M, Zouine M, Latché A, Bouzayen M, Pech JC (2010) Characteristics of the tomato chromoplast revealed by proteomic analysis. J Exp Bot 61:2413–2431
Baumgartner BJ, Rapp JC, Mullet JE (1989) Plastid transcription activity and DNA copy number increase early in barley chloroplast development. Plant Physiol 89:1011–1018
Baumgartner BJ, Rapp JC, Mullet JE (1993) Plastid genes encoding the transcription/translation apparatus are differentially transcribed early in barley (*Hordeum vulgare*) chloroplast development (evidence for selective stabilization of *psbA* mRNA). Plant Physiol 101:781–791
Benne R, Van den Burg J, Brakenhoff J, Sloof P, Van Boom JH, Tromp MC (1986) Major transcript of the frameshift *coxII* from trypanosome mitochondria contains four nucleotides that are not encoded in the DNA. Cell 46:819–826
Bentolila S, Elliott LE, Hanson MR (2008) Genetic architecture of mitochondrial editing in *Arabidopsis thaliana*. Genetics 178:1693–1708
Biehl A, Richly E, Noutsos C, Salamini F, Leister D (2005) Analysis of 101 nuclear transcriptomes reveals 23 distinct regulons and their relationship to metabolism, chromosomal gene distribution and co-ordination of nuclear and plastid gene expression. Gene 344:33–41
Binder S, Brennicke A (1993) A transfer RNA gene transcription initiation site is similar to messenger RNA and rRNA promoters in plant mitochondria. Nucleic Acids Res 21:5012–5019
Binder S, Marchfelder A, Brennicke A, Wissinger B (1992) RNA editing in *trans*-splicing intron sequences of *nad2* mRNAs in *Oenothera* mitochondria. J Biol Chem 267:7615–7623
Bock R, Koop HU (1997) Extraplastidic site-specific factors mediate RNA editing in chloroplasts. EMBO J 16:3282–3288
Bock R, Hagemann R, Kössel H, Kudla J (1993) Tissue- and stage-specific modulation of RNA editing of the psbF and psbL transcript from spinach plastids–a new regulatory mechanism? Mol Gen Genet 240:238–244
Bock H, Brennicke A, Schuster W (1994a) *Rps3* and *rpl16* genes do not overlap in *Oenothera* mitochondria: GTG as a potential translation initiation codon in plant mitochondria? Plant Mol Biol 24:811–818
Bock R, Kössel H, Maliga P (1994b) Introduction of a heterologous editing site into the tobacco plastid genome: the lack of RNA editing leads to a mutant phenotype. EMBO J 13:4623–4628
Bohne AV, Irihimovitch V, Weihe A, Stern DB (2006) *Chlamydomonas reinhardtii* encodes a single sigma70-like factor which likely functions in chloroplast transcription. Curr Genet 49:333–340
Bollenbach T, Schuster G, Portnoy V, Stern D (2007) Processing, degradation, and polyadenylation of chloroplast transcripts. Top Curr Genet 19:175–211
Bonen L, Vogel J (2001) The ins and outs of group II introns. Trends Genet 17:322–331
Bouchez D, Höfte H (1998) Functional genomics in plants. Plant Physiol 118:725–732
Bräutigam K, Dietzel L, Pfannschmidt T (2007) Plastid-nucleus communication: anterograde and retrograde signalling in the development and function of plastids. Top Curr Genet 19:409–456
Brosch M, Krause K, Falk J, Krupinska K (2007) Analysis of gene expression in amyloplasts of potato tubers. Planta 227:91–99
Buchanan BB, Gruissem W, Jones RL (eds) (2000) Biochemistry & molecular biology of plants. American Society of Plant Physiologists, Rockwille
Cahoon AB, Harris FM, Stern DB (2004) Analysis of developing maize plastids reveals two mRNA stability classes correlating with RNA polymerase type. EMBO Rep 5:801–806
Cahoon AB, Takacs EM, Sharpe RM, Stern DB (2008) Nuclear, chloroplast, and mitochondrial transcript abundance along a maize leaf developmental gradient. Plant Mol Biol 66:33–46
Canino G, Bocian E, Barbezier N, Echeverría M, Forner J, Binder S, Marchfelder A (2009) Arabidopsis encodes four tRNase Z enzymes. Plant Physiol 150:1494–1502
Carrillo C, Bonen L (1997) RNA editing status of *nad7* intron domains in wheat mitochondria. Nucleic Acids Res 25:403–409
Carter ML, Smith AC, Kobayashi H, Purton S, Herrin DL (2004) Structure, circadian regulation and bioinformatic analysis of the unique sigma factor gene in *Chlamydomonas reinhardtii*. Photosynth Res 82:339–349

Chang CC, Stern DB (1999) DNA-binding factors assemble in a sequence-specific manner on the maize mitochondrial atpA promoter. Curr Genet 35: 506–511

Chapdelaine Y, Bonen L (1991) The wheat mitochondrial gene for subunit-I of the NADH dehydrogenase complex – a *trans*-splicing model for this gene-in-pieces. Cell 65:465–472

Chateigner-Boutin AL, Hanson MR (2003) Developmental co-variation of RNA editing extent of plastid editing sites exhibiting similar *cis*-elements. Nucleic Acids Res 31:2586–2594

Chateigner-Boutin AL, Small I (2007) A rapid high-throughput method for the detection and quantification of RNA editing based on high-resolution melting of amplicons. Nucleic Acids Res 35:e114

Chateigner-Boutin AL, Ramos-Vega M, Guevara-García A, Andrés C, de la Luz Gutiérrez-Nava M, Cantero A, Delannoy E, Jiménez LF, Lurin C, Small I, León P (2008) CLB19, a pentatricopeptide repeat protein required for editing of *rpoA* and *clpP* chloroplast transcripts. Plant J 56:590–602

Chaudhuri S, Carrer H, Maliga P (1995) Site-specific factor involved in the editing of the *psbL* mRNA in tobacco plastids. EMBO J 14:2951–2957

Chelm BK, Hallick RB (1976) Changes in the expression of the chloroplast genome of *Euglena gracilis* during chloroplast development. Biochemistry 15:593–599

Chelm BK, Hallick RB, Gray PW (1979) Transcription program of the chloroplast genome of *Euglena gracilis* during chloroplast development. Proc Natl Acad Sci USA 76:2258–2262

Cheung F, Haas BJ, Goldberg SM, May GD, Xiao Y, Town CD (2006) Sequencing *Medicago truncatula* expressed sequenced tags using 454 Life Sciences technology. BMC Genomics 7:272

Cho Y, Qiu Y-L, Kuhlman P, Palmer JD (1998) Explosive invasion of plant mitochondria by a group I intron. Proc Natl Acad Sci USA 95:14244–14249

Cho WK, Geimer S, Meurer J (2009) Cluster analysis and comparison of various chloroplast transcriptomes and genes in *Arabidopsis thaliana*. DNA Res 16:31–44

Clarke AK, MacDonald TM, Sjögren LLE (2005) The ATP-dependent Clp protease in chloroplasts of higher plants. Physiol Plant 123:406–412

Corneille S, Lutz K, Maliga P (2000) Conservation of RNA editing between rice and maize plastids: are most editing events dispensable? Mol Gen Genet 264:419–424

Covello PS, Gray MW (1989) RNA editing in plant mitochondria. Nature 341:662–666

Daher Z, Recorbet G, Valot B, Robert F, Balliau T, Potin S, Schoefs B, Dumas-Gaudot E (2010) Proteomic analysis of Medicago truncatula root plastids. Proteomics 10:2123–2137

Daniell H, Lee S-B, Grevich J, Saski C, Quesada-Vargas T, Guda C, Tomkins J, Jansen R (2006) Complete chloroplast genome sequences of *Solanum bulbocastanum*, *Solanum lycopersicum* and comparative analyses with other *Solanaceae* genomes. Theor Appl Genet 112:1503–1518

Danon A (1997) Translational regulation in the chloroplast. Plant Physiol 115:1293–1298

De Santis-Maciossek G, Kofer W, Bock A, Schoch S, Maier RM, Wanner G, Rüdiger W, Koop HU, Herrmann RG (1999) Targeted disruption of the plastid RNA polymerase genes *rpoA, B* and *C1*: molecular biology, biochemistry and ultrastructure. Plant J 18:477–489

Demarsy E, Courtois F, Azevedo J, Buhot L, Lerbs-Mache S (2006) Building up of the plastid transcriptional machinery during germination and early plant development. Plant Physiol 142:993–1003

Deng XW, Gruissem W (1988) Constitutive transcription and regulation of gene expression in non-photosynthetic plastids of higher plants. EMBO J 7:3301–3308

Dhingra A, Bies DH, Lehner KR, Folta KM (2006) Green light adjusts the plastid transcriptome during early photomorphogenic development. Plant Physiol 142:1256–1266

Dix KP, Rawson JRY (1983) In vivo transcriptional products of the chloroplast DNA of *Euglena gracilis*. Curr Genet 7:265–272

Dombrowski S, Brennicke A, Binder S (1997) 3′-Inverted repeats in plant mitochondrial mRNAs are processing signals rather than transcription terminators. EMBO J 16:5069–5076

Dong FG, Wilson KG, Makaroff CA (1998) The radish (*Raphanus sativus* L.) mitochondrial *cox2* gene contains an ACG at the predicted translation initiation site. Curr Genet 34:79–87

Dreyfus M, Régnier P (2002) The poly(A) tail of mRNAs: bodyguard in eukaryotes, scavenger in bacteria. Cell 111:611–613

Duchêne AM, Giritch A, Hoffmann B, Cognat V, Lancelin D, Peeters NM, Zaepfel M, Maréchal-Drouard L, Small ID (2005) Dual targeting is the rule for organellar aminoacyl-tRNA synthetases in *Arabidopsis thaliana*. Proc Natl Acad Sci USA 102:16484–16489

Duff RJ, Moore FB (2005) Pervasive RNA editing among hornwort *rbcL* transcripts except *Leiosporoceros*. J Mol Evol 61:571–578

Eberhard S, Drapier D, Wollman FA (2002) Searching limiting steps in the expression of chloroplast-encoded proteins: relations between gene copy number, transcription, transcript abundance and

translation rate in the chloroplast of *Chlamydomonas reinhardtii*. Plant J 31:149–160
Emrich SJ, Barbazuk WB, Li L, Schnable PS (2007) Gene discovery and annotation using LCM-454 transcriptome sequencing. Genome Res 17:69–73
Erickson B, Stern DB, Higgs DC (2005) Microarray analysis confirms the specificity of a *Chlamydomonas reinhardtii* chloroplast RNA stability mutant. Plant Physiol 137:534–544
Fedorova O, Zingler N (2007) Group II introns: structure, folding and splicing mechanism. Biol Chem 388:665–678
Fey J, Maréchal-Drouard L (1999) Compilation and analysis of plant mitochondrial promoter sequences: an illustration of a divergent evolution between monocot and dicot mitochondria. Biochem Biophys Res Commun 256:409–414
Forner J, Hölzle A, Jonietz C, Thuss S, Schwarzländer M, Weber B, Meyer RC, Binder S (2008) Mitochondrial mRNA polymorphisms in different Arabidopsis accessions. Plant Physiol 148:1106–1116
Freyer R, Kiefer-Meyer MC, Kössel H (1997) Occurrence of plastid RNA editing in all major lineages of land plants. Proc Natl Acad Sci USA 94:6285–6290
Fujiwara M, Nagashima A, Kanamaru K, Tanaka K, Takahashi H (2000) Three new nuclear genes, *sigD, sigE and sigF*, encoding putative plastid RNA polymerase sigma factors in *Arabidopsis thaliana*. FEBS Lett 481:47–52
Gagliardi D, Perrin R, Maréchal-Drouard L, Grienenberger JM, Leaver CJ (2001) Plant mitochondrial polyadenylated mRNAs are degraded by a 3′- to 5′- exoribonuclease activity, which proceeds unimpeded by stable secondary structures. J Biol Chem 276:43541–43547
Gargano D, Vezzi A, Scotti N, Gray JC, Valle G, Grillo S, Cardi T (2005) The complete nucleotide sequence genome of potato (Solanum tuberosum cv Désirée) chloroplast DNA. In: Proceedings of the 2nd Solanaceae Genome Workshop 2005, Ischia, 25–29 Sept 2005
Geimer S, Belicová A, Legen J, Slávikovà S, Herrmann RG, Krajcovic J (2009) Transcriptome analysis of the *Euglena gracilis* plastid chromosome. Curr Genet 55:425–438
Giegé P, Brennicke A (1999) RNA editing in *Arabidopsis* mitochondria effects 441 C to U changes in ORFs. Proc Natl Acad Sci USA 96:15324–15329
Giegé P, Hoffmann M, Binder S, Brennicke A (2000) RNA degradation buffers asymmetries of transcription in *Arabidopsis* mitochondria. EMBO Rep 1:164–170
Giegé P, Rayapuram N, Meyer EH, Grienenberger JM, Bonnard G (2004) CcmF(C) involved in cytochrome c maturation is present in a large sized complex in wheat mitochondria. FEBS Lett 563:165–169
Giegé P, Sweetlove LJ, Cognat V, Leaver CJ (2005) Coordination of nuclear and mitochondrial genome expression during mitochondrial biogenesis in *Arabidopsis*. Plant Cell 17:1497–1512
Glanz S, Kück U (2009) *Trans*-splicing of organelle introns–a detour to continuous RNAs. Bioessays 31:921–934
Gobert A, Gutmann B, Taschner A, Gössringer M, Holzmann J, Hartmann RK, Rossmanith W, Giegé P (2010) A single *Arabidopsis* organellar protein has RNase P activity. Nat Struct Mol Biol 17:740–744
Gonzalez DH, Welchen E, Attallah CV, Comelli RN, Mufarrege EF (2007) Transcriptional coordination of the biogenesis of the oxidative phosphorylation machinery in plants. Plant J 51:105–116
Grewe F, Viehoever P, Weisshaar B, Knoop V (2009) A *trans*-splicing group I intron and tRNA-hyperediting in the mitochondrial genome of the lycophyte *Isoetes engelmannii*. Nucleic Acids Res 37:5093–5104
Grewe F, Herres S, Viehover P, Polsakiewicz M, Weisshaar B, Knoop V (2010) A unique transcriptome: 1782 positions of RNA editing alter 1406 codon identities in mitochondrial mRNAs of the lycophyte *Isoetes engelmannii*. Nucleic Acids Res 39:2890–2902
Grohmann L, Thieck O, Herz U, Schröder W, Brennicke A (1994) Translation of *nad9* mRNAs in mitochondria from *Solanum tuberosum* is restricted to completely edited transcripts. Nucleic Acids Res 22:3304–3311
Gualberto JM, Lamattina L, Bonnard G, Weil JH, Grienenberger JM (1989) RNA editing in wheat mitochondria results in the conservation of protein sequences. Nature 341:660–662
Hajdukiewicz PT, Allison LA, Maliga P (1997) The two RNA polymerases encoded by the nuclear and the plastid compartments transcribe distinct groups of genes in tobacco plastids. EMBO J 16:4041–4048
Hallick RB, Hong L, Drager RG, Favreau MR, Monfort A, Orsat B, Spielmann A, Stutz E (1993) Complete sequence of *Euglena gracilis* chloroplast DNA. Nucleic Acids Res 21:3537–3544
Hammani K, Okuda K, Tanz SK, Chateigner-Boutin AL, Shikanai T, Small I (2009) A study of new Arabidopsis chloroplast RNA editing mutants reveals general features of editing factors and their target sites. Plant Cell 21:3686–3699
Han C-d, Patrie W, Polacco M, Coe EHJ (1993) Aberrations in plastid transcripts and deficiency of plastid DNA in striped and albino mutants in maize. Planta 191:552–563
Handa H (2003) The complete nucleotide sequence and RNA editing content of the mitochondrial genome

of rapeseed (*Brassica napus* L.): comparative analysis of the mitochondrial genomes of rapeseed and *Arabidopsis thaliana*. Nucleic Acids Res 31:5907–5916

Hayes R, Kudla J, Gruissem W (1999) Degrading chloroplast mRNA: the role of polyadenylation. Trends Biochem Sci 24:199–202

Hedtke B, Börner T, Weihe A (1997) Mitochondrial and chloroplast phage-type RNA polymerases in *Arabidopsis*. Science 277:809–811

Hedtke B, Wagner I, Börner T, Hess WR (1999) Inter-organellar crosstalk in higher plants: impaired chloroplast development affects mitochondrial gene and transcript levels. Plant J 19:635–643

Hedtke B, Börner T, Weihe A (2000) One RNA polymerase serving two genomes. EMBO Rep 1:435–440

Hess WR, Prombona A, Fieder B, Subramanian AR, Börner T (1993) Chloroplast *rps15* and the *rpo*B/C1/C2 gene cluster are strongly transcribed in ribosome-deficient plastids: evidence for a functioning non-chloroplast-encoded RNA polymerase. EMBO J 12:563–571

Hiesel R, Wissinger B, Schuster W, Brennicke A (1989) RNA editing in plant mitochondria. Science 246:1632–1634

Hildebrand M, Hallick RB, Passavant CW, Bourque DP (1988) *Trans*-splicing in chloroplasts: the *rps12* loci of *Nicotiana tabacum*. Proc Natl Acad Sci USA 85:372–376

Hirose T, Sugiura M (1997) Both RNA editing and RNA cleavage are required for translation of tobacco chloroplast ndhD mRNA: a possible regulatory mechanism for the expression of a chloroplast operon consisting of functionally unrelated genes. EMBO J 16:6804–6811

Hirose T, Sugiura M (2001) Involvement of a site-specific *trans*-acting factor and a common RNA-binding protein in the editing of chloroplast mRNAs: development of a chloroplast in vitro RNA editing system. EMBO J 20:1144–1152

Hirose T, Fan H, Suzuki JY, Wakasugi T, Tsudzuki T, Kössel H, Sugiura M (1996) Occurrence of silent RNA editing in chloroplasts: its species specificity and the influence of environmental and developmental conditions. Plant Mol Biol 30:667–672

Hirose T, Kusumegi T, Tsudzuki T, Sugiura M (1999) RNA editing sites in tobacco chloroplast transcripts: editing as a possible regulator of chloroplast RNA polymerase activity. Mol Gen Genet 262:462–467

Hoch B, Maier RM, Appel K, Igloi GL, Kössel H (1991) Editing of a chloroplast mRNA by creation of an initiation codon. Nature 353:178–180

Hoffmann M, Dombrowski S, Guha C, Binder S (1999) Cotranscription of the *rpl5-rps14-cob* gene cluster in pea mitochondria. Mol Gen Genet 261:537–545

Holec S, Lange H, Kühn K, Alioua M, Börner T, Gagliardi D (2006) Relaxed transcription in *Arabidopsis* mitochondria is counterbalanced by RNA stability control mediated by polyadenylation and polynucleotide phosphorylase. Mol Cell Biol 26:2869–2876

Hollingsworth MJ, Johanningmeier U, Karabin GD, Stiegler GL, Hallick RB (1984) Detection of multiple, unspliced precursor mRNA transcripts for the Mr 32,000 thylakoid membrane protein from *Euglena gracilis* chloroplasts. Nucleic Acids Res 12:2001–2017

Howell KA, Millar AH, Whelan J (2006) Ordered assembly of mitochondria during rice germination begins with promitochondrial structures rich in components of the protein import apparatus. Plant Mol Biol 60:201–223

Inada M, Sasaki T, Yukawa M, Tsudzuki T, Sugiura M (2004) A systematic search for RNA editing sites in pea chloroplasts: an editing event causes diversification from the evolutionarily conserved amino acid sequence. Plant Cell Physiol 45:1615–1622

Isono K, Niwa Y, Satoh K, Kobayashi H (1997a) Evidence for transcriptional regulation of plastid photosynthesis genes in *Arabidopsis thaliana* roots. Plant Physiol 114:623–630

Isono K, Shimizu M, Yoshimoto K, Niwa Y, Satoh K, Yokota A, Kobayashi H (1997b) Leaf-specifically expressed genes for polypeptides destined for chloroplasts with domains of sigma70 factors of bacterial RNA polymerases in *Arabidopsis thaliana*. Proc Natl Acad Sci USA 94:14948–14953

Jenkins BD, Kulhanek DJ, Barkan A (1997) Nuclear mutations that block group II RNA splicing in maize chloroplasts reveal several intron classes with distinct requirements for splicing factors. Plant Cell 9:283–296

Jonietz C, Forner J, Holzle A, Thuss S, Binder S (2010) RNA PROCESSING FACTOR2 is required for 5′ end processing of *nad9* and *cox3* mRNAs in mitochondria of *Arabidopsis thaliana*. Plant Cell 22:443–453

Kahlau S, Bock R (2008) Plastid transcriptomics and translatomics of tomato fruit development and chloroplast-to-chromoplast differentiation: chromoplast gene expression largely serves the production of a single protein. Plant Cell 20:856–874

Kahlau S, Aspinall S, Gray JC, Bock R (2006) Sequence of the tomato chloroplast DNA and evolutionary comparison of solanaceous plastid genomes. J Mol Evol 63:194–207

Kanamaru K, Fujiwara M, Seki M, Katagiri T, Nakamura M, Mochizuki N, Nagatani A, Shinozaki K, Tanaka K, Takahashi H (1999) Plastidic RNA polymerase sigma factors in *Arabidopsis*. Plant Cell Physiol 40:832–842

Kanamaru K, Nagashima A, Fujiwara M, Shimada H, Shirano Y, Nakabayashi K, Shibata D, Tanaka K, Takahashi H (2001) An *Arabidopsis* sigma factor (SIG2)-dependent expression of plastid-encoded tRNAs in chloroplasts. Plant Cell Physiol 42:1034–1043

Kaneko T, Sato S, Kotani H, Tanaka A, Asamizu E, Nakamura Y, Miyajima N, Hirosawa M, Sugiura M, Sasamoto S, Kimura T, Hosouchi T, Matsuno A, Muraki A, Nakazaki N, Naruo K, Okumura S, Shimpo S, Takeuchi C, Wada T, Watanabe A, Yamada M, Yasuda M, Tabata S (1996) Sequence analysis of the genome of the unicellular cyanobacterium *Synechocystis* sp. strain PCC6803. II. Sequence determination of the entire genome and assignment of potential protein-coding regions. DNA Res 3:109–136

Karcher D, Bock R (1998) Site-selective inhibition of plastid RNA editing by heat shock and antibiotics: a role for plastid translation in RNA editing. Nucleic Acids Res 26:1185–1190

Karcher D, Bock R (2002a) The amino acid sequence of a plastid protein is developmentally regulated by RNA editing. J Biol Chem 277:5570–5574

Karcher D, Bock R (2002b) Temperature sensitivity of RNA editing and intron splicing reactions in the plastid *ndhB* transcript. Curr Genet 41:48–52

Karcher D, Kahlau S, Bock R (2008) Faithful editing of a tomato-specific mRNA editing site in transgenic tobacco chloroplasts. RNA 14:217–224

Keren I, Klipcan L, Bezawork-Geleta A, Kolton M, Shaya F, Ostersetzer-Biran O (2008) Characterization of the molecular basis of group II intron RNA recognition by CRS1-CRM domains. J Biol Chem 283:23333–23342

Khanam SM, Naydenov NG, Kadowaki K, Nakamura C (2007) Mitochondrial biogenesis as revealed by mitochondrial transcript profiles during germination and early seedling growth in wheat. Genes Genet Syst 82:409–420

Kim M, Christopher DA, Mullet JE (1993) Direct evidence for selective modulation of *psbA, rpoA, rbcL* and 16S RNA stability during barley chloroplast development. Plant Mol Biol 22:447–463

Kim SR, Yang JI, Moon S, Ryu CH, An K, Kim KM, Yim J, An G (2009) Rice OGR1 encodes a pentatricopeptide repeat-DYW protein and is essential for RNA editing in mitochondria. Plant J 59:738–749

Klaff P, Gruissem W (1991) Changes in chloroplast mRNA stability during leaf development. Plant Cell 3:517–529

Klein RR, Mullet JE (1987) Control of gene expression during higher plant chloroplast biogenesis. Protein synthesis and transcript levels of *psbA, psaA-psaB*, and *rbcL* in dark-grown and illuminated barley seedlings. J Biol Chem 262:4341–4348

Knoop V (2004) The mitochondrial DNA of land plants: peculiarities in phylogenetic perspective. Curr Genet 46:123–139

Knoop V, Schuster W, Wissinger B, Brennicke A (1991) *Trans*-splicing integrates an exon of 22 nucleotides into the *nad5* messenger RNA in higher plant mitochondria. EMBO J 10:3483–3493

Kobayashi Y, Dokiya Y, Sugita M (2001) Dual targeting of phage-type RNA polymerase to both mitochondria and plastids is due to alternative translation initiation in single transcripts. Biochem Biophys Res Commun 289:1106–1113

Kode V, Mudd EA, Iamtham S, Day A (2005) The tobacco plastid *accD* gene is essential and is required for leaf development. Plant J 44:237–244

Kotera E, Tasaka M, Shikanai T (2005) A pentatricopeptide repeat protein is essential for RNA editing in chloroplasts. Nature 433:326–330

Kozak M (2005) Regulation of translation via mRNA structure in prokaryotes and eukaryotes. Gene 361: 13–37

Krause K, Maier RM, Kofer W, Krupinska K, Herrmann RG (2000) Disruption of plastid-encoded RNA polymerase genes in tobacco: expression of only a distinct set of genes is not based on selective transcription of the plastid chromosome. Mol Gen Genet 263:1022–1030

Krupinska K, Apel K (1989) Light-induced transformation of etioplasts to chloroplasts of barley without transcriptional control of plastid gene expression. Mol Gen Genet 219:467–473

Kubicki A, Steinmüller K, Westhoff P (1994) Differential transcription of plastome-encoded genes in the mesophyll and bundle-sheath chloroplasts of the monocotyledonous NADP-malic enzyme-type C4 plants maize and sorghum. Plant Mol Biol 25:669–679

Kudla J, Hayes R, Gruissem W (1996) Polyadenylation accelerates degradation of chloroplast mRNA. EMBO J 15:7137–7146

Kugita M, Yamamoto Y, Fujikawa T, Matsumoto T, Yoshinaga K (2003) RNA editing in hornwort chloroplasts makes more than half the genes functional. Nucleic Acids Res 31:2417–2423

Kühn K, Weihe A, Börner T (2005) Multiple promoters are a common feature of mitochondrial genes in *Arabidopsis*. Nucleic Acids Res 33:337–346

Kühn K, Bohne AV, Liere K, Weihe A, Börner T (2007) Arabidopsis phage-type RNA polymerases: accurate in vitro transcription of organellar genes. Plant Cell 19:959–971

Kühn K, Richter U, Meyer EH, Delannoy E, de Longevialle AF, O'Toole N, Börner T, Millar AH, Small ID, Whelan J (2009) Phage-type RNA polymerase *RPOTmp* performs gene-specific transcription in mitochondria of *Arabidopsis thaliana*. Plant Cell 21:2762–2779

Kuntz M, Evrard J-L, d'Harlingue A, Weil JH, Camara B (1989) Expression of plastid and nuclear genes during chromoplast differentiation in bell pepper (*Capsicum annuum*) and sunflower (*Helianthus annuus*). Mol Gen Genet 216:156–163

Kuntz M, Camara B, Weil JH, Schantz R (1992) The *psbL* gene from bell pepper (*Capsicum annuum*): plastid RNA editing also occurs in non-photosynthetic chromoplasts. Plant Mol Biol 20:1185–1188

Kunzmann A, Brennicke A, Marchfelder A (1998) 5′ end maturation and RNA editing have to precede tRNA 3′ processing in plant mitochondria. Proc Natl Acad Sci USA 95:108–113

Kuroda H, Maliga P (2003) The plastid *clpP1* protease gene is essential for plant development. Nature 425:86–89

Kurth J, Varotto C, Pesaresi P, Biehl A, Richly E, Salamini F, Leister D (2002) Gene-sequence-tag expression analyses of 1,800 genes related to chloroplast functions. Planta 215:101–109

Lambowitz AM, Zimmerly S (2004) Mobile group II introns. Annu Rev Genet 38:1–35

Legen J, Kemp S, Krause K, Profanter B, Herrmann RG, Maier RM (2002) Comparative analysis of plastid transcription profiles of entire plastid chromosomes from tobacco attributed to wild-type and PEP-deficient transcription machineries. Plant J 31:171–188

Leino M, Teixeira R, Landgren M, Glimelius K (2003) *Brassica napus* lines with rearranged *Arabidopsis* mitochondria display CMS and a range of developmental aberrations. Theor Appl Genet 106:1156–1163

Leino M, Thyselius S, Landgren M, Glimelius K (2004) *Arabidopsis thaliana* chromosome III restores fertility in a cytoplasmic male-sterile *Brassica napus* line with *A. thaliana* mitochondrial DNA. Theor Appl Genet 109:272–279

Leino M, Landgren M, Glimelius K (2005) Alloplasmic effects on mitochondrial transcriptional activity and RNA turnover result in accumulated transcripts of *Arabidopsis* orfs in cytoplasmic male-sterile *Brassica napus*. Plant J 42:469–480

Lemieux B, Aharoni A, Schena M (1998) Overview of DNA chip technology. Mol Breed 4:277–289

Lerbs-Mache S (1993) The 110-kDa polypeptide of spinach plastid DNA-dependent RNA polymerase: single-subunit enzyme or catalytic core of multimeric enzyme complexes? Proc Natl Acad Sci USA 90:5509–5513

Liere K, Börner T (2007) Transcription and transcriptional regulation in plastids. Top Curr Genet 19:121–174

Liere K, Link G (1995) RNA-binding activity of the matK protein encoded by the chloroplast *trnK* intron from mustard (*Sinapis alba* L.). Nucleic Acids Res 23:917–921

Liere K, Maliga P (1999) In vitro characterization of the tobacco *rpoB* promoter reveals a core sequence motif conserved between phage-type plastid and plant mitochondrial promoters. EMBO J 18:249–257

Li-Pook-Than J, Carrillo C, Bonen L (2004) Variation in mitochondrial transcript profiles of protein-coding genes during early germination and seedling development in wheat. Curr Genet 46:374–380

Lu B, Hanson MR (1994) A single homogeneous form of ATP6 protein accumulates in petunia mitochondria despite the presence of differentially edited *atp6* transcripts. Plant Cell 6:1955–1968

Lupold DS, Caoile AG, Stern DB (1999) Genomic context influences the activity of maize mitochondrial *cox2* promoters. Proc Natl Acad Sci USA 96:11670–11675

MacLean D, Jerome CA, Brown AP, Gray JC (2008) Co-regulation of nuclear genes encoding plastid ribosomal proteins by light and plastid signals during seedling development in tobacco and *Arabidopsis*. Plant Mol Biol 66:475–490

Magee AM, Kavanagh TA (2002) Plastid genes transcribed by the nucleus-encoded plastid RNA polymerase show increased transcript accumulation in transgenic plants expressing a chloroplast-localized phage T7 RNA polymerase. J Exp Bot 53:2341–2349

Magee AM, Coyne S, Murphy D, Horvath EM, Medgyesy P, Kavanagh TA (2004) T7 RNA polymerase-directed expression of an antibody fragment transgene in plastids causes a semi-lethal pale-green seedling phenotype. Transgenic Res 13:325–337

Magee AM, MacLean D, Gray JC, Kavanagh TA (2007) Disruption of essential plastid gene expression caused by T7 RNA polymerase-mediated transcription of plastid transgenes during early seedling development. Transgenic Res 16:415–428

Maier RM, Neckermann K, Igloi GL, Kössel H (1995) Complete sequence of the maize chloroplast genome: gene content, hotspots of divergence and fine tuning of genetic information by transcript editing. J Mol Biol 251:614–628

Malek O, Lättig K, Hiesel R, Brennicke A, Knoop V (1996) RNA editing in bryophytes and a molecular phylogeny of land plants. EMBO J 15:1403–1411

Marano MR, Carrillo N (1992) Constitutive transcription and stable RNA accumulation in plastids during the conversion of chloroplasts to chromoplasts in ripening tomato fruits. Plant Physiol 100:1103–1113

Maréchal-Drouard L, Kumar R, Remacle C, Small I (1996a) RNA editing of larch mitochondrial tRNA(His) precursors is a prerequisite for processing. Nucleic Acids Res 24:3229–3234

Maréchal-Drouard L, Cosset A, Remacle C, Ramamonjisoa D, Dietrich A (1996b) A single editing event is a prerequisite for efficient processing of potato mitochondrial phenylalanine tRNA. Mol Cell Biol 16:3504–3510

Mayfield SP, Yohn CB, Cohen A, Danon A (1995) Regulation of chloroplast gene expression. Annu Rev Plant Physiol Plant Mol Biol 46:147–166

Meng Q, Wang Y, Liu XQ (2005) An intron-encoded protein assists RNA splicing of multiple similar introns of different bacterial genes. J Biol Chem 280:35085–35088

Mereschkowsky C (1905) Ueber Natur und Ursprung der Chromatophoren im Pflanzenreiche. Biol Centralbl 25:593–604

Meyers BC, Galbraith DW, Nelson T, Agrawal V (2004) Methods for transcriptional profiling in plants. Be fruitful and replicate. Plant Physiol 135:637–652

Miller ME, Jurgenson JE, Reardon EM, Price CA (1983) Plastid translation in organello and in vitro during light-induced development in *Euglena*. J Biol Chem 258:14478–14484

Minoda A, Nagasawa K, Hanaoka M, Horiuchi M, Takahashi H, Tanaka K (2005) Microarray profiling of plastid gene expression in a unicellular red alga, *Cyanidioschyzon merolae*. Plant Mol Biol 59:375–385

Miyata Y, Sugita M (2004) Tissue- and stage-specific RNA editing of rps14 transcripts in moss (*Physcomitrella patens*) chloroplasts. J Plant Physiol 161:113–115

Mohr G, Lambowitz AM (2003) Putative proteins related to group II intron reverse transcriptase/maturases are encoded by nuclear genes in higher plants. Nucleic Acids Res 31:647–652

Mower J, Palmer J (2006) Patterns of partial RNA editing in mitochondrial genes of *Beta vulgaris*. Mol Genet Genomics 276:285–293

Muise RC, Hauswirth WW (1992) Transcription in maize mitochondria: effects of tissue and mitochondrial genotype. Curr Genet 22:235–242

Mullet JE (1993) Dynamic regulation of chloroplast transcription. Plant Physiol 103:309–313

Nagashima A, Hanaoka M, Motohashi R, Seki M, Shinozaki K, Kanamaru K, Takahashi H, Tanaka K (2004) DNA microarray analysis of plastid gene expression in an *Arabidopsis* mutant deficient in a plastid transcription factor sigma, SIG2. Biosci Biotechnol Biochem 68:694–704

Nakagawa N, Sakurai N (2006) A mutation in At-nMat1a, which encodes a nuclear gene having high similarity to group II intron maturase, causes impaired splicing of mitochondrial NAD4 transcript and altered carbon metabolism in *Arabidopsis thaliana*. Plant Cell Physiol 47:772–783

Nakamura T, Furuhashi Y, Hasegawa K, Hashimoto H, Watanabe K, Obokata J, Sugita M, Sugiura M (2003) Array-based analysis on tobacco plastid transcripts: preparation of a genomic microarray containing all genes and all intergenic regions. Plant Cell Physiol 44:861–867

Nakamura T, Sugiura C, Kobayashi Y, Sugita M (2005) Transcript profiling in plastid arginine tRNA-CCG gene knockout moss: construction of *Physcomitrella patens* plastid DNA microarray. Plant Biol 7:258–265

Naydenov NG, Khanam SM, Atanassov A, Nakamura C (2008) Expression profiles of respiratory components associated with mitochondrial biogenesis during germination and seedling growth under normal and restricted conditions in wheat. Genes Genet Syst 83:31–41

Naydenov NG, Khanam S, Siniauskaya M, Nakamura C (2010) Profiling of mitochondrial transcriptome in germinating wheat embryos and seedlings subjected to cold, salinity and osmotic stresses. Genes Genet Syst 85:31–42

Neckermann K, Zeltz P, Igloi GL, Kössel H, Maier RM (1994) The role of RNA editing in conservation of start codons in chloroplast genomes. Gene 146:177–182

Notsu Y, Masood S, Nishikawa T, Kubo N, Akiduki G, Nakazono M, Hirai A, Kadowaki K (2002) The complete sequence of the rice (*Oryza sativa* L.) mitochondrial genome: frequent DNA sequence acquisition and loss during the evolution of flowering plants. Mol Genet Genomics 268:434–445

Obukosia SD, Richards CM, Boyer CD (2003) Expression of plastid-encoded photosynthetic genes during chloroplast or chromoplast differentiation in *Cucurbitae pepo* L. fruits. Phytochemistry 64:1213–1221

Ohyama K, Fukuzawa H, Kohchi T, Shirai H, Sano T, Sano S, Umesono K, Shiki Y, Takeuchi M, Chang Z, Aota S, Inokuchi H, Ozeki H (1986) Chloroplast gene organization deduced from complete sequence of liverwort *Marchantia polymorpha* chloroplast DNA. Nature 322:572–574

Okada S, Brennicke A (2006) Transcript levels in plant mitochondria show a tight homeostasis during day and night. Mol Genet Genomics 276:71–78

Okuda K, Nakamura T, Sugita M, Shimizu T, Shikanai T (2006) A pentatricopeptide repeat protein is a site recognition factor in chloroplast RNA editing. J Biol Chem 281:37661–37667

Okuda K, Myouga F, Motohashi R, Shinozaki K, Shikanai T (2007) Conserved domain structure of pentatricopeptide repeat proteins involved in chloroplast RNA editing. Proc Natl Acad Sci USA 104:8178–8183

Peeters NM, Hanson MR (2002) Transcript abundance supercedes editing efficiency as a factor in developmental variation of chloroplast gene expression. RNA 8:497–511

Peled-Zehavi H, Danon A (2007) Translation and translational regulation in chloroplasts. Top Curr Genet 19:249–281

Perrin R, Lange H, Grienenberger JM, Gagliardi D (2004a) AtmtPNPase is required for multiple aspects of the 18S rRNA metabolism in *Arabidopsis thaliana* mitochondria. Nucleic Acids Res 32:5174–5182

Perrin R, Meyer EH, Zaepfel M, Kim YJ, Mache R, Grienenberger JM, Gualberto JM, Gagliardi D (2004b) Two exoribonucleases act sequentially to process mature 3′-ends of *atp9* mRNAs in *Arabidopsis* mitochondria. J Biol Chem 279:25440–25446

Pfalz J, Bayraktar OA, Prikryl J, Barkan A (2009) Site-specific binding of a PPR protein defines and stabilizes 5′ and 3′ mRNA termini in chloroplasts. EMBO J 28:2042–2052

Phreaner CG, Williams MA, Mulligan RM (1996) Incomplete editing of rps12 transcripts results in the synthesis of polymorphic polypeptides in plant mitochondria. Plant Cell 8:107–117

Piechulla B, Imlay KRC, Gruissem W (1985) Plastid gene expression during fruit ripening in tomato (*Lycopersicon esculentum*). Plant Mol Biol 5: 373–384

Pring DR, Mullen JA, Kempken F (1992) Conserved sequence blocks 5′ to start codons of plant mitochondrial genes. Plant Mol Biol 19:313–317

Pyke KA (2007) Plastid biogenesis and differentiation. Top Curr Genet 19:1–28

Raczynska KD, Le Ret M, Rurek M, Bonnard G, Augustyniak H, Gualberto JM (2006) Plant mitochondrial genes can be expressed from mRNAs lacking stop codons. FEBS Lett 580:5641–5646

Rawson JR, Boerma CL (1976) A measurement of the fraction of chloroplast DNA transcribed during chloroplast development in *Euglena gracilis*. Biochemistry 15:588–592

Remacle C, Maréchal-Drouard L (1996) Characterization of the potato mitochondrial transcription unit containing 'native' trnS (GCU), trnF (GAA) and trnP (UGG). Plant Mol Biol 30: 553–563

Ribas-Carbo M, Taylor NL, Giles L, Busquets S, Finnegan PM, Day DA, Lambers H, Medrano H, Berry JA, Flexas J (2005) Effects of water stress on respiration in soybean leaves. Plant Physiol 139:466–473

Richly E, Dietzmann A, Biehl A, Kurth J, Laloi C, Apel K, Salamini F, Leister D (2003) Covariations in the nuclear chloroplast transcriptome reveal a regulatory master-switch. EMBO Rep 4:491–498

Rüdinger M, Funk HT, Rensing SA, Maier UG, Knoop V (2009) RNA editing: only eleven sites are present in the *Physcomitrella patens* mitochondrial transcriptome and a universal nomenclature proposal. Mol Genet Genomics 281:473–481

Ruf S, Kössel H (1997) Tissue-specific and differential editing of the two *ycf3* editing sites in maize plastids. Curr Genet 32:19–23

Sakai A, Kawano S, Kuroiwa T (1992) Conversion of proplastids to amyloplasts in tobacco cultured cells is accompanied by changes in the transcriptional activities of plastid genes. Plant Physiol 100: 1062–1066

Saldanha R, Mohr G, Belfort M, Lambowitz AM (1993) Group I and group II introns. FASEB J 7:15–24

Salinas T, Duchêne AM, Maréchal-Drouard L (2008) Recent advances in tRNA mitochondrial import. Trends Biochem Sci 33:320–329

Schmitz-Linneweber C, Barkan A (2007) RNA splicing and RNA editing in chloroplasts. Top Curr Genet 19:213–248

Schmitz-Linneweber C, Regel R, Du TG, Hupfer H, Herrmann RG, Maier RM (2002) The plastid chromosome of *Atropa belladonna* and its comparison with that of *Nicotiana tabacum*: the role of RNA editing in generating divergence in the process of plant speciation. Mol Biol Evol 19:1602–1612

Schmitz-Linneweber C, Williams-Carrier R, Barkan A (2005) RNA Immunoprecipitation and microarray analysis show a chloroplast pentatricopeptide repeat protein to be associated with the 5′ region of mRNAs whose translation it activates. Plant Cell 17:2791–2804

Schnable PS, Ware D, Fulton RS, Stein JC, Wei F, Pasternak S, Liang C, Zhang J, Fulton L, Graves TA, Minx P, Reily AD, Courtney L, Kruchowski SS, Tomlinson C, Strong C, Delehaunty K, Fronick C, Courtney B, Rock SM, Belter E, Du F, Kim K, Abbott RM, Cotton M, Levy A, Marchetto P, Ochoa K, Jackson SM, Gillam B, Chen W, Yan L, Higginbotham J, Cardenas M, Waligorski J, Applebaum E, Phelps L, Falcone J, Kanchi K, Thane T, Scimone A, Thane N, Henke J, Wang T, Ruppert J, Shah N, Rotter K, Hodges J, Ingenthron E, Cordes M, Kohlberg S, Sgro J, Delgado B, Mead K, Chinwalla A, Leonard S, Crouse K, Collura K, Kudrna D, Currie J, He R, Angelova A, Rajasekar S, Mueller T, Lomeli R, Scara G, Ko A, Delaney

K, Wissotski M, Lopez G, Campos D, Braidotti M, Ashley E, Golser W, Kim H, Lee S, Lin J, Dujmic Z, Kim W, Talag J, Zuccolo A, Fan C, Sebastian A, Kramer M, Spiegel L, Nascimento L, Zutavern T, Miller B, Ambroise C, Muller S, Spooner W, Narechania A, Ren L, Wei S, Kumari S, Faga B, Levy MJ, McMahan L, Van Buren P, Vaughn MW, Ying K, Yeh CT, Emrich SJ, Jia Y, Kalyanaraman A, Hsia AP, Barbazuk WB, Baucom RS, Brutnell TP, Carpita NC, Chaparro C, Chia JM, Deragon JM, Estill JC, Fu Y, Jeddeloh JA, Han Y, Lee H, Li P, Lisch DR, Liu S, Liu Z, Nagel DH, McCann MC, SanMiguel P, Myers AM, Nettleton D, Nguyen J, Penning BW, Ponnala L, Schneider KL, Schwartz DC, Sharma A, Soderlund C, Springer NM, Sun Q, Wang H, Waterman M, Westerman R, Wolfgruber TK, Yang L, Yu Y, Zhang L, Zhou S, Zhu Q, Bennetzen JL, Dawe RK, Jiang J, Jiang N, Presting GG, Wessler SR, Aluru S, Martienssen RA, Clifton SW, McCombie WR, Wing RA, Wilson RK (2009) The B73 maize genome: complexity, diversity, and dynamics. Science 326:1112–1115

Schulze A, Downward J (2001) Navigating gene expression using microarrays-a technology review. Nat Cell Biol 3:E190–E195

Schuster W (1993) Ribosomal protein gene *rpl5* is cotranscribed with the *nad3* gene in *Oenothera* mitochondria. Mol Gen Genet 240:445–449

Schweer J, Loschelder H, Link G (2006) A promoter switch that can rescue a plant sigma factor mutant. FEBS Lett 580:6617–6622

Serino G, Maliga P (1998) RNA polymerase subunits encoded by the plastid *rpo* genes are not shared with the nucleus-encoded plastid enzyme. Plant Physiol 117:1165–1170

Shiina T, Tsunoyama Y, Nakahira Y, Khan MS (2005) Plastid RNA polymerases, promoters, and transcription regulators in higher plants. In: Jeon KW (ed) International review of cytology, vol 244. Academic, Amsterdam, pp 1–68

Shikanai T (2006) RNA editing in plant organelles: machinery, physiological function and evolution. Cell Mol Life Sci 63:698–708

Shikanai T, Shimizu K, Ueda K, Nishimura Y, Kuroiwa T, Hashimoto T (2001) The chloroplast *clpP* gene, encoding a proteolytic subunit of ATP-dependent protease, is indispensable for chloroplast development in tobacco. Plant Cell Physiol 42:264–273

Shinozaki K, Ohme M, Tanaka M, Wakasugi T, Hayashida N, Matsubayashi T, Zaita N, Chunwongse J, Obokata J, Yamaguchi-Shinozaki K, Ohto C, Torazawa K, Meng BY, Sugita M, Deno H, Kamogashira T, Yamada K, Kusuda J, Takaiwa F, Kato A, Tohdoh N, Shimida H, Sugiura M (1986) The complete nucleotide sequence of the tobacco chloroplast genome: its gene organization and expression. EMBO J 5:2043–2049

Siddique MA, Grossmann J, Gruissem W, Baginsky S (2006) Proteome analysis of bell pepper (*Capsicum annuum* L.) chromoplasts. Plant Cell Physiol 47:1663–1673

Silhavy D, Maliga P (1998) Mapping of promoters for the nucleus-encoded plastid RNA polymerase (NEP) in the iojap maize mutant. Curr Genet 33:340–344

Siniauskaya M, Naydenov N, Davydenko O, Nakamura C (2008) Macroarray for studying chloroplast gene expression profiles associated with the initial development of wheat. In: Proceedings of the 11th International Wheat Genetics Symposium, Sidney University Press, Sidney

Small ID, Peeters N (2000) The PPR motif – a TPR-related motif prevalent in plant organellar proteins. Trends Biochem Sci 25:46–47

Stears RL, Martinsky T, Schena M (2003) Trends in microarray analysis. Nat Med 9:140–145

Stern DB, Kindle KL (1993) 3¢end maturation of the *Chlamydomonas reinhardtii* chloroplast *atpB* mRNA is a two-step process. Mol Cell Biol 13:2277–2285

Stern DB, Goldschmidt-Clermont M, Hanson MR (2010) Chloroplast RNA metabolism. Annu Rev Plant Biol 61:125–155

Sugiura M, Hirose T, Sugita M (1998) Evolution and mechanism of translation in chloroplasts. Annu Rev Genet 32:437–459

Svensson AS, Rasmusson AG (2001) Light-dependent gene expression for proteins in the respiratory chain of potato leaves. Plant J 28:73–82

Takenaka M, Verbitskiy D, van der Merwe JA, Zehrmann A, Brennicke A (2008) The process of RNA editing in plant mitochondria. Mitochondrion 8:35–46

Tanaka K, Tozawa Y, Mochizuki N, Shinozaki K, Nagatani A, Wakasa K, Takahashi H (1997) Characterization of three cDNA species encoding plastid RNA polymerase sigma factors in *Arabidopsis thaliana*: evidence for the sigma factor heterogeneity in higher plant plastids. FEBS Lett 413:309–313

Tasaki E, Hattori M, Sugita M (2010) The moss pentatricopeptide repeat protein with a DYW domain is responsible for RNA editing of mitochondrial *ccmFc* transcript. Plant J 62:560–570

Tavares-Carreón F, Camacho-Villasana Y, Zamudio-Ochoa A, Shingú-Vázquez M, Torres-Larios A, Pérez-Martínez X (2008) The pentatricopeptide repeats present in Pet309 are necessary for translation but not for stability of the mitochondrial *cox1* mRNA in yeast. J Biol Chem 283:1472–1479

Tillich M, Funk HT, Schmitz-Linneweber C, Poltnigg P, Sabater B, Martin M, Maier RM (2005) Editing of

plastid RNA in *Arabidopsis thaliana* ecotypes. Plant J 43:708–715
Tillich M, Lehwark P, Morton BR, Maier UG (2006) The evolution of chloroplast RNA editing. Mol Biol Evol 23:1912–1921
Tracy RL, Stern DB (1995) Mitochondrial transcription initiation: promoter structures and RNA polymerases. Curr Genet 28:205–216
Tsudzuki T, Wakasugi T, Sugiura M (2001) Comparative analysis of RNA editing sites in higher plant chloroplasts. J Mol Evol 53:327–332
Unseld M, Marienfeld JR, Brandt P, Brennicke A (1997) The mitochondrial genome of *Arabidopsis thaliana* contains 57 genes in 366,924 nucleotides. Nat Genet 15:57–61
Uyttewaal M, Mireau H, Rurek M, Hammani K, Arnal N, Quadrado M, Giegé P (2008) PPR336 is associated with polysomes in plant mitochondria. J Mol Biol 375:626–636
Valkov VT, Scotti N, Kahlau S, Maclean D, Grillo S, Gray JC, Bock R, Cardi T (2009) Genome-wide analysis of plastid gene expression in potato leaf chloroplasts and tuber amyloplasts: transcriptional and posttranscriptional control. Plant Physiol 150:2030–2044
Vaughn JC, Mason MT, Sper-Whitis GL, Kuhlman P, Palmer JD (1995) Fungal origin by horizontal transfer of a plant mitochondrial group I intron in the chimeric *coxI* gene of Peperomia. J Mol Evol 41:563–572
Verbitskiy D, Zehrmann A, Brennicke A, Takenaka M (2010) A truncated MEF11 protein shows site-specific effects on mitochondrial RNA editing. Plant Signal Behav 5:558–560
Vogel A, Schilling O, Späth B, Marchfelder A (2005) The tRNase Z family of proteins: physiological functions, substrate specificity and structural properties. Biol Chem 386:1253–1264
Wakasugi T, Hirose T, Horihata M, Tsudzuki T, Kössel H, Sugiura M (1996) Creation of a novel protein-coding region at the RNA level in black pine chloroplasts: the pattern of RNA editing in the gymnosperm chloroplast is different from that in angiosperms. Proc Natl Acad Sci USA 93:8766–8770
Wang L, Li P, Brutnell TP (2010) Exploring plant transcriptomes using ultra high-throughput sequencing. Brief Funct Genomics 9:118–128
Wank H, SanFilippo J, Singh RN, Matsuura M, Lambowitz AM (1999) A reverse transcriptase/maturase promotes splicing by binding at its own coding segment in a group II intron RNA. Mol Cell 4:239–250
Weber AP, Weber KL, Carr K, Wilkerson C, Ohlrogge JB (2007) Sampling the *Arabidopsis* transcriptome with massively parallel pyrosequencing. Plant Physiol 144:32–42
Welchen E, Gonzalez DH (2006) Overrepresentation of elements recognized by TCP-domain transcription factors in the upstream regions of nuclear genes encoding components of the mitochondrial oxidative phosphorylation machinery. Plant Physiol 141:540–545
Wissinger B, Schuster W, Brennicke A (1991) Trans splicing in *Oenothera* mitochondria: *nad1* mRNAs are edited in exon and *trans*-splicing group-II intron sequences. Cell 65:473–482
Wolf PG, Rowe CA, Hasebe M (2004) High levels of RNA editing in a vascular plant chloroplast genome: analysis of transcripts from the fern *Adiantum capillus-veneris*. Gene 339:89–97
Wood CK, Dudley P, Albury MS, Affourtit C, Leach GR, Pratt JR, Whitehouse DG, Moore AL (1996) Developmental regulation of respiratory activity and protein import in plant mitochondria. Biochem Soc Trans 24:746–749
Yamaguchi K, Subramanian AR (2000) The plastid ribosomal proteins. Identification of all the proteins in the 50 S subunit of an organelle ribosome (chloroplast). J Biol Chem 275:28466–28482
Yamaguchi K, Subramanian AR (2003) Proteomic identification of all plastid-specific ribosomal proteins in higher plant chloroplast 30S ribosomal subunit. Eur J Biochem 270:190–205
Yamaguchi K, von Knoblauch K, Subramanian AR (2000) The plastid ribosomal proteins. Identification of all the proteins in the 30 S subunit of an organelle ribosome (chloroplast). J Biol Chem 275:28455–28465
Yamaguchi K, Prieto S, Beligni MV, Haynes PA, McDonald WH, Yates JR 3rd, Mayfield SP (2002) Proteomic characterization of the small subunit of *Chlamydomonas reinhardtii* chloroplast ribosome: identification of a novel S1 domain-containing protein and unusually large orthologs of bacterial S2, S3, and S5. Plant Cell 14:2957–2974
Yamaguchi K, Beligni MV, Prieto S, Haynes PA, McDonald WH, Yates JR 3rd, Mayfield SP (2003) Proteomic characterization of the *Chlamydomonas reinhardtii* chloroplast ribosome. Identification of proteins unique to the 70S ribosome. J Biol Chem 278:33774–33785
Yukawa M, Tsudzuki T, Sugiura M (2005) The 2005 version of the chloroplast DNA sequence from tobacco (*Nicotiana tabacum*). Plant Mol Biol Rep 23:359–365
Zandueta-Criado A, Bock R (2004) Surprising features of plastid *ndhD* transcripts: addition of non-encoded nucleotides and polysome association of mRNAs

with an unedited start codon. Nucleic Acids Res 32:542–550

Zehrmann A, van der Merwe JA, Verbitskiy D, Brennicke A, Takenaka M (2008) Seven large variations in the extent of RNA editing in plant mitochondria between three ecotypes of *Arabidopsis thaliana*. Mitochondrion 8:319–327

Zehrmann A, Verbitskiy D, van der Merwe JA, Brennicke A, Takenaka M (2009) A DYW domain-containing pentatricopeptide repeat protein is required for RNA editing at multiple sites in mitochondria of *Arabidopsis thaliana*. Plant Cell 21:558–567

Zerges W (2000) Translation in chloroplasts. Biochimie 82:583–601

Zybailov B, Friso G, Kim J, Rudella A, Rodriguez VR, Asakura Y, Sun Q, van Wijk KJ (2009) Large scale comparative proteomics of a chloroplast Clp protease mutant reveals folding stress, altered protein homeostasis, and feedback regulation of metabolism. Mol Cell Proteomics 8:1789–1810

Chapter 15

Organellar Proteomics: Close Insights into the Spatial Breakdown and Functional Dynamics of Plant Primary Metabolism

Hans-Peter Braun and Holger Eubel*
Department V: Plant Proteomics, Institute for Plant Genetics, Leibniz Universität Hannover, Herrenhäuser Str. 2, D-30419 Hannover, Germany

Summary

The proteomes of cells, cellular compartments or fluids are mirrors of their functions and the changes detectable in these proteomes have direct impact on their physiology. Proteomics aims at identifying the components of a target proteome as well as characterizing its dynamics, but it is not restricted to the mere identification and quantitation of proteins. Because of their ability to modulate enzymatic activity, the analysis of post-translational modifications (PTMs) produces information on the regulation of cell physiology beyond the level of protein abundance. In the past, plant proteome analyses were dominated by studies performed on Arabidopsis thaliana, while other plants such as spinach, tobacco and rice were only of minor importance. The focus on Arabidopsis is mostly due to the high quality of publicly available genome data, a major pillar of mass spectrometry based identification of proteins.

*Author for correspondence, e-mail: heubel@genetik.uni-hannover.de

R. Bock and V. Knoop (eds.), Genomics of Chloroplasts and Mitochondria, Advances in Photosynthesis and Respiration 35, pp. 357–378,
DOI 10.1007/978-94-007-2920-9_15, © Springer Science+Business Media B.V. 2012

However, the flow of information (from genomics to proteomics) is not a one way alley. Proteogenomic mapping contributes to the process of gene annotation by the discovery of proteins which do not fit predicted gene sequences, but instead require re-annotation of predicted open reading frames (ORFs). Since it also leads to the discovery of so far unrecognized ORFs, it may additionally be helpful in the identification of genes in newly sequenced genomes.

In plants, proteomic analysis of the three major energy-related organelles, plastids, mitochondria and peroxisomes, gives evidence to the spatial distribution of the bulk of enzymes involved in energy metabolism. Furthermore, proteomics, often in combination with other 'omic'-techniques, has been used to investigate a wide spectrum of questions in organelle biology. These include, among others, responses to environmental conditions, stress situations and gene knock-outs or knock-downs. Proteomics has also been employed to investigate organelle biogenesis and protein-protein interactions and forms the data basis for computational approaches ranging from protein targeting prediction to the modeling of metabolic networks. Finally, although more commonly used to monitor the changes in an experimental system, proteomics also serves to provide targets for more detailed studies, for example involving reverse genetics. Focusing on the gain in biological understanding generated by proteomics, the following paragraphs provide a short history of plant organelle proteomics and try to give an outlook on future directions of this 'omic' platform.

Abbreviations: 1D – one-dimensional; 2D – two-dimensional; ACCase – acetyl-CoA carboxylase complex; AMT – accurate mass and time; BN – blue native; CCS – capsanthin/capsorubin synthase; cTP – chloroplast transit peptide; FFE – free flow electrophoresis; HNE – 4-hydroxy-2-nonenal; HSP – heat shock proteins; IEF – Isoelectric focusing; IMAC – immobilized metal affinity chromatography; LC-ESI-LTQ-Orbitrap – liquid chromatography electrospray ionization linear ion-trap quadrupole orbitrap (mass spectrometer); LC-ESI-Q-TOF – liquid chromatography electrospray ionization quadrupole time-of-flight (mass spectrometer); LOPIT – localization of organelle proteins by isotope tagging; lTP – luminal transit peptide; MCO – metal catalyzed oxidation; MRM – multiple reaction monitoring; MS – mass spectrometry; MS/MS – tandem mass spectrometry; MuDPIT – multi dimensional protein identification technology; NEP – nucleus encoded plastid RNA-polymerase; OPPP – oxidative pentose phosphate pathway; ORF – open reading frame; PAGE – polyacrylamide gel electrophoresis; PDC – pyruvate dehydrogenase complex; PEP – plastid encoded RNA polymerase; pETC – plastid electron transfer chain; PG – plastoglobuli; PTM – post-translational modification; PTS – peroxisomal targeting sequence; RNP – ribonucleoprotein; ROS – reactive oxygen species; RuBisCO – ribulose-1,5-bisphosphate-carboxylase/oxygenase; SDS – sodium dodecyl sulfate; SEC – size exclusion chromatography; SOD – superoxide dismutase; SpC – spectral counting; TAT – twin arginine protein translocation; WT – wild type

I. Introduction

Proteins are the main effectors of cellular metabolism. Therefore, the analysis of the protein content of cells, cellular compartments or biological fluids provides broad overviews of their functions in the same way the purpose of a craftsman's workshop is revealed by its equipment: car maintenance requires a different set of tools than carpentry or stone masonry. The global analysis of proteins and their abundance, usually referred to as 'proteomics', has been fostered by the growing numbers of species for which the respective genomic sequences are known and will continue to benefit from the genomic sequences, which will become known in the near future. Although proteomics has been done in the 'pre-genomic era', it is the information produced by such genome sequencing efforts, which enables the quick and reliable identification of proteins by means of mass spectrometry. However, proteome analysis of whole cells is usually difficult for three major reasons:

1. They contain highly complex mixtures of different protein species.
2. The dynamic range of protein abundance in proteomes is high, ranging between 10^5 and 10^7.

3. Being assembled from unique combinations of 20 amino acids, proteins display a wide spectrum of chemical diversity.

For technical reasons, these three parameters inevitably compromise the quality of any proteome analysis. While the chemical composition and the dynamic range of the proteins are fixtures, a helpful reduction in complexity can be achieved by splitting the cells into their building blocks prior to the analysis. These building blocks can then be analyzed individually, which delivers better overall proteome coverage. In plant cells, plastids, mitochondria, and (to a lesser degree) also peroxisomes are popular targets for proteomic investigations at the subcellular level. This is not only because of their impact on primary metabolism, but also due to the ease with which such organellar fractions can be produced from plant material (the impact of the isolation procedure on the outcome of the analyses will be discussed in the Sect. II). Apart from the increased proteome coverage, dissecting cells into their distinct compartments has an additional advantage. It delivers information about the intracellular whereabouts of the identified proteins, which can then be used to create a model of the spatial breakdown of cellular metabolism.

However, the benefits of the static investigations of organelle protein content are not limited to the assessment of the internal cellular architecture. Our understanding of the dynamic processes taking place within the cell in general and within these cellular organelles in particular also benefits from such studies. For example, metabolic regulation often is achieved by modulating protein abundance and such changes often correlate with alterations of enzymatic activity. Also, plant genomes posses large numbers of coding regions for paralogues which are flanked by different regulatory sequences. Often, these paralogues are targeted to different locations within the cell. Knowledge of the mechanisms regulating gene expression combined with that of the intracellular use of paralogues, therefore, provides insight into key aspects of metabolic regulation.

Even though the assessment of the protein content of cellular compartments produces results, which have direct or indirect implications for functional investigations of organelle biology, comparative proteome studies are usually better suited to serve this purpose. The effects of development, environment, stress, gene knock-outs etc. on organellar metabolism can be monitored by quantitative proteome analysis. Alterations in the proteomes of organelles isolated from differently treated plants, knockout versus WT plants or over different stages of development can be detected by direct comparison. Changes detected in such studies may be strong indicators of modified organelle functions. Compared to global transcript analysis (transcriptomics), proteomics often delivers results which are more meaningful in the context of protein abundance given that transcript and protein levels are frequently not directly correlated. Moreover, the same protein is often found in several versions within the target compartment due to maturation, modification or breakdown of the polypeptide chain. Such observations may serve as an indication for regulatory events, but since it is often unclear which protein version is the (most) active one, enzymatic assays or the quantitative comparison of metabolite abundance is imperative to complement proteome data in functional studies.

However, the potential advantage of proteomics over global transcript analysis comes at a price. Proteins are chemically more diverse than ribonucleic acids, and their global assessment therefore requires highly sophisticated technical setups to cover the broad spectrum of proteins present in any proteome comprehensively. As a result, only abundant proteins are covered by the analysis while less abundant ones (often those with regulatory functions) are hidden below the detection threshold.

II. Concepts and General Technical Challenges of Organelle Proteomics

This chapter aims to focus on the benefits of proteomics for the fields of plant organelle biochemistry and physiology. But since proteomics is a field which is strongly driven by technical improvements, we deem it necessary to first discuss the technical background as well as strengths and pitfalls of proteomics in order to prepare the non-expert reader for the following sections.

Recent developments in mass spectrometry have had a strong impact on the quality of proteome studies. As already mentioned, earlier technical shortcomings favored the detection of prevalent proteins while ignoring those of low abundance. However, instruments with ever increasing sensitivity and speed are responsible for improved proteome coverage and depth of proteome studies. While mass spectrometry is an interesting topic in itself, it is far too complex to be dealt with at length in this article. In the context of plant organelle proteomics, it is more important to point out that improvements in mass spectrometry require concomitant upgrades in the steps upstream of MS in order to ensure good quality results. Identifications of proteins (obviously) stemming from other cellular compartments than the targeted one have frequently troubled organelle proteome studies. This problem became all the more apparent with the increase in performance of modern mass spectrometers. Hence, measures to reestablish confidence in organelle proteome data have become increasingly important. Two approaches improving data quality have successfully been employed in the past:

1. Purification of the organelles to higher degrees of homogeneity prior to MS
2. Comparative quantitation of protein abundance.

Mass spectrometry performed on organelles with a higher level of purity yields lower numbers of foreign proteins than heavily contaminated samples. State-of-the-art in organelle preparation is differential centrifugation followed by one, two or even three isopycnic centrifugations (Fig. 15.1). Although enzymatic assays of marker enzymes often suggest higher purity levels, organelles can be purified to a maximum of approximately 90% homogeneity this way. However, such assays have to be treated with caution (Baginsky et al. 2004). Often, contaminations do not stem from intact organelles but from broken ones. The activities of soluble marker proteins are therefore not representative for the presence of non-soluble contaminations and vice versa. Moreover, measuring very low levels of enzymatic activities, for example that of a foreign marker enzyme in the target fraction, is difficult and the margins for errors are high.

Biological variation is the main culprit in the isolation of cell organelles, especially that in the density of the organelles. While density is the main parameter of separation in isopycnic centrifugation, it is particle size in differential centrifugation. However, also in this step, density affects the sedimentation speed of the organelles. Consequently, these two centrifugation based techniques cannot be regarded as completely independent dimensions in the separation process and for this reason will never yield plant organelles of ultra-high purity, regardless of the number of centrifugation steps. To further improve the purity of organelle isolates, it is therefore necessary to employ techniques which are not based on centrifugation but employ other physical properties of the organelles. One such technique is free flow electrophoresis (FFE) which has successfully been used on yeast mitochondria, plant mitochondria and peroxisomes, and rat peroxisomes (Eubel et al. 2007, 2008; Volkl et al. 1997, 1999; Zischka et al. 2006). FFE separates particles with different surface charges and therefore constitutes a bona fide second dimension to centrifugation. Mitochondria isolated by FFE have a higher level of homogeneity than those isolated by centrifugation. Other approaches, for example flow cytometry, are potentially capable of supplementing centrifugation in the quest for improved quality of the isolates, but such techniques still need to prove their potential.

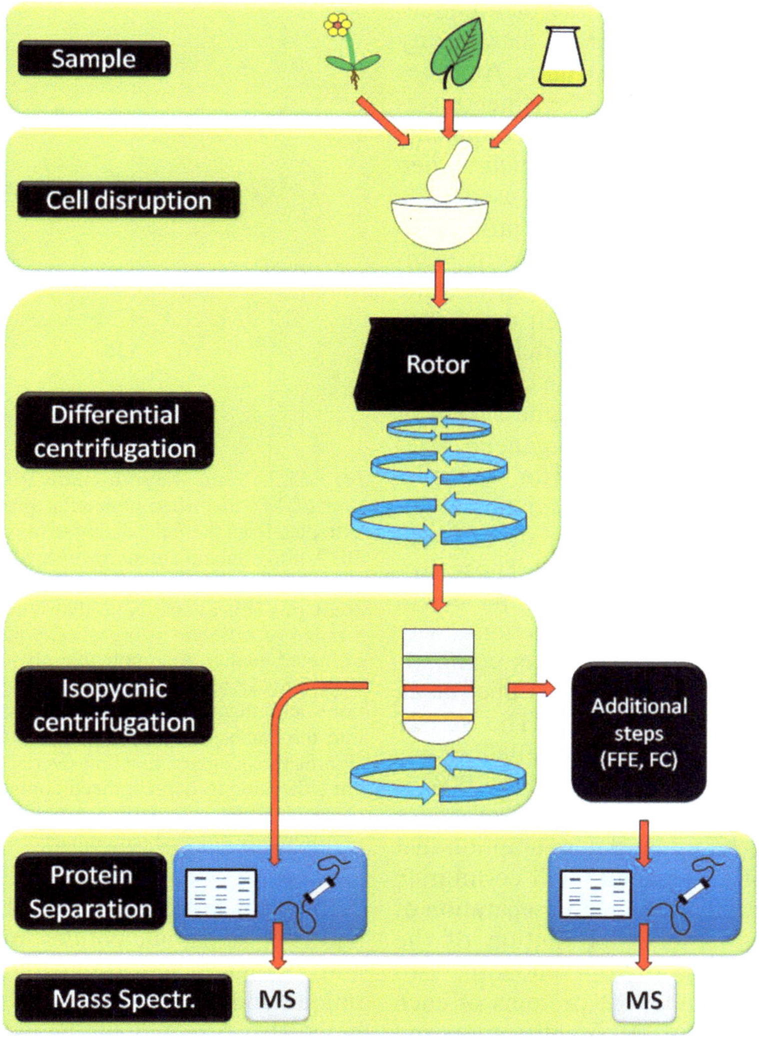

Fig. 15.1. Flowchart of organellar proteomics. Samples (whole plants, plant organs or cell cultures) are homogenized to break up the cells. Organelles are then isolated by differential centrifugation followed by isopycnic centrifugation. The latter step is often repeated under different conditions to reduce contamination. The target organelle fraction is removed from the gradient and washed. Organellar proteins are then separated by gel electrophoresis or chromatographic means before they are analyzed in a mass spectrometer. Alternatively, organelles can be further purified prior to protein separation using more elaborate techniques, such as free flow electrophoresis or flow cytometry. *FC* flow cytometry, *FFE* free flow electrophoresis, *MS* mass spectrometry.

A less efficient but simple strategy to reduce organelle contamination is the subfractionation of plant organelles. Gentle solubilization steps are effective means in this respect, since organelle membranes possess different lipid compositions and therefore react differently to the same detergent. An additional density gradient can then be employed to separate intact organelles from the content of lysed ones.

However, regardless of the methods employed in the organelle purification steps, it seems doubtful that purity levels >99% are possible. Therefore, even if additional steps can improve confidence in the data noticeably, other methods to improve data quality will become necessary in the future. Analyses of the protein abundances of the target organelle and of the most prominent contaminants will yield quantitative data which, when compared, can for example be used as a means to allocate most identified proteins to plastids or mitochondria. It is worth mentioning here that this technique delivers the most unambiguous results when the spread in protein abundance is high, that is, when the isolates of the target and the contaminating fractions each have the highest level of purity. If purifying certain organelle fractions proves to be difficult (for example, Arabidopsis peroxisomes), protein abundances across target fractions of different levels of purity can be monitored. Those proteins belonging to the target organelle will become more abundant in fractions with higher purity while those of other compartments will be less prominent (Eubel et al. 2007; Heinemeyer et al. 2009). The LOPIT approach (Localization of Organelle Proteins by Isotope Tagging, Kleffmann et al. 2004) is a variation of the comparative proteomics approach. It is based on the assumption that the proteins of an organelle will co-migrate in a density gradient during the separation of cell organelles. After fractionation of the gradient, quantifiable MS-compatible isotopes are used to label the proteins of each fraction, followed by MS identification and quantitation. By statistical analysis, the identified proteins can be grouped and their subcellular localizations can be deduced. Figure 15.2 gives an overview of the locations of the majority of proteins identified in proteomic studies of plastids, mitochondria and peroxisomes.

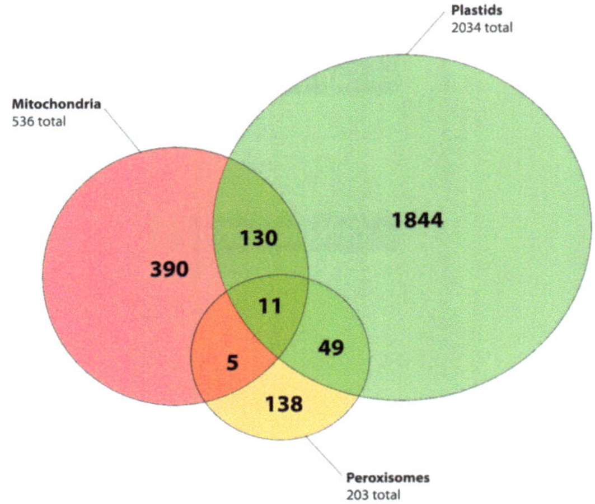

Fig. 15.2. Venn diagram indicating the localization of MS-identified organellar proteins. Data were extracted from the Suba database (as of December 15th 2010, http://suba.plantenergy.uwa.edu.au, Heazlewood et al. 2007), which contains proteome data from hundreds of publications. Note that it contains data from a range of different approaches, which inevitably are of mixed quality. Also note that obvious contaminants were already removed by the authors of the publications and that, if only raw data were incorporated into the database, the overlaps can be expected to be significantly higher. Existing overlaps in the diagram are either due to unrecognized contamination, dual or triple targeting or, possibly, ambiguous identification of differently targeted paralogues.

III. Plastid Proteomics

A. Proteomics of Autotrophic Plastids

Chloroplasts are of central importance to the physiology of the plant. They are the sites of photosynthesis and produce essential cellular metabolites, such as amino acids and fatty acids, nucleotides, plant hormones and secondary metabolites. They possess a unique and highly organized internal membrane structure and can be easily broken up into four major fractions: envelope (consisting of outer and inner envelope membrane as well as the plastid intermembrane space), thylakoid membranes (harboring the protein complexes of the chloroplast electron transfer chain in the light-driven reactions of photosynthesis), thylakoid lumen and chloroplast stroma. Due to the nature of these compartments, the proteomes of envelope and thylakoid membranes are dominated by hydrophobic proteins whereas soluble proteins dominate the stroma and thylakoid lumen proteomes. Kleffmann and co-workers were able to identify 690 proteins from Arabidopsis chloroplasts by a two-track strategy, focusing on soluble proteins in one approach and membrane proteins in the other one (Kleffmann et al. 2004). Functional characterization of 141 unknown proteins based on the in silico prediction of functional domains in the polypeptides was successful in the case of 59 proteins. In addition, to assess the prevalence of biochemical pathways, the coverage of a pathway was assessed by dividing the number of detected proteins by the number of expected proteins. To further substantiate the results, correlations between protein detection and RNA abundance (which has also been analyzed in this study using DNA chip technology) were produced at the level of complete pathways. Although there was a positive overall correlation between protein detection and high transcript levels, there were marked differences between the different metabolic pathways. For example, while the Calvin cycle and the amino acid metabolism showed a good correlation, this was not the case for the tetrapyrrole pathway and the proteins of the photosystems. Overall, this indicates a transcriptional control for the Calvin cycle and amino acid metabolism, in contrast to the tetrapyrrole pathway and the photosystems, whose abundances are probably controlled to a high degree by translational and posttranslational mechanisms. At the level of individual genes, it was found that protein abundance (by means of sequence coverage) and RNA abundance were moderately positively correlated, as indicated by a Spearman rank correlation of 0.53 (0–0.5=weak; 0.5–0.7=moderate; 0.7–0.9=strong and 0.9–1.0=very strong correlation, respectively). Using the same dataset, the correlation between protein and RNA abundance was investigated in more detail later (Baginsky et al. 2005). Essentially, the following conclusions were drawn:

1. Compared to genome-wide expression levels and also to the expression levels of all proteins predicted to be imported into plastids, RNA expression levels are higher for the proteins identified in the study. As expected, this suggests a bias towards the detection of highly abundant proteins, at least in proteomic studies using the shotgun approach.
2. Against this trend, some proteins with rather low mRNA levels were detected nevertheless. A common feature of these proteins is an above-average size, indicative of a bias towards the detection of high molecular weight proteins.
3. RNA/protein abundance correlation can be used to increase confidence in the localization of an identified protein. Since proteins contaminating the target organelle fraction are usually the most abundant ones in their 'home-compartment', they can be expected to correlate with an equally high RNA abundance. Proteins with low sequence coverage combined with high RNA levels are therefore indicative of 'foreign' proteins.

While, in general, the mRNA-to-protein correlation provides insight into those aspects of metabolic regulation which ultimately use protein abundance as the key factor, other mechanisms are employed by chloroplasts as well. One of these was investigated by Ströher and Dietz, who focused on redox-regulated proteins of the Arabidopsis chloroplast stroma, thylakoid lumen and the thylakoid periphery (Ströher and Dietz 2008). Using diagonal 2D redox SDS-PAGE, 49 potentially redox-regulated proteins were identified. Formation and breakage of cystein disulfide bonds induce conformational changes in the affected proteins and are considered the main redox-regulatory events. Such changes result in altered electrophoretic mobility of the proteins between the first and second gel dimensions of the employed 2D system. Proteins affected by conformational changes due to disulfide bond breakage or formation migrate above or below the diagonal line on which most (non-redox regulated) proteins are found. For two of these proteins, redox-regulation was demonstrated in functional assays. Proteins involved in photosynthesis, Calvin cycle, photorespiration, response to oxidative stress, protein folding and turnover, transcription as well as mRNA degradation were identified.

Based on targeting prediction of proteins it is estimated that plastid functions are performed by up to 3,450 proteins (Peltier et al. 2006) or up to 3,800 proteins (Kleffmann et al. 2007), which would account for approximately 14–15% of all Arabidopsis proteins. Although not all of these proteins are expected to be present in the organelles at the same time, the complexity of the plastid proteome can best be described as high and, as a consequence, in most cases the plastids are fractionated into subcompartments prior to proteome analysis.

Recently, Zybailov and co-workers published a plastid proteome list consisting of 1,325 accession numbers (Zybailov et al. 2008). Besides protein identities, relative abundances as well as posttranslational modifications are reported for a subset of the identified proteins. The identification of this, from a technical viewpoint, astonishing number of proteins was achieved by first subfractionating the isolated chloroplasts into stroma and thylakoid membrane fractions, each of which was then separated on 1D gels. Gel lanes were cut into several pieces followed by in-gel trypsination and peptide extraction. MS was performed on LC-ESI-Q-TOF or LC-ESI-LTQ-Orbitrap instruments. In a similar approach, the identification of nearly exactly the same number of proteins (1,323) was achieved from thylakoid membranes, envelope membranes and stroma subfractions of Arabidopsis chloroplasts (Ferro et al. 2010). The generation of quantitative data by means of spectral counting (SpC) enabled the assignment of proteins to one of the three plastid compartments investigated in this study. Furthermore, accurate mass and time tags (AMT tags) for the detected peptides of identified proteins were generated and publicly archived in order to facilitate a simpler means of quantitation. A detailed description of the multiple features and many results of these two studies is outside the scope of this chapter. Interested readers are therefore referred to the original publications. However, these two 'combined' sub-plastid proteome studies demonstrate the state-of-the-art in organelle proteomics. Further examples of targeted analyses of plastid suborganellar proteomes are described below. Where possible, we attempted to follow the chronological order of publication to provide an insight into the impressive progress the scientific field of plant organelle proteomics has made over the last decade.

1. Thylakoid Lumen Proteomics

Although it has been known that extrinsic subunits of PSII are attached to the luminal side of the thylakoid membrane (Andersson and Akerlund 1978), it was only discovered in 1991 that these proteins can also be found in the soluble part of the thylakoid lumen (Ettinger and Theg 1991). The proteomic investigation of the thylakoid lumen started with the study of Kieselbach and co-workers (Kieselbach et al. 1998) using spinach. An isolation procedure for the thylakoid lumen was developed and 25 proteins from 1D SDS gels were identified by N-terminal sequencing and immunoblotting. Later, using Edman degradation and MS for the identification of 2D IEF/SDS-PAGE separated proteins, Peltier and colleagues analyzed 400 spots of soluble and peripheral (i.e. of the lumen-side of the thylakoid membrane) proteins of this compartment, which resulted in the identification of 61 proteins (Peltier et al. 2000). It is noteworthy that this study was performed using pea, which compromised the identification of proteins by MS markedly by the absence of genomic sequence information. In a second attempt using the same approach, the authors made use of the recently annotated Arabidopsis genome by changing over to this model system. Eighty-one proteins were identified, of which 30 were localized in the thylakoid lumen (Peltier et al. 2002). Thirty-two proteins were of non-luminal but chloroplast origin and six were of other cellular compartments. Apart from the sheer identification of luminal proteins, the authors also increased the value of their data by using them for the re-annotation of some Arabidopsis genes and a detailed analysis of the chloroplast transit peptide (cTP) and the luminal transit peptide (lTP). In addition to the genome-wide prediction of the luminal proteome and its functional properties, Schubert and co-workers directly compared the luminal proteome of Arabidopsis with that of spinach and found that the two overlap well (Schubert et al. 2002). Furthermore, they estimated the Arabidopsis luminal proteome to consist of approximately 80 proteins, which is close to the 93 proteins predicted by Friso and colleagues, albeit their prediction only covered those proteins of the twin arginine protein translocation (TAT) pathway (Friso et al. 2004). Both studies clearly report the presence of proteins involved in protein folding and protein fate in the thylakoid lumen. Especially interesting among these are the members of the immunophilin family, which are characterized by their peptidyl-prolyl cis-trans isomerase activity. The function of these proteins is not entirely clear in that they also have been hypothesized to be involved in redox-regulation (Gopalan et al. 2004). Functional aspects of the thylakoid lumen proteome were first investigated in a study attempting to unravel the mechanisms of plastid cold acclimation (Goulas et al. 2006). The results indicate a substitution of nuclear-encoded paralogues within the PsbO and PsbP protein families in response to cold stress. These proteins are lumen-localized extrinsic subunits of the oxygen-evolving complex of PSII and are substituted by respective paralogues in order to facilitate adaptation to reduced temperatures. At the same time, some members of the immunophilins were found to decrease in abundance in cold acclimated plants whereas a potentially major agent of protein folding in the thylakoid lumen, a cyclophilin-type isomerase, increased in abundance. Due to its close resemblance with a stroma-localized protein, which has been shown to be redox-dependently regulated by thioredoxin (Romano et al. 2005), it has been suggested that this may also happen to the lumen-localized isomerase. However, this notion was not further supported in a study of potential thioredoxin targets in the thylakoid lumen, since it was not among the 19 proteins identified (Hall et al. 2010). Instead, the identity of the thioredoxin regulated proteins suggests a strong impact on PSII under changing light conditions.

2. Thylakoid Membrane Proteomics

The thylakoid membrane is tightly packed with proteins involved in photosynthesis. This poses one of the major challenges to the proteomic analysis of this compartment. Compared to the subunits of the four photosynthetic protein complexes, the abundance of most other thylakoid membrane proteins is one or more orders of magnitude lower. As a result, it can be expected that their mass-spectrometric identification is severely hindered. This problem comes on top of the difficulty inherent to all membrane proteomics: the solubilization and extraction of hydrophobic membrane proteins.

Kugler and colleagues resolved 50 subunits of the plastid electron transfer chain and the plastid ATP synthase of spinach and tobacco by applying blue-native/SDS-PAGE (BN-SDS/PAGE, Schägger and von Jagow 1991), which had successfully been employed for the analysis of the mitochondrial electron transfer chain (Kugler et al. 1997; for references on BN-PAGE of mitochondrial protein complexes see below). Heinemeyer and co-workers were able to investigate the stoichiometry of supercomplexes of Arabidopsis photosynthetic and photosynthesis-associated protein complexes (Heinemeyer et al. 2004). Later, the protein composition of the barley photosynthetic complexes was evaluated in more detail using the same technical approach (Granvogl et al. 2006).

In an attempt to increase the depth in the proteome analysis of thylakoid membranes, Friso and colleagues established a three-step extraction protocol to circumvent the issues associated with solubilization of membrane proteins and identified 154 proteins. Seventy-six of these were considered to be integral membrane proteins due to the presence of α-helical domains (Friso et al. 2004). As expected, proteins of the photosynthetic apparatus were prevalent, but the differential extraction seemed to have facilitated also the identification of other, lower abundant polypeptides. An upgraded protocol that involved more steps led to further improvements in proteome coverage, while retaining a similar percentage of membrane proteins (Peltier et al. 2004). In total, 242 proteins were identified by Peltier and co-workers. However, while such a strategy seems worthwhile for the purpose of increasing the coverage of membrane proteins, it most probably is too complicated to generate reproducible results in comparative approaches, like the one reported by Giacomelli and colleagues (2006). Monitoring the thylakoid membrane proteome of WT and the ascorbate deficient *vic2-2* mutant under high light stress across four time points, the abundance of proteins was compared using 72 2D gels in total. For both the WT and the *vic2-2* mutant, changes in protein abundances in response to light stress were most pronounced in proteins localized in thylakoid-associated plastoglobules (PGs, for a proteomic study of PGs see below), which suggests an important role of PGs in the defense against high light stress. One protein with an increased abundance in the light-stressed lines was YCF37, which had been implicated in the assembly or oligomerization of PSI. The data may suggest a higher turnover rate (or repair rate) of this photosynthetic protein complex under high light conditions. Surprisingly, the differences between the WT and the ascorbate-deficient mutant were rather small and included general stress-responsive proteins, like superoxide dismutases (SODs) and (stroma-localized) heat shock proteins (HSPs), along with PsbS (whose suggested role lies in the protection of PSII from excess light, see Li et al. 2000) and glyoxylase I. The low intensity of the response to lowered ascorbate levels in the mutant under high light conditions was interpreted as the result of a sufficient level of redundancy in the defense systems against oxidative stress in plastids, which can compensate for the loss of one of its components.

3. Envelope Proteomics

The proteomic analysis of the plastid envelope is not hampered to the same degree as that of thylakoid membranes, since it does not contain the all-overshadowing photosynthetic protein complexes. At the same time, exactly this is most probably the reason for the envelope not receiving as much attention as the thylakoids. However, the investigation of the boundary membranes of plastids is far from being unimportant. The envelope membranes connect the plastids with the cytosol and therefore with the rest of the cell. Every molecule which is imported into the plastid or exported into the cytosol has to pass these barriers, including nuclear-encoded plastid proteins synthesized on cytoplasmic ribosomes. Therefore, a diverse range of transporters is expected to be present in the envelope membranes. Identification and characterization of these transporters gives useful insights into the interlocking of plastid and cytosolic biochemistry. Furthermore, the envelope membranes themselves are sites of biochemical activity, since they harbor enzymes involved in the synthesis and breakdown of lipids and lipid-derived products. Additionally, the chloroplast envelope is also engaged in the degradation of chlorophyll. The large-scale quest for plastid transporters began in 2002. Proteins of highly enriched envelope fractions from spinach chloroplasts were extracted by organic solvents, separated on 1D SDS PAGE and subjected to MS/MS (Ferro et al. 2002). Altogether, 54 proteins were identified, half of which were of unknown function. Shortly afterwards, more than 100 envelope proteins from Arabidopsis were identified using a different extraction protocol and gel-free shotgun proteomics (Ferro et al. 2003). By using a combination of 1D SDS-PAGE and offline multi-dimensional LC followed by MS/MS (MuDPIT), an astonishing 392 proteins were identified from mixed envelope membranes (inner and outer membranes) by Fröhlich and co-workers (Froehlich et al. 2003). Twenty-six percent of the proteins identified in this study had been reported before as being members of the envelope proteome, but the functions of the majority of proteins were unknown. Interestingly, from a technical point of view, the set of MuDPIT-identified proteins contained a higher percentage of proteins with at least one TM domain than the set derived from 1D SDS-PAGE (46% for MuDPIT, 10% for 1D SDS and 25% for the proteins identified by both techniques).

4. Stroma Proteomics

The chloroplast stroma is a hotspot of enzymatic activity. Enzymes of the Calvin cycle, gluconeogenesis, the pentose phosphate pathway and glycolysis are located in the stroma as well as enzymes involved in nitrogen and sulfur assimilation, to name but a few. Due to the high content of soluble proteins, the identification of proteins from the stroma is straightforward, once a fraction of sufficient quantity and purity has been obtained. By concentrating on proteins that form part of small complexes (<1 MDa), Peltier and co-workers did not only identify 241 stromal proteins, but also assessed their relative abundance (Peltier et al. 2006). This additional layer of information is able to scrutinize experimental and predicted protein localization data. It also allows to draw conclusions about the importance of biochemical pathways within the stroma and gives insights into the usage of paralogues. It was estimated that the 241 proteins identified by Peltier et al. comprise about 99% of the stromal proteome. Since, based on targeting prediction software, the stroma is expected to comprise far more proteins, the abundance of the so far undetected proteins can be expected to be extremely low, several orders of magnitude below the most abundant stromal proteins. In a follow-up study, Olinares and co-workers employed size exclusion chromatography (SEC) in order to detect stromal protein complexes of up to 5 MDa (Olinares et al. 2010). The higher mass complexes consist of protein-protein and protein-nucleic acid complexes and comprise the plastid pyruvate dehydrogenase complex (PDC), acetyl-CoA carboxylase complex (ACCase), the plastid-encoded RNA polymerase complex (PEP) and ribosome particles. Composition of protein complexes was deduced by grouping of proteins with similar SEC elution characteristics through hierarchical clustering. Several proteins of low abundance, especially those involved in RNA-processing, were identified for the first time. The proteins of these high-molecular mass complexes were estimated to comprise 10–13% of the overall stromal proteome, which does not seem to match well with the calculated 99% coverage stated earlier for the low-molecular mass complexes (Peltier et al. 2006). However, considering a certain amount of overlap between the studies (especially in Calvin cycle components) as well as the unavoidable inaccuracy inherent to quantitation by gel spot volume (in Peltier et al. 2006) and spectral counting in shotgun proteomics (in Olinares et al. 2010), the estimations in both two studies complement each other reasonably well.

In an attempt to shed new light on the modulation of plastid mRNA stability (a major regulatory level of gene expression), Baginsky and co-workers investigated the proteins responsible for mRNA processing (Baginsky et al. 2007). Plastid extracts, enriched in RNA-processing proteins, which are capable of processing mRNAs in vitro, were produced by chromatographic means and analyzed for their protein content. A range of lowly abundant RNA-binding proteins were identified. In addition, substances of known influence on the stability of RNA were tested in in vitro assays of stromal extracts prepared in the light (light protein fraction, LPF, with long mRNA half-life) and the dark (dark protein fraction, DPF, with short mRNA half-life) suggesting that the degradation of mRNA in the dark is a two-step process. In summary, the proteomic study first supplied a list of potential candidate enzymes involved in the modulation of RNA stability, which subsequently was reduced by intelligently designed functional assays. The authors concluded that a $MgCl_2$-independent endoribonuclease initiates the process and is superseded by an unidentified nuclease, which is $MgCl_2$ dependent. As for the thylakoid lumen proteome, in a study investigating the influence of cold acclimation, the stromal proteome (as well as the thylakoid lumen proteome) was found to contain proteins responsive to this environmental change (Goulas et al. 2006). In the stroma, these proteins are mainly enzymes of the Calvin cycle. While the small and large RuBisCO subunits increased in abundance as a result of the cold acclimation, six other enzymes of this pathway decreased in abundance, which is in agreement with reported results of enzymatic assays (Strand et al. 1999). At the same time, the abundance of subunits of the extrinsic CF_1 part of the thylakoid ATP synthase (which protrudes into the stroma) increased. This may indicate either damage of this protein complex under cold stress or a higher level of turn-over of this enzyme during the acclimation process. Levels of CF_1 subunits were back to normal values in those leaves that had grown after the onset of the cold treatment (Goulas et al. 2006). Furthermore, certain proteins known to react to oxidative stress, like glyoxylase I, a glutathione S-transferase (GSTF8), peroxiredoxins of the 2-Cys type and a fibrillin increased in abundance. The latter one is known to be up-regulated during abiotic stress and is suggested to play a role in the protection of PSII. Single protein subunits of PSI, the b_6f complex and an extrinsic subunit of PSII were also reported to increase in abundance in the stroma during the acclimation process, potentially due to a compromised assembly rate of the photosynthetic protein complexes, which may be a direct result of a not yet cold-adjusted and therefore sub-optimal membrane composition.

B. Proteomics of Heterotrophic Plastids

Autotrophic chloroplasts represent the type of plastids that has been studied most intensively. However, studies of non-green (heterotrophic) plastid types such as etioplasts, amyloplasts, chromoplasts and undifferentiated proplastids are necessary in order to elucidate the mechanisms of plastid biogenesis and interconversion and to identify proteins involved in non-photosynthetic aspects of plastid differentiation. Due to the shortage of tissues rich in non-green plastids in Arabidopsis, none of the studies outlined below used this model plant. Instead, they relied on other systems, such as, rice, wheat, tobacco or bell pepper.

Investigation of the proteome of undifferentiated plastids is probably the ideal starting point for a survey of the proteomes of heterotrophic plastid proteomes. By employing a sequential extraction strategy, it was found that in cultured tobacco BY-2 cells (harboring non-green undifferentiated plastids), integral membrane proteins could be solubilized using milder detergents than needed for the solubilization of these proteins in chloroplasts (Baginsky et al. 2004). This was attributed to the lack of an extensive internal membrane system in undifferentiated plastids and a potentially different membrane composition. A total of 168 proteins were identified, 124 of which could be assigned to plastids with good confidence. The functions of the identified proteins suggest that they are mainly involved in amino acid synthesis. These amino acids may then be exported from the plastids to support growth of the fast dividing tobacco suspension culture cells. Another abundant functional class comprises proteins involved in protein folding and turnover, suggesting that the turnover of proteins is high in undifferentiated plastids. As expected, in the absence of the photosynthetic machinery, the high abundance of the ATP/ADP antiporter indicates a strong demand of these plastids for externally supplied energy. Glucose-6-phosphate imported into the undifferentiated plastids by the glucose-6-phosphate translocator is potentially fed into the oxidative pentose phosphate pathway, which in turn produces reduction equivalents necessary for the fixation of nitrogen. This process requires oxoglutarate, which can be imported into the organelles by the oxoglutarate/malate translocator. The latter has also been found at high abundance in the organelle membrane fraction. It is noteworthy that, in contrast to chloroplasts, plastid-encoded proteins are under-represented in undifferentiated plastids and that nearly all of the few detected plastid-encoded proteins are products of genes which are transcribed by the nucleus-encoded plastid RNA polymerase (NEP). Furthermore, comparison of the proteome of undifferentiated plastids with that of amyloplasts (Andon et al. 2002) and chloroplasts

(Peltier et al. 2002; Schubert et al. 2002; Ferro et al. 2003; Froehlich et al. 2003; Kleffmann et al. 2004) suggest distinct differences in the proteomes of these plastid varieties. On a similar notion, after the rice etioplast proteome had been investigated by von Zychlinsky and colleagues (2005), the differences between non-photosynthetic and photosynthetic plastids became also apparent when the transition of rice etioplasts into chloroplasts was investigated by 2D IEF/SDS-PAGE (Kleffmann et al. 2007). In total, 369 reproducible protein spots were identified and compared to the list generated earlier by shotgun proteomics (von Zychlinski et al. 2005). The 2D approach also allowed the relative quantitation of protein abundances by means of spot volume. It was found that the majority of the proteins (237) were identified exclusively from spots in the gel-based approach, whereas 118 proteins were detected by shotgun proteomics only. This leaves an overlap of just 24 proteins and shows that the different technical approaches complement each other nicely. The most significant changes happened 4 h after illumination (which triggers the transition from etioplasts into chloroplasts) and are marked by a reduction in enzymes of amino acid and fatty acid metabolism as well as an increase in proteins involved in carbohydrate metabolism and photosynthesis. It is suggested that the energy required for the transition is generated by the oxidative branch of the pentose phosphate pathway in conjunction with a major part of the glycolytic pathway. The amino acids required for plastid gene expression are expected to stem from recycled etioplast proteins, facilitated by an enlarged protein degradation machinery involving Clp protease isoforms. Furthermore, it has been found that an RNA-binding protein (RNP29), which is suspected to be involved in plastid mRNA degradation, changes its phosphorylation pattern after illumination. Since general transcription rates of plastid- encoded genes do not change to a large amount during plastid development, RNP29 may be a major regulator of gene expression in the light-induced transition of etioplasts to chloroplasts. Phosphorylation of this protein might stabilize mRNAs of genes coding for proteins of the photosynthetic apparatus, especially since other identified proteins known to influence mRNA stability did not change in abundance. The regulation of protein abundance at the post-transcriptional level is also consistent with the finding that elongation factors Tu and P increased in abundance during the transition phase.

Another type of heterotrophic plastid has been analyzed by Andon and co-workers (2002). Filling of wheat grains with starch is performed by the amyloplasts of the wheat endosperm. Analysis of the amyloplast proteome therefore is a promising approach to gain deeper insight into this agronomically and nutritionally important process. Genomic sequence information was scarce in 2002 and the number of publically available ESTs for wheat was also low. Despite these circumstances, 171 amyloplast proteins from narrow- and medium-range 2D gels as well as 1D SDS-PAGE were identified. However, sequence coverage was generally low and many proteins were identified by only a single cross species matched peptide. Nevertheless, potential key proteins involved in wheat grain filling were identified. With an increase in the number of publically available ESTs, the identification of wheat proteins by MS became more and more efficient over the years. While it was less than 25,000 ESTs in 2000, the number rose to over 1,000,000 in 2005 (source: International Triticeae EST Cooperative (*ITEC*), http://avena.pw.usda.gov/genome, 06.01.2011). This may have been one factor contributing to the identification of as many as 289 proteins from the amyloplasts of developing wheat endosperm in 2006 (Balmer et al. 2006). Another factor may have been the slightly different fractionation strategy. While Andon and colleagues used whole amyloplasts and amyloplast membranes in their study, Balmer et al. generated soluble and membrane amyloplast fractions and relied completely on 2D gels for protein separation (2D and 1D-PAGE in Andon et al. 2002).

The pigment-containing chromoplasts represent another differentiation form of plastids. The protein content of bell pepper (*Capsicum annuum*) chromoplasts has been studied by Siddique and co-workers, who were able to identify 151 proteins (Siddique et al. 2006). The availability of sequence information that can be used to identify bell pepper proteins was (and still is) even more sparse than in wheat. No genome data and only a limited number of ESTs were available at that point in time, which necessitated an identification strategy supported by de novo sequencing. Capsorubin and capsanthin are the prevailing carotinoid pigments in the fibrillar chromoplasts of bell pepper. It is therefore not surprising that capsanthin/capsorubin synthase (CCS), an enzyme of the carotenoid synthesis pathway, was found to be a major component of the chromoplast proteome, along with fibrillin. Enzymes involved in carbon metabolism, especially of the OPPP were also found in high abundance. Since chromoplasts develop from fully functional chloroplasts, residual amounts of Calvin cycle enzymes were still detectable. These proteins are probably recycled by the organelle and used for the synthesis of new polypeptides. Hence, there is a lower requirement for amino acid synthesis, which might be the reason for the reduced amino acid synthesis machinery in the bell pepper chromoplasts. Comparison of the chromoplast proteome with that of other plastid types revealed that a set of 79 proteins were identified in all types of plastids. This set is, therefore, expected to form part of the standard plastid equipment. Only nine chromoplast proteins had not been identified previously, among them CCS and a protein potentially involved in chlorophyll degradation. Chromoplasts were also compared to chloroplasts at the sub-organellar level. Plastoglobuli (PGs), thylakoid membrane associated structures with not well-defined functions, from bell pepper chromoplasts and Arabidopsis chloroplasts were analyzed at the proteome level (Ytterberg et al. 2006). PGs are known to serve as storage repositories for compounds, such as, α-tocopherol, triacylglycerols, plastoquinone, and especially in chromoplasts, carotenoids. The most abundant proteins of PGs are members of the fibrillin family, which primarily serve structural purposes. However, it has been found that their overexpression improves plant growth under light stress (Rey et al. 2000). At the same time, knockdown of a stromal protease subunit (ClpR2) increases the amount of PGs in the plastids (Ytterberg et al. 2006). However, the underlying mechanisms and the general enzymatic properties of PGs are unclear. Therefore, the proteome of Arabidopsis PGs from plants grown under normal light conditions were compared with the proteome of plants grown under high light and in darkness. It was found that the Arabidopsis PG proteome consists of >30, mainly soluble, proteins, many of which are associated with isoprenoid metabolism. The most striking difference in the PG proteomes of normally and alternatively illuminated plants was an increased abundance of an esterase/lipase/thioesterase which may serve a function in lipid catabolism during periods of stress. No striking differences of clpr2 and WT PGs were found. Chromoplast PGs were found to contain additional enzymes of the carotenoid biosynthesis pathway, which is consistent with their high carotenoid content.

IV. Mitochondrial Proteomics

A. Assessment of Proteome Composition

Proteomic studies of plant mitochondria are nearly entirely focused on the model plant Arabidopsis, with rice being the most common exception. Since isolating mitochondria from green tissue bears the risk of chloroplast co-purification, often non-green, heterotrophically grown Arabidopsis cell cultures were used as the starting material (Kruft et al. 2001; Millar et al. 2001; Sweetlove et al. 2002; Heazlewood et al. 2004). In other studies, rice (Heazlewood et al. 2003c) or pea (Bardel et al. 2002) mitochondria were investigated with respect to their proteome composition. Initial studies used gel based (IEF/SDS-PAGE) approaches to separate the mitochondrial proteome and (with the exception of the rice mitochondrial proteome) relatively low numbers of proteins (<100) were identified (Kruft et al. 2001; Millar et al. 2001). Mostly relying on 2D IEF/SDS-PAGE, these studies were not only limited by the total amount of identified proteins, but were also characterized by an overrepresentation of hydrophilic proteins. In order to achieve better coverage of the hydrophobic proteome of plant mitochondria, second-generation studies very much focused on other separation strategies. Shotgun proteomics was able to raise the detection limit with respect to both overall coverage and detection of hydrophobic proteins considerably. More than 400 proteins were identified by Heazlewood and co-workers (Heazlewood et al. 2004). In an alternative approach, Brugiere and colleagues successfully used a differential extraction strategy coupled to 1D SDS-PAGE (Brugiere et al. 2004). More than 100 proteins were identified by this approach, about half of which were membrane proteins. Although not used with the primary intention to identify hydrophobic proteins, BN-PAGE of mitochondrial proteins in combination with SDS-PAGE as second dimension also proved to be an effective strategy in this regard. Apart from Arabidopsis, rice, spinach and the green alga Chlamydomonas were also investigated using this technique (Eubel et al. 2003; Heazlewood et al. 2003a, b; van Lis et al. 2003; Krause et al. 2004; Millar et al. 2004a; Klodmann and Braun 2011; Klodmann et al. 2011). In a 3D approach, BN-PAGE was also used as a first dimension to investigate the protein composition of the plant mitochondrial NADH dehydrogenase complex (complex I). Tricine-SDS PAGE under different conditions served as second and third dimensions (Meyer et al. 2008). By doing so, the resolution was increased in comparison to conventional BN/SDS-PAGE and new subunits, mainly of complex I, were identified. Surprisingly, Klodmann and Braun (2011) were able to identify even more complex I proteins from conventional BN/SDS gels. In summary, these studies not only revealed the composition of the plant mitochondrial electron transfer chain and the ATP synthase complex, they also enabled the detection of a range of additional proteins in the respiratory complexes, the majority of which are plant-specific subunits.

B. Functional Proteomics

At the functional level, mitochondria have been studied extensively using proteomic approaches. The response of mitochondria isolated from heterotrophically grown Arabidopsis cell suspension cultures to externally applied oxidative stress has been assessed relatively early following the initial stock-taking of the mitochondrial proteome (Sweetlove et al. 2002). Interestingly, a large proportion of proteins which were found to change in abundance in the tested mitochondrial fractions were of non-mitochondrial origin and some of them were later shown to associate with the cytosolic side of the outer envelope of mitochondria for biological reasons (Giegé et al. 2003; Graham et al. 2007). Later, Taylor and colleagues assessed the effect of cold, drought and herbicides on the mitochondrial proteome of pea plants (Taylor et al. 2005). All three types of stress were found to impose oxidative stress on mitochondria, with the herbicide paraquat causing the most severe effects, in view of the fact that it was the only stress treatment inducing lipid peroxidation. Oxidative damage to mitochondrial proteins did occur in all treatments and proteins found in the matrix where the most susceptible ones. Recently, defense against oxidative stress has also been linked to salt tolerance in a proteome analysis of Australian wheat varieties (Jacoby et al. 2010), indicating that improved mitochondrial ROS protection might also provide tolerance to other types of abiotic stress.

The questions of oxidative stress and oxidative damage of proteins have fostered more detailed studies in this area of research. Oxidative damage is expected to be more severe in proteins associating with or containing metal ions, a process termed metal-catalyzed oxidation (MCO). Using immobilized metal affinity chromatography (IMAC), the metallo-proteome of plant mitochondria has been investigated (Tan et al. 2010). More than 140 proteins were found to bind Cu^{2+}, Zn^{2+} or Co^{2+} (or combinations thereof) and thus represent potential targets for oxidative damage in the mitochondrial proteome.

Proteins are not only damaged by MCO, but also by the products of oxidative damage of the fatty acids in membrane lipids, of which trans-4-hydroxy-2-nonenal (HNE) is probably the most prominent example. Winger and co-workers found that many of the identified HNE-susceptible mitochondrial proteins interact with the mitochondrial ubiquinone pool or are subunits of lipoic acid-containing protein complexes (Winger et al. 2007). Apart from the dehydrogenases found in the lipoic acid containing protein complexes, a limited number of other dehydrogenases were also found to be susceptible to HNE modification. Furthermore, the authors were able to establish a connection of HNE adducts with the occurrence of protein species of low molecular weight, which suggests some influence of HNE on protein degradation.

While HNE-adduction can be considered an accidental form of PTM, phosphorylation of proteins is a common mechanism employed to regulate enzymatic activity. Analysis of the plant mitochondrial phosphoproteome can therefore provide insights into the regulation of mitochondrial functions. In a first step, Ito and co-workers were able to identify potential phosphorylation targets by analyzing ATP binding proteins (Ito et al. 2006). Using an enrichment strategy, 34 proteins were considered to be ATP binding and 13 of these were identified in mitochondrial isolates for the first time. Since ATP affinity can only be taken as a weak evidence for phosphorylation, the Arabidopsis mitochondrial phosphoproteome was subsequently investigated more directly by using a phosphorylation-specific fluorescent dye. Additionally, enrichment of already phosphorylated proteins by Titanium dioxide affinity chromatography was achieved, which enabled Ito and colleagues to directly identify seven phosphorylation sites in the Arabidopsis mitochondrial proteome, a considerable increase from only two that had been known previously (Ito et al. 2009).

Proteomics has also had an impact on the investigation of the physiology of plant mitochondria. Three studies focused on the physiological differences between mitochondria in light-exposed plant tissues in comparison to those which were not illuminated. During the diurnal cycle and the associated alterations in photosynthetic activity, mitochondria face a changing supply of substrates for respiration. At the same time, major parts of cellular metabolism are redirected, requiring different metabolites to be exported from mitochondria. In a recent study by Lee et al. (2010), changes in the mitochondrial proteome accompanying these processes have been analyzed. Albeit small, alterations in the abundances of 55 protein spots were found repeatedly. They affect components of the TCA cycle, nitrogen metabolism, sulfur metabolism and ROS defense. In general, these results match those obtained earlier by comparing green and non-green tissues in pea (Bardel et al. 2002) and Arabidopsis (Lee et al. 2008). In a two-step approach, Bardel and co-workers first thoroughly characterized the pea mitochondrial proteome (see above) and then investigated the differences between the mitochondrial proteomes of green and non-green tissues (such as etiolated pea leaves, roots and seeds) in a second step (Bardel et al. 2002). In a similar approach, Lee and colleagues compared the proteome of mitochondria isolated from an Arabidopsis cell culture with that of Arabidopsis leaves from hydroponic cultures (Lee et al. 2008). By using an established model system, the latter study also allowed the comparison of the changes in transcript and protein abundance. A positive but weak correlation between transcript ratios and protein abundance ratios is indicated by a Spearman rank correlation of 0.49. Therefore, similar to the results obtained in plastids, the abundance of mitochondrial proteins is only partly controlled by transcript levels. Proteins whose abundance is not controlled transcriptionally include enzymes of the citric acid cycle, the branched chain amino acid metabolism and stress defense.

Further noteworthy is the work which has been done on rice mitochondria. Not only is the importance of this staple-food species undisputed, its ability to germinate under hypoxia enabled the investigation of some aspects of mitochondrial biology, which could not be analyzed in Arabidopsis. Especially the connection between respiration and mitochondrial protein import has been highlighted in studies that used proteomics as an integral part of their experimental setup (Millar et al. 2004b; Howell et al. 2006, 2007).

V. Peroxisome Proteomics

Although not directly involved in photosynthesis or respiration, peroxisomes link chloroplast and mitochondrial metabolism by taking part in the process of photorespiration. Therefore, these cellular organelles rightfully deserve a place in this survey of plant organelle proteomics. Again, Arabidopsis has been the model system of choice for studies of the proteome of peroxisomes. This is despite the fact that the isolation of peroxisomes from this plant seems to be more difficult than from other species. The relatively low number of studies conducted on peroxisomes and the usually high levels of non-peroxisomal contaminants experienced by researchers are indicators of the limited accessibility of Arabidopsis peroxisomes to proteomics. Therefore, compared to plastids and mitochondria, peroxisomal proteomics can still be considered to be in its infancy. Hence, the characterization of the protein composition rather than the investigation of functional aspects of the peroxisomal proteome has been the primary focus so far. Analysis of leaf peroxisomes of greening Arabidopsis cotyledons (Fukao et al. 2002) and glyoxysomes of etiolated cotelydons (Fukao et al. 2003) laid the foundation for subsequent studies of the plant peroxisomal proteome. Twenty-nine and 19 major peroxisomal proteins, respectively, were identified from 2D IEF/SDS-PAGE in these studies along with some potential contaminations. The next big step forward came in 2007, when Reumann and co-workers were able to identify 78 proteins from Arabidopsis leaf peroxisomes by a combination of gel-based and shotgun proteomics (Reumann et al. 2007). A large proportion of these proteins had previously not been assigned to this organelle. Results of the proteome analysis were supported by *in silico* predictions and protein localization analysis employing chimeric reporter gene fusions. New peroxisomal targeting sequences (PTSs) were recognized and the presence of proteins involved in the protection from herbivores and pathogens in the peroxisomal compartment was deduced from the data generated in this study. Most of these data were confirmed in a later study on Arabidopsis cell suspension cultures (Eubel et al. 2008) with the exception of the latter group of proteins, which were not found in peroxisomes prepared from this type of material. The study by Eubel et al. differed from the work of Reumann et al. and Fukao et al. in two aspects. Firstly, it not only relied on density gradients for the isolation of peroxisomal fractions, but also employed FFE as an additional step. The isolation strategy also enabled a wide-scale comparison in protein abundance of the peroxisomal samples with the dominating source of contamination, mitochondria. Quantitative data for a large proportion of the identified proteins were generated, which enabled the high-confidence assignment of proteins to either peroxisomes or mitochondria. Eighty-nine proteins were identified from cell culture peroxisomes, many of which have so far unknown functions. Metabolic network analysis provided potential candidates for substrates entering and exiting the peroxisomal compartment. Knowledge of the Arabidopsis peroxisomal proteome was further increased by a follow-up study by Reumann and co-workers, which provided further insights into the protein composition of this compartment (Reumann et al. 2009). The herbivore and pathogen defense proteins were identified for a second time in leaf peroxisomes, this time along with a multitude of proteins from other cell organelles.

Soybean cotyledons and spinach leaves are the only non-Arabidopsis sources of peroxisomes that have been investigated for their protein composition to date. In soybean, Arai and colleagues were able to identify 70 proteins, 30 of which could be assigned to peroxisomes (Arai et al. 2008). This amounts for approximately the same number of non-redundant proteins identified from spinach mitochondria (Babujee et al. 2010), among which enzymes involved in the production of vitamin K (phylloquinone) were found. Production of phylloquinone was believed to occur exclusively in plastids but new studies of fluorescent fusion proteins show that some enzymes of this pathway might be dual targeted to plastids and peroxisomes, whereas one protein (MenB/NS) is believed to be located in peroxisomes exclusively (Babujee et al. 2010).

VI. The General Impact of Proteomics on Organelle Research

Proteomic data provide broad insights into the presence, abundance, dynamics and modifications of the ultimate gene products, the proteins. Proteomics, therefore, can contribute significantly to our understanding of the mechanisms that execute and regulate cellular metabolism. Proteomics is heavily interlocked with genomics and relies on high quality genomic (or transcriptomic) data. However, the flow of information from genomics to proteomics is not a one-way alley. Proteogenomic mapping contributes to the process of gene annotation by the discovery of proteins that do not fit predicted gene sequences and thus require re-annotation of predicted open reading frames (ORFs) or lead to the discovery of so far unrecognized genes (Baerenfaller et al. 2008; Castellana et al. 2008). This may also help with the identification of genes in newly sequenced genomes. Proteomics can also reveal potential candidates for reverse genetics approaches. All these general attributes also apply to organelle proteomics, but in addition, subcellular proteomics is able to provide some added benefits.

1. By dividing the cell into units of manageable complexity and performing MS on these instead of whole cells, the localization of proteins is revealed in addition to their identity. Although the approach may be troubled by false positive localization (due to contamination with non-target compartments as discussed above), organelle proteomics can be considered as a high-throughput approach to reveal the intracellular distribution of proteins.
2. Since most of the mitochondrial and plastid proteins as well as all peroxisomal proteins are synthesized in the cytosol, they need to be imported into the organelles. The assignment of proteins to their individual organelle by proteomics allows studying the characteristics of the target sequences that direct the import into specific organelles. This knowledge can then be used to improve the performance of in silico targeting prediction tools.
3. Plant mitochondria and chloroplasts change single nucleotides in transcripts of organelle-encoded genes by mRNA editing, which affects the amino acid sequence of the resulting polypeptides. Using specialized databases, proteomics may become an independent way of identifying editing sites.

VII. Outlook

Already now, proteomics is a versatile field of research and it will probably become even more diverse in the future. This is partly due to the increasingly better integration of the proteomics platform with other approaches, and partly due to technical developments in the proteomic field itself. Although it may never achieve the depth of transcriptomics analyses, proteomics most likely will become better applicable to high-throughput applications and, in combination with an increased coverage, will be able to deliver a more complete picture of organelle biochemistry and physiology. In our view, future developments in proteomics will include:

1. In-depth characterization of proteomes: The process of discovering new organellar proteins is not slowing down. Modern MS instruments, especially of the Orbitrap type, enable the characterization of complex protein mixtures

and require less extensive pre-fractionation steps upstream of MS.
2. Focus on specialized subsets of proteins, for example, in the area of posttranslational modifications: This is facilitated by technical developments in the fields of chromatography (shotgun proteomics) and specialized dyes (gel-based proteomics) as well as mass spectrometry itself.
3. A more functional definition of the proteomes of plant cell organelles: For example, distinguishing between contaminants and proteins sticking to the outer envelope in a functionally significant manner remains a major challenge.
4. In the future, quantitation of proteins will play an ever increasing role in organelle proteomics. Multiple Reaction Monitoring (MRM) will generate quantitative data faster and more easily. Based on results obtained on the location of proteins by proteomics or other means, functional organellar proteomics can be done without the need for demanding organelle isolating procedures. This will allow the assessment of proteomic changes under more physiological conditions and will be very helpful for the functional characterization of organelles which are notoriously hard to isolate, such as peroxisomes. Where necessary, MRM can also be employed to assess the purity of organellar fractions. Hundreds of proteins from organelles potentially contaminating the target fraction can be tested in a single MRM run, which will generate far more meaningful results than measuring the activity of a few selected marker proteins.

Acknowledgements

Research in our laboratory is supported by the Deutsche Forschungsgemeinschaft (grants Br1829/8-2 and Br1829/10-1)

References

Andersson B, Akerlund HE (1978) Inside-out membrane vesicles isolated from spinach thylakoids. Biochim Biophys Acta 503:462–472

Andon NL, Hollingworth S, Koller A, Greenland AJ, Yates JR 3rd, Haynes PA (2002) Proteomic characterization of wheat amyloplasts using identification of proteins by tandem mass spectrometry. Proteomics 2:1156–1168

Arai Y, Hayashi M, Nishimura M (2008) Proteomic analysis of highly purified peroxisomes from etiolated soybean cotyledons. Plant Cell Physiol 49:526–539

Babujee L, Wurtz V, Ma C, Lueder F, Soni P, van Dorsselaer A, Reumann S (2010) The proteome map of spinach leaf peroxisomes indicates partial compartmentalization of phylloquinone (vitamin K) biosynthesis in plant peroxisomes. J Exp Bot 61:1441–1453

Baerenfaller K, Grossmann J, Grobei MA, Hull R, Hirsch-Hoffmann M, Yalovsky S, Zimmermann P, Grossniklaus U, Gruissem W, Baginsky S (2008) Genome-scale proteomics reveals *Arabidopsis thaliana* gene models and proteome dynamics. Science 320:938–941

Baginsky S, Siddique A, Gruissem W (2004) Proteome analysis of tobacco bright yellow-2 (BY-2) cell culture plastids as a model for undifferentiated heterotrophic plastids. J Proteome Res 3:1128–1137

Baginsky S, Kleffmann T, von Zychlinski A, Gruissem W (2005) Analysis of shotgun proteomics and RNA profiling data from Arabidopsis thaliana chloroplasts. J Proteome Res 4:637–640

Baginsky S, Grossmann J, Gruissem W (2007) Proteome analysis of chloroplast mRNA processing and degradation. J Proteome Res 6:809–820

Balmer Y, Vensel WH, DuPont FM, Buchanan BB, Hurkman WJ (2006) Proteome of amyloplasts isolated from developing wheat endosperm presents evidence of broad metabolic capability. J Exp Bot 57:1591–1602

Bardel J, Louwagie M, Jaquinod M, Jourdain A, Luche S, Rabilloud T, Macherel D, Garin J, Bourguignon J (2002) A survey of the plant mitochondrial proteome in relation to development. Proteomics 2:880–898

Brugiere S, Kowalski S, Ferro M, Seigneurin-Berny D, Miras S, Salvi D, Ravanel S, d'Herin P, Garin J, Bourguignon J, Joyard J, Rolland N (2004) The hydrophobic proteome of mitochondrial membranes from Arabidopsis cell suspensions. Phytochemistry 65:1693–1707

Castellana NE, Payne SH, Shen Z, Stanke M, Bafna V, Briggs SP (2008) Discovery and revision of Arabidopsis genes by proteogenomics. PNAS 105:21034–21038

Ettinger WF, Theg SM (1991) Physiologically active chloroplasts contain pools of unassembled extrinsic proteins of the photosynthetic oxygen-evolving enzyme complex in the thylakoid lumen. J Cell Biol 115:321–328

Eubel H, Jänsch L, Braun H-P (2003) New insights into the respiratory chain of plant mitochondria. Supercomplexes and a unique composition of Complex II. Plant Phys 133:274–286

Eubel H, Lee CP, Kou J, Meyer EH, Taylor NL, Millar AH (2007) Free flow electrophoresis for purification of plant mitochondria by surface charge. Plant J 52:583–594

Eubel H, Meyer EH, Taylor NL, Bussell JD, O'Toole N, Heazlewood JL, Castleden I, Small ID, Smith SM, Millar AH (2008) Novel proteins, putative membrane transporters, and an integrated metabolic network are revealed by quantitative proteomic analysis of Arabidopsis cell culture peroxisomes. Plant Physiol 148:1809–1829

Ferro M, Salvi D, Riviere-Rolland H, Vermat T, Seigneurin-Berny D, Grunwald D, Garin J, Joyard J, Rolland N (2002) Integral membrane proteins of the chloroplast envelope: identification and subcellular localization of new transporters. Proc Natl Acad Sci USA 99:11487–11492

Ferro M, Salvi D, Brugiere S, Miras S, Kowalski S, Louwagie M, Garin J, Joyard J, Rolland N (2003) Proteomics of the chloroplast envelope membranes from *Arabidopsis thaliana*. Mol Cell Proteomics 2:325–345

Ferro M, Brugiere S, Salvi D, Seigneurin-Berny D, Court M, Moyet L, Ramus C, Miras S, Mellal M, Le Gall S, Kieffer-Jaquinod S, Bruley C, Garin J, Joyard J, Masselon C, Rolland N (2010) AT_CHLORO, a comprehensive chloroplast proteome database with subplastidial localization and curated information on envelope proteins. Mol Cell Proteomics 9:1063–1084

Friso G, Giacomelli L, Ytterberg AJ, Peltier JB, Rudella A, Sun Q, Wijk KJ (2004) In-depth analysis of the thylakoid membrane proteome of Arabidopsis thaliana chloroplasts: new proteins, new functions, and a plastid proteome database. Plant Cell 16:478–499

Froehlich JE, Wilkerson CG, Ray WK, McAndrew RS, Osteryoung KW, Gage DA, Phinney BS (2003) Proteomic study of the Arabidopsis thaliana chloroplastic envelope membrane utilizing alternatives to traditional two-dimensional electrophoresis. J Proteome Res 2:413–425

Fukao Y, Hayashi M, Nishimura M (2002) Proteomic analysis of leaf peroxisomal proteins in greening cotyledons of Arabidopsis thaliana. Plant Cell Physiol 43:689–696

Fukao Y, Hayashi M, Hara-Nishimura I, Nishimura M (2003) Novel glyoxysomal protein kinase, GPK1, identified by proteomic analysis of Glyoxysomes in etiolated cotyledons of Arabidopsis thaliana. Plant Cell Physiol 44:1002–1012

Giacomelli L, Rudella A, van Wijk KJ (2006) High light response of the thylakoid proteome in Arabidopsis wild type and the ascorbate-deficient mutant vtc2-2. A comparative proteomics study. Plant Physiol 141:685–701

Giegé P, Heazlewood JL, Roessner-Tunali U, Millar AH, Fernie AR, Leaver CJ, Sweetlove LJ (2003) Enzymes of glycolysis are functionally associated with the mitochondrion in Arabidopsis cells. Plant Cell 15:2140–2151

Gopalan G, He Z, Balmer Y, Romano P, Gupta R, Heroux A, Buchanan BB, Swaminathan K, Luan S (2004) Structural analysis uncovers a role for redox in regulating FKBP13, an immunophilin of the chloroplast thylakoid lumen. Proc Natl Acad Sci USA 101:13945–13950

Goulas E, Schubert M, Kieselbach T, Kleczkowski LA, Gardestrom P, Schroder W, Hurry V (2006) The chloroplast lumen and stromal proteomes of Arabidopsis thaliana show differential sensitivity to short- and long-term exposure to low temperature. Plant J 47:720–734

Graham JWA, Williams TCR, Morgan M, Fernie AR, Ratcliffe RG, Sweetlove LJ (2007) Glycolytic enzymes associate dynamically with mitochondria in response to respiratory demand and support substrate channeling. Plant Cell 19:3723–3738

Granvogl B, Reisinger V, Eichacker LA (2006) Mapping the proteome of thylakoid membranes by de novo sequencing of intermembrane peptide domains. Proteomics 6:3681–3695

Hall M, Mata-Cabana A, Akerlund HE, Florencio FJ, Schroder WP, Lindahl M, Kieselbach T (2010) Thioredoxin targets of the plant chloroplast lumen and their implications for plastid function. Proteomics 10:987–1001

Heazlewood JL, Howell KA, Millar AH (2003a) Mitochondrial complex I from Arabidopsis and rice: orthologs of mammalian and fungal components coupled with plant-specific subunits. Biochim Biophys Acta Bioenergetics 1604:159–169

Heazlewood JL, Whelan J, Millar AH (2003b) The products of the mitochondrial orf25 and orfB genes are F-o components in the plant F1Fo ATP synthase. FEBS Lett 540:201–205

Heazlewood JL, Howell KA, Whelan J, Millar AH (2003c) Towards an analysis of the rice mitochondrial proteome. Plant Physiol 132:230–242

Heazlewood JL, Tonti-Filippini JS, Gout AM, Day DA, Whelan J, Millar AH (2004) Experimental analysis of the Arabidopsis mitochondrial proteome highlights signaling and regulatory components, provides assessment of targeting prediction programs, and indicates plant-specific mitochondrial proteins. Plant Cell 16:241–256

Heazlewood JL, Verboom RE, Tonti-Filippini J, Small I, Millar AH (2007) SUBA: the Arabidopsis subcellular database. Nucleic Acids Res 35:D213–D8218

Heinemeyer J, Eubel H, Wehmhoner D, Jänsch L, Braun HP (2004) Proteomic approach to characterize the supramolecular organization of photosystems in higher plants. Phytochemistry 65:1683–1692

Heinemeyer J, Scheibe B, Schmitz UK, Braun HP (2009) Blue native DIGE as a tool for comparative analyses of protein complexes. J Proteomics 72:539–544

Howell KA, Millar AH, Whelan J (2006) Ordered assembly of mitochondria during rice germination begins with promitochondrial structures rich in components of the protein import apparatus. Plant Mol Biol 60:201–223

Howell KA, Cheng K, Murcha MW, Jenkin LE, Millar AH, Whelan J (2007) Oxygen initiation of respiration and mitochondrial biogenesis in rice. J Biol Chem 282:15619–15631

Ito J, Heazlewood JL, Millar AH (2006) Analysis of the soluble ATP-binding proteome of plant mitochondria identifies new proteins and nucleotide triphosphate interactions within the matrix. J Proteome Res 5:3459–3469

Ito J, Taylor NL, Castleden I, Weckwerth W, Millar AH, Heazlewood JL (2009) A survey of the Arabidopsis thaliana mitochondrial phosphoproteome. Proteomics 9:4229–4240

Jacoby RP, Millar AH, Taylor NL (2010) Wheat mitochondrial proteomes provide new links between antioxidant defense and plant salinity tolerance. J Proteome Res 9:6595–6604

Kieselbach T, Hagman AB, Schroder WP (1998) The thylakoid lumen of chloroplasts. Isolation and characterization. J Biol Chem 273:6710–6716

Kleffmann T, Russenberger D, von Zychlinski A, Christopher W, Sjolander K, Gruissem W, Baginsky S (2004) The Arabidopsis thaliana chloroplast proteome reveals pathway abundance and novel protein functions. Curr Biol 14:354–362

Kleffmann T, von Zychlinski A, Russenberger D, Hirsch-Hoffmann M, Gehrig P, Gruissem W, Baginsky S (2007) Proteome dynamics during plastid differentiation in rice. Plant Physiol 143:912–923

Klodmann J, Braun HP (2010) Proteomic approach to characterize mitochondrial complex I from plants. Phytochemistry 72:1071–1080

Klodmann J, Sunderhaus S, Nimtz M, Jänsch L, Braun HP (2010) Internal architecture of mitochondrial complex I from *Arabidopsis thaliana*. Plant Cell 22: 797–810

Krause F, Reifschneider NH, Vocke D, Seelert H, Rexroth S, Dencher NA (2004) "Respirasome"-like supercomplexes in green leaf mitochondria of spinach. J Biol Chem 279:48369–48375

Kruft V, Eubel H, Jänsch L, Werhahn W, Braun HP (2001) Proteomic approach to identify novel mitochondrial proteins in *Arabidopsis*. Plant Physiol 127:1694–1710

Kugler M, Jänsch L, Kruft V, Schmitz UK, Braun HP (1997) Analysis of the chloroplast protein complexes by blue-native polyacrylamide gel electrophoresis (BN-PAGE). Photosynth Res 53:35–44

Lee CP, Eubel H, O'Toole N, Millar AH (2008) Heterogeneity of the mitochondrial proteome for photosynthetic and non-photosynthetic Arabidopsis metabolism. Mol Cell Proteomics 7:1297–1316

Lee CP, Eubel H, Millar AH (2010) Diurnal changes in mitochondrial function reveal daily optimization of light and dark respiratory metabolism in Arabidopsis. Mol Cell Proteomics 9:2125–2139

Li XP, Bjorkman O, Shih C, Grossman AR, Rosenquist M, Jansson S, Niyogi KK (2000) A pigment-binding protein essential for regulation of photosynthetic light harvesting. Nature 403:391–395

Meyer EH, Taylor NL, Millar AH (2008) Resolving and identifying protein components of plant mitochondrial respiratory complexes using three dimensions of gel electrophoresis. J Proteome Res 7:786–794

Millar AH, Sweetlove LJ, Giegé P, Leaver CJ (2001) Analysis of the Arabidopsis mitochondrial proteome. Plant Physiol 127:1711–1727

Millar AH, Eubel H, Jänsch L, Kruft V, Heazlewood JL, Braun HP (2004a) Mitochondrial cytochrome c oxidase and succinate dehydrogenase complexes contain plant specific subunits. Plant Mol Biol 56:77–90

Millar AH, Trend AE, Heazlewood JL (2004b) Changes in the mitochondrial proteome during the anoxia to air transition in rice focus around cytochrome-containing respiratory complexes. J Biol Chem 279:39471–39478

Olinares PD, Ponnala L, van Wijk KJ (2010) Megadalton complexes in the chloroplast stroma of Arabidopsis thaliana characterized by size exclusion chromatography, mass spectrometry, and hierarchical clustering. Mol Cell Proteomics 9:1594–1615

Peltier JB, Friso G, Kalume DE, Roepstorff P, Nilsson F, Adamska I, van Wijk KJ (2000) Proteomics of the chloroplast: systematic identification and targeting analysis of lumenal and peripheral thylakoid proteins. Plant Cell 12:319–341

Peltier JB, Emanuelsson O, Kalume DE, Ytterberg J, Friso G, Rudella A, Liberles DA, Soderberg L, Roepstorff P, von Heijne G, van Wijk KJ (2002) Central functions of the lumenal and peripheral thylakoid proteome of Arabidopsis determined by experimentation and genome-wide prediction. Plant Cell 14:211–236

Peltier J-B, Ytterberg J, Sun Q, van Wijk KJ (2004) New functions of the thylakoid membrane proteome of *Arabidopsis thaliana* revealed by a simple, fast, and versatile fractionation strategy. J Biol Chem 279:49367–49383

Peltier JB, Cai Y, Sun Q, Zabrouskov V, Giacomelli L, Rudella A, Ytterberg AJ, Rutschow H, van Wijk KJ (2006) The oligomeric stromal proteome of *Arabidopsis thaliana* chloroplasts. Mol Cell Proteomics 5:114–133

Reumann S, Babujee L, Ma C, Wienkoop S, Siemsen T, Antonicelli GE, Rasche N, Luder F, Weckwerth W, Jahn O (2007) Proteome analysis of Arabidopsis leaf peroxisomes reveals novel targeting peptides, metabolic pathways, and defense mechanisms. Plant Cell 19:3170–3193

Reumann S, Quan S, Aung K, Yang P, Manandhar-Shrestha K, Holbrook D, Linka N, Switzenberg R, Wilkerson CG, Weber APM, Olsen LJ, Hu J (2009) In-depth proteome analysis of Arabidopsis leaf peroxisomes combined with in vivo subcellular targeting verification indicates novel metabolic and regulatory functions of peroxisomes. Plant Physiol 150:125–143

Rey P, Gillet B, Romer S, Eymery F, Massimino J, Peltier G, Kuntz M (2000) Over-expression of a pepper plastid lipid-associated protein in tobacco leads to changes in plastid ultrastructure and plant development upon stress. Plant J 21:483–494

Romano P, Gray J, Horton P, Luan S (2005) Plant immunophilins: functional versatility beyond protein maturation. New Phytol 166:753–769

Schagger H, von Jagow G (1991) Blue native electrophoresis for isolation of membrane-protein complexes in enzymatically active form. Anal Biochem 199:223–231

Schubert M, Petersson UA, Haas BJ, Funk C, Schroder WP, Kieselbach T (2002) Proteome map of the chloroplast lumen of *Arabidopsis thaliana*. J Biol Chem 277:8354–8365

Siddique MA, Grossmann J, Gruissem W, Baginsky S (2006) Proteome analysis of bell pepper (*Capsicum annuum* L.) chromoplasts. Plant Cell Physiol 47: 1663–1673

Strand A, Hurry V, Henkes S, Huner N, Gustafsson P, Gardestrom P, Stitt M (1999) Acclimation of Arabidopsis leaves developing at low temperatures. Increasing cytoplasmic volume accompanies increased activities of enzymes in the Calvin cycle and in the sucrose-biosynthesis pathway. Plant Physiol 119:1387–1398

Stroher E, Dietz KJ (2008) The dynamic thiol-disulphide redox proteome of the *Arabidopsis thaliana* chloroplast as revealed by differential electrophoretic mobility. Physiol Plant 133:566–583

Sweetlove LJ, Heazlewood JL, Herald V, Holtzapffel R, Day DA, Leaver CJ, Millar AH (2002) The impact of oxidative stress on Arabidopsis mitochondria. Plant J 32:891–904

Tan YF, O'Toole N, Taylor NL, Millar AH (2010) Divalent metal ions in plant mitochondria and their role in interactions with proteins and oxidative stress-induced damage to respiratory function. Plant Physiol 152:747–761

Taylor NL, Heazlewood JL, Day DA, Millar AH (2005) Differential impact of environmental stresses on the pea mitochondrial proteome. Mol Cell Proteomics 4:1122–1133

van Lis R, Atteia A, Mendoza-Hernández G, González-Halphen D (2003) Identification of novel mitochondrial protein components of *Chlamydomonas reinhardtii*. A proteomic approach. Plant Physiol 132:318–330

Volkl A, Mohr H, Weber G, Fahimi HD (1997) Isolation of rat hepatic peroxisomes by means of immune free flow electrophoresis. Electrophoresis 18:774–780

Volkl A, Mohr H, Fahimi HD (1999) Peroxisome subpopulations of the rat liver. Isolation by immune free flow electrophoresis. J Histochem Cytochem 47:1111–1118

von Zychlinski A, Kleffmann T, Krishnamurthy N, Sjolander K, Baginsky S, Gruissem W (2005) Proteome analysis of the rice etioplast – metabolic and regulatory networks and novel protein functions. Mol Cell Proteomics 4:1072–1084

Winger AM, Taylor NL, Heazlewood JL, Day DA, Millar AH (2007) The cytotoxic lipid peroxidation product 4-hydroxy-2-nonenal covalently modifies a selective range of proteins linked to respiratory function in plant mitochondria. J Biol Chem 282:37436–37447

Ytterberg AJ, Peltier JB, van Wijk KJ (2006) Protein profiling of plastoglobules in chloroplasts and chromoplasts. A surprising site for differential accumulation of metabolic enzymes. Plant Physiol 140: 984–997

Zischka H, Braun RJ, Marantidis EP, Buringer D, Bornhovd C, Hauck SM, Demmer OCJG, Reichert AS, Madeo F, Ueffing M (2006) Differential analysis of *Saccharomyces cerevisiae* mitochondria by free flow electrophoresis. Mol Cell Proteomics 5:2185–2200

Zybailov B, Rutschow H, Friso G, Rudella A, Emanuelsson O, Sun Q, van Wijk KJ (2008) Sorting signals, N-terminal modifications and abundance of the chloroplast proteome. PLoS One 3:e1994

Chapter 16

Plastid Transformation in Algae

Jörg Nickelsen and Alexandra-Viola Bohne
Molekulare Pflanzenwissenschaften, Biozentrum LMU München, Großhaderner Str. 2-4, D-82152 Planegg-Martinsried, Germany

Summary

Pioneering work from the late 1980s using the green alga *Chlamydomonas reinhardtii* has paved the way for biolistic chloroplast transformation in general. Since then, the continuous development of a molecular toolkit has made this chlorophyte alga the prime organism for algal transplastomic biotechnological applications. However, comparatively little progress has been made with the stable genetic manipulation of members of other algal groups with the red alga *Porphyridium* UTEX 637 representing a rare exception. In this chapter, we summarize the basic molecular principles of chloroplast transformation in algae as well as current approaches to optimize foreign gene expression in Chlamydomonas.

I. Introduction

Algae represent a diverse group of photosynthetic eukaryotes which are of fundamental ecological importance as primary producers of ca. 50% of the total organic carbon produced on earth per year and consequently a fundamental basis of the food chain. Algae arose during evolution by the uptake of a cyanobacterium by a heterotrophic protist

*Author for correspondence, e-mail: joerg.nickelsen@lrz.uni-muenchen.de

R. Bock and V. Knoop (eds.), Genomics of Chloroplasts and Mitochondria, Advances in Photosynthesis and Respiration 35, pp. 379–392,
DOI 10.1007/978-94-007-2920-9_16,

host approximately 1.5 billion years ago, a process referred to as endosymbiosis (Gross and Bhattacharya 2009). The engulfed cyanobacterium, probably related to extant *Anabaena* species, then transformed into a cellular organelle, the chloroplast (Deusch et al. 2008). This gave rise to at least three different lineages: the Glaucophyta (glaucophyte algae), the Rhodophyta (red algae) and the Viridiplantae (green algae and land plants). However, the evolutionary spectrum of algal life-forms was further increased due to secondary and even tertiary endosymbiotic events which involved the uptake of green or red eukaryotic algae by other heterotrophic eukaryotic hosts (see Chap. 2). As a consequence, the resulting algal groups contain "complex" plastids which are surrounded by three or four membranes. In contrast, plastids derived from primary endosymbiosis possess only two membranes forming the chloroplast envelope.

Algae have a long-standing tradition as food for humans and animal feed, especially in Asian countries. Todays biotechnological applications involve mainly non-transgenic approaches including the production of polyunsaturated fatty acids, polysaccharides and carotenoids (for review see Hallmann 2007). More recently, microalgae have attracted more attention as a source for the production of renewable energy like biodiesel and hydrogen (Mata et al. 2010; Stephens et al. 2010). Nevertheless with accumulating sequence information from various algal genomes and the parallel development of transformation techniques, a "transgenic century" for algal biotechnology has been initiated. To date ca. 25 algal species have been stably genetically manipulated in their nuclear genomes including green, red and brown algae as well as diatoms and dinoflagellates (for an overview see Walker et al. 2005). Most of them represent unicellular microalgae but also macroalgae like *Laminaria japonica*, *Porphyra miniata* and *Ulva lactuca* have successfully been subjected to stable genetic transformation (Qin et al. 2005). The transformation methods in use mainly include bombardment of algal cells with DNA-coated particles, agitation of algae with glass beads or silicate whiskers in the presence of DNA, electroporation and in rare cases *Agrobacterium tumefaciens* mediated transformation (Walker et al. 2005).

As outlined below, biolistic transformation is the method of choice for generating chloroplast transformants in both algae and land plants. However, with the exception of the green alga *Chlamydomonas reinhardtii* only a very limited number of algal species has been genetically transformed in their plastid genomes (see Sect. IV). As such, *C. reinhardtii* currently still represents a more-or-less stand-alone model system for algal chloroplast transformation.

II. Chloroplast Biology of *C. reinhardtii*

C. reinhardtii is a flagellated unicellular green alga which has a size of 10 μm in diameter. It contains a single cup-shaped chloroplast which accounts for 40% of the total cellular volume. This relatively "big" chloroplast has significantly facilitated chloroplast transformation attempts by using the biolistic approach described below (see Sect. III.A). The chloroplast is surrounded by two envelope membranes and inside the organelle one can identify the organization of the thylakoid membrane system by electron microscopy. In *C. reinhardtii*, as well as in other green algae, plant-like differentiated thylakoid grana regions are lacking but instead thylakoid membranes can be arranged in multiple stacks of 2–10 discs (Harris 2009). Like many algae, *C. reinhardtii* contains a basally located chloroplast pyrenoid whose function is mainly dedicated to carbon concentrating mechanisms and which primarily consists of the CO_2-fixing enzyme ribulose-1,5-bisphosphate carboxylase/

Abbreviations: AHAS – Acetohydroxyacid synthase; CAI – Codon adaption index; CES – Control by epistasy of synthesis; DCMU – 3-(3′ 4′-di-chlorphenyl)-1,1-dimethylurea

oxygenase (Rubisco). Interestingly, the pyrenoid has recently been shown to also serve as the site of thylakoid membrane biogenesis and to play a pivotal role in the spatial organization of chloroplast gene expression (Uniacke and Zerges 2007). Furthermore, motile algae like *C. reinhardtii* can contain an "eye" allowing phototactic movements. One prominent eye constituent is the so-called eyespot, which is located at the periphery of the chloroplast where it associates with the inner chloroplast envelope and thylakoids.

The chloroplast genome of *C. reinhardtii* (203 kbp) is present in ca. 80 copies per cell (Koop et al. 2007). It encodes 109 genes and harbours a high number of short dispersed repeat regions (see Chap. 3). During sexual reproduction, the chloroplast DNA is uniparentally inherited from the mating type + (mt^+) parent. Thus, resembling the situation in vascular plants, an outcrossing of chloroplast transgenes from a mating type – (mt^-) parent is unlikely, increasing the ecological safeness of transplastomic algal lines for biotechnological applications (see also Chap. 18).

The chloroplast gene expression machinery has been intensively studied at the molecular level and crucial regulatory factors and elements regulating this machinery have been dismantled by both genetic and biochemical means (see Sect. V). One remarkable feature of some Chlamydomonas species, including *C. reinhardtii*, is their capacity to grow heterotrophically by consumption of acetate as reduced carbon source. The precise acetate assimilation pathway is still unclear but it is generally assumed that acetate is converted into acetyl-CoA by either acetyl-CoA synthetase activity or by a two-step process catalyzed by acetate kinase and phosphate acetyltransferase (Spalding 2009). Subsequently, most of the acetate assimilation is proposed to occur through the glyoxylate cycle whose subcellular compartmentation is still uncertain (Spalding 2009). However despite these open questions, acetate-containing media allow the maintenance of photosynthetic mutants which still retain their fertility and, thus, can be used for downstream genetic analyses. More important in the course of chloroplast transformation events, these mutants can serve as recipient strains for chloroplast DNA manipulations (Nickelsen and Kück 2000; see Sect. III.D).

III. Transformation Procedures

The principal challenge for chloroplast transformation of both plant and algal cells is the delivery of DNA across the cell wall and at least three membranes, i.e., the plasma membrane and the inner and outer chloroplast envelope membranes. In case of algae harbouring complex plastids, the situation is even more complicated due to the presence of one or two additional chloroplast envelope membranes. Through the development of a so-called particle gun this problem has been overcome. Stable transplastomic lines have been generated for algae and land plants mainly by using this biolistic transformation technique (see below). However, for vascular plants also PEG-treatment of protoplasts has frequently been applied to deliver DNA into the chloroplast compartment of the cell (Koop et al. 2007; see Chap. 18). For *C. reinhardtii*, an alternative, very simple method has been described which is based on the agitation of cell-wall-less strains with glass beads and, therefore, requires no specialized equipment like a particle-gun (Kindle et al. 1991). Usually, this method yields high levels of nuclear transformants, but also chloroplast transformants can be generated to some extent. In contrast to protoplast transformation of plant cells, addition of PEG had no positive effect on chloroplast transformation rates using the glass bead method (Kindle et al. 1991). When this simple method was directly compared to the biolistic technique, however, a drastically (more than tenfold) lower chloroplast transformation efficiency was observed. Therefore nowadays, biolistic transformation is the routinely used method for generating transplastomic lines in *C. reinhardtii*.

A. Biolistic Gene Transfer

In 1988, Boynton and co-workers reported on the first successful transformation of a chloroplast genome. They used *C. reinhardtii* and microparticle bombardment which in parallel had successfully been applied to the transformation of mitochondria from yeast cells (Johnston et al. 1988). The basic steps of the method involve the coating of microparticles with DNA which are then used to bombard an algal cell lawn on an agar plate under vacuum. For acceleration of particles to sufficiently high velocity, initially home-built powder explosion devices were used but these have nowadays been replaced by commercially available helium-powered guns. Alternatively, so-called particle inflow guns have very successfully been used (Finer et al. 1992). Particles are accelerated in a helium stream which is controlled by a timer relay driven solenoid working at moderate helium pressure of eight bar and, thus, causing less damage to algal cells (Nickelsen and Kück 2000). As microprojectiles, both tungsten and gold particles are used with gold being more expensive but also more inert and uniform in size. For chloroplast transformation in *C. reinhardtii*, tungsten particles work sufficiently well.

Once DNA-coated particles hit a cell, they first penetrate through the cell wall/plasma membrane and then are supposed to penetrate the organelle's envelope and deposit the transforming DNA into the chloroplast stroma. How punctured membranes reseal afterwards is completely unknown. Alternatively, one might envisage that DNA is delivered to the cytosol and an unknown mechanism would subsequently transfer this DNA into the organelle (Koop et al. 2007). Irrespective of the incomplete picture of the entire transformation process, the feasibility of the approach is well established and, therefore, the development of the biolistic transformation procedure clearly marks the breakthrough for organelle transgenics.

B. Stable Transformation

Upon reaching the chloroplast, foreign DNA molecules then can integrate into the chloroplast genome. Due to the evolutionary history of chloroplasts as former cyanobacteria, a bacterial recombination system still exists in plastids that mediates integration of DNA via homologous recombination. Therefore, a prerequisite for plastome transgene integration is the presence of flanking homologous regions in the transforming DNA, which is thought to recombine via a double crossover event into the chloroplast genome. Usually, foreign recombinant DNAs should contain homologous flanking regions comprising ca. 1 kbp. However, successful high-frequency integration of DNA at the *psbA* locus from *C. reinhardtii* was also obtained with non-purified PCR fragments containing only 51 bp upstream and 121 bp of homology downstream of the integration site (Dauvillee et al. 2004). Furthermore, recombination between artificially introduced direct repeats of 483 bp – but not of 230 bp – was demonstrated in *C. reinhardtii* chloroplasts (Fischer et al. 1996) suggesting that the size of minimal "recombination platforms" may depend on the structural characteristics of the involved chloroplast genome regions.

As in bacteria, single crossover events due to only one homologous flanking region lead to the integration of the entire plasmid DNA and the generation of direct repeats at the plasmid sequence ends. This, however, creates an unstable situation since the plasmid tends to immediately recombine out after selective pressure is released (Purton 2007).

C. Heteroplasmy and Episomal Maintenance

The recombination event between a copy of the chloroplast genome and the foreign DNA results in a state, termed "heteroplasmic", where only one or few of the 80 genome copies have been altered. However, when transformants are repeatedly transferred to fresh selective medium, eventually, a "homoplasmic" state is accomplished at which all copies of the chloroplast genome contain the transgenic manipulation. In *C. reinhardtii*, homoplasmy is usually achieved after 3–4 weeks with a weekly transfer of transformant colonies. At this point, the transgenic state is stable and selective pressure can be released (Koop et al. 2007).

A heteroplasmic state is maintained, however, when the genetic alteration affects an essential gene. In this case, about 50% of the cp-genome copies remain wild-type to guarantee survival of cells. As a consequence, subsequent removal of selective conditions leads to a rapid reversion to the homoplasmic wild-type state. Thus, for reverse genetic approaches, a persisting heteroplasmic state is indicative of essential gene functions being compromised (see Chap. 19).

As outlined by Purton (2007), heteroplasmy might also cause problems when recessive mutations are introduced into the chloroplast genome of *C. reinhardtii* via co-integration of a selectable marker. The natural selection against such mutations, for instance in photosynthetic genes, could eventually lead to a low recovery of transformants containing both the site-directed mutation and the selectable marker. This might be due to low incorporation or maintenance of site-directed alterations and/or "copy correction" mechanisms acting on them. Strategies to overcome these problems include the pre-treatment of cells with FUdR (5-fluorodeoxyuridine), an inhibitor of chloroplast DNA replication that leads to reduced chloroplast-genome copy number. Apparently, this reduced copy number facilitates the subsequent segregation process during subculturing of transformants (Goldschmidt-Clermont 1998). Secondly, the use of strains containing chloroplast deletions of the target site avoids the problem of elimination via copy correction by wild-type gene versions (Guergova-Kuras et al. 2001).

Although homologous recombination usually results in stable integration of foreign DNA into the chloroplast genome, exceptions from this rule have been observed. In 1994, Kindle et al. reported on the detection of plasmid-like structures in the chloroplast of *C. reinhardtii* after chloroplast transformation. The episomal elements contained mutated versions of the *atpB* gene and accumulated to ca. 2,000 copies per chloroplast. Genetic crossings revealed that they were uniparentally inherited from the mt^+ parent indicating a chloroplast location. However, subsequent attempts to generate autonomously replicating systems based on these elements failed probably due to a very specialized *atpB*-specific effect (Suzuki et al. 1997). Episomal maintenance of transforming DNA was also observed during transformation of the unicellular alga *Euglena gracilis*, but in this case, copy numbers were drastically reduced as compared to chloroplast genome copy number (Doetsch et al. 2001; see Sect. IV). Taken together, the apparent possibility of creating high-copy-number plasmids within chloroplasts sounds appealing but available data suggest that it will be very difficult to generate chloroplast high-expression systems for transgenic biotechnological applications based on this system.

D. Chloroplast Markers and Marker Recycling

Biolistic chloroplast transformation in *C. reinhardtii* is usually performed by bombardment of a lawn of ca. 1×10^8 algal cells. Usually hundreds of transformants can be generated with one "shot" which are then selected directly on the plate. Alternatively, after bombardment, the cell lawn is transferred to a fresh plate containing the selective medium. In principle, three different selection strategies for transgenic lines have been applied to date. In their first successful attempts, Boynton et al. (1988) complemented an *atpB* deletion mutant by using a wild-type version of the chloroplast *atpB* gene encoding the β subunit of the chloroplast ATP synthase. Consequently, selection for restored photosynthetic activity was performed on minimal medium containing no acetate as reduced carbon source. Similarly, the chloroplast *tscA* gene involved in group II intron trans-splicing of the photosystem I subunit *psaA* mRNA restored photoautotrophic growth of transformants upon introduction into the chloroplast genome of the *tscA* deletion mutant *H13* (Goldschmidt-Clermont et al. 1991). Despite the fact that this approach is limited by the availability of appropriate chloroplast mutants, it has the main advantage that problems of heteroplasmic states (see Sect. III.C) are minimized. Moreover, no selectable bacterial markers must be co-introduced during transformation avoiding risks of marker spreading into the environment via horizontal gene transfer.

A second strategy described for *C. reinhardtii* is based on the introduction of point mutations into the chloroplast genome that confer resistance to either antibiotics like spectinomycin or herbicides like DCMU (for an overview see Goldschmidt-Clermont 1998). In the red alga *Porphyridium* spec., a mutated version of the chloroplast acetohydroxyacid synthase (AHAS) has successfully been used to select chloroplast transformants based on their resistance against the herbicide sulfometuron methyl (SMM; Lapidot et al. 2002). The advantage of such an approach is that basically any strain can be used as recipient for transformation and that no bacterial marker sequences are involved. However, since resistances occur spontaneously, a background of pseudo-transformants must always be considered which have to be sorted out by molecular analyses.

Most convenient for many applications in basic research is the use of dominant bacterial marker genes fused to regulatory chloroplast 5′ and 3′ regions. The most frequently used one is the *aadA* cassette from *Escherichia coli* which confers resistance to both spectinomycin and streptomycin (Goldschmidt-Clermont 1991). The second marker that is available for chloroplast transformation in *C. reinhardtii* is based on the *aphA-6* gene from *Acinetobacter baumannii* conferring resistance to kanamycin or amikacin (Bateman and Purton 2000). Marker cassettes can be introduced at any site of the chloroplast genome and have been used to inactivate or modify a number of chloroplast genes in *C. reinhardtii* (see Chap. 19). Furthermore, cassette co-integration is the method of choice for the establishment of foreign gene expression in chloroplasts (see Sect. V). During chloroplast transformation of *C. reinhardtii*, high frequencies (ca. 80%) of co-transformation events have been observed when two different markers on separate vectors where transformed (Boynton and Gillham 1993). This offers the possibility of efficient strategies for site-directed mutagenesis even if the marker cassette cannot be integrated close to the mutated site.

Obviously, the number of different markers and, consequently, selection strategies for algal chloroplast transformants are limited. This prevents the manipulation of multiple sites of the chloroplast genome in successive rounds of transformation. One solution to this problem is the use of markers that can be recycled. Fischer et al. (1996) reported on two different approaches for the use of the *aadA* marker for transformant selection and its subsequent removal from the chloroplast genome. The first strategy requires a marker flanked by direct repeats of 483 bp from bacterial plasmid DNA. After co-integration into the chloroplast genome, transformants are selected until homoplasmy is reached and, subsequently, selective pressure is released by cultivation of cells on appropriate (antibiotic-free) medium. Under non-selective conditions, recombination events between the direct repeats result in the excision of the marker located between the repeats. Afterwards, the marker-free strains can be applied to a next round of transformation via the *aadA* cassette (Fischer et al. 1996; Redding et al. 1998). Alternatively, the *aadA* cassette is introduced into an essential gene leading to a heteroplasmic state with regard to the selectable marker. A co-transformed construct creating a mutation of interest in a non-essential gene will reach homoplasmy during the selection period. Upon release of selective pressure, the marker cassette will then be eliminated from its heteroplasmic integration site via copy correction mechanisms or lost by random genome sorting. In summary, a complete molecular toolkit for the genetic manipulation of the chloroplast genome from *C. reinhardtii* is nowadays available and has successfully been applied in both basic and applied science.

IV. Transformed Algae Species

As already mentioned in the introductionary section, the transformation of algal species other than *C. reinhardtii* is still in its infancy. To date, only two additional algae species have been reported to have been successfully transformed. Doetsch and coworkers (2001) subjected the complex plastid of *Euglena gracilis* to transformation with the *aadA* cassette driven by *E. gracilis* control elements from the *psbA* 5′ and 3′ regions. In contrast to *C. reinhardtii,* for which recipient cells are transformed directly on agar plates, Euglena cells had to be spread on filter membranes in form of a mono-layer before bombardment. Apparently, this procedure results in a stabilization of cells against a semirigid backbone during microprojectile entry and, thus, allows the penetration through the protein pellicle surrounding *E. gracilis* cells (Doetsch et al. 2001). Molecular characterization of spectinomycin/streptomycin resistant colonies revealed that the transforming DNA did not integrate into the chloroplast genome, but was maintained as an extrachromosomal copy. Although the copy number was quite low with only 1–2 copies per chloroplast compared to ca. 100–300 copies of the chloroplast genome, this episomal element was maintained during at least 2 years of cultivation on solid medium. Apparently, long-term maintenance was sequence- or gene-dependent, respectively, since constructs containing a complete *psbK* operon were lost after only a few weeks on solid medium, probably due to overexpression problems caused by the introduced genes (Doetsch et al. 2001). Nevertheless despite several unsolved problems with *E. gracilis* chloroplast transformation, these first steps hopefully pave the way for the genetic manipulation of other algae containing complex plastids like brown algae or diatoms (see Chap. 2).

Initial attempts to stably transform the chloroplast genome of the diatom *Phaeodactylum tricornutum* proved to be difficult. Materna et al. (2009) reported on the generation of site-directed plastid mutants of the *psbA* gene encoding the D1 protein of the photosystem II reaction center. Using a commercial particle gun, constructs with mutant variants of codon 264 of the D1 protein leading to herbicide resistance against DCMU were introduced into to *P. tricornutum* cells (see also Chap. 12). However, molecular analyses of resulting DCMU resistant strains revealed that, apparently, the delivered DNA induced elevated mutation rates at the *psbA* locus but no real transformation events (Materna et al. 2009; P. Kroth, Konstanz, personal communication). The molecular basis for this phenomenon is unclear and, therefore, further efforts are required before a reliable chloroplast transformation protocol for diatoms will be available.

Similar to *C. reinhardtii*, the unicellular marine red alga *Porphyridium* UTEX 637 contains primary chloroplasts which are surrounded by only two envelope membranes. Biolistic transformation of this species resulted in stable transformants due to single crossover events that mediated the homologous recombination into the alga's plastid AHAS locus (Lapidot et al. 2002; for selection strategy see Sect. III.D). This resulted in the presence of two AHAS gene copies, one wild-type form and the SMM resistance conferring form. Thus, release of selection pressure is likely to lead to an immediate loss of the introduced DNA via recombination between the two AHAS repeats. Under continuous selective conditions however, the introduced DNA was stable for at least 1 year of cultivation. During the course of their work, the authors noticed that the transformation efficiency drastically increased when cells from dark/light synchronized cultures were used immediately after the dark cycle. As speculated by the authors, this is likely to be an effect of reduced amounts of cell wall polysaccharides at this time point and, consequently, a less solid barrier for the DNA-loaded microprojectiles (Lapidot et al. 2002). Therefore, careful evaluation of chloroplast transformation rates of cells from different time points of synchronized liquid cultures might represent one promising approach to overcome problems of genetic manipulation of recalcitrant algal species including especially those harbouring complex plastids.

V. Expression of Foreign Genes and Algal Chloroplast Biotechnology

Recent years have seen an increasing interest in using transplastomic approaches for the commercial production of recombinant therapeutic proteins. In algae – mostly in *C. reinhardtii* – substantial progress has been made with regard to the number and yields of recombinant protein production in chloroplasts. As reviewed by Specht et al. (2010), algae have distinct advantages as compared to vascular plants for biotechnological transgenic applications. Usually, they grow in contained bioreactors limiting risks of contaminations of production cultures on the one hand and the environment on the other hand. Transformation protocols are fast and microalgal cells are relatively uniform in size and differentiation status thereby facilitating downstream processing.

One major goal of transplastomic biotechnological applications is the optimization of product yields during the production process, i.e., the increase in stable accumulation of the foreign recombinant protein which is usually measured in relation to total protein amount. Current optimization attempts significantly benefit from the comprehensive knowledge of the molecular principles underlying endogenous chloroplast gene expression in *C. reinhardtii*. Chloroplast gene expression has been shown to be controlled at almost all levels including transcription, RNA metabolism and translation, with the latter one representing in most cases the rate-limiting step in expression of a chloroplast gene (Eberhard et al. 2002). Reverse genetic approaches involving site-directed mutagenesis of the flanking regions of chloroplast genes dramatically accelerated the identification of crucial *cis*-acting elements which are directly involved in the control of gene expression processes (see Chap. 19). The critical elements include promoter structures, which usually are of the bacterial sigma 70 type and contain so-called −10 and −35 elements. However, highly expressed algal genes like *psbD* encoding the D2 protein of photosystem II possess promoters containing only a −10 element (Klinkert et al. 2005). Furthermore, the analysis of plastid reporter gene constructs has demonstrated that the 5′ UTRs of plastid mRNAs play critical roles in RNA stabilization and for translation initiation. These 5′ UTRs serve as recognition sites for trans-acting regulatory protein factors or form structural RNA elements which influence posttranscriptional processes by the formation of barriers against nucleolytic attack or by the control of ribosomal access to mRNAs (Nickelsen et al. 1994; Drager et al. 1998; Bruick and Mayfield 1998; Vaistij et al. 2000; Suay et al. 2005; Klinkert et al. 2006). Complementary biochemical and genetic analyses in *C. reinhardtii* have revealed the nature of many trans-acting factors which exert their function via these cis-elements and, thus, represent the pacemakers for the expression of chloroplast genes and, consequently, also for transgenes (for a recent review see Bohne et al. 2009; Stern et al. 2010). One remarkable general regulatory principle which underlies chloroplast gene expression has been named "control by epistasy of synthesis" (CES). It is valid for the synthesis/assembly of various photosynthetic complexes, i.e., Cytb$_6$f, PSII, PSI, as well as the ATPase in *C. reinhardtii* and the Rubisco enzyme in tobacco (for a recent overview see Choquet and Wollman 2009). Basically, unassembled subunits of multisubunit protein complexes exert a feedback-loop inhibition on their own synthesis via the 5′ untranslated regions of the respective mRNAs. It appears likely that at least some of the abovementioned trans-acting factors are involved in these control circuits.

A. Determinants for the Efficiency of Chloroplast Transgene Expression in C. reinhardtii

Based on the extensive knowledge on the control of endogenous chloroplast gene expression in *C. reinhardtii*, several transgenic lines have been generated which express foreign reporter genes like *gfp* and *uidA* (encoding the green fluorescent protein and the β-D-glucuronidase, respectively) in the chloroplast under the control of plastid regulatory elements. In a combinatorial approach, several 5′ and 3′ flanking regions from algal chloroplast protein-coding genes were tested (Ishikura et al. 1999; Barnes et al. 2005; Fletcher et al. 2007). Generally, the 5′ regions from the *atpA* and *psbD* genes were found to confer the highest expression rates on transgenes whereas different 3′ regions had only minor impacts. More recently, the *psaA*-exon1 5′ UTR was added to the list of tested regulatory regions and found to confer the highest expression rates amongst the known 5′ regions (Michelet et al. 2010). Nevertheless, all transgene expression rates were found to be significantly lower than those of the highly abundant endogenous algal chloroplast proteins. Moreover, transgene expression in *C. reinhardtii* is usually an order of magnitude lower than that in chloroplasts of higher plants (see Chap. 18). In a systematic evaluation, Surzcycki et al. (2009) defined four main determinants which affect chloroplast transgene expression in *C. reinhardtii*. These include (1) codon optimization, (2) protein toxicity, (3) protease activity and (4) genotypic modifications. Recently, this list was extended by Coragliotti et al. (2010), who showed that translation of recombinant mRNA molecules (5) also affects the accumulation of heterologous proteins.

1. Codon Optimization: Initial work on *C. reinhardtii* had shown that adaption of the codon-usage of a transplastomic *gfp* gene increases the accumulation of its product ca. 80-fold (Franklin et al. 2002). Since then, chloroplast transgenes are routinely designed according to the CAI (Codon Adaption Index) which provides a quantitative method for the prediction of protein expression levels. While codon optimization is usually calculated against a chloroplast codon usage which is derived from a list of all chloroplast genes, Surzcycki et al. (2009) pointed out that it is important to include only highly expressed chloroplast genes into the reference list. When following this rule, the authors obtained the highest expression level for a foreign protein in *C. reinhardtii* chloroplasts, i.e., the VP28 protein of the white spot syndrome virus accumulated to 21% of total cellular protein (TCP).
2. Protein toxicity: Another severe problem of algal transgene expression in chloroplasts is the toxicity of some foreign proteins which per se is not predictable (Surzcycki et al. 2009). However, like in other systems, inducible gene expression systems could solve this problem by expressing the toxic protein only after an appropriate biomass of transgenic lines has been generated (Koop et al. 2007). In *C. reinhardtii*, the *psbA* 5′ region mediates light-dependent regulation of D1 synthesis and, thus, can be used for a controlled onset of the translation of recombinant mRNAs via light (Barnes et al. 2005; Mayfield et al. 2007). However, even in the dark, the *psbA* 5′ region promotes a substantial level of basic gene expression which could lead to accumulation of toxic proteins. A similar, incomplete repression of transgene expression was observed when an artificial *lac* regulation system from *E. coli* was introduced into the chloroplast *rbcL* promoter region of *C. reinhardtii* (Kato et al. 2007).

A tighter induction system has recently been developed based on the nucleus-encoded Nac2 factor controlling the stabilization of the chloroplast *psbD* mRNA encoding the D2 protein of photosystem II (Boudreau et al. 2000; Surzcycki et al. 2007). Nac2 has been shown to exert its function via the *psbD* 5′ UTR in a concerted manner together with the translational regulator RBP40 (Ossenbühl and Nickelsen 2000; Klinkert et al. 2006; Schwarz et al. 2007). As depicted in Fig. 16.1a, the Nac2 gene has been placed under the control of the copper-sensitive cytochrome c_6 promoter in the nuclear genome of *C. reinhardtii* (Surzcycki et al. 2007). Therefore, the expression of any gene via the *psbD* 5′ UTR is strictly copper-regulated, i.e., under copper-depleted conditions, Nac2 and, consequently, recombinant proteins accumulate while copper-repleted

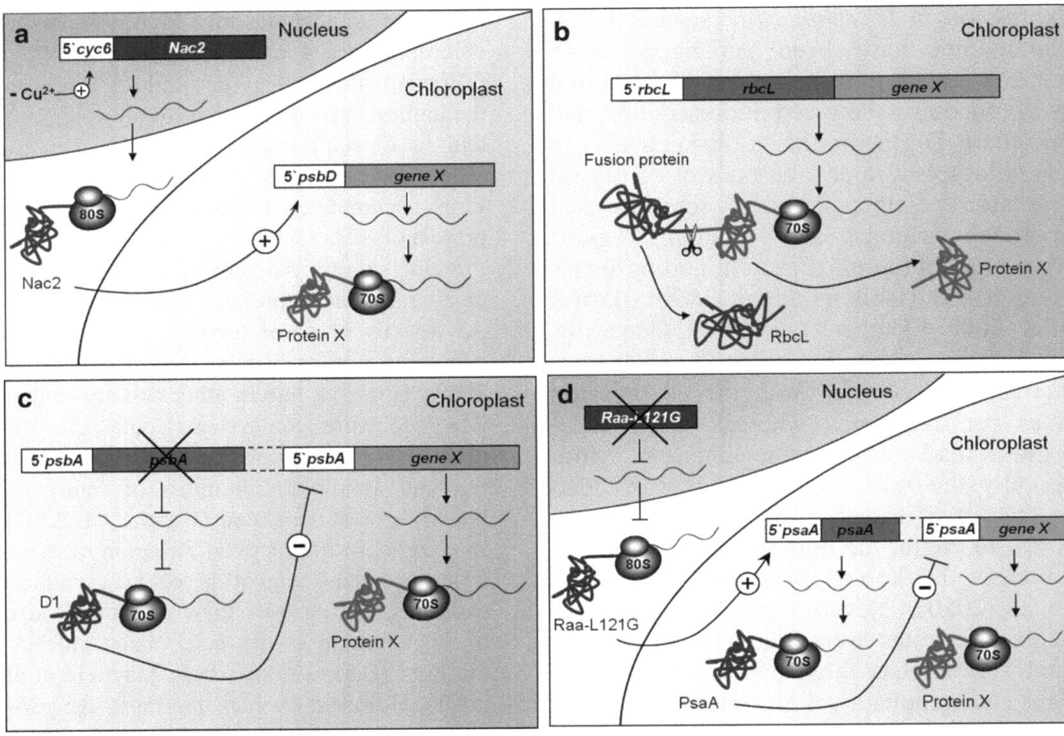

Fig. 16.1. Optimization of foreign gene expression in algal chloroplasts. (**a**) Chloroplast transgene expression is induced by copper depletion-induced expression of the nucleus-encoded Nac2 factor, which is required for stabilization of the transgene mRNA via its *psbD*-derived 5′ UTR (Surzcycki et al. 2007). (**b**) Combination of a foreign gene and the endogenous *rbcL* gene leads to the synthesis of a fusion protein which is processed to yield active RbcL and the recombinant protein (Muto et al. 2009). (**c**) Inactivation of the endogenous chloroplast gene enhances transgene expression driven by the same 5′ UTR due to inactivated negative feedback loops (Manuell et al. 2007). (**d**) Inactivation of negative feedback control by a nucleus-encoded mutation in a gene which is required for the expression of the plastid gene providing the 5′ UTR for transgene expression (Michelet et al. 2010). For further explanations, see text.

medium results in the complete loss of expression from the *psbD* 5′ UTR. To bypass the requirement of endogenous *psbD* gene expression for Nac2 function, the 5′ UTR of the chloroplast-encoded *psbD* gene was replaced with that of the *petA* 5′ UTR. This manipulation then allowed to synthesize PS II and, thus, enabled photoautotrophic growth for optimal biomass production in a Nac2-independent manner in the presence of copper (Surzcycki et al. 2007, 2009). By using this inducible system, the synthesis and accumulation of the otherwise toxic growth promoter DILP-2 in chloroplasts of *C. reinhardtii* was achieved indicating the viability of the system for biotechnological applications (Surzcycki et al. 2009).

3. Protease activity: The stability of recombinant proteins within the chloroplast is an obvious yield-affecting parameter (Mayfield et al. 2007). Nevertheless, relatively limited data are available on the influence of protein degradation on net accumulation of recombinant chloroplast proteins in alga. Recently, a first evaluation revealed a threefold higher recombinant protein stability in *C. reinhardtii* cells which had been treated with the energy uncoupler cyanide m-chlorophenylhydrazone (CCCP) as compared to those which had not

been treated (Surzcycki et al. 2009). This suggests that ATP-dependent proteolytic activities can significantly diminish the levels of foreign proteins in chloroplasts of *C. reinhardtii*. Another recently applied strategy to enhance recombinant protein accumulation in algal chloroplasts – probably due to protein stabilization effects –is the translational fusion of the foreign protein to the large subunit of the Rubisco enzyme, RbcL. Muto et al. (2009) achieved a 33-fold increase of luciferase activity when it was expressed together with RbcL and posttranslationally liberated from the fusion protein via an artificially introduced protease cleavage site from pre-ferredoxin (Fig. 16.1b).

4. Genotypic background/modification: The genetic background of recipient strains for chloroplast genetic manipulation has a significant impact on transgene expression rates. When a *C. reinhardtii* chloroplast transgene is expressed via the *psbA* 5′ region, a tenfold increase in accumulation of recombinant protein is observed in strains lacking the endogenous *psbA* gene (Fig. 16.1c). This phenomenon was attributed to less competition for trans-acting activators of translation and/or less negative feedback control by the abovementioned CES system (Manuell et al. 2007; Rasala et al. 2010; Minai et al. 2006). A similar enhancement of transgene expression driven by the *rbcL* 5′ region has been observed upon deletion of the endogenous *rbcL* gene suggesting that deletion of endogenous chloroplast regulatory regions represents a fruitful general strategy for transplastomic algal biotechnology.

   Besides the chloroplast genome also the nuclear genome has a significant impact on chloroplast gene expression via the abovementioned trans-acting regulatory factors. Michelet et al. (2010) recently showed that transgene expression driven by the *psaA* 5′ region is enhanced in a nuclear mutant background of a factor involved in splicing of the *psaA* mRNA. This phenomenon is likely to be due to two effects, i.e., increased RNA accumulation in the splicing deficient mutant background and the bypass of negative feedback loops caused by unassembled PsaA protein (Fig. 16.1d). However, one major drawback of this approach is the non-photosynthetic phenotype of the producing strain which limits photoautotrophic growth rates.

5. Translation: Recently, the role of translational activity on recombinant chloroplast mRNA templates was analysed in detail. The results suggest that protein synthesis on the level of ribosome association and even more important during translation elongation has severe impacts on heterologous protein accumulation (Coragliotti et al. 2010). Taken together, further understanding of the regulatory principles of chloroplast gene expression will clearly help to optimize biotechnological recombinant protein production strategies.

B. Expressed Transgenes

By using the abovementioned strategies for chloroplast transgene expression in *C. reinhardtii* several recombinant proteins have been produced to date. Early attempts mainly focussed on the expression of marker and reporter genes including the *aadA*, *uidA*, luciferase and *gfp* genes (for a review see Koop et al. 2007). In addition, the *E. coli* RecA protein as well as allophycocyanin from *Spirulina maxima* were successfully expressed in *C. reinhardtii* chloroplasts (Cerutti et al. 1995; Su et al. 2005). The first chloroplast-expressed proteins of pharmaceutical relevance were a human large single-chain antibody, a fusion protein of cholera toxin B subunit and foot and mouth disease VP1 protein (Mayfield et al. 2003; Sun et al. 2003). Since then, several other vaccines and therapeutics have been produced in transplastomic *C. reinhardtii* cells including, for instance, a correctly assembled human monoclonal antibody (Surzcycki et al. 2009; Dreesen et al. 2010; Rasala et al. 2010; Tran et al. 2009; for a comprehensive overview see Specht et al. (2010) and further references therein). Taken together, the available data clearly demonstrate that algal chloroplasts can provide an efficient platform for the production of high value recombinant proteins for human and animal biopharmaceuticals.

VI. Future Perspectives

Recent advances in the genetic manipulation of the chloroplast genome of *C. reinhardtii* promise a bright future for algal biotechnology. The development of a complete molecular toolkit basically allows any alteration of interest to be introduced into the chloroplast genome. Especially, foreign gene expression has been substantially optimized by following various approaches like codon optimization of foreign genes, use of inducible systems, defining optimal genetic backgrounds for gene expression cassettes and stabilization of recombinant proteins via translational fusions to endogenous chloroplast proteins. It is foreseeable, that on-going systematic evaluation of combinations of these and new strategies will further increase the yields of recombinant therapeutic production in chloroplasts into the range of economic sustainability.

The development of transformation protocols for other algae than *C. reinhardtii*, including those harbouring complex plastids, remains a major challenge for algal chloroplast transformation. Especially, some ecologically and/or economically highly relevant groups, like diatoms or brown algae, should be major targets of research efforts in that direction.

Acknowledgments

Work in the authors' laboratory is supported by the Deutsche Forschungsgemeinschaft (Ni390/4-2).

References

Barnes D, Franklin S, Schultz J, Henry R, Brown E, Coragliotti A, Mayfield SP (2005) Contribution of 5′- and 3′-untranslated regions of plastid mRNAs to the expression of *Chlamydomonas reinhardtii* chloroplast genes. Mol Gen Genomics 274:625–636

Bateman JM, Purton S (2000) Tools for chloroplast transformation in *Chlamydomonas*: expression vectors and new dominant selectable marker. Mol Gen Genet 263:404–410

Bohne A-V, Schwarz C, Jalal A, Ossenbühl F, Nickelsen J (2009) Control of organellar gene expression in Chlamydomonas – future perspectives -. Endocyt Cell Res 19:70–80

Boudreau E, Nickelsen J, Lemaire S, Ossenbühl F, Rochaix J-D (2000) The *Nac2* gene of *Chlamydomonas reinhardtii* encodes a chloroplast TPR-like protein involved in *psbD* mRNA stability. EMBO J 19:3366–3376

Boynton JE, Gillham NW, Harris EH, Hosler JP, Johnson AM, Jones AR, Randolph-Andersen BL, Robertson D, Klein TM, Shark KB, Sanford JC (1988) Chloroplast transformation in Chlamydomonas with high velocity microprojectiles. Science 240:1534–1538

Boynton JE, Gillham NW (1993) Chloroplast transformation in Chlamydomonas. Methods Enzymol 217:510–536

Bruick RK, Mayfield SP (1998) Processing of the *psbA* 5′ untranslated region in *Chlamydomonas reinhardtii* depends upon factors mediating ribosome association. J Cell Biol 143:1145–1153

Cerutti H, Johnson AM, Boynton JE, Gillham NW (1995) Inhibition of chloroplast DNA recombination and repair by dominant negative mutants of *Escherichia coli* RecA. Mol Cell Biol 15:3003–3011

Choquet Y, Wollman FA (2009) The CES process. In: Harris EH, Stern DB (eds) The Chlamydomonas sourcebook, vol 2, Organellar and metabolic processes. Academic, Oxford

Coragliotti AT, Beligni MV, Franklin SE, Mayfield SP (2010) Molecular factors affecting the accumulation of recombinant proteins in the Chlamydomonas reinhardtii chloroplast. Mol Biotechnol 48:60–75

Dauvillee D, Hilbig L, Preiss S, Johanningmeier U (2004) Minimal extent of sequence homology required for homologous recombination at the *psb*A locus in *Chlamydomonas reinhardtii* chloroplasts using PCR-generated DNA fragments. Photosyn Res 79:219–224

Deusch O, Landan G, Roettger M, Gruenheit N, Kowallik KV, Allen JF, Martin W, Dagan T (2008) Genes of cyanobacterial origin in plant nuclear genomes point to a heterocyst-forming plastid ancestor. Mol Biol Evol 25:748–761

Doetsch N, Favreau M, Kuscuoglu N, Thompson M, Hallick RB (2001) Chloroplast transformation in *Euglena gracilis*: splicing of a groupIII twintron transcribed from a transgenic *psbK* operon. Curr Genet 39:49–60

Drager RG, Girard-Bascou J, Choquet Y, Kindle KL, Stern DB (1998) In vivo evidence for 5′–>3′ exoribonuclease degradation of an unstable chloroplast mRNA. Plant J 13:85–96

Dreesen IAJ, Charpin-El Hamri G, Fussenegger M (2010) Heat-stable oral alga-based vaccine protects mice from *Staphylococcus aureus* infection. J Biotechnol 145:273–280

Eberhard S, Drapier D, Wollman FA (2002) Searching limiting steps in the expression of chloroplast-encoded proteins: relations between gene copy number, transcription, transcript abundance and translation rate in the chloroplast of *Chlamydomonas reinhardtii*. Plant J 31:149–160

Finer JJ, Vain P, Jones MW, McMullen MD (1992) Development of the particle inflow gun for DNA delivery to plant cells. Plant Cell Rep 11:323–328

Fischer N, Stampacchia O, Redding K, Rochaix JD (1996) Selectable marker recycling in the chloroplast. Mol Gen Genet 251:373–380

Fletcher SP, Muto M, Mayfield SP (2007) Optimization of recombinant protein expression in the chloroplasts of green algae. Adv Exp Med Biol 616:90–98

Franklin S, Ngo B, Efuet E, Mayfield SP (2002) Development of a GFP reporter gene for *Chlamydomonas reinhardtii* chloroplast. Plant J 30: 733–744

Goldschmidt-Clermont M (1991) Transgenic expression of aminoglycoside adenine trans-ferase in the chloroplast: a selectable marker for site-directed transformation of Chlamydomonas. Nucleic Acids Res 19:4083–4089

Goldschmidt-Clermont M, Choquet Y, Girard-Bascou J, Michel F, Schirmer-Rahire M, Rochaix J-D (1991) A small chloroplast RNA may be required for trans-splicing in *Chlamydomonas reinhardtii*. Cell 65:135–143

Goldschmidt-Clermont M (1998) Chloroplast transformation. In: Rochaix J-D, Goldschmidt-Clermont M, Merchant S (eds) The molecular biology of chloroplasts and mitochondria in chlamydomonas. Kluwer Academic Publishers, Dordrecht, pp 139–149

Gross J, Bhattacharya D (2009) Mitochondrial and plastid evolution in eukaryotes: an outsiders perspective. Nat Rev Genet 10:495–505

Guergova-Kuras M, Boudreaux B, Joliot A, Joliot P, Redding K (2001) Evidence for two active branches for electron transfer in photosystem I. Proc Natl Acad Sci USA 98:4437–4442

Hallmann A (2007) Algal transgenics and biotechnology. Transgenic Plant J 1:81–98

Harris EH (ed) (2009) The Chlamydomonas sourcebook, vol 2, Introduction to Chlamydomonas and its laboratory use. Academic, Oxford

Ishikura K, Takaoka Y, Kato K, Sekine M, Yoshida K, Shinmyo A (1999) Expression of a foreign gene in *Chlamydomonas reinhardtii* chloroplast. J Biosci Bioengin 87:307–314

Johnston SA, Anziano PQ, Shark K, Sanford JC, Butow RA (1988) Mitochondrial transformation in yeast by bombardment with microprojectiles. Science 240:1538–1541

Kato K, Marui T, Kasai S, Shinmyo A (2007) Artificial control of transgene expression in *Chlamydomonas reinhardtii* chloroplast using the *lac* regulation system from *Escherichia coli*. J Biosci Bioeng 104:207–213

Kindle KL, Richards KL, Stern DB (1991) Engineering the chloroplast genome: techniques and capabilities for chloroplast transformation in *Chlamydomonas reinhardtii*. Proc Natl Acad Sci USA 88:1721–1725

Kindle KL, Suzuki H, Stern DB (1994) Gene amplification can correct a photosynthetic growth defect caused by mRNA instability in Chlamydomonas chloroplasts. Plant Cell 6:187–200

Klinkert B, Schwarz C, Pohlmann S, Nickelsen J (2005) Correlation of mRNA levels and protein accumulation in a *psbD* promotor mutant of *Chlamydomonas reinhardtii*. Mol Gen Genom 274:637–643

Klinkert B, Elles I, Nickelsen J (2006) Translation of chloroplast *psbD* mRNA in Chlamydomonas depends on a structural RNA element blocking the AUG start codon. Nucleic Acids Res 34:386–394

Koop HU, Herz S, Golds T, Nickelsen J (2007) The genetic transformation of plastids. In: Bock R (ed) Cell and molecular biology of plastids, vol 19, Topics in current genetics. Springer, Berlin/Heidelberg, pp 457–510

Lapidot M, Raveh D, Sivan A, Shoshana A, Shapira M (2002) Stable chloroplast transfromation of the unicellular red alga *Porphyridium spec*. Plant Physiol 129:7–12

Manuell AL, Beligni MV, Elder JH, Siefker DT, Tran M, Weber A, McDonald TL, Mayfield SP (2007) Robust expression of a bioactive mammalian protein in Chlamydomonas chloroplast. Plant Biotechnol J 5:402–412

Mata TM, Martins AA, Caetano NS (2010) Microalgae for biodiesel production and other applications: a review. Renew Sustain Energy Rev 14:217–232

Materna AC, Sturm S, Kroth PG, Lavaud J (2009) First induced plastid genome mutations in an alga with secondary plastid: *psbA* mutations in the diatom *Phaeodactylum tricornutum* reveal consequences on the regulation of photosynthesis. J Phycol 45:838–846

Mayfield SP, Franklin SE, Lerner RA (2003) Expression and assembly of a fully active antibody in algae. Proc Natl Acad Sci USA 100:438–442

Mayfield SP, Manuell AL, Chen S, Wu J, Tran M, Siefker D, Muto M, Marin-Navarro J (2007) *Chlamydomonas reinhardtii* chloroplasts as protein factories. Curr Opin Biotechnol 18:126–133

Michelet L, Lefebvre-Legendre L, Burr SE, Rochaix J-D, Goldschmidt-Clermont M (2010) Enhanced chloroplast transgene expression in a nuclear mutant of Chlamydomonas. Plant Biotechnol J 9:565–574

Minai L, Wostrikoff K, Wollman FA, Choquet Y (2006) Chloroplast biogenesis of photosystem II cores involves a series of assembly-controlled steps that regulate translation. Plant Cell 18:159–175

Muto M, Henry RE, Mayfield SP (2009) Accumulation and processing of a recombinant protein designed as a cleavable fusion to the endogenous Rubisco LSU protein in Chlamydomonas chloroplast. BMC Biotechnol 9:26

Nickelsen J, Dillewyn J, Rahire M, Rochaix J-D (1994) Determinants for stability of the chloroplast *psbD* mRNA are located within its short leader region in *Chlamydomonas reinhardtii*. EMBO J 13:3182–3191

Nickelsen J, Kück U (2000) The unicellular green alga *Chlamydomonas reinhardtii* as an experimental system to study chloroplast RNA metabolism. Naturwiss 87:97–107

Ossenbühl F, Nickelsen J (2000) *Cis* and *trans*-acting determinants for the translation of the *psbD* mRNA in *Chlamydomonas reinhardtii*. Mol Cell Biol 20:8134–8142

Purton S (2007) Tools and techniques for chloroplast transformation of Chlamydomonas. In: Leon R, Galvan A, Fernandez E (eds) Transgenic microalgae as green cell factories, vol 616, Advances in experimental medicine and biology. Springer Science + Business Media, New York, pp 34–45

Qin S, Jiang P, Tseng C (2005) Transforming kelp into a marine bioreactor. Trends Biotech 23:264–268

Rasala BA, Muto M, Lee PA, Jager M, Cardoso RMF, Behnke CA, Kirk P, Hokanson CA, Crea R, Mendez M, Mayfield SP (2010) Production of therapeutic proteins in algae, analysis of expression of seven human proteins in the chloroplast of *Chlamydomonas reinhardtii*. Plant Biotechnol J 8:719–733

Redding K, MacMillan F, Leibl W, Brettel K, Hanley J, Rutherford AW, Breton J, Rochaix JD (1998) A systematic survey of conserved histidines in the core subunits of Photosystem I by site-directed mutagenesis reveals the likely axial ligands of P700. EMBO J 17:50–60

Schwarz C, Elles I, Kortmann J, Piotrowski M, Nickelsen J (2007) Synthesis of the D2 protein of photosystem II in *Chlamydomonas* is controlled by a high-molecular-weight complex containing the RNA stabilization factor Nac2 and the translational activator RBP40. Plant Cell 19:3627–3639

Spalding MH (2009) The CO_2-concentrating mechanism and carbon assimilation. In: Harris EH, Stern DB (eds) The Chlamydomonas sourcebook, vol 2, Organellar and metabolic processes. Academic, Oxford

Specht E, Miyake-Stoner S, Mayfield S (2010) Microalgae come of age as a platform for recombinant protein production. Biotechnol Lett 32:1373–1383

Stephens E, Ross IL, Mussgnug JH, Wagner LD, Borowitzka MA, Posten C, Kruse O, Hankamer B (2010) Future prospects of microalgal biofuel production systems. Trends Plant Sci 15:554–564

Stern DB, Goldschmidt-Clermont M, Hanson RH (2010) Chloroplast RNA metabolism. Ann Rev Plant Biol 61:125–155

Su ZL, Qian KX, Tan CP, Meng CX, Qin S (2005) Recombination and heterologous expression of allophycyanin gene in the chloroplast of *Chlamydomonas reinhardtii*. Acta Biochim Biophys Sinica 37:709–712

Suay L, Salvador ML, Abesha E, Klein U (2005) Specific roles of 5′ RNA secondary structures in stabilizing transcripts in chloroplasts. Nucleic Acid Res 33:4754–4761

Sun M, Qian K, Su N, Chang H, Liu J, Chen G (2003) Foot-and-mouth disease virus VP1 protein fused with cholera toxin B subunit expressed in *Chlamydomonas reinhardtii* chloroplast. Biotechnol Lett 25:1087–1092

Surzcycki R, Cournac L, Peltier G, Rochaix J-D (2007) Potential for hydrogen production with inducible chloroplast gene expression in Chlamydomonas. Proc Natl Acad Sci USA 104:17548–17553

Surzcycki R, Greenham K, Kityama K, Dibal F, Wagner R, Rochaix J-D, Ajam T, Surzcycki S (2009) Factors effecting expression of vaccines in microalgae. Biologicals 37:133–138

Suzuki H, Ingersoll J, Stern DB, Kindle KL (1997) Generation and maintenance of tandemly repeated extrachromosomal plasmid DNA in Chlamydomonas chloroplasts. Plant J 11:635–648

Tran M, Zhou B, Petterson PL, Gonzalez MJ, Mayfield SP (2009) Synthesis and assembly of a full-length human monoclonal antibody in algal chloroplasts. Biotechnol Bioeng 104:663–673

Uniacke J, Zerges W (2007) Photosystem II assembly and repair are differentially localized in Chlamydomonas. Plant Cell 19:3640–3654

Vaistij FE, Goldschmidt-Clermont M, Wostrikoff K, Rochaix JD (2000) Stability determinants in the chloroplast psbB/T/H mRNAs of *Chlamydomonas reinhardtii*. Plant J 21:469–481

Walker TL, Collet C, Purton S (2005) Algal transgenics in the genomic era. J Phycol 41:1077–1093

Chapter 17

Plastid Transformation in Flowering Plants

Pal Maliga*
Waksman Institute, Rutgers University, 190 Frelinghuysen Road, Piscataway, NJ 08854, USA
and Department of Plant Biology, Rutgers University, 59 Dudley Road, New Brunswick, NJ 08901, USA

* Author for correspondence, e-mail: maliga@waksman.rutgers.edu

R. Bock and V. Knoop (eds.), Genomics of Chloroplasts and Mitochondria, Advances in Photosynthesis and Respiration 35, pp. 393–414,
DOI 10.1007/978-94-007-2920-9_17,

Summary

The plastid genome of higher plants is relatively small, 120–230-kb in size, and present in up to 10,000 copies per cell. Standard protocols for the introduction of transforming DNA employ biolistic DNA delivery or polyethylene glycol treatment. Genetically stable, transgenic plants are obtained by modification of the plastid genome by homologous recombination, followed by selection for the transformed genome copy by the expression of marker genes that protect the cells from selective agents. Commonly used selective agents are antibiotics, including spectinomycin, streptomycin, kanamycin and chloramphenicol. Selection for resistance to amino acid analogues has also been successful. The types of plastid genome manipulations include gene deletion, gene insertion, and gene replacement, facilitated by specially designed transformation vectors. Methods are also available for post-transformation removal of marker genes. The model species for plastid genetic manipulation is Nicotiana tabacum, in which most protocols have been tested. Plastid transformation is also available in several solanaceous crops (tomato, potato, eggplant) and ornamental species (petunia, Nicotiana sylvestris). Significant progress has been made with Brasssicaceae including cabbage, oilseed rape and Arabidopsis. Recent additions to the crops in which plastid transformation is reproducibly obtained are lettuce, soybean and sugar beet. The monocots are a taxonomic group recalcitrant to plastid transformation; initial inroads have been made only in rice.

I. Introduction

Plastids are semi-autonomous plant organelles containing their own genome (plastid DNA; ptDNA). The compact 120–230-kb plastid genome encodes less than 100 proteins (Sugiura 1989; Raubeson and Jansen 2005); the majority of plastid functions is carried out by proteins encoded in ~3,000 nuclear genes (Leister 2003). Plastid genes, transcription and translation have many conserved prokaryotic features (Barkan 2011).

Transformation of the plastid genome was first achieved in 1988 in the unicellular green alga *Chlamydomonas reinhardtii* (Boynton et al. 1988). Transformation of the plastid genome in tobacco (*Nicotiana tabacum*), a flowering plant species, followed in 1990 (Svab et al. 1990). Progress in Chlamydomonas plastome engineering has been the source of continued inspiration for researchers working with flowering plants. Shared features between the algal and flowering plant plastids are a polyploid genetic system, and reliance on nuclear genes for plastid function. However, the evolutionary distance is reflected in many mechanistic differences, and there is no expectation that protocols developed in either system would be

Abbreviations: AAD – Aminoglycoside 3″-adenylyltransferase; AS – Anthranilate synthase; ASA2 – Anthranilate synthase alpha-subunit; BA – Betaine aldehyde; BADH – Betaine aldehyde dehydrogenase enzyme; CAT – Chloramphenicol acetyltransferase; GFP – Green fluorescent protein; GUS – β-glucuronidase; NPTII – Neomycin phosphotransferase II; PEG – Polyethylene glycol; PIG – Particle inflow gun; PPT – Phosphinothricin herbicide; ptDNA – Plastid DNA, plastid genome

interchangeable. The principal difference in the methodology can be traced back to engineering of the plastid genome of algal cells in photoautotrophic cultures and manipulation of the plastid genome in higher plants in heterotrophically grown tissue culture cells.

Since 1990 plastid transformation has been implemented in numerous flowering plant species. This review will focus on the methods for engineering the plastid genome of flowering plants and gives an overview of the progress made in implementing plastid transformation in different taxonomic groups. For information on the applications of plastid transformation in basic science and biotechnology, the reader is referred to recent reviews (Daniell et al. 2009; Cardi et al. 2010; Day and Goldschmidt-Clermont 2011; Maliga and Bock 2011; Whitney et al. 2011).

II. Methods for DNA Introduction

There are two practical methods of DNA introduction into plastids: biolistic DNA delivery and polyethylene glycol (PEG)-mediated DNA uptake.

A. Biolistic DNA Delivery

Protocols for biolistic delivery of RNA and DNA into living cells were developed by John Sanford's laboratory. In the first experiments, delivery of tobacco mosaic virus RNA was confirmed by formation of viral inclusion bodies in onion cells (Klein et al. 1987) and transient expression of introduced nuclear reporter genes was confirmed by measuring CAT and GUS reporter enzymes in bombarded onion and maize tissue (Klein et al. 1987, 1988a). Stable genetic transformation of the tobacco nucleus (Klein et al. 1988b), yeast mitochondria (Johnston et al. 1988) and the chloroplasts in Chlamydomonas (Boynton et al. 1988; Blowers et al. 1989) and higher plants (Svab et al. 1990) followed in rapid succession. Early protocols for biolistic DNA delivery involved precipitation of the transforming DNA with $CaCl_2$ and spermidine free base on the surface of microscopic (0.6–1.0 μm) tungsten or gold particles, and accelerating the particles using a gunpowder-charge driven device to speeds that enable penetration of multiple cell layers. Acceleration of particles was carried out in vacuum in the PDS-1000 gun and solid support of the bombarded cells was provided in the form of a filter paper facilitating particle penetration. All these important elements for success were identified early on (Klein et al. 1987, 1988a). A cleaner, more efficient device is PDS-1000/He in which helium replaces the role of the gunpowder charge (Ye et al. 1990). A useful recent addition to the PDS-1000/He device is the hepta adaptor enabling simultaneous bombardment with seven macrocarriers.

An alternative particle gun design is the Particle Inflow Gun (PIG) that also uses pressurized helium in combination with a partial vacuum to accelerate DNA-coated tungsten or gold particles (Finer et al. 1992). The particles in the PIG are accelerated directly in a helium stream rather than being supported by a macrocarrier, as in the PDS1000/He gun. Because the PIG is not available commercially, it is relatively rarely used. However, it appears to be as efficient as the PDS1000/He gun for plastid transformation (Dufourmantel et al. 2004, 2007).

The targets for plastid transformation by biolistic DNA delivery most often are plastids in leaves (Svab et al. 1990; Svab and Maliga 1993) or less frequently in tissue culture cells (Langbecker et al. 2004). Osmotic stabilizers in some instances are used to protect tissue culture cells during bombardment, although the efficiency of protection has not been rigorously proven (Langbecker et al. 2004).

Historically, biolistic DNA delivery to plastids was optimized using transient expression of GUS and CAT reporter enzymes expressed from plastid signals (Daniell et al. 1990; Ye et al. 1990); for review see (Sanford et al. 1993). Only a small fraction of overall activity detected in these experiments is likely to derive from plastids because genes in plastid cassettes are also expressed in the nucleus (Cornelissen and Vandewiele 1989). The nucleus is transformed 20–40-times more efficiently than plastids (Langbecker et al. 2004) and initial plastid expression from a few transformed ptDNA copies is only a fraction of protein levels measured at the homoplastomic state. Therefore, these experiments likely determined conditions for DNA delivery to the plant nucleus rather than to plastids. Protocols detecting DNA delivery to the nucleocytosolic compartment are still useful to identify conditions for plastid transformation, because delivery of DNA into the cell is sufficient to obtain plastid transformation (see PEG-mediated plastid transformation below). Only one systematic study of biolistic DNA delivery was carried out that measured the success of DNA delivery by the number of transplastomic clones (Langbecker et al. 2004). The number of transplastomic clones obtained with 0.6 or 1.0 μm particles in tissue culture cells and in leaves was comparable, ~1 per bombarded sample. However, plastid transformation in tobacco tissue culture cells with the smaller 0.4 μm particles was 3–4-times more efficient that with the standard 0.6–1.0 μm particles, yielding ~4 transplastomic clones per bombarded sample. Detailed protocols are available for biolistic transformation of tobacco leaf cells (Bock 2001; Lutz et al. 2006b; Lutz and Maliga 2007a; Maliga and Svab 2011) and tissue culture cells (Langbecker et al. 2004).

B. Polyethylene Glycol Treatment

Plastid transformation by polyethylene glycol (PEG) treatment of protoplasts utilizes the empiric DNA uptake process developed for nuclear gene transformation (Paszkowski et al. 1984). PEG treatment was first used to demonstrate transient expression of the introduced GUS reporter gene in isolated tobacco chloroplasts (Sporlein et al. 1991), followed by stable genetic transformation of the plastid genome in *Nicotiana tabacum* (Golds et al. 1993) and *Nicotiana plumbaginifolia* (O'Neill et al. 1993). A detailed protocol for PEG-mediated transformation of plastids in tobacco protoplasts is available (Koop et al. 1996).

Because of its ease of application, biolistic DNA delivery is by far the most frequently used method for plastid transformation. Protoplast isolation, PEG treatment and plant regeneration from protoplasts require more training and are more laborious and time-consuming. However, plastid transformation by PEG treatment is in the public domain and does not require expensive equipment, thus it may be preferable to biolistic DNA delivery in some applications (Dix and Kavanagh 1995).

III. Marker Genes

The challenge of plastid transformation has been to uniformly alter the hundreds to thousands of plastid genome copies localized in ten to hundreds of organelles in a plant cell. DNA delivery produces only a few transformed ptDNA copies, which are then selectively amplified while the cells are grown in tissue culture. Selection for transformed plastid genomes is essential to recover genetically uniform transplastomic plants. Tobacco shoots regenerated from a bombarded leaf are always chimeric. Two cycles of plant regeneration on a selective medium, coupled with probing total cellular DNA for the uniformity of ptDNA, is typically sufficient to obtain genetically stable plants. Repeated cycles of plant regeneration are necessary, because cells in different developmental layers in a shoot apex may differ in their segregation patterns of the two plastid genome types. Regeneration of a new shoot apex from a small group of cells on a selective medium is used to obtain genetically uniform, homoplastomic plants (Lutz and Maliga 2008). Alternatively, visual-selective markers may track progress toward the homoplastomic state (Tungsuchat-Huang et al. 2011). Below is a review of the selectable marker genes that are available for the construction of transplastomic clones.

A. Primary Positive Selection

Detoxifying enzymes that enable the growth of cells on a normally toxic medium provide selective advantage to plastids so that they gradually outnumber non-transformed plastids in cells grown in culture. If the cellular target of antibiotic action is known, genes encoding insensitive forms of the cellular target may also be used as selective markers. The selective plastid markers fall in two classes: primary selective markers that confer a selective advantage early on, when only a few ptDNA copies are amplified; and secondary selective markers that confer protection only when a significant portion of ptDNA copies already carry the marker (see below).

Most primary selective agents are selective inhibitors of plastid protein synthesis on the prokaryotic type (70S) ribosomes, which do not affect mRNA translation on the eukaryotic 80S ribosomes in the cytoplasm. The group of antibiotics that can be used as a primary selective agent includes spectinomycin, streptomycin, kanamycin and chloramphenicol. These antibiotics inhibit greening, cell division and shoot formation in culture on a shoot regeneration medium. Transplastomic clones can be identified by the absence of phenotypes associated with antibiotic treatment of wild-type cells, that is, they show greening, faster proliferation and shoot formation on an antibiotic-containing plant regeneration medium. The first transplastomic clones were obtained by spectinomycin selection for mutant forms of the 16S rRNA, which do not bind the antibiotic (Svab et al. 1990; Staub and Maliga 1992, 1993). The mutant *rrn16* genes in the plastid transformation vectors were soon replaced with the more efficient *aadA* gene encoding aminoglycoside 3 -adenylyltransferase or AAD (Svab and Maliga 1993). AAD inactivates both spectinomycin and streptomycin. Resistance to both antibiotics is exploited to distinguish relatively frequent spontaneous spectinomycin resistant mutants from transplastomic clones, because only transplastomic clones, but not plastid rRNA mutants, are resistant to both antibiotics.

Kanamycin resistance has also been suitable to recover transplastomic clones. The first plastid-engineered kanamycin resistance (*neo*) genes were relatively inefficient (Carrer et al. 1993), but increasing expression of the encoded enzyme neomycin phosphotransferase II (NPTII) yielded marker gene variants that are as efficient as *aadA*, yielding about one transplastomic clone per bombarded sample (Lutz et al. 2004). Kanamycin resistant clones were also recovered by selection for the *aph(3′)IIa* gene (Huang et al. 2002).

There are two recent additions to the primary selective plastid markers, both of which were tested in tobacco. One of the new selective agents is chloramphenicol, inhibiting translation on plastid ribosomes as do spectinomycin and kanamycin. Chloramphenicol resistance appears to be less robust than spectinomycin or kanamycin resistance, because selection in tobacco should be carried out in low light and the color change is more subtle (Li et al. 2011). A distinct advantage of the marker is the absence of spontaneous chloramphenicol resistance mutants. The second marker system explored selection for the feedback-insensitive anthranilate synthase (AS) alpha-subunit gene of tobacco (ASA2) that confers resistance to the indole analogue 4-methylindole (4MI) or the tryptophan analogue 7-methyl-DL-tryptophan (7MT) (Barone et al. 2009). Testing of the new markers in additional plant species will be necessary to fully assess their utility.

Selection for betaine aldehyde (BA) resistance after transformation with a vector carrying a spinach betaine aldehyde dehydrogenase (*badh*) gene was reported to be efficient for the recovery of transplastomic clones (Daniell et al. 2001; Verma and Daniell 2007). The betaine aldehyde dehydrogenase enzyme (BADH) converts toxic BA to betaine, an osmoprotectant accumulating in some plants in dry or saline environments. Attempts to duplicate the selection protocol in other laboratories were unsuccessful, as discussed in a recent review (Maliga 2004). Because no plants were described in the literature that carry *badh* as the only selective marker (*badh* was always combined with *aadA*), for the time being, *badh* should be considered a putative marker only.

B. Secondary Positive Selection

Protection conferred to plant cells by secondary selective markers is dose dependent. These markers are not suitable to enrich for transplastomic plastids when only a few ptDNA copies are transformed, but will confer a selective advantage when many or most genome copies carry the marker gene. Examples for secondary selective marker genes are those that confer resistance to the herbicides phosphinothricin (PPT; (Lutz et al. 2001; Ye et al. 2003)), glyphosate (Ye et al. 2003), sulfonylurea, pyrimidinylcarboxylate (Shimizu et al. 2008) and diketonitrile (Dufourmantel et al. 2007). Low level expression of the protective enzyme from the few initially transformed ptDNA copies, as opposed to full expression from a nuclear transgene may explain why these markers are suitable to directly recover nuclear transformants, but require enrichment to recover transplastomic clones. Subcellular localization of the protective enzymatic activity may also be a contributing factor.

Actinonin is a selective and potent inhibitor of plant peptide deformylases (Fernandez-San Millan et al. 2011). Expression of the *Arabidopsis thaliana* peptide deformylase PDF1B (linked to spectinomycin resistance) in tobacco chloroplasts conferred actinonin resistance to the transformed plants. However, when the combination of the PDF1B gene and actinonin was used as the primary selective marker system for chloroplast transformation, all developed shoots were escapes. Therefore, the use of this system would be limited to the role of a secondary selective marker (Fernandez-San Millan et al. 2011).

C. Negative Selection

Negative selection is also available in plastids. It selects for the loss of a conditionally toxic gene. Negative selection in plastids is based on the expression of the cytosine deaminase enzyme making the cells sensitive to 5-fluorocytosine. The loss of the bacterial *codA* gene (encoding cytosine deaminase) could be detected by cellular proliferation on 5-fluorocytosine-containing medium (Serino and Maliga 1997; Corneille et al. 2001).

D. Visual Plastid Marker Systems

Because the plants that are expressing selectable marker gene have no visual phenotype, the uniform transformation of plastid genomes (= homoplastomic state) can be verified only by DNA gel blot analyses and the absence of segregation in the seed progeny. Since deletion of most plastid genes causes a dramatic change in leaf color, changes in chlorophyll content have been utilized as a marker system to facilitate rapid identification of plastid genotypes. The Koop laboratory (Klaus et al. 2003) developed a system for the rapid identification of transplastomic sectors using pigment-deficient tobacco knockout plants as recipients. In the knockout plants, the first plastid marker (*aadA*, encoding spectinomycin resistance) replaces a plastid gene that causes chlorophyll deficiency. The second transformation vector carries the photosynthetic gene to restore green pigmentation linked to a second marker (*aphA-6*, encoding kanamycin resistance). Homoplastomic sectors and plants can be readily identified by the restoration of green pigmentation among plants selected for kanamycin resistance.

Variants of this protocol have been developed that require only one selectable marker and are directed towards manipulation of *rbcL*, the plastid-encoded Rubisco large subunit gene in tobacco. In one approach (Kode et al. 2006), deletion of the plastid *rbcL* gene was obtained by homology-based deletion using a two-step protocol. First, selection for spectinomycin resistance (*aadA*) was used to duplicate the *rbcL* flanking sequence. Subsequently, deletion of *rbcL* and the linked *aadA* by a (spontaneously occurring) homologous recombination event was recognized in the seed progeny by appearance of the pigment-deficient phenotype. The *rbcL* deletion line could subsequently be transformed with a functional *rbcL* allele linked to *aadA*. The homoplastomic sectors (plants) could be readily identified by their green pigmentation. In a variant approach (Whitney and Sharwood 2008), the tobacco *rbcL* gene was replaced with a heterologous *rbcL* sequence using *aadA* as a selective marker. The *aadA* gene was subsequently removed by the Cre site-specific recombinase, so the master line was ready to be transformed with *rbcL* variants using *aadA* as a selective marker.

The visual marker system discussed above relies on pigment deficiency caused by a missing or defective plastid gene. Our novel visual marker system relies on interference of a plastid transgene with the expression of the *clpP* plastid gene. The transgene acts as a "poison pill" because it contains a *clpP* segment that interferes with the maturation of the native *clpP* mRNA (Kuroda and Maliga 2002). So far, two variants of the visual marker have been tested: the aurea *bar* (*bar*^{au}) (Kittiwongwattana et al. 2007; Lutz and Maliga 2008) and *aadA*^{au} (Tungsuchat-Huang et al. 2011) transgenes conferring a golden leaf phenotype to plants. Because the *bar*^{au} gene is not a primary selectable marker, its deployment requires two genes: the aurea *bar* (*bar*^{au}) gene that confers a golden leaf phenotype and a spectinomycin resistance (*aadA*) gene that is necessary for the introduction of the *bar*^{au} gene in the plastid genome. The *aadA*^{au} transgene fulfills both functions: it is a conventional selectable *aadA* gene in culture, and allows detection of transplastomic sectors in the greenhouse by leaf color. Because the aurea plants are viable, the aurea plastid genes are useful to query rare events in large populations (Tungsuchat-Huang et al. 2010).

E. Reporter Genes

The *E. coli* β-glucuronidase (GUS) reporter enzyme facilitates the monitoring of gene expression, but does not confer a selective advantage or disadvantage to plastids. GUS enzymatic activity expressed in chloroplasts has been measured using fluorogenic assays (Staub and Maliga 1993, 1994; Eibl et al. 1999; Zou et al. 2003) and visualized by histochemical staining (Staub and Maliga 1993; Iamtham and Day 2000; Zubko et al. 2004; Sheppard et al. 2008).

The *Aequorea victoria* green fluorescent protein (GFP) is a visual marker, allowing direct imaging of the fluorescent gene product in living cells. Its chromophore forms autocatalytically in the presence of oxygen and fluoresces green when absorbing blue or UV light. GFP has been used to detect transient gene expression (Hibberd et al. 1998) and stable transformation events (Sidorov et al. 1999; Shiina et al. 2000; Reed et al. 2001) in chloroplasts. GFP-expressing chloroplasts in tissue grafts facilitated demonstration of the transfer of genetic material between cells (Stegemann and Bock 2009).

GFP was fused with AAD, the *aadA* gene product that confers spectinomycin resistance, to be used as a bifunctional visual and selective (spectinomycin resistance) marker gene (Khan and Maliga 1999). Transformation vectors carrying the *aadA-gfp* marker gene were used to recover stable transplastomic clones in *N. tabacum* (Khan and Maliga 1999), *N. sylvestris* (Maliga and Svab 2011) and *Lesquerella fendleri* (Skarjinskaia et al. 2003).

Luciferases are enzymes that emit light in the presence of oxygen and a substrate (luciferin) and which have been used for real-time, low-light imaging of gene expression in cell cultures, individual cells, whole organisms, and transgenic organisms. Luciferases have served as reporters in a number of promoter search and targeted gene expression experiments over the last two decades (Greer and Szalay 2002). Until now, expression of various luciferases in plants has required exogenous application of luciferins – frequently toxic and high-cost compounds – to achieve only temporary and relatively low light emission levels from live plant tissues. Evolutionary conservation of the prokaryotic gene expression machinery enabled expression of the six genes of the *lux* operon in chloroplasts yielding plants that are capable of autonomous light emission (Krichevsky et al. 2010). This system now can be modified for gene expression studies and for genetic screens.

IV. Vectors

Plastid transformation vectors consist of a vector backbone for cloning and propagation in *E. coli*, a plastid targeting region with a selectable plastid marker to facilitate integration of the gene-of-interest into the plastid genome, and optional sequences to facilitate marker gene excision. The vector backbones are pUC or pBluescript plasmid derivatives carrying a ColE1 replication origin that ensures plasmid replication in *E. coli* but not in plastids. Because the ColE1 replication origin does not function in plastids, the plastid marker is expressed in the plant cell only if it integrates into the plastid genome. The pUC and pBluescript vectors encode ampicillin resistance as the selectable marker in *E. coli*, which is not a suitable selectable marker in plastids. Spectinomycin, kanamycin or chloramphenicol resistance genes engineered for expression in plastids are also selectable in *E. coli*, therefore dual selection for the bacterial ampicillin resistance and the plastid marker ensures maintenance of intact (deletion-free) copies of plastid vectors.

The plastid-targeting region is a ~0.5–2.0-kb ptDNA fragment flanking the marker gene (and gene of interest) to facilitate integration of the marker gene (and the gene-of-interest) into the ptDNA by two homologous recombination events. The vector design is dependent on the desired ptDNA manipulation that can be insertion of foreign genes, replacement of native plastid genes with mutant forms, gene deletion or cotransformation.

A. Insertion Vectors

Expression of transgenes requires plastid insertion vectors that enable convenient DNA manipulation in *E. coli* and targeted insertion of the gene-of-interest into the plastid genome. Because the insertion vectors are repeatedly used for the insertion of different genes, significant effort has been invested to characterize the insertion site in the plastid genome and endow the vectors with convenient features. Characterization of the insertion site includes, for example, ensuring that there is no interference with the expression of adjacent plastid genes and identification of read-through transcripts that may enhance or reduce transgene expression. Vector convenience features are, for example, convenient restriction sites for cloning, alternative selection markers, and sequences to facilitate post-transformation removal of marker genes. Because vector development requires a significant effort, only a few vectors are used routinely. The pRB94/95 vectors (Ruf et al. 2001) and our pSS24/25 vectors (Sinagawa-Garcia et al. 2009) target transgenes in the single-copy region of the plastid genome, whereas our pPRV vector series (Zoubenko et al. 1994; Lutz et al. 2007) and the pSBL-CTV2 vectors (Daniell et al. 1998) target insertions in the repeated region of the plastid genome. Insertion of transgenes in the repeated region yields ptDNA with two transgene copies per genome.

When choosing plastid-targeting sequences for vector construction, DNA sequence variation within species and between species is a concern. Ideally, vectors should contain sequences identical to the target ptDNA for optimal recombination. Targeting regions with point mutations in synthetic DNA behave as homologous sequences; the recombination sites are at either ends of the targeting region (Sinagawa-Garcia et al. 2009). Some degree of sequence variation is tolerated as long as sufficiently extensive regions of homology are present. In a now classic study, transformation of *N. tabacum* plastids with *Solanum nigrum* vectors has shown that transformation with 97.6% similar (homeologous) sequences (sequence divergence 2.4%) is as efficient as with identical sequences (Kavanagh et al. 1999). Vectors with *N. tabacum* targeting sequences are used to transform plastids in potato (Sidorov et al. 1999), tomato (Ruf et al. 2001), petunia (Zubko et al. 2004) and *N. sylvestris* (Maliga and Svab 2011). The plastid genomes of the amphiploid species *Nicotiana tabacum* and its maternal progenitor *N. sylvestris* differ only by seven sites: three in introns, two in spacer regions and two in coding regions (Yukawa et al. 2006). None of the known differences are within the plastid targeting regions of our standard pPRV or pSS24/25 vectors and, even if they were, the point mutations and insertions/deletions (affecting one or two nucleotides) would not significantly affect transformation efficiency. However, replacement of tobacco-specific vectors (sequence divergence 4.6%) with potato-specific vectors increased potato plastid transformation efficiency 10-fold (Valkov et al. 2011). Thus, construction of species-specific, or even line-specific, vectors is advisable, if there is significant intraspecific variation in the ptDNA. Sequencing the plastid genomes of two tomato cultivars (IPA-6 and Ailsa Craig) revealed that they are identical to the nucleotide (Kahlau et al. 2006); thus, one vector for tomato should be sufficient. However, significant sequence variation in the ptDNAs of rice subspecies (Tang et al. 2004) may justify construction of multiple plastid transformation vectors for rice.

There is only limited information on the importance of choosing homologous expression signals for transgene expression. In most plastid transformation vectors the marker genes are driven by the "heterologous" tobacco *rrn* operon PEP promoter. Because the *rrn* PEP promoter elements are conserved between dicots and monocots (with the only know exception being spinach; (Sriraman et al. 1998; Suzuki et al. 2003)), this promoter is not really heterologous. However, the efficiency of expressing recombinant proteins from the *psbA* promoter appears species specific (Ruhlman et al. 2010). Systematic testing of the utility of expression signals in heterologous systems will be an important area for future research.

The general insertion vectors have only a marker gene and a linked multicloning site. Specialized vectors, in addition, have a gene of interest on which one element, for example the promoter, can be readily exchanged to create a series of constructs. Such specialized vectors are the vectors developed to study plastid RNA editing. Three approaches were used. Conceptually the simplest design was construction of minigenes that were obtained by inserting in a plastid expression cassette a DNA fragment that contains (an) editing site(s) (Reed and Hanson 1997). The second approach, translational fusion with a reporter gene was used to study the *psbL* and *ndhD* editing events that create an AUG translation initiation codon by editing of an ACG codon at the mRNA level (Chaudhuri and Maliga 1996). The third approach was incorporation of editing segments in the 3′UTR of the *aadA* marker gene where the editing status of the segment does not affect expression of the marker gene (Bock et al. 1996). For a review of plastid editing vectors, see (Lutz and Maliga 2007a).

B. Replacement Vectors

Replacement vectors are variants of insertion vectors, when the sequence to be inserted is already present in the ptDNA and the intent is to replace the native sequence with a variant gene (mutant allele) incorporated in the vector targeting region. Replacement vectors are individually tailored to engineer specific genes. Replacement vectors have been developed for engineering *rbcL*, the gene encoding the large subunit of the Rubisco enzyme. Significant similarity between the native sequence and the variant, such as the tobacco and sunflower *rbcL* genes allowed undesirable recombination within the *rbcL* gene (Kanevski et al. 1999). To avoid this, the target gene sequence was either deleted (Klaus et al. 2003) or replaced with a dissimilar sequence (Whitney and Sharwood 2008), and the knockout/engineered plant is then used a master recipient for gene replacement. Efficient recovery of transplastomic clones was facilitated by restoration of green pigmentation, as discussed in Sect. III.D.

C. Deletion Vectors

Deletion vectors are designed to create knockout lines lacking specific plastid genes by replacing the target gene with a selectable marker gene by homologous recombination via the flanking ptDNA sequences. Knockout lines could be obtained for most plastid genes. For example, deletion of the plastid *rbcL* or *rpoB* genes makes the plants pigment deficient, but the knockout plants can be maintained on sucrose-containing medium or by grafting onto wild-type plants. In some instances, for example in the case of the plastid *ndh* genes, the knockout phenotype does not significantly interfere with photosynthesis and viability, while in other cases, for example *clpP1*, the plastid genes are essential for viability even on sucrose-containing medium. For reviews see (Bock 2001; Maliga 2004) and Chap. 18 in this volume.

D. Cotransformation

Cotransformation is a process when transformation is carried out with two (or more) vectors, targeting multiple regions of the plastid genome. At least one of the vectors carries a selectable marker gene so that transplastomic clones can be recovered by selection. Because bombardment is carried out with mixed plasmids and integration of both plasmids is efficient, ~20% of the clones selected by the antibiotic resistance encoded in one vector will carry integrated copies of the second vector lacking a selectable marker (Carrer and Maliga 1995). Cotransformation has been exploited to tag an unlinked *ndh* gene (Rumeau et al. 2005) and to obtain marker-free herbicide resistance plants (Sect. V.D, Ye et al. 2003).

V. Marker Excision

The marker genes are essential for the selective enrichment of rare transformed ptDNA copies. However, when uniform transformation of ptDNA copies is achieved, the marker gene is no longer necessary to maintain the transplastomic state. Reasons for posttransformation removal of marker genes are: the shortage of primary selectable markers (spectinomycin selection for *aadA* is by far the most convenient), high-level expression of the marker genes imposing a metabolic burden on the plant, and consumer acceptance. There are four principal protocols for marker excision, each of which requires a special vector design discussed below. For reviews, see (Lutz and Maliga 2007b; Day and Goldschmidt-Clermont 2011).

A. Repeat-Mediated Excision

Repeat-mediated marker excision, developed in Anil Day's laboratory, requires flanking the sequence targeted for deletion by a duplicated segment of at least a few hundred base pairs. The duplicated structure is unstable, and homologous recombination will eventually result in deletion of the sequence between the repeats. The advantage of homology-based marker excision is that it is seamless, leaving behind no extraneous sequence. However, repeat-mediated marker excision is difficult to control, because deletion may take place in *E. coli* during cloning or during transformation before reaching the homoplastomic state (Iamtham and Day 2000; Day and Goldschmidt-Clermont 2011). Homology-based marker excision has been used in soybean to obtain marker-free herbicide-resistant plants (Dufourmantel et al. 2007).

B. Excision by Phage Recombinases

Marker excision by phage site-specific recombinases is a two-step process: first, transplastomic plants are obtained in the absence of recombinases and, when marker excision is desired, plastid-targeted recombinases are expressed in the cells (Lutz and Maliga 2007b). To set up the lines for marker excision, the P1 phage *loxP* site (Corneille et al. 2001; Hajdukiewicz et al. 2001) or the phiC31 phage *attP*/*attB* sites (Kittiwongwattana et al. 2007) flank the marker genes in the plastid transformation vectors. The plastid genomes carrying target site-flanked marker genes are stable in the absence of recombinases (Tungsuchat-Huang et al. 2010). However, excision of the marker genes is very efficient when the gene of the plastid-targeted recombinase is introduced into the nuclear genome by transformation or crossing (Corneille et al. 2001; Hajdukiewicz et al. 2001; Kittiwongwattana et al. 2007), or transiently from Agrobacterium T-DNA (Lutz et al. 2006a). When using phage site-specific recombinases, a copy of the recombinant target site is left behind in the plastid genome.

C. Transient Cointegration

The third approach is the so-called transient cointegration protocol, in which the marker gene is outside the plastid targeting region of the transformation vector (Klaus et al. 2004). Placing the marker gene outside the targeting region enables selection for a cointegrate structure that forms by recombination between the ptDNA and the transformation vector *via* only one of the plastid targeting regions. As the result, the entire vector is incorporated in the ptDNA. When selection for the antibiotic resistance marker is stopped, recombination *via* the second targeting region can take place and the marker gene is excised. This marker excision system is also seamless, and antibiotic selection provides a degree of control.

D. Cotransformation and Segregation

Marker-free herbicide resistance plants have been obtained after transformation with mixed plasmids and a consecutive two-step selection process (Ye et al. 2003). The transformed plastids were first selected on spectinomycin-containing medium to identify clones, which were grown from cells bombarded with mixed plastids. A significant fraction of plastid genome copies in these cells carried integrated herbicide resistance genes targeted to a second integration site. Glyphosate or phosphinothricin are not suitable for the recovery of transplastomic clones when present in only a few copies in a cell, as discussed in Sect. III.B. However, spectinomycin resistance enabled propagation of integrated herbicide-resistance genes so that they could be directly selected for during a second cycle of plant regeneration. Some of the ptDNA copies carrying integrated herbicide resistance genes do not have integrated copies of the spectinomycin resistance gene, thus enabling segregation of spectinomycin marker-free plants (Ye et al. 2003).

VI. Flowering Plant Species with Systems for Plastid Transformation

Identification of transplastomic tobacco lines is based on two general criteria: greening of transplastomic cells (chlorophyll accumulation) on the selective medium that normally inhibits growth and chlorophyll accumulation, and capacity for regeneration from cultured cells so that homoplastomic cells can be obtained during repeated cycles of plant regeneration. The key to extending plastid transformation to new species has been combining a species-specific regeneration protocol with antibiotic treatment that blocks greening and tissue proliferation. Below is a brief review of the state of the art of plastid transformation in the different taxonomic groups. Highlighted in the crop species section will be (1) the laboratories making significant contributions to technology development, (2) the choice of methods for DNA introduction, (3) the marker genes used for selection, (4) the cultivars or accession in which the methods have been tested, (5) the salient features of the system and (6) its main uses.

A. Tobacco: Nicotiana tabacum and Other Species in the Genus Nicotiana

N. tabacum cv. Petit Havana was the first tobacco cultivar in which we reported plastid transformation with a mutant *rrn16* gene in 1990 (Svab et al. 1990). The recessive *rrn16* gene was soon replaced with the dominant *aadA* gene that is more efficient yielding about one transplastomic clone per bombarded sample (Svab and Maliga 1993). To date, virtually all tools and protocols for plastid transformation have been developed using this cultivar (for details, see sections above). The most commonly used protocols employ shoot regeneration from bombarded leaf tissue (Lutz et al. 2006b; Lutz and Maliga 2007a), although a protocol for transforming proplastids in tissue culture cells was also described (Langbecker et al. 2004). Plastid transformation in other Nicotiana species with a similar tissue culture response could be readily duplicated using *N. tabacum* cv. Petit Havana protocols, including *Nicotiana plumbaginifolia* (O'Neill et al. 1993), *Nicotiana benthamiana* (Davarpanah et al. 2009) and *Nicotiana sylvestris* TW137 (Maliga and Svab 2011). The cv. Petit Havana plants are relatively small and flower early. To obtain plants with a larger biomass, plastid transformation has been extended to additional tobacco cultivars, including Wisconsin 38 (Iamtham and Day 2000), Xanthi, Burley (Lee et al. 2003), Samsun, K327 (22X-1; (Yu et al. 2007)) and Maryland Mammoth (McCabe et al. 2008). *N. tabacum* is the model species of plastome engineering and is widely used in basic science studies and for biotechnological applications (Daniell et al. 2009; Cardi et al. 2010; Day and Goldschmidt-Clermont 2011; Maliga and Bock 2011; Whitney et al. 2011).

B. Potato: Solanum tuberosum

Plastid transformation in potato was reported by the Monsanto group (Sidorov et al. 1999) in FL1607, a highly regenerable, non-commercial potato line. Transformation was carried out with tobacco-specific vectors, which carried tobacco ptDNA fragments to target insertions into the potato ptDNA. The vectors carried *aadA* as a selectable marker and shoot regeneration was carried out in the presence of spectinomycin (300 mg/L). The yield of transplastomic clones was lower than in tobacco, one transplastomic clone in 15–30 bombarded leaf samples. Comparably low plastid transformation efficiency was obtained with the *aadA* marker gene, spectinomycin selection (300 mg/L) and tobacco-specific vectors in *Solanum tuberosum* cv. Desiree, a commercial cultivar (Nguyen et al. 2005). Transplastomic clones in FL1607 were recovered in a single-step regeneration protocol as in tobacco. In cv. Desiree, a two-step procedure was adopted: selection was first carried out on a callus-induction medium, then on shoot-induction medium. A dramatic, ~10-fold increase in transformation efficiency was obtained when the tobacco-specific targeting sequences were replaced with potato-specific targeting sequences in cv. Desiree, using an improved two-step procedure yielding about one transplastomic clone per bombarded sample (Valkov et al. 2011). Leaf bombardment was carried out on a medium containing 0.1 M sorbitol and 0.1 M mannitol as osmoticum. Because the transplastomic clones were grown for a long time (3–4 months) as callus before plant regeneration, almost all (92%) of the regenerated plants were homoplastomic. GFP in transplastomic leaves accumulated up to 3–5% of total soluble protein as compared to 0.02–0.05% in tubers (Sidorov et al. 1999; Valkov et al. 2011) indicating that optimization of protein expression is required if expression of recombinant proteins in potato tuber amyloplasts is the goal.

C. Tomato: Solanum lycopersicum

Plastid transformation in tomato has been developed in Ralph Bock's laboratory using biolistic DNA delivery, tobacco-specific vectors carrying the *aadA* marker gene and spectinomycin selection (500 mg/L; (Ruf et al. 2001)). Transformation has been carried out in two South American varieties: Santa Clara and IPA-6 (Wurbs et al. 2007; Zhou et al. 2008; Apel and Bock 2009). Plastid transformation in tomato has also been obtained by PEG-treatment of protoplasts, using tobacco (*N. tabacum*) or *Solanum nigrum*-specific vectors carrying binding-type spectinomycin and streptomycin resistance markers in the *rrn16* genes and selection for spectinomycin resistance (300 mg/L). Transformation was carried out in the tomato processing cultivar T1783 (Nugent et al. 2005). Although plastid transformation in tomato has been significantly improved over time (Wurbs et al. 2007; Zhou et al. 2008), initial construct optimization in the well-established tobacco system is advisable.

Applications of tomato plastid transformation include engineering the carotenoid metabolic pathway and expression of antigens for subunit vaccines (Wurbs et al. 2007; Zhou et al. 2008; Apel and Bock 2009). Some of the recombinant proteins (p24-Nef) accumulated to up to 40% of the total soluble cellular protein in tomato leaves, but no significant protein accumulation was detected in ripe tomato fruits suggesting that protein expression in chromoplasts will require a specialized expression system (Zhou et al. 2008). The presumably relatively low enzyme levels were sufficient for successful metabolic pathway engineering (Wurbs et al. 2007; Apel and Bock 2009).

D. Petunia: Petunia hybrida

Plastid transformation in petunia has been reported from Anil Day's laboratory (Zubko et al. 2004). Tobacco-specific transformation vectors carrying an *aadA* gene were introduced into petunia leaves by the biolistic process, and transplastomic shoots were regenerated on a medium containing spectinomycin (200 mg/L) and streptomycin (200 mg/L). Transformation was carried out in the Pink Waive commercial cultivar. *Petunia hybrida* is a diploid species that is suitable to study the biology of flowering plants using transgenic approaches (Gerats and Vandenbussche 2005; Gillman et al. 2009). Thus, applications of plastid transformation in Petunia are expected to follow.

E. Eggplant: Solanum melongena

Plastid transformation in eggplant was developed in K.C. Bansal's laboratory (Singh et al. 2010). Tobacco-specific vectors carrying the *aadA* marker gene were introduced into green stem segments by the biolistic process and transplastomic shoots were regenerated on spectinomycin-containing medium (300 mg/L) using a one-step protocol. Initial selection on spectinomycin was followed up by selection for spectinomycin and streptomycin (300 mg/L each). Plastid transformation was essentially carried out as in tobacco, except that the transforming DNA was introduced into green stem segments instead of leaves.

F. Soybean: Gycine max

Soybean was the first major agronomic crop in which plastid transformation was implemented by a group of researchers at Bayer Crop Science (Dufourmantel et al. 2004). Plastid transformation was achieved by biolistic delivery of soybean-specific vectors carrying an *aadA* gene and the transplastomic clones were identified by their green color on spectinomycin medium in cv. Jack. The Bayer group used the particle inflow gun (PIG), rather than the DuPont biolistic gun. Noteworthy about the soybean system is that the transforming DNA was introduced into green embryogenic calli. The green embryogenic callus bleached in the presence of 200 or 300 mg/L spectinomycin, so that the resistant clones could be identified by their green color. The green embryogenic calli were then converted into embryos on a suitable medium in the presence of spectinomycin (150 mg/L). After 2 months on the embryo induction medium, the embryos were transferred to an embryo-germination medium containing spectinomycin (150 mg/L). Interestingly, soybean is naturally resistant to high concentrations (800 mg/L) of streptomycin. Plastid transformation in soybean is a good example for combining a crop-specific plant regeneration protocol with spectinomycin color selection. Another salient feature of the soybean system is the absence of spontaneous spectinomycin-resistant mutants. This may be the case because the mutations that would confer spectinomycin resistance are not compatible with ribosome function. The third salient feature of the soybean system is the absence of wild-type ptDNA copies in the regenerated plants, as in potato. The uniform population of transformed ptDNA copies in the regenerated plants is likely to be due to protracted cultivation on the selective medium prior to plant regeneration. Construction of insect resistant (Dufourmantel et al. 2005) and herbicide resistant (Dufourmantel et al. 2007) transplastomic soybean plants confirmed the utility of plastid transformation in soybean. Soybean is the most important agronomic crop in which reproducible plastid transformation is currently available.

G. Alfalfa: Medicago sativa

Plastid transformation of alfalfa has been accomplished in Shaochen Xing's laboratory using biolistic delivery of a homologous, *aadA*-carrying vector to leaves (Wei et al. 2011). The tissue culture system for cv. Longmu 803 used a typical multi-stage medium for embryo induction, multiplication, germination and rooting. Selection was carried out in the presence of 500 mg/L spectinomycin. Because alfalfa is edible and is used as feedstuff to livestock, it is a suitable crop for oral delivery of vaccines and therapeutic proteins.

H. Lettuce: Lactuca sativa

Two groups reported plastid transformation in lettuce at about the same time. Cilia Lelivelt, Jackie Nugent and a group of collaborating researchers from Rijk Zwaan Breeding, B.V., Fijnaart, The Netherlands and The National University of Ireland, Maynooth, reported plastid transformation in cv. Flora. Transformation was carried out with a homologous lettuce vector carrying an *aadA* gene that was introduced into protoplasts by PEG treatment. Transplastomic clones were identified on a medium containing 500 mg/L spectinomycin.

Kanamoto and colleagues (Kanamoto et al. 2006) described plastid transformation in cv. Cisco after biolistic delivery of a lettuce-specific vector into leaves carrying an *aadA* gene, and shoot regeneration on spectinomycin-containing medium. The levels of selective spectinomycin concentrations were 10× lower than in tobacco, 50 mg/L. The efficiency of plastid transformation was comparable to tobacco (one transplastomic clone per bombarded sample), and the level of GFP was very high, ~36% of total soluble cellular protein. In the meantime, the same group (Lim et al. 2011) has extended plastid transformation to a different lettuce cultivar, Romana.

By 2010, the Daniell laboratory developed an efficient transformation and regeneration protocol for cv. Simpson Elite (Ruhlman et al. 2010). Contributing to the success were (1) adoption of native targeting sequences and regulatory sequences, and (2) cultivar-specific optimization of the regeneration medium to produce transplastomic shoots by direct organogenesis. The Daniell group successfully used the lettuce system for the expression of various recombinant proteins (Ruhlman et al. 2007; Boyhan and Daniell 2011; Kanagaraj et al. 2011). The advantage of the system is that lettuce is edible raw, thus it is suitable for oral delivery of therapeutic proteins and vaccines.

I. Cabbage: Brassica oleracea and Other Species in the Brassicacae Family

The plastids in several species in the mustard (Brassicaceae) family have been the targets of plastome engineering. The plastid transformation vectors carried *aadA* genes and identification of transplastomic clones was based on spectinomycin resistance. Selective concentrations of spectinomycin, in most cases, were lower (10–60 mg/L) than the concentrations used in tobacco (500 mg/L).

Plastid transformation of oilseed rape (*Brassica napus*) was carried out by bombardment of green cotyledon petioles (Hou et al. 2003) or cotyledons of cv. FY-4; (Cheng et al. 2010) and selected on a medium containing 10 mg/L spectinomycin. The regenerated plants were heteroplastomic, a problem that can be addressed by repeated cycles of plant regeneration, segregating away the wild-type copies in the seed progeny, or choosing alternate insertion sites to avoid interference with flanking genes.

Genetically stable, homoplastomic lines have been described in two other species in the Brassicaceae family. Plastid transformation in cauliflower (*Brassica oleracea* var. *botrytis*) by PEG treatment and selection on 20–60 mg/L spectinomycin yielded a single homoplastomic plant (Nugent et al. 2006). Plastid transformation has also been reported in *Lesquerella fendleri* (Gray) Wats A14581, a species with a desirable seed oil composition and a high capacity for plant regeneration from leaves (Skarjinskaia et al. 2003). Leaf bombardment with *aadA* vectors and selection on spectinomycin (400 mg/L) yielded fertile, homoplastomic plants. Plastid transformation was relatively inefficient: in 51 bombarded leaf samples, only two transplastomic clones were obtained, possibly due to the use of heterologous vectors. Surprising was the large number (110) of spontaneous mutants in the experiment.

Systematic research in the laboratory of Menq-Jiau Tseng led to the establishment of a reproducible system for plastid transformation in cabbage (*Brassica oleracea* L. var. *capitata* L.) (Liu et al. 2007). The protocols have been implemented in cultivars K-Y cross and Summer Summit. Biolistic DNA delivery with homologous, *aadA*-containing vectors was followed by initial selection on 50 mg/L spectinomycin, followed by cultivation on 200 mg/L spectinomycin. The utility of plastid transformation in cabbage was demonstrated by the expression of insecticidal *cry1Ab* protein gene in chloroplasts (Liu et al. 2008).

J. Thale Cress: *Arabidopsis thaliana*

Arabidopsis thaliana is also a member of the mustard family (Brassicaceae). We obtained plastid transformation in Arabidopsis by combining the tobacco leaf transformation protocol with the two-step (callus induction, plant regeneration) Arabidopsis tissue culture and plant regeneration protocols (Sikdar et al. 1998). Because the leaf cells in Arabidopsis are polyploid, we obtained sterile plants. However, the meristematic cells in a shoot apex or cells of a developing embryo are diploid. To maintain the diploid state in our culture, we developed an embryogenic culture system for plastid transformation in Arabidopsis by regulated expression of the BABY BOOM transcription factor (Lutz et al. 2011). This investment has yet to yield fertile transplastomic plants.

K. Sugar Beet: *Beta vulgaris*

Plastid transformation in sugar beet was reported from Michele Bellucci's laboratory following biolistic DNA delivery to leaf petioles using a homologous vector, and selection in the presence of 50 mg/L spectinomycin (De Marchis et al. 2009). Interestingly, like soybean, sugar beet is also naturally resistant to high concentrations (1,000 mg/L) of streptomycin. The transplastomic clones appeared after 5 months of selection in the Z025 line. Plant regeneration was obtained only after spectinomycin was removed from the medium. The regenerated plants were heteroplastomic; however, two additional rounds of shoot regeneration in the presence of low (12.5 mg/L) spectinomycin concentrations yielded homoplastomic plants. Overall, in this first experimental series, it took 14 months to obtain transplastomic sugar beet plants. Sugar beet is an important industrial crop of the temperate zone in which chloroplast DNA is not transmitted through pollen, like in most flowering plant species. Plastid localization of transgenes could alleviate concerns about gene flow in the field due to the well documented cross-compatibility of sugar beet with its wild relative sea beet (*B. vulgaris* ssp. *maritima*; (De Marchis et al. 2009)).

L. Carrot: *Daucus carota*

Transplastomic carrot (*Daucus carota* cv. Half long) was reported from the Daniell laboratory (Kumar et al. 2004b). Transplastomic carrot was obtained after biolistic DNA delivery of a homologous vector carrying an *aadA* gene, and selection for increasing concentrations (150, 350 and 500 mg/L) of spectinomycin. The transgenic calli had a green colour, attributed to the expression of *badh* transgene introduced by linkage to the *aadA* gene (see Sect. III.A).

M. Poplar: Populus alba

Okamura and colleagues (Okumura et al. 2006) reported plastid transformation in poplar after biolistic DNA delivery of a homologous vector carrying an *aadA* marker gene. The vector DNA was introduced into leaves, and the transplastomic shoots were recovered by selection for spectinomycin resistance (30 mg/L). A significant number of spontaneous plastid-encoded spectinomycin resistant mutants were also obtained. Poplar is a potential biofuel crop, in which plastid transformation may be useful to improve the value of the crop by co-expression of value-added products.

N. Cotton: Gossypium hirsutum

Plastid transformation in cotton was reported from the Daniell laboratory using cv. Coker 310FR (Kumar et al. 2004a). Because spectinomycin was reportedly toxic, after biolistic DNA delivery, the selection of transplastomic clones was carried out on kanamycin. The selective concentration of kanamycin was initially 50 mg/L and then increased to 100 mg/L in subsequent cycles. The Double Gene/Single Selection vector carried the *aphA-6* and *nptII* genes. Because both genes confer resistance to kanamycin and expression of neither of the genes alone has been tested, the rationale behind the approach remains unclear. Although of significant potential economic interest, plastid transformation in cotton has not yet been duplicated nor have transplastomic seeds been distributed for analyses.

O. Cereals: Rice (Oryza sativa) and Wheat (Triticum aestivum)

Cereal plastids are naturally resistant to spectinomycin due to having the 16S rRNA nucleotide substitution that confers spectinomycin resistance to sensitive ribosomes (Fromm et al. 1987). Therefore, we attempted selection for streptomycin resistance with homologous rice vectors carrying an *aadA* gene (Khan and Maliga 1999). Biolistic delivery of the transformation vector into cultured embryogenic cells was followed by selection on streptomycin-containing plant regeneration medium. Because AAD, the *aadA* gene product, was fused with GFP, chloroplast localization of the fusion protein could be detected by fluorescence microscopy. However, in the absence of repeated cycles of plant regeneration, we did not obtain homoplastomic plants. Lee and colleagues (Lee et al. 2006) duplicated the experiment and carried it a step further by demonstrating that the transformed plastids can be transmitted into the next generation. However, they could not find a solution to the problem of obtaining homoplastomic plants from the cultured rice cells.

Chloroplast transformation in wheat (*Triticum aestivum* L.) was reported recently (Cui et al. 2011). The transformation vector was introduced into immature scutella and inflorescences by the biolistic process, and transplastomic clones were selected by resistance to 30, 40 and 50 mg/L of G418 in the first, second and third selection cycles, respectively. Bombardment of ~2,500 scutella and ~600 immature inflorescence sections yielded one homoplastomic and two heteroplastomic plants, a relatively low frequency. Two facts cast doubt on the validity of the claims. (1) Transformation with the vector, as described, results in the deletion of the *atpB* coding region N-terminus. Deletion of *atpB* would results in pigment deficiency in other plants. The transplastomic wheat plants were reported to have a green, wild type phenotype. (2) Probing of total plant cellular DNA with the *rbcL-atpB* targeting region was reported to detect a 2.5-kb *Bam*HI fragment. In the wild type wheat plastid genome (AB042240), the *rbcL-atpB* region is contained in a 9.5-kb BamHI fragment. Because the artificial *Bam*HI cloning sites from the transformation vector are not incorporated in the transplastomic wheat ptDNA, the 2.5-kb signal in Fig. 4 suggests probing plasmid, rather than total plant cellular DNA. If the reported data are true, we shall soon see confirmation of these findings from multiple laboratories.

VII. Perspectives

Plastid transgene expression offers many advantages, but it is still barely utilized. The obvious reason is that, more than 20 years after its first implementation, the technology is still not available in most crops. What needs to be done to accelerate progress?

Relative uniformity of plastid genomes and the acceptance of heterologous vectors within the Solanaceae led us to believe that we did not necessarily need species-specific vectors. The recent example of efficient potato plastid transformation being dependent on homologous vectors is a wake-up call (Valkov et al. 2011). If significant intraspecific sequence diversity turns out to be the rule, we may need to develop multiple vectors for each species, dependent on the tolerance of its recombination system for sequence variation. This calls for more plastid genome sequencing. Fortunately, next generation sequencing provides the tool to rapidly determine the plastid genome sequence from total plant DNA of the cultivar we intend to transform (Nock et al. 2011) and, based on the sequence, we can decide if construction of line-specific vectors is justified. If construction of homologous vectors is required, we can replace cloning by purchasing synthetic targeting regions.

Spectinomycin selection was useful to recover transplastomic clones in many species. Even if spectinomycin selection is feasible in a crop, we need at least one additional marker for multistep engineering. However, some of the crops, such as the cereals, are naturally resistant to spectinomycin. Just finding the right antibiotic or selectable marker gene may solve the problem of obtaining homoplastomic plants in monocots. That is why testing a wider array of antibiotics, for example G418 to which the *neo* (*nptII*) and *aph(3′)IIa* genes confer resistance (Sect. III.A), and new selectable marker genes may be important for extending plastid transformation to new crops.

Plastid transformation is available for the expression of recombinant proteins in the well-established tobacco system and the newly developed, edible hosts lettuce and alfalfa. Extension of the technology to new crops would significantly enhance its utility. The most desirable agronomic application would be containment of herbicide-resistance transgenes and disease resistance traits in the wind-pollinated cereal crops.

Acknowledgments

Research in the author's laboratory was supported by grants from the USDA National Institute of Food and Agriculture Biotechnology Risk Assessment Research Grant Program Award No. 2005-33120-16524, 2008-03012 and 2010-2716.

References

Apel W, Bock R (2009) Enhancement of carotenoid biosynthesis in transplastomic tomatoes by induced lycopene-to-provitamin A conversion. Plant Physiol 151:59–66

Barkan A (2011) Expression of plastid genes: organelle-specific elaborations on a prokaryotic scaffold. Plant Physiol 155:1520–1532

Barone P, Zhang XH, Widholm JM (2009) Tobacco plastid transformation using the feedback-insensitive anthranilate synthase [alpha]-subunit of tobacco (ASA2) as a new selectable marker. J Exp Bot 60:3195–3202

Blowers AD, Bogorad L, Shark KB, Sanford JC (1989) Studies on Chlamydomonas chloroplast transformation: foreign DNA can be stably maintained in the chromosome. Plant Cell 1:123–132

Bock R (2001) Transgenic plastids in basic research and plant biotechnology. J Mol Biol 312:425–438

Bock R, Hermann M, Kössel H (1996) In vivo dissection of *cis*-acting determinants for plastid RNA editing. EMBO J 15:5052–5059

Boyhan D, Daniell H (2011) Low-cost production of proinsulin in tobacco and lettuce chloroplasts for injectable or oral delivery of functional insulin and C-peptide. Plant Biotechnol J 9:585–598

Boynton JE, Gillham NW, Harris EH, Hosler JP, Johnson AM, Jones AR, Randolph-Anderson BL, Robertson D, Klein TM, Shark KB, Sanford JC (1988) Chloroplast transformation in *Chlamydomonas* with high velocity microprojectiles. Science 240:1534–1538

Cardi T, Lenzi P, Maliga P (2010) Chloroplasts as expression platforms for plant-produced vaccines. Expert Rev Vaccines 9:893–911

Carrer H, Maliga P (1995) Targeted insertion of foreign genes into the tobacco plastid genome without physical linkage to the selectable marker gene. Biotechnology 13:791–794

Carrer H, Hockenberry TN, Svab Z, Maliga P (1993) Kanamycin resistance as a selectable marker for plastid transformation in tobacco. Mol Gen Genet 241:49–56

Chaudhuri S, Maliga P (1996) Sequences directing C to U editing of the plastid *psbL* mRNA are located within a 22 nucleotide segment spanning the editing site. EMBO J 15:5958–5964

Cheng L, Li HP, Qu B, Huang T, Tu JX, Fu TD, Liao YC (2010) Chloroplast transformation of rapeseed (*Brassica napus*) by particle bombardment of cotyledons. Plant Cell Rep 29:371–381

Corneille S, Lutz K, Svab Z, Maliga P (2001) Efficient elimination of selectable marker genes from the plastid genome by the CRE-*lox* site-specific recombination system. Plant J 72:171–178

Cornelissen M, Vandewiele M (1989) Nuclear transcriptional activity of the tobacco plastid *psbA* promoter. Nucleic Acids Res 17:19–29

Cui C, Song F, Tan Y, Zhou X, Zhao W, Ma F, Liu Y, Hussain J, Wang Y, Yang G, He G (2011) Stable chloroplast transformation of immature scutella and inflorescences in wheat (*Triticum aestivum* L.). Acta Biochim Biophys Sin (Shanghai) 43:284–291

Daniell H, Vivekananda J, Nielsen BL, Ye GN, Tewari KK, Sanford JC (1990) Transient foreign gene expression in chloroplasts of cultured tobacco cells after biolistic delivery of chloroplast vectors. Proc Natl Acad Sci USA 87:88–92

Daniell H, Datta R, Varma S, Gray S, Lee SB (1998) Containment of herbicide resistance through genetic engineering of the chloroplast genome. Nat Biotechnol 16:345–348

Daniell H, Muthukumar B, Lee SB (2001) Marker free transgenic plants: engineering the chloroplast genome without the use of antibiotic selection. Curr Genet 39:109–116

Daniell H, Singh ND, Mason H, Streatfield SJ (2009) Plant-made vaccine antigens and biopharmaceuticals. Trends Plant Sci 14:669–679

Davarpanah SJ, Jung SH, Kim YJ, Park YI, Min SR, Liu JR, Jeong WJ (2009) Stable plastid transformation in *Nicotiana benthamiana*. J Plant Biol 52:244–250

Day A, Goldschmidt-Clermont M (2011) The chloroplast transformation toolbox: selectable markers and marker removal. Plant Biotechnol J 9:540–553

De Marchis F, Wang Y, Stevanato P, Arcioni S, Bellucci M (2009) Genetic transformation of the sugar beet plastome. Transgenic Res 18:17–30

Dix PJ, Kavanagh TA (1995) Transforming the plastome: genetic markers and DNA delivery systems. Euphytica 85:29–34

Dufourmantel N, Pelissier B, Garcon F, Peltier G, Ferullo JM, Tissot G (2004) Generation of fertile transplastomic soybean. Plant Mol Biol 55: 479–489

Dufourmantel N, Tissot G, Goutorbe F, Garcon F, Muhr C, Jansens S, Pelissier B, Peltier G, Dubald M (2005) Generation and analysis of soybean plastid transformants expressing *Bacillus thuringiensis* Cry1Ab protoxin. Plant Mol Biol 58:659–668

Dufourmantel N, Dubald M, Matringe M, Canard H, Garcon F, Job C, Kay E, Wisniewski JP, Ferullo JM, Pelissier B, Sailland A, Tissot G (2007) Generation and characterization of soybean and marker-free tobacco plastid transformants over-expressing a bacterial 4-hydroxyphenylpyruvate dioxygenase which provides strong herbicide tolerance. Plant Biotechnol J 5:118–133

Eibl C, Zou Z, Beck A, Kim M, Mullet J, Koop HU (1999) In vivo analysis of plastid *psbA*, *rbcL* and *rpl32* UTR elements by chloroplast transformation: tobacco plastid gene expression is controlled by modulation of transcript levels and translation efficiency. Plant J 19:333–345

Fernandez-San Millan A, Obregon P, Veramendi J (2011) Over-expression of peptide deformylase in chloroplasts confers actinonin resistance, but is not a suitable selective marker system for plastid transformation. Transgenic Res 20:613–624

Finer JJ, Vain P, Jones MW, McMullen MD (1992) Development of the particle inflow gun for DNA delivery to plant cells. Plant Cell Rep 11:323–328

Fromm H, Edelman M, Aviv D, Galun E (1987) The molecular basis for rDNA-dependent spectinomycin resistance in *Nicotiana* chloroplasts. EMBO J 6:3233–3237

Gerats T, Vandenbussche M (2005) A model system comparative for research: Petunia. Trends Plant Sci 10:251–256

Gillman JD, Bentolila S, Hanson MR (2009) Cytoplasmic male sterility in Petunia. In: Gerats T, Strommer J (eds) Petunia. Springer Science+Business Media, LLC, New York, pp 107–129

Golds T, Maliga P, Koop HU (1993) Stable plastid transformation in PEG-treated protoplasts of *Nicotiana tabacum*. Biotechnology 11:95–97

Greer LF 3rd, Szalay AA (2002) Imaging of light emission from the expression of luciferases in living cells and organisms: a review. Luminescence 17:43–74

Hajdukiewicz PTJ, Gilbertson L, Staub JM (2001) Multiple pathways for Cre/*lox*-mediated recombination in plastids. Plant J 27:161–170

Hibberd JM, Linley PJ, Khan MS, Gray JC (1998) Transient expression of green fluorescent protein in various plastid types following microprojectile bombardment. Plant J 16:627–632

Hou BK, Zhou YH, Wan LH, Zhang ZL, Shen GF, Chen ZH, Hu ZM (2003) Chloroplast transformation in oilseed rape. Transgenic Res 12:111–114

Huang FC, Klaus SMJ, Herz S, Zuo Z, Koop HU, Golds TJ (2002) Efficient plastid transformation in tobacco using the *aphA-6* gene and kanamycin selection. Mol Genet Genomics 268:19–27

Iamtham S, Day A (2000) Removal of antibiotic resistance genes from transgenic tobacco plastids. Nat Biotechnol 18:1172–1176

Johnston SA, Anziano PQ, Shark K, Sanford JC, Butow RA (1988) Mitochondrial transformation in yeast by bombardment with microprojectiles. Science 240:1538–1541

Kahlau S, Aspinall S, Gray JC, Bock R (2006) Sequence of the tomato chloroplast DNA and evolutionary comparison of solanaceous plastid genomes. J Mol Evol 63:194–207

Kanagaraj AP, Verma D, Daniell H (2011) Expression of dengue-3 premembrane and envelope polyprotein in lettuce chloroplasts. Plant Mol Biol 76:323–333

Kanamoto H, Yamashita A, Asao H, Okumura S, Takase H, Hattori M, Yokota A, Tomizawa K (2006) Efficient and stable transformation of *Lactuca sativa* L. cv. Cisco (lettuce) plastids. Transgenic Res 15: 205–217

Kanevski I, Maliga P, Rhoades DF, Gutteridge S (1999) Plastome engineering of ribulose-1,5-bisphosphate carboxylase/oxygenase in tobacco to form a sunflower large subunit and a tobacco small subunit hybrid. Plant Physiol 119:133–141

Kavanagh TA, Thanh ND, Lao NT, McGrath N, Peter SO, Horváth EM, Dix PJ, Medgyesy P (1999) Homeologous plastid DNA transformation in tobacco is mediated by multiple recombination events. Genetics 152:1111–1122

Khan MS, Maliga P (1999) Fluorescent antibiotic resistance marker to track plastid transformation in higher plants. Nat Biotechnol 17:910–915

Kittiwongwattana C, Lutz KA, Clark M, Maliga P (2007) Plastid marker gene excision by the phiC31 phage site-specific recombinase. Plant Mol Biol 64:137–143

Klaus SMJ, Huang FC, Eibl C, Koop HU, Golds TJ (2003) Rapid and proven production of transplastomic tobacco plants by restoration of pigmentation and photosynthesis. Plant J 35:811–821

Klaus SMJ, Huang FC, Golds TJ, Koop H-U (2004) Generation of marker-free plastid transformants using a transiently cointegrated selection gene. Nat Biotechnol 22:225–229

Klein TM, Wolf ED, Wu R, Sanford JC (1987) High-velocity microprojectiles for delivering nucleic acids in living cells. Nature 327:70–73

Klein TM, Gradziel T, Fromm ME, Sanford JM (1988a) Factors influencing gene delivery into *Zea* mays cell by high-velocity microprojectiles. Biotechnology 6:559–563

Klein TM, Harper EC, Svab Z, Sanford JC, Fromm ME, Maliga P (1988b) Stable genetic transformation of intact Nicotiana cells by the particle bombardment process. Proc Natl Acad Sci USA 85: 8502–8505

Kode V, Mudd E, Iamtham S, Day A (2006) Isolation of precise plastid deletion mutants by homology-based excision: a resource for site-directed mutagenesis, multi-gene changes and high-throughput plastid transformation. Plant J 46:901–909

Koop HU, Steinmüller K, Wagner H, Rössler C, Eibl C, Sacher L (1996) Integration of foreign sequences into the tobacco plastome via PEG-mediated protoplast transformation. Planta 199:193–201

Krichevsky A, Meyers B, Vainstein A, Maliga P, Citovsky V (2010) Autoluminescent plants. PLoS One 5:e15461

Kumar S, Dhingra A, Daniell H (2004a) Stable transformation of the cotton plastid genome and maternal inheritance of transgenes. Plant Mol Biol 56: 203–216

Kumar S, Dhingra A, Daniell H (2004b) Plastid-expressed betaine aldehyde deydrogenase gene in carrot cultured cells, roots and leaves confers enhanced salt tolerance. Plant Physiol 136: 2843–2854

Kuroda H, Maliga P (2002) Over-expression of the *clpP* 5′-UTR in a chimeric context causes a mutant phenotype suggesting competition for a *clpP*-specific RNA maturation factor in tobacco chloroplasts. Plant Physiol 129:1600–1606

Langbecker CL, Ye GN, Broyles DL, Duggan LL, Xu CW, Hajdukiewicz PT, Armstrong CL, Staub JM (2004) High-frequency transformation of undeveloped plastids in tobacco suspension cells. Plant Physiol 135:39–46

Lee SB, Kwon HB, Kwon SJ, Park SC, Jeong MJ, Han SE, Byun MO, Daniell H (2003) Accumulation of trehalose within transgenic chloroplasts confers drought tolerance. Mol Breed 11:1–13

Lee SM, Kang KS, Chung H, Yoo SH, Xu XM, Lee SB, Cheong JJ, Daniell H, Kim M (2006) Plastid transformation in the monocotyledonous cereal crop, rice (*Oryza sativa*) and transmission of transgenes to their progeny. Mol Cells 21:401–410

Leister D (2003) Chloroplast research in the genomic age. Trends Genet 19:47–56

Li W, Ruf S, Bock R (2011) Chloramphenicol acetyltransferase as selectable marker for plastid transformation. Plant Mol Biol 76:443–451

Lim S, Ashida H, Watanabe R, Inai K, Kim YS, Mukougawa K, Fukuda H, Tomizawa K, Ushiyama K, Asao H, Tamoi M, Masutani H, Shigeoka S, Yodoi J, Yokota A (2011) Production of biologically active human thioredoxin 1 protein in lettuce chloroplasts. Plant Mol Biol 76:335–344

Liu CW, Lin CC, Chen JJ, Tseng MJ (2007) Stable chloroplast transformation in cabbage (*Brassica oleracea* L. var. *capitata* L.) by particle bombardment. Plant Cell Rep 26:1733–1744

Liu CW, Lin CC, Yiu JC, Chen JJ, Tseng MJ (2008) Expression of a *Bacillus thuringiensis* toxin (*cry1Ab*) gene in cabbage (*Brassica oleracea* L. var. *capitata* L.) chloroplasts confers high insecticidal efficacy against *Plutella xylostella*. Theor Appl Genet 117:75–88

Lutz KA, Maliga P (2007a) Transformation of the plastid genome to study RNA editing. Methods Enzymol 424:501–518

Lutz KA, Maliga P (2007b) Construction of marker-free transplastomic plants. Curr Opin Biotechnol 18:107–114

Lutz KA, Maliga P (2008) Plastid genomes in a regenerating tobacco shoot derive from a small number of copies selected through a stochastic process. Plant J 56:975–983

Lutz KA, Knapp JE, Maliga P (2001) Expression of *bar* in the plastid genome confers herbicide resistance. Plant Physiol 125:1585–1590

Lutz K, Corneille S, Azhagiri AK, Svab Z, Maliga P (2004) A novel approach to plastid transformation utilizes the phiC31 phage integrase. Plant J 37: 906–913

Lutz KA, Bosacchi MH, Maliga P (2006a) Plastid marker gene excision by transiently expressed CRE recombinase. Plant J 45:447–456

Lutz KA, Svab Z, Maliga P (2006b) Construction of marker-free transplastomic tobacco using the Cre-*loxP* site-specific recombination system. Nat Protoc 1:900–910

Lutz KA, Azhagiri AK, Tungsuchat-Huang T, Maliga P (2007) A guide to choosing vectors for transformation of the plastid genome of higher plants. Plant Physiol 145:1201–1210

Lutz KA, Azhagiri A, Maliga P (2011) Transplastomics in Arabidopsis: progress towards developing an efficient method. In: Jarvis RP (ed) Chloroplast research in arabidopsis. Springer Science+Business Media, LLC, New York

Maliga P (2004) Plastid transformation in higher plants. Annu Rev Plant Biol 55:289–313

Maliga P, Bock R (2011) Plastid biotechnology: food, fuel and medicine for the 21st century. Plant Physiol 155:1501–1510

Maliga P, Svab Z (2011) Engineering the plastid genome of *Nicotiana sylvestris*, a diploid model species for plastid genetics. In: Birchler JJ (ed) Plant chromosome engineering: methods and protocols. Springer Science+Business Media, LLC, New York, pp 37–50

McCabe MS, Klaas M, Gonzalez-Rabade N, Poage M, Badillo-Corona JA, Zhou F, Karcher D, Bock R, Gray JC, Dix PJ (2008) Plastid transformation of high-biomass tobacco variety Maryland Mammoth for production of human immunodeficiency virus type 1 (HIV-1) p24 antigen. Plant Biotechnol J 6: 914–929

Nguyen TT, Nugent GD, Cardi T, Dix PJ (2005) Generation of homoplasmic plastid transformants of a commercial cultivar of potato (*Solanum tuberosum* L.). Plant Sci 168:1495–1500

Nock CJ, Waters DLE, Edwards MA, Bowen SG, Rice N, Cordeiro GM, Henry RJ (2011) Chloroplast genome sequences from total DNA for plant identification. Plant Biotechnol J 9:328–333

Nugent GD, Ten Have M, van der Gulik A, Dix PJ, Uijtewaal BA, Mordhorst AP (2005) Plastid transformants of tomato selected using mutations affecting ribosome structure. Plant Cell Rep 24:341–349

Nugent GD, Coyne S, Nguyen TT, Kavanagh TA, Dix PJ (2006) Nuclear and plastid transformation of *Brassica oleracea* var. *botrytis* (cauliflower) using PEG-mediated uptake into protoplasts. Plant Sci 170:135–142

O'Neill C, Horvath GV, Horvath E, Dix PJ, Medgyesy P (1993) Chloroplast transformation in plants: polyethylene glycol (PEG) treatment of protoplasts is an alternative to biolistic delivery systems. Plant J 3:729–738

Okumura S, Sawada M, Park YW, Hayashi T, Shimamura M, Takase H, Tomizawa K (2006) Transformation of poplar (*Populus alba*) plastids and expression of foreign proteins in tree chloroplasts. Transgenic Res 15:637–646

Paszkowski J, Shillito RD, Saul M, Mandak V, Hohn T, Hohn B, Potrykus I (1984) Direct gene transfer to plants. EMBO J 3:2717–2722

Raubeson LA, Jansen RK (2005) Chloroplast genomes of plants. In: Henry RJ (ed) Diversity and evolution of plants – genotypic and phenotypic variation in higher plants. CABI, Wallingford, pp 45–68

Reed ML, Hanson MR (1997) A heterologous maize *rpoB* editing site is recognized by transgenic tobacco chloroplasts. Mol Cell Biol 17:6948–6952

Reed ML, Wilson SK, Sutton CA, Hanson MR (2001) High-level expression of a synthetic red-shifted GFP coding region incorporated into the chloroplasts. Plant J 27:257–265

Ruf S, Hermann M, Berger IJ, Carrer H, Bock R (2001) Stable genetic transformation of tomato plastids: foreign protein expression in fruit. Nat Biotechnol 19:870–875

Ruhlman T, Ahangari R, Devine A, Samsam M, Daniell H (2007) Expression of cholera toxin B-proinsulin fusion protein in lettuce and tobacco chloroplasts – oral administration protects against development of insulitis in non-obese diabetic mice. Plant Biotechnol J 5:495–510

Ruhlman T, Verma D, Samson N, Daniell H (2010) The role of heterologous chloroplast sequence elements in transgene integration and expression. Plant Physiol 152:2088–2104

Rumeau D, Becuwe-Linka N, Beyly A, Louwagie M, Garin J, Peltier G (2005) New subunits NDH-M, -N, and -O, encoded by nuclear genes, are essential for plastid NDH complex functioning in higher plants. Plant Cell 17:219–232

Sanford JC, Smith FD, Russell JA (1993) Optimizing the biolistic process for different biological applications. Methods Enzymol 217:483–509

Serino G, Maliga P (1997) A negative selection scheme based on the expression of cytosine deaminase in plastids. Plant J 12:697–701

Sheppard AE, Ayliffe MA, Blatch L, Day A, Delaney SK, Khairul-Fahmy N, Li Y, Madesis P, Pryor AJ, Timmis JN (2008) Transfer of plastid DNA to the nucleus is elevated during male gametogenesis in tobacco. Plant Physiol 148:328–336

Shiina T, Hayashi K, Ishii N, Morikawa K, Toyoshima Y (2000) Chloroplast tubules visualized in transplastomic plants expressing green fluorescent protein. Plant Cell Physiol 41:367–371

Shimizu M, Goto M, Hanai M, Shimizu T, Izawa N, Kanamoto H, Tomizawa K, Yokota A, Kobayashi H (2008) Selectable tolerance to herbicides by mutated acetolactate synthase genes integrated into the chloroplast genome of tobacco. Plant Physiol 147: 1976–1983

Sidorov VA, Kasten D, Pang SZ, Hajdukiewicz PTJ, Staub JM, Nehra NS (1999) Stable chloroplast transformation in potato: use of green fluorescent protein as a plastid marker. Plant J 19:209–216

Sikdar SR, Serino G, Chaudhuri S, Maliga P (1998) Plastid transformation in *Arabidopsis thaliana*. Plant Cell Rep 18:20–24

Sinagawa-Garcia SR, Tungsuchat-Huang T, Paredes-Lopez O, Maliga P (2009) Next generation synthetic vectors for transformation of the plastid genome of higher plants. Plant Mol Biol 70:487–498

Singh AK, Verma SS, Bansal KC (2010) Plastid transformation in eggplant (*Solanum melongena* L.). Transgenic Res 19:113–119

Skarjinskaia M, Svab Z, Maliga P (2003) Plastid transformation in *Lesquerella fendleri*, an oilseed Brassicacea. Transgenic Res 12:115–122

Sporlein B, Streubel M, Dahlfeld G, Westhoff P, Koop HU (1991) PEG-mediated plastid transformation: a new system for transient gene expression assays in chloroplasts. Theor Appl Genet 82:717–722

Sriraman P, Silhavy D, Maliga P (1998) Transcription from heterologous rRNA operon promoters in chloroplasts reveals requirement for specific activating factors. Plant Physiol 117:1495–1499

Staub JM, Maliga P (1992) Long regions of homologous DNA are incorporated into the tobacco plastid genome by transformation. Plant Cell 4:39–45

Staub JM, Maliga P (1993) Accumulation of D1 polypeptide in tobacco plastids is regulated via the untranslated region of the *psbA* mRNA. EMBO J 12:601–606

Staub JM, Maliga P (1994) Translation of *psbA* mRNA is regulated by light *via* the 5′-untranslated region in tobacco plastids. Plant J 6:547–553

Stegemann S, Bock R (2009) Exchange of genetic material between cells in plant tissue grafts. Science 324:649–651

Sugiura M (1989) The chloroplast chromosomes in land plants. Annu Rev Cell Biol 5:51–70

Suzuki JY, Sriraman P, Svab Z, Maliga P (2003) Unique architecture of the plastid ribosomal RNA operon promoter recognized by the multisubunit RNA polymerase (PEP) in tobacco and other higher plants. Plant Cell 15:195–205

Svab Z, Maliga P (1993) High-frequency plastid transformation in tobacco by selection for a chimeric *aadA* gene. Proc Natl Acad Sci USA 90:913–917

Svab Z, Hajdukiewicz P, Maliga P (1990) Stable transformation of plastids in higher plants. Proc Natl Acad Sci USA 87:8526–8530

Tang J, Xia H, Cao M, Zhang X, Zeng W, Hu S, Tong W, Wang J, Wang J, Yu J, Yang H, Zhu L (2004) A comparison of rice chloroplast genomes. Plant Physiol 135:412–420

Tungsuchat-Huang T, Sinagawa-Garcia SR, Paredes-Lopez O, Maliga P (2010) Study of plastid genome stability in tobacco reveals that the loss of marker genes is more likely by gene conversion than by recombination between 34-bp *loxP* repeats. Plant Physiol 153:252–259

Tungsuchat-Huang T, Slivinski KM, Sinagawa-Garcia SR, Maliga P (2011) Visual spectinomycin resistance gene for facile identification of transplastomic sectors in tobacco leaves. Plant Mol Biol 76:453–461

Valkov VT, Gargano D, Manna C, Formisano G, Dix PJ, Gray JC, Scotti N, Cardi T (2011) High efficiency plastid transformation in potato and regulation of transgene expression in leaves and tubers by alternative 5′ and 3′ regulatory sequences. Transgenic Res 20:137–151

Verma D, Daniell H (2007) Chloroplast vector systems for biotechnology applications. Plant Physiol 145:1129–1143

Wei Z, Liu Y, Lin C, Wang Y, Cai Q, Dong Y, Xing S (2011) Transformation of alfalfa chloroplasts and expression of green fluorescent protein in a forage crop. Biotechnol Lett 33:2487–2494

Whitney SM, Sharwood RE (2008) Construction of a tobacco master line to improve Rubisco engineering in chloroplasts. J Exp Bot 59(7):1909–1921

Whitney SM, Houtz RL, Alonso H (2011) Advancing our understanding and capacity to engineer nature's CO2 sequestering enzyme, Rubisco. Plant Physiol 155:27–35

Wurbs D, Ruf S, Bock R (2007) Contained metabolic engineering in tomatoes by expression of carotenoid biosynthesis genes from the plastid genome. Plant J 49:276–288

Ye GN, Daniell H, Sanford JC (1990) Optimization of delivery of foreign DNA into higher-plant chloroplasts. Plant Mol Biol 15:809–819

Ye GN, Colburn S, Xu CW, Hajdukiewicz PTJ, Staub JM (2003) Persistence of unselected transgenic DNA during a plastid transformation and segregation approach to herbicide resistance. Plant Physiol 133:402–410

Yu LX, Gray BN, Rutzke CJ, Walsker LP, Wilson DB, Hanson MR (2007) Expression of thermostable microbial cellulases in the chloroplast of nicotine-free tobacco. J Biotechnol 131:362–369

Yukawa M, Tsudzuki T, Sugiura M (2006) The chloroplast genome of *Nicotiana sylvestris* and *Nicotiana tomentosiformis*: complete sequencing confirms that the *Nicotiana sylvestris* progenitor is the maternal genome donor of *Nicotiana tabacum*. Mol Genet Genomics 275:367–373

Zhou F, Badillo-Corona JA, Karcher D, Gonzalez-Rabade N, Piepenburg K, Borchers AM, Maloney AP, Kavanagh TA, Gray JC, Bock R (2008) High-level expression of human immunodeficiency virus antigens from the tobacco and tomato plastid genomes. Plant Biotechnol J 6:897–913

Zou Z, Eibl C, Koop HU (2003) The stem-loop structure of the tobacco *psbA* 5′UTR is an important determinant of mRNA stability and translation efficiency. Mol Genet Genomics 269:340–349

Zoubenko OV, Allison LA, Svab Z, Maliga P (1994) Efficient targeting of foreign genes into the tobacco plastid genome. Nucleic Acids Res 22:3819–3824

Zubko MK, Zubko EI, van Zuilen K, Mayer P, Day A (2004) Stable transformation of petunia plastids. Transgenic Res 13:523–530

Chapter 18

Reverse Genetics in Flowering Plant Plastids

Anil Day
Faculty of Life Sciences, The University of Manchester, Michael Smith Building, Oxford Road, Manchester, M13 9PT, UK

Summary

Plastid reverse genetics exploits the predominance of homologous DNA recombination in this organelle, which allows targeted mutations to be introduced into plastid genes. Most studies have used tobacco and involve replacement of wild-type plastid genes with mutant alleles. Mutant alleles are either disrupted by the marker gene or lie adjacent to the marker gene. Marker selection with antibiotics is required to remove wild-type plastid genomes and reveal the phenotype of homoplasmic mutant plants. Targeted knock-outs have shown that tobacco plastid genes are either dispensable or essential. Dispensable plastid genes include

*Author for correspondence, e-mail: anil.day@manchester.ac.uk

R. Bock and V. Knoop (eds.), Genomics of Chloroplasts and Mitochondria, Advances in Photosynthesis and Respiration 35, pp. 415–441,
DOI 10.1007/978-94-007-2920-9_18,

those encoding photosynthesis-related proteins, subunits of the plastid-encoded RNA polymerase, ribosomal proteins *rps15, rpl33 and rpl36*, valyl transfer RNA(GAC), glycyl transfer RNA(GCC) and putative origins of DNA replication. Loss-of-photosynthesis is dispensable if mutant plants are propagated on sucrose-containing medium. Knock-outs were particularly useful for elucidating the roles of conserved but dispensable hypothetical reading frames (*ycf* genes) in photosynthesis. Site-directed mutations allow structure-function studies on the products of plastid genes. Marker-free plants containing deletions of dispensable plastid genes, e.g. the large subunit of RuBisCO gene, facilitate the rapid isolation of plants containing site-directed mutant alleles. Knock-outs of essential tobacco plastid genes (*accD, clpP, ycf1, ycf2, rps2, rps3, rps4, rps16, rps18, rpl20, rpl22, rpl23, rpl32, trnC-GCA, trnN-GUU, trnG-UCC*) persist as heteroplasmic mixtures with the wild-type allele under antibiotic selection; removal of selection results in loss of the knock out allele. Homoplasmic cells containing knock out alleles of essential genes would not be viable and this explains the leaf-lamina-loss phenotype of mutant plants. Strong selection for the wild-type gene may hinder the isolation of partial-function alleles of essential plastid genes containing site-directed mutations. New methods are required to study essential plastid genes involving regulated expression or inducible excision mediated by site-specific recombinases. Progress may require the use of angiosperm species, in which homologues of essential tobacco plastid genes are dispensable.

I. Introduction

Advances in DNA sequencing technologies combined with the relatively small sizes of plastid genomes (Chaps. 3, 4, 5) have led to a steady rise in the number of plastid genomes sequenced. The number of species for which complete sequences are available in the National Centre for Biotechnology Information (NCBI) database in the USA (http://www.ncbi.nlm.nih.gov/) now exceeds 150 plastid genomes. The availability of a plastid genome sequence is the first step required to elucidate its coding content. Plastid genomes are relatively well conserved allowing rapid annotation of plastid genes based on previously studied genomes. The functions of plastid genes can in many cases be deduced by comparisons with well-characterised homologues in other plastids or in bacteria. Plastids are likely to be descended from ancient cyanobacteria (Martin et al. 2002) and encode proteins that are homologous to relatively well characterised *Escherichia coli* proteins such as the β-carboxyl transferase subunit of acetyl-CoA carboxylase (Li and Cronan 1992); the enzyme that catalyses the first committed step of fatty acid synthesis. Abundant products of plastid genes are amenable to biochemical and molecular analyses providing information on their intra-plastidic location, activities, solubility and assembly into complexes. The majority of plastid genes can be divided into two groups (see Bock 2007): those related to photosynthesis (photosynthesis-related genes) and those involved in gene expression (genetic-system genes). A number of plastid genes lie outside these two categories including the plastid *accD* gene, which encodes the β-carboxyl transferase subunit of acetyl CoA carboxylase. In addition, plastid genomes contain open reading frames encoding proteins that cannot be ascribed a function from sequence alone, because their products have not been characterised and homologues of known function have not been found. Open reading frames (ORFs) that are conserved in plastid genomes from a range of plant species indicate that their retention is due to functions needed in plastids. These conserved plastid ORFs are

Abbreviations: LS RuBisCO – Large subunit of Ribulose Bisphosphate Carboxylase/Oxygenase; NEP – Nucleus-encoded plastid RNA polymerase; ORF – Open reading frame; *ori* – Origin of DNA replication; PEP – Plastid-encoded plastid RNA polymerase; pt DNA – Plastid DNA; WT – Wild-type; *ycf* – Hypothetical chloroplast open reading frame

known as hypothetical chloroplast open reading frames (*ycf*) and are ideal subjects for reverse genetics to elucidate their functions.

Reverse genetics has revolutionised the study of plastid genes and was made possible by the development of plastid transformation in microalgae and plants (Chaps. 16, 17). The term reverse genetics arises from the order of investigation, starting from gene sequence→targeted mutation→phenotype, which is the reverse of the order in classical forward genetics screens, where phenotype→mutation→gene sequence is the order of investigation. The procedure involves the introduction of targeted mutations in the plastid gene under investigation followed by an analysis of the consequences of these mutations on phenotype at the molecular, physiological, cellular and whole plant levels. The majority of reverse genetics studies on the plastid genome have been carried out in the green alga *Chlamydomonas reinhardtii* and *Nicotiana tabacum* (tobacco), which are the current species of microalgae and flowering plant most amenable to plastid transformation (Chaps. 16, 17). Reverse genetics is also applicable to bryophyte plastids (Nakamura et al. 2005). It is important to recognise that the reverse genetics approaches used in plastids differ from those used in the nucleus of flowering plants. Homologous recombination predominates in plastids allowing precise targeted mutations to be made in plastid genes. The precise replacement of a wild-type gene (WT allele) with mutant versions of the gene (mutant alleles) using homologous recombination is an effective tool for reverse genetics. Illegitimate DNA recombination pathways predominate in the nucleus of flowering plants. The intrinsic frequency of gene targeting in flowering plant nuclei is in the order of 1×10^{-3} to 1×10^{-4} per insertion event (Hohn and Puchta 2003), which is too low to be useful for reverse genetics. Nuclear genome projects utilising reverse genetics involving insertional mutagenesis or site-directed mutations often require relatively large resources and number of personnel. These resources and personnel are required to isolate and catalogue insertion mutations, resulting from 'random' integration of T-DNA or transposons, or point mutations involving targeting induced local lesions in genomes ('TILLING'), in the >30,000 genes present in the nucleus (Colbert et al. 2001; Gilchrist and Haughn 2010; Hardy et al. 2010). In contrast, targeted mutations in plastid genomes can be made relatively easily by small research groups capable of constructing DNA vectors and proficient in plastid transformation. Knockdown of gene expression using RNA interference (RNAi) or anti-sense RNA approaches that are applicable to nuclear genes are not used in plastids. This is because plastids appear to lack RNAi–mediated gene silencing pathways. Over-expression of a natural plastid antisense RNA to 5S rRNA and tRNA(Arg) led to an increase rather than a decrease in the accumulation of tRNA(Arg) hinting at a complexity of regulation that requires further investigation (Hotto et al. 2010). Dominant-negative mutations have not been used to study the functions of plastid genes (Herskowitz 1987). However, this approach has been used to interfere with the expression of a plastid-targeted RecA protein, which is nucleus encoded, by over-expressing a dominant-negative mutant bacterial RecA protein in Chlamydomonas chloroplasts (Cerutti et al. 1995). The use of homologous recombination to introduce mutations into plastid genes is the reverse genetics method of choice for studying the function of plastid genes.

The high copy number of plastid genomes per cell is a major difference with other systems such as the nuclear genomes of yeast or the moss *Physcomitrella patens*, in which homologous recombination is the method of choice to replace wild-type genes with mutant alleles (Rothstein 1991; Schaefer 2002). A single copy of a nuclear gene is usually present in the haploid cells of yeast and moss, whereas ~1,000–5,000 plastid genomes are distributed amongst the 25–100 chloroplasts present in a leaf mesophyll cell (Pyke et al. 2000; Zoschke et al. 2007; Rauwolf et al. 2010). This means that in the early stages of gene replacement, a mutant gene transformed into plastids will be masked by the many copies of the original wild-type allele present in the cell. Reverse genetics methods must therefore overcome the multiple copies of plastid genomes present per cell in order to reveal the phenotype of a mutant plastid gene. In this chapter, I review the use of reverse genetics to study the coding content of plastid genomes. This includes the analyses of genes encoding RNA and protein products, as well as DNA sequences involved in plastid genome maintenance, such as origins of replication. In these studies, wild-type (WT) plastid genes have been replaced by knock-out alleles or alleles containing site-directed mutations. Analyses of the resulting mutant plants provide information on the role of the plastid gene under investigation.

II. Principles of Plastid Reverse Genetics Methods

A. Targeted Insertional Mutagenesis with the aadA Marker Gene

The most common procedure to inactivate plastid genes is illustrated in Fig. 18.1. The example shown uses the *rbcL* gene encoding the large subunit of RuBisCO, which was the first photosynthesis-related plastid gene to be mutagenised by reverse genetics in flowering plants (Kanevski and Maliga 1994). The *rbcL* gene was disrupted by insertion of the *aadA* marker gene. The *aadA* gene is the marker gene of choice for plastid transformation and encodes the enzyme aminoglycoside 3″-adenylyltransferase. The antibiotics spectinomycin and streptomycin inhibit plastid protein synthesis and are inactivated by adenylylation mediated by aminoglycoside 3″-adenylyltransferase. The *aadA* gene is expressed using regulatory elements (e.g. promoters, 5′ UTRs containing ribosome-binding sites and 3′ UTRs for RNA stability) that function in plastids. The plastid transformation vector containing the mutant *rbcL* knock-out allele was assembled in the bacterium *Escherichia coli* using standard cloning techniques. The procedure involves inserting *aadA* into the coding region of the *rbcL* gene present on a previously cloned sequence of chloroplast DNA. When inactivating genes it is common practice to remove coding sequences in the regions

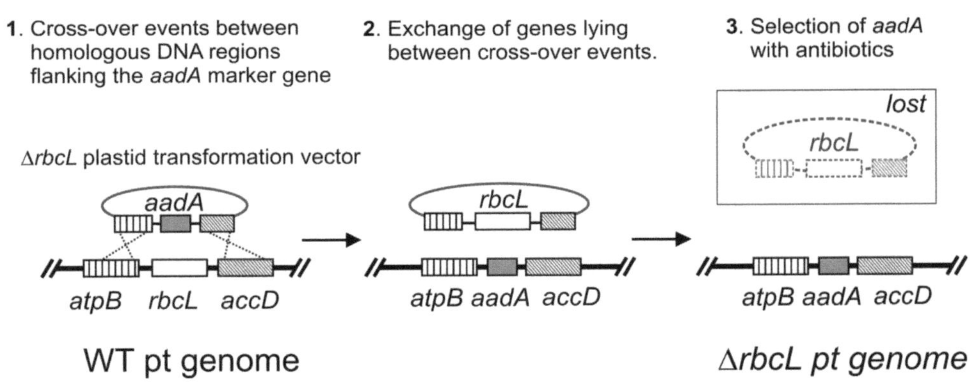

Fig. 18.1. Scheme showing targeted knock-out of the tobacco plastid *rbcL* gene with an *aadA*-disrupted *rbcL* allele (Kanevski and Maliga 1994). The plastid transformation vector containing the *aadA*-disrupted *rbcL* allele is introduced into plastids most commonly by particle bombardment where the recombinant DNA integrates into plastid DNA by homologous recombination. The consequence of a reciprocal recombination event is shown. (**1**) The *aadA* gene is flanked by a *left arm* (*atpB* gene) and *a right arm* (*accD* gene), which contain DNA sequences common to the resident plastid genome. Homologous recombination events between common DNA sequences involve exchange of DNA strands between vector and plastid genome and are represented as crosses flanking the *aadA* gene. (**2**) Recombination events on either side of *aadA* exchange this marker gene for *rbcL*. (**3**) Integration of *aadA* into the plastid genome is essential to ensure its replication in plastids. Plastid transformation vectors are unstable and are not maintained in plastids. Antibiotic selection ensures multiplication of plastids containing *aadA*.

flanking *aadA*. Loss-of-function results from deletion of gene sequences combined with disruption of the coding region by *aadA*. The vector is transformed into plastids most frequently by particle-bombardment with DNA-coated microprojectiles (Klein et al. 1987). DNA cross-over events between DNA sequences common to the vector and resident plastid genome, to either side of *aadA* (Fig. 18.1, step 1), exchange the wild-type plastid gene for the mutant allele (Fig. 18.1, step 2). This results in integration of the *aadA*-knock-out allele into the plastid genome allowing its propagation within plastids. Whilst a double recombination event is shown for simplicity, the actual mechanism most probably proceeds in two steps involving the insertion of the entire plastid transformation vector into the plastid genome by a single recombination event in one of the targeting arms to produce a co-integrate that is then resolved by a second recombination event in the other arm, which is duplicated in the co-integrate (Ahlert et al. 2003; Klaus et al. 2004).

B. Selection for Transformed Plastids and the Attainment of Homoplasmy

The plastid genome lacking *rbcL* (Δ*rbcL* plastid genome) contains *aadA* and is selected using antibiotics. Spectinomycin can either be used on its own or in combination with streptomycin. Double selection with both antibiotics increases the time required to isolate resistant shoots but has the advantage of selecting against spontaneous resistant mutants to the single antibiotics. The plastid transformation vector is eventually lost because it lacks DNA sequences required for stable propagation in plastids (Fig. 18.1, step 3). This is an example of a reciprocal DNA recombination event where sequences are exchanged between the transforming vector and resident plastid genome. Non-reciprocal DNA recombination or gene conversion is an alternative mechanism for introducing the disrupted *rbcL* allele into the plastid genome. In the simplified scheme shown in Fig. 18.2, DNA replication is shown as dotted arrows

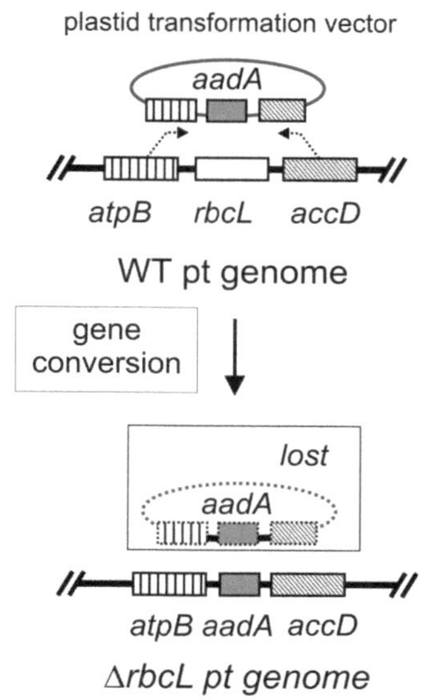

Fig. 18.2. Gene conversion between the transforming vector and the resident WT plastid genome replaces the WT *rbcL* gene with the *aadA*-disrupted allele. In the simplified scheme, DNA strands from the resident plastid genome copy the region between *atpB* and *accD* genes in the plastid transformation vector.

copying sequences from the plastid transformation vector resulting in the introduction of the copied region into the plastid genome.

Early in the transformation process, two plastid genomes will be present in cells: the original wild-type plastid genome and the second mutant genome containing the *aadA* knock-out allele (Fig. 18.3, top). The presence of two or more different plastid genomes within a plant is called heteroplasmy. Heteroplasmy is normally an unstable state. Following cell division, the different plastid genome types present in a heteroplasmic cell segregate away from each other to form cells with a uniform population of plastids (homoplasmic cells). In the absence of selection that favours one plastid form, the proportion of homoplasmic cells with each plastid genome type appears to reflect the starting dosage of each plastid type in heteroplasmic cells and a random segregation process (Birky 2001). Antibiotics enable the preferential multiplication of plastids expressing *aadA* resulting in the enrichment of homoplasmic cells containing a uniform

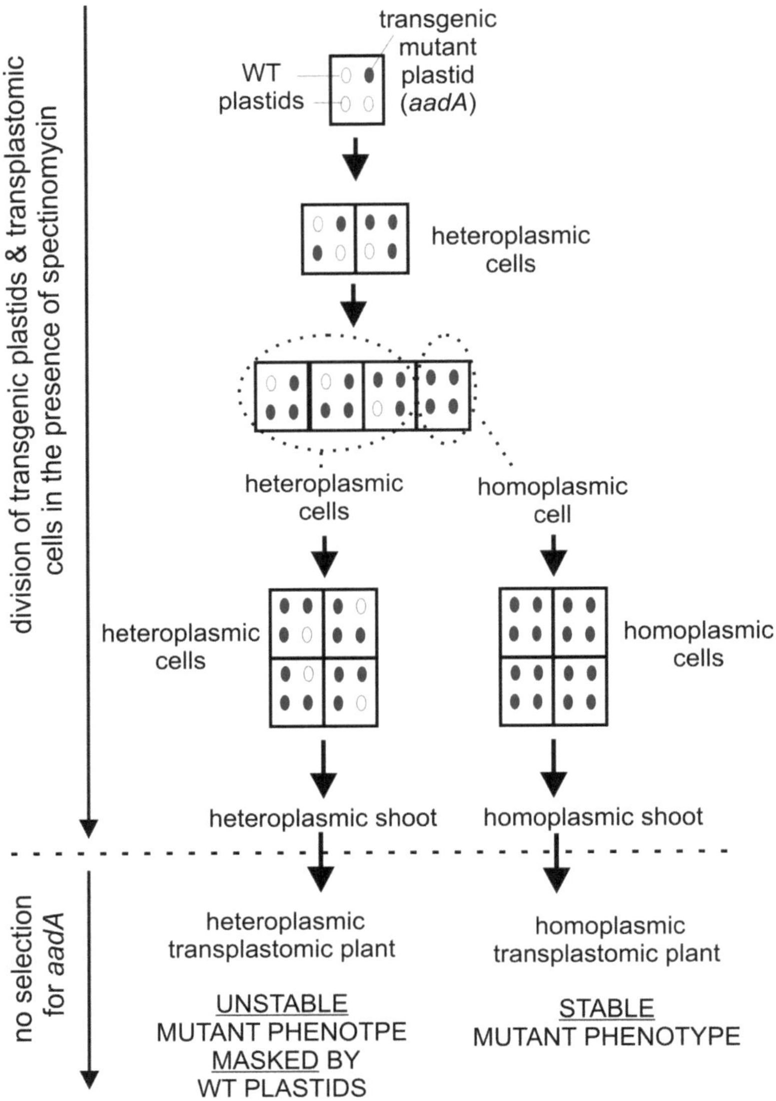

Fig. 18.3. Steps involved in isolating homoplasmic mutant plants. In the scheme, a single transgenic plastid containing *aadA* divides under spectinomycin selection to replace WT plastids during cell divisions. Regeneration of a plant from a homoplasmic cell gives rise to a homoplasmic plant with a stable mutant phenotype. Plants regenerated from heteroplasmic cells will have an unstable mutant phenotype.

population of transgenic plastid genomes. To obtain homoplasmic cells with the knock-out *aadA* allele, cells with transformed plastids are maintained on antibiotics until all wild-type plastid genomes have been removed. Attainment of homoplasmy will be influenced by DNA replication and repair pathways that ensure a homogeneous population of DNA molecules, including gene conversion events that change WT genomes to mutant Δ*rbcL* genomes (Fig. 18.4). Division of homoplasmic cells with mutant plastids will give rise to homoplasmic shoots and plants with a stable mutant phenotype. This allows mutant plants to be grown in soil in the absence of selection. In contrast, heteroplasmy will give rise to an unstable phenotype due to segregation of WT and mutant plastids. Moreover, the presence of WT plastids in plants could mask the mutant phenotype (Fig. 18.3). Removal of WT plastid genomes is critical for the analysis of the mutant phenotype.

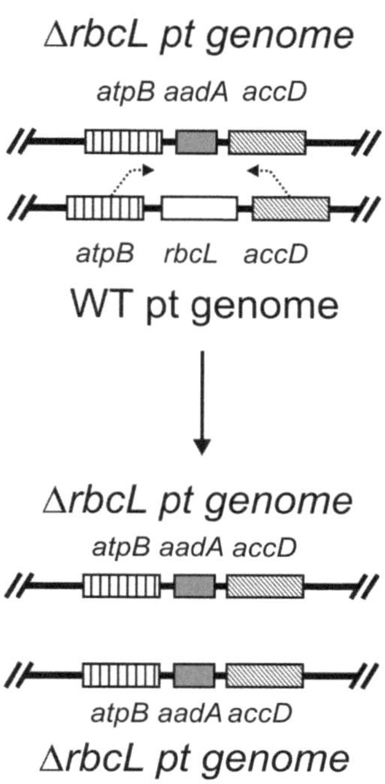

Fig. 18.4. Scheme showing non-reciprocal recombination between plastid genomes resulting in conversion of the WT *rbcL* gene to the Δ*rbcL* knock out allele.

In practice, homoplasmy is achieved by prolonged cell divisions in the presence of antibiotics. This usually involves two to three cycles of shoot formation from leaf explants placed on regeneration medium containing antibiotics. Homoplasmy is not attainable for knock-outs of essential plastid genes (discussed below).

C. Deletion of Plastid Genes Linked to an Adjacent Marker Gene

The scheme shown in Fig. 18.1 is applicable to genes that are expressed as single coding regions in a monocistronic transcript such as *rbcL*. A scheme for deleting a gene that is co-transcribed with other genes in an operon is shown in Fig. 18.5. In the example shown the *aadA* marker gene was located outside the *psbEFLJ* genes to minimise any impact on the expression of genes within the operon (Hager et al. 2002). To knock out expression of *psbJ*, it was deleted from the operon. Physical distance between marker and operon means that insertion of *aadA* is not always linked to insertion of the deleted operon. Cross-over events that take place outside the deletion in the *psbEFL* operon and *aadA* will insert both into the plastid genome (Fig. 18.5a). Any cross-over event between

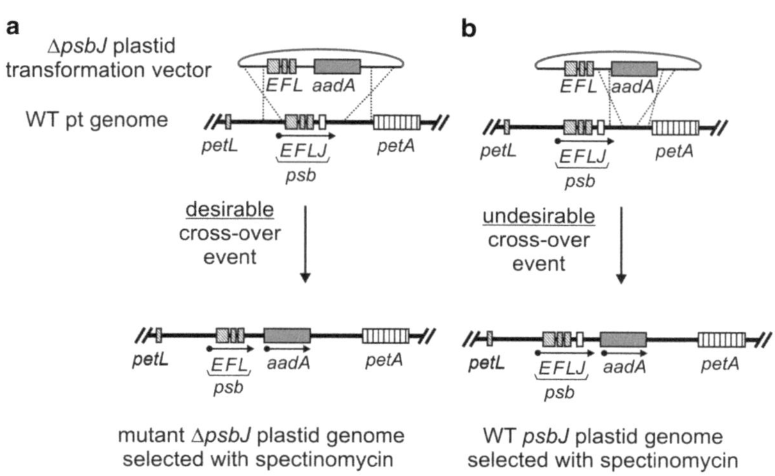

Fig. 18.5. Deletion of the tobacco *psbJ* plastid gene that is located some distance away from the selectable *aadA* marker gene (Hager et al. 2002). (**a**) Cross-over events that flank *aadA* and the Δ *psbJ* allele introduce the mutant allele into the plastid gene. (**b**) Cross-over events that introduce *aadA* but not the Δ *psbJ* allele into the plastid genome.

the deleted operon and *aadA* will only insert *aadA* into the plastid genome (Fig. 18.5b). This means a proportion of antibiotic resistant plants will not contain the desired *psbJ* knock-out mutation. In this example, of the four transplastomic lines isolated from independent transformation events, two lines contained the deletion in the *psbEFL* operon (Fig. 18.5a) and the other two lines retained the WT *psbEFLJ* operon (Fig. 18.5b). This example illustrates the need to minimise the distance between marker gene and knock-out allele to reduce the frequency of transplastomic plants containing *aadA* but not the desired mutation.

Most reverse genetics studies on plastid genes have utilised *aadA* and delivery of mutant alleles into plastids by particle bombardment. An alternative scheme has used mutant alleles of the plastid 16S rRNA gene (*16S rrn*) and the *rps12*(3′) gene, which confer resistance to spectinomycin and streptomycin, respectively. The proximity of *ndhB* to these native plastid genes allowed insertion of a mutant allele of *ndhB* into the plastid genome when cross-over events flanked both *rps12*(3′) and the nearby mutant *ndhB* allele (Horváth et al. 2000). In this case, vector DNA was delivered into plastids by treating protoplasts with solutions containing polyethylene glycol (PEG). These methods are not widely used because *16S rrn* and *rps12(3′)* genes are relatively inefficient markers compared to *aadA* (Svab and Maliga 1993) and PEG-mediated transformation of protoplasts is technically more demanding than particle bombardment of leaves.

III. Verifying Homoplasmy of Mutant Plastid Genomes

Attainment of homoplasmy is critical to the analysis of the phenotype of a mutant plastid gene. The presence of a uniform population of mutant plastid genomes is normally verified by DNA blot analysis on leaf DNA from mutant plants. The technique verifies replacement of a wild-type plastid gene with a mutant allele. This is visualised as DNA bands of different sizes corresponding to WT and mutant alleles, which result from cutting DNA with restriction enzymes. Even after prolonged selection on antibiotics, a number of studies have detected faint bands corresponding in size to WT alleles in restriction enzyme digests of DNA from transplastomic plants with clear and stable mutant phenotypes (Kofer et al. 1998; Hager et al. 1999; Ruf et al. 2000; Baena-González et al. 2001). The apparent presence of WT plastid genomes led to the qualification of mutant phenotypes by stating the level of apparent heteroplasmy of WT plastid genomes. This apparent residual presence of WT plastid genomes was noted to be below 5% (Kofer et al. 1998; Baena-González et al. 2001). Persistent heteroplasmy of WT genes complicates the analysis of plastid mutants (Koop et al. 1998; Maliga and Nixon 1998). Whilst our understanding of plastid genome maintenance is far from complete, cells with mixed populations of plastids appear to be unstable and following cell division segregate into cells with pure populations of the different plastid types present in the original heteroplasmic cells. Given this observation, WT plastids would not be expected to persist with mutant plastids in a stable ratio arguing against heteroplasmy as the basis for the presence of faint WT bands in blot analyses of mutant DNA.

In those cases where the faint WT bands detected in DNA from plastid mutants have been examined in detail, they appear to originate from DNA located outside plastids. Plastid DNA sequences are known to be present in nuclei and mitochondria and result from transfer of DNA from plastids to these organelles (Bock and Timmis 2008). Purification of mutant chloroplasts removes faint WT bands on DNA blots indicating their derivation from extra-plastidic DNA (Hager et al. 1999; Ruf et al. 2000). Absence of WT bands in DNA blot analyses of purified chloroplast DNA confirms homoplasmy of mutant plastid genomes. Fractionation of mutant plastid genomes away from extra-plastidic DNA by pulsed field gel electrophoresis is an alternative method to verify homoplasmy (Swiatek et al. 2003a). The plastid-derived DNA sequences located outside plastids are likely to be non-functional. Polymerase chain reaction (PCR) analyses using *rbcL* primers did not detect any WT *rbcL* sequences in a Δ*rbcL* plastid mutant. Rather the *rbcL* sequences present in the amplified PCR products contained base-substitutions, insertions and deletions indicating they were non-functional and most probably extra-plastidic in origin (Kode et al. 2006). A further line of evidence to support the absence of WT plastid genomes in plastid mutants comes from inheritance data. If any WT plastid genomes were present in plastid knock out mutants they would be expected to segregate to individual eggs. On fertilisation the resulting seeds would lack *aadA* and give rise to seedlings that are sensitive and bleach on antibiotic medium allowing them to be identified amongst the thousands of green resistant seedlings containing *aadA* (Ruf et al. 2000). The absence of bleached leaves, or bleached sectors in leaves, in seedlings is consistent with the absence of WT plastid genomes in *aadA*-knock-out mutants. All these lines of evidence support the absence of wt plastid genomes in plastid mutants.

IV. Loss-of-Function Mutations in Tobacco Plastid Genes

The methods and principles described in Figs. 18.1, 18.2, 18.3, 18.4, and 18.5 apply to the analysis of all plastid genes. Reverse genetics studies using *aadA* to inactivate plastid genes have shown that tobacco plastid genes can be divided into two groups. Plastid genes are either dispensable or essential for growth and development of tobacco plants. Dispensable plastid genes include those required for photoautotrophic growth. Mutations in these genes give rise to plastid mutants that are propagated in vitro on plant growth medium containing sucrose. Whilst these genes are essential for photosynthesis and growth in soil they are categorised as dispensable because they can be rescued by supplying sucrose in the growth medium. Dispensable plastid genes include photosynthesis-related genes and a small number of non-photosynthesis-related genes.

V. Deletion of Dispensable Tobacco Plastid Genes

A. Deletion of Dispensable Photosynthesis-Related Genes

Table 18.1 provides a list of photosynthesis-related genes that have been inactivated by replacing the WT plastid genes with loss-of-function *aadA*-disrupted mutant alleles. The knock-outs are listed in chronological order. The first plastid gene to be inactivated was the *rbcL* gene (Kanevski and Maliga 1994). The mutant plants lacking *rbcL* function (Δ*rbcL* plants) were pale-green and non-photosynthetic and were propagated in vitro on plant growth medium containing sucrose. These Δ*rbcL* plants were used as recipients for nuclear transformation to show that a plastid-targeted product of a nuclear-localised *rbcL* gene could rescue the phenotype. Since this pioneering study, the list of knock-outs in photosynthesis-related genes has risen steadily to 29 genes (Table 18.1). In a few cases, knock-out mutations in the same plastid genes have been isolated independently in different laboratories, e.g. the *ndh* genes encoding subunits of NAD(P)H dehydrogenase (Burrows et al. 1998; Kofer et al. 1998; Shikanai et al. 1998), and these are listed in Table 18.1.

The overall impact of a knock-out on general plant phenotype varies for different photosynthesis-related genes. Loss-of-function mutations in a number of genes (*petA, petB, petD, psaA, psbA, psbE, psbF, psbL, psbJ, rbcL, ycf3, ycf5, ycf6*) result in pale pigmentation of leaves and loss of phototrophic growth, requiring propagation of mutants on sucrose-containing media (Table 18.1). Sucrose-dependent growth indicates that these genes play an essential role in photosynthesis and that there are no redundant pathways in chloroplasts, which could rescue

Table 18.1. Reports of deletions of photosynthesis-related plastid genes using *aadA*-based plastid transformation in tobacco. List ordered by date published

			Mutant phenotype			
	Gene name	Gene product	Pigmentation	Growth	Comments	Reference
1	*rbcL*	Ribulose bisphosphate carboxylase/oxygenase	Pale-green	Sucrose-dependent	Absence of RuBisCO	Kanevski and Maliga (1994)
2	*ycf3*	20 kDa protein	Pale-green	Sucrose-dependent	Required for assembly of PSI	Ruf et al. (1997)
3	*ndhB*	Subunit of NAD(P)H dehydrogenase	Normal	Photoautotrophic	Required for cyclic electron flow around PSI; chlororespiration	Shikanai et al. (1998)
4	*ndhI* *ndhJ* *ndhK*	Subunits of NAD(P)H dehydrogenase	Normal	Photoautotrophic	Required for cyclic electron flow around PSI	Burrows et al. (1998)
5	*ndhA* *ndhC* *ndhH* *ndhI* *ndhJ* *ndhK*	Subunits of NAD(P)H dehydrogenase	Normal	Photoautotrophic	Required to remove excess reducing equivalents; 5% heteroplasmy	Kofer et al. (1998)
6	*petN (ycf6)*	29 aa subunit VIII of the cytochrome b_6/f complex	Light sensitive: white leaves in normal light	Sucrose-dependent	Required for assembly or stability of cytochrome b_6f complex	Hager et al. (1999)
7	*petA* *petB* *petD*	cytf, cytb_6 & subunit IV components of the cytochrome b_6/f complex	Pale-green	Sucrose-dependent	Loss of cytochrome b_6f complex	Monde et al. (2000)
8	*psbZ (ycf9)*	62 aa PSII subunit	Normal	Photoautotrophic	Links LHC complex CP26 to PSII; reduced growth in low light	Ruf et al. (2000)
9	*psbZ (ycf9)*	62 aa PSII subunit	Normal	Photoautotrophic	Genuine subunit of PSII; pale-green on sucrose medium	Swiatek et al. (2001)
10	*psbZ (ycf9)*	62 aa PSII subunit	Normal	Photoautotrophic	Reduction in CP26; abnormal electron transport	Baena-González et al. (2001)
11	*psbJ*	40 aa PSII subunit	Light sensitive: leaves turn white in normal light	Sucrose-dependent	Assembly oxygen evolving complex	Hager et al. (2002)
12	*psbA*	D1 protein of PSII	Pale-green leaves	Sucrose-dependent	PSII null; upregulation of plastid terminal oxidase & NAD(P)H dehydrogenase	Baena-González et al. (2003)
13	*ycf10 (cemA)*	27 kDa protein	n.r.	Photoautotrophic	Extra-plastidic *ycf10* genes resolved from pt DNA by pulsed-field gels	Swiatek et al. (2003a), Świątek (2002)

14		PSII low MW subunits	Light sensitive: leaves turn white in normal light	Sucrose-dependent	*psbE* & *psbF* mutants lack PSII; *psbL* & *psbJ* mutants cxhibit low PSII activity in young leaves	Swiatek et al. (2003b), Ohad et al. (2004)
	psbE	83 aa subunit V				
	psbF	39 aa subunit VI				
	psbL	38 aa protein				
	psbJ	40 aa protein				
15	*petL*	31 amino aa subunit VI cytochromeb$_6$/f complex	Normal	Photoautotrophic	Cytochrome b$_6$f complex less stable; evolution of plastid editing sites	Fiebig et al. (2004), Schöttler et al. (2007a)
16	*psbI*	36 amino acid PSII protein	Normal	Photoautotrophic	50% reduction in PSII core; light sensitive	Schwenkert et al. (2006)
17	*ycf5 (ccsA)*	36 kDa protein (involved in heme attachment to c-type cytochome)	Pale-green	Sucrose-dependent	Heterotrophic growth indicated a role related to photosynthesis	Tsuruya et al. (2006)
18	*psaJ*	44 amino acid subunit IX of PSI	Light sensitive: reduced chlorophyll in low light	Photoautotrophic	Similar growth and pigmentation to WT at 600 μE m^{-2} s^{-1}; involved in PSI excitation	Schöttler et al. (2007b)
19	*petG* *petN (ycf6)*	Cytochrome b$_6$f complex: 37 aa subunit V 29 aa subunit VIII	Bleached in tissue culture 100 μE m^{-2} s^{-1}	Sucrose-dependent	Essential for stability of the cytochrome b$_6$f complex	Schwenkert et al. (2007)
20	*psbM*	34 aa PSII protein	Bleached if light intensity exceeded 200 μE m^{-2} s^{-1}	Photoautotrophic	Involved in interaction of redox components for electron flow to/ from PSII	Umate et al. (2007)
21	*psbTc*	34 aa PSII protein	Normal	Photoautotrophic	Moderate increase in light sensitivity of PSII	Umate et al. (2008)
22	*psaA*	Subunit PSI reaction centre	Pale green, bleached in high light	Sucrose-dependent	Altered expression of nuclear & plastid genes	Leelavathi et al. (2011)

aa amino acid, *n.r.* not reported

the photosynthetic functions encoded by the deleted plastid genes. Light intensity is known to influence the level of pigmentation of a number of these mutants (Table 18.1). Often, the pale-green phenotype is more apparent at higher light intensities, an effect we have also observed with Δ*rbcL* plants (Kode et al. 2006).

Flowering, fertilisation and setting seeds are problematic for tobacco plants grown *in vitro*. This limitation can be overcome by collecting seeds from mutant shoots grafted onto wild-type plants (Kanevski and Maliga 1994). Alternatively, seeds collected from heteroplasmic plants grown in soil can be screened to isolate mutant seedlings (Klaus et al. 2003). Mutations in a number of photosynthesis-related plastid genes (*ndh* genes, *petL, psaJ, psbI, psbM, psbTc, psbZ-ycf9, ycf10*) are compatible with photoautotrophic growth, indicating that these genes are not essential for photosynthesis. Photoautotrophic growth allows mutant plants to be grown in soil for the collection of seeds that can be stored or disseminated amongst the scientific community.

Knock-outs were pivotal to elucidate the roles of the plastid *ndh* genes, *ycf* genes, and genes encoding small protein subunits of photosystem I, photosystem II and the cytochrome b_6/f complex (Table 18.1). Analysis of plants with mutations in *ndh* genes supported a role in cyclic electron flow around photosystem I and chlororespiration to remove excess reducing equivalents (Burrows et al. 1998; Kofer et al. 1998; Shikanai et al. 1998). These *aadA*-based insertional knock-outs and targeted point mutations (discussed below) that result in loss of function (Horváth et al. 2000) continue to provide a useful resource for studying the NAD(P)H dehydrogenase complex. Systematic deletion of *ycf* genes, including those encoding very small polypeptides, e.g. the 29 amino acid product of *ycf6*, identified roles in photosynthesis (Table 18.1). These include *ycf3*, *ycf5* (*ccsA*), *ycf6* (*petN*), *ycf9* (*psbZ*) and *ycf10* (*cemA*). Elucidation of gene function was facilitated by analysis of knock-outs in homologous genes present in Chlamydomonas chloroplasts (see Chap. 11). For example, *cemA* is likely to be involved in chloroplast uptake of inorganic carbon (Rolland et al. 1997) and *ycf5* (*ccsA*) is likely to be required for attachment of heme to chloroplast c-type cytochromes (Xie and Merchant 1996).

B. Deletion of Dispensable Non-photosynthesis Related Genes

The inactivation of genetic-system genes encoding subunits of the bacterial-like plastid-encoded-plastid (PEP) RNA polymerase (*rpo*) was instrumental in confirming an alternative bacteriophage-like nucleus-encoded-plastid (NEP) RNA polymerase in chloroplasts (Allison et al. 1996), which had been suggested from earlier studies on ribosome deficient barley plastids (Hess et al. 1993). Further work on *rpo*-deficient plants facilitated the analysis of genes transcribed by NEP and PEP RNA polymerases (Hajdukiewicz et al. 1997; Serino and Maliga 1998; De Santis-Maciossek et al. 1999). Isolation of *rpo* mutants provides an interesting case study of the isolation of knock-out mutations in genes that are required to express the *aadA* marker gene. This is illustrated by Fig. 18.6. PEP RNA polymerase recognises the *rrn* promoter and is required to express the *aadA* marker gene. This means that in the presence of antibiotics, the WT *rpo* genes are required to express the *aadA* gene disrupting the *rpo* gene. Therefore, both knock-out and WT *rpo* alleles are required to confer resistance to antibiotics. Homoplasmic cells containing the knock-out *aadA* allele are obtained by removing antibiotics. This allows the segregation of green WT chloroplasts from white *rpo*-deficient plastids during plant growth and development. Using this approach pure albino *rpo*-deficient lines were isolated.

Other dispensable genetic-system genes (Table 18.2) include *rpl33*, *trnV-GAC* and *trnG-GCC*. Excision of *trnV-GAC* was noted during a study on marker excision using the *Cre/LoxP* system (Hajdukiewicz et al. 2001). The presence of Cre recombinase in plastids appeared to promote homologous

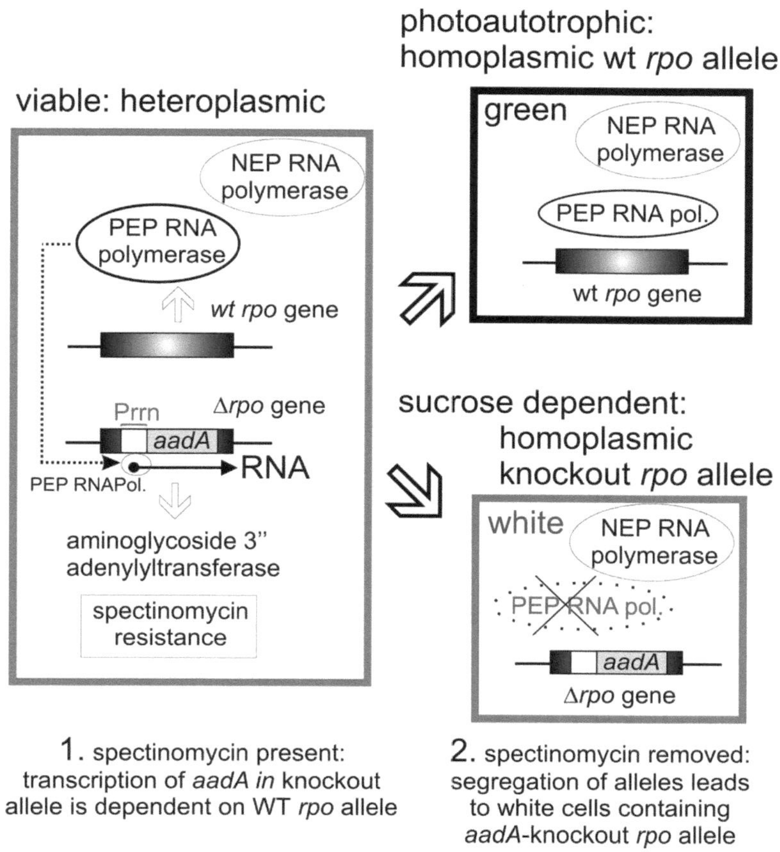

Fig. 18.6. Scheme used to isolate stable targeted knock-outs in tobacco plastid genes encoding subunits of the PEP RNA polymerase (Allison et al. 1996). PEP RNA polymerase is required to express the *aadA* marker gene. In the presence of antibiotics, both WT and knock-out *rpo* genes are required. In the absence of antibiotics, segregation of WT and knock-out alleles gives rise to albino shoots homoplasmic for mutant *rpo* alleles. Essential plastid genes are transcribed by the NEP RNA polymerase.

recombination between short direct *rrn* repeats of ~120 bp that flanked and excised the intervening *trnV-GAC*, *aadA* and green fluorescent protein genes (Hajdukiewicz et al. 2001). The plants lacking *trnV-GAC* appeared normal, grew in soil but were infertile. The authors (Hajdukiewicz et al. 2001) speculated that the apparent normal growth phenotype may result from a redundant gene product that could be either the plastid *trnV-UAC* gene (with extended codon recognition) or a nuclear-encoded *trnV-GAC* imported from the cytosol. Loss of function of the *rpl33* gene reduces tolerance to chilling (Rogalski et al. 2008b). Loss of *trnG-GCC* supported the idea that its function could be compensated by extended codon recognition (superwobbling) by the essential plastid *trnG-UCC* gene (Table 18.3). This was shown to be the case but *trnG-GCC* knock-outs grew more slowly than WT plants (Rogalski et al. 2008a). Disruption of the *sprA* gene encoding the 218 base small plastid RNA did not reveal a phenotype and its function remains unclear (Sugita et al. 1997a). Disruption of a potential origin of DNA replication mapped by electron microscopy did not have an impact on plastid DNA maintenance (Mühlbauer et al. 2002) indicating the presence of multiple DNA

Table 18.2. Reports of deletions of plastid genes not encoding photosynthesis-related proteins using *aadA* based plastid transformation in tobacco. Ordered by date published

	Gene name	Gene product	Mutant phenotype			Reference
			Pigmentation	Growth	Comments	
1	*rpoB*	β subunit of plastid- encoded plastid (PEP) RNA polymerase	White	Sucrose-dependent	Evidence for a nuclear encoded plastid (NEP) RNA polymerase	Allison et al. (1996), Hajdukiewicz et al. (1997)
2	*sprA*	218 nucleotide RNA	Normal	Photo-autotrophic	Dispensable with no clear mutant phenotype	Sugita et al. (1997b)
3	*rpoA* *rpoC1* *rpoC2*	α-, β′-, β″-subunits of plastid- encoded plastid (PEP) RNA polymerase	White	Sucrose-dependent	Redundant plastid targeted rpo-like proteins are not encoded by the nucleus	Serino and Maliga (1998)
4	*rpoA* *rpoB* *rpoC1*	α-, β-, β′subunits of plastid- encoded plastid (PEP) RNA polymerase	White	Sucrose-dependent	Functional ribosomes present	De Santis-Maciossek et al. (1999)
5	*trnV-GAC*	Valyl-transfer RNA	Normal	Photo-autotrophic	Infertile plants	Hajdukiewicz et al. (2001)
6	*oriA*	Origin of DNA replication mapped by EM	Normal	Photo-autotrophic	*oriA* is dispensable for plastid DNA replication	Mühlbauer et al. (2002)
7	*rpl33*	Ribosomal protein large subunit no. 33	Normal	Photo-autotrophic	Increased sensitivity to 4°C chilling stress	Rogalski et al. (2008b)
8	*trnG*-GCC	Glycyl-transfer RNA	Pale-green	Photo-autotrophic	Slow growth rate due to reduced translation efficiency	Rogalski et al. (2008a)

n.r. not reported, *EM* electron microscopy

replication origins or alternative pathways of DNA replication.

VI. Identification and Analysis of Essential Plastid Genes in Tobacco

A. Persistent Heteroplasmy of aadA-knock Outs Under Selection

The first genes to be characterized as indispensable in the tobacco plastid genome were the *ycf1* and *ycf2* genes, which encode large proteins of 226 and 267 kDa, respectively (Drescher et al. 2000). The functions of the *ycf* genes were not known, thus making them ideal targets for reverse genetics studies. However, unlike the dispensable plastid genes listed in Tables 18.1 and 18.2, it was not possible to isolate homoplasmic plants containing the mutant alleles. Despite propagation of plants for long periods of selection on antibiotic-containing medium, it was not possible to replace the WT allele with the strongly selected *aadA*-knock-out allele. The retention of the WT allele indicates that it is required for the survival of cells. Essential genes are also present in the Chlamydomonas chloroplast genome (Fischer et al. 1996). Persistent heteroplasmy of wild-type and *aadA*-knock-out alleles in the presence of antibiotics is a diagnostic feature of essential plastid genes. In this situation, the WT allele provides the essential function and the *aadA* knock-out allele is required to confer resistance to antibiotics (Fig. 18.7). Any homoplasmic cells containing the *aadA*-knock out allele resulting from segregation of plastid genomes

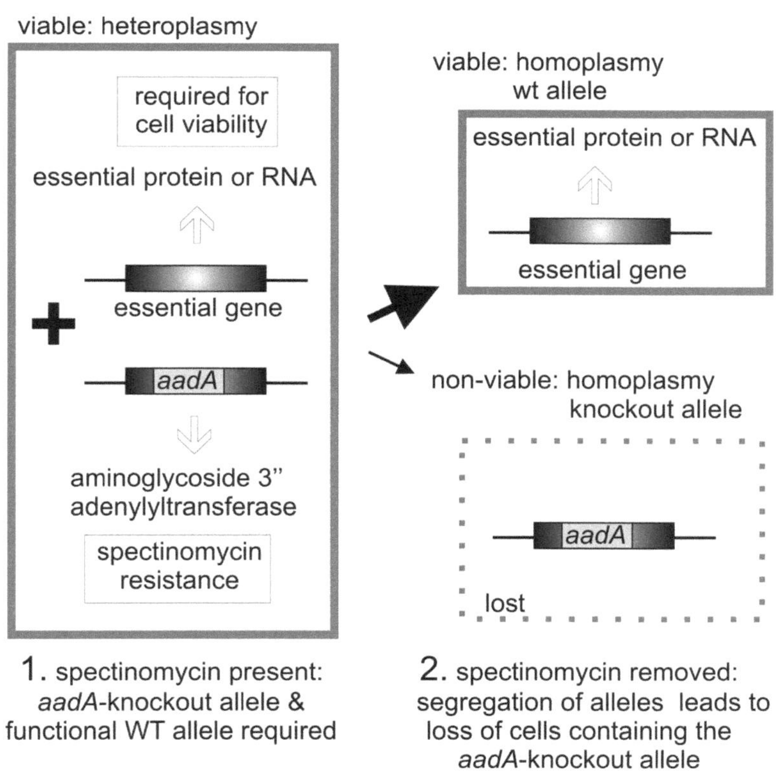

Fig. 18.7. Essential plastid genes are identified by persistent heteroplasmy in the presence of antibiotics. The knock-out allele is required for antibiotic resistance and the WT allele to provide the essential gene product needed for cell viability. Removal of antibiotics results in the loss of the knock-out allele. Homoplasmic cells containing the knock-out allele would not be viable and this explains the leaf-lamina-loss phenotype of some essential plastid genes (see Fig. 18.8).

Fig. 18.8. Phenotype of heteroplasmic tobacco plants containing WT and knock-out alleles of the essential plastid *accD* gene (Kode et al. 2005). (**a**) Irregular leaves resulting from variable loss of the leaf lamina. (**b**) Variegated leaves showing areas enriched in WT (*dark*) and mutant (*pale*) plastid genomes.

will not be viable (Fig. 18.7). Absence of the WT allele either arrests cell division or results in cell death. This is readily visualised in leaves, where large sections of the leaf lamina are lost presumably due to the production of non-viable cells containing the *aadA*-knock-out allele in the homoplasmic state. Leaf-lamina-loss is observed in the presence and absence of antibiotics. In the presence of antibiotics, the phenotype would also be affected by bleaching or arrest of division of any WT cells produced by cytoplasmic sorting of WT and *aadA*-knock-out alleles. Growth in the absence of antibiotics allows visualisation of the leaf-lamina-loss phenotype resulting from segregation of WT and knock-out alleles (Fig. 18.8).

Table 18.3 lists the essential genes identified in the tobacco plastid genome. The functions of *ycf1* and *ycf2* remain unclear. The list includes the *accD* gene, which encodes the β-carboxyl transferase subunit of acetyl CoA carboxylase. Acetyl CoA carboxylase is considered to be the regulatory enzyme of fatty acid synthesis. The essential *clpP1* gene encodes a subunit of a protease, which must act on an essential plastid polypeptide encoded by the nuclear or plastid genomes. The essentiality of the replication origin *oriB1* may be a consequence of its location in the essential gene *ycf1* rather than an indispensable role in DNA replication (Mühlbauer et al. 2002). The remaining genes listed in Table 18.3 are genetic system genes, which are presumably essential, because they are required to express the *ycf1*, *ycf2*, *clpP1* and *accD* genes. An alternative nomenclature distinguishes 'structural genes' coding for structural proteins and enzymes from 'regulatory genes', which are involved in gene expression. However, the terminology can lead to misunderstanding because a regulatory gene such as the plastid *clpP1* gene involved in protein turnover encodes the proteolytic rather than the regulatory subunits of Clp protease, which are encoded by nuclear genes. Essential genetic system genes encode subunits of plastid ribosomes as well as plastid transfer RNA molecules (Legen et al. 2007; Rogalski et al. 2008a). Moreover, this indicates that alternative gene products encoded by the nuclear or plastid genomes cannot rescue loss-of-function mutations in these genetic system genes. The majority of these studies (Table 18.3) recorded variable loss of leaf lamina (discussed above, see example in Fig. 18.8). Once selection was removed, variable rates of persistence of the *aadA*-knock-out allele were reported (Drescher et al. 2000), possibly reflecting varying levels of selection for restoration of the essential function encoded by the genes. Persistence of the knock-out allele during growth and development of plants allowed its transmission to the next generation in some cases (Table 18.3). The presence of essential genes in tobacco plastids means that plastid protein synthesis is indispensable for growth

Table 18.3. Essential genes that cannot be deleted from the tobacco plastid genome. In the presence of antibiotics both the *aadA*-knock-out and WT alleles are maintained. Sexual transmission of the knock-out allele to progeny was at low frequency (less than 10% and often ~1% or less). List ordered by date published

	Gene name	Gene product	Comments	Reference
1	*ycf1*	226 kDa protein	Persistence of knock-out allele in the absence of selection allows sexual transmission	Drescher et al. (2000)
2	*ycf2*	267 kDa protein	Spectinomycin removal resulted in rapid loss of the knock-out allele	Drescher et al. (2000)
3	*clpP1*	Subunit of the ATP-dependent clpP protease	Variable loss of leaf lamina	Shikanai et al. (2001)
4	*oriB1 (ycf1)*	Origin of DNA replication in *ycf1*	*oriB1* is located in the large inverted repeat. *oriB1* in *ycf1* cannot be deleted. *oriB2* in orf350 can be deleted	Mühlbauer et al. (2002)
5	*accD*	β-carboxyl transferase subunit of acetyl-CoA carboxylase	Variable loss of leaf lamina; Variegated leaves with pale-green sectors; Knock-out allele transmitted to progeny	Kode et al. (2005)
6	*rps18*	Ribosomal protein of the small subunit no. 18	Variable loss of leaf lamina Knock-out allele transmitted to progeny	Rogalski et al. (2006)
7	*trnC*-GCA	Cysteinyl-transfer RNA	Variable loss of leaf lamina	Legen et al. (2007)
8	*trnN*-GUU	Asparagyl- transfer RNA(GUU)	Variable loss of leaf lamina; Variegated leaves with white sectors; Knock-out allele transmitted to progeny	Legen et al. (2007)
9	*trnG*-UCC	Glycyl-transfer RNA(UCC)	Variable loss of leaf lamina	Rogalski et al. (2008a)
10	*rps2* *rps4* *rpl20*	Ribosomal proteins: small subunit nos. 2 & 4, large subunit no. 20	Variable loss of leaf lamina Knock-out allele transmitted to progeny	Rogalski et al. (2008b)

and development of tobacco (Ahlert et al. 2003). Plastid protein synthesis is dispensable in Brassicas (Zubko and Day 1998) and cereals (Hess et al. 1993; Zubko and Day 2002), indicating the absence of essential plastid-encoded proteins in these species. The presence of an alternative plastid-targeted acetyl CoA carboxylase encoded by a nuclear gene in Brassicas and cereals, that is not present in tobacco, has been suggested to account for the dispensability of plastid protein synthesis in these species (Kode et al. 2005).

B. Deleting Essential Genes Using Site-Specific Recombinases

Site-specific recombinases provide an alternative method to delete essential plastid genes such as the *clpP1* gene (Kuroda and Maliga 2003). This is illustrated in Fig. 18.9. The *loxP* target sites of the Cre site specific recombinase were introduced into a cloned *clpP1* gene. The construct with *loxP* sites was transformed into tobacco plastids using an adjacent *aadA* marker gene to select antibiotic-resistant transplastomic plants. The resulting *clpP1* gene is functional. Introduction of a nucleus-localised gene encoding a plastid targeted Cre recombinase by sexual crosses resulted in the simultaneous excision of sequences flanked by *loxP* sites from the majority of plastid genomes in a cell. This results in excision of *clpP1* exons 2 and 3 from the plastid genome giving rise to a deleted *clpP1* gene in the majority of plastid genomes. The result was an arrest of seedling growth confirming that the *clpP1* gene is an essential plastid gene in tobacco (Kuroda and

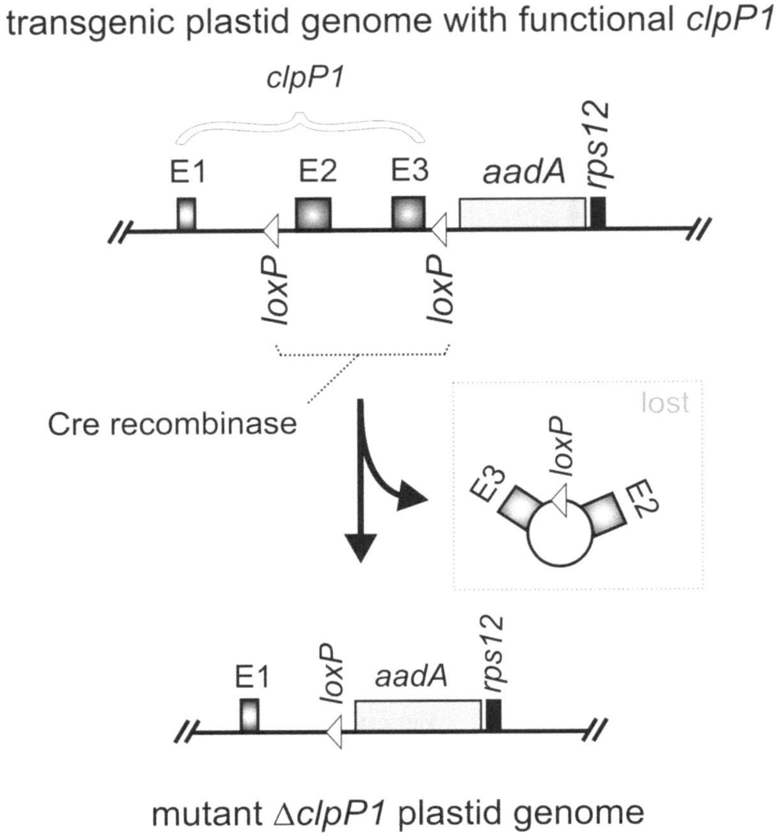

Fig. 18.9. Scheme showing excision of the essential *clpP1* plastid gene using the site-specific recombinase Cre acting on its *loxP* DNA recognition sites (Kuroda and Maliga 2003). The plastid *clpP1* gene is expressed giving rise to viable tobacco plants. Introduction of plastid targeted Cre recombinase results in the simultaneous deletion of *clpP1 from* the majority of plastids and arrests seedling growth.

Maliga 2003). The production of a uniform population of deleted plastid genomes is an advantage of this approach for studying essential plastid genes. Use of a chemically inducible nuclear promoter (Zuo and Chua 2000) to regulate Cre expression would allow the impact of gene deletion at different growth stages to be evaluated. The approach would allow the study of essential plastid genes encoding both RNA and protein products.

C. New Approaches to Study the Function of Essential Plastid Genes

The demonstration of a translational riboswitch that functions in tobacco chloroplasts provides an opportunity to regulate the expression of essential plastid genes encoding protein products (Verhounig et al. 2010). This is illustrated in Fig. 18.10. The theophylline riboswitch is active in the presence of theophylline. Adding this riboswitch to the 5′ UTR of an essential plastid gene requires replacement of the cognate wt essential gene by plastid transformation. In the presence of theophylline, the RNA from the riboswitch version of the essential plastid gene would be translated providing the essential function required for cell viability. When theophylline is removed translation would no longer take place resulting in loss of the essential protein. This would allow control over the timing of loss of an essential gene product enabling the consequences of the deficiency to be followed at different developmental stages. Kinetic analysis that follows

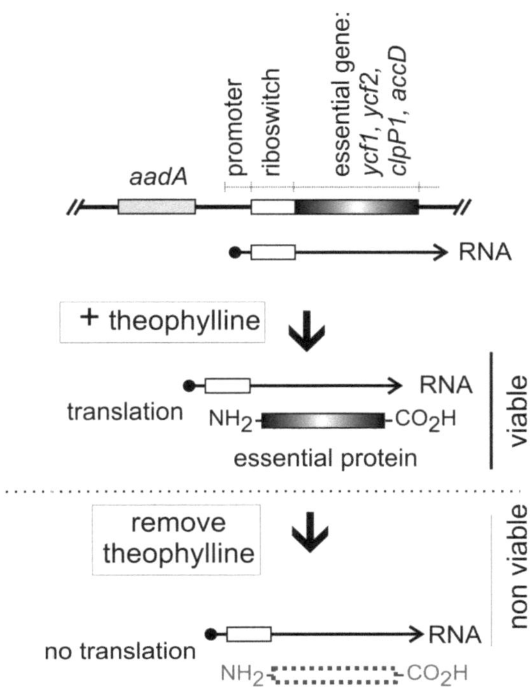

Fig. 18.10. Inducible gene expression with a theophylline 'translational-on' riboswitch (Verhounig et al. 2010) would allow the isolation of homoplasmic plants in which the expression of essential genes can be regulated. This allows analysis at different developmental stages and at different time points after gene expression is reduced to investigate potential molecular targets affected by loss of the essential gene product.

the impact of the progressive loss of an essential gene product following removal of theophylline would provide information on the plastid pathways affected. Accompanying changes in the biochemistry, cell biology, molecular biology and physiology of affected plants would shed light on the function(s) of the essential gene.

VII. Introducing Site-Directed Mutations into Plastid Genes

A. Replacing WT Plastid Genes with Mutant Alleles

Targeted replacement of WT plastid genes with alleles containing site-directed mutations involves the same homologous recombination pathway and selection used to isolate knock-out mutants. The marker gene is located in close proximity to the mutant allele on the same vector (Fig. 18.11) or co-transformed on a physically separated vector (Rumeau et al. 2004). The scheme shown in Fig. 18.11 is similar to that used to delete a gene within an operon (Fig. 18.5). Integration of the mutant allele requires cross-over events that flank both the mutant allele and the *aadA* marker gene (Fig. 18.11a). A cross-over event between the mutant allele and marker gene only inserts the marker gene (Fig. 18.11b). Once the mutant allele is inserted into the plastid genome, homoplasmy of transgenic mutant genomes is attained by selection for the marker gene. Even when the mutant allele is integrated (Fig. 18.11a) in a fraction of plastid genomes, it can be lost through cross-over or gene conversion events with WT plastid genomes in the early heteroplasmic stages of plastid transformation. Therefore only a proportion of antibiotic resistant mutant plants will contain the mutant allele. Examples of replacement of WT alleles with mutant alleles are listed in Table 18.4. The method has been used successfully: to change codons to study RNA editing (Bock et al. 1994), to mutate LS RuBisCO (Whitney et al. 1999), to study the functions of introns (Petersen et al. 2011) and to analyse the role of protein phosphorylation (Martin et al. 2009). Targeted replacements have also been used to introduce His-tags (Rumeau et al. 2004; Martin et al. 2009) or other epitope tags (Albus et al. 2010) into protein products of plastid genes to aid their purification. The introduction of a His-tag into LS RuBisCO altered zinc levels in leaves when tobacco plants were grown on media with raised zinc content (Rumeau et al. 2004). To study the role of a potential replication origin (*oriB1*) located in the essential *ycf1* gene, site directed mutations were introduced into the replication origin to preserve the *ycf1* open reading frame (Scharff and Koop 2007). The results indicate that *oriB* is probably dispensable for plastid DNA replication.

The scheme shown in Fig. 18.11a can also be used to replace tobacco plastid genes with homologous genes from other species to

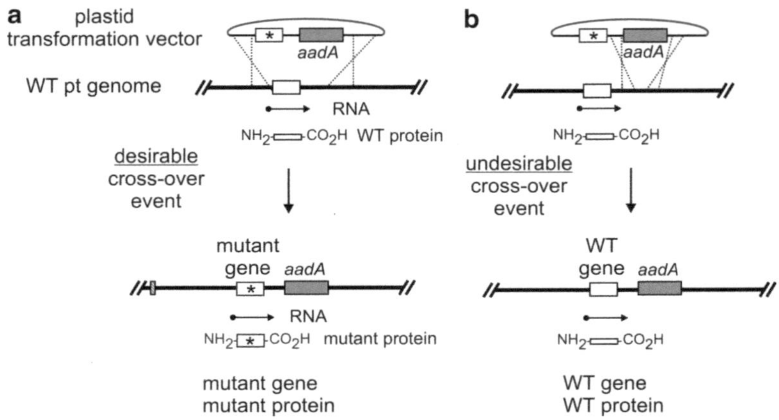

Fig. 18.11. Scheme for replacing WT plastid genes with a mutant allele containing a site-directed mutation (*). (**a**) Cross-over events flanking the mutation and *aadA* marker gene introduce the mutation into the plastid genome. (**b**) Cross-over events incorporating the *aadA* marker gene but not the mutation into the plastid genome.

determine the impact of sequence divergence on plastid gene function. For example, the native *rbcL* gene was replaced with the sunflower r*bcL* gene in tobacco chloroplasts (Kanevski et al. 1999). Thirty-two amino acid differences are found between the 477 amino acid tobacco and 485 amino acid sunflower LS RuBisCO polypeptides. Plants containing a hybrid RuBisCO holoenzyme comprised of sunflower large subunits and tobacco small subunits possessed 12% of the RuBisCO activity of wild-type tobacco plants (Kanevski et al. 1999) and required grafting onto wild-type tobacco rootstocks to flower and set seed in a greenhouse.

B. Introducing Mutations into Essential Plastid Genes

Targeted knock-outs resulting in persistent heteroplasmy in the presence of antibiotics have identified essential genes in the tobacco plastid genome (Table 18.3 and Sect. VI above). Heteroplasmy hinders analyses of the functions of essential plastid genes. Point mutations in essential genes that retain some function may allow the isolation of homoplasmic mutant plants. Analyses of these plants enables studies on the molecular processes affected and would provide information on gene function. Mutant plants with partial-function alleles of essential plastid genes have not been isolated. When mutations changing the ATG start codons of the *ycf1* and *ycf2* to GTG or ACG were introduced into plastids the mutant alleles did not persist (Khakhlova and Bock 2006). Strong selection pressure converted the mutant alleles back to the WT sequence. A change from ATG to ATT or ATA in the *ycf2* initiator codon persisted longer in transformed plastids but did not give rise to homoplasmic mutant plants (Khakhlova and Bock 2006). Whilst these studies were useful for providing information on biased gene conversion in plastids they did not further our understanding of the functions of the *ycf1* and *ycf2* genes. In summary, introduction of partial-function mutant alleles of essential plastid genes can be difficult due to strong selection for the WT gene resulting in conversion of the mutant alleles back to the original WT sequence.

VIII. Multiple Rounds of Plastid Transformation: Double Mutants; Site Directed Mutations

Multiple rounds of plastid transformation allow targeted changes to more than one plastid gene or the replacement of knock-out alleles with partial function alleles. This

Table 18.4. Introduction of site-directed mutations into tobacco plastid genes. Experiments involved *aadA*-based transformation using the particle gun unless otherwise indicated

	Gene name	Gene product	Mutation	Comments	Reference
1	*psbF*	β-subunit of cytochrome b559	UUU → UCU spinach editing site	Lack of editing results in slow growth & pale green leaves	Bock et al. (1994)
2	*rbcL*	LS RuBisCO	Leucine codon 335 changed to Valine	Growth required 0.3% CO_2	Whitney et al. (1999), Whitney and Sharwood (2008)
3[a]	*rbcL*	LS RuBisCO	32 amino acid changes	Replaced with sunflower *rbcL* RuBisCO activity 12% of WT	Kanevski et al. (1999)
3[b]	*ndhB*	Subunit of NAD(P)H dehydrogenase	C inserted into codon 206 causes frameshift and termination at stop codons	Reduced photosynthesis under humidity stress	Horváth et al. (2000)
4[c]	*rbcL*	LS RuBisCO	C-terminal His_6	Phenotype similar to WT; elevated zinc in leaves when zinc content in media was raised	Rumeau et al. (2004)
5	*ycf1*	226 kDa protein	Start ATG → GTG or ACG	Mutation corrected back to ATG by gene conversion	Khakhlova and Bock (2006)
6	*ycf2*	267 kDa protein	Start ATG → GTG, ACG, ATT or ATA	Gene conversion back to ATG is more rapid for GTG & ACG than ATT or ATA	Khakhlova and Bock (2006)
7	*oriB1*	Origin of DNA replication mapped by EM	Bases changed to destabilise *oriB* stem-loop	*oriB* is probably dispensable for plastid DNA replication	Scharff and Koop (2007)
8	*psbE*	α-subunit cytochrome b559	N-terminal His_6 or His_{10}- tags	Oxygen evolution reduced 10–30%	Fey et al. (2008)
9	*ndhF*	Subunit of NAD(P)H dehydrogenase	ACT → GCT or TCT or GAT	Thr 181 mutated to Ala, Ser or Asp to study role of phosphorylation	Martin et al. (2009)
10	*ycf3*	Assembly of PSI	C-terminal FLAG peptide	Purification of Ycf3 complex	Albus et al. (2010)
11	*ycf3*	Assembly of PSI	Removal of one or both introns	Intron I removal causes a slow growth phenotype in low light & chilling sensitivity	Petersen et al. (2011)

[a]Replacement with a homologous gene

[b]Mutations in *ndhB* were linked to streptomycin-resistant *rps12* and spectinomycin-resistant 16S *rrn* alleles in a 7.8 kbp DNA sequence transformed using PEG and protoplasts

[c]The His-tagged LS RuBisCO gene was introduced by PEG-based co-transformation of protoplasts with a mutant plastid 16S *rrn* gene conferring spectinomycin resistance

requires the use of different selectable markers at each round of transformation. Alternatively, marker excision following the isolation of mutant plants allows the re-use of the highly efficient *aadA* marker gene to introduce mutations in other plastid genes.

A. Use of Different Marker Genes

To study the role of origins of replication (ori) in tobacco plastid DNA, plants in which *oriA* was deleted using the *aadA* marker gene were subject to a second round of transformation using a *nptII* marker gene and kanamycin selection to mutate *oriB* (Scharff and Koop 2007). Plastid transformation using *nptII* involved transforming micro-colonies derived from protoplasts rather than leaves (Scharff and Koop 2007). Other marker gene/selection agent combinations used to isolate plastid transformants include the *aphA6* gene/kanamycin (Huang et al. 2002), the anthranilate synthase gene/7-methyl-DL-tryptophan or 4-methylindole (Barone et al. 2009) and the *cat* gene/chloramphenicol (Li et al. 2010). The use of multiple rounds of plastid transformation based on different selection agents will depend on the efficiencies of these alternative markers compared to the widely used *aadA* marker gene.

B. Marker Excision and Re-transformation Using the Same Marker Gene

The use of site-directed mutations for detailed investigations on the structure-function relationships of a plastid encoded protein is facilitated by the isolation of plastid deletion mutants lacking marker genes. This allows the rapid introduction of mutant alleles using the *aadA* marker gene without undesirable recombination events. The isolation and use of marker-free plastid deletion mutants is illustrated in Fig. 18.12. Figure 18.12a illustrates a recombination event between 649 bp direct repeats that deletes *rbcL* and the foreign *aadA* and *gusA* genes. This excision event deletes the region between the direct repeats to give rise to a marker free Δ*rbcL* plastid genome (Kode et al. 2006). The marker free Δ*rbcL* plants are pale green and heterotrophic. Figure 18.12b shows excision of *aadA* using the Cre/*loxP* system to give rise to a marker-free genome containing the *Rhodospirillum rubrum rbcM* gene (Whitney and Sharwood 2008). In this mutant, the

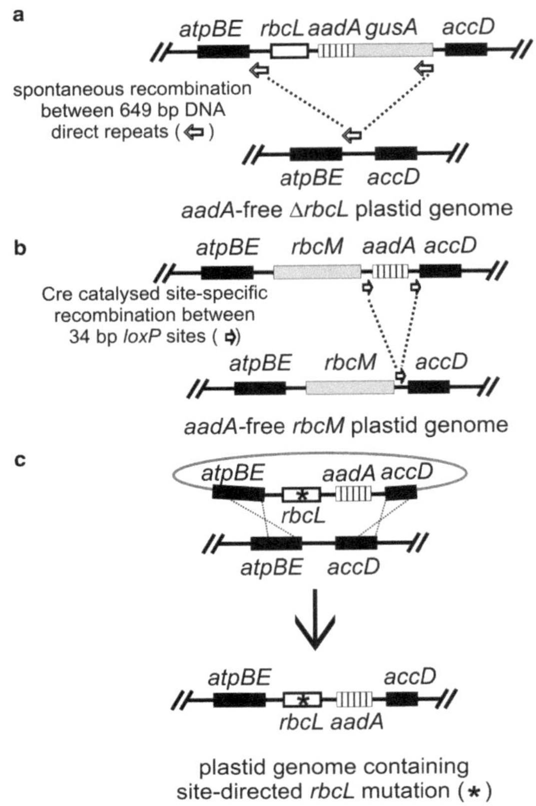

Fig. 18.12. Isolation of *aadA*-free tobacco plastid deletion mutants. (**a**) Map of transgenic plastid DNA with foreign *aadA* and *gusA* genes containing a 649 bp duplication that flanked the foreign genes and the *rbcL* gene. Spontaneous recombination between these direct repeats deleted the intervening genes and one copy of the repeat giving rise to a marker-free plastid genome with a 1,894 bp deletion (Δ*rbcL*) removing the entire *rbcL* gene (Kode et al. 2006). (**b**) A transplastomic plant in which the plastid *rbcL* gene was replaced by a bacterial *rbcM* gene. The Cre site-specific recombinase recognises the *loxP* sites and excises the *aadA* marker gene (Whitney and Sharwood 2008). (**c**) Mutant *rbcL* genes are incorporated into the plastid genome lacking *rbcL* without any risk of undesirable cross-over events that would insert the *aadA* marker but not the mutation (see Fig. 18.11b).

tobacco *rbcL* gene is replaced by the *rbcM* which allows growth in high CO_2 but not under normal atmospheric CO_2 concentration. Therefore both marker-free Δ*rbcL* (Kode et al. 2006) and *rbcM* (Whitney and Sharwood 2008) tobacco plants can be rescued by transformation with the WT *rbcL* gene. Importantly, when mutant *rbcL* alleles are transformed into Δ*rbcL* or *rbcM* plants, all plastid transformants will contain the mutant allele because a cross-over event between *aadA* and *rbcL* cannot take place (Fig. 18.12c); the cross-over event shown in Fig. 18.11b is ruled out. This is because the DNA sequences between *aadA* and the mutation in the vector have been deleted from the resident plastid genome preventing a cross-over in this region. Use of Δ*rbcL* plants as recipients for plastid transformation has the added advantage that the phenotype of site-directed mutant *rbcL* genes will not be masked by WT plastid genomes and will be revealed early in the transformation process.

IX. Perspective

Reverse genetics is a routine procedure in tobacco and has been an indispensable tool for studying the functions of plastid genes. The field has been dominated by single rounds of transformation to isolate knockout plastid genes using *aadA*-based plastid transformation. Methods to isolate plastid deletion mutants lacking marker genes facilitate the introduction of site-directed mutations to study the structure-function relationships of plastid gene products. New methods involving regulated gene expression or inducible gene excision are required to study the functions of essential plastid genes. The successful replacement of an essential plastid gene with partial function alleles would further our understanding of the roles of these genes in plastids. Reverse genetics studies on the plastid genomes in flowering plants other than tobacco would allow comparative analysis of mutant phenotypes. The essential genes encoding polypeptides in tobacco (listed in Table 18.3) may be dispensable in *Brassica napus*. Use of alternative angiosperm species such as *B. napus* for reverse genetics may further our understanding of the essential genes that are difficult to study in tobacco. However, the application of routine plastid reverse genetics methods to a wider range of angiosperm species requires improving their efficiencies of plastid transformation to the frequencies achievable in tobacco.

Acknowledgements

I thank: Dr Elisabeth Mudd and the editors for improving the manuscript, and Prof Ulrich Koop (München) for information on the *ycf10* knockout. This work was supported by research grants BB/E020445 and BB/I011552 from the Biotechnology and Biological Sciences Research Council (UK).

Note Added in Proof

Articles published after going to press: Fleischmann et al. (2011) used *aadA*-knockouts to show that the plastid genes encoding ribosomal proteins *rpl22, rpl23, rpl32, rps3* and *rps16* were essential whereas ribosomal proteins *rps15* and *rpl36* were nonessential. Whitney et al. (2011) introduced *rbcL* genes from *Flaveria* C3 and C4 species into *aadA*-free *rbcM* tobacco plants (Fig. 18.12) to identify amino acids affecting the carboxylation rate and CO_2 affinity of RuBisCO.

References

Ahlert D, Ruf S, Bock R (2003) Plastid protein synthesis is required for plant development in tobacco. Proc Natl Acad Sci USA 100:15730–15735

Albus C, Ruf S, Schöttler MA, Lein W, Kehr J, Bock R (2010) Y3IP1, a nucleus-encoded thylakoid protein, co-operates with the plastid-encoded Ycf3 protein in photosystem I assembly. Plant Cell 22: 2838–2855

Allison LA, Simon LD, Maliga P (1996) Deletion of *rpoB* reveals a second distinct transcription system in plastids of higher plants. EMBO J 15:2802–2809

Baena-González E, Gray JC, Tyystjarvi E, Aro EM, Mäenpää P (2001) Abnormal regulation of photosynthetic electron transport in a chloroplast *ycf9* inactivation mutant. J Biol Chem 276:20795–20802

Baena-González E, Allahverdiyeva Y, Svab Z, Maliga P, Josse EM, Kuntz M, Mäenpää P, Aro EM (2003) Deletion of the tobacco plastid *psbA* gene triggers an upregulation of the thylakoid-associated NAD(P)H dehydrogenase complex and the plastid terminal oxidase (PTOX). Plant J 35:704–716

Barone P, Zhang XH, Widholm JM (2009) Tobacco plastid transformation using the feedback-insensitive anthranilate synthase [alpha]-subunit of tobacco (ASA2) as a new selectable marker. J Exp Bot 60:3195–3202

Birky CW (2001) The inheritance of genes in mitochondria and chloroplasts: laws, mechanisms, and models. Ann Rev Genet 35:125–148

Bock R (2007) Structure, function, and inheritance of plastid genomes. Cell and molecular biology of plastids. In: Bock R (ed) Topics in current genetics, vol 19. Springer, Berlin/Heidelberg, pp 29–63

Bock R, Timmis JN (2008) Reconstructing evolution: gene transfer from plastids to the nucleus. Bioessays 30:556–566

Bock R, Kössel H, Maliga P (1994) Introduction of a heterologous editing site into the tobacco plastid genome: the lack of RNA editing leads to a mutant phenotype. EMBO J 13:4623–4628

Burrows PA, Sazanov LA, Svab Z, Maliga P, Nixon PJ (1998) Identification of a functional respiratory complex in chloroplasts through analysis of tobacco mutants containing disrupted plastid *ndh* genes. EMBO J 17:868–876

Cerutti H, Johnson AM, Boynton JE, Gillham NW (1995) Inhibition of chloroplast DNA recombination and repair by dominant negative mutants of *Escherichia coli* RecA. Mol Cell Biol 15: 3003–3011

Colbert T, Till BJ, Tompa R, Reynolds S, Steine MN, Yeung AT, McCallum CM, Comai L, Henikoff S (2001) High-throughput screening for induced point mutations. Plant Physiol 126:480–484

De Santis-Maciossek G, Kofer W, Bock A, Schoch S, Maier RM, Wanner G, Rüdiger W, Koop HU, Herrmann RG (1999) Targeted disruption of the plastid RNA polymerase genes *rpoA, B* and *C1*: molecular biology, biochemistry and ultrastructure. Plant J 18:477–489

Drescher A, Ruf S, Calsa T, Carrer H, Bock R (2000) The two largest chloroplast genome-encoded open reading frames of higher plants are essential genes. Plant J 22:97–104

Fey H, Piano D, Horn R, Fischer D, Schmidt M, Ruf S, Schröder WP, Bock R, Büchel C (2008) Isolation of highly active photosystem II core complexes with a His-tagged Cyt b(559) subunit from transplastomic tobacco plants. Biochim Biophys Acta Bioenerg 1777:1501–1509

Fiebig A, Stegemann S, Bock R (2004) Rapid evolution of RNA editing sites in a small non-essential plastid gene. Nucleic Acids Res 32:3615–3622

Fischer N, Stampacchia O, Redding K, Rochaix JD (1996) Selectable marker recycling in the chloroplast. Mol Gen Genet 251:373–380

Fleischmann TT, Scharff LB, Alkatib S, Hasdorf S, Schoettler MA, Bock R (2011) Nonessential plastid-encoded ribosomal proteins in tobacco: a developmental role for plastid translation and implications for reductive genome evolution. Plant Cell 23:3137–3155

Gilchrist E, Haughn G (2010) Reverse genetics techniques: engineering loss and gain of gene function in plants. Brief Funct Genomics 9:103–110

Hager M, Biehler K, Illerhaus J, Ruf S, Bock R (1999) Targeted inactivation of the smallest plastid genome-encoded open reading frame reveals a novel and essential subunit of the cytochrome b(6)f complex. EMBO J 18:5834–5842

Hager M, Hermann M, Biehler A, Krieger-Liszkay A, Bock R (2002) Lack of the small plastid-encoded *PsbJ* polypeptide results in a defective water-splitting apparatus of photosystem II, reduced photosystem I levels, and hypersensitivity to light. J Biol Chem 277:14031–14039

Hajdukiewicz PTJ, Allison LA, Maliga P (1997) The two RNA polymerases encoded by the nuclear and the plastid compartments transcribe distinct groups of genes in tobacco plastids. EMBO J 16:4041–4048

Hajdukiewicz PTJ, Gilbertson L, Staub JM (2001) Multiple pathways for Cre/*lox*-mediated recombination in plastids. Plant J 27:161–170

Hardy S, Legagneux V, Audic Y, Paillard L (2010) Reverse genetics in eukaryotes. Biol Cell 102:561–580

Herskowitz I (1987) Functional inactivation of genes by dominant negative mutations. Nature 329:219–222

Hess WR, Prombona A, Fieder B, Subramanian AR, Börner T (1993) Chloroplast rps15 and the *rpoB/C1/C2* gene cluster are strongly transcribed in ribosome deficient plastids: evidence for a functioning non chloroplast encoded RNA polymerase. EMBO J 12:563–571

Hohn B, Puchta H (2003) Some like it sticky: targeting of the rice gene Waxy. Trends Plant Sci 8:51–53

Horváth EM, Peter SO, Joët T, Rumeau D, Cournac L, Horváth GV, Kavanagh TA, Schäfer C, Peltier G, Medgyesy P (2000) Targeted inactivation of the plastid *ndhB* gene in tobacco results in an enhanced sensitivity of photosynthesis to moderate stomatal closure. Plant Physiol 123:1337–1349

Hotto AM, Huston ZE, Stern DB (2010) Overexpression of a natural chloroplast-encoded antisense RNA in tobacco destabilizes 5S rRNA and retards plant growth. BMC Plant Biol 10:213

Huang FC, Klaus SM, Herz S, Zou Z, Koop HU, Golds TJ (2002) Efficient plastid transformation in tobacco using the *aphA-6* gene and kanamycin selection. Mol Genet Genomics 268:19–27

Kanevski I, Maliga P (1994) Relocation of the plastid *rbcL* gene to the nucleus yields functional ribulose-1,5-bisphosphate carboxylase in tobacco chloroplasts. Proc Natl Acad Sci USA 91:1969–1973

Kanevski I, Maliga P, Rhoades DF, Gutteridge S (1999) Plastome engineering of ribulose-1,5-bisphosphate carboxylase/oxygenase in tobacco to form a sunflower large subunit and tobacco small subunit hybrid. Plant Physiol 119:133–141

Khakhlova O, Bock R (2006) Elimination of deleterious mutations in plastid genomes by gene conversion. Plant J 46:85–94

Klaus SMJ, Huang FC, Golds TJ, Koop HU (2004) Generation of marker-free plastid transformants using a transiently cointegrated selection gene. Nat Biotechnol 22:225–229

Klaus SMJ, Huang FC, Eibl C, Koop HU, Golds TJ (2003) Rapid and proven production of transplastomic tobacco plants by restoration of pigmentation and photosynthesis. Plant J 35:811–821

Klein TM, Wolf ED, Wu R, Sanford JC (1987) High velocity microprojectiles for delivering nucleic acids into living cells. Nature 327:70–73

Kode V, Mudd EA, Iamtham S, Day A (2005) The tobacco plastid *accD* gene is essential and is required for leaf development. Plant J 44:237–244

Kode V, Mudd EA, Iamtham S, Day A (2006) Isolation of precise plastid deletion mutants by homology-based excision: a resource for site-directed mutagenesis, multi-gene changes and high-throughput plastid transformation. Plant J 46:901–909

Kofer W, Koop HU, Wanner G, Steinmüller K (1998) Mutagenesis of the genes encoding subunits A, C, H, I, J and K of the plastid NAD(P)H-plastoquinone-oxidoreductase in tobacco by polyethylene glycol-mediated plastome transformation. Mol Gen Genet 258:166–173

Koop HU, Kofer W, Steinmüller K (1998) Judging the homoplastomic state of plastid transformants – reply. Trends Plant Sci 3:377

Kuroda H, Maliga P (2003) The plastid *clpP1* protease gene is essential for plant development. Nature 425:86–89

Legen J, Wanner G, Herrmann RG, Small I, Schmitz-Linneweber C (2007) Plastid tRNA genes *trnC-GCA* and *trnN-GUU* are essential for plant cell development. Plant J 51:751–762

Leelavathi S, Bhardwaj A, Kumar S, Dass A, Pathak R, Pandey S, Tripathy B, Padmalatha K, Dhandapani G, Kanakachari M, Kumar P, Cella R, Siva Reddy V (2011) Genome-wide transcriptome and proteome analyses of tobacco *psaA* and *psbA* deletion mutants. Plant Mol Biol 76:407–423

Li SJ, Cronan JE (1992) Putative zinc finger protein encoded by a conserved chloroplast gene is very likely a subunit of a biotin dependent carboxylase. Plant Mol Biol 20:759–761

Li W, Ruf S, Bock R (2010) Chloramphenicol acetyltransferase as selectable marker for plastid transformation. Plant Mol Biol 72:443–451

Maliga P, Nixon PJ (1998) Judging the homoplastomic state of plastid transformants. Trends Plant Sci 3: 376–377

Martin M, Funk HT, Serrot PH, Poltnigg P, Sabater B (2009) Functional characterization of the thylakoid Ndh complex phosphorylation by site-directed mutations in the ndhF gene. Biochim Biophys Acta Bioenerg 1787:920–928

Martin W, Rujan T, Richly E, Hansen A, Cornelsen S, Lins T, Leister D, Stoebe B, Hasegawa M, Penny D (2002) Evolutionary analysis of Arabidopsis, cyanobacterial, and chloroplast genomes reveals plastid phylogeny and thousands of cyanobacterial genes in the nucleus. Proc Natl Acad Sci USA 99:12246–12251

Monde RA, Zito F, Olive J, Wollman FA, Stern DB (2000) Post-transcriptional defects in tobacco chloroplast mutants lacking the cytochrome b(6)/f complex. Plant J 21:61–72

Mühlbauer SK, Lossl A, Tzekova L, Zou ZR, Koop HU (2002) Functional analysis of plastid DNA replication origins in tobacco by targeted inactivation. Plant J 32:175–184

Nakamura T, Sugiura C, Kobayashi Y, Sugita M (2005) Transcript profiling in plastid arginine tRNA-CCG gene knock-out moss: construction of *Physcomitrella patens* plastid DNA microarray. Plant Biol 7:258–265

Ohad I, Dal Bosco C, Herrmann RG, Meurer J (2004) Photosystem II proteins *PsbL* and *PsbJ* regulate electron flow to the plastoquinone pool. Biochemistry 43:2297–2308

Petersen K, Schöttler MA, Karcher D, Thiele W, Bock R (2011) Elimination of a group II intron from a plastid gene causes a mutant phenotype. Nucleic Acids Res 39:5181–5192

Pyke K, Zubko MK, Day A (2000) Marking cell layers with spectinomycin provides a new tool for monitoring cell fate during leaf development. J Exp Bot 51:1713–1720

Rauwolf U, Golczyk H, Greiner S, Herrmann RG (2010) Variable amounts of DNA related to the size of chloroplasts III. Biochemical determinations of DNA amounts per organelle. Mol Genet Genomics 283:35–47

Rogalski M, Ruf S, Bock R (2006) Tobacco plastid ribosomal protein S18 is essential for cell survival. Nucleic Acids Res 34:4537–4545

Rogalski M, Karcher D, Bock R (2008a) Superwobbling facilitates translation with reduced tRNA sets. Nat Struct Mol Biol 15:192–198

Rogalski M, Schöttler MA, Thiele W, Schulze WX, Bock R (2008b) Rpl33, a nonessential plastid-encoded ribosomal protein in tobacco, is required under cold stress conditions. Plant Cell 20:2221–2237

Rolland N, Dorne AJ, Amoroso G, Sultemeyer DF, Joyard J, Rochaix JD (1997) Disruption of the plastid *ycf10* open reading frame affects uptake of inorganic carbon in the chloroplast of Chlamydomonas. EMBO J 16:6713–6726

Rothstein R (1991) Targeting, disruption, replacement, and allele rescue: integrative DNA transformation in yeast. Methods Enzymol 194:281–301

Ruf S, Kössel H, Bock R (1997) Targeted inactivation of a tobacco intron-containing open reading frame reveals a novel chloroplast-encoded photosystem I-related gene. J Cell Biol 139:95–102

Ruf S, Biehler K, Bock R (2000) A small chloroplast-encoded protein as a novel architectural component of the light-harvesting antenna. J Cell Biol 149: 369–377

Rumeau D, Becuwe-Linka N, Beyly A, Carrier P, Cuine S, Genty B, Medgyesy P, Horváth E, Peltier G (2004) Increased zinc content in transplastomic tobacco plants expressing a polyhistidine-tagged Rubisco large subunit. Plant Biotechnol J 2:389–399

Schaefer DG (2002) A new moss genetics: targeted mutagenesis in *Physcomitrella patens*. Annu Rev Plant Biol 53:477–501

Scharff LB, Koop HU (2007) Targeted inactivation of the tobacco plastome origins of replication A and B. Plant J 50:782–794

Schöttler MA, Flugel C, Thiele W, Bock R (2007a) Knock-out of the plastid-encoded *PetL* subunit results in reduced stability and accelerated leaf age-dependent loss of the cytochrome b(6)f complex. J Biol Chem 282:976–985

Schöttler MA, Flugel C, Thiele W, Stegemann S, Bock R (2007b) The plastome-encoded PsaJ subunit is required for efficient Photosystem I excitation, but not for plastocyanin oxidation in tobacco. Biochem J 403:251–260

Schwenkert S, Legen J, Takami T, Shikanai T, Herrmann RG, Meurer J (2007) Role of the low-molecular-weight subunits *PetL*, *PetG*, and *PetN* in assembly, stability, and dimerization of the cytochrome b(6)f complex in tobacco(1[C]). Plant Physiol 144:1924–1935

Schwenkert S, Umate P, Dal Bosco C, Volz S, Mlcochova L, Zoryan M, Eichacker LA, Ohad I, Herrmann RG, Meurer J (2006) *PsbI* affects the stability, function, and phosphorylation patterns of photosystem II assemblies in tobacco. J Biol Chem 281:34227–34238

Serino G, Maliga P (1998) RNA polymerase subunits encoded by the plastid *rpo* genes are not shared with the nucleus-encoded plastid enzyme. Plant Physiol 117:1165–1170

Shikanai T, Endo T, Hashimoto T, Yamada Y, Asada K, Yokota A (1998) Directed disruption of the tobacco *ndhB* gene impairs cyclic electron flow around photosystem I. Proc Natl Acad Sci USA 95:9705–9709

Shikanai T, Shimizu K, Ueda K, Nishimura Y, Kuroiwa T, Hashimoto T (2001) The chloroplast *clpP* gene, encoding a proteolytic subunit of ATP-dependent protease, is indispensable for chloroplast development in tobacco. Plant Cell Physiol 42:264–273

Sugita M, Svab Z, Maliga P, Sugiura M (1997a) Targeted deletion of *sprA* from the tobacco plastid genome indicates that the encoded small RNA is not essential for pre-16S rRNA maturation in plastids. Mol Gen Genet 257:23–27

Sugita M, Sugiura M, Svab Z, Maliga P (1997b) Gene-disruption of *sprA* in tobacco plastid. Plant Physiol 114:887

Svab Z, Maliga P (1993) High frequency plastid transformation in tobacco by selection for a chimeric *aadA* gene. Proc Natl Acad Sci USA 90:913–917

Świątek M (2002) Functional analysis of plastid-encoded genes. Application of reverse genetics on *Nicotiana tabacum*. PhD dissertation, der Fakultät für Biologie der Ludwig-Maximilian-Universität München, München

Swiatek M, Greiner S, Kemp S, Drescher A, Koop HU, Herrmann RG, Maier RM (2003a) PCR analysis of pulsed-field gel electrophoresis-purified plastid DNA, a sensitive tool to judge the hetero-/homoplastomic status of plastid transformants. Curr Genet 43:45–53

Swiatek M, Regel RE, Meurer J, Wanner G, Pakrasi HB, Ohad I, Herrmann RG (2003b) Effects of selective inactivation of individual genes for low-molecular-mass subunits on the assembly of photosystem II, as revealed by chloroplast transformation: the *psbEFLJ* operon in *Nicotiana tabacum*. Mol Genet Genomics 268:699–710

Swiatek M, Kuras R, Sokolenko A, Higgs D, Olive J, Cinque G, Muller B, Eichacker LA, Stern DB, Bassi R, Herrmann RG, Wollman FA (2001) The chloroplast gene *ycf9* encodes a photosystem II (PSII) core subunit, *PsbZ*, that participates in PSII supramolecular architecture. Plant Cell 13:1347–1367

Tsuruya K, Suzuki M, Plader W, Sugita C, Sugita M (2006) Chloroplast transformation reveals that tobacco *ycf5* is involved in photosynthesis. Acta Physiol Plantarum 28:365–371

Umate P, Fellerer C, Schwenkert S, Zoryan M, Eichacker LA, Sadanandam A, Ohad I, Herrmann RG, Meurer J (2008) Impact of *PsbTc* on forward and back electron flow, assembly, and phosphorylation patterns of photosystem II in tobacco. Plant Physiol 148:1342–1353

Umate P, Schwenkert S, Karbat I, Dal Bosco C, Mlčochová L, Volz S, Zer H, Herrmann RG, Ohad I, Meurer J (2007) Deletion of *PsbM* in tobacco alters the Q(B) site properties and the electron flow within photosystem II. J Biol Chem 282: 9758–9767

Verhounig A, Karcher D, Bock R (2010) Inducible gene expression from the plastid genome by a synthetic riboswitch. Proc Natl Acad Sci USA 107: 6204–6209

Whitney SM, Sharwood RE (2008) Construction of a tobacco master line to improve RuBisCO engineering in chloroplasts. J Exp Bot 59:1909–1921

Whitney SM, von Caemmerer S, Hudson GS, Andrews TJ (1999) Directed mutation of the Rubisco large subunit of tobacco influences photorespiration and growth. Plant Physiol 121:579–588

Whitney SM, Sharwood RE, Orr D, White SJ, Alonso H, Galmes J (2011) Isoleucine 309 acts as a C(4) catalytic switch that increases ribulose-1,5-bisphosphate carboxylase/oxygenase (rubisco) carboxylation rate in Flaveria. Proc Natl Acad Sci USA 108:14688–14693

Xie ZY, Merchant S (1996) The plastid-encoded *ccsA* gene is required for heme attachment to chloroplast c-type cytochromes. J Biol Chem 271:4632–4639

Zoschke R, Liere K, Börner T (2007) From seedling to mature plant: arabidopsis plastidial genome copy number, RNA accumulation and transcription are differentially regulated during leaf development. Plant J 50:710–722

Zubko MK, Day A (1998) Stable albinism induced without mutagenesis: a model for ribosome-free plastid inheritance. Plant J 15:265–271

Zubko MK, Day A (2002) Differential regulation of genes transcribed by nucleus-encoded plastid RNA polymerase, and DNA amplification, within ribosome- deficient plastids in stable phenocopies of cereal albino mutants. Mol Gen Genomics 267:27–37

Zuo JR, Chua NH (2000) Chemical-inducible systems for regulated expression of plant genes. Curr Opin Biotechnol 11:146–151

Chapter 19

Transformation and Nucleic Acid Delivery to Mitochondria

Claire Remacle*, Veronique Larosa, and Thalia Salinas
Laboratory of Genetics of Microorganisms, Institute of Botany B22, University of Liège, B-4000 Liège, Belgium

Patrice Hamel, and Nitya Subrahmanian
Department of Molecular Genetics and Department of Molecular and Cellular Biochemistry, 500 Aronoff Laboratory, 318 W. Ave, Columbus, OH 43210, USA and Plant Cellular and Molecular Biology Graduate Program, The Ohio State University, Columbus, OH USA

Nathalie Bonnefoy
Centre National de la Recherche Scientifique, Centre de Génétique Moléculaire FRE3144, FRC3115, Avenue de la Terrasse, 91198 Gif-sur-Yvette Cedex, France

Frank Kempken*
Abteilung Botanische Genetik und Molekularbiologie, Botanisches Institut, Christian-Albrechts-Universität zu Kiel, Olshausenstr. 40, D-24098 Kiel, Germany

* Authors for correspondence, e-mail:c.remacle@ulg.ac.be; fkempken@bot.uni-kiel.de

R. Bock and V. Knoop (eds.), Genomics of Chloroplasts and Mitochondria, Advances in Photosynthesis and Respiration 35, pp. 443–458,
DOI 10.1007/978-94-007-2920-9_19,

Summary

Genomic, transcriptomic and proteomic approaches have yielded considerable information, which impacted our understanding of the interactions between the nucleus and the mitochondria. Plant mitochondrial (mt) genomes are very large (220–2,000 kb) and often occur as complex pools of recombined molecules whose stoichiometry is tightly controlled by the nucleus. Unlike their mammalian and fungal counterparts, plant mt transcripts undergo complex post-transcriptional modifications such as editing and trans-splicing. Due to the impossibility to stably transform plant mitochondria and hence to manipulate mt gene expression, the genetic regulation of plant mt genomes has remained poorly understood. In this chapter, we will review the experimental data concerning the unicellular green alga *Chlamydomonas reinhardtii*, the only photosynthetic organism for which mt transformation has been achieved. Although *Chlamydomonas* harbors an extremely compact linear mt genome (15.8 kb) that differs from the one typically found in vascular plants, this system could bring novel insights on the role of the few subunits of the respiratory chain that are encoded in the mt genome. This is particularly relevant for the nd genes, which encode subunits of complex I since the yeast *Saccharomyces cerevisiae*, the other unicellular organism where mt transformation is performed nearly at will, is deprived of complex I. Moreover, because the *Chlamydomonas* mt genome only encodes three tRNAs, genetic manipulation of the organellar genome is a promising avenue to dissect the import of cytosolic tRNAs, a process that is now known to take place in plant and also human mitochondria. We also present alternative approaches such as the in vitro import of DNA or RNA and electroporation of isolated mitochondria followed by *in organello* synthesis that have been developed. These approaches have generated fruitful information about transcription and post-transcriptional processing of plant mt RNAs.

I. Mt Transformation

A. Introduction

Two types of mt DNA are present in the embryophytes (land plants). The mitochondria of the three clades of bryophytes (liverworts, hornworts and mosses, see chapter "Promiscuous Organellar DNA") possess circular genomes of around 105–185 kb resembling that of their green algal relatives like *Chara* and even the early-branching nongreen eukaryote, *Reclinomonas americana* (Li et al. 2009). They are therefore considered to be of the ancestral type despite the acquisition of several additional features such as moderate size increase, intron gain and RNA editing, which are generally absent in algal genomes (Wang et al. 2009). On the contrary, the mitochondria of seed plants (see Chap. 10 "Horizontal Gene Transfer in Eukaryotes: Fungi-to-Plant and Plant-to-Plant Transfers of organellar DNA") harbor

Abbreviations: mt – Mitochondrial; tRNA – Transfer RNA; VDAC – Voltage-dependent anion channel

much larger genomes, from 221 kb for the smallest mt genome sequenced to date (Handa 2003) to more than 2,000 kb in the Cucurbitaceae family (Ward et al. 1981). The size increase does not reflect an increase of the gene content but addition of noncoding sequences such as introns, DNA repeat motifs and insertion of nuclear and chloroplast fragments (Knoop 2004). In addition seed plants possess a complex pool of frequently recombining molecules, the stoichiometry of which is controlled by nuclear genes (Abdelnoor et al. 2003; Arrieta-Montiel et al. 2009). Plant mt transcription is also complex and plant mt RNA transcripts undergo unique post-transcriptional modifications such as editing and trans-splicing (see section "DNA and RNA Delivery into Plant Mitochondria" of this chapter). Investigations of such processes remain difficult due to the impossibility to stably transform plant mitochondria and to manipulate mt gene expression. A plant mt transformation system would therefore be of great value.

Several significant seed plant mtDNA mutations (see chapter "Expression Profiling of Organellar Genes") were identified. Among them, some are found in subunits of the respiratory-chain complexes and are responsible for visible phenotypes such as the cytoplasmic male sterility in tobacco (Pla et al. 1995; Gutierres et al. 1997; Pineau et al. 2005) or the nonchromosomal stripe phenotype in maize (Karpova and Newton 1999; Kubo and Newton 2008).

In principle, and similarly to the situation in other unicellular organisms like the yeast *Saccharomyces cerevisiae* (Fox et al. 1988; Johnston et al. 1988) and in the green alga *Chlamydomonas* (see below for details), such respiratory deficient mutants could serve as recipient strains for mt transformation. A few years ago, cucumber was described as a promising system for mt transformation for several reasons: (1) the mt genome is inherited by the paternal parent and microspores possess a few huge mitochondria that could be transformed before ovule fertilization, (2) mutants with mt deletions, responsible for a mosaic phenotype, exist and could serve as non-reverting recipient strains for mt transformation (Havey et al. 2002). However, despite several attempts, no successful transformation of mitochondria in cucumber or any plant system has been reported to date. The reasons for the failure are probably multiple: (1) the lack of a method to transform either microspores or pollen cells or the ovule for plants with maternal transmission of mitochondria; (2) the fact that plant mt genomes are extremely complex and unstable and can rapidly evolve via substoichiometric shifting due to recombination (Small et al. 1987, 1989; Mackenzie 2007); (3) the lack of a selection that is maintained throughout the formation of the adult plant after zygote formation. Auxotrophic markers corresponding to nuclear genes encoding enzymes targeted to mitochondria are attractive for the development of a selection method. In yeast, expression of the synthetic gene $ARG8^m$ from the mt genome allows nuclear *arg8* mutants to grow without arginine (Steele et al. 1996). The Arg8p protein is normally imported into mitochondria from the cytoplasm, but also functions when synthesized within the organelle in the mt transformants. Thus, arginine prototrophy can become a phenotype dependent on mt gene expression.

Unfortunately, there are very few auxotrophic markers that could function within mitochondria of plants. For example, the *Chlamydomonas* or *Arabidopsis* ortholog of *ARG8* cannot be used as a mitochondrial marker since it actually encodes a plastid protein (Remacle et al. 2009). One notable exception is the BIO2 protein, an iron-cluster enzyme responsible for the last step of biotin synthesis inside the mitochondria. The *bio2* mutants defective for the BIO2 protein have been characterized in *Arabidopsis* and they survive if supplemented with biotin (Patton et al. 1998; Arnal et al. 2006). The *bio2* mutants may represent a promising system for mitochondrial transformation, by transforming either protoplasts with subsequent regeneration of plants or the ovule before pollen fertilization, using as selection a medium devoid of biotin.

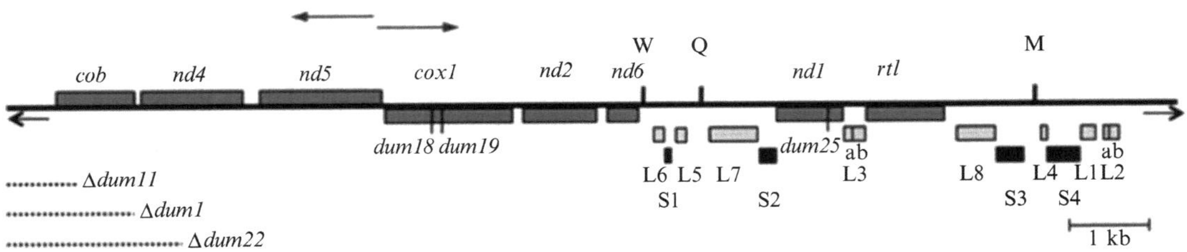

Fig. 19.1. Physical map of the 15.8 kb mt genome of *C. reinhardtii*. The *rectangles* represent protein-coding genes: *cob*, gene encoding *apocytochrome b* of complex III; *nd1, 2, 4, 5*, and *6*, genes encoding the corresponding subunits of complex I; *cox1*, gene encoding subunit 1 of complex IV, *rtl*: reverse transcriptase-like protein. L and S represent modules encoding segments of rRNAs of the large and the small ribosomal subunits, respectively. W, Q, and M represent tRNAs for Trp, Gln, and Met, respectively. The inverted telomeric ends are represented by *short arrows* and the bidirectional origin of transcription between *nd5* and *cox1* by *longer arrows*. Positions of the *dum1*, *dum11* and *dum22* deletions and of the *dum18*, *dum19* and *dum25* mutations are indicated.

B. Mt Transformation in the Unicellular Green Alga Chlamydomonas

1. The Mt Genome of Chlamydomonas

The mt genome of *C. reinhardtii* is a 15.8-kb linear molecule containing at each extremity telomeres corresponding to inverted repeats of about 500 bp, with 40-bp single-stranded extensions (Vahrenholz et al. 1993). Thirteen genes that encode five subunits of the NADH:ubiquinone oxidoreductase or complex I (*nd1*, *nd2*, *nd4*, *nd5* and *nd6*), apocytochrome *b* of the bc_1 complex or complex III (*cob*), subunit 1 of cytochrome *c* oxidase or complex IV (*cox1*), a reverse transcriptase-like protein (*rtl*), three tRNAs (*trnW*, *trnQ*, *trnM*) and the ribosomal RNAs (rRNAs) are present in the mt genome (Fig. 1). The rRNA genes are discontinuous and split into mini-sequences encoding four small subunit (*S*) and eight large subunit (*L*) rRNA modules, interspersed with one another and with protein and tRNA genes. The small rRNA segments of the two ribosomal subunits are believed to interact by way of extensive intermolecular pairing between one another to form conventional rRNA molecules (Boer and Gray 1988).

The mt genome is a multicopy system of around 50–100 copies organized into about 20–30 nucleoids (Nishimura et al. 1998; Hiramatsu et al. 2006). The nucleoids and the mitochondria seem extremely dynamic and undergo changes in their size and shape during the cell cycle (Ehara et al. 1995; Hiramatsu et al. 2006).

2. The Mt Mutants of Chlamydomonas

Several mutations altering the mt *cob*, *cox1* and *nd* genes have been isolated following random mutagenesis with acriflavine (Remacle et al. 2001b). Phenotypically, the mutants in the *cob* and *cox1* genes have lost the capacity to grow under heterotrophic conditions i.e. in the dark, with acetate as carbon source because they lack the cytochrome pathway of respiration. In contrast, mutants altered in the *nd* genes, which encode subunits of complex I, are able to grow in the dark, but considerably more slowly than the wild-type strain (Remacle et al. 2001a; Cardol et al. 2002). Most of the mutations located in the *cob* gene are deletions covering not only the coding sequence but also the left telomere, whereas mutations in the *cox1* and *nd* genes are usually frameshifts caused by deletion or insertion of one thymidine (Remacle et al. 2001b). In addition, the *mud2* mutation at codon 129 (Phe TTC → Leu CTC) of the *cob* gene confers resistance to myxothiazol and mucidine, inhibitors of the cytochrome bc_1 complex (Bennoun et al. 1991).

All the mt point mutants studied so far were found to be homoplasmic, i.e. they contain only mutated mt genomes. However, the mt DNA present in the deletion mutants exists as a mixture of both deleted monomers and dimers arising from head-to-head fusions between deleted monomers (Matagne et al. 1989; Dorthu et al. 1992; Duby and Matagne 1999). The total amount of mt DNA in such mutants is generally lower than in the wild type. The deletion mutants do not revert and are ideally suited as recipients for mt transformation experiments.

3. Recombination and Segregation of mt DNA

In *Chlamydomonas*, homologous mt DNA recombination is only detected after crosses between mt^+ and mt^- strains in mitotic zygotes that do not undergo meiosis. In such zygotes, mt DNA is transmitted by both parents and recombination between the parental genomes is frequent (Remacle et al. 1990; Remacle and Matagne 1993). This demonstrates that enzymes involved in homologous recombination are active in *Chlamydomonas* mitochondria, an important feature for mt transformation as stable integration of the transforming DNA relies on recombination. Segregation of the mt genomes occurs in the zygotes and their mitotic progeny and after 15–20 divisions, most of the diploid cells are homoplasmic for a mt genome, either recombined or parental.

4. Mt Transformation

a. Deletion Mutants as Recipient for mt Transformation

In the first report of mt transformation in *Chlamydomonas*, a mutant (*dum1*) deleted for the left telomere and *cob* gene (1.5 kb deletion) was successfully transformed to respiratory competence with partially purified mt DNA from *C. reinhardtii* or *Chlamydomonas smithii* using a biolistic device (Randolph-Anderson et al. 1993). *C. reinhardtii* and *C. smithii* are two interfertile species, which harbor identical mt genomes with the exception of a 1-kb group I intron located in the *cob* gene that is present in *C. smithii* but absent in *C. reinhardtii* (Boynton et al. 1987; Remacle et al. 1990). Later, biolistics was again used successfully to transform the same recipient strain, with purified mt DNA or cloned mt DNA fragments (Yamasaki et al. 2005). In both cases, the wild-type mt sequence of the transforming DNA had replaced the deleted genome in the transformants selected under heterotrophic conditions (dark+acetate). Transformation efficiency was low (0.4–3 transformants/μg DNA). This precluded any genetic manipulation of the mt genome, since isolation of transformants with the desired genotype usually requires the screening of many colonies, as this is also the case in *Saccharomyces* mt transformation (Bonnefoy et al. 2007).

Subsequently, biolistic transformation was optimized using cloned mt DNA or PCR fragments as transforming molecules (Remacle et al. 2006). Another deletion mutant carrying a 1.2 kb deletion including the left telomere and part of the *cob* gene (*dum11*) could be rescued after selection in the dark using a mt DNA fragment covering the deletion and the *cob* gene as donor DNA (Fig. 1). Homologous recombination occurred between the introduced DNA and the endogenous mt genome and homologous sequences as short as 28 nucleotides could direct recombination (Remacle et al. 2006). Mt transformants were homoplasmic for the 15.8 kb wild-type genome and did not exhibit the dimeric forms of the mt genome that were present in the recipient strain. Moreover, a high transformation efficiency was achieved (100–250 transformants/μg DNA), the best results being obtained with linearized plasmid DNA.

Interestingly, a strain lacking the *cob* gene and the left telomere could be rescued, although at a very low rate, when the transforming DNA is nearly completely devoid of the left telomere (Remacle et al. 2006). This indicated that the right telomere can be copied to reconstruct the left telomere by recombination. Using the strategy described above, we were able to introduce non-deleterious mutations and also loss-of-function molecular lesions in the mt genome. Myxothiazol-resistant transformants were generated by introducing the nucleotide substitution that is present in the *cob* gene of the strains displaying myxothiazol resistance (Remacle et al. 2006). Similarly, an in-frame deletion of 23 codons was reconstructed in the *nd4* gene with a frequency of one homoplasmic *nd4* transformant among 90 transformants analyzed (Remacle et al. 2006). During selection in the dark, recombination events resulted in the co-integration of the *cob* gene and the deletion in *nd4* in some molecules of the mt genome despite the negative effect of the *nd4* mutation on complex I assembly and activity as well as on whole cell respiration (Fig. 1) (Remacle et al. 2006). These results open the way to reverse genetics in *Chlamydomonas* mitochondria and more specifically, to site-directed mutagenesis of mitochondrially encoded subunits of complex I (ND subunits). This is of special interest because the yeast *S. cerevisiae*, whose mt genome can be manipulated virtually at will, lacks complex I.

After a 2 month selection in the dark, heteroplasmy could still be detected in some transformants, suggesting that the segregation process of the mt molecules was extremely slow. To circumvent the problem of heteroplasmy, we recently used a deletion mutant with a deletion extending up to *nd4* (the *dum22* mutant) (Remacle et al. 2001b) to force the insertion of mutations into *nd4* by recombination. The *dum22* mutant is absolutely dependent on glycolysis and the chloroplast for ATP formation since it lacks the three key protein complexes of the respiratory chain (complex I, complex III and IV). This mutant has proven to be a good recipient strain for isolation of homoplasmic *nd4* transformants, although transformation efficiency is very low (1–2 transformants/μg of DNA). For example, using the *dum22* mutant as recipient strain, we have been recently able to isolate two homoplasmic transformants affected in *nd4*. One transformant bears a point mutation (Leu TTG → Pro CCA) corresponding to a human disease (Chronic Progressive External Ophthalmoplegia) at codon 158 of *nd4* (V. Larosa, unpublished). The other one has been created to study mt codon usage and bears a set of 11 codons, which differ from the usual mt codon usage (T. Salinas and C. Remacle, unpublished). It is worth mentioning that a similar transformant remained heteroplasmic when using the *dum11* mutant only affected in *cob* as a recipient strain (T. Salinas, unpublished).

b. Frameshift Mutants as Recipients for mt Transformation

Point mutants can also be used for mt transformation. A double frameshift mutant in both the *cox1* and *nd1* genes (*dum19 dum25*) (Fig. 1) could be rescued for heterotrophic growth and the 23 codon in-frame deletion in *nd4* cited above could be reconstructed (Remacle et al. 2006). Another frameshift mutant (*dum18*) in the *cox1* gene was also employed as a recipient strain for transformation (Colin et al. 1995). This mutant seems best suited for site-directed mutagenesis of the *nd* genes since the *cox1* gene lies between the *nd4* and *nd5* genes on one end and the *nd2* and *nd6* genes on the other end of the genome (Fig. 1). Unfortunately, the high frequency of reversion of this frameshift mutation precluded the use of this strain as a recipient for transformation experiments.

5. Toward a Selection Independent of the Restoration of Heterotrophic Growth

An obvious limitation to mt transformation in *Chlamydomonas* is the selection process that requires a 2-month incubation period in the dark before any molecular characterization can be performed. The development of a more rapid selection using phototrophic growth for the generation of mt transformants is therefore a high priority. We first tested a co-transformation strategy using a primary selection in the light, by bombarding the recipient strain with a plasmid carrying a nuclear marker and another construct containing the mt DNA of interest. Nuclear transformants were first selected in the light and then tested for their respiratory competence to detect mt transformation events

(Remacle et al. 2006). This type of selection, commonly used for yeast transformation (Bonnefoy et al. 2007), was not successful when applied to *Chlamydomonas*, presumably because simultaneous transformation of the nuclear and mitochondrial genomes is a extremely rare event (Remacle et al. 2006). A novel method of selection for mt transformation based on the use of the maize URF13 protein is currently being tested. URF13 is a chimeric mt inner membrane protein that arose from the recombination of mt molecules (Hanson 1991). The presence of URF13 in the mitochondria causes cytoplasmic male sterility and susceptibility to methomyl, an insecticide commonly used for crops (Hanson 1991; Levings and Siedow 1992; Rhoads et al. 1995). Interestingly, the expression of a mitochondrially targeted URF13 in the fungi *S. cerevisiae* and *Pichia pastoris* also confers methomyl sensitivity (Glab et al. 1990; Huang et al. 1990; Soderholm et al. 2001). A *Chlamydomonas* recipient strain carrying a codon-optimized version of URF13 in place of the *nd4* gene will be created by biolistic transformation using dark selection (*nd4::URF13*). We expect the *nd4::URF13* strain to be methomyl sensitive (and deficient for complex I) based on the fact that expression of URF13 in the mitochondria of *Saccharomyces*, *Pichia* and tobacco confers this trait (Glab et al. 1990; Huang et al. 1990; von Allmen et al. 1991; Chaumont et al. 1995; Soderholm et al. 2001). As a proof of concept, the methomyl sensitive *nd4::URF13* recipient strain will be transformed with a construct containing the wild-type *nd4* gene and selection of methomyl resistant transformants will be attempted in the light. Such transformants are expected to arise from homologous recombination between the *nd4::URF13* region in the mt DNA and the transforming DNA carrying the wild-type *nd4* gene. If such a selection is successfully established, mutations in the *nd4* gene resulting in complex I deficiency in humans will be reconstructed in the *Chlamydomonas* mt genome of the *nd4::URF13* strain using the methomyl resistance selection. The same methodology can be applied to manipulate other mt *nd* genes.

II. DNA and RNA Delivery into Plant Mitochondria

A. Cytosolic tRNA Import into Plant Mitochondria

Mitochondria perform protein biosynthesis. Therefore, they require a complete set of transfer RNAs (tRNAs). The availability of several complete mt genomes of land plants and algae allowed the identification of mt *trn* genes encoding tRNAs (O'Brien et al. 2009). In algae (with the exception of *C. reinhardtii* and *Polytomella capuana*) and in bryophytes, the number of *trn* genes seems sufficient or nearly sufficient for mt translation (O'Brien et al. 2009). In contrast, in seed plants the number of *trn* genes is clearly insufficient as *trn* genes for tRNAs corresponding to 5–7 amino acids are absent from the mt genomes. Experimental studies in a number of these organisms showed that this lack is compensated by the import of the corresponding cytosolic tRNAs (Glover et al. 2001; Vinogradova et al. 2009). Transport of tRNAs from the cytosol to mitochondria is not restricted to plants but is a widespread process that also occurs in organisms that would not need to import tRNAs to sustain mt translation (Salinas et al. 2008; Alfonzo and Soll 2009; Lithgow and Schneider 2010). An important feature in plants is that the number and the identity of imported tRNAs vary from one species to another and are not always consistent with the assigned phylogenetic position. The significance of this observation is currently unclear. Interestingly, with a few notable exceptions, the mt population of nuclear-encoded tRNAs was found to be primarily complementary to those encoded in the mt genome. Therefore, tRNA import in plant mitochondria represents a highly specific process as only a subset of cytosolic tRNAs appears to be routed to the mitochondria. In vivo studies in tobacco demonstrated the involvement of aminoacyl-tRNA synthetases in tRNA import (Dietrich et al. 1996). These studies also revealed the presence of different import determinants within tRNAs depending on the tRNA studied, showing the complexity and selectivity of the import process (Delage et al. 2003b; Salinas et al. 2005). The development of an in vitro tRNA import system (Delage et al. 2003a) together with biochemical approaches in potato mitochondria allowed the identification of some components of the translocation machinery of tRNAs through the mt membranes. These investigations implicated the Voltage Dependent Anion Channel (VDAC), a known player in metabolite transport, as the major component of the tRNA transport system through the outer mitochondrial membrane. Moreover, TOM20 and TOM40, two major components of the TOM (Translocase of the Outer mitochondrial Membrane) complex, are likely to be important for tRNA binding at the surface of mitochondria (Salinas et al. 2006).

B. In Vitro Import of DNA

Isolated plant mitochondria are able to take up double-stranded DNA without sequence specificity. This was first documented using a 2.3 kb linear DNA plasmid originally described in maize mitochondria (Leon et al. 1989). Additional experiments indicate the ability of mitochondria to take up DNA molecules of more than 10 kb. This uptake appeared to be an active transmembrane potential-dependent mechanism and DNA was found to be transcribed *in organello* after import in the mt matrix (Koulintchenko et al. 2003). Mitochondria from mammalian and yeast cells also display the ability to take up DNA, which subsequently is transcribed *in organello* (Koulintchenko et al. 2006; Weber-Lotfi et al. 2009). The authors concluded that the process of DNA uptake may involve VDAC and the adenine nucleotide translocator, which are core components of the mt permeability transition pore complex in animal cells (Zamzami and Kroemer 2001). At the same time the authors ruled out mt membrane permeabilization as a possible mechanism for the DNA uptake in the organelle. In a yeast system, it was shown that DNA import is inhibited by VDAC effectors. Yeast strains deleted for the VDAC-1 or VDAC-2 gene are severely reduced in mt DNA import (Weber-Lotfi et al. 2009), supporting the role of this component in DNA import into mitochondria. Understanding the mechanism that mitochondria use to import DNA in vitro may be useful for the genetic engineering of plant or animal mitochondria in vivo. This has not yet been possible, even with vectors designed for mt gene expression. If VDACs are able to import DNA into mitochondria in living cells, this would greatly increase the chances to establish mt transformation (Weber-Lotfi et al. 2009).

C. Electroporation of Isolated Mitochondria with DNA and RNA

Electroporation changes the conductance and molecular permeability of cell membranes and has been used for delivery of nucleic acids into a broad spectrum of cells (see literature in Rao et al. 2009). Delivery of small RNAs into plant mitochondria has been reported upon electroporation of protoplasts with tRNAs, which subsequently were imported into mitochondria (Wintz and Dietrich 1996). The first reports of successful electroporation of isolated organelles were published some 20 years ago. One group demonstrated the introduction of RNA into isolated chloroplasts and used this system to study the control of mRNA stability in chloroplasts by 3′ inverted repeats (Adams and Stern 1990), while another group used mitochondria from maize and rice (Mulligan et al. 1989). These attempts were not continued with plant organelles at first, but some work was done using mitochondria isolated from mice liver cells (Collombet et al. 1997). Electroporation of maize, *Sorghum* and wheat mitochondria was successfully established a few years later (Farré and Araya 2001; Staudinger and Kempken 2003). Recently, isolated mitochondria from *Arabidopsis* and cauliflower were also used for electroporation (Bolle and Kempken 2006; Kempken et al. 2009) and the details of the methodology have been published (Farré et al. 2007; Kempken et al. 2007).

The data from the mt electroporation systems differ with respect to efficiency. While in animal and wheat mitochondria, electroporation at 1.4 kV/cm (Collombet et al. 1997) and 1.3 kV/cm (Farré and Araya 2001), respectively, gave optimal results, in maize, 1.8–2.0 kV/cm appeared to be most efficient (Staudinger and Kempken 2003). In contrast, Mulligan et al. (1989) reported 8.8 kV/cm to be optimal. It is possible that these older data may be due to differences in the type of apparatus used, while the deviation of the newer data may be due to differences in the amount of mitochondria and plasmid DNA used.

There appears to be a size limit for the DNA to be introduced. While plasmid DNA of up to 11 kb could be introduced into mitochondria via electroporation, larger DNA molecules of about 30 kb were not (Staudinger and Kempken 2003). Moreover, while both linear and covalently closed circular (ccc) plasmid DNA can be imported into the mt matrix, the ccc DNA is mostly converted to open circular and linear DNA during the process (Collombet et al. 1997; Staudinger and Kempken 2003). Furthermore, there is no evidence for integration of the introduced DNA into the mt chromosome (Staudinger and Kempken 2003). The use of a mt plasmid from *Chenopodium album* for electroporation (Backert et al. 1997) gave some indication that replication of mt plasmids may be possible in isolated mitochondria (Kempken, unpublished data, see Fig. 2).

In addition to DNA, RNA can also be transformed into mitochondria using electroporation. This was demonstrated recently for maize and cauliflower mitochondria (Hinrichsen et al. 2009). The data presented demonstrate that plant mt RNA processing appears to be independent of both transcription and respiratory regulation. Moreover, introduction of RNA offers interesting experimental strategies to investigate RNA processing events.

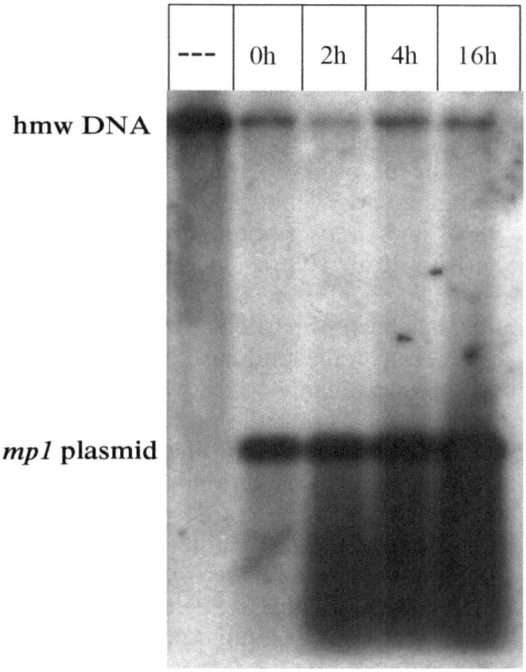

Fig. 19.2. Introduction of *mp1* plasmid DNA (Backert et al. 1997) in maize mitochondria via electroporation. DNA was introduced in mitochondria as described in (Staudinger and Kempken 2003), and *in organello* incubation was carried out in the presence of α-^{32}P-dCTP for up to 16 h. DNA was isolated after incubation and subjected to agarose gel electrophoresis. Over time, an increase in the *mp1* signal is observed in the autoradiogram.

Electroporation has been used for transformation of isolated organelles only. However, it may be possible to employ electroporation for genetic engineering of organelles in intact cells as well. Recently, a mathematical cell model was established, which indicates the possibility to use electric fields for intracellular manipulations (Esser et al. 2010). It was indeed demonstrated that high-intensity electric pulses induce mitochondria-dependent apoptosis in mice cells (Li et al. 2008). Therefore electroporation may be useful in the future for the in vivo transformation of mitochondria. This certainly will have applications not only in plant science but also in the design of therapies for human mt disorders (Cwerman-Thibault et al. 2010).

D. In organello Analysis

1. DNA Replication

Plant and fungal mt *in organello* systems have been used to analyze replication (Bedinger and Walbot 1986; Kempken et al. 1989) and transcription of mt genomes (Carlson et al. 1986). One specific type of genetic elements, the so-called linear plasmids, has been of particular interest in these studies. Linear plasmids are several thousand bases in size, characterized by long terminal inverted repeats and the presence of 5′-end bound terminal proteins. They are believed to be remnants of bacteriophages with similar structures (Meinhardt et al. 1990; Kempken et al. 1992). Studies on plant mitochondria undergoing *in organello* replication while using radiolabeled nucleotides support the existence of DNA repair and DNA replication activities within the organelle. Linear plasmids showed much higher incorporation of radiolabeled nucleotides than the high molecular weight DNA, which might be due to the activity of DNA polymerases encoded by linear plasmids. Similar results were obtained in a fungal *in organello* system employing mitochondria from *Ascobolus immersus*. End fragments from the plasmid were found to be more heavily labeled than internal fragments, indicating a potential role of the terminal proteins in DNA replication. It is assumed that terminal proteins function as primers for the replication of linear plasmids, as is the case for certain viruses with linear plasmids (Bedinger and Walbot 1986; Kempken et al. 1989). However, the fungal system provided stronger evidence for true DNA replication versus DNA repair, as the use of strand specific probes gave evidence for protein-primed DNA replication (Kempken et al. 1989).

2. Transcription and RNA Processing

Molecular analysis of plant mt transcription and RNA processing was long hampered due to the inability to transform plant mitochondria. Alternatively, in vitro systems have successfully been employed (e.g. Takenaka et al. 2004; Kühn et al. 2005) and *in organello* systems provide another substitute for *in planta* experiments (e.g. Farré and Araya 2001; Staudinger and Kempken 2003).

Two processes have mainly been studied using *in organello* systems: RNA editing and splicing. RNA editing is a post-transcriptional process that alters the information content of RNA (see chapter "Organellar Proteomics: Close Insights into the Spatial Breakdown and Functional Dynamics of Plant Primary Metabolism"). Several functional types are known (Knoop 2010), one of which occurs mostly by conversion of specific cytidine residues to uridine residues in the RNAs of higher plant mitochondria (Covello and Gray 1989; Gualberto et al. 1989; Hiesel et al. 1989). In addition, in several plant clades such as hornworts or ferns, many U-to-C changes have also been observed to occur in mitochondria (for reviews see Shikanai 2006; Takenaka et al. 2008). RNA editing is also observed in mosses, some liverworts and in gymnosperms (Knoop 2010). The number of RNA editing sites differs, but there are up to maybe more than 500 in higher plant mitochondria (Takenaka et al. 2008; Bruhs and Kempken 2010; Knoop 2010), and even more than 1,500 in lycophyte mitochondria (Grewe et al. 2011).

In attempts to elucidate higher plant mt RNA editing and the mechanism by which the C-to-U transition is achieved, *in organello* systems based on mitochondria from pea seedlings, potato tubers (Yu and Schuster 1995) or maize seedlings (Rajasekhar and Mulligan 1993) were developed. Taken together, the studies have excluded nucleotide excision and base exchange as possible editing reactions, while deamination or transamination reactions both are possible mechanisms. However, so far no conclusive evidence for either reaction has been obtained (Takenaka et al. 2008).

In organello assays in combination with uptake of DNA via electroporation have been successfully employed to analyze RNA processing (Farré et al. 2001; Staudinger and

Kempken 2003). Using biotinylated Uridine Tri-Phosphate (UTP), de novo transcription and RNA processing were observed *in organello*. The *in organello* systems were shown to faithfully reflect the *in planta* situation. Transcripts that are fully edited *in planta* are fully edited *in organello*, while those partially edited *in planta* are also partially edited in the *in organello* system (Staudinger and Kempken 2003, 2004). Using a wheat *in organello* system, a core sequence including 16 nucleotides upstream and six nucleotides downstream of the edited nucleotide was defined for recognition of two editing sites (Farré et al. 2001; Choury et al. 2004). However, for efficient editing, an upstream sequence of 40 nucleotides appeared to be required as shown by in vitro experiments (Takenaka et al. 2004). While it is now established that pentatricopeptide proteins (see chapter "Organellar Proteomics: Close Insights into the Spatial Breakdown and Functional Dynamics of Plant Primary Metabolism") participate in the recognition of mt RNA editing sites (e.g. Verbitskiy et al. 2009; Zehrmann et al. 2009), experiments using cauliflower and maize *in organello* systems also suggest an influence of a RNA secondary or tertiary structure (Bolle and Kempken 2006).

The way splicing and editing function in cross-species analyses is not fully predictable. While the *cox2* RNA is spliced and edited in mono- and di-cotyledonous mitochondria regardless of the mono- or dicot origin of the gene (Staudinger and Kempken 2003), the *rps10* RNA from potato (dicot) is neither spliced nor edited in wheat (monocot) mitochondria (Choury et al. 2005). The *atp6* mRNA from sorghum (monocot) is not even edited in maize (monocot) mitochondria despite very high sequence similarity. However, a chimeric *atp6* transcript composed of maize and sorghum DNA sequences gave rise to partial RNA editing, which may be due to presence of a RNA editing recognition sequence in the *atp6* transcript (Staudinger and Kempken 2003). An *in organello* system has also demonstrated the inability of plant mitochondria to recognize chloroplast editing sites (Bolle et al. 2007).

RNA editing can be essential for subsequent RNA splicing (Farré and Araya 2002; Castandet et al. 2010). However, both RNA splicing and editing are independent of the transcription machinery, as demonstrated using the *cox2* mRNA electroporated into mitochondria (Hinrichsen et al. 2009).

3. Translation

In organello translation of proteins employing ^{35}S labeled amino acids has long been established (Grohmann 1995). It has been employed for analysis of cytoplasmic male sterility (e.g. Horn et al. 1991; Moneger et al. 1994) and for the analysis of translation of partially edited transcripts. Amino acid sequencing of immunoprecipitated NAD9 protein indicated that only proteins from fully edited *nad9* mRNAs accumulate *in organello* (Grohmann et al. 1994).

While transcription and RNA processing can occur from DNA electroporated into isolated mitochondria and subsequent *in organello* incubation, there is no experimental evidence supporting that introduced RNA can be translated. In animal mitochondria, RNAs from a mt reporter gene (modified luciferase) could be stably maintained in the presence of functioning mitochondrial protein synthesis. However, the RNAs were not translated (McGregor et al. 2001). Likewise there is no proof for translation of RNA transcribed from introduced DNA into higher plant mitochondria (Kempken, unpublished data). As the introduced mRNA is faithfully processed (Hinrichsen et al. 2009), the reason for why translation cannot occur remains unclear.

III. Conclusion

Mt transformation of the unicellular alga *C. reinhardtii* can be achieved with surprisingly high efficiency and is of great value for the analysis of mt gene functions. Unfortunately comparable attempts to transform vascular plant mitochondria have been unsuccessful so far. While plant mt transformation has not yet been achieved, mt *in organello* systems and the ability of mitochondria to uptake DNA provide an attractive alternative to study transcription and RNA processing mechanisms, such as RNA splicing and RNA editing. However, although *in organello* translation has long been established, the translation of RNA either directly introduced into isolated mitochondria, or transcribed from DNA electroporated into mitochondria, has not been reported yet. It is conceivable that a close coupling between transcription and translation as well as the proximity of the mt membrane could be necessary to synthesize polypeptides within mitochondria, but other parameters could be important, too. In addition, there could be a connection between the failure to translate imported nucleic acids within mitochondria and the inability to transform plant mitochondria. Therefore, understanding what might be the reason(s) for this lack of translation represents a major challenge that will provide clues to successful transformation of plant mitochondria.

Acknowledgements

Work in the lab of C. Remacle is supported by 'Fonds National de la Recherche Scientifique' from Belgium (F.R.S.-FNRS) 1.5.255.08 and 2.4601.08 and by Action de Recherche Concertée ARC07/12 04 and a joint United Mitochondrial Disease Foundation Grant (UMDF) to C. Remacle and P. Hamel. T. Salinas is supported by the European Community's Seventh Framework Programme (FP7/2007–2013) under grant agreement n° 220808 (Marie Curie Fellowship). V. Larosa is supported by 'Fonds pour la Formation à la Recherche dans l'Industrie et dans l'Agriculture' (FRIA). Work in the lab of F. Kempken its supported by the German Research Foundation (DFG). N. Bonnefoy is the recipient of an ANR contract (JC-JC-06-0163). We warmly thank Dr R. Lamb (Ohio State University, OH) for careful reading of the manuscript.

References

Abdelnoor RV, Yule R, Elo A, Christensen AC, Meyer-Gauen G, Mackenzie SA (2003) Substoichiometric shifting in the plant mitochondrial genome is influenced by a gene homologous to MutS. Proc Natl Acad Sci USA 100:5968–5973

Adams CC, Stern DB (1990) Control of mRNA stability in chloroplasts by 3′ inverted repeats: effects of stem and loop mutations on degradation of psbA mRNA in vitro. Nucleic Acids Res 18:6003–6010

Alfonzo JD, Soll D (2009) Mitochondrial tRNA import – the challenge to understand has just begun. Biol Chem 390:717–722

Arnal N, Alban C, Quadrado M, Grandjean O, Mireau H (2006) The *Arabidopsis* Bio2 protein requires mitochondrial targeting for activity. Plant Mol Biol 62:471–479

Arrieta-Montiel MP, Shedge V, Davila J, Christensen AC, Mackenzie SA (2009) Diversity of the *Arabidopsis* mitochondrial genome occurs via nuclear-controlled recombination activity. Genetics 183:1261–1268

Backert S, Meißen K, Börner T (1997) Unique features of the mitochondrial rolling circle-plasmid *mp1* from the higher plant *Chenopodium album* (L.). Nucleic Acid Res 25:582–589

Bedinger P, Walbot V (1986) DNA synthesis in purified maize mitochondria. Curr Genet 10:631–637

Bennoun P, Delosme M, Kuck U (1991) Mitochondrial genetics of *Chlamydomonas reinhardtii*: resistance mutations marking the cytochrome b gene. Genetics 127:335–343

Boer PH, Gray MW (1988) Scrambled ribosomal RNA gene pieces in *Chlamydomonas reinhardtii* mitochondrial DNA. Cell 55:399–411

Bolle N, Kempken F (2006) Mono- and dicotyledonous plant-specific RNA editing sites are correctly edited in both *in organello* systems. FEBS Lett 580:4443–4448

Bolle N, Hinrichsen I, Kempken F (2007) Plastid mRNAs are neither spliced nor edited in maize and cauliflower mitochondrial *in organello* systems. RNA 13:2061–2065

Bonnefoy N, Remacle C, Fox TD (2007) Genetic transformation of *Saccharomyces cerevisiae* and *Chlamydomonas reinhardtii* mitochondria. Methods Cell Biol 80:525–548

Boynton JE, Harris EH, Burkhart BD, Lamerson PM, Gillham NW (1987) Transmission of mitochondrial and chloroplast genomes in crosses of *Chlamydomonas*. Proc Natl Acad Sci USA 84:2391–2395

Bruhs A, Kempken F (2010) RNA editing in higher plant mitochondria. In: Kempken F (ed) Plant mitochondria. Springer, New York/Dordrecht/Heidelberg/London, pp 157–176

Cardol P, Matagne RF, Remacle C (2002) Impact of mutations affecting ND mitochondria-encoded subunits on the activity and assembly of complex I in *Chlamydomonas*. Implication for the structural organization of the enzyme. J Mol Biol 319:1211–1221

Carlson JE, Brown GL, Kemble RJ (1986) *In organello* mitochondrial DNA and RNA synthesis in fertile and cytoplasmic sterile *Zea mays* L. Curr Genet 11:151–160

Castandet B, Choury D, Bégu D, Jordana X, Araya A (2010) Intron RNA editing is essential for splicing in plant mitochondria. Nucleic Acids Res 38:7112–7121

Chaumont F, Bernier B, Buxant R, Williams ME, Levings CS III, Boutry M (1995) Targeting the maize T-urf13 product into tobacco mitochondria confers methomyl sensitivity to mitochondrial respiration. Proc Natl Acad Sci USA 92:1167–1171

Choury D, Farre JC, Jordana X, Araya A (2004) Different patterns in the recognition of editing sites in plant mitochondria. Nucleic Acids Res 32:6397–6406

Choury D, Farré JC, Jordana X, Araya A (2005) Gene expression studies in isolated mitochondria: *Solanum tuberosum rps10* is recognized by cognate potato but not by the transcription, splicing and editing machinery of wheat mitochondria. Nucleic Acids Res 33:7058–7065

Colin M, Dorthu MP, Duby F, Remacle C, Dinant M, Wolwertz MR, Duyckaerts C, Sluse F, Matagne RF (1995) Mutations affecting the mitochondrial genes encoding the cytochrome oxidase subunit I and apocytochrome *b* of *Chlamydomonas reinhardtii*. Mol Gen Genet 249:179–184

Collombet J-M, Wheeler VC, Vogel F, Coutelle C (1997) Introduction of plasmid DNA into isolated mitochondria by electroporation. J Biol Chem 272:5342–5347

Covello PS, Gray MW (1989) RNA editing in plant mitochondria. Nature 341:662–666

Cwerman-Thibault H, Sahel JA, Corral-Debrinski M (2010) Mitochondrial medicine: to a new era of gene therapy for mitochondrial DNA mutations. J Inherit Metab Dis 34:327–344

Delage L, Dietrich A, Cosset A, Maréchal-Drouard L (2003a) In vitro import of a nuclearly encoded tRNA into mitochondria of *Solanum tuberosum*. Mol Cell Biol 23:4000–4012

Delage L, Duchene AM, Zaepfel M, Maréchal-Drouard L (2003b) The anticodon and the D-domain sequences are essential determinants for plant cytosolic tRNA(Val) import into mitochondria. Plant J 34:623–633

Dietrich A, Maréchal-Drouard L, Carneiro V, Cosset A, Small I (1996) A single base change prevents import of cytosolic tRNA(Ala) into mitochondria in transgenic plants. Plant J 10:913–918

Dorthu MP, Remy S, Michel-Wolwertz MR, Colleaux L, Breyer D, Beckers MC, Englebert S, Duyckaerts C, Sluse FE, Matagne RF (1992) Biochemical, genetic and molecular characterization of new respiratory-deficient mutants in *Chlamydomonas reinhardtii*. Plant Mol Biol 18:759–772

Duby F, Matagne RF (1999) Alteration of dark respiration and reduction of phototrophic growth in a mitochondrial DNA deletion mutant of *Chlamydomonas* lacking *cob*, *nd4*, and the 3′ end of *nd5*. Plant Cell 11:115–125

Ehara T, Osafune T, Hase E (1995) Behavior of mitochondria in synchronized cells of *Chlamydomonas reinhardtii* (Chlorophyta). J Cell Sci 108(Pt 2): 499–507

Esser AT, Smith KC, Gowrishankar TR, Vasilkoski Z, Weaver JC (2010) Mechanisms for the intracellular manipulation of organelles by conventional electroporation. Biophys J 98:2506–2514

Farré J-C, Araya A (2001) Gene expression in isolated plant mitochondria: high fidelity of transcription, splicing and editing of a transgene product in electroporated organelles. Nucleic Acids Res 29:2484–2491

Farré JC, Araya A (2002) RNA splicing in higher plant mitochondria: determination of functional elements in group II intron from a chimeric *cox II* gene in electroporated wheat mitochondria. Plant J 29:203–213

Farré JC, Leon G, Jordana X, Araya A (2001) *cis* recognition elements in plant mitochondrion RNA editing. Mol Cell Biol 21:6731–6737

Farré JC, Choury D, Araya A (2007) *In organello* gene expression and RNA editing studies by electroporation-mediated transformation of isolated plant mitochondria. Methods Enzymol 424:483–500

Fox TD, Sanford JC, McMullin TW (1988) Plasmids can stably transform yeast mitochondria lacking endogenous mtDNA. Proc Natl Acad Sci USA 85: 7288–7292

Glab N, Wise RP, Pring DR, Jacq C, Slonimski P (1990) Expression in *Saccharomyces cerevisiae* of a gene associated with cytoplasmic male sterility from maize: respiratory dysfunction and uncoupling of yeast mitochondria. Mol Gen Genet 223:24–32

Glover KE, Spencer DF, Gray MW (2001) Identification and structural characterization of nucleus-encoded transfer RNAs imported into wheat mitochondria. J Biol Chem 276:639–648

Grewe F, Herres S, Viehöver P, Polsakiewicz M, Weisshaar B, Knoop V (2011) A unique transcriptome: 1782 positions of RNA editing alter 1406 codon identities in mitochondrial mRNAs of the lycophyte *Isoetes engelmannii*. Nucleic Acids Res 39:2890–2902

Grohmann L (1995) *In organello* protein synthesis. Methods Mol Biol 49:391–397

Grohmann L, Thieck O, Herz U, Schröder W, Brennicke A (1994) Translation of *nad9* mRNAs in mitochondria from *Solanum tuberosum* is restricted to completely edited transcripts. Nucleic Acids Res 22:3304–3311

Gualberto JM, Lamattina L, Bonnard G, Weil JH, Grienenberger JM (1989) RNA editing in wheat mitochondria results in the conservation of protein sequences. Nature 341:660–662

Gutierres S, Sabar M, Lelandais C, Chetrit P, Diolez P, Degand H, Boutry M, Vedel F, de Kouchkovsky Y, De Paepe R (1997) Lack of mitochondrial and nuclear-encoded subunits of complex I and alteration of the respiratory chain in *Nicotiana sylvestris* mitochondrial deletion mutants. Proc Natl Acad Sci USA 94:3436–3441

Handa H (2003) The complete nucleotide sequence and RNA editing content of the mitochondrial genome of rapeseed (*Brassica napus* L.): comparative analysis of the mitochondrial genomes of rapeseed and *Arabidopsis thaliana*. Nucleic Acids Res 31:5907–5916

Hanson MR (1991) Plant mitochondrial mutations and male sterility. Annu Rev Genet 25:461–486

Havey MJ, Lilly JW, Bohanec B, Bartoszewski G, Malepszy S (2002) Cucumber: a model angiosperm for mitochondrial transformation? J Appl Genet 43:1–17

Hiesel R, Wissinger B, Schuster W, Brennicke A (1989) RNA editing in plant mitochondria. Science 246:1632–1634

Hinrichsen I, Bolle N, Paun L, Kempken F (2009) RNA processing in plant mitochondria is independent of transcription. Plant Mol Biol 70:663–668

Hiramatsu T, Nakamura S, Misumi O, Kuroiwa T, Nakamura S (2006) Morphological changes in mitochondrial and chloroplast nucleoids and mitochondria during the *Chlamydomonas reinhardtii* (Chlorophyceae) cell cycle. J Phycol 42:1048–1058

Horn R, Kohler RH, Zetsche K (1991) A mitochondrial 16 kDa protein is associated with cytoplasmic male sterility in sunflower. Plant Mol Biol 17:29–36

Huang J, Lee SH, Lin C, Medici R, Hack E, Myers AM (1990) Expression in yeast of the T-urf13 protein from Texas male-sterile maize mitochondria confers sensitivity to methomyl and to Texas-cytoplasm-specific fungal toxins. EMBO J 9:339–347

Johnston SA, Anziano PQ, Shark K, Sanford JC, Butow RA (1988) Mitochondrial transformation in yeast by bombardment with microprojectiles. Science 240:1538–1541

Karpova OV, Newton KJ (1999) A partially assembled complex I in ND4-deficient mitochondria of maize. Plant J 17:511–521

Kempken F, Meinhardt F, Esser K (1989) *In organello* replication and viral affinity of linear, extrachromosomal DNA of the ascomycete *Ascobolus immersus*. Mol Gen Genet 218:523–530

Kempken F, Hermanns J, Osiewacz HD (1992) Evolution of linear plasmids. J Mol Evol 35:502–513

Kempken F, Bolle N, Forner J, Binder S (2007) Transcript end mapping and analysis of RNA editing in plant mitochondria. Methods Mol Biol 372:177–192

Kempken F, Bolle N, Bruhs A (2009) Higher plant *in organello* systems as a model for RNA editing. Endocyt Cell Res 19:1–10

Knoop V (2004) The mitochondrial DNA of land plants: peculiarities in phylogenetic perspective. Curr Genet 46:123–139

Knoop V (2010) When you can't trust the DNA: RNA editing changes transcript sequences. Cell Mol Life Sci 68:567–586

Koulintchenko M, Konstantinov Y, Dietrich A (2003) Plant mitochondria actively import DNA via the permeability transition pore complex. EMBO J 22: 1245–1254

Koulintchenko M, Temperley RJ, Mason PA, Dietrich A, Lightowlers RN (2006) Natural competence of mammalian mitochondria allows the molecular investigation of mitochondrial gene expression. Hum Mol Genet 15:143–154

Kubo T, Newton KJ (2008) Angiosperm mitochondrial genomes and mutations. Mitochondrion 8:5–14

Kühn K, Weihe A, Börner T (2005) Multiple promoters are a common feature of mitochondrial genes in *Arabidopsis*. Nucleic Acids Res 33:337–346

Leon P, Walbot V, Bedinger P (1989) Molecular analysis of the linear 2.3 kb plasmid of maize mitochondria: apparent capture of tRNA genes. Nucleic Acids Res 17:4089–4099

Levings CS III, Siedow JN (1992) Molecular basis of disease susceptibility in the Texas cytoplasm of maize. Plant Mol Biol 19:135–147

Li C, Hu LN, Dong XJ, Sun CX, Mi Y (2008) High-intensity electric pulses induce mitochondria-dependent apoptosis in ovarian cancer xenograft mice. Int J Gynecol Cancer 18:1258–1261

Li L, Wang B, Liu Y, Qiu YL (2009) The complete mitochondrial genome sequence of the hornwort *Megaceros aenigmaticus* shows a mixed mode of conservative yet dynamic evolution in early land plant mitochondrial genomes. J Mol Evol 68:665–678

Lithgow T, Schneider A (2010) A Evolution of macromolecular import pathways in mitochondria, hydrogenosomes and mitosomes. Philos Trans R Soc Lond B Biol Sci 365:799–817

Mackenzie SA (2007) The unique biology of mitochondrial genome instability in plants. In: Logan DC (ed) Plant mitochondria. Blackwell Publishing, Singapore

Matagne RF, Michel-Wolwertz MR, Munaut C, Duyckaerts C, Sluse F (1989) Induction and characterization of mitochondrial DNA mutants in *Chlamydomonas reinhardtii*. J Cell Biol 108: 1221–1226

McGregor A, Temperley R, Chrzanowska-Lightowlers ZM, Lightowlers RN (2001) Absence of expression from RNA internalised into electroporated mammalian mitochondria. Mol Genet Genomics 265:721–729

Meinhardt F, Kempken F, Kämper J, Esser K (1990) Linear plasmids among eukaryotes: fundamentals and application. Curr Genet 17:89–95

Moneger F, Smart CJ, Leaver CJ (1994) Nuclear restoration of cytoplasmic male sterility in sunflower is associated with the tissue-specific regulation of a novel mitochondrial gene. EMBO J 13:8–17

Mulligan RM, Leon P, Calvin N, Walbot V (1989) Introduction of DNA into maize and rice mitochondria by electroporation. Maydica 34:207–216

Nishimura Y, Higashiyama T, Suzuki L, Misumi O, Kuroiwa T (1998) The biparental transmission of the mitochondrial genome in *Chlamydomonas reinhardtii* visualized in living cells. Eur J Cell Biol 77:124–133

O'Brien EA, Zhang Y, Wang E, Marie V, Badejoko W, Lang BF, Burger G (2009) GOBASE: an organelle genome database. Nucleic Acids Res 37:D946–D950

Patton DA, Schetter AL, Franzmann LH, Nelson K, Ward ER, Meinke DW (1998) An embryo-defective mutant of *arabidopsis* disrupted in the final step of biotin synthesis. Plant Physiol 116:935–946

Pineau B, Mathieu C, Gerard-Hirne C, De Paepe R, Chetrit P (2005) Targeting the NAD7 subunit to mitochondria restores a functional complex I and a wild type phenotype in the *Nicotiana sylvestris* CMS II mutant lacking *nad7*. J Biol Chem 280:25994–26001

Pla M, Mathieu C, De Paepe R, Chetrit P, Vedel F (1995) Deletion of the last two exons of the mitochondrial *nad7* gene results in lack of the NAD7 polypeptide in a *Nicotiana sylvestris* CMS mutant. Mol Gen Genet 248:79–88

Rajasekhar VK, Mulligan RM (1993) RNA editing in plant mitochondria: α-phosphate is retained during C-to-U conversion in mRNAs. Plant Cell 5:1843–1852

Randolph-Anderson BL, Boynton JE, Gillham NW, Harris EH, Johnson AM, Dorthu MP, Matagne RF (1993) Further characterization of the respiratory deficient *dum-1* mutation of *Chlamydomonas reinhardtii* and its use as a recipient for mitochondrial transformation. Mol Gen Genet 236:235–244

Rao AQ, Bakhsh A, Kiani S, Shahzad K, Shahid AA, Husnain T, Riazuddin S (2009) The myth of plant transformation. Biotechnol Adv 27:753–763

Remacle C, Matagne RF (1993) Transmission, recombination and conversion of mitochondrial markers in relation to the mobility of a group I intron in *Chlamydomonas*. Curr Genet 23:518–525

Remacle C, Bovie C, Michel-Wolwertz MR, Loppes R, Matagne RF (1990) Mitochondrial genome transmission in *Chlamydomonas* diploids obtained by sexual crosses and artificial fusions: role of the mating type and of a 1 kb intron. Mol Gen Genet 223:180–184

Remacle C, Baurain D, Cardol P, Matagne RF (2001a) Mutants of *Chlamydomonas reinhardtii* deficient in mitochondrial complex I: characterization of two mutations affecting the *nd1* coding sequence. Genetics 158:1051–1060

Remacle C, Duby F, Cardol P, Matagne RF (2001b) Mutations inactivating mitochondrial genes in *Chlamydomonas reinhardtii*. Biochem Soc Trans 29:442–446

Remacle C, Cardol P, Coosemans N, Gaisne M, Bonnefoy N (2006) High-efficiency biolistic transformation of *Chlamydomonas* mitochondria can be used to insert mutations in complex I genes. Proc Natl Acad Sci USA 103:4771–4776

Remacle C, Cline S, Boutaffala L, Gabilly S, Larosa V, Barbieri RM, Coosemans N, Hamel PP (2009) The *ARG9* gene encodes the plastid-resident *N*-acetyl ornithine aminotransferase in the green alga *Chlamydomonas reinhardtii*. Eukaryot Cell 8:1460–1463

Rhoads DM, Levings CS III, Siedow JN (1995) URF13, a ligand-gated, pore-forming receptor for T-toxin in the inner membrane of cms-T mitochondria. J Bioenerg Biomembr 27:437–445

Salinas T, Schaeffer C, Maréchal-Drouard L, Duchene AM (2005) Sequence dependence of tRNA(Gly) import into tobacco mitochondria. Biochimie 87:863–872

Salinas T, Duchene AM, Delage L, Nilsson S, Glaser E, Zaepfel M, Maréchal-Drouard L (2006) The voltage-dependent anion channel, a major component of the tRNA import machinery in plant mitochondria. Proc Natl Acad Sci USA 103:18362–18367

Salinas T, Duchene AM, Maréchal-Drouard L (2008) Recent advances in tRNA mitochondrial import. Trends Biochem Sci 33:320–329

Shikanai T (2006) RNA editing in plant organelles: machinery, physiological function and evolution. Cell Mol Life Sci 63:698–708

Small ID, Isaac PG, Leaver CJ (1987) Stoichiometric differences in DNA molecules containing the *atpA* gene suggest mechanisms for the generation of mitochondrial genome diversity in maize. EMBO J 6:865–869

Small I, Suffolk R, Leaver CJ (1989) Evolution of plant mitochondrial genomes via substoichiometric intermediates. Cell 58:69–76

Soderholm J, Bevis BJ, Glick BS (2001) Vector for pop-in/pop-out gene replacement in *Pichia pastoris*. Biotechniques 31:306–310

Staudinger M, Kempken F (2003) Electroporation of isolated higher-plant mitochondria: transcripts of an introduced *cox2* gene, but not an *atp6* gene, are edited *in organello*. Mol Genet Genomics 269:553–561

Staudinger M, Kempken F (2004) *In organello* editing of mitochondrial *atp9*, *cox2*, and *nad9* transcripts. Endocyt Cell Res 15:551–560

Steele DF, Butler CA, Fox TD (1996) Expression of a recoded nuclear gene inserted into yeast mitochondrial DNA is limited by mRNA-specific translational activation. Proc Natl Acad Sci USA 93:5253–5257

Takenaka M, Neuwirt J, Brennicke A (2004) Complex *cis*-elements determine an RNA editing site in pea mitochondria. Nucleic Acids Res 32:4137–4144

Takenaka M, Verbitskiy D, van der Merwe JA, Zehrmann A, Brennicke A (2008) The process of RNA editing in plant mitochondria. Mitochondrion 8:35–46

Vahrenholz C, Riemen G, Pratje E, Dujon B, Michaelis G (1993) Mitochondrial DNA of *Chlamydomonas reinhardtii*: the structure of the ends of the linear 15.8-kb genome suggests mechanisms for DNA replication. Curr Genet 24:241–247

Verbitskiy D, Zehrmann A, van der Merwe JA, Brennicke A, Takenaka M (2009) The PPR protein encoded by the LOVASTATIN INSENSITIVE 1 gene is involved in RNA editing at three sites in mitochondria of *Arabidopsis thaliana*. Plant J 61:446–455

Vinogradova E, Salinas T, Cognat V, Remacle C, Maréchal-Drouard L (2009) Steady-state levels of imported tRNAs in *Chlamydomonas* mitochondria are correlated with both cytosolic and mitochondrial codon usages. Nucleic Acids Res 37:1521–1528

von Allmen JM, Rottmann WH, Gengenbach BG, Harvey AJ, Lonsdale DM (1991) Transfer of methomyl and HmT-toxin sensitivity from T-cytoplasm maize to tobacco. Mol Gen Genet 229:405–412

Wang B, Xue J, Li L, Liu Y, Qiu YL (2009) The complete mitochondrial genome sequence of the liverwort *Pleurozia purpurea* reveals extremely conservative mitochondrial genome evolution in liverworts. Curr Genet 55:601–609

Ward BL, Anderson RS, Bendich AJ (1981) The mitochondrial genome is large and variable in a family of plants (Cucurbitaceae). Cell 25:793–803

Weber-Lotfi F, Ibrahim N, Boesch P, Cosset A, Konstantinov Y, Lightowlers RN, Dietrich A (2009) Developing a genetic approach to investigate the mechanism of mitochondrial competence for DNA import. Biochim Biophys Acta 1787:320–327

Wintz H, Dietrich A (1996) Electroporation of small RNAs into plant protoplasts: mitochondrial uptake of transfer RNAs. Biochem Biophys Res Commun 223:204–210

Yamasaki T, Kurokawa S, Watanabe KI, Ikuta K, Ohama T (2005) Shared molecular characteristics of successfully transformed mitochondrial genomes in *Chlamydomonas reinhardtii*. Plant Mol Biol 58:515–527

Yu W, Schuster W (1995) Evidence for a site-specific cytidine deamination reaction involved in C-to-U RNA editing of plant mitochondria. J Biol Chem 270:18227–18233

Zamzami N, Kroemer G (2001) The mitochondrion in apoptosis: how Pandora's box opens. Nat Rev Mol Cell Biol 2:67–71

Zehrmann A, Verbitskiy D, van der Merwe JA, Brennicke A, Takenaka M (2009) A DYW domain-containing pentatricopeptide repeat protein is required for RNA editing at multiple sites in mitochondria of *Arabidopsis thaliana*. Plant Cell 21:558–567

Subject Index

R. Bock and V. Knoop (eds.), Genomics of Chloroplasts and Mitochondria, Advances in Photosynthesis and Respiration 35, pp. 459–475,
DOI 10.1007/978-94-007-2920-9, © Springer Science+Business Media B.V. 2012

D

S

MIX
Papier aus verantwortungsvollen Quellen
Paper from responsible sources
FSC® C105338

If you have any concerns about our products,
you can contact us on
ProductSafety@springernature.com

In case Publisher is established outside the EU,
the EU authorized representative is:
Springer Nature Customer Service Center GmbH
Europaplatz 3, 69115 Heidelberg, Germany

Printed by Libri Plureos GmbH
in Hamburg, Germany